12–7
A mutant red delicious apple. The mutant allele that is responsible for the golden color arose in a cell of the flower's ovary wall, which subsequently developed into the fleshy part of the apple. The seeds of this apple would produce red apples because they were unaffected by the mutation. Golden delicious apples, however, were discovered on a mutated branch of a tree that produced red delicious apples. The seeds of these golden apples, which carried the mutation, produced trees bearing golden delicious apples.

mutations occur without cause but rather that the events triggering them are independent of their subsequent effects. Although the rate of mutation can be influenced by environmental factors, the specific mutations produced are independent of the environment—and independent of their potential for subsequent benefit or harm to the organism and its offspring.

Although the rate of spontaneous mutation is generally low, mutations provide the raw material for evolutionary change because they provide the variation on which other evolutionary forces act. However, because mutation rates are so low, mutations probably never determine the direction of evolutionary change.

Gene Flow Is the Movement of Alleles into or out of a Population

Gene flow can occur as a result of immigration or emigration of individuals of reproductive age. In the case of plants, it can also occur through the movement of gametes via pollen between populations.

Gene flow can introduce new alleles into a population or it can change existing allele frequencies. Its overall effect is to decrease the difference between populations. Natural selection, by contrast, is more likely to increase differences, producing populations more suited for local conditions. Thus, gene flow often counteracts natural selection.

The possibilit lations of most p tance. Although pollen can be dispersed great distances at times, the chances of its falling on a receptive stigma at any great distance are slight. For several kinds of insect-pollinated plants that grow in temperate regions, a gap of only 300 meters may effectively isolate two populations. Rarely will more than 1 percent of the pollen that reaches a given individual come from this far away. In plants whose pollen is spread by wind, very little falls more than 50 meters from the parent plant under normal circumstances.

Genetic Drift Refers to Changes That Occur Due to Chance

As we stated previously, the Hardy–Weinberg equilibrium holds true only if the population is large. This qualification is necessary because the equilibrium depends on the laws of probability. Consider, for example, an allele, say *a*, that has a frequency of 1 percent. In a population of 1 million individuals, 20,000 *a* alleles would be present in the gene pool. (Remember that each diploid individual carries two alleles for any given gene. In the gene pool of this population there are 2 million alleles for this particular gene, of which 1 percent, or 20,000, are allele *a*.) If a few individuals in this population were destroyed by chance before leaving offspring, the effect on the frequency of allele *a* would be negligible.

In a population of 50 individuals, however, the situation would be quite different. In this population, it is likely that only one copy of allele *a* would be present. If the lone individual carrying this allele failed to reproduce or were destroyed by chance before leaving offspring, allele *a* would be completely lost. Similarly, if 10 of the 49 individuals homozygous for allele *A* were lost, the frequency of *a* would jump from 1 in 100 to 1 in 80.

This phenomenon, a change in the gene pool that takes place as a result of chance, is **genetic drift.** Population geneticists and other evolutionary biologists generally agree that genetic drift plays a role in determining the evolutionary course of small populations. Its relative importance, however, as compared with that of natural selection, is a matter of debate. There are at least two situations—the founder effect and the bottleneck effect—in which genetic drift has been shown to be important.

The Founder Effect Occurs When a Small Population Colonizes a New Area A small population that becomes separated from a larger one may or may not be genetically representative of the larger population from which it was derived (Figure 12–8). Some rare alleles may be overrepresented or, conversely, may be completely absent in the small population. An extreme case would be

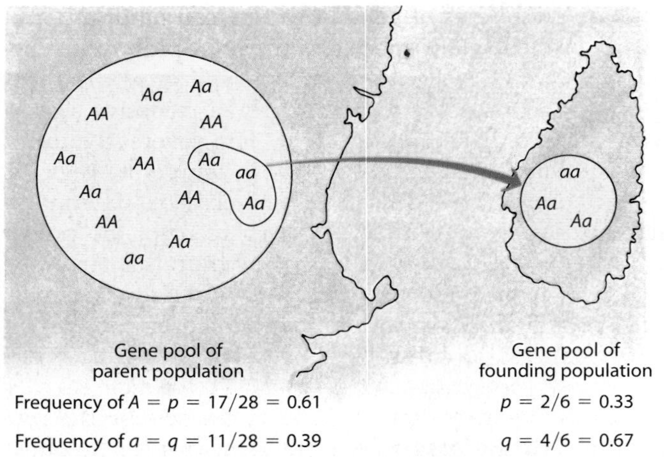

12–8
The founder effect. When a small subset of a population founds a new colony (for example, on a previously uninhabited island), the allele frequencies within the founding group may be different from those within the parent population. Thus the gene pool of the new population will have a different composition than the gene pool of the parent population.

the initiation of a new population by a single plant seed. As a consequence, when and if the small population increases in size, it will continue to have a different genetic composition—a different gene pool—from that of the parent group. This phenomenon, a type of genetic drift, is known as the **founder effect.**

The Bottleneck Effect Occurs When Environmental Factors Suddenly Decrease Population Size The **bottleneck effect** is another type of situation that can lead to genetic drift. It occurs when a population is drastically reduced in numbers by an event, such as an earthquake, flood, or fire, that may have little or nothing to do with the usual forces of natural selection. A population bottleneck is likely not only to eliminate some alleles entirely but also to cause others to become overrepresented in the gene pool.

Nonrandom Mating Decreases the Frequency of Heterozygotes

Disruption of the Hardy–Weinberg equilibrium can also be produced by nonrandom mating. Typically, the members of a population mate more often with close neighbors than with more distant ones. Thus, within the large population, neighboring individuals tend to be closely related. Such nonrandom mating promotes **inbreeding,** the mating of closely related individuals. An extreme form of nonrandom mating that is particularly important in plants is self-pollination (as in the pea plants Mendel studied).

Inbreeding and self-pollination tend to increase the frequencies of homozygotes in a population at the expense of the heterozygotes. Take, for example, Mendel's pea plants in which only two alleles are involved in flower color, *W* (purple) and *w* (white). When *WW* plants and *ww* plants self-pollinate, all of their progeny will be homozygous. When *Ww* plants self-pollinate, however, only half of their progeny will be heterozygous. With succeeding generations, there will be a decrease in the frequency of heterozygotes, with a corresponding increase in the frequencies of the two homozygotes. Note that although nonrandom mating, as exemplified by the pea plants, can change the ratio of genotypes and phenotypes in a population, the frequencies of the alleles in question remain the same.

Preservation and Promotion of Variability

Sexual Reproduction Produces New Genetic Combinations

By far the most important method by which eukaryotic organisms promote variation in their offspring is sexual reproduction. Sexual reproduction produces new genetic combinations in three ways: (1) by independent assortment at the time of meiosis (see Figure 9–10), (2) by crossing-over with genetic recombination (see Figure 9–4), and (3) by the combination of two different parental genomes at fertilization. At every generation, alleles are assorted into new combinations.

In contrast, consider organisms that reproduce only asexually—that is, by processes that involve mitosis and cytokinesis but not meiosis. Except when a mutation has occurred in the duplication process, the new organism will exactly resemble its only parent. In the course of time, various clones may form, each carrying one or more mutations, but potentially favorable combinations are unlikely to accumulate in one genotype. The only advantage to the organism of sexual reproduction, from a strictly scientific viewpoint, appears to be the promotion of variation by the production of new combinations of alleles among the offspring. Why such variation is advantageous to the individual organism is a matter of controversy.

Various Mechanisms Promote Outbreeding

Many means have evolved by which new genetic combinations are promoted in sexually reproducing populations. Among plants, various mechanisms ensure that the sperm-bearing pollen from flowers on one plant is

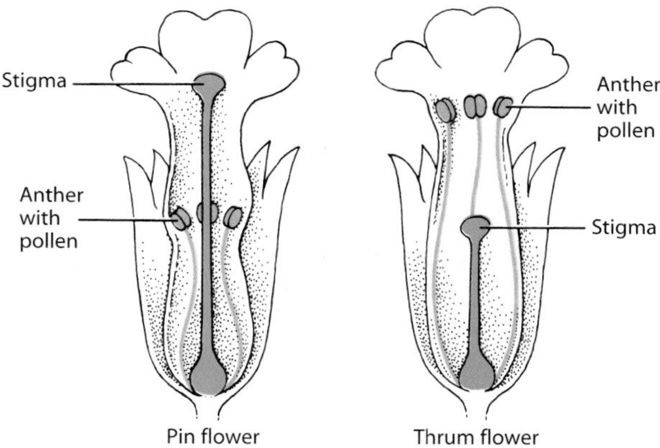

12–9
Diagrams of two types of flowers ("pin" and "thrum") of the same species of primrose. Notice that the pollen-bearing anthers of the pin flower and the pollen-receiving stigma of the thrum flower are both situated about halfway up the length of the flower and that the pin stigma is level with the thrum anthers. An insect foraging for nectar deep inside these flowers would collect pollen on different areas of its body, so that thrum pollen would be deposited on pin stigmas, and vice versa. Self-pollination is therefore inhibited.

delivered to the stigmas of flowers on a different plant. In some plants, such as the holly and the date palm, male flowers are on one tree and female flowers on another. In others, such as the avocado, the pollen of a particular plant matures at a time when its own stigma is not receptive. In some species, anatomical arrangements inhibit self-pollination (Figure 12–9).

Some plants have genes for self-sterility. Typically, such a gene has multiple alleles—s^1, s^2, s^3, and so on. A plant carrying the s^1 allele cannot fertilize a plant with an s^1 allele. One with an s^1s^2 genotype cannot fertilize any plant with either of those alleles, and so forth. In one population of about 500 evening primrose plants, 37 different self-sterility alleles were found, and it has been estimated that there may be hundreds of alleles for self-sterility in red clover. A plant with a rare self-sterility allele is more likely to be able to fertilize another plant than is a plant with a common self-sterility allele. As a consequence, the self-sterility system strongly encourages variability in a population. Selection for the rare allele makes it more common, whereas more common alleles become rarer.

Diploidy Enables Recessive Alleles to Be Stored

Another factor in the preservation of variability in eukaryotes is diploidy. In a haploid organism, genetic variations are immediately expressed in the phenotype and are therefore exposed to the selection process. In a diploid organism, however, such variations may be stored as recessives, as with the allele for white flowers in Mendel's pea plants (Table 12–1). As the table reveals, the lower the frequency of allele *w*, the smaller the proportion of it exposed in the *ww* homozygote becomes. The removal of the allele by natural selection slows down accordingly.

TABLE 12–1 Protection of Recessive Alleles by Diploidy

Frequency of Allele *w* in Gene Pool	Genotype Frequencies *WW*	*Ww*	*ww*	Percentage of Allele *w* in Heterozygotes
0.9	0.01	0.18	0.81	10
0.1	0.81	0.18	0.01	90
0.01	0.9801	0.0198	0.0001	99

Heterozygotes May Have a Selective Advantage over Homozygotes

Recessive alleles, even ones that are harmful in the homozygous state, may not only be sheltered in the heterozygous state but sometimes may actually be selected for. This phenomenon, in which the heterozygotes produce more offspring than either type of homozygote, is known as **heterozygote advantage.** It is another way that genetic variability is preserved.

Probably the best example of heterozygote advantage in plants is found in the crossing of inbred lines of crop plants, with the resultant increase in vigor and productivity of the **hybrids**—by definition, the offspring of genetically dissimilar parents. Hybrids exhibiting such **hybrid vigor** have played an important role in increasing the yield of crop plants throughout the world, most notably of maize. Hybridization was the first systematic application of genetic principles to crop breeding. The discovery of hybrid vigor in maize dates to 1908, when the American plant breeder G. H. Shull discovered that crossing inbred lines produced hybrids that quadrupled the yield (from about 20 to 80 bushels per acre!) of the inbred lines.

Responses to Selection

In the course of the controversies that led to the synthesis of evolutionary theory with Mendelian genetics, some biologists argued that natural selection could serve only to eliminate the "less fit." As a consequence, it would tend to reduce the genetic variation in a population and thus reduce the potential for further evolution. Modern population genetics has demonstrated that this is not the case. Natural selection can, in fact, be a critical factor in preserving and promoting genetic variability in a population.

In general, only the phenotype is being selected, in the sense that it is the relationship between the phenotype and the environment that determines how many offspring an individual will contribute to the next generation. Because nearly all characteristics of individual organisms are determined by the interactions of many genes, phenotypically similar individuals can have many different genotypes.

When some feature, such as tallness, is strongly selected for, there is an accumulation of alleles that contribute to this feature and an elimination of those alleles that work in the opposite direction. But selection for a polygenic characteristic is not simply the accumulation of one set of alleles and the elimination of others. Gene interactions, such as epistasis and pleiotropy (pages 195–196), are of fundamental importance in determining the course of selection in a population.

Bear in mind, also, that the phenotype is not determined solely by the interactions of the multitude of alleles making up the genotype. The phenotype is also a product of the interaction of the genotype with the environment in the course of the individual's life (Figure 12–10).

(a)

(b)

12–10
Jeffrey pines (Pinus jeffreyi) *usually grow tall and straight, as in* **(a)**. *Environmental forces, however, can alter the normal growth patterns, as shown in* **(b)**. *This tree is growing on a mountaintop in Yosemite National Park, California, where it is exposed to strong, constant winds.*

Evolutionary Changes in Natural Populations May Occur Rapidly

Under certain circumstances, the characteristics of populations may change rapidly, often in response to a rapidly changing environment. Particularly during the past few centuries, the influence of human beings in many areas has been so great that some populations of other organisms have had to adjust rapidly in order to survive. Evolutionary biologists have been particularly interested in examples of such rapid changes—"evolution in action"—because the principles involved are presumed to be the same as those that govern changes (even though much slower changes) in populations generally.

In plants, strong selective forces have been observed to produce rapid changes in natural populations. For example, plants in the grazed part of an experimental pasture in Maryland were much shorter than those in the ungrazed part, and it was thought that this might be due to the direct action of grazing. This hypothesis was tested by digging up some of the plants from both parts of the pasture and growing them together in a garden. It was assumed that if the short plants were short merely because they had been grazed, they would soon become tall in the absence of grazing. Some of them did, but plants of white clover *(Trifolium repens)*, Kentucky bluegrass *(Poa pratensis)*, and orchard grass *(Dactylis glomerata)* remained short, indicating that these populations had been modified genetically by the selective force of grazing, over a period of only two or three centuries. A similar example is illustrated in Figure 12–11.

In Wales, the tailings around a number of abandoned lead mines are rich in lead (up to 1 percent) and zinc (up to 0.03 percent)—substances that are toxic to most plants at these concentrations. Because of the presence of these metals, the tailings are often nearly devoid of plant life. Observing that one species of grass, *Agrostis tenuis*, was colonizing these areas, scientists took some of the mine plants and other *Agrostis* from nearby pastures and grew them together either in normal soil or in mine soil. In the normal soil, the mine plants were slower growing and smaller than the pasture plants. On the mine soil, how-

(a)

(b)

12–11
Prunella vulgaris *is a common herb of the mint family; it is widespread in woods, meadows, and lawns in temperate regions of the world. Most populations consist of erect plants, such as those shown in* **(a),** *which grow in open, often somewhat moist, grassy places throughout the cooler regions of the world. Populations found in lawns, however, always consist of prostrate plants, such as those shown in* **(b),** *growing in Berkeley, California. Erect plants of* P. vulgaris *cannot survive in lawns because they are damaged by mowing and do not have the capability for resprouting low branches from the base, which would be necessary for survival. When lawn plants are grown in an experimental garden, some remain prostrate whereas others grow erect. The prostrate habit is determined genetically in the first group and environmentally in the second group.*

ever, the mine plants grew normally but the pasture plants did not grow at all. Half the pasture plants in mine soil were dead in three months and had misshapen roots that were rarely more than 2 millimeters long. But a few of the pasture plants (3 of 60) showed some resistance to the effects of the metal-rich soil. They were doubtless genetically similar to the plants originally selected in the development of the lead-resistant strain of *Agrostis*. The mines were no more than 100 years old and so the lead-resistant strain had developed in a relatively short period of time. The resistant plants had been selected from the genetically variable pasture plants, some of whose seeds had fallen on the mine soil, germinated, and produced seeds that could survive there. Then, due to natural selection, a distinct strain of *Agrostis* was constituted (Figure 12–12).

12–12
Bent grass (Agrostis tenuis) *growing in the foreground on the tailings of abandoned lead mines rich in lead, photographed in Wales. Such tailings are often devoid of plant life because the lead is present at levels toxic to plants. The lead-tolerant bent grass plants growing here are descendants of a population of plants that gradually adapted to the lead in the tailings.*

The Result of Natural Selection: Adaptation

Natural selection results in **adaptation,** a term with several meanings in biology. First, it can mean a state of being adjusted to the environment. Every living organism is adapted in this sense, just as Abraham Lincoln's legs were, as he remarked, "just long enough to reach the ground." Second, adaptation can refer to a particular characteristic that aids in the adjustment of an organism to its environment. Third, adaptation can mean the evolutionary process, occurring over the course of many generations, that produces organisms better suited to their environment.

Natural selection involves interactions between individual organisms, their physical environment, and their biological environment—that is, other organisms. In many cases, the adaptations that result from natural selection can be clearly correlated with environmental factors or with the selective forces exerted by other organisms.

Clines and Ecotypes Are Reflections of Adaptation to the Physical Environment

Developmental plasticity is the tendency of individuals to vary over time in response to different environmental conditions, or for genetically identical organisms to differ in response to different environmental stimuli. Such plasticity is much greater in plants than in animals because the indeterminate (unrestricted) growth pattern that is characteristic of plants can be modified more easily to produce strikingly different expressions of a particular genotype. Environmental factors, as every gardener knows, can cause profound differences in the phenotypes of many species of plants. Leaves that develop in the shade, for example, may be thinner and larger than those that develop in the sun.

Sometimes phenotypic variation within the same species follows a geographic distribution and can be correlated with gradual changes in temperature, humidity, or some other environmental condition. A gradual change of this kind in the characteristics of populations of an organism is called a **cline.** Many species exhibit north-south clines of various traits. Plants, for example, growing in the south often have slightly different requirements for flowering or for ending dormancy than the same kind of plants growing in the north, although they may all belong to the same species.

Clines are frequently encountered in organisms that live in the sea, where the temperature often rises or falls very gradually with changes in latitude. Clines are also characteristic of organisms that occur in such areas as the eastern United States, where rainfall gradients may extend over thousands of kilometers. When populations of plants are sampled along a cline, the differences are often proportional to the distance between the populations.

A species that occupies many different habitats may appear to be slightly different in each one. Each group of distinct phenotypes is known as an **ecotype.** Are the differences among ecotypes determined entirely by the environment, or do these differences represent adaptation resulting from the action of natural selection on genetic variation?

Of particular interest in the study of ecotypes is the work of Jens Clausen, David Keck, and William Hiesey with the perennial herb *Potentilla glandulosa,* which ranges through a wide variety of climatic zones in California. These scientists established experimental gardens at three localities in California at sites where native populations of *P. glandulosa* occur: (1) Stanford, located between the inner and outer Coast Ranges, at 30 meters elevation, with warm temperate weather and predominant winter rainfall; (2) Mather, on the western slope of the Sierra Nevada, at 1400 meters elevation, with long, cold, snowy winters and hot, mostly dry summers; and (3) Timberline, east of the crest of the Sierra Nevada at roughly the same latitude as the two other stations but at 3050 meters elevation, with very long and cold, snowy winters and short, cool, rather dry summers (Figure 12–13).

When *P. glandulosa* plants from numerous locations were grown side-by-side in the gardens set up at the three stations, four distinct ecotypes became apparent. The morphological, or structural, characteristics of each ecotype were correlated with their physiological responses, which in turn were critical to the survival of each ecotype in its native environment.

For example, the Coast Ranges ecotype consists of plants that grew actively in both winter and summer when cultivated at Stanford, which lies within their native range. Plants of this ecotype decreased in size but survived at Mather, outside their native range, even though they were subjected to about five months of cold winter weather. At Mather they became winter-dormant, but they stored enough food during their growing season to carry them through the long, unfavorable winter. At Timberline, plants of the Coast Ranges ecotype failed to survive, almost invariably dying during the first winter. The short growing season at this high elevation did not permit them to store enough food to survive the long winter. Other species that occur in California's Coast Ranges produce ecotypes that have physiological responses comparable to those of *P. glandulosa.* Indeed, strains of unrelated plant species that occur together naturally in a given location are often more similar to one another physiologically than they are to other populations of their own species.

The physiological and morphological characteristics of ecotypes, as in *P. glandulosa,* usually have a complex genetic basis, involving dozens (or, in some cases, perhaps even hundreds) of genes. Sharply defined ecotypes are characteristic of regions, like western North America, where the breaks between adjacent habitats are sharply defined. On the other hand, when the environment changes more gradually from one habitat to another, the characteristics of the plants that occur in that region may do likewise.

Ecotypes Differ Physiologically In order to understand why ecotypes flourish where they do, we must understand the physiological basis for their ecotypic differentiation. For example, when Scandinavian strains of goldenrod *(Solidago virgaurea)* from shaded habitats and from exposed habitats were grown under different light intensities, they showed differences in their photosynthetic response. The plants from shaded environments grew rapidly under low light intensities, whereas their growth rate was markedly retarded under high light intensities. In contrast, plants from exposed habitats grew rapidly under conditions of high light intensity, but much less well at low light levels.

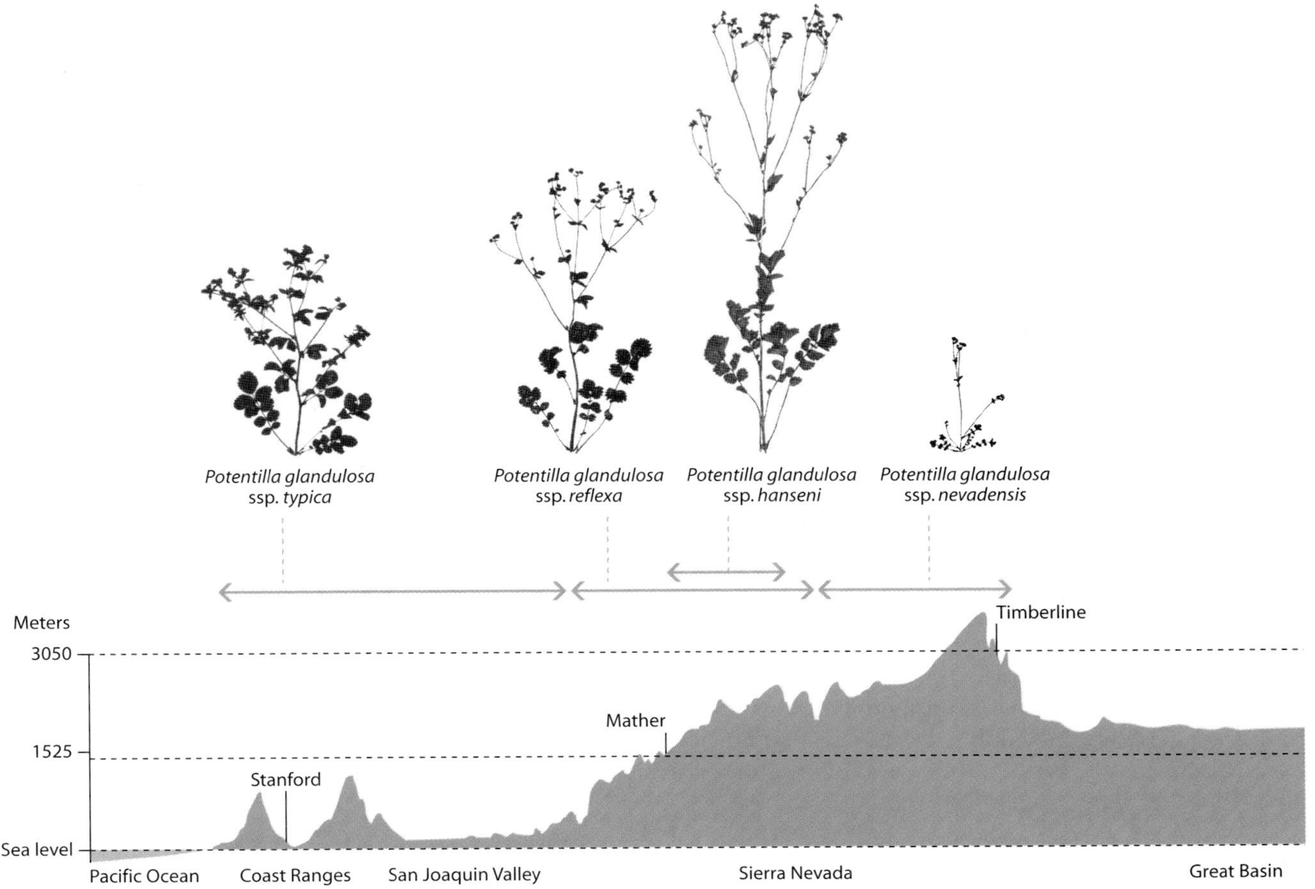

12–13

Ecotypes of Potentilla glandulosa, *a relative of the strawberry. Plants from a number of populations of* P. glandulosa *were collected at 38° north latitude from the Pacific Ocean to Timberline and transplanted to experimental gardens at Stanford, Mather, and Timberline, all also at about 38° north latitude. The plants were propagated asexually so that genetically alike individuals could be grown at the three climatically unlike gardens. When grown side-by-side, four distinct ecotypes became apparent. The four ecologically distinct ecotypes were correlated with differences in their morphology, especially flower and leaf characteristics. Many of these characteristics were passed on to the next generation, indicating that these differences are genotypic as well as phenotypic.*

Each ecotype is distributed over a range of altitudes. Where the ranges of two ecotypes overlap, the two ecotypes grow in different environments. Representatives of the four ecotypes at the flowering stage and their approximate ranges are shown. Each of the four ecotypes has been given a subspecies name.

In another experiment, arctic and alpine populations of the widespread herb *Oxyria digyna* were studied using strains from an enormous latitudinal range that extended southward from Greenland and Alaska to the mountains of California and Colorado. Plants of northern populations had more chlorophyll in their leaves, as well as higher respiration rates at all temperatures, compared with plants from farther south. High-elevation plants from near the southern limits of the species' range carried out photosynthesis more efficiently at high light intensities than did low-elevation plants from farther north. Thus the respective ecotypes could function better in their own habitats, characterized, for example, by high light intensities in high mountain habitats and lower intensities in the far north. The existence of *O. digyna* over such a wide area and such a wide range of ecological conditions is made possible, in part, by differences in metabolic potential among its various populations.

Coevolution Results from Adaptation to the Biological Environment

When populations of two or more species interact so closely that each exerts a strong selective force on the other, simultaneous adjustments occur that result in **coevolution**. One of the most important, in terms of sheer number of species and individuals involved, is the coevolution of flowers and their pollinators, which is described in Chapter 22. Another example of coevolution involves the monarch butterfly and the milkweed plants (page 34).

The Origin of Species

Although Darwin titled his monumental book *On the Origin of Species,* he was never really able to explain how species might originate. However, an enormous body of work, mostly in the twentieth century, has provided many insights into the process. In addition, a great amount of time and discussion has been spent attempting to develop a clear definition of the term "species."

What Is a Species?

In Latin, **species** simply means "kind," and so species are, in the simplest sense, different kinds of organisms. More precisely, a species is a group of natural populations whose members can interbreed with one another but cannot (or at least usually do not) interbreed with members of other such groups. This concept of species is called the *biological species concept.* The key criterion of this definition is **genetic isolation:** if members of one species freely exchanged genes with members of another species, they could no longer retain those unique characteristics that identify them as different kinds of organisms. The biological species concept does not work in all situations, and it is difficult to apply to actual data in nature. Several alternative species concepts have therefore been proposed. In practice, "biological species" are generally identified purely on an assessment of their morphological, or structural, distinctness. In fact, most species recognized by taxonomists have been designated as such based on anatomical and morphological criteria. This practical approach has been referred to as the *morphological species concept.*

The inability to form fertile hybrids has often been used as a basis for defining species. However, this criterion is not generally applicable. In some groups of plants—particularly such long-lived woody plants as trees and shrubs—very morphologically distinct species often can form fertile hybrids with one another. Take the case of the sycamores *Platanus orientalis* and *Platanus occidentalis,* which have been isolated from one another in nature for at least 50 million years. *P. orientalis* is native from the eastern Mediterranean region to the Himalayas, while *P. occidentalis* is native to eastern North America (Figure 12–14). *P. orientalis* has been widely cultivated in southern Europe since Roman times, but it cannot be grown in northern Europe away from the moderating influence of the sea. After the European discovery of the New World, *P. occidentalis* was brought into cultivation in the colder portions of northern Europe, where it flourished. About 1670, these two very distinct trees produced intermediate and fully fertile hybrids when they were cultivated together in England. Called the London plane, the hybrid *(Platanus × hybrida)* is capable of growing in regions with cold winters and is now grown as a street tree in New York City and throughout the temperate regions of the world.

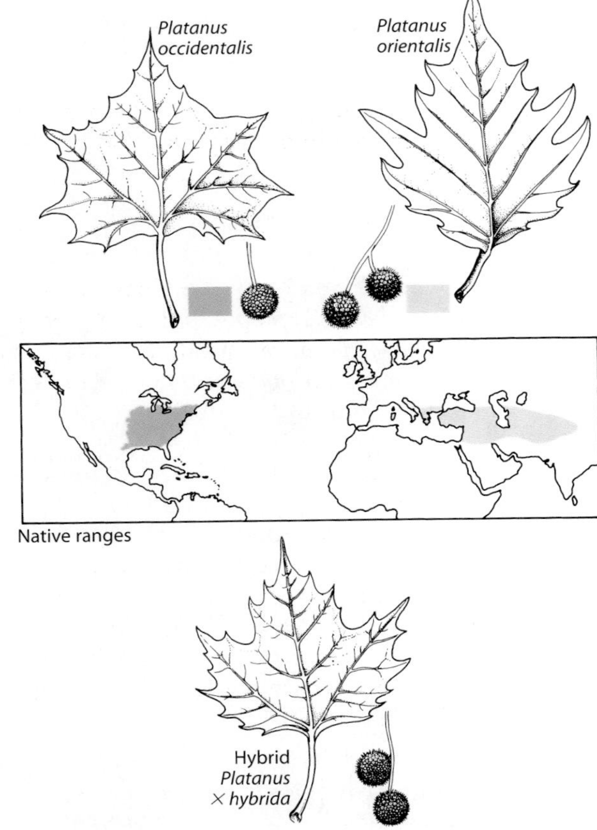

12–14
The distribution of two species of sycamore, Platanus orientalis, *which is native to areas from the eastern Mediterranean region to the Himalayas, and* P. occidentalis, *which is native to North America. The fully fertile hybrid,* Platanus × hybrida, *is the London plane, a sturdy tree suitable for growing along city streets.*

One might argue that *P. orientalis* and *P. occidentalis* had no potential to interbreed in nature. But, then, there are many groups, such as the birches, oaks, and willows, in which many species freely interbreed in nature. Does that mean that each of these groups of interbreeding species should be considered a single species, and that what appear as morphologically distinct species should be considered subspecies (subdivisions of a species)? It is unlikely that most plant taxonomists would be willing to go that far.

From an evolutionary perspective, a species is a population or group of organisms reproductively united but very probably changing as it moves through space and time. Splinter groups reproductively isolated from the population as a whole can undergo sufficient change that they become new species. This process is known as **speciation.** Occurring repeatedly in the course of more than 3.5 billion years, it has given rise to the diversity of organisms that have lived in the past and that live today.

How Does Speciation Occur?

By definition, the members of a species share a common gene pool that is effectively separated from the gene pools of other species. A central question, then, is, how does one pool of genes split off from another to begin a separate evolutionary journey? A subsidiary question is, how do two species, often very similar to one another, inhabit the same place at the same time and yet remain reproductively isolated?

According to current thinking, speciation is most commonly the result of the geographic separation of a population of organisms; this process is known as **allopatric** ("other country") **speciation.** Under certain circumstances, speciation may also occur without geographic isolation, in which case it is known as **sympatric** ("same country") **speciation.**

Allopatric Speciation Involves the Geographic Separation of Populations

Every widespread species that has been carefully studied has been found to contain geographically representative populations that differ from each other to a greater or lesser extent. Examples are the ecotypes of *Potentilla glandulosa* and the strains of *Oxyria digyna.* A species composed of such geographic variants is particularly susceptible to speciation if geographic barriers arise, preventing gene flow.

Geographic barriers are of many different types (Figure 12–15). Islands are frequently sites for the development of new species, and set the stage for the sudden (in geologic time) diversification of a group of organisms that share a common ancestor, often itself newly evolved. The sudden diversification of such a group of organisms is called **adaptive radiation.** It is associated with the opening up of a new biological frontier that may be as vast as the land or the air, or, as in the case of the Galápagos tortoises, as small as an archipelago. Adaptive radiation results in the almost simultaneous formation of many new species in a wide range of habitats. (See "Adaptive Radiation in Hawaiian Tarweeds" on pages 250–251.)

Differentiation on islands is particularly striking because, in the absence of competition, organisms seem more likely to produce highly unusual forms than do related species on the continents. In island localities, the characteristics of plants and animals may change more rapidly than on the mainland, and features that are never encountered elsewhere may arise. Similar clusters of species may also arise in mainland areas, of course, and may involve spectacular differentiation.

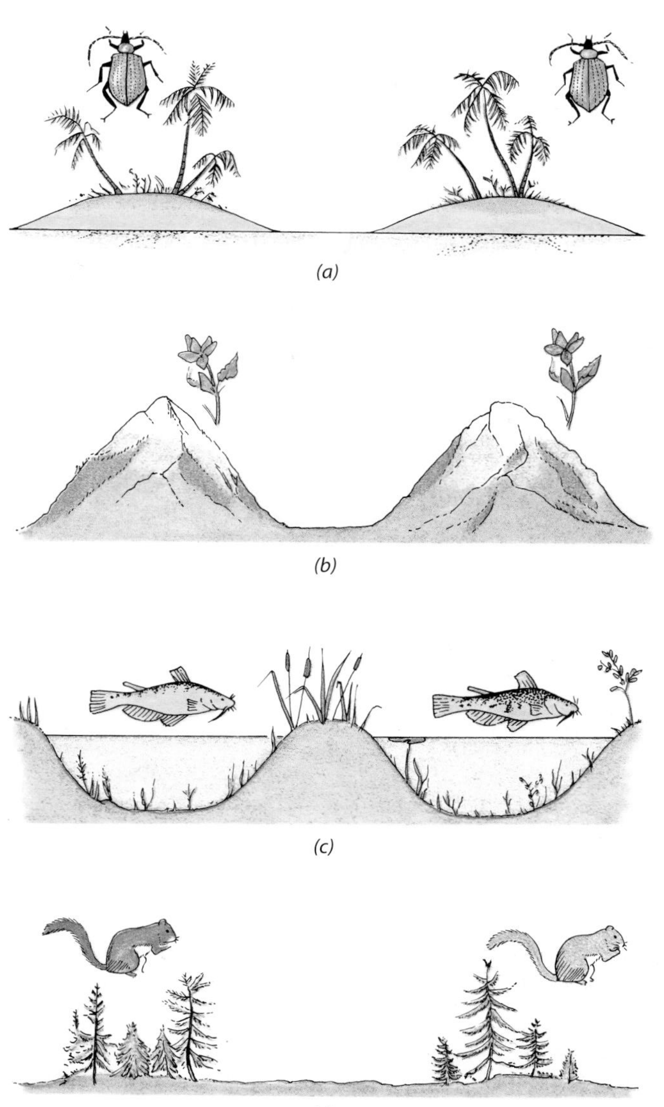

12–15
Four different geographic barriers that can lead to speciation may be ***(a)*** *islands;* ***(b)*** *mountaintops;* ***(c)*** *ponds, lakes, or even oceans; and* ***(d)*** *isolated clumps of vegetation.*

Sympatric Speciation Occurs without Geographic Separation

A well-documented mechanism by which new species are produced through sympatric speciation—that is, when there is no geographic isolation—is **polyploidy.** By definition, polyploidy is an increase in the number of chromosomes beyond the typical diploid ($2n$) complement. Polyploidy may arise as a result of nondisjunction (page 173) during mitosis or meiosis, or it may be generated when the chromosomes divide properly during mitosis or meiosis but cytokinesis does not subsequently occur. Polyploid individuals can be produced deliberately in the laboratory by use of the drug colchicine, which disrupts microtubule formation and hence prevents separation of chromosomes during mitosis (page 161).

Adaptive Radiation in Hawaiian Tarweeds

Some spectacular groups of native plants and animals occur in the Hawaiian Islands. On this archipelago, whose major islands arose from the sea in isolation over millions of years, the habitats are exceedingly diverse, and the distances to mainland areas from which plants and animals migrated to colonize the islands are great. Those organisms that did reach the Hawaiian Islands often changed greatly in their characteristics as they occupied the varied habitats that were available to them. The process by which such changes occur is called adaptive radiation.

The striking Hawaiian silversword alliance, which includes 28 species in three closely related genera of the sunflower family, Asteraceae, *is one of the most remarkable examples of adaptive radiation in plants. These species belong to the tarweed subtribe* Madiinae, *most of whose members are found in California and adjacent regions. The 28 species belong to the genus* Argyroxiphium *(silverswords) and to two other genera also found only in Hawaii,* Dubautia *and* Wilkesia. *The species of these genera range in habit from small, mat-forming shrubs and rosette plants to large trees and climbing vines. They grow in habitats as diverse as exposed lava, dry scrub, dry woodland, moist forest, wet forest, and bogs. In these habitats, the annual precipitation ranges from less than 40 centimeters to more than 1230 centimeters. The rainiest of these habitats is among the wettest places on Earth.*

This enormous variation in habitats is paralleled by significant variation in leaf sizes and shapes among these plants. For example, species of Dubautia *that grow in bright, dry habitats usually have very small leaves, while those that inhabit the shaded understories of wet forests have much larger ones. The silversword,* Argyroxiphium sandwicense, *which grows on the dry alpine slopes of Haleakala Crater on the island of Maui, has leaves that are covered with a dense, silvery mat of hairs. These hairs apparently provide protection from intense solar radiation and aid in the conservation of moisture. The leaves of the closely related greensword,* Argyroxiphium grayanum, *which also occurs on Maui but in wet forest and bog habitats, lack these hairs.*

Important physiological differences likewise characterize the species of

The silversword, Argyroxiphium sandwicense, *a remarkable plant that grows on the exposed, upper cinder slopes of Haleakala Crater on the island of Maui, where the plants are exposed to high levels of solar radiation and low humidities.*

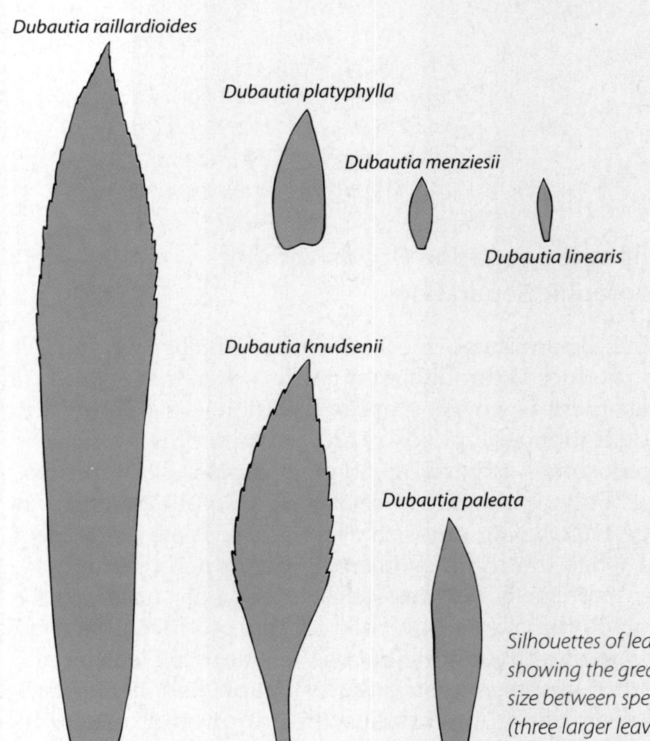

Silhouettes of leaves of six species of Dubautia, *showing the great differences in leaf shape and size between species that grow in wet habitats (three larger leaves) and those that grow in dry ones (three smaller leaves).*

Dubautia reticulata. *Species of the genus* Dubautia *(which includes 21 of the estimated 28 species of Hawaiian tarweeds) range from trees and shrubs to lianas (woody vines) and small, matted plants that are scarcely woody.* Dubautia reticulata, *shown here, grows in the wet forests of Maui, where it may reach a height of 8 meters or more and develop a trunk with a diameter of nearly 0.5 meter.*

Wilkesia gymnoxiphium. *This bizarre, yuccalike plant occurs only on the island of Kauai, where it is restricted to dry scrub vegetation along the fringes of Waimea Canyon. Kauai is the oldest of the main islands of the Hawaiian Archipelago. Robert Robichaux, shown here, is studying the physiological ecology of this fascinating group of plants.*

Dubautia scabra. *This low, matted, herbaceous member of the genus is found in moist to wet habitats on several of the islands in the Hawaiian Archipelago. It is believed to be the progenitor of several additional species on the younger islands of the group.*

Hawaiian tarweeds, which must meet very different kinds of environmental challenges. For example, species of* Dubautia *that occur in dry habitats have a much greater tolerance to water stress than those that occur in moist to wet habitats. The species that grow in dry habitats have leaves with more elastic cell walls and under dry conditions are able to maintain higher turgor pressures than their relatives.

***Despite their great diversity in appearance and the very different habitats in which they grow, all 28 species of these three genera are very closely related to one another. Any two of them can be hybridized, as far as we know, and all of the hybrids are at least partly fertile. Additionally, experimental hybrids between Hawaiian and Californian tarweeds have been produced, underscoring their very close relationship. The entire group of three genera appears to have evolved in isolation on the Hawaiian Islands following the arrival of a single, original colonist from western North America. Molecular studies have revealed that this ancestral species was closely related (and similar) to modern Californian species of the tarweed genera* Madia *and* Raillardiopsis.**

Evolutionary patterns, reconstructed from DNA sequence data, indicate that the most recent common ancestor of the modern silversword alliance species occurred on the oldest high island of Kauai. The many species of this group that occur outside Kauai owe their existence to inter-island dispersal of founders from older to progressively younger islands, followed by speciation that involved major ecological shifts—the hallmarks of adaptive radiation. Similar patterns have been confirmed recently for several other endemic lineages of Hawaiian organisms.

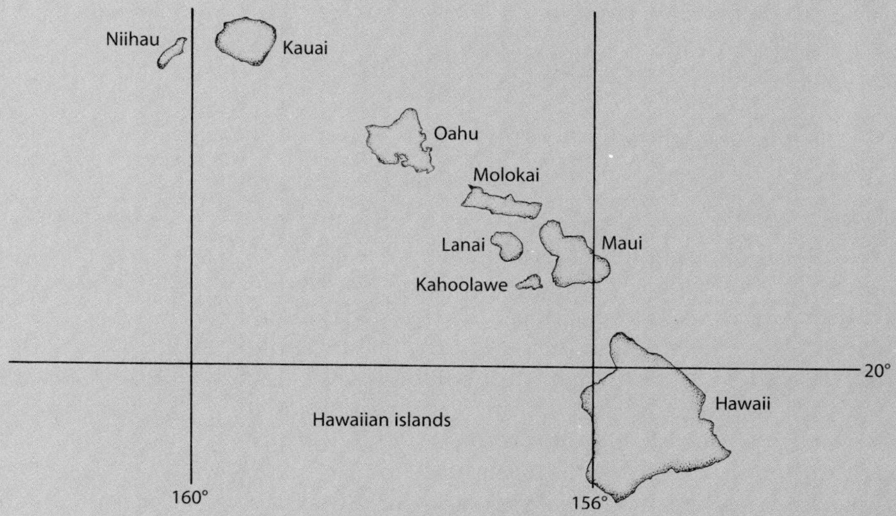

The Hawaiian Islands. The oldest of the main islands, Kauai, includes some rocks that are as much as 6 million years old; the youngest, Hawaii, is still being formed. The Hawaiian Archipelago is gradually moving northwest with the Pacific Plate, the older islands gradually being eroded below the surface of the sea and the younger ones continuously being formed, apparently as they pass over a "hot spot," which is a thin spot in the Earth's crust through which lava erupts. Thus we know that there were islands in approximately the present position of Hawaii much more than 6 million years ago.

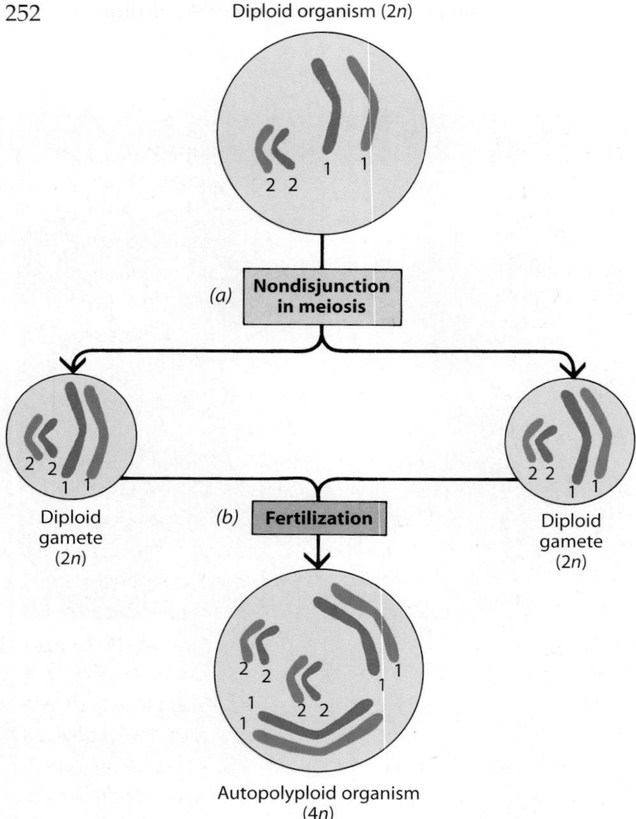

12–16
Autopolyploidy. Polyploidy within individual organisms can lead to the formation of new species. ***(a)*** *If the chromosomes of a diploid organism do not separate during meiosis (nondisjunction), diploid (2n) gametes may result.* ***(b)*** *Union of two such gametes, produced either by the same individual or by different individuals of the same species, will produce an autopolyploid, or tetraploid (4n), individual. Although this individual may be capable of sexual reproduction, it will be reproductively isolated from the diploid parent species.*

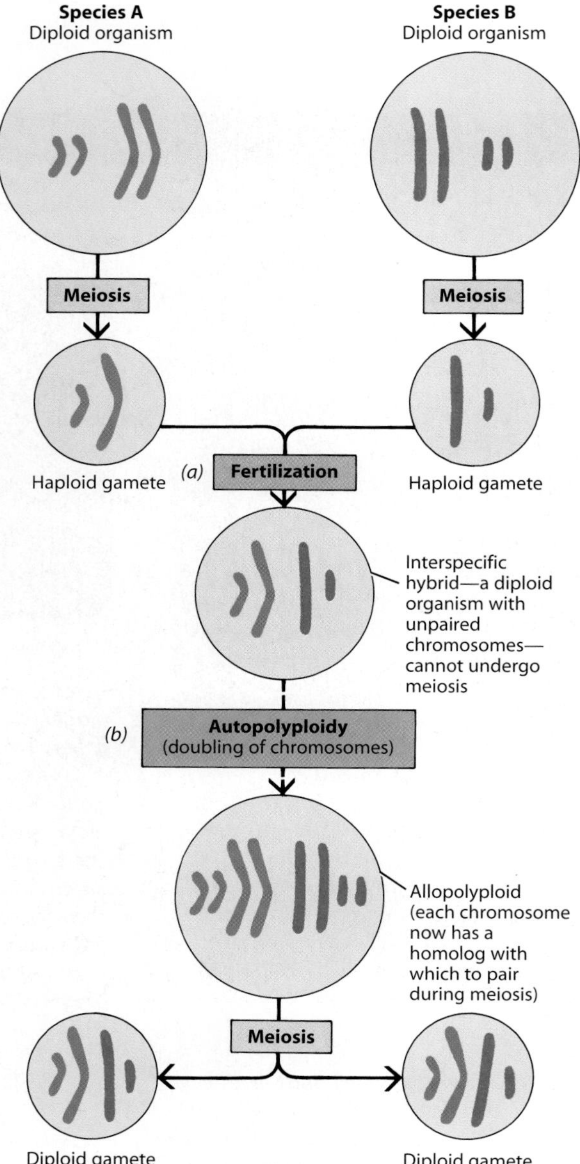

12–17
Allopolyploidy. ***(a)*** *An organism that is a hybrid between two different species—an interspecific hybrid—and is produced from two haploid (n) gametes can grow normally because mitosis is normal. It cannot reproduce sexually, however, because the chromosomes cannot pair at meiosis.* ***(b)*** *If autopolyploidy subsequently occurs and the chromosome number doubles, the chromosomes can pair at meiosis. As a result, the hybrid—an allopolyploid—can produce viable diploid (2n) gametes and is now a new species capable of reproducing sexually.*

Polyploidy leading to the formation of new species as a result of a doubling of chromosome number within individual organisms of a species is called **autopolyploidy** (Figure 12–16). Such individuals are called **autopolyploids.** Sympatric speciation by autopolyploidy was first discovered by Hugo de Vries, during his investigation of the genetics of the evening primrose *(Oenothera lamarckiana),* a diploid species with 14 chromosomes. Among his plants appeared an unusual variant, which, upon microscopic examination, was found to be tetraploid (4*n*), with 28 chromosomes. De Vries was unable to breed the tetraploid primrose with the diploid primrose, because of problems with pairing of the chromosomes during meiosis. The tetraploid (an autopolyploid) was a new species, which de Vries named *Oenothera gigas* (*gigas*, meaning "giant").

A much more common mode of polyploidy is **allopolyploidy,** which results from a cross between two different species, producing an **interspecific hybrid** (Figure 12–17). Such hybrids are usually sterile because the chromosomes cannot pair at meiosis (having no ho-

12–18

Flowering heads of Tragopogon *(goat's beard). Three highly fertile diploid* (2n = 12) *species introduced from Europe,* ***(a)*** Tragopogon dubius, ***(b)*** Tragopogon porrifolius, *and* ***(c)*** Tragopogon pratensis, *were all well established in southeastern Washington and adjacent Idaho by 1930. These species hybridized easily, resulting in highly sterile diploid* (F_1) *interspecific hybrids:* ***(d)*** Tragopogon dubius × porrifolius, ***(e)*** Tragopogon porrifolius × pratensis, *and* ***(f)*** Tragopogon dubius × pratensis. *In 1949, four small populations of* Tragopogon *were discovered that were clearly different from the diploid hybrids and were immediately suspected of being two newly originated polyploid species. The suspected polyploids* ***(g), (h)*** *differed in none of their characters from the diploid hybrids* ***(d), (f)*** *except that they were much larger in every way and were obviously fertile, with flower heads containing many developing fruits. It was soon confirmed that these populations were tetraploid* (2n = 24) *species and were named* ***(g)*** Tragopogon mirus *and* ***(h)*** Tragopogon miscellus. *These tetraploid species are among the few allopolyploids whose time of origin is known with a high degree of certainty.* ***(i)*** *Note that this inflorescence of* Tragopogon porrifolius × pratensis *is a highly sterile diploid hybrid of a generation later than the* F_1 *seen here as* ***(e)***.

mologs), a necessary step for producing viable gametes. If, however, autopolyploidy then occurs in such a sterile hybrid and the resulting cells divide by future mitosis and cytokinesis, they eventually produce a new individual asexually. That individual—an allopolyploid—will have twice as many chromosomes as its parent. As a consequence, it is reproductively isolated from its parental line. However, its chromosomes—now duplicated—can pair, meiosis can occur normally, and fertility is restored. It is a new species capable of sexual reproduction.

Hybridization and sympatric speciation through polyploidy are important, well-established phenomena in plants. It is clear that they have played significant roles in the evolution of flowering plants. Recent estimates are that from 47 percent to over 70 percent of flowering plants are polyploid. Furthermore, 80 percent of the species in the grass family are estimated to be polyploid, and a large number of the major food crops are polyploid, including wheat, sugarcane, potatoes, sweet potatoes, and bananas.

Some polyploids that originated as weeds in habitats associated with the activities of human beings have been spectacularly successful. Probably the best-documented examples are two species of goat's beard, *Tragopogon mirus* and *Tragopogon miscellus*, which are products of allopolyploid speciation (Figure 12–18). Both arose during the last hundred years in the Palouse region of southeastern Washington and adjacent Idaho, following the introduction and naturalization of their Old World progenitors, *Tragopogon dubius*, *Tragopogon porrifolius*, and *Tragopogon pratensis*. All three of the Old World species are diploid ($2n = 12$), and each crosses readily with both of the others, forming F_1 hybrids that are highly sterile. In 1949, two tetraploid ($2n = 24$) hybrids were discovered that are fairly fertile and have increased substantially in the Palouse region since their discovery. Both *T. mirus* and *T. miscellus* have been reported in Arizona, and *T. miscellus* also occurs in Montana and Wyoming. The progenitors of *T. mirus* are *T. dubius* and *T. porrifolius*, and those of *T. miscellus* are *T. dubius* and *T. pratensis*. The origin of the allopolyploids of *Tragopogon* has recently been confirmed by molecular analysis of ribosomal RNA genes.

(a)

(b)

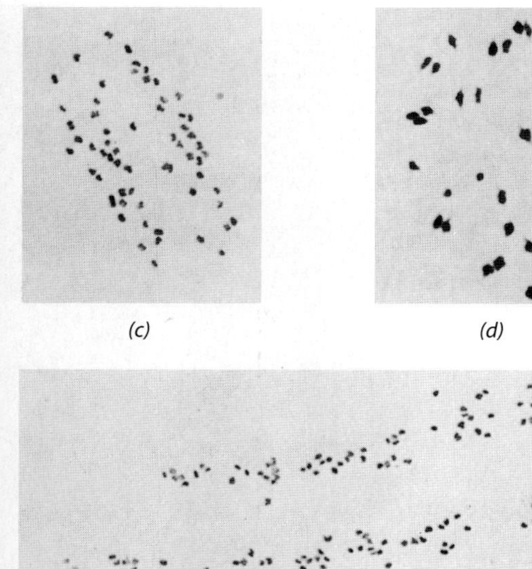

(c) (d)

(e)

12–19

Polyploidy has been investigated extensively among grasses of the genus Spartina, *which grow in salt-marsh habitats along the coasts of North America and Europe.* ***(a)*** *This salt marsh is on the coast of Great Britain.* ***(b)*** *A* Spartina *hybrid.* ***(c)*** Spartina maritima, *the native European species of salt-marsh grass, has 2*n *= 60 chromosomes, shown here from a cell in meiotic anaphase I.* ***(d)*** Spartina alterniflora *is a North American species with 2*n *= 62 chromosomes (there are 30 bivalents and 2 unpaired chromosomes), shown here from a cell in meiotic metaphase I.* ***(e)*** *A vigorous polyploid,* S. anglica, *arose spontaneously from a hybrid between these species and was first collected in the early 1890s. This polyploid, which has 2*n *= 122 chromosomes, shown here from a cell in meiotic anaphase I, is now extending its range through the salt marshes of Great Britain and other temperate countries.*

One of the best known polyploids, the origin of which is associated with human activity, is a salt-marsh grass of the genus *Spartina* (Figure 12–19). One native species, *S. maritima,* occurs in marshes along the coasts of Europe and Africa. A second species, *S. alterniflora,* was introduced into Great Britain from eastern North America in about 1800, spreading from where it was first planted and forming large but local colonies.

In Britain, the native *S. maritima* is short in stature whereas *S. alterniflora* is much taller, frequently growing to 0.5 meter and occasionally to 1 meter or even more in height. Near the harbor at Southampton, in southern England, both the native species and the introduced species existed side by side throughout the nineteenth century. In 1870, botanists discovered a sterile hybrid between these two species that reproduced vigorously by rhizomes. Of the two parental species, *S. maritima* has a somatic chromosome number of $2n = 60$ and *S. alterniflora* has $2n = 62$; the hybrid, owing perhaps to some minor meiotic misdivision, also has $2n = 62$ chromosomes. This sterile hybrid, which was named *Spartina* × *townsendii,* still persists. About 1890, a vigorous seed-producing polyploid, named *S. anglica,* was derived naturally from it. This fertile polyploid, which had a diploid chromosome number of $2n = 122$ (one chromosome pair was evidently lost), spread rapidly along the coasts of Great Britain and northwestern France. It is often planted to bind mud flats, and such use has contributed to its further spread.

One of the most important polyploid groups of plants is the genus *Triticum,* the wheats. The most commonly cultivated crop in the world, bread wheat, *T. aestivum,* has $2n = 42$ chromosomes. Bread wheat originated at least 8000 years ago, probably in central Europe, following the natural hybridization of a cultivated wheat with $2n = 28$ chromosomes and a wild grass of the same genus with $2n = 14$ chromosomes. The wild grass probably occurred spontaneously as a weed in the fields where wheat was being cultivated. The hybridization that gave rise to bread wheat probably occurred between polyploids that arose from time to time within the populations of the two ancestral species.

It is likely that the desirable characteristics of the new, fertile, 42-chromosome wheat were easily recognized, and it was selected for cultivation by the early farmers of Europe when it appeared in their fields. One of its parents, the 28-chromosome cultivated wheat, had itself originated following hybridization between two wild 14-chromosome species in the Near East. Species of wheat with $2n = 28$ chromosomes are still cultivated, along with their 42-chromosome derivative. Such 28-chromosome wheats are the chief grains used in macaroni products because of the desirable agglutinating properties of their proteins.

12–20
Sympatric speciation. Helianthus anomalus *is the result of a cross between two other distinct species of sunflower,* Helianthus annuus *and* Helianthus petiolaris. *Individually, the three species are easily distinguished from one another.* H. anomalus, *for example, has few petals, which are wider than those in* H. petiolaris *and* H. annuus, *and has small leaves with short petioles (leaf stalks).* H. petiolaris *is named for its long, thin petioles, and* H. annuus *has large leaves with thick petioles.*

An example of sympatric speciation not involving polyploidy is provided by the anomalous sunflower *(Helianthus anomalus),* the product of interbreeding between two other distinct species of sunflower, the common sunflower *(Helianthus annuus)* and the petioled sunflower *(Helianthus petiolaris)* (Figure 12–20). All three species occur widely in the western United States. Molecular evidence indicates that *H. anomalus* arose by **recombination speciation,** a process in which two distinct species hybridize, the mixed genome of the hybrid becoming a third species that is genetically (reproductively) isolated from its ancestors.

First-generation hybrids of *H. annuus* and *H. petiolaris* are semisterile, a condition apparently due to unfavorable interactions between the genomes of the parental species that result in difficulties during meiosis in the hybrid. Over several generations, however, full fertility is achieved in the hybrids as the gene combinations become rearranged. Individuals with the newly arranged genomes are compatible with one another—that is, *H. anomalus* with *H. anomalus*—but are incompatible with *H. annuus* and *H. petiolaris.*

In a recent study undertaken by Loren H. Rieseberg and coworkers, the genomic composition of three experimentally produced hybrid lineages involving crosses between *H. annuus* and *H. petiolaris* was compared with that of the naturally occurring *H. anomalus.* Surprisingly, by the fifth generation, the genomic composition of all three lineages was remarkably similar to that of the naturally occurring *H. anomalus.* In addition, in all three experimentally produced hybrid lineages, the fertility was uniformly high (greater than 90 percent). It has been hypothesized that certain combinations of genes from *H. annuus* and *H. petiolaris* consistently work better together and, hence, are always found together in surviving hybrids. This study has been characterized as "a first-ever re-creation of a new species."

Sterile Hybrids May Become Widespread If They Are Able to Reproduce Asexually

Even if hybrids are sterile, as in the horsetail hybrid *Equisetum* × *ferrissii,* they may become widespread providing they are able to reproduce asexually (Figure 12–21). In some groups of plants, sexual reproduction is combined with frequent asexual reproduction, so that recombination occurs but successful genotypes can be multiplied exactly (see Figure 9–12).

An outstanding example of such a system is the extremely variable Kentucky bluegrass *(Poa pratensis),* which in one form or another occurs throughout the cooler portions of the Northern Hemisphere. Occasional hybridization with a whole series of related species has produced hundreds of distinct races of this grass, each characterized by a form of asexual reproduction called **apomixis,** in which seeds are formed but they contain embryos that are produced independent of fertilization. Consequently, the embryos are genetically identical to

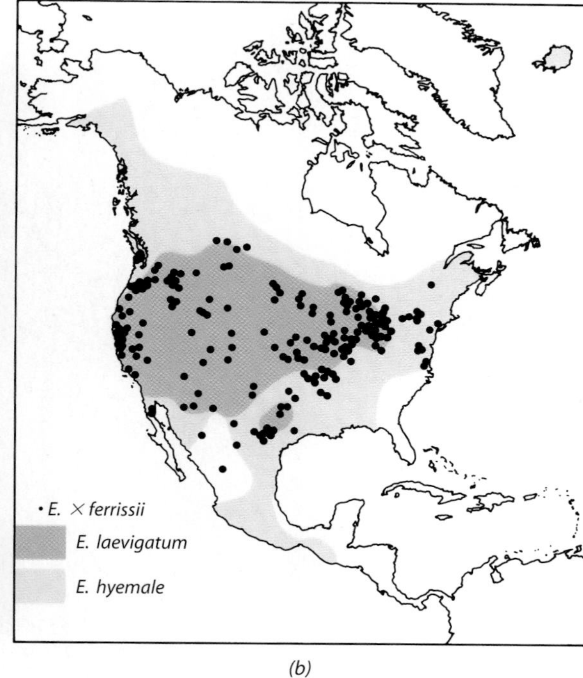

12–21
One of the most abundant and vigorous of the horsetails (see Figure 19–25) found in North America is **(a)** Equisetum × ferrissii, *a completely sterile hybrid of* Equisetum hyemale *and* Equisetum laevigatum. *Horsetails propagate readily from small fragments of underground stems, and the hybrid maintains itself over its wide range through such vegetative propagation.* **(b)** *Range of* E. × ferrissii *and those of its parental species.*

the parent. Apomixis occurs in the ovule or immature seed, with the apomictic embryo being formed by one of two different pathways, depending upon the species. Of the angiosperms, over 300 species from more than 35 families have been described as apomictic. Among them are the *Poaceae* (grasses), *Asteraceae* (composites), and *Rosaceae* (rose family).

In apomictic species, or in species with well-developed vegetative reproduction, the individual strains may be particularly successful in specific habitats. In addition, such asexually propagated strains do not require **outcrossing** (cross-pollination between individuals of the same species) and therefore often do well in environments such as high mountains, where pollination by insects may be uncertain.

Maintaining Reproductive Isolation

Once speciation has occurred, the now-separate species can live together without interbreeding, despite the fact that some are so similar phenotypically that only an expert can tell them apart. What factors operate to maintain this **reproductive,** or **genetic, isolation** of closely related species?

Isolating mechanisms may be conveniently divided into two categories: **prezygotic mechanisms** (premating mechanisms), which prevent the formation of hybrid zygotes, and **postzygotic mechanisms** (postmating mechanisms), which prevent or limit gene exchange even after hybrid zygotes have been formed. One of the most significant things about postzygotic isolating mechanisms is that, in nature, they are rarely tested. The prezygotic mechanisms alone usually prevent any hybridization.

One type of prezygotic mechanism in plants is microhabitat isolation. For example, both scarlet oak *(Quercus coccinea)* and black oak *(Quercus velutina)* occur throughout the eastern United States (Figure 12–22), and the two species, which are wind-pollinated, form fertile hybrids readily in cultivation. Nevertheless, hybrids between these two species are rare in nature. In general, scarlet oaks are found in relatively moist, low areas with acidic soil, whereas black oaks are found in drier, well-drained habitats. Although these two species occur in the same general area, their separation in these two different habitats virtually precludes cross-fertilization. Only where the environment has been disturbed—as by burning or cutting of the trees—do hybrids become more common.

Among other prezygotic mechanisms that can prevent the formation of hybrids between plant species that occur together are seasonal differences in time of flowering. If two species do not flower together, they will not hybridize in nature even when they grow side by side. Alternatively, they may be pollinated by different kinds of insects or other animals, and pollen may rarely be transferred between them. They will hybridize only when a pollinator mistakenly visits the "wrong" flower.

The Origin of Major Groups of Organisms

As knowledge about the ways in which species may originate has become better developed, evolutionary biologists have turned their attention to the origin of genera and higher taxonomic groups of organisms. As suggested by the essay on adaptive radiation (pages 250–251), genera may originate through the same kinds of evolutionary processes that are responsible for the origin of species. If a particular species has a distinctive adaptation to a habitat that is very different from the habitat in which its progenitor grows, the adapted species may come to be very different from that progeni-

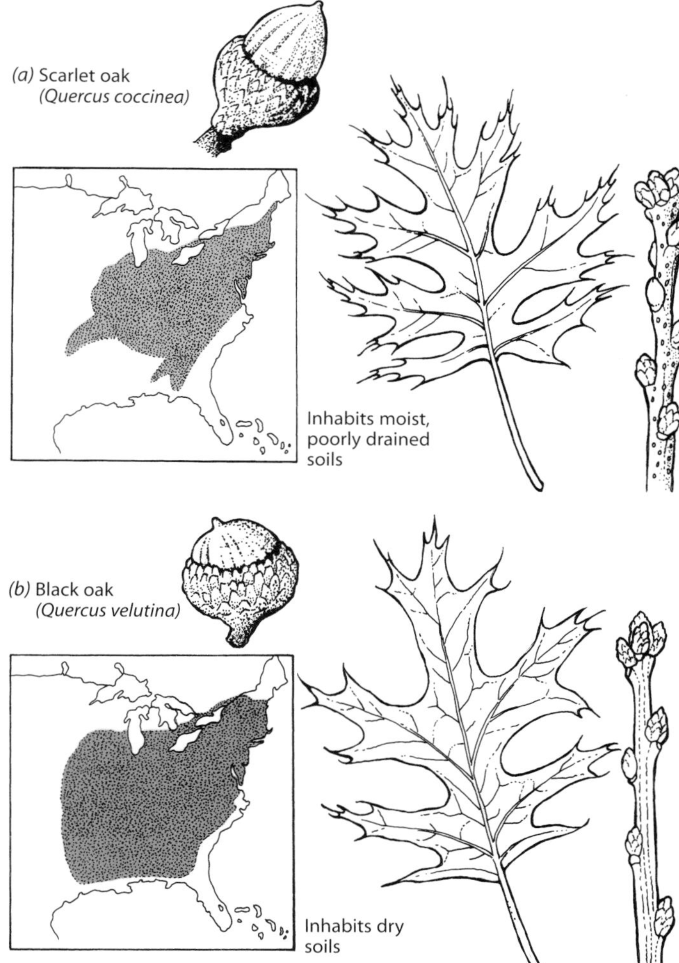

12–22

Microhabitat isolation, a prezygotic mechanism that prevents inbreeding. Although these two species of oaks occur throughout the eastern United States, they are isolated reproductively, largely by their specific habitat requirements. ***(a)*** *Scarlet oak* (Quercus coccinea) *inhabits relatively moist, poorly drained acidic soils, and* ***(b)*** *black oak* (Quercus velutina) *inhabits drier, well-drained soils.*

tor. It may gradually give rise to many new species and come to constitute a novel evolutionary line, a process called phyletic change. Gradually, it may become so distinct that this line may be classified as a new genus, family, or even class of organisms. (The levels of classification, or taxonomic groups, are discussed in Chapter 13.) Accordingly, no special mechanisms would be necessary to account for the origin of taxonomic groups above the level of species, that is, for **macroevolution**—only discontinuity in habitat, range, or way of life, and the accumulation of many small changes in the frequencies of alleles in gene pools. This is the **gradualism model** of evolution.

Although the fossil record documents many important stages in evolutionary history, there are numerous gaps, and gradual transitions of fossil forms are rarely found. Instead, new forms representing new species appear rather suddenly (in geologic terms) in strata, apparently persist unchanged for their tenure on Earth, and then disappear from the rocks as suddenly as they appeared. For many years this discrepancy between the model of slow phyletic change and the poor documentation of such change in much of the fossil record was attributed to the imperfection of the record itself. Darwin, in *On the Origin of Species*, noted that the geologic record is ". . . a history of the world imperfectly kept, and written in a changing dialect; of this history we possess the last volume alone, relating only to two or three countries. Of this volume, only here and there a short chapter has been preserved; and of each page, only here and there a few lines."

In 1972, two young scientists, Niles Eldredge and Stephen Jay Gould, ventured the proposal that perhaps the fossil record is not so imperfect after all. Both Eldredge and Gould had backgrounds in geology and invertebrate paleontology, and both were impressed with the fact that there was very little evidence for a gradual phyletic change in the fossil species they studied.

Typically a species would appear "abruptly" in fossil-bearing strata, last 5 million to 10 million years, and disappear, apparently not much different from when it first appeared. Another species, related but distinctly different, would take its place, persist with little change, and disappear "abruptly." Suppose, Eldredge and Gould argued, that these long periods of little or no change, followed by what appear to be gaps in the fossil record, are not flaws in the record but *are* the record, the evidence of what really happens.

Eldredge and Gould proposed that species undergo most of their morphological modification as they first diverge from their progenitors, and then change little even as they give rise to additional species. In other words, long periods of gradual change or of no change at all (periods of equilibrium) are punctuated by periods of rapid change, that is, of rapid speciation. This theory is called the **punctuated equilibrium model** of evolution.

How could new species make such "sudden" appearances? Eldredge and Gould turned to allopatric speciation for the answer. If new species formed principally in small populations isolated from parent populations and occurred rapidly (in thousands rather than millions of years), and if the new species then outcompeted the old species, taking over their geographic range, the resulting fossil pattern would be the one observed.

The punctuated equilibrium model has stimulated a vigorous and continuing debate among biologists, a reexamination of evolutionary mechanisms as currently understood, and a reappraisal of the evidence. Perhaps populations change more rapidly at some times than at others, particularly in periods of environmental stress.

Regularly, new studies are published in support of gradualism or of punctuated equilibrium, but a consensus on either model has not been achieved. These activities have at times been misinterpreted as a sign that Darwin's theory is "in trouble." In fact, they indicate that evolutionary biology is alive and well and that scientists are doing what they are supposed to be doing—asking questions. Darwin, we think, would have been delighted.

Summary

Darwin Proposed a Theory of Evolution by Natural Selection

Charles Darwin was not the first to propose a theory of evolution, but his theory differed from others in that it envisioned evolution as a two-part process, depending upon (1) the existence in nature of inheritable variations among organisms and (2) the process of natural selection by which some organisms, by virtue of their inheritable variations, leave more surviving progeny than others. Darwin's theory is regarded as the greatest unifying principle in biology.

Population Genetics Is the Study of Gene Pools

Population genetics is a synthesis of the Darwinian theory of evolution with the principles of Mendelian genetics. For the population geneticist, a population is an interbreeding group of organisms, defined and united by its gene pool (the sum of all the alleles of all the genes of all the individuals in the population). Evolution is the result of accumulated changes in the composition of the gene pool.

The Hardy–Weinberg Law States That in an Ideal Population, the Frequency of Alleles Will Not Change over Time

The Hardy–Weinberg law describes the steady state in allele and genotype frequencies that would exist in an ideal, nonevolving population, in which five conditions are met: (1) no mutation, (2) isolation from other populations, (3) large population size, (4) random mating, and (5) no natural selection. The Hardy–Weinberg equilibrium demonstrates that the genetic recombination that results from meiosis and fertilization cannot, in itself, change the frequencies of alleles in the gene pool. The mathematical expression of the Hardy–Weinberg equilibrium provides a quantitative method for determining the extent and direction of change in allele and genotype frequencies.

Five Agents Cause Gene Frequencies in a Gene Pool to Change

The principal agent of change in the composition of the gene pool is natural selection. Other agents of change include mutation, gene flow, genetic drift, and nonrandom mating. Mutations provide the raw material for change, but mutation rates are usually so low that mutations, in themselves, do not determine the direction of evolutionary change. Gene flow, the movement of alleles into or out of the gene pool, may introduce new alleles or alter the proportions of alleles already present. It often has the effect of counteracting natural selection. Genetic drift is the phenomenon in which certain alleles increase or decrease in frequency, and sometimes even disappear, as a result of chance events. Circumstances that can lead to genetic drift, which is most likely to occur in small populations, include the founder effect and the bottleneck effect. Nonrandom mating causes changes in the proportions of genotypes but may or may not affect allele frequencies.

Various Processes Preserve and Promote Variability

Sexual reproduction is the most important factor promoting genetic variability in populations. Mechanisms that favor outbreeding further promote variability, including self-sterility alleles in plants. Variability is preserved by diploidy, which shelters rare, recessive alleles from selection. In cases of heterozygote advantage, the heterozygote is selected over either homozygote, thus maintaining both the recessive and the dominant alleles in the population. One of the best examples of heterozygote advantage in plants is hybrid vigor.

Natural Selection Acts on the Phenotype, Not the Genotype

Only the phenotype is accessible to selection. Similar phenotypes can result from very different combinations of alleles. Because of epistasis and pleiotropy, single alleles cannot be selected in isolation. Selection affects the whole genotype.

The Result of Natural Selection Is the Adaptation of Populations to Their Environment

Evidence of adaptation to the physical environment can be seen in gradual variations that follow a geographic distribution (cline) and in distinct groups of phenotypes (ecotypes) of the same species occupying different habitats. Adaptation to the biological environment results from the selective forces exerted by interacting species of organisms on each other (coevolution).

Species Are Usually Defined on the Basis of Genetic Isolation

A species is commonly defined as a group of natural populations whose members can interbreed with one another but cannot (or at least usually do not) interbreed with members of other such groups. In practice, most species are generally identified purely on an assessment of their morphological, or structural, distinctness.

In order for speciation—the formation of new species—to occur, populations that formerly shared a common gene pool must be reproductively isolated from one another and subsequently subjected to different selection pressures.

Allopatric Speciation Involves the Geographic Separation of Populations, while Sympatric Speciation Occurs among Organisms Living Together

Two principal modes of speciation are recognized, allopatric ("other country") and sympatric ("same country"). Allopatric speciation occurs in geographically isolated populations. Islands are frequently sites for sudden diversification and the development of new species from a common ancestor, a pattern of speciation that is called adaptive radiation. Sympatric speciation, which does not require geographic isolation, occurs principally in plants through polyploidy, often coupled with hybridization. Hybrid populations derived from two species are common in plants, especially trees and shrubs. Even if hybrids are sterile, they may become widespread by asexual means of reproduction, including apomixis, in which seeds are formed but with embryos that are produced independent of fertilization.

The Gradualism and Punctuated Equilibrium Models Are Used to Explain Evolution of Major Groups of Organisms

Carried out over time, the same processes responsible for the evolution of species may give rise to genera and other major groups. This is the gradualism model of evolution. Paleontologists have presented evidence for an additional pattern of evolution known as punctuated equilibrium. They propose that new species are formed during bursts of rapid speciation among small isolated populations, that the new species outcompete many of the existing species (which become extinct), and that, in turn, the new species abruptly become extinct.

Selected Key Terms

adaptation p. 245
adaptive radiation p. 249
allopatric speciation p. 249
allopolyploidy p. 252
apomixis p. 255
artificial selection p. 238
autopolyploidy p. 252
bottleneck effect p. 242
cline p. 246
coevolution p. 247
Darwin's theory p. 237
developmental plasticity p. 246
ecotype p. 246
founder effect p. 242
gene flow p. 241
gene pool p. 239
genetic drift p. 241
gradualism model of evolution p. 257
Hardy–Weinberg law p. 240
hybrid vigor p. 243
macroevolution p. 257
microevolution p. 240
mutations p. 240
natural selection p. 236
polyploidy p. 249
population genetics p. 239
punctuated equilibrium model of evolution p. 257
recombination speciation p. 255
reproductive (genetic) isolation p. 256
speciation p. 248
species p. 248
sympatric speciation p. 249

Questions

1. Explain the influence of Thomas Malthus and Charles Lyell on the development of Darwin's theory of evolution.
2. How did Darwin's concept of evolution differ primarily from that of his predecessors? What was the major weakness in Darwin's theory?
3. What is meant by developmental plasticity? Why is developmental plasticity much greater in plants than in animals?
4. Distinguish between each of the following: cline/ecotype; microevolution/macroevolution; allopatric speciation/sympatric speciation; autopolyploidy/allopolyploidy.
5. Define genetic isolation. Why is it such an important factor in speciation?
6. When two distinct species of plants hybridize, their hybrid offspring usually are sterile. Why? Explain how such a sterile hybrid can give rise to a new species capable of sexual reproduction.

SECTION 4

Diversity

Yellow lady's slipper (Cypripedium calceolus), *an orchid, blooming in early summer in a Wisconsin wood.* Orchidaceae *is the largest family of flowering plants, with some 20,000 species, the majority of which grow in the tropics. As is true of most orchids, especially those of temperate climates, the lady's slipper is threatened with extinction.*

Chapter 13 Systematics: The Science of Biological Diversity

OVERVIEW

When you pause to examine a flower, shrub, or tree, you may very well wonder: "What's the name of that plant?" Such a question—arising out of a simple curiosity to identify organisms in the world around us—has intrigued people as far back as Aristotle, and no doubt before. In this chapter, you will find that the seemingly trivial process of naming an organism is, in fact, part of a highly organized system for establishing genetic relationships and identifying evolutionary trends.

Because people commonly name plants and other organisms in the language of their country, there will be almost as many names for the same organism as there are languages. For botanists and other biologists, this multitude of names represents a significant barrier to the sharing of information. Therefore, in addition to the "common names" that vary from country to country, each organism also has a "scientific name"—a two-word Latin name that identifies it precisely to the world at large.

Not only does a scientific name provide a universal "identity card" for an organism, but it also gives clues about the relationships of one organism to another. Thus, after describing the rules and rationale behind the scientific naming of organisms, we broaden the scope to discuss the different characteristics used for classifying organisms into groups. This is followed by an overview of the major groups of organisms and the hypothesized mechanism by which eukaryotic organisms evolved from prokaryotes.

CHECKPOINTS

By the time you finish reading this chapter, you should be able to answer the following questions:

1. What is the binomial system of nomenclature?
2. Why is the term "hierarchical" used to describe taxonomic categories, and what are the principal categories between the levels of species and kingdom?
3. What is cladistic analysis, and how is a cladogram constructed?
4. What evidence is there for the existence of the three major domains, or groups, of living organisms?
5. What are the four kingdoms of eukaryotes, and what are the major identifying characteristics of each?

In the previous section, we considered the mechanisms by which evolutionary change occurs. Now we turn our attention to the products of evolution, that is, to the multitude of different kinds, or species, of living organisms—estimated at over 30 million—that share our biosphere today. The scientific study of biological diversity and its evolutionary history is called **systematics.**

Taxonomy and Hierarchical Classification

An important aspect of systematics is **taxonomy**—the identifying, naming, and classifying of species. Modern biological classification began with the eighteenth-century Swedish naturalist Carl Linnaeus (Figure 13–1), whose ambition was to name and describe all of the known kinds of plants, animals, and minerals. In 1753, Linnaeus published a two-volume work entitled *Species Plantarum* ("The Kinds of Plants") in which he described each species in Latin by a sentence limited to twelve words. He regarded these descriptive Latin phrase names, or **polynomials,** as the proper names for the species, but in adding an important innovation devised earlier by Caspar Bauhin (1560–1624), Linnaeus made permanent the **binomial** ("two-term") **system** of nomenclature. In the margin of the *Species Plantarum,* next to the "proper" polynomial name of each species, he wrote a single word. This word, when combined with the first word of the polynomial—the **genus** (plural: genera)—formed a convenient "shorthand" designation for the species. For example, for catnip, which was formally named *Nepeta floribus interrupte spicatus pedunculatis* (meaning "*Nepeta* with flowers in an interrupted pedunculate spike"), Linnaeus wrote the word "cataria" (meaning "cat-associated") in the margin of the text, thus calling attention to a familiar attribute of the plant. He and his contemporaries soon began calling this species *Nepeta cataria,* and this Latin name is still used for this species today.

13–1
The eighteenth-century Swedish professor, physician, and naturalist Carl Linnaeus (1707–1778), who devised the binomial system for naming species of organisms and established the major categories that are used in the hierarchical system of biological classification. When he was 25, Linnaeus spent five months exploring Lapland for the Swedish Academy of Sciences; he is shown here wearing his Lapland collector's outfit.

The convenience of this new system was obvious, and the cumbersome polynomial names were soon replaced by binomial names. The earliest binomial name applied to a particular species has priority over other names applied to the same species later. The rules governing the application of scientific names to plants are embodied in the *International Code of Botanical Nomenclature.* Codes also exist for animals *(International Code of Zoological Nomenclature)* and microbes *(International Code of Nomenclature of Bacteria).*

The Species Name Consists of the Genus Name Plus the Specific Epithet

A species name consists of two parts. The first is the name of the genus—also called the generic name—and the second is the **specific epithet.** For catnip, the generic name is *Nepeta,* the specific epithet is *cataria,* and the species name is *Nepeta cataria.*

A generic name may be written alone when one is referring to the entire group of species making up that genus. For instance, Figure 13–2 shows three species of the violet genus, *Viola.* A specific epithet is meaningless, however, when written alone. The specific epithet *biennis,* for example, is used in conjunction with dozens of different generic names. *Artemisia biennis,* a kind of wormwood, and *Lactuca biennis,* a species of wild lettuce, are two very different members of the sunflower family, and *Oenothera biennis,* an evening primrose, belongs to a different family altogether. Because of the danger of confusing names, a specific epithet is always preceded by the name or the initial letter of the genus that includes it:

(a)

(b)

(c)

13–2
Three members of the violet genus. ***(a)*** *The common blue violet,* Viola papilionaceae, *which grows in temperate regions of eastern North America as far west as the Great Lakes.* ***(b)*** Viola tricolor, *a yellow-flowered violet.* ***(c)*** *Pansy,* Viola tricolor *var.* hortensis, *an annual, cultivated strain of a mostly perennial species that is native to western Europe. These photographs indicate the kinds of differences in flower color and size, leaf shape and margin, and other features that distinguish the species of this genus, even though there is an overall similarity between all of them. There are about 500 species of the genus* Viola; *most of them are found in temperate regions of the Northern Hemisphere.*

for example, *Oenothera biennis* or *O. biennis*. Names of genera and species are printed in italics or are underlined when written or typed.

If a species is discovered to have been placed in the wrong genus initially and must then be transferred to another genus, the specific epithet moves with it to the new genus. If there is already a species in that genus that has that particular specific epithet, however, an alternative name must be found.

Each species has a **type specimen,** usually a dried plant specimen housed in a museum or herbarium, which is designated either by the person who originally named that species or by a subsequent author if the original author failed to do so (Figure 13–3). The type specimen serves as a basis for comparison with other specimens in determining whether they are members of the same species or not.

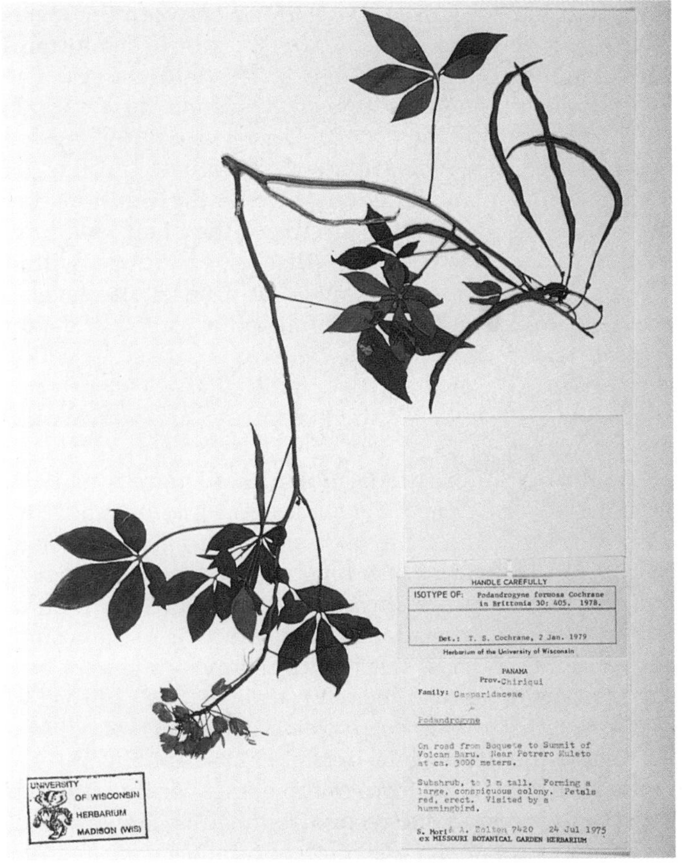

13–3
Type specimen of the angiosperm Podandrogyne formosa *(family* Capparidaceae*), found in Costa Rica and western Panama. This specimen was collected by Theodore S. Cochrane and described by him in a paper published in the journal* Britonnia *(volume 30, pages 405–410, 1978).*

The Members of a Species May Be Grouped into Subspecies or Varieties

Certain species consist of two or more subspecies or varieties (some botanists consider varieties to be subcategories of subspecies, and others regard them as equivalent). All of the members of a subspecies or variety of a given species resemble one another and share one or more features not present in other subspecies or varieties of that species. As a result of these subdivisions, although the binomial name is still the basis of classification, the names of some plants and animals may consist of three parts. The subspecies or variety that includes the type specimen of the species repeats the name of the species, and all the names are written in italics or underlined. Thus the peach tree is *Prunus persica* var. *persica,* whereas the nectarine is *Prunus persica* var. *nectarina.* The repeated *persica* in the name of the peach tree tells us that the type specimen of the species *P. persica* belongs to this variety, abbreviated "var."

Organisms Are Grouped into Broader Taxonomic Categories Arranged in a Hierarchy

Linnaeus (and earlier scientists) recognized three kingdoms—plant, animal, and mineral—and until very recently the **kingdom** was the most inclusive unit used in biological classification. In addition, several hierarchical taxonomic categories were added between the levels of genus and kingdom: genera were grouped into **families,** families into **orders,** and orders into **classes.** The Swiss-French botanist Augustin-Pyramus de Candolle (1778–1841), who invented the word "taxonomy," added another category—**division**—to designate groups of classes in the plant kingdom. Hence, the divisions became the largest inclusive groups of the plant kingdom. In this hierarchical system—that is, of groups within groups, with each group ranked at a particular level—the taxonomic group at any level is called a **taxon** (plural: taxa). The level at which it is ranked is called a **category.** For example, genus and species are categories, and *Prunus* and *Prunus persica* are taxa within those categories.

At the XV International Botanical Congress in 1993, the *International Code of Botanical Nomenclature* made the term **phylum** (plural: phyla) nomenclaturally equivalent to division. "Phylum" has long been used by zoologists for groups of classes. In addition, the Editorial Committee of the Code adopted the practice of italicizing all taxonomic names, not just the names of genera and species. We have adopted that practice and the use of the term "phylum" instead of "division" in this edition. All taxa at the genus level or higher are capitalized.

Regularities in the form of the names for the different taxa make it possible to recognize them as names at that level. For example, names of plant families end in *-aceae,* with a very few exceptions. Older names are allowed as alternatives for a few families, such as *Fabaceae,* the pea family, which may also be called by the older name, *Leguminosae; Apiaceae,* the parsley family, also known as *Umbelliferae;* and *Asteraceae,* the sunflower family, also known as *Compositae.* Names of plant orders end in *-ales.*

Sample classifications of maize *(Zea mays)* and the commonly cultivated edible mushroom *(Agaricus bisporus)* are given in Table 13–1.

Classification and Phylogeny

As mentioned previously, taxonomy is only one aspect of systematics. For Linnaeus and his immediate successors, the goal of taxonomy was the revelation of the grand, unchanging design of creation. After publication of Darwin's *On the Origin of Species* in 1859, however, differences and similarities among organisms came to be

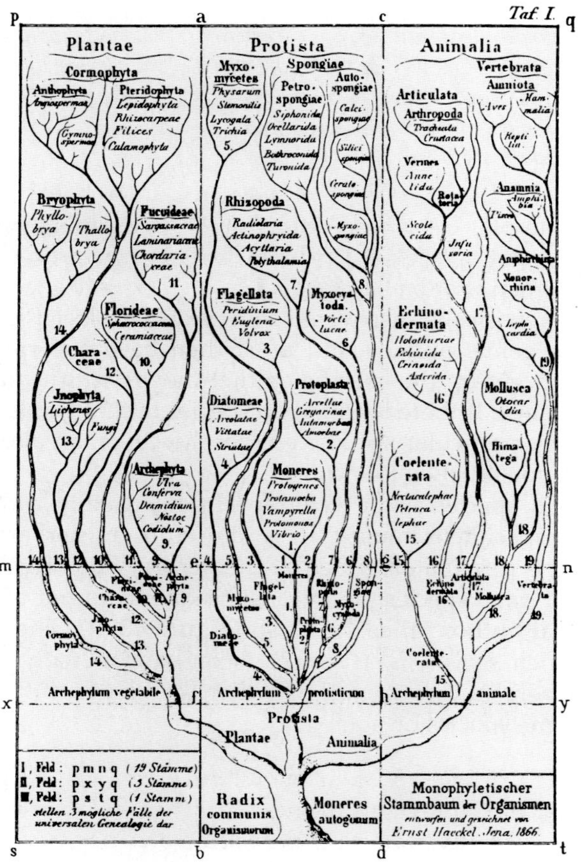

13–4

One of the earliest phylogenetic trees appeared in The History of Creation, *by the German naturalist Ernst Haeckel, published in 1866. The terminology reflects the system of classification in use at the time.*

TABLE 13–1 **Biological Classification. Notice how much you can tell about an organism when you know its place in the system. The descriptions here do not define the various categories but tell you something about their characteristics. The kingdoms *Plantae* and *Fungi* belong to the domain *Eukarya*.**

Category	Taxon	Description
Maize		
Kingdom	*Plantae*	Organisms that are primarily terrestrial, with chlorophylls *a* and *b* contained in chloroplasts, spores enclosed in sporopollenin (a tough wall substance), and nutritionally dependent multicellular embryos.
Phylum	*Anthophyta*	Vascular plants with seeds and flowers; ovules enclosed in an ovary, pollination indirect; the angiosperms.
Class	*Monocotyledones*	Embryo with one cotyledon; flower parts usually in threes; many scattered vascular bundles in the stem; the monocots.
Order	*Commelinales*	Monocots with fibrous leaves; reduction and fusion in flower parts.
Family	*Poaceae*	Hollow-stemmed monocots with reduced greenish flowers; fruit a specialized achene (caryopsis); the grasses.
Genus	*Zea*	Robust grasses with separate staminate and carpellate flower clusters; caryopsis fleshy.
Species	*Zea mays*	Maize, or corn.
Edible Mushroom		
Kingdom	*Fungi*	Nonmotile, multinucleate, heterotrophic, absorptive organisms in which chitin predominates in the cell walls.
Phylum	*Basidiomycota*	Dikaryotic fungi that form a basidium bearing four spores (basidiospores); the *Basidiomycetes, Teliomycetes,* and *Ustomycetes.*
Class	*Basidiomycetes*	Fungi that produce basidiomata, or "fruiting bodies," and club-shaped, aseptate basidia that line gills or pores; the hymenomycetes.
Order	*Agaricales*	Fleshy fungi with radiating gills or pores.
Family	*Agaricaceae*	*Agaricales* with gills.
Genus	*Agaricus*	Dark-spored soft fungi with a central stalk and gills free from the stalk.
Species	*Agaricus bisporus*	The common edible mushroom.

seen as products of their evolutionary history, or **phylogeny.** Biologists now wanted classifications to be not only informative and useful but also an accurate reflection of the evolutionary relationships among organisms. The evolutionary relationships among organisms have often been diagrammed as **phylogenetic trees,** which depict the genealogic relationships between taxa as hypothesized by a particular investigator (Figure 13–4). Like other hypotheses, certain aspects of phylogenetic trees can be tested and revised as necessary. Examples of such testing would be detailed study of the fossil record and examination of the structural and molecular characteristics of living organisms.

In a classification scheme that accurately reflects phylogeny, every taxon is, ideally, **monophyletic.** This means that the members of a taxon, at whatever categorical level, be it genus, family, or order, should all be descendants of a single common ancestral species. Thus, a genus should consist of all species descended from the most recent common ancestor—and only of species de-

Convergent Evolution

Comparable selective forces, acting on plants growing in similar habitats but different parts of the world, often cause totally unrelated species to assume a similar appearance. The process by which this happens is known as **convergent evolution.**

Let us consider some of the adaptive characteristics of plants growing in desert environments—fleshy, columnar stems (which provide the capacity for water storage), protective spines, and reduced leaves. Three fundamentally different families of flowering plants—the spurge family (Euphorbiaceae), *the cactus family* (Cactaceae), *and the milkweed family* (Asclepiadaceae)*—have members that have evolved these features. The cactuslike representatives of the spurge and milkweed families shown here evolved from leafy plants that look quite different from one another.*

Native cacti occur (with the exception of one species) exclusively in the New World. The comparably fleshy members of the spurge and milkweed families occur mainly in desert regions in Asia and especially Africa, where they play an ecological role similar to that of the New World cacti.

Although the plants shown here—(a) Euphorbia, *a member of the spurge family;* (b) Echinocereus, *a cactus;* (c) Hoodia, *a fleshy milkweed—have CAM photosynthesis, all three are related to and derived from plants that have only C_3 photosynthesis (page 147). This indicates that the physiological adaptations involved in CAM photosynthesis also arose as a result of convergent evolution.*

(a)

(b)

(c)

scended from that ancestor. Similarly, a family should consist of all genera descended from a more distant common ancestor—and only of genera descended from that ancestor.

Although this ideal, which results in **natural taxa,** sounds relatively straightforward, it is often difficult to attain. In many cases, biologists do not know enough about the evolutionary history of the organisms to establish taxa that are, with a reasonable degree of certainty, monophyletic. However, where relationships are unknown or uncertain, it may be more practical to create an **artificial taxon.** Thus, as we shall see, some widely accepted taxa contain members descended from more than one ancestral line. Such taxa are said to be **polyphyletic.** Other taxa exclude one or more descendants of a common ancestor. These taxa are said to be **paraphyletic.**

Homologous Features Have a Common Origin, and Analogous Features Have a Common Function but Different Evolutionary Origins

Systematics is, to a great extent, a comparative science. It groups organisms into taxa from the categorical levels of genus through phylum based on similarities in structure and other characters. From Aristotle on, however, biologists have recognized that superficial similarities are not useful criteria for taxonomic decisions. To take a simple example, birds and insects should not be grouped together simply because both have wings. A wingless insect (such as an ant) is still an insect, and a flightless bird (such as the kiwi) is still a bird.

A key question in systematics is the origin of a similarity or difference. Does the similarity of a particular feature reflect inheritance from a common ancestor, or does it reflect adaptation to similar environments by organisms that do not share a common ancestor? A related question arises concerning differences between organisms: Does a difference reflect separate evolutionary histories, or does it reflect instead the adaptations of closely related organisms to very different environments? As we shall see in later chapters, foliage leaves, cotyledons, bud scales, and floral parts have quite different functions and appearances, but all are modifications of the same type of organ, namely, the leaf. Such structures, which have a common origin but not necessarily a common function, are said to be **homologous** (from the Greek *homologia,* meaning "agreement"). These are the features upon which evolutionary classification systems are ideally constructed.

By contrast, other structures, which may have a similar function and superficial appearance, have an entirely different evolutionary background. Such structures are said to be **analogous** and are the result of convergent evolution (see the essay on the facing page). Thus the wings of a bird and those of an insect are analogous, not homologous. Similarly, the spine of a cactus (a modified leaf) and the thorn of a hawthorn (a modified stem) are analogous, not homologous. Distinguishing between homology and analogy is seldom so simple, and generally requires detailed comparison as well as evidence from other features of the organisms under study.

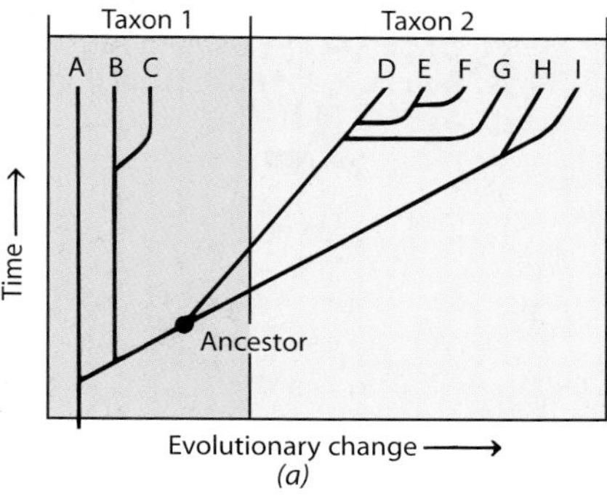

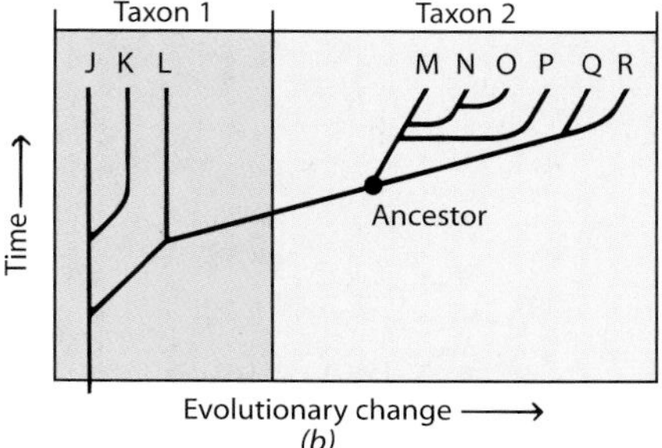

13–5
The evolutionary history of a group of related organisms can be represented by a phylogenetic tree constructed by using traditional methods. The vertical locations of the branching points indicate when particular taxa diverged from one another. The horizontal distances indicate how much the taxa have diverged, taking into account a number of different characteristics.

The two diagrams shown here represent the evolutionary histories of two different groups of taxa, labeled A through I in **(a)** *and J through R in* **(b)**. *In* **(a)**, *the ancestor of taxa D through I is included in taxon 1 because of its close resemblance to taxa B and C. In* **(b)**, *the ancestor of taxa M through R is placed in taxon 2, because of its close resemblance to taxon M. In each case, taxa 1 and 2 would themselves be members of a taxon at a higher categorical level, which would probably include other taxa as well.*

Methods of Classification

The Traditional Method Is Based on a Comparison of Outward Similarities

Traditionally, the classification of a recently discovered organism and its phylogenetic relationship to other organisms has been assessed on the basis of its overall outward similarities to other members of that taxon. Phylogenetic trees constructed by traditional methods rarely include detailed considerations of comparative information. Instead, they reflect the relatively intuitive consideration and weighing of a large number of factors. The resulting trees often contain information about both the sequence in which branchings occurred and the extent of the subsequent biological changes (Figure 13–5). Although this approach has produced many useful results, it is based to a large extent on the investigator's opinion of the relative importance of the various factors being taken into account in determining the classification. Therefore, it is not surprising that very different classifications have sometimes been proposed for the same groups of organisms.

The Cladistic Method Is Based on Phylogeny

The most widely used method of classifying organisms today is known as **cladistics,** or **phylogenetic analysis,** because it explicitly seeks to understand phylogenetic relationships. The approach focuses on the branching of one lineage from another in the course of evolution. It attempts to identify monophyletic groups, or **clades,** which can be defined by the possession of unique features (sometimes called *shared derived character states*), as opposed to the possession of more widespread features. These widespread features can be interpreted as *preexisting*, or *ancestral, character states*. The two types of character states usually are distinguished from one another by comparison with one or more **outgroups,** that is, with closely related taxa outside the group that is being analyzed.

The result of cladistic analysis is a **cladogram,** which provides a graphical representation of a working model, or hypothesis, of branching sequences. These hypotheses can then be tested by attempting to incorporate additional plants or characters that may or may not conform to the predictions of the model.

To see how a cladogram is constructed, let us consider four different groups of plants: mosses, ferns, pines, and oaks. For each of the plant groups, we have selected four homologous characters to be analyzed

TABLE 13–2 Selected Characters Used in Analyzing the Phylogenetic Relationships of Four Plant Taxa

Taxon	Characters*			
	Xylem and Phloem	Wood	Seeds	Flowers
Mosses	−	−	−	−
Ferns	+	−	−	−
Pines	+	+	+	−
Oaks	+	+	+	+

*The character state "present" (+) is the derived condition; the character state "absent" (−) is the ancestral condition.

(Table 13–2). To keep matters simple, the characters are considered to have only two different states: present (+) and absent (−).

Through their possession of embryos, mosses are known to be related to the other three plant groups, which also have embryos. However, mosses lack many features that the other three plants share—for example, xylem and phloem, and many characters not shown here. The mosses can be used as the outgroup and can be considered to have diverged earlier than the other taxa from a common ancestor. Accordingly, the mosses can be used to determine whether features shared among ferns, pines, and oaks can potentially be used to define a clade. For example, seeds are not present in mosses and can therefore be hypothesized as a potential shared derived feature that would support uniting pines and oaks as a monophyletic group. Applying this argument to our few characters results in the "absent" character state being consistently recognized as the ancestral condition and a "present" character state as the derived condition.

Figure 13–6a shows how one might sketch a cladogram based on the presence or absence of the vascular tissues xylem and phloem. Inasmuch as the ferns, pines, and oaks all have xylem and phloem, they can be hypothesized to form a monophyletic group. Figure 13–6b shows how further resolution is obtained as information about other features is added.

How does one interpret the cladogram in Figure 13–6b? To begin, note that cladograms do not indicate that one group gave rise to another, as in many phylogenetic trees constructed by the traditional method. Rather, they imply that groups terminating adjacent branches (the branch points are called **nodes**) shared a common ancestor. The cladogram of Figure 13–6b tells us that oaks shared a more recent common ancestor with pines than with ferns, and are more closely related to pines than to ferns. The relative positions of plants on the cladogram indicate their relative times of divergence.

A fundamental principle of cladistics is that a cladogram should be constructed in the simplest, least complicated, and most efficient way. This principle is called the **principle of parsimony.** When conflicting cladograms are constructed from the data at hand, the one with the greatest number of statements of homology and the fewest of analogy is preferred.

Molecular Systematics

Until relatively recently, classification by any methodology was based largely on comparative morphology and anatomy. During the past decade, however, plant systematics has been revolutionized by the application of molecular techniques. The techniques most widely used are those for determining both the sequence of amino acids in proteins and of nucleotides in nucleic acids—sequences that are genetically determined. Molecular data are different from data obtained from traditional sources in several important ways: in particular, they are easier to quantify, potentially provide many more characters for phylogenetic analysis, and allow comparison of organisms that are morphologically very different. With the development of molecular techniques, it has become possible to compare organisms at the most basic level—the gene. The drawbacks with molecular data are that they can rarely be obtained from fossils and that homologies are sometimes very difficult to assess.

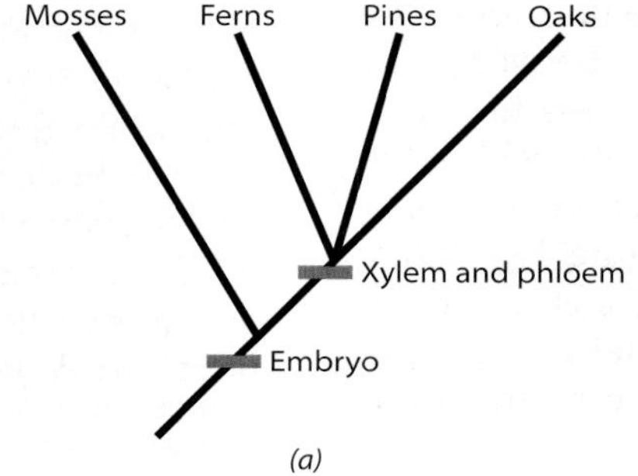

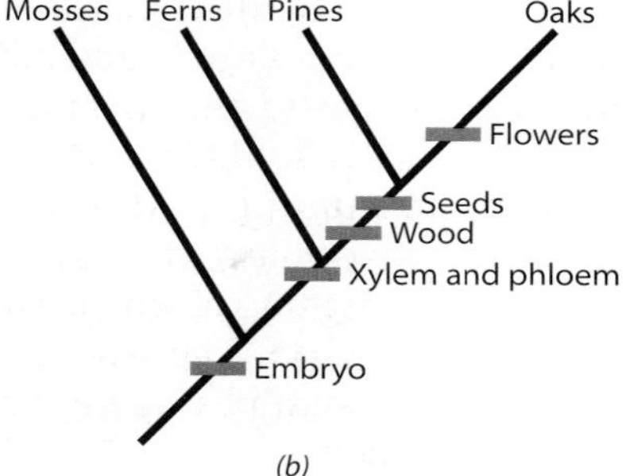

13–6 Cladograms showing phylogenetic relationships between ferns, pines, and oaks, indicating the shared characters that support the patterns of relationships. ***(a)*** *A cladogram based on the presence or absence of xylem and phloem.* ***(b)*** *Further resolution of the relationships, based on additional information regarding the presence or absence of wood, seeds, and flowers.*

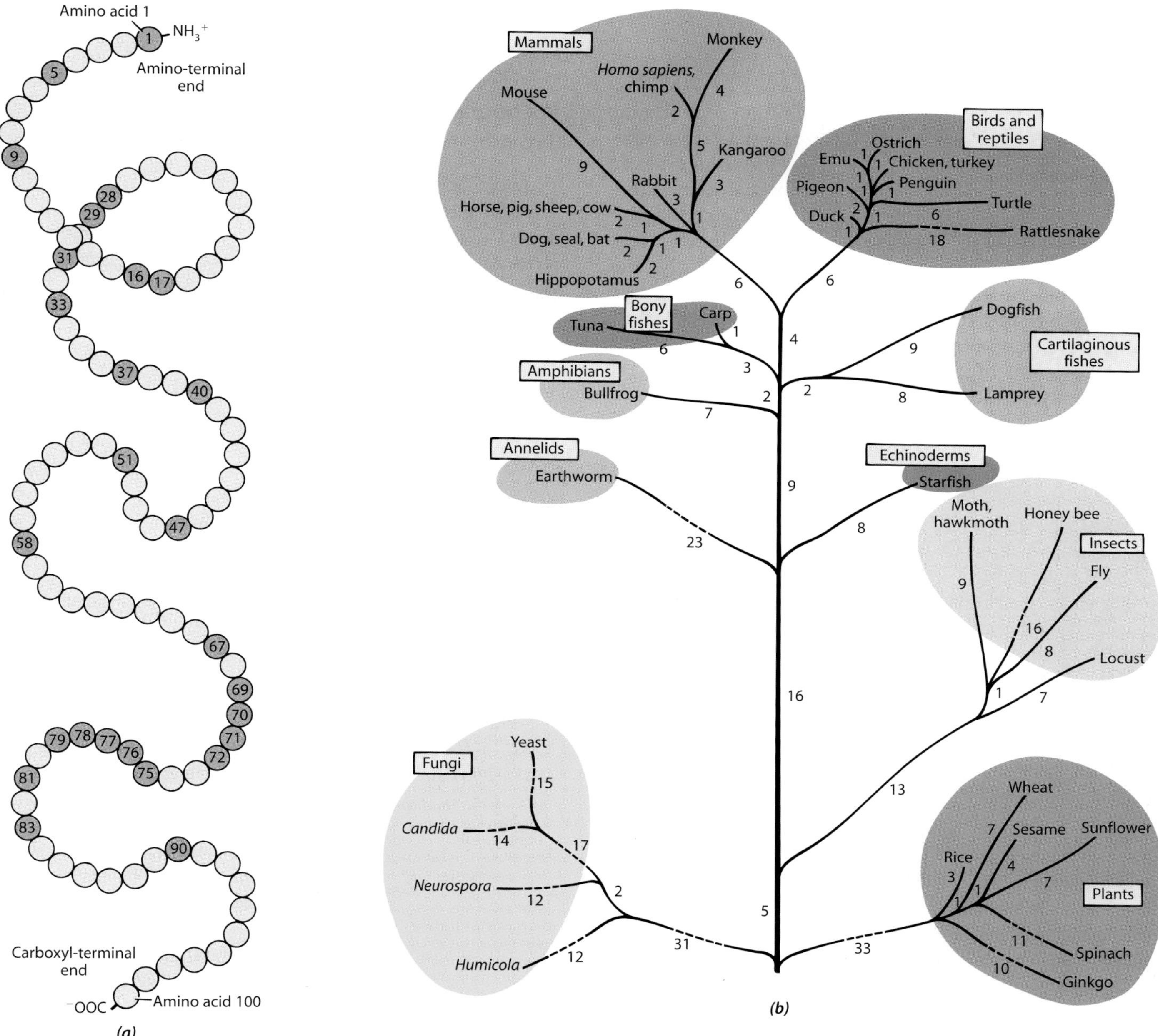

13–7

The use of amino acid sequences of homologous proteins to determine evolutionary relationships. This method assumes that the greater the number of amino acid differences between homologous proteins of any two organisms, the more distant their evolutionary relationship. Conversely, the fewer the differences, the closer their relationship.

One of the most frequently sequenced proteins is cytochrome c, a carrier in the electron transport chain. ***(a)*** *In cytochrome c molecules from the more than 60 species that have been studied thus far, 27 of the amino acids are identical (dark blue-green).* ***(b)*** *The main branches of an evolutionary tree based on comparisons of amino acid sequences of cytochrome c molecules. The numbers indicate the number of amino acids by which each cytochrome c differs from the cytochrome c at the nearest branch point. Dashes indicate that a line has been shortened and is therefore not to scale. Although based on comparisons of a single type of protein molecule, this tree is in fairly good agreement with evolutionary trees constructed by more conventional means, which take into account a variety of data.*

A Comparison of Amino Acid Sequences Provides a Molecular Clock

Among the first proteins to be analyzed in taxonomic studies was cytochrome *c*, one of the carriers of the electron transport chain (page 116). Cytochrome *c* molecules from a great variety of organisms were sequenced, making it possible to determine the number of amino acids by which the molecules of various organisms differ. The number of similarities and differences between the amino acid constituents of different organisms was then used to evaluate their evolutionary relationship: the smaller the number of differences, the closer the relationship between any two organisms. Figure 13–7 illustrates a phylogeny of eukaryotic organisms based on cytochrome *c* data. The results conform fairly well, but not perfectly, with phylogenies constructed by more traditional methods.

As data on protein variations accumulated, it became clear that although protein structure is a useful parameter of evolutionary relationships, there are difficulties in interpreting the results. Some biologists maintained that differences in protein structure resulted in functional differences among the molecules. Other biologists contended that amino acid changes occur regularly and randomly as the result of mutation—and that they do not represent the result of a selection process. These changes are seen as merely marking off the passage of time, like grains of sand trickling through an hourglass, or the decay of radioisotopes, or the ticking of a clock. From this viewpoint, the amino acid differences in the homologous proteins of different groups of organisms do not represent functional differences. Instead, they represent the differences in the number of amino acid substitutions that have occurred in the homologous proteins since the lineages branched apart. This viewpoint led to the concept of a **molecular clock,** which, simply stated, uses the rate at which proteins (or nucleic acids) shared by different groups of organisms changed over time as an indication of when those groups diverged from a common ancestor.

Although the use of homologous proteins to estimate evolutionary relationships has been largely abandoned, amino acid sequences from 57 different enzymes have recently been used to determine the divergence times of prokaryotes and eukaryotes. The results of this study indicate that prokaryotic organisms and eukaryotic organisms last shared a common ancestor about 2 billion years ago. These results conflict with an earlier assessment, based on ribosomal RNA sequences, that prokaryotes and eukaryotes evolved from a common ancestor over a relatively short time in the planet's history and were already in existence 3.5 billion years ago.

A Comparison of Nucleotide Sequences Provides Evidence for Three Domains of Life

Nucleic acid sequencing is technically far easier than protein sequencing, dealing as it does with only four different nucleotides, as compared with 20 different amino acids. Also, it assesses similarities and differences more sensitively, since changes in nucleotides may not be reflected in amino acids because of the many synonyms in the genetic code (page 209).

As the sequences of nucleic acids from a variety of species have been determined, the information has been entered into computer banks. It is therefore possible to make detailed comparisons among large numbers of taxa. Such comparisons have demonstrated the value of nucleic acid sequences in systematic studies. For example, analyzing the sequences of the small-subunit ribosomal RNA provided the first evidence that the living world is divided into three major groups, or domains—*Bacteria, Archaea,* and *Eukarya* (Figure 13–8). As discussed further in Chapter 14, the ***Bacteria*** are the prokaryotes considered to be true bacteria, and the ***Archaea*** are prokaryotes able to live in extreme environments. The ***Eukarya*** include all eukaryotes.

The first full sequencing of a genome—that is, the DNA—for a representative of the *Archaea, Methanococcus jannaschii,* has recently been completed (Figure 13–9). The results of this DNA sequencing further support the presence of three domains of life and indicate that the *Archaea* and *Eukarya* shared a common evolutionary pathway independent of the lineage of *Bacteria.* Direct sequencing of ribosomal RNA has also provided data for studying plant phylogeny and for testing hypotheses on the origin of the angiosperms (see Chapter 22).

The most comprehensive studies of seed plant phy-

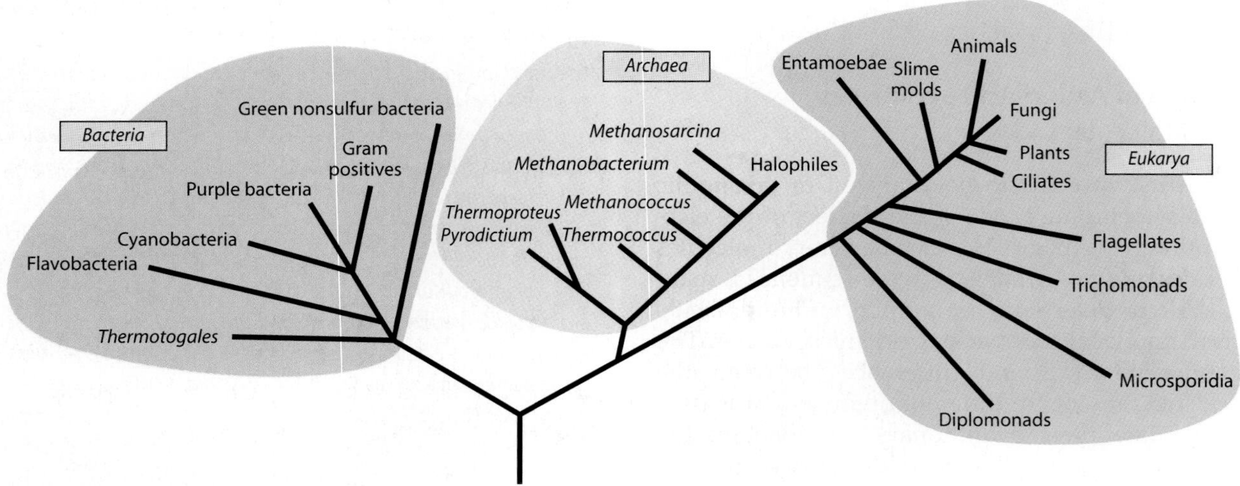

13–8
*A universal evolutionary tree as determined by comparing sequences of ribosomal RNA. The data support the division of the living world into three domains, two of which consist of prokaryotic organisms (*Bacteria *and* Archaea*) and one of eukaryotic organisms* (Eukarya)*. The universal ancestor of all cells is found at the base of the tree. Animals, fungi, and plants each constitute a separate kingdom in the domain* Eukarya*. The remaining groups in the* Eukarya *belong to the kingdom* Protista.

logeny have been based on the variation in nucleotide sequences of the *rbc*L gene found in the DNA of the chloroplast. The *rbc*L gene, which codes for the large subunit of the Rubisco enzyme of the Calvin cycle (page 140), is especially well suited for analysis of such a broad group of plants. Not only is it a slow-evolving, single-copy gene, but it lacks introns and is large enough (about 1500 base pairs) to preserve a significant number of phylogenetically informative characters.

Molecular data alone may not provide the most accurate account of phylogenetic relationships. Some systematists think that all available evidence—molecular, morphological, anatomical, ultrastructural, developmental, and fossil—should be taken into consideration in assessing phylogenetic relationships.

The Major Groups of Organisms: *Bacteria, Archaea,* and *Eukarya*

In Linnaeus's time, as we mentioned earlier, three kingdoms were recognized—animals, plants, and minerals—and until fairly recently it was common to classify every living thing as either an animal or a plant. Kingdom *Animalia* included those organisms that moved and ate things and whose bodies grew to a certain size and then stopped growing. Kingdom *Plantae* comprised all living things that did not move or eat and that grew indefinitely. Thus the fungi, algae, and bacteria, or prokaryotes, were grouped with the plants, and the protozoa—the one-celled organisms that ate and moved—were classified as animals. Lamarck, Cuvier, and most other eighteenth- and nineteenth-century biologists continued to place all organisms in one or the other of these kingdoms. This old division into plants and animals is still widely reflected in the organization of college textbooks, including this one. That is why, in addition to plants, we include algae, fungi, and prokaryotes in this text.

In the twentieth century, new data began to emerge. This was partly a result of improvements in the light microscope and, subsequently, the development of the electron microscope. It was also due, in part, to the application of biochemical techniques to studies of differences and similarities among organisms. As a result, the number of groups recognized as constituting different kingdoms increased. The new techniques revealed, for example, the fundamental differences between prokaryotic and eukaryotic cells. These differences were sufficiently great to warrant placing the prokaryotic organisms in a separate kingdom, *Monera.* Since then, as we have just noted, studies of ribosomal RNA sequences have revealed the two distinct lineages of prokaryotic organisms—*Bacteria* and *Archaea*—in addition to the distinct lineage of eukaryotic organisms—*Eukarya* (Figure 13–10).

Utilizing this molecular evidence, the system of classification adopted in this book recognizes the three **domains,** *Bacteria, Archaea,* and *Eukarya.* The protists, fungi, plants, and animals are now regarded as kingdoms within the domain *Eukarya.* Thus, the domain, not the kingdom, is the highest level of taxonomic categories. Table 13–3 summarizes some of the major differences that distinguish the three domains.

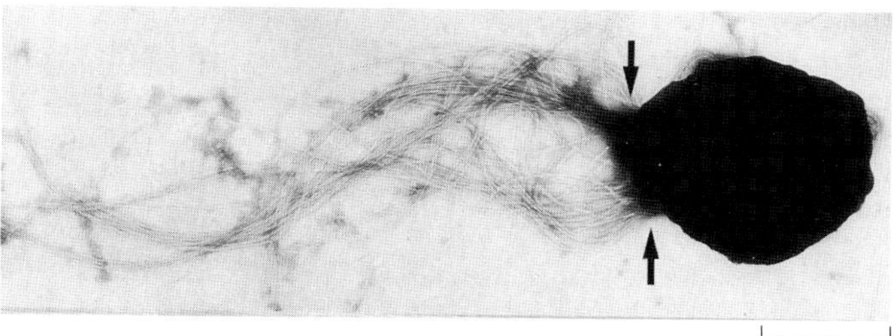

13–9
Micrograph of Methanococcus jannaschii, *the first* Archaea *whose entire genome has been fully sequenced. Its genome consists of a circular chromosome made up of DNA. Completed in 1996, the sequencing showed that two-thirds of the organism's genes consist of sequences never seen before by researchers. Arrows indicate where the cell has ruptured, spilling its contents.*

TABLE 13–3 Some Major Distinguishing Features of the Three Domains of Life*

Characteristic	*Bacteria*	*Archaea*	*Eukarya*
Cell type	Prokaryotic	Prokaryotic	Eukaryotic
Nuclear envelope	Absent	Absent	Present
Number of chromosomes	1	1	More than 1
Chromosome configuration	Circular	Circular	Linear
Organelles (mitochondria and plastids)	Absent	Absent	Present (in all but a few)
Cytoskeleton	Absent	Absent	Present
Chlorophyll-based photosynthesis	Yes	No	Yes

*Note that some features listed apply to only certain representatives of a certain domain.

13–10
Representatives of the three domains. Electron micrographs of **(a)** *a prokaryote, the cyanobacterium* Anabaena *(domain* Bacteria*);* **(b)** *a prokaryote, the archaea* Methanothermus fervidus *(domain* Archaea*); and* **(c)** *a eukaryotic cell, from the leaf of a sugarbeet* (Beta vulgaris) *(domain* Eukarya*). The cyanobacterium is a common inhabitant of ponds, while* Methanothermus, *which is adapted to high temperatures, grows optimally at 83° to 88°C. Note the greater complexity of the eukaryotic cell, with its conspicuous nucleus and chloroplasts, and its much larger size (note the scale markers).*

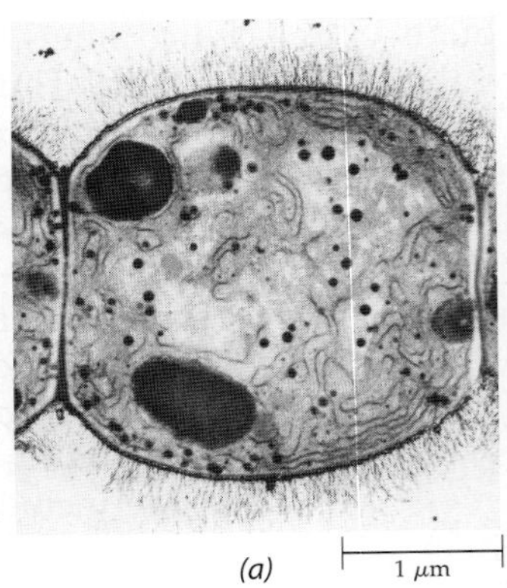

(a) 1 μm

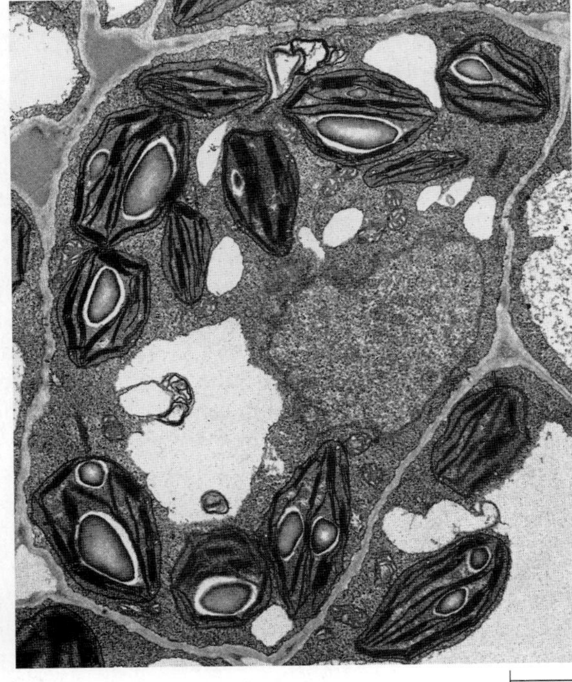

(c) 10 μm

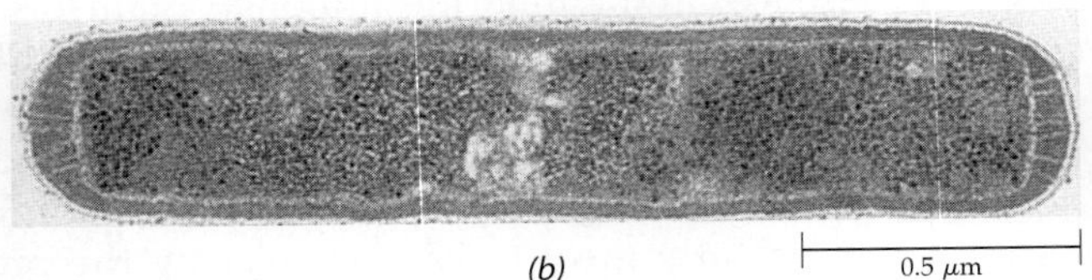

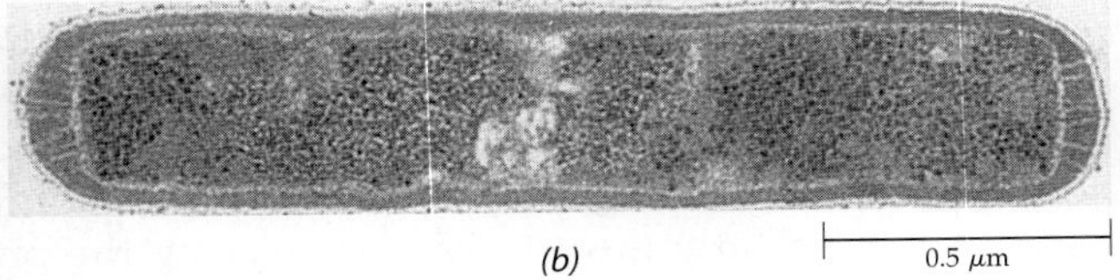

(b) 0.5 μm

Origin of the Eukaryotes

One of the most momentous series of events to take place in the evolution of life on Earth was the transformation of relatively simple prokaryotic cells into intricately organized eukaryotic cells. You will recall from Chapter 3 that eukaryotic cells are typically much larger than prokaryotes, and their DNA, which is more highly structured, is enclosed within a nuclear envelope. In addition to an internal cytoskeleton, eukaryotic cells differ further from prokaryotic cells in their possession of mitochondria and, in plants and algae, chloroplasts that are about the size of a prokaryotic cell.

The Serial Endosymbiotic Theory Provides a Hypothesis for the Origin of Mitochondria and Chloroplasts

Both mitochondria and chloroplasts are believed to be the descendants of bacteria that were taken up and adopted by some ancient **host cell.** This concept for the origin of mitochondria and chloroplasts is known as the **serial endosymbiotic theory,** with the prokaryotic ancestors of mitochondria and chloroplasts as the **endosymbionts.** An endosymbiont is an organism that lives within another, dissimilar organism. The process by which eukaryotic cells originated is termed *serial* endosymbiosis because the events did not occur simultaneously—mitochondria are believed to pre-date chloroplasts.

The Endomembrane System Is Thought to Have Evolved from Portions of the Plasma Membrane Endosymbiosis has had a profound influence upon diversification of the eukaryotes. Most experts believe that the process of establishing an endosymbiotic relationship was preceded by the evolution of some prokaryotic host cell into a primitive **phagocyte** (meaning "eating cell")—a cell capable of engulfing large particles such as bacteria (Figure 13–11). It is likely that the ancestral host cell was a wall-less heterotroph living in an environment that provided it with food. Such cells would need a flexible plasma membrane capable of enveloping bulky extracellular objects by folding inward. This endocytosis would be followed by the breakdown of the food particles within vacuoles derived from the plasma membrane. The plasma membrane would have been rendered flexible by the incorporation of sterols, and the development of a

13–11
Diagrammatic representation of the origin of a photosynthetic eukaryotic cell from a heterotrophic prokaryote. **(a)** *Most prokaryotes contain a rigid cell wall, so it is likely that an initial step in the transformation of a prokaryote to a eukaryotic cell was loss of the prokaryote's ability to form a cell wall.* **(b), (c)** *This free-living, naked form now had the*

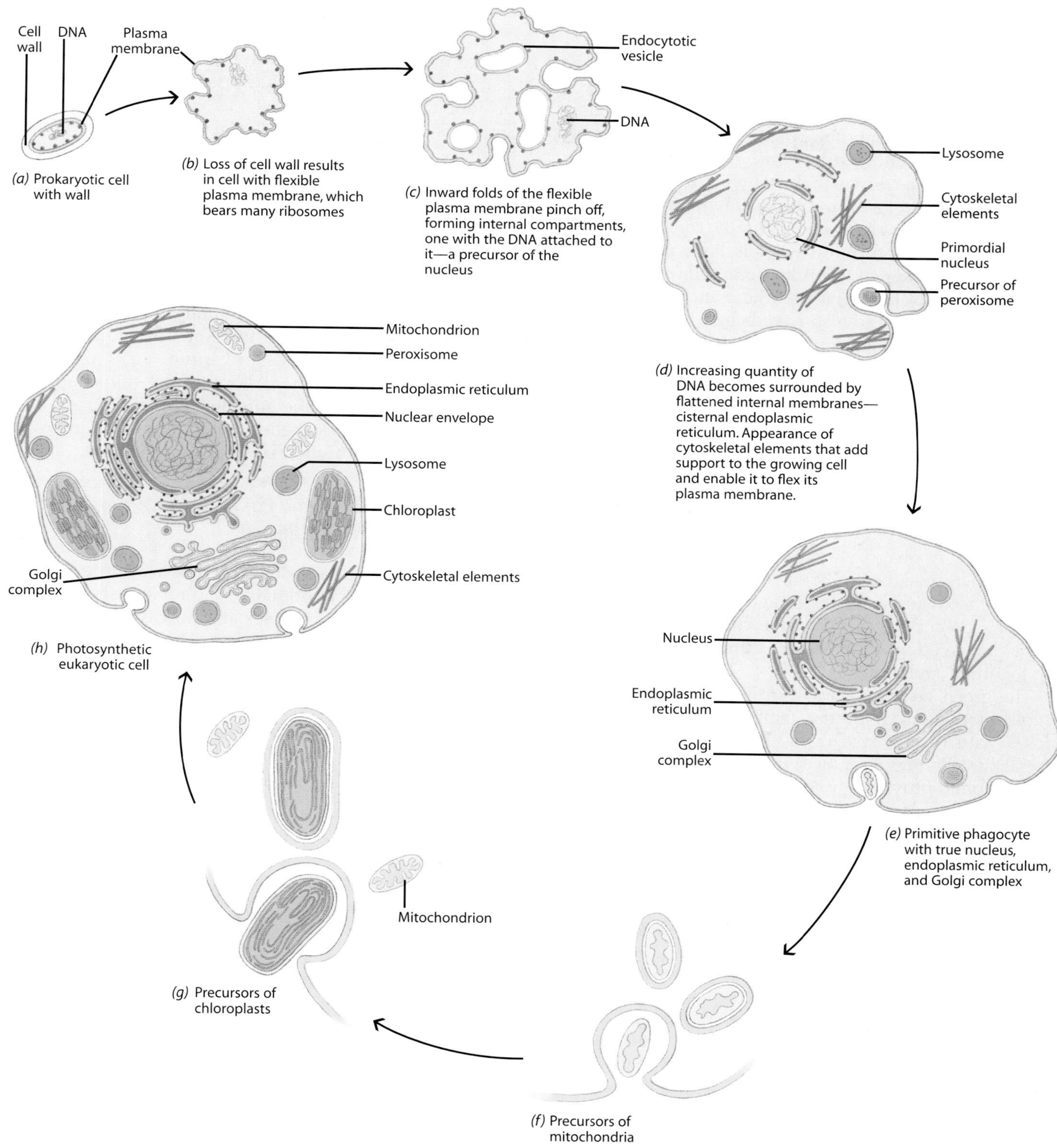

ability to increase in size, change shape, and engulf extracellular objects by infolding of the plasma membrane (endocytosis). ***(d), (e)*** *Internalization of a patch of the plasma membrane to which DNA was attached was the probable precursor of the nucleus. The primitive phagocyte eventually acquired a true nucleus containing an increased quantity of DNA. A cytoskeleton must also have developed to provide inner support for the wall-less cell and to play a role in movement, of both the cell itself and its internal components.* ***(f)*** *The mitochondria of the eukaryotic cell had their origin as bacterial endosymbionts, which ultimately transferred most of their DNA to the host's nucleus.* ***(g)*** *Chloroplasts also are believed to be the descendants of bacteria. They ultimately transferred most of their DNA to the host's nucleus.* ***(h)*** *The photosynthetic eukaryotic cell contains a complex endomembrane system and a variety of other internal structures such as the peroxisomes, mitochondria, and chloroplasts depicted here.*

cytoskeleton (especially of microtubules) would have provided the mechanism necessary for capturing food or prey and carrying it inward by endocytosis. The host cell's lysosomes (membrane-bounded vesicles that contain degradative enzymes) would fuse with the food vacuoles, breaking down their contents into usable organic compounds. The intracellular membranes derived from the plasma membrane gradually compartmentalized the host cells, forming what is known as the endomembrane system of the eukaryotic cell (page 58).

The genesis of the nucleus—the principal feature of eukaryotic cells—could also have begun by infolding of the plasma membrane. You will recall that in prokaryotes the circular DNA molecule, or chromosome, is attached to the plasma membrane. Infolding of this portion of the plasma membrane could have resulted in enclosure of the DNA within an intracellular sac, the primordial nucleus (Figure 13–11).

Mitochondria and Chloroplasts Are Thought to Have Evolved from Bacteria That Were Phagocytized The stage is set. A phagocyte now exists that can prey on bacteria. The phagocyte, however, still lacks mitochondria. The next step is for the phagocyte not to digest the bacterial precursors of the mitochondria (or chloroplasts) but to adopt them, establishing a symbiotic ("living together") relationship.

The green *Vorticella* shown in Figure 13–12 is an example of a modern protist that establishes endosymbioses with certain species of the green alga *Chlorella.* The algal cells remain intact within the host cells as endosymbionts, providing photosynthetic products useful to the heterotrophic host. In return, the algae receive essential mineral nutrients from the host. There are many examples of prokaryotic (bacterial) and eukaryotic endosymbionts in other protists, as well as in the cells of some 150 genera of freshwater and marine invertebrate animals. Algal endosymbionts, including those occurring in the polyps of reef-building corals, increase host productivity and survival (see Chapter 16).

Transformation of an endosymbiont into an organelle usually involved loss of the endosymbiont's cell wall (if any existed) and other unneeded structures. In the course of evolution, the DNA of the endosymbiont and many of its functions were gradually transferred to the host's nucleus. Hence, the genomes of modern mitochondria and chloroplasts are quite small compared with the nuclear genome. Although the mitochondrion or chloroplast cannot live outside a eukaryotic cell, both are self-replicating organelles that have retained many of the characteristics of their prokaryotic ancestors.

Relationships within the Eukaryotes Are Not Always Well Defined

In contrast to the sharp distinction between the *Bacteria, Archaea,* and *Eukarya* domains, the relationships within the *Eukarya* are more complex and the divisions between eukaryotic kingdoms much less clear-cut. For convenience, the predominantly unicellular phyla and some of the multicellular lines associated with them are grouped in the kingdom *Protista,* from which the three kingdoms consisting essentially of multicellular organisms—*Plantae, Animalia,* and *Fungi*—have evolved. While plants, animals, and fungi are probably monophyletic groups, *Protista* is paraphyletic.

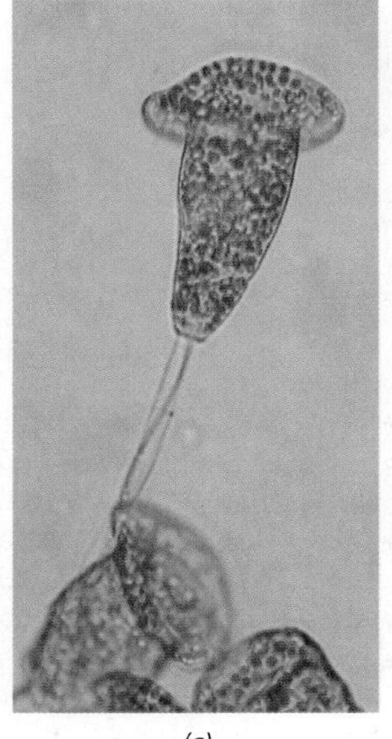

(a)

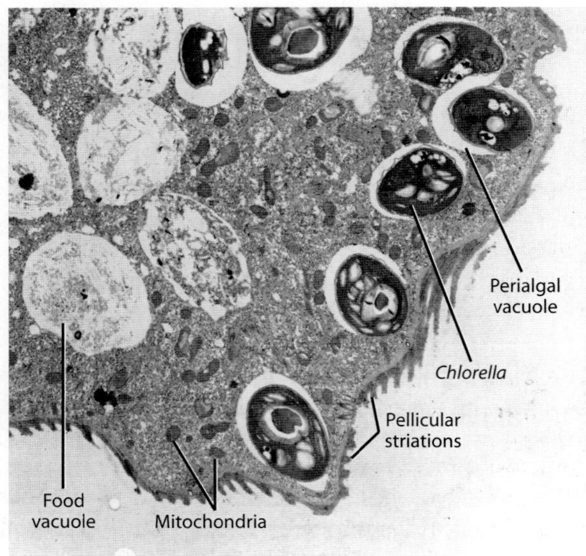

(b)

13–12
Endosymbiosis in Vorticella. ***(a)*** *Each bell-shaped cell of the protozoan* Vorticella *contains numerous cells of the autotrophic, endosymbiotic alga* Chlorella. ***(b)*** *Electron micrograph of a* Vorticella *containing cells of* Chlorella. *Each algal cell is found in a separate vacuole (perialgal vacuole) bounded by a single membrane. The protozoan provides the algae with protection and mineral nutrients, while the algae produce carbohydrates that serve as nourishment for their heterotrophic host cell.*

TABLE 13–4 Classification of Living Organisms Included in This Book. (See Appendix D for summary descriptions of these groups.)

Prokaryotic Domains		
Bacteria	Bacteria	
Archaea	Archaea	
Eukaryotic Domain		
Eukarya		
Kingdom *Fungi*	Fungi	Phylum *Chytridiomycota* (chytrids) Phylum *Zygomycota* (zygomycetes) Phylum *Ascomycota* (ascomycetes) Phylum *Basidiomycota (Basidiomycetes, Teliomycetes,* and *Ustomycetes)*
Kingdom *Protista*	Heterotrophic protists	Phylum *Myxomycota* (plasmodial slime molds) Phylum *Dictyosteliomycota* (cellular slime molds) Phylum *Oomycota* (water molds)
	Photosynthetic protists ("algae")	Phylum *Euglenophyta* (euglenoids) Phylum *Cryptophyta* (cryptomonads) Phylum *Rhodophyta* (red algae) Phylum *Dinophyta* (dinoflagellates) Phylum *Haptophyta* (haptophytes) Phylum *Bacillariophyta* (diatoms) Phylum *Chrysophyta* (chrysophytes) Phylum *Phaeophyta* (brown algae) Phylum *Chlorophyta* (green algae)
Kingdom *Plantae*	Bryophytes	Phylum *Hepatophyta* (liverworts) Phylum *Anthocerophyta* (hornworts) Phylum *Bryophyta* (mosses)
	Vascular plants Seedless vascular plants	Phylum *Psilotophyta* (psilotophytes) Phylum *Lycophyta* (lycophytes) Phylum *Sphenophyta* (horsetails) Phylum *Pterophyta* (ferns)
	Seed plants	Phylum *Cycadophyta* (cycads) Phylum *Ginkgophyta* (ginkgo) Phylum *Coniferophyta* (conifers) Phylum *Gnetophyta* (gnetophytes) Phylum *Anthophyta* (angiosperms)

The Eukaryotic Kingdoms

The following is a synopsis of the four kingdoms included in the domain *Eukarya* (see Table 13–4, which does not include the kingdom *Animalia*).

The Kingdom *Protista* Includes Unicellular, Colonial, and Simple Multicellular Eukaryotes

Protista (Figure 13–13) is here considered to comprise all organisms traditionally regarded as protozoa (one-celled "animals"), which are heterotrophic, as well as all algae, which are autotrophic. Also included in the kingdom *Protista* are some heterotrophic groups of organisms that have traditionally been placed with the fungi—including the water molds and their relatives (phylum *Oomycota*), the cellular slime molds (phylum *Dictyosteliomycota*), and the plasmodial slime molds (phylum *Myxomycota*).

The reproductive cycles of protists are varied but typically involve both cell division and sexual reproduction. Protists may be motile by 9-plus-2 flagella or cilia or by amoeboid movement, or they may be nonmotile. Red algae and some predominantly unicellular groups of protists included in this book are discussed in Chapter 16, and two major groups of algae—green algae and brown algae (and related organisms)—are discussed in Chapter 17. Green algae are clearly very closely related to plants.

(a)

(b)

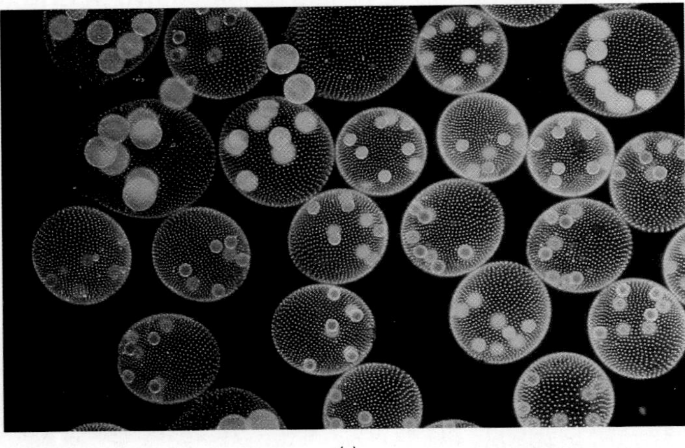

(c)

(d)

(e)

13–13
Protists. ***(a)*** *Plasmodium of a plasmodial slime mold,* Physarum *(phylum* Myxomycota*), growing on a leaf.* ***(b)*** Postelsia palmiformis, *the "sea palm" (phylum* Phaeophyta*), growing on exposed intertidal rocks off Vancouver Island, British Columbia.* ***(c)*** Volvox, *a motile colonial green alga (phylum* Chlorophyta*).* ***(d)*** *A red alga (phylum* Rhodophyta*).* ***(e)*** *A pennate diatom (phylum* Bacillariophyta*) showing the intricately marked shell characteristic of this group.*

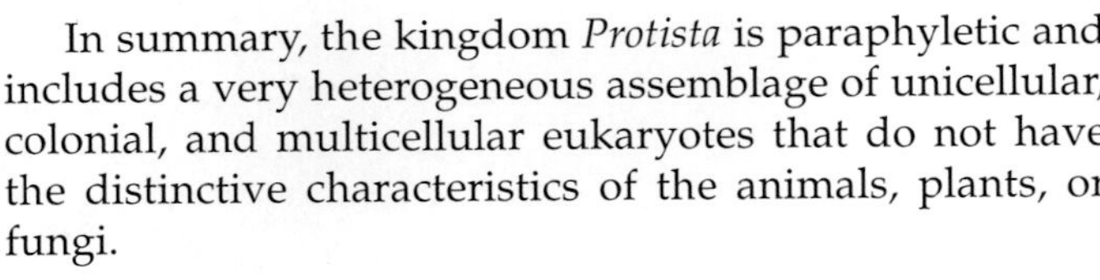

In summary, the kingdom *Protista* is paraphyletic and includes a very heterogeneous assemblage of unicellular, colonial, and multicellular eukaryotes that do not have the distinctive characteristics of the animals, plants, or fungi.

The Kingdom *Animalia* Includes Eukaryotic, Multicellular Ingesters

The animals are multicellular organisms with eukaryotic cells lacking cell walls, plastids, and photosynthetic pigments. Nutrition is primarily ingestive—food is taken in through a mouth or other opening—with digestion occurring in an internal cavity. In some forms, however, nutrition is absorptive, and a number of groups lack an internal digestive cavity. The level of organization and tissue differentiation in complex animals far exceeds that of the other kingdoms, particularly with the evolution of complex sensory and neuromotor systems. The motility of the organism (or, in sessile forms, of its component parts) is based on contractile fibrils. Reproduction is predominantly sexual. Animals are not discussed in this book, except in relation to some of their interactions with plants and other organisms.

The Kingdom *Fungi* Includes Eukaryotic, Multicellular Absorbers

Members of the kingdom *Fungi,* which are nonmotile, filamentous eukaryotes that lack plastids and photosynthetic pigments, absorb their nutrients from either dead or living organisms (Figure 13–14). The fungi have traditionally been grouped with plants, but there is no longer any doubt that the fungi are an independent evolutionary line. Moreover, comparisons of ribosomal RNA sequences indicate that the fungi are more closely related to animals than to plants. Apparently animals and fungi share a unique evolutionary history, their most recent common ancestor being a flagellated protist similar to modern choanoflagellates. Choanoflagellates are characterized by flattened mitochondrial cristae and a posterior flagellum. Aside from their filamentous growth habit, fungi have little in common with any of the protist groups that have been classified as algae. Typically, for example, the cell walls of fungi include a matrix of chitin. The structures in which fungi form their spores are often complex. Fungal reproductive cycles, which can also be quite complex, typically involve both sexual and asexual processes. Fungi are discussed in Chapter 15.

The Kingdom *Plantae* Includes Eukaryotic, Multicellular Photosynthesizers

Plants—with three phyla of bryophytes (mosses, liverworts, and hornworts) and the nine living, or extant, phyla of vascular plants—constitute a kingdom of photosynthetic organisms adapted for a life on the land (Figure 13–15). Their ancestors were specialized green algae. All plants are multicellular and are composed of

13–14
Fungi. ***(a)*** *Red blanket lichen* (Herpothallon sanguineum) *growing on a tree trunk.* ***(b)*** *A white coral fungus (family* Clavariaceae*).* ***(c)*** *Mushrooms (genus probably* Mycena*), with dew droplets, growing in a rainforest in Peru.* ***(d)*** *An earthball,* Scleroderma aurantium.

(a) *(b)* *(c)* *(d)*

(a)

(b)

(c)

(d)

(e)

13–15

Plants. ***(a)*** Sphagnum, *the peat moss (phylum* Bryophyta*), forms extensive bogs in cold and temperate regions of the world.* ***(b)*** Marchantia *is by far the most familiar of the thallose liverworts (phylum* Hepatophyta*). It is a widespread, terrestrial genus that grows on moist soil and rocks.* ***(c)*** Diphasiastrum digitatum, *a "club moss" (phylum* Lycophyta*).* ***(d)*** *Wood horsetail,* Equisetum sylvaticum *(phylum* Sphenophyta*).* ***(e)*** *Bulblet fern,* Cystopteris bulbifera *(phylum* Pterophyta*).* ***(f)*** *The dandelion,* Taraxacum officinale, *and* ***(g)*** *fishhook cactus,* Mammillaria microcarpa, *are eudicots (phylum* Anthophyta*).* ***(h)*** *Foxtail barley,* Hordeum jubatum, *and* ***(i)*** Cymbidium *orchids are monocots (phylum* Anthophyta*).* ***(j)*** *Sugar pine,* Pinus lambertiana, *and incense cedar,* Calocedrus decurrens, *are both conifers (phylum* Coniferophyta*).*

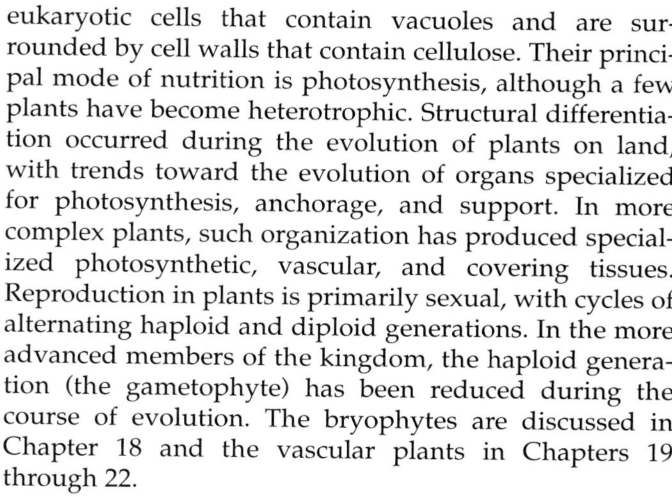

eukaryotic cells that contain vacuoles and are surrounded by cell walls that contain cellulose. Their principal mode of nutrition is photosynthesis, although a few plants have become heterotrophic. Structural differentiation occurred during the evolution of plants on land, with trends toward the evolution of organs specialized for photosynthesis, anchorage, and support. In more complex plants, such organization has produced specialized photosynthetic, vascular, and covering tissues. Reproduction in plants is primarily sexual, with cycles of alternating haploid and diploid generations. In the more advanced members of the kingdom, the haploid generation (the gametophyte) has been reduced during the course of evolution. The bryophytes are discussed in Chapter 18 and the vascular plants in Chapters 19 through 22.

(f)

(g)

Summary

Systematics, the scientific study of biological diversity, encompasses taxonomy—the identifying, naming, and classifying of species—and phylogenetics—discerning the evolutionary interrelationships among organisms.

Organisms Are Named with a Binomial and Grouped into Taxonomic Categories Arranged in a Hierarchy

Organisms are designated scientifically by a name that consists of two words—a binomial. The first word in the binomial is the name of the genus, and the second word,

(h)

(i)

(j)

the specific epithet, combined with the name of the genus, completes the name of the species. Species are sometimes subdivided into subspecies or varieties. Genera are grouped into families, families into orders, orders into classes, classes into phyla, and phyla into kingdoms. Based on recent taxonomic studies, kingdoms are further grouped into domains.

Organisms Are Classified Phylogenetically Using Homologous, Rather than Analogous, Features

In classifying organisms into the categories of genus through domain, systematists seek to group the organisms in ways that reflect their phylogeny (evolutionary history). In a phylogenetic system, every taxon should be, ideally, monophyletic—that is, every taxon should consist only of organisms descended from a common ancestor. A major principle of such classification is that the similarities should be homologous—that is, the result of common ancestry—rather than the result of convergent evolution.

The frequently intuitive traditional methods of classifying organisms have been largely replaced by more explicit cladistic methods, which attempt to understand branching sequences (genealogy) based on the possession of shared derived character traits.

A Comparison of the Molecular Composition of Organisms Can Be Used to Predict Their Evolutionary Relationships

New techniques in molecular systematics are providing a relatively objective and explicit method of comparing organisms at the most basic level of all, the gene. A large body of evidence indicates that under certain circumstances protein and nucleic acid molecules are molecular clocks, with changes in composition reflecting the time that has elapsed since different groups of organisms diverged from one another. Both amino acid and nucleotide sequencing are making valuable contributions to more accurate classification schemes that reflect an improved understanding of biological diversity and its evolutionary history.

Organisms Are Classified into Two Prokaryotic Domains and One Eukaryotic Domain, Which Consists of Four Kingdoms

In this text—based largely on data obtained from the sequencing of the small-subunit ribosomal RNA—living organisms are grouped into three domains, *Bacteria, Archaea,* and *Eukarya*. The *Bacteria* and *Archaea* are two distinct lineages of prokaryotic organisms. The *Archaea* are more closely related to the *Eukarya* than they are to the *Bacteria*. *Eukarya* consist entirely of eukaryotes. The *Protista, Fungi, Plantae,* and *Animalia* are kingdoms within the *Eukarya*.

Selected Key Terms

analogy p. 267
Archaea p. 270
artificial taxon p. 266
Bacteria p. 270
binomial system p. 262
category p. 264
clades p. 267
cladistics p. 267
cladogram p. 267
convergent evolution p. 266
domains p. 271
endosymbiosis p. 272
Eukarya p. 270
genus p. 262
homology p. 266
molecular clock p. 270
molecular systematics p. 268
monophyletic p. 265
outgroup p. 267
paraphyletic p. 266
phylogenetic tree p. 265
phylogeny p. 265
polyphyletic p. 266
principle of parsimony p. 268
serial endosymbiotic theory p. 272
specific epithet p. 262
systematics p. 262
taxon p. 264
taxonomy p. 262
type specimen p. 263

Questions

1. Distinguish between or among the following: category/taxon; monophyletic/polyphyletic/paraphyletic; host/endosymbiont.
2. Identify which of the following are categories and which are taxa: undergraduates; the faculty of The Pennsylvania State University; the Green Bay Packers; major league baseball teams; the U.S. Marine Corps; the family Robinson.
3. A key question in systematics is the origin of a similarity or difference. Explain.
4. How does a cladogram differ from a phylogenetic tree constructed by the traditional method?
5. It is generally thought that similarities and differences in homologous nucleotide sequences provide a more sensitive indicator of the evolutionary distance between organisms than similarities and differences in the amino acid sequences of homologous proteins. Explain why this should be so.
6. Explain the role of endosymbiosis in the origin of eukaryotic cells.

Chapter 14

Prokaryotes and Viruses

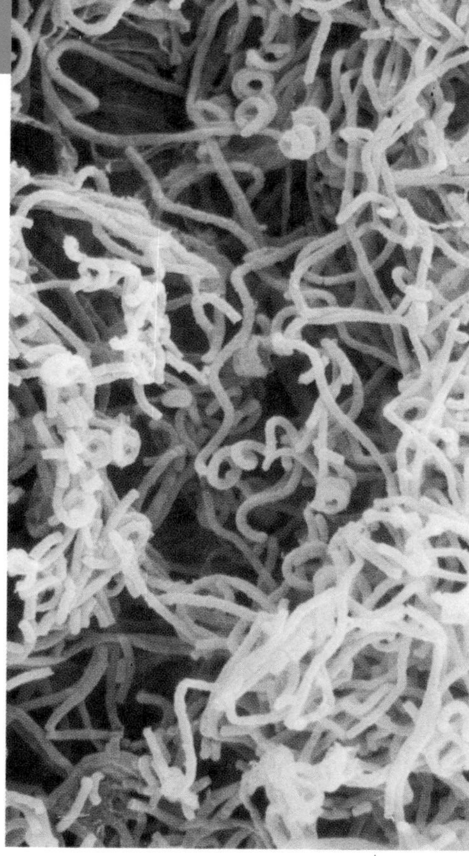

14–1
A common prokaryote, the filamentous actinomycete known as Streptomyces scabies. *Actinomycetes are abundant in soil, where they are largely responsible for the "moldy" odor of damp soil and decaying material.* Streptomyces scabies *is the bacterium that causes potato scab disease.*

OVERVIEW

The most abundant organisms on Earth—that is, the prokaryotes, which make up the domains Bacteria *and* Archaea*—are usually invisible to the naked eye. But small does not mean insignificant. Prokaryotes, for example, brought oxygen to the atmosphere of the early Earth and are responsible for decomposing dead organisms and recycling their chemical elements. In addition, prokaryotes supply commercially important products such as antibiotics, and they provide scientists with model systems for studying the ways in which organisms survive and even thrive in hostile environments. On the other hand, certain prokaryotes attain notoriety because of the problems they cause, most notably diseases of plants and animals.*

Viruses—other microscopic agents of disease discussed in this chapter—do not belong to any domain or kingdom and thus are something of a paradox. By themselves, viruses are metabolically inert entities. Yet, once inside a living cell, a virus can reproduce by taking over the cell's genetic machinery.

CHECKPOINTS

By the time you finish reading this chapter, you should be able to answer the following questions:

1. What is the basic structure of a prokaryotic cell?
2. How do prokaryotes reproduce, and in what ways does genetic recombination take place in prokaryotes?
3. In what ways are the cyanobacteria ecologically important?
4. Metabolically, what are the principal differences between the cyanobacteria and the purple and green bacteria?
5. How do the mycoplasmas differ from all other *Bacteria*?
6. Physiologically, what are the three large groups of *Archaea*?
7. What is the basic structure of a virus? How do viruses reproduce?

Of all organisms, the prokaryotes are the simplest structurally, smallest physically, and the most abundant worldwide. Even though individuals are microscopically small, the total weight of prokaryotes in the world is estimated to exceed that of all other living organisms combined. In the sea, for example, prokaryotes make up an estimated 90 percent or more of the total weight of living organisms. In a single gram (about $\frac{1}{28}$ of an ounce) of fertile agricultural soil, there may be 2.5 billion individuals. About 2700 species of prokaryotes are currently recognized, but many more await discovery.

The prokaryotes are, in evolutionary terms, the oldest organisms on Earth. The oldest known fossils are chainlike prokaryotes found in rocks from Western Australia, dated at about 3.5 billion years old (see Figure 1–2). Although some present-day prokaryotes phenotypically resemble these ancient organisms, none of the prokaryotes living today are primitive. Rather, they are modern organisms that have succeeded in adapting to their particular environments.

The prokaryotes are, in fact, the most dominant and successful forms of life on Earth. Their success is undoubtedly due to their great metabolic diversity and their rapid rate of cell division. Growing under optimal conditions, a population of *Escherichia coli*, probably the best-known prokaryote, can double in size and divide every 20 minutes. Prokaryotes can survive in many environments that support no other form of life. They occur in the icy wastes of Antarctica, in the dark depths of the ocean, in the near-boiling waters of natural hot springs (Figure 14–2), and in the superheated water found near undersea vents. Some prokaryotes are among the very few modern organisms that can survive without free oxygen, obtaining their energy by anaerobic processes (page 122). Oxygen is lethal to some types, whereas others can exist with or without oxygen.

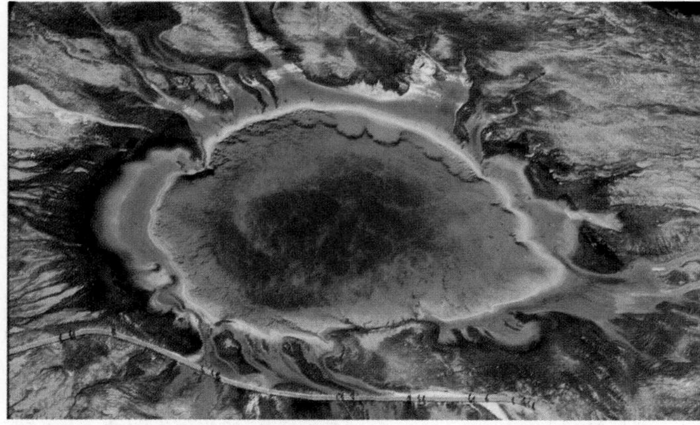

14–2
Aerial view of a very large boiling hot spring, Grand Prismatic Spring, in Yellowstone National Park in Wyoming. Thermophilic ("heat-loving") bacteria thrive in such hot springs. Carotenoid pigments of thickly growing thermophilic bacteria, including cyanobacteria, color the runoff channels a brownish orange.

In Chapter 13, we emphasized that there are two distinct lineages of prokaryotes, the *Bacteria* and the *Archaea*. At the molecular level, these two domains, while both prokaryotic, are as evolutionarily distinct from one another as either is from the *Eukarya*. We begin by considering some features largely shared by prokaryotes, while noting any differences that may exist between the two domains (see Table 13–3). We then turn our attention specifically to *Bacteria* and then to *Archaea*. Finally, we look briefly at the viruses. Viruses are not cells and, hence, they have no metabolism of their own. A virus consists primarily of a genome (either DNA or RNA) that replicates itself within a living host cell by directing the genetic machinery of that cell to synthesize viral nucleic acids and proteins.

Characteristics of the Prokaryotic Cell

Prokaryotes lack an organized nucleus bounded by a nuclear envelope (see page 41). Instead, a single circular, or continuous, molecule of DNA, associated with nonhistone proteins, is localized in a region of the cell called the **nucleoid.** In addition to its so-called "chromosome," a prokaryotic cell may also contain one or more smaller extrachromosomal pieces of circular DNA, called **plasmids,** that replicate independently of the cell's chromosome.

The cytoplasm of most prokaryotes is relatively unstructured, although it often has a fine granular appearance due to its many ribosomes—as many as 10,000 in a single cell. These are smaller (70S in size) than the cytoplasmic ribosomes (80S in size) of eukaryotes (see Table 3–2). Prokaryotes occasionally contain **inclusions,** distinct granules consisting of storage material. They lack a cytoskeleton. The cytoplasm does not contain any membrane-bounded organelles, nor is it generally divided or compartmentalized by membranes. The principal exception occurs in the cyanobacteria, which contain an extensive system of membranes (thylakoids) bearing chlorophyll and other photosynthetic pigments (see Figure 14–10). Purple bacteria also have extensive membranes for their photosynthetic apparatus.

The Plasma Membrane Serves as a Site for the Attachment of Various Molecular Components

The plasma membrane of a prokaryotic cell is formed from a lipid bilayer and is similar in chemical composition to that of a eukaryotic cell. However, except in the mycoplasmas (the smallest free-living cells known), the plasma membranes of prokaryotes lack cholesterol or other sterols. In aerobic prokaryotes, the plasma membrane incorporates the electron transport chain found in the mitochondrial membrane of eukaryotic cells, lending further support to the serial endosymbiotic theory (page 272). In the photosynthetic purple and green bacteria (but not the cyanobacteria), the sites of photosynthesis are found in the plasma membrane. In the photosynthetic purple bacteria and aerobes with large energy requirements, the membrane is often extensively convoluted, greatly increasing its working surface. Also, the membrane apparently contains specific attachment sites for the DNA molecules, ensuring separation of the replicated chromosomes at cell division (page 156).

The Cell Wall of Most Prokaryotes Contains Peptidoglycans

The protoplasts of almost all prokaryotes are surrounded by a cell wall, which gives the different types their characteristic shapes. Many prokaryotes have rigid walls, some have flexible walls, and only the mycoplasmas have no cell walls at all.

The cell walls of prokaryotes are complex and contain many kinds of molecules not present in eukaryotes. Except for the *Archaea,* the walls of prokaryotes contain complex polymers known as **peptidoglycans,** which are primarily responsible for the mechanical strength of the wall. Thus, peptidoglycan has been dubbed a "signature molecule" for differentiating species of *Bacteria* from species of *Archaea.*

Bacteria can be divided into two major groups on the basis of the capacity of their cell walls to retain the dye known as crystal violet. Those whose cell walls retain the dye are called **gram-positive,** whereas those that do not are called **gram-negative,** after Hans Christian Gram, the Danish microbiologist who discovered the distinction. Gram-positive and gram-negative bacteria differ markedly in the structure of their cell walls. In gram-positive bacteria, the wall, which ranges from 10 to 80 nanometers in thickness, has a homogeneous appearance and consists of as much as 90 percent peptidoglycan (Figure 14–3a). In gram-negative bacteria, the wall consists of two layers: an inner peptidoglycan layer, only 2 to 3 nanometers thick, and an outer layer of lipopolysaccharides and proteins (Figure 14–3b). The molecules of the outer layer are arranged in the form of a bilayer, about 7 to 8 nanometers in thickness, similar in structure to the plasma membrane. Gram staining is widely used to identify and classify bacteria, because it reflects a fundamental difference in the architecture of the cell wall.

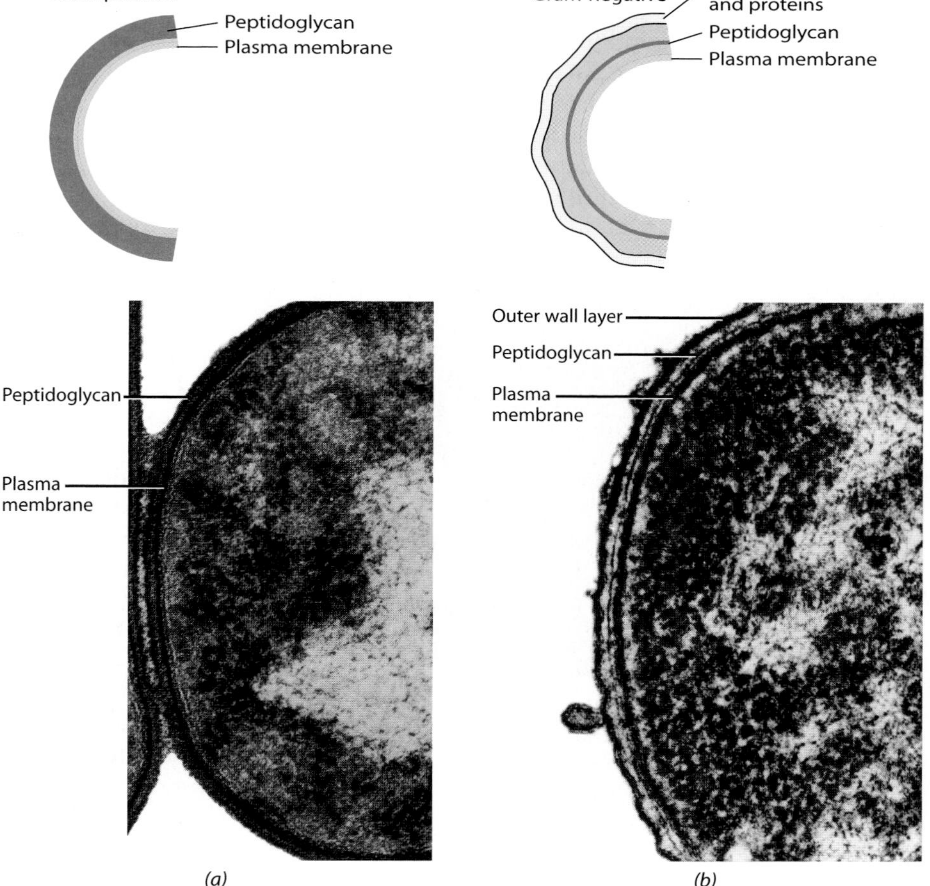

14–3
Bacterial cell walls. ***(a)*** *The wall of a gram-positive bacterium consists of a homogeneous layer of mostly peptidoglycan, seen here as the inner dark band. The outer dark layer in this electron micrograph represents a layer of surface proteins.* ***(b)*** *In a gram-negative bacterium, a layer of peptidoglycan is sandwiched between the plasma membrane and an outer layer of lipopolysaccharides and proteins, similar in composition to the plasma membrane.*

Many prokaryotes secrete slimy or gummy substances on the surface of their walls. Most of these substances consist of polysaccharides; a few consist of proteins. Commonly known as a "capsule," the general term for these layers is **glycocalyx.** The glycocalyx plays an important role in infection by allowing certain pathogenic bacteria to attach to specific host tissues. It is believed that the glycocalyx may also protect the bacteria from desiccation.

Prokaryotes Store Various Compounds in Granules

A wide variety of prokaryotes—both *Bacteria* and *Archaea*—contain inclusion bodies, or storage granules, consisting of **poly-β-hydroxybutyric acid,** a lipidlike compound, which serves as a depository for carbon and energy. Another storage compound of prokaryotes is **glycogen,** a starchlike substance. Glycogen granules are usually smaller than poly-β-hydroxybutyric acid granules.

Prokaryotes Have Distinctive Flagella

Many prokaryotes are motile, and their ability to move independently is usually due to long, slender appendages known as **flagella** (singular: flagellum) (Figure 14–4). Lacking microtubules and a plasma membrane, these flagella differ greatly from those of eukaryotes (page 60). Each prokaryotic flagellum is composed of subunits of a protein called flagellin, which are arranged into chains that are wound in a triple helix (three chains) with a hollow central core. Bacterial flagella grow at the tip. Flagellin molecules formed in the cell pass up through the hollow core and are added at the far end of the chains. In some species, the flagella are distributed over the entire cell surface; in others, they occur in tufts at one end of the cell.

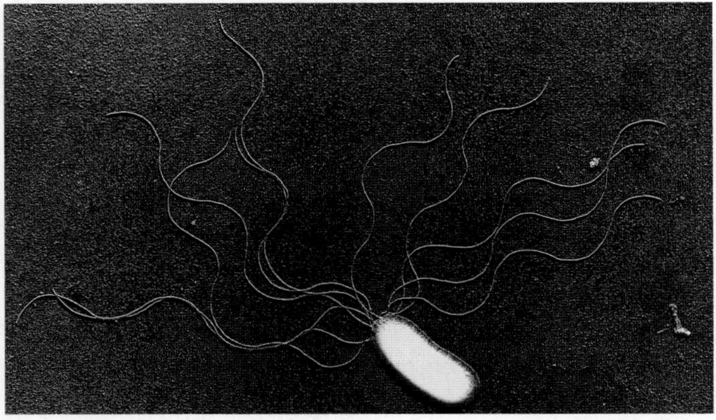

14–4
Flagella on Pseudomonas marginalis, *a common bacterium that is widespread in soils. It causes a soft-rot disease found mainly in fleshy and leafy vegetables.*

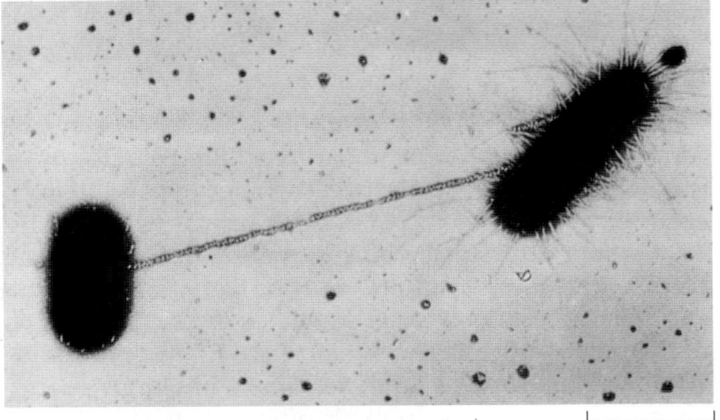

14–5
Electron micrograph of conjugating E. coli *cells. The elongated donor cell at the right of the micrograph is connected to the more rotund recipient cell by a long pilus, which is necessary for conjugation. Numerous short fimbriae are visible on the donor cell.*

Fimbriae and Pili Are Involved in Attachment

Fimbriae and pili—the two terms are often used interchangeably—are filamentous structures assembled from protein subunits in much the same way as the filaments of flagella. **Fimbriae** (singular: fimbria) are much shorter, more rigid, and typically more numerous than flagella (Figure 14–5). The functions of fimbriae are not known for certain, but they may serve to attach the organism to a food source or other surfaces.

Pili (singular: pilus) are generally longer than fimbriae, and only one or a few are present on the surface of an individual (Figure 14–5). Some pili are involved in the process of conjugation between prokaryotes, serving first to connect the two cells and then, by retracting, to draw them together for the actual transfer of DNA.

Diversity of Form

The oldest method of identifying prokaryotes is by their physical appearance. Prokaryotes exhibit considerable diversity of form, but many of the most familiar species fall into one of three categories (Figure 14–6). A prokaryote with a cylindrical shape is called a **rod,** or **bacillus** (plural: bacilli); spherical ones are called **cocci** (singular: coccus); and long curved, or spiral, rods are called **spirilla** (singular: spirillum). Cell shape is a relatively constant feature in most species of prokaryotes.

In many prokaryotes, the cells remain together, producing filaments, clusters, or colonies that also have a distinctive shape. For example, cocci and rods may adhere to form chains, and such behavior is characteristic of particular genera. Bacilli usually separate after cell division. When they do remain together, they form long, thin chains of cells, as are found in the filamentous acti-

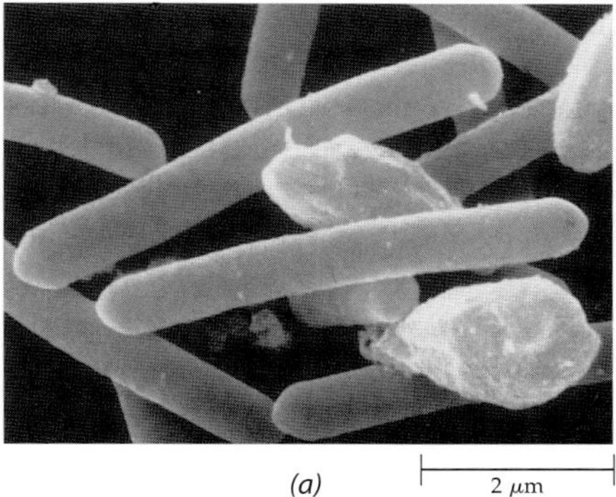

(a)

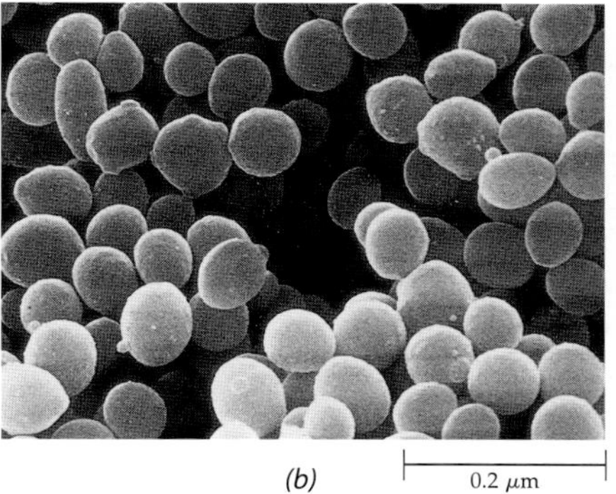

(b)

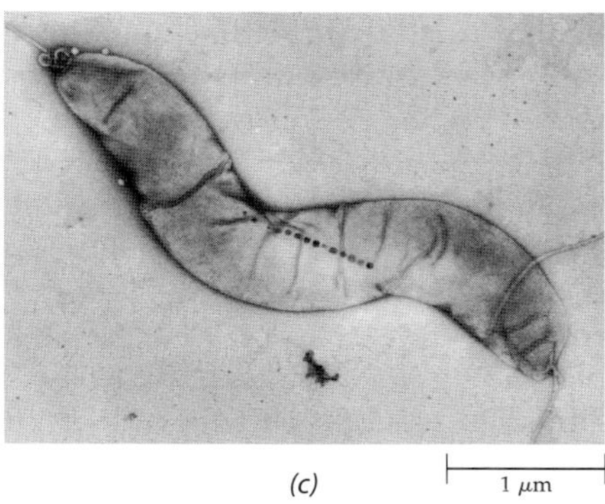

(c)

14–6
The three major forms of prokaryotes are: bacilli, cocci, and spirilla. ***(a)*** Clostridium botulinum, *the source of the toxin that causes deadly food poisoning, or botulism, is a bacillus, or rod-shaped bacterium. The saclike structures are endospores, which are resistant to heat and cannot easily be destroyed. Bacilli are responsible for many plant diseases, including fire blight of apples and pears (caused by* Erwinia amylovora*) and bacterial wilt of tomatoes, potatoes, and bananas (caused by* Pseudomonas solanacearum*).* ***(b)*** *Many prokaryotes, such as* Micrococcus luteus, *shown here, take the shape of spheres. Among the cocci are* Streptococcus lactis, *a common milk-souring agent and* Nitrosococcus nitrosus, *a soil bacterium that oxidizes ammonia to nitrites.* ***(c)*** *Spirilla, such as* Magnetospirillum magnetotacticum, *are less common than bacilli and cocci. Flagella can be seen at either end of this cell, which was isolated from a swamp. The string of dark magnetic particles orient the cell in the Earth's magnetic field.*

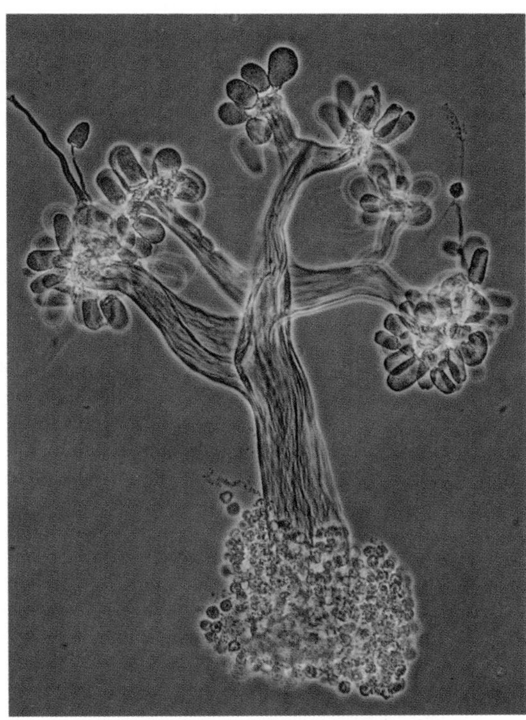

14–7
One of the diverse types of prokaryotes. This is a fruiting body of Chondromyces crocatus, *a myxobacterium, or gliding bacterium. Each fruiting body, which may contain as many as 1 million cells, consists of a central stalk that branches to form clusters of myxospores. Normally, the myxobacteria are rods that glide together along slime tracks and eventually form the kinds of bodies shown here.*

nomycetes (Figure 14–1). The gram-negative rods of the myxobacteria aggregate and construct complex fruiting bodies, within which some cells become converted into resting cells called myxospores (Figure 14–7). The myxospores are more resistant to drying, UV radiation, and heat than the vegetative cells and are of survival value to the bacterium.

Reproduction and Gene Exchange

Most prokaryotes reproduce by a simple type of cell division called **binary fission** (page 156). In some forms, reproduction is by budding or by fragmentation of filaments of cells. As they multiply, prokaryotes, barring mutation, produce clones of genetically identical cells. Mutations do occur, however. It has been estimated that in a culture of *E. coli* that has divided 30 times, about 1.5 percent of the cells are mutants. Mutations, combined with a rapid generation time, are responsible for the extraordinary adaptability of prokaryotes.

Further adaptability is provided by the genetic recombinations that take place as a result of conjugation, transformation, and transduction. Evidence is rapidly accumulating that such genetic recombinations are quite common in nature. All three mechanisms of gene transfer occur together in certain bacteria as well as archaea.

Conjugation has been characterized as the prokaryotic version of sex. This form of mating takes place when a pilus produced by the donor cell comes in contact with the recipient cell (Figure 14–5). This "sex pilus" then retracts, pulling the two cells together so that a tube, called a conjugation bridge, can form between them. A portion of the donor chromosome then passes through this bridge to the recipient cell. Conjugation is the mechanism used by plasmids to transfer copies of themselves to a new host. Conjugation can transfer genetic information between distantly related organisms; for example, plasmids can be transferred between bacteria and fungi as well as between bacteria and plants. **Transformation** occurs when a prokaryote takes up free, or naked, DNA from the environment. The free DNA may have been left behind by an organism that died. Because DNA is not chemically stable outside cells, transformation is probably less important than conjugation. **Transduction** occurs when viruses that attack bacteria—viruses known as bacteriophages (see page 297)—bring with them DNA they have acquired from their previous host.

Endospores

Certain species of bacteria have the capacity to form **endospores,** which are dormant resting cells (Figure 14–8). This process, called sporulation, has been studied extensively in the genera *Bacillus* and *Clostridium.* It characteristically occurs when a population of cells begins to use up its food supply.

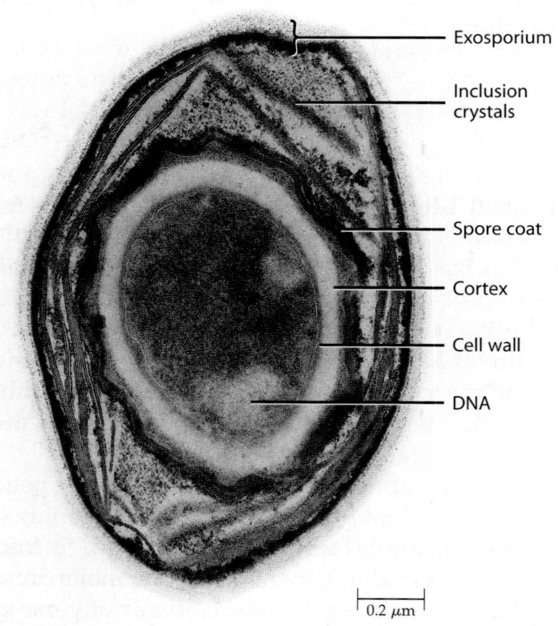

14–8
Mature free spore of Bacillus megaterium. *The outermost layer is the enveloping exosporium, made up of a faint peripheral layer and a dark basal layer. Underlying the exosporium are large inclusion crystals. The spore proper is covered with a protein spore coat. Beneath the spore coat is a thick peptidoglycan cortex, which is essential for the unique resistance properties of bacterial spores. Interior to the cortex is a thin cell wall, also of peptidoglycan, that covers a dehydrated spore protoplast.*

Endospore formation greatly increases the capacity of the prokaryotic cell to survive. Endospores are exceedingly resistant to heat, radiation, and chemical disinfectants, primarily due to their dehydrated protoplasts. The endospores of *Clostridium botulinum,* the organism that causes often-fatal blood poisoning, are not destroyed by boiling for several hours. In addition, endospores can remain viable (that is, they can germinate and develop into vegetative cells) for a very long time. For example, endospores recovered in 7000-year-old fractions of sediment cores from a Minnesota lake proved to be viable. More remarkably, ancient endospores preserved in the gut of an extinct bee trapped in amber were also reported to be revivable. The amber—and hence the endospores—was estimated to be 25 to 40 million years old.

Metabolic Diversity

Prokaryotes Are Autotrophs or Heterotrophs

Prokaryotes exhibit tremendous metabolic diversity. Although some are autotrophs (meaning literally "self-feeding")—organisms that use carbon dioxide as their sole source of carbon—most are **heterotrophs**—organisms requiring organic compounds as a carbon source. The vast majority of heterotrophs are **saprophytes** (Gk. *sapros,* "rotten," or "putrid"), which obtain their carbon from dead organic matter. Saprophytic bacteria and fungi are responsible for the decay and recycling of organic material in the soil; indeed, they are the recyclers of the biosphere.

14–9
Thiothrix, *a colorless sulfur bacterium.* Thiothrix *obtains energy by the oxidation of hydrogen sulfide,* H_2S. *The chains of cells, each filled with particles of sulfur, are attached to the substrate at the base (center of micrograph) and so form a characteristic rosette.*

Among the autotrophs are those that obtain their energy from light. These organisms are referred to as **photosynthetic autotrophs.** Some autotrophs, known as **chemosynthetic autotrophs,** are able to use inorganic compounds, rather than light, as an energy source (Figure 14–9). The energy is obtained from the oxidation of reduced inorganic compounds containing nitrogen, sulfur, or iron or from the oxidation of gaseous hydrogen.

Prokaryotes Vary in Their Tolerance of Oxygen and Temperature

Prokaryotes vary in their need for, or tolerance of, oxygen. Some species, called **aerobes,** require oxygen for respiration. Others, called **anaerobes,** lack an aerobic pathway and therefore cannot use oxygen as a terminal electron acceptor. There are two kinds of anaerobes: **strict anaerobes,** which are killed by oxygen and, hence, can live only in the absence of oxygen, and **facultative anaerobes,** which can grow in the presence or absence of oxygen.

Prokaryotes also vary with regard to the range of temperatures under which they can grow. Some have a low optimum temperature (that is, a temperature at which growth is most rapid). These organisms, called **psychrophiles,** can grow at 0°C or lower and can survive indefinitely when it is much colder. At the other extreme are the **thermophiles** and **extreme thermophiles,** which have high and very high temperature optima, respectively. Thermophilic prokaryotes, with growth optima below 80°C, are common inhabitants of hot springs. Some extreme thermophiles have optimal growth temperatures greater than 100°C and have even been found growing in the 140°C water near deep-sea vents. Because their heat-stable enzymes are capable of catalyzing biochemical reactions at high temperatures, thermophiles and extreme thermophiles are being intensively investigated for use in industrial and biotechnological processes.

Prokaryotes Play a Vital Role in the Functioning of the World Ecosystem

The autotrophic bacteria make major contributions to the global carbon balance. The role of certain bacteria in fixing atmospheric nitrogen—that is, in incorporating nitrogen gas into nitrogen compounds—is likewise of major biological significance (see page 290 and Chapter 30). Through the action of decomposers, materials incorporated into the bodies of once-living organisms are degraded and released and made available for successive generations. More than 90 percent of the CO_2 production in the biosphere, other than that associated with human activities, results from the metabolic activity of bacteria and fungi. The abilities of certain bacteria to decompose toxic natural and synthetic substances such as petroleum, pesticides, and dyes may lead to their widespread use in cleaning up dangerous spills and toxic dumps, when the techniques of using these bacteria are better developed.

Some Prokaryotes Cause Disease

In addition to their ecological roles, bacteria are important as agents of disease in both animals and plants. Human diseases caused by bacteria include tuberculosis, cholera, anthrax, gonorrhea, whooping cough, bacterial pneumonia, Legionnaires' disease, typhoid fever, botulism, syphilis, diphtheria, and tetanus. A clear link has now been established between stomach ulcers and infection with *Helicobacter pylori* and between coronary heart disease and *Chlamydia pneumoniae,* which have been found growing in arteries. About 80 species of bacteria, including many strains that appear to be identical but differ in the species of plant they infect, have been found to cause diseases of plants. Many of these diseases are highly destructive; some of them are described later in this chapter.

Some Prokaryotes Are Used Commercially

Industrially, bacteria are sources of a number of important antibiotics: streptomycin, aureomycin, neomycin, and tetracycline, for example, are produced by actinomycetes. Bacteria are also widely used commercially for the production of drugs and other substances, such as vinegar, various amino acids, and enzymes. The production of almost all cheeses involves bacterial fermentation of the sugar lactose into lactic acid, which coagulates milk proteins. The same kinds of bacteria that are responsible for the production of cheese are also responsible for the production of yogurt and for the lactic acid that preserves sauerkraut and pickles.

Bacteria

Phylogenetic analysis, employing sequencing of ribosomal RNA, reveals that there are twelve major lineages, or kingdoms, of *Bacteria*. The *Bacteria* are a diverse group of organisms. They range from the most ancient lineage of extreme thermophilic chemosynthetic autotrophs that oxidize gaseous hydrogen or reduce sulfur compounds to the lineages of photosynthetic autotrophs, represented by the cyanobacteria and the purple and green bacteria. The bacteria selected for individual treatment here are those we consider to be of special evolutionary and ecological importance.

Cyanobacteria Are Important from Ecological and Evolutionary Perspectives

The **cyanobacteria** deserve special emphasis because of their great ecological importance, especially in the global carbon and nitrogen cycles, as well as their evolutionary significance. They represent one of the major evolutionary lines of *Bacteria*. Photosynthetic cyanobacteria have chlorophyll *a*, together with carotenoids and other, unusual accessory pigments known as **phycobilins.** There are two kinds of phycobilins: **phycocyanin,** a blue pigment, and **phycoerythrin,** a red one. Within the cells of cyanobacteria are numerous layers of membranes, often parallel to one another (Figure 14–10). These membranes are photosynthetic thylakoids that resemble those found in chloroplasts—which, in fact, correspond in size to the entire cyanobacterial cell. The main storage product of the cyanobacteria is glycogen. Cyanobacteria are believed to have given rise through symbiosis to at least some eukaryotic chloroplasts. In biochemical and structural detail, cyanobacteria are especially similar to the chloroplasts of red algae (see page 358).

Many cyanobacteria produce a mucilaginous envelope, or sheath, that binds groups of cells or filaments together. The sheath is often deeply pigmented, particularly in species that sometimes occur in terrestrial habitats. The colors of the sheaths in different species include light gold, yellow, brown, red, emerald green, blue, violet, and blue-black. Despite their former name ("blue-green algae"), therefore, only about half of the species of cyanobacteria are actually blue-green in color.

Cyanobacteria often form filaments and may grow in large masses 1 meter or more in length. Some cyanobacteria are unicellular, a few form branched filaments, and a very few form plates or irregular colonies (Figure 14–11). Following division of the cyanobacterial cell, the resulting subunits may then separate to form new colonies. In addition, the filaments may break into fragments called **hormogonia** (singular: hormogonium). As in other filamentous or colonial bacteria, the cells of cyanobacteria are usually joined only by their walls or by mucilaginous sheaths, so that each cell leads an independent life.

Some filamentous cyanobacteria are motile, gliding and rotating around the longitudinal axis. Short segments that break off from a cyanobacterial colony can glide away from their parent colony at rates as rapid as 10 micrometers per second. This movement may be connected with the extrusion of mucilage through small pores in the cell wall, together with the production of contractile waves in one of the surface layers of the wall. Some cyanobacteria exhibit intermittent jerky movements.

Cyanobacteria Can Live in a Wide Variety of Environments

Although more than 7500 species of cyanobacteria have been described and given names, there may actually be as few as 200 distinct, free-living nonsymbiotic species. Like other bacteria, cyanobacteria sometimes grow under extremely inhospitable conditions, from the water of hot springs to the frigid lakes of Antarctica, where they sometimes form luxuriant mats 2 to 4 centimeters thick in water beneath more than 5 meters of permanent ice. However, cyanobacteria are absent in acidic waters, where eukaryotic algae are often abundant. The greenish color of some polar bears in zoos is due to the presence of colonies of cyanobacteria that develop within the hollow hairs of their fur.

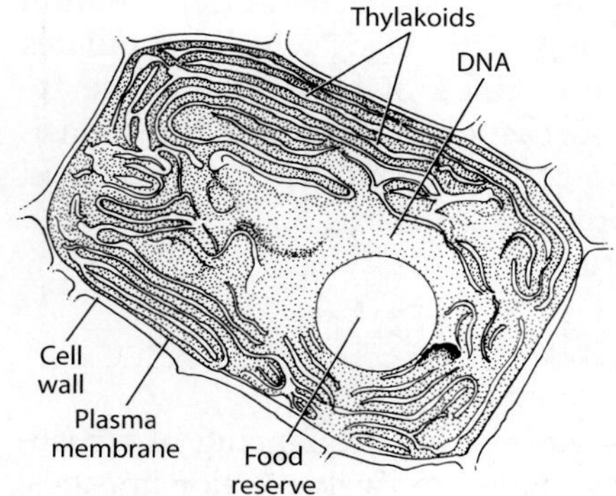

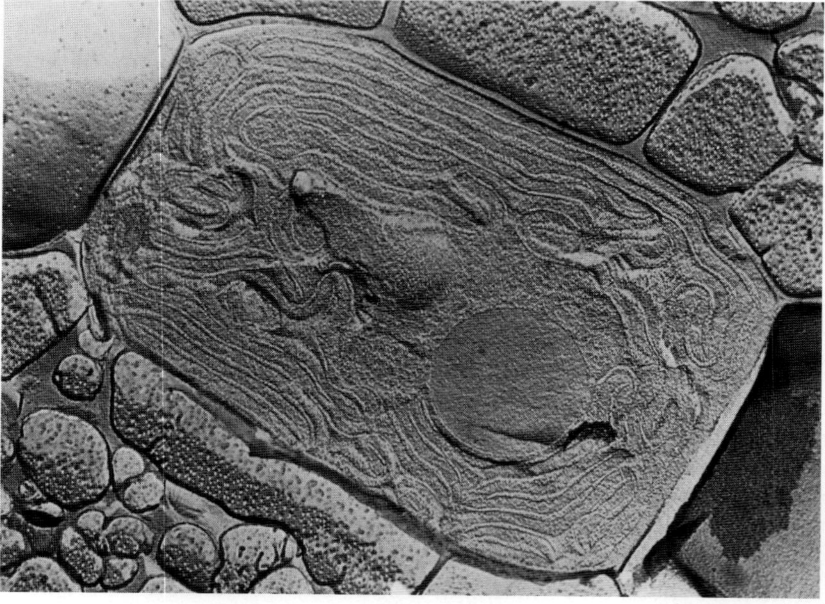

14–10
A cell of the cyanobacterium Anabaena cylindrica. *Photosynthesis takes place in the chlorophyll-containing membranes—the thylakoids—within the cell. The three-dimensional quality of this electron micrograph is due to the freeze-fracturing technique used in the preparation of the tissue.*

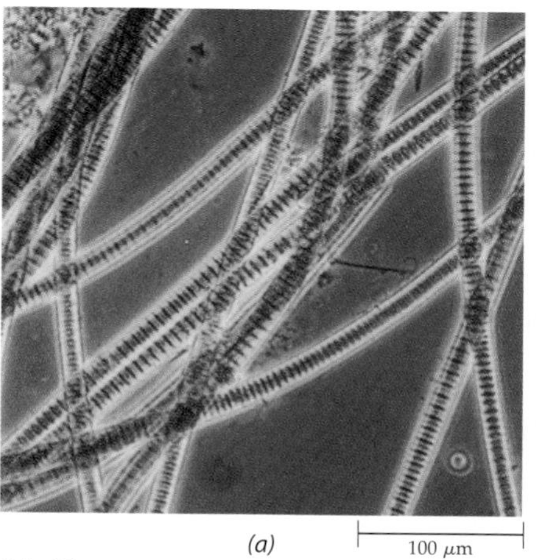

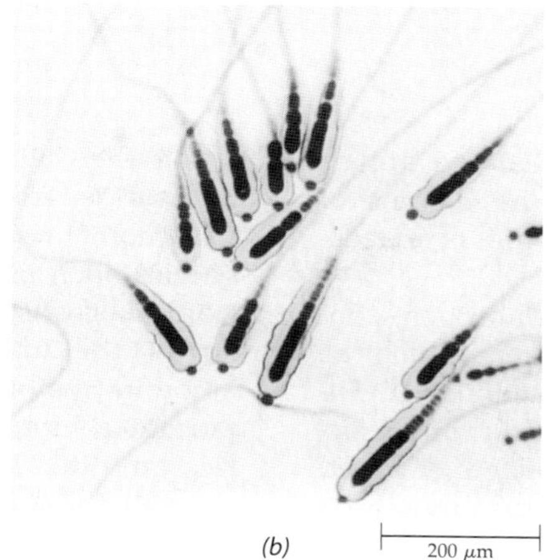

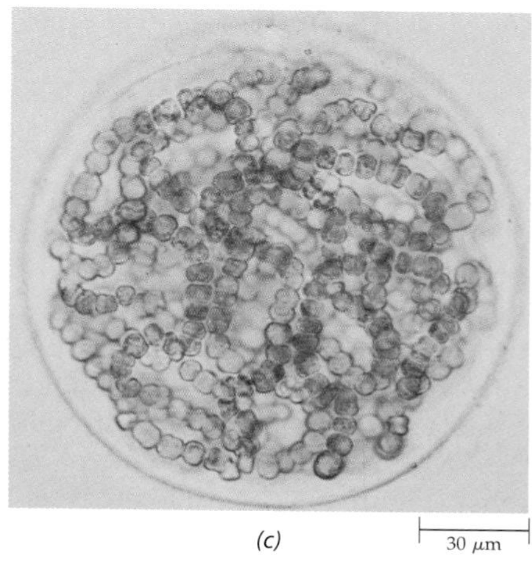

14–11
Three common genera of cyanobacteria. ***(a)*** Oscillatoria, *in which the only form of reproduction is by means of fragmentation of the filament.* ***(b)*** Calothrix, *a filamentous form with a basal heterocyst (see page 290).* Calothrix *is capable of forming akinetes—enlarged cells that develop a resistant outer envelope—just above the heterocysts.* ***(c)*** *A gelatinous "ball" of* Nostoc commune, *containing numerous filaments. These cyanobacteria occur frequently in freshwater habitats.*

Layered chalk deposits called **stromatolites** (Figure 14–12), which have a continuous geologic record covering 2.7 billion years, are produced when colonies of cyanobacteria bind calcium-rich sediments. Today, stromatolites are formed only in a few places, such as shallow pools in hot, dry climates. Their abundance in the fossil record is evidence that such environmental conditions were prevalent in the past, when cyanobacteria played the decisive role in elevating the level of free oxygen in the atmosphere of the early Earth. More ancient stromatolites (3 billion years or older), produced in an oxygen-free environment, probably were made by purple and green bacteria. Recent evidence suggests that some stromatolites may have been produced solely by geologic processes.

Many marine cyanobacteria occur in limestone (calcium carbonate) or lime-rich substrates, such as coralline algae (see page 359) and the shells of mollusks. Some freshwater species, particularly those that grow in hot springs, often deposit thick layers of lime in their colonies.

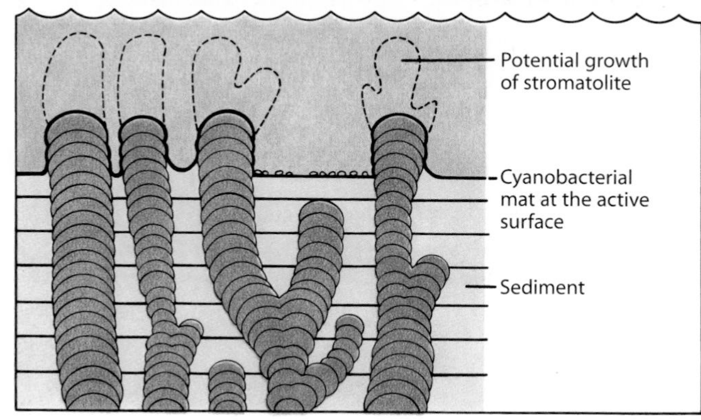

14–12
Stromatolites are produced when flourishing colonies of cyanobacteria bind calcium carbonate into domed structures, like the ones shown in the diagram and photograph, or others with a more intricate form. Such structures are abundant in the fossil record but today are being formed in only a few, highly suitable environments, such as the tidal flats of Hamlin Pool in Western Australia shown here.

Cyanobacteria Form Gas Vesicles, Heterocysts, and Akinetes The cells of cyanobacteria living in freshwater or marine habitats—especially those that inhabit the surface layers of the water, in the community of microscopic organisms known as the **plankton**—commonly contain bright, irregularly shaped structures called **gas vesicles.** These vesicles provide and regulate the buoyancy of the organisms, thus allowing them to float at certain levels in the water. When numerous cyanobacteria become unable to regulate their gas vesicles properly—for example, because of extreme fluctuations of temperature or oxygen supply—they may float to the surface of the body of water and form visible masses called "blooms." Some cyanobacteria that form blooms secrete chemical substances that are toxic to other organisms, causing large numbers of deaths. The Red Sea apparently was given its name because of the blooms of planktonic species of *Trichodesmium*, a red cyanobacterium.

Many genera of cyanobacteria can fix nitrogen, converting nitrogen gas to ammonium, a form in which the nitrogen is available for biological reactions. In filamentous cyanobacteria, **nitrogen fixation** often occurs within **heterocysts,** which are specialized, enlarged cells (Figure 14–13). Heterocysts are surrounded by thickened cell walls containing large amounts of glycolipid, which serves to impede the diffusion of oxygen into the cell. Within a heterocyst, the cell's internal membranes are reorganized into a concentric or reticulate pattern. Heterocysts are low in phycobilins, and they lack Photosystem II, so the cyclic photophosphorylation that occurs in these cells does not result in the evolution of oxygen (page 139). The oxygen that is present is either rapidly reduced by hydrogen, a by-product of nitrogen fixation, or expelled through the wall of the heterocyst. Nitrogenase, the enzyme that catalyzes the nitrogen-fixing reactions, is sensitive to the presence of oxygen, and nitrogen fixation therefore is an anaerobic process. Heterocysts have small plasmodesmatal connections—microplasmodesmata—with adjacent vegetative cells. It is via these microplasmodesmata that the products of nitrogen fixation are transported from the heterocyst to the vegetative cells and the products of photosynthesis move from the vegetative cells to the heterocyst.

Among cyanobacteria that fix nitrogen are free-living species such as *Trichodesmium,* which occurs in certain tropical oceans. *Trichodesmium* accounts for about a quarter of the total nitrogen fixed there, an enormous amount. Symbiotic cyanobacteria are likewise very important in nitrogen fixation. In the warmer parts of Asia, rice can often be grown continuously on the same land without the addition of fertilizers because of the presence of nitrogen-fixing cyanobacteria in the rice paddies (Figure 14–14). Here the cyanobacteria, especially members of the genus *Anabaena* (Figure 14–13), often occur in association with the small, floating water fern *Azolla,* which forms masses on the paddies.

Cyanobacteria occur as symbionts within the bodies of some sponges, amoebas, flagellated protozoa, diatoms, green algae that lack chlorophyll, other cyanobacteria, mosses, liverworts, vascular plants, and oomycetes, in addition to their familiar role as the photosynthetic partner in many lichens (see Chapter 15). Some symbiotic cyanobacteria lack a cell wall, in which case they function as chloroplasts. The symbiotic cyanobacterium divides at the same time as the host cell by a process similar to chloroplast division. Bacteria similar to cyanobacteria appear to have given rise to at least some chloroplasts.

In addition to the heterocysts, some cyanobacteria form resistant spores called **akinetes,** which are enlarged cells surrounded by thickened envelopes (Figure 14–13b). Like the endospores formed by other bacteria,

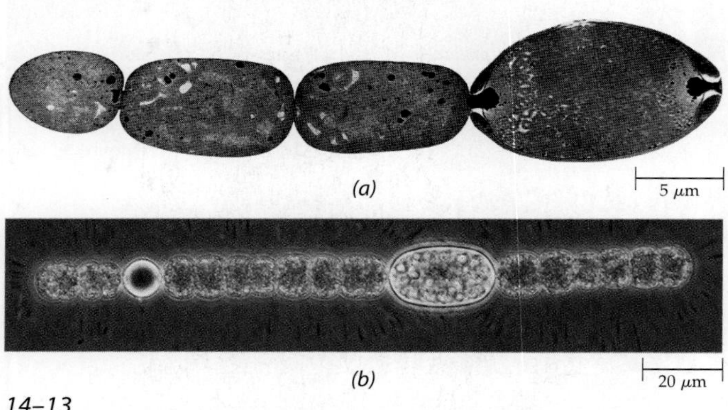

14–13
Filament of Anabaena. ***(a)*** *This electron micrograph shows a chain of cells held together along incompletely separated walls. The first cell on the right end of the chain is a heterocyst, in which nitrogen fixation takes place. The gelatinous matrix of this filament was destroyed during preparation of the specimen for electron microscopy.* ***(b)*** *In this preparation, the gelatinous matrix is barely discernible as striations extending outward from the cell surfaces. The third cell from the left is a heterocyst.* Anabaena, *like the* Calothrix *shown in Figure 14–11b, forms akinetes (large oval body toward the right).*

14–14
Women planting rice in a paddy in Perak, Malaysia. In Southeast Asia, rice can often be grown on the same land continuously without the addition of fertilizers because of nitrogen fixation by Anabaena azollae, *which grows in the tissues of the water fern* Azolla *growing in the rice paddies.*

akinetes are resistant to heat and drought and thus allow the cyanobacterium to survive during unfavorable periods.

Purple and Green Bacteria Have Unique Types of Photosynthesis

The purple and green bacteria together represent the second major group of photosynthetic bacteria. The overall photosynthetic process and the photosynthetic pigments utilized by these bacteria differ from those used by the cyanobacteria. Whereas cyanobacteria produce oxygen during photosynthesis, the purple and green bacteria do not. In fact, purple and green bacteria can grow in light only under anaerobic conditions because pigment synthesis in these organisms is repressed by oxygen. Cyanobacteria employ chlorophyll *a* and two photosystems in their photosynthetic process. By contrast, purple and green bacteria employ several different types of bacteriochlorophyll, which differ in certain respects from chlorophyll, and only one photosystem (Figure 14–15). The photosystems present in purple and green bacteria appear to be ancestors of individual photosystems—Photosystem II and Photosystem I, respectively—present in photosynthetic autotrophs, including plants, algae, and cyanobacteria.

The colors characteristic of photosynthetic bacteria are associated with the presence of several accessory pigments that function in photosynthesis. In two groups of purple bacteria, these pigments are yellow and red carotenoids. In the cyanobacteria, as we have seen, the pigments are the red and blue phycobilins, which are not found in purple and green bacteria.

The purple and green bacteria are subdivided into those that use mostly sulfur compounds as electron donors and those that do not. In the purple sulfur and green sulfur bacteria, sulfur compounds play the same role in photosynthesis that water plays in organisms containing chlorophyll *a* (page 128).

Purple or green sulfur bacterium:

$$\underset{\text{Carbon dioxide}}{CO_2} + \underset{\text{Hydrogen sulfide}}{2H_2S} \xrightarrow{\text{Light}} \underset{\text{Carbohydrate}}{(CH_2O)} + \underset{\text{Water}}{H_2O} + \underset{\text{Sulfur}}{2S}$$

Cyanobacterium, alga, or plant:

$$\underset{\text{Carbon dioxide}}{CO_2} + \underset{\text{Water}}{2H_2O} \xrightarrow{\text{Light}} \underset{\text{Carbohydrate}}{(CH_2O)} + \underset{\text{Water}}{H_2O} + \underset{\text{Oxygen}}{O_2}$$

The purple nonsulfur and green nonsulfur bacteria, which are able to use sulfide at only low levels, also use organic compounds as electron donors. These compounds include alcohols, fatty acids, and a variety of other organic substances.

The prokaryote that became an endosymbiont of eukaryotes and evolved into mitochondria is thought to have been a relative of the purple nonsulfur bacteria. This conclusion is based both on the similar metabolic features of mitochondria and purple nonsulfur bacteria and on comparisons of the base sequences in their small-subunit ribosomal RNAs.

Because of their requirement for H_2S or a similar substrate, the purple and green sulfur bacteria are able to grow only in habitats that contain large amounts of decaying organic material, recognizable by the sulfurous odor. In these bacteria, as in the closely related colorless sulfur bacterium *Thiothrix*, elemental sulfur may accumulate as deposits within the cell (Figure 14–9).

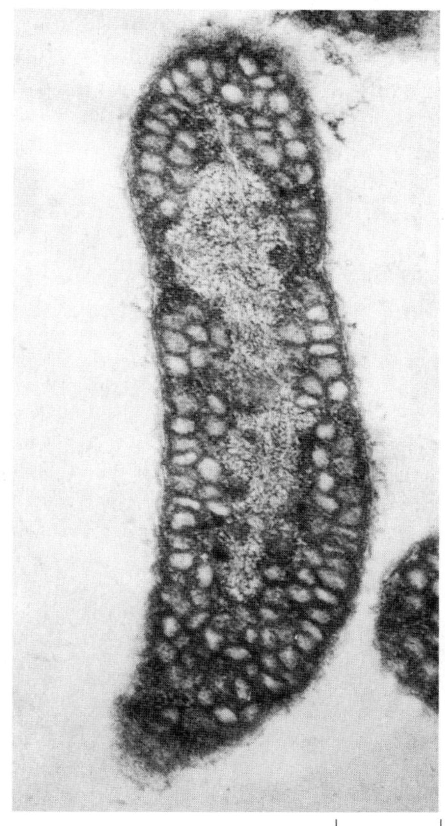

14–15

A purple nonsulfur bacterium, Rhodospirillum rubrum. *The structures that resemble vesicles are intrusions of the plasma membrane, which contains the photosynthetic pigments. This cell, with its many membrane intrusions, has a very high content of bacteriochlorophyll. It is from a culture that was grown in dim light. In cells grown in bright light, the membrane intrusions are less extensive because there is a decreased need for photosynthetic pigments.*

Prochlorophytes Contain Chlorophylls *a* and *b* and Carotenoids

The **prochlorophytes** are a group of photosynthetic prokaryotes that contain chlorophylls *a* and *b*, as well as carotenoids, but do not contain phycobilins. Thus far, only three genera of prochlorophytes have been discovered. The first of these is *Prochloron,* which lives only along tropical seashores, as a symbiont within colonial sea squirts (ascidians). The cells of *Prochloron* are roughly spherical in shape and contain an extensive system of thylakoids (Figure 14–16).

The two other known prochlorophyte genera are *Prochlorothrix* and *Prochlorococcus*. *Prochlorothrix,* which is filamentous, has been found growing in several shallow lakes in the Netherlands. *Prochlorococcus* has been found deep in the euphotic zone—the zone into which sufficient light penetrates for photosynthesis to occur—of the open oceans. Prochlorococci are extremely small cocci.

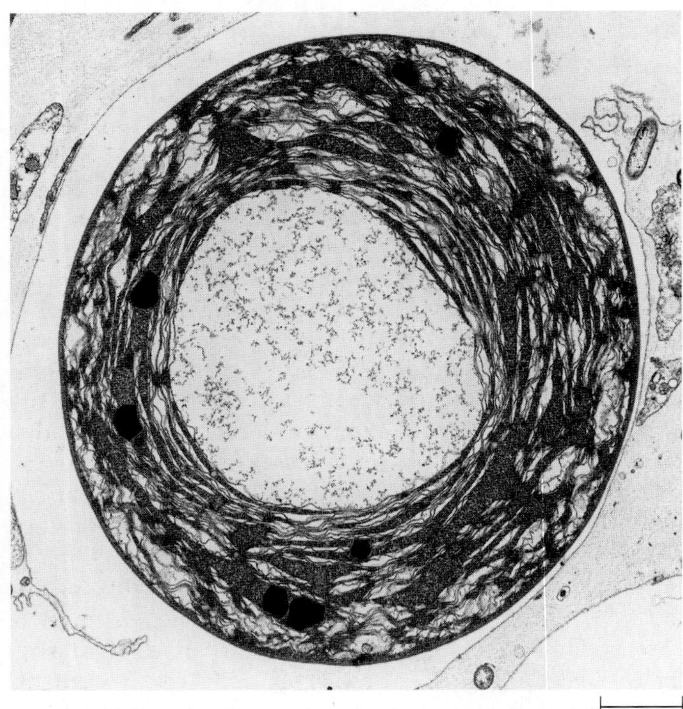

14–16
A single cell of the bacterium Prochloron, *showing the extensive system of thylakoids.* Prochloron *is a photosynthetic bacterium with chlorophylls* a *and* b *and carotenoids, the same pigments found in the green algae and plants. The prochlorophytes resemble both cyanobacteria (because they are prokaryotic and contain chlorophyll* a*) and the chloroplasts of green algae and plants (because they contain chlorophyll* b *instead of phycobilins).*

Mycoplasmas Are Wall-less Organisms That Live in a Variety of Environments

Mycoplasmas and mycoplasmalike organisms are bacteria that lack cell walls. Usually about 0.2 to 0.3 micrometer in diameter, they are probably the smallest organisms capable of independent growth. The genome of the mycoplasmas is likewise small, only one-fifth that of *E. coli* and other common prokaryotes. Because they lack a cell wall and, consequently, rigidity, mycoplasmas can assume various forms. In a single culture, a mycoplasma may range from small rods to highly branched filamentous forms.

Mycoplasmas may be free-living in soil and sewage, parasites of the mouth or urinary tract of humans, or pathogens in animals and plants. Among the plant-pathogenic mycoplasmas are the **spiroplasmas,** long spiral or corkscrew-shaped cells less than 0.2 micrometer in diameter, which are motile even though they lack flagella (Figure 14–17). They are motile by means of a rotary motion or a slow undulation. Some spiroplasmas have been cultured on artificial media, including *Spiroplasma citri,* which causes citrus stubborn disease. Symptoms of stubborn disease, such as bunchy upright growth of twigs and branches, are slow to develop and difficult to detect. The disease, which is widespread, is difficult to control. In California and some Mediterranean countries, stubborn is probably the greatest threat to the production of grapefruit and sweet oranges. *Spiroplasma citri* has also been isolated from corn (maize) plants suffering from corn stunt disease.

Mycoplasmalike Organisms Cause Diseases in Plants

Mycoplasmalike organisms (MLOs), or phytoplasmas, have been identified in more than 200 distinct plant dis-

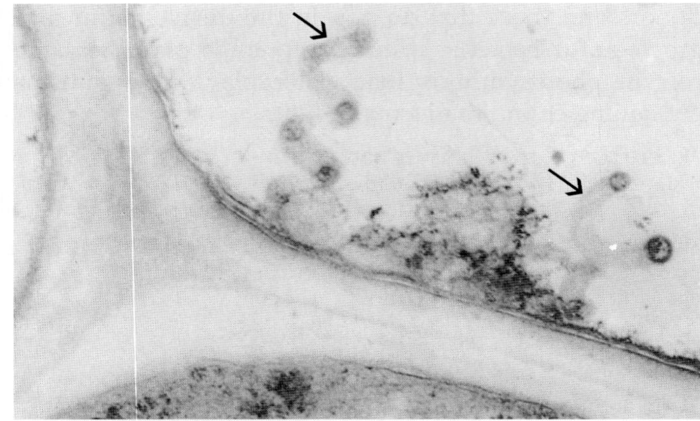

14–17
Spiroplasmas (arrows) in a sieve tube of a maize (Zea mays) *plant that was suffering from corn stunt disease.* Spiroplasma citri *causes both corn stunt disease and citrus stubborn disease.*

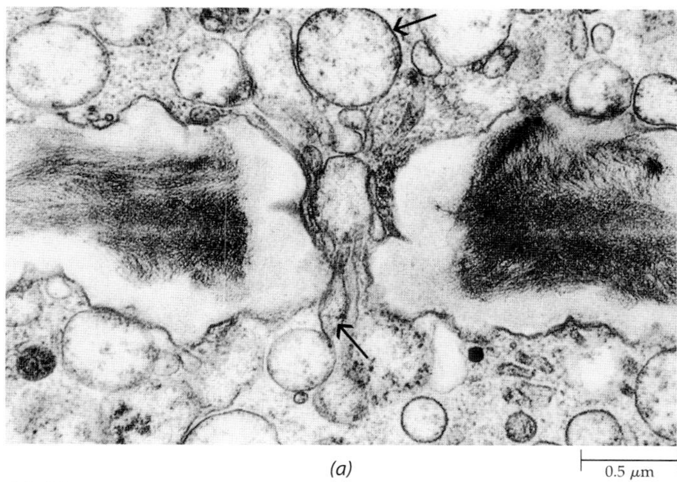

14–18
(a) Mycoplasmalike organisms (arrows) apparently traversing a sieve-plate pore in a young inflorescence of a coconut palm (Cocos nucifera) *affected by lethal yellowing disease. (b) A devastated grove of coconut palms—now looking like telephone poles—in Jamaica. Lethal yellowing has been responsible for the death of palms of many genera in southern Florida and elsewhere.*

eases affecting several hundred genera of plants. Some of these diseases are very destructive, for example, X-disease of peach, which can render a tree commercially worthless in 2 to 4 years, and pear decline, so-called because it commonly causes a slow, progressive weakening and ultimate death of pear trees. Aster yellows, another MLO-caused disease, results in a general yellowing (chlorosis) of the foliage and infects a wide variety of host crops, ornamentals, and weeds. Carrots are among the host crops suffering the greatest losses, commonly 10 to 25 percent and as high as 90 percent. Lethal yellowing of coconut is also caused by an MLO (Figure 14–18).

In flowering plants, MLOs are generally confined to the conducting elements of the phloem known as sieve tubes. Most MLOs are believed to move passively from one sieve-tube member to another through the sieve-plate pores, as the sugar solution is transported in the phloem. Motile spiroplasmas, which also are found in sieve tubes, may be able to move actively in phloem tissue. Most MLOs and spiroplasmas are transmitted from plant to plant by insect vectors that acquire the pathogen while feeding on an infected plant.

Plant-Pathogenic Bacteria Cause a Wide Variety of Diseases

In addition to those already mentioned, many more economically important diseases of plants are caused by bacteria, contributing substantially to the one-eighth of crops worldwide that are lost to disease annually. Almost all plants can be affected by bacterial diseases, and many of these diseases can be extremely destructive.

Virtually all plant-pathogenic bacteria are gram-negative, and all but *Streptomyces,* which is filamentous, are rod-shaped. They are **parasites**—symbionts that are harmful to their hosts. The symptoms caused by plant-pathogenic bacteria are quite varied, with the most common appearing as spots of various sizes on stems, leaves, flowers, and fruits (Figure 14–19). Almost all such bacterial spots are caused by members of two closely related genera, *Pseudomonas* and *Xanthomonas.*

Some of the most destructive diseases of plants—such as blights, soft rots, and wilts—are also caused by bacteria. Blights are characterized by rapidly developing necroses (dead, discolored areas) on stems, leaves, and flowers. Fire blight in apples and pears, caused by *Erwinia amylovora,* is a widespread, economically important disease that can kill young trees within a single season. Fire blight was so destructive to pear trees that by the 1930s their cultivation on a commercial scale had been eliminated from almost all parts of the United States except the Northwest. Bacterial soft rots occur most commonly in the fleshy storage tissues of vegetables, such as potatoes or carrots, as well as in fleshy fruits, such as tomatoes and eggplants, and succulent stems or leaves, such as cabbage or lettuce. The most destructive soft rots are caused by bacteria of the genus *Erwinia,* with heavy losses occurring in the postharvest period.

Bacterial vascular wilts affect mainly herbaceous plants. The bacteria invade the vessels of the xylem, where they multiply. They interfere with the movement of water and inorganic nutrients by producing high-molecular-weight polysaccharides, which results in the wilting and death of the plants. The bacteria commonly degrade portions of the vessel walls and can even cause the vessels to rupture. Once the walls have ruptured, the bacteria then spread to the adjacent parenchyma tissues,

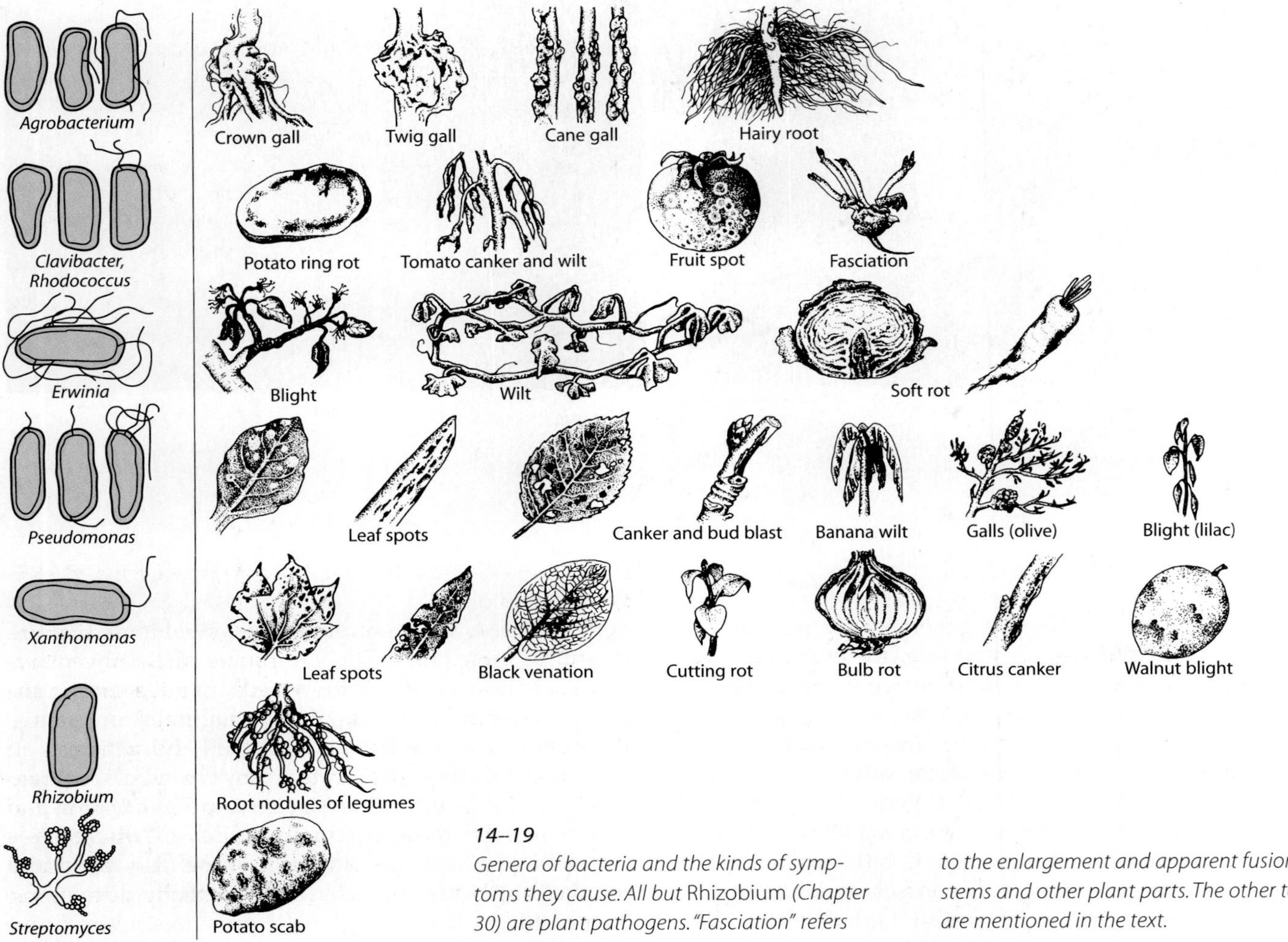

14–19
Genera of bacteria and the kinds of symptoms they cause. All but Rhizobium *(Chapter 30) are plant pathogens. "Fasciation" refers to the enlargement and apparent fusion of stems and other plant parts. The other terms are mentioned in the text.*

where they continue to multiply. In some bacterial wilts, the bacteria ooze to the surface of the stems or leaves through cracks formed over cavities filled with cellular debris, gums, and bacterial cells. More commonly, however, the bacteria do not reach the surface of the plant until the plant has been killed by the disease.

Among the most important examples of wilts are bacterial wilt of alfalfa, tomato, and bean plants (each caused by different species of *Clavibacter*); bacterial wilt of cucurbits, such as squashes and watermelons (caused by *Erwinia tracheiphila*); and black vein of crucifers, such as cabbage (caused by *Xanthomonas campestris*). The most economically important wilt disease of plants, however, is caused by *Pseudomonas solanacearum*. It affects more than 40 different genera of plants, including such major crops as bananas, peanuts, tomatoes, potatoes, eggplants, and tobacco, to name a few. This disease occurs worldwide in tropical, subtropical, and warm temperate areas.

The members of another genus of *Bacteria*, known as *Agrobacterium*, cause a cancer-like disease known as crown gall in plants, "gall" being a general term for swelling. Because of their role in genetic engineering, these *Bacteria* will be discussed in Chapter 28.

Archaea

There are four phylogenetically distinct lineages of *Archaea*, exhibiting enormous physiological diversity. On the basis of that diversity, the *Archaea* can be divided into three large groups—extreme halophiles, methanogens, and extreme thermophiles—and one small group represented by a thermophile that lacks a cell wall. Until very recently, the archaea were generally thought of as noncompetitive relics inhabiting hostile environments that were of little importance to global ecology. It is now known, however, that the archaea are present in less hostile environments such as soil. Archaea also constitute a major component of the oceanic picoplankton (organisms smaller than 1 micrometer), possibly outnumbering all other oceanic organisms.

14–20
Extreme halophilic archaea growing in evaporating ponds of seawater, as seen in an aerial view taken near San Francisco Bay, California. The ponds yield table salt, as well as other salts of commercial value. As the water evaporates and the salinity increases, the halophiles multiply harmlessly, forming massive growths, or "blooms," that turn the seawater pink or reddish purple.

The Extreme Halophiles Are the "Salt-Loving" Archaea

The **extreme halophilic archaea** are a diverse group of prokaryotes that occur everywhere in nature where the salt concentration is very high—in places such as Great Salt Lake and the Dead Sea, as well as in ponds where seawater is left to evaporate, yielding table salt (Figure 14–20). Extreme halophiles have a very high requirement for salt, with most of them requiring 12 to 23 percent salt (sodium chloride, NaCl) for optimal growth. Their cell walls, ribosomes, and enzymes are stabilized by the sodium ion, Na^+.

All extreme halophiles are chemoorganotrophs (heterotrophs that obtain their energy from the oxidation of organic compounds), and most species require oxygen. In addition, certain species of extreme halophiles exhibit a light-mediated synthesis of ATP that does not involve any chlorophyll pigments. Among them is *Halobacterium halobium,* the prevalent species of *Archaea* in the Great Salt Lake. Although the high concentration of salt in their environment limits the availability of O_2 for respiration, such extreme halophiles are able to supplement their ATP-producing capacity by converting light energy into ATP using a protein called bacteriorhodopsin, which is found in the plasma membrane.

The Methanogens Are the Methane-Producing Archaea

The **methanogens** are a unique group of prokaryotes—the only ones that produce methane (Figure 14–21). All are strictly anaerobic and will not tolerate even the slightest exposure to oxygen. Methanogens can produce methane (CH_4) from hydrogen (H_2) and carbon dioxide (CO_2), with the needed electrons being derived from the H_2 and the CO_2 serving as both carbon source and electron acceptor.

All methanogens use ammonium (NH_4^+) as a nitrogen source, and a few can fix nitrogen. Methanogens are common in sewage-treatment plants, in bogs, and in the ocean depths. In fact, most of the natural gas reserves now used as fuel were produced by the activities of methane-producing prokaryotes in the past. Methanogens are also found in the digestive tracts of cattle and other ruminant (cud-chewing) animals, where they make possible the digestion of cellulose. It is estimated that a cow belches about 50 liters of methane a day while chewing the cud.

14–21
Scanning electron micrograph of a methane-producing archaea from the digestive tract of a cud-chewing animal. Cells such as those shown here produce methane and carbon dioxide. Methanogens are strict anaerobes and can therefore live only in the absence of oxygen—a condition prevailing on the young Earth but occurring today only in isolated environments.

14–22
A steaming hot spring in Yellowstone National Park, with hydrogen sulfide–rich steam rising to the surface of the Earth. This is a typical habitat for extreme thermophilic archaea, where, because of the heat and acidity, higher forms of life cannot develop. The archaea form a mat around the spring.

The Extreme Thermophiles Are the "Heat-Loving" Archaea

The **extreme thermophilic archaea** contain representatives of the most heat-loving of all known prokaryotes. The membranes and enzymes of these archaea are unusually stable at high temperatures: all have temperature optima above 80°C, and some grow at temperatures over 110°C. Most species of extreme thermophiles metabolize sulfur in some way, and, with only a few exceptions, they are strict anaerobes. These archaea are inhabitants of hot, sulfur-rich environments, such as the hot springs and geysers found in Iceland, Italy, New Zealand, and Yellowstone National Park (Figure 14–22). As noted, extreme thermophilic archaea also grow near deep-sea hydrothermal vents and cracks in the ocean floor from which geothermally superheated water is emitted (page 287).

Thermoplasma Is an Archaea Lacking a Cell Wall

A fourth group of *Archaea* consists of a single known genus, *Thermoplasma,* containing a single species, *T. acidophilum. Thermoplasma* resembles the mycoplasmas (page 292) in lacking a cell wall and being very small; individuals vary from spherical (0.3 to 2 micrometers in diameter) to filamentous. *Thermoplasma* is only known to occur in acidic, self-heating coal refuse piles in southern Indiana and western Pennsylvania, in sites within the piles where the temperatures range from 32° to 80°C—the sort of very unusual habitat where archaea seem to thrive.

Because archaea often grow in places that resemble habitats characteristic of the early Earth, it has been suggested that the methanogens might have originated more than 3 billion years ago, when there was an atmosphere rich in CO_2 and H_2, and that they might have persisted in certain habitats similar to those in which they first evolved.

Viruses

The existence of viruses was first recognized just over 100 years ago when Dutch, German, and Russian scientists showed that tobacco mosaic disease could be produced on plants with a liquid that had passed through a filter used to trap bacteria. Until the 1930s, viruses were considered to be extremely small bacteria or even enzymes or proteins. Evidence against this point of view began to accumulate in 1933, when Wendell Stanley, a biochemist working at the Rockefeller Institute for Medical Research, prepared an extract from diseased tobacco plants that contained the tobacco mosaic virus (TMV) and purified it. In the presence of high salt concentrations, the purified virus precipitated in the form of crystals, thus behaving more like a chemical than an organism. When these needlelike crystals were redissolved and the solution applied to a tobacco leaf, the characteristic symptoms of tobacco mosaic disease appeared. Thus the virus had retained its ability to infect tobacco plants even after it had been crystallized, unlike any known organism.

Practically every kind of organism is now known to be infected by distinct viruses, and it is clear that a tremendous diversity of viruses exists. A virus typically is associated with a specific type of host and is usually discovered and studied because it causes a disease in that host.

In humans, viruses are responsible for many infectious diseases, including chicken pox, measles, mumps, influenza, colds (often complicated by secondary bacterial infections), infectious hepatitis, polio, rabies, herpes, AIDS, and deadly hemorrhagic fevers (such as those caused by Ebola virus and Hantaan virus). Viruses also

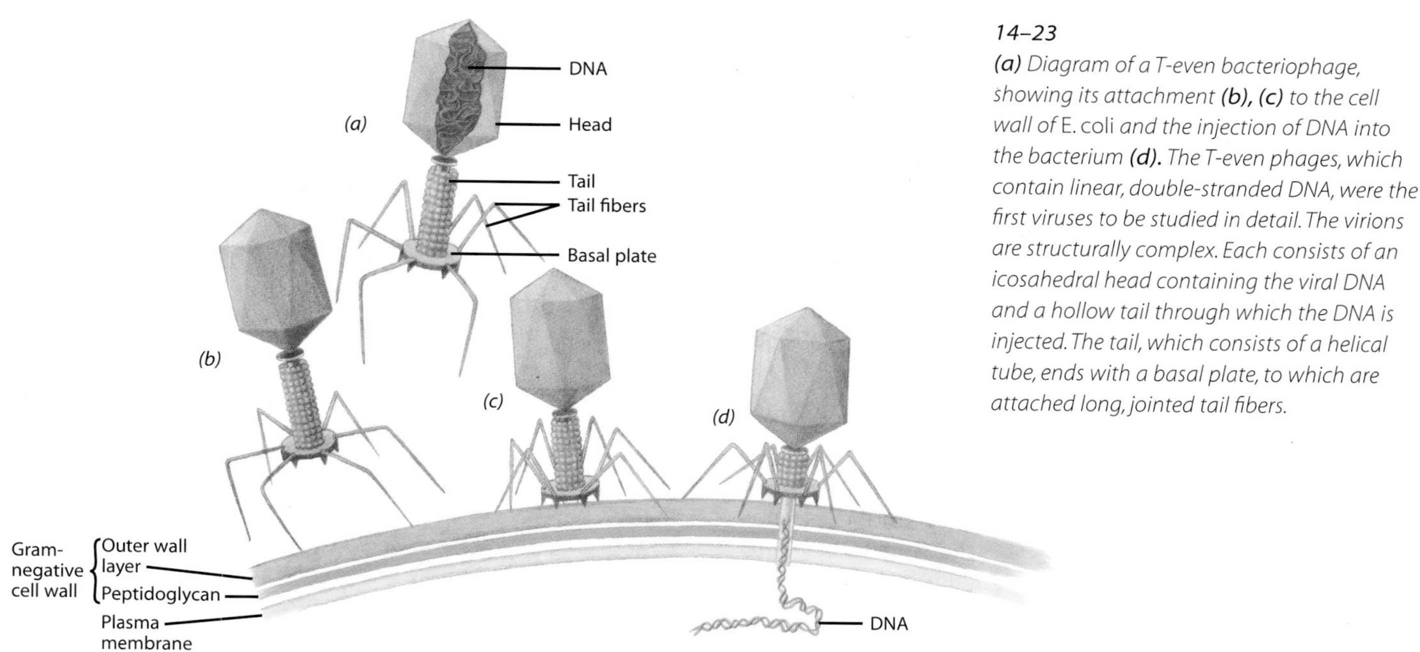

14–23
*(**a**) Diagram of a T-even bacteriophage, showing its attachment (**b**), (**c**) to the cell wall of* E. coli *and the injection of DNA into the bacterium (**d**). The T-even phages, which contain linear, double-stranded DNA, were the first viruses to be studied in detail. The virions are structurally complex. Each consists of an icosahedral head containing the viral DNA and a hollow tail through which the DNA is injected. The tail, which consists of a helical tube, ends with a basal plate, to which are attached long, jointed tail fibers.*

attack other animals, as well as prokaryotes (including mycoplasmas), protists, fungi, and plants. Viruses cause many plant diseases. In fact, about one-fourth of all known viruses attack and cause diseases of plants. Some of the viruses infecting plants are quite similar to those infecting animals. The main differences are in the proteins on the cell surface, which restrict a virus to a narrow ecological niche. Once inside a cell, viral gene expression and replication require specific interactions with other host components. Subsequent spread of viruses between plant cells requires yet other virus-host interactions. Any of these steps can be blocked by incompatibilities between virus and host components.

Viruses that infect bacteria are called **bacteriophages** ("bacteria eaters") or simply **phages** (Figure 14–23). Studies of bacteriophages were very important in the early development of molecular biology. Increasingly powerful approaches have also been developed for studying animal and plant viruses.

Experiments with bacteriophages have revealed many important and previously unappreciated features of eukaryotic cell biology, including RNA splicing and many aspects of DNA replication, transcription, and translation. Thus viruses have been and continue to be very important sources of insight into the general biology of cells. Viruses also serve as gene-transfer systems and have become indispensable tools of modern biotechnology, as discussed in Chapter 28.

Viruses Are Noncellular Structures Consisting of Protein and DNA or RNA

Viruses consist of a nucleic acid genome surrounded by protein, which protects the genome in an external environment and helps the virus attach to the next cell or host. Viruses may survive on their own outside a living cell, but increase in number only by replication in appropriate susceptible host cells. Outside the cell a virus is a submicroscopic particle that is metabolically inert. The virus particle, commonly called the **virion,** is the structure by which the virus genome is carried from one cell to another.

Viral genomes are composed of either RNA or DNA—double-stranded or single-stranded—depending on the virus. Viruses lack plasma membranes, cytoplasm with ribosomes, and enzymes necessary for protein synthesis and energy production. Viral genes, however, encode the information for the production of enzymes involved in nucleic acid replication and protein processing. Viral genomes are highly organized sequences of nucleotides with distinct genes and regulatory sequences. The regulatory sequences in DNA viruses often function in ways similar to those that regulate genes in cellular DNA. In RNA viruses, however, the mechanisms of regulation of gene expression often differ from the DNA-based pathways common in cells. The entire nucleotide sequence has been determined for hundreds of viruses and has shown the similarity of viruses infecting very different hosts.

The Viral Capsid Is Composed of Protein Subunits

All viruses have one or more proteins, called capsid or coat proteins, which assemble in a precise symmetrical manner to form the **capsid,** a shell-like covering that protects the nucleic acid. Some viruses also have an envelope of lipid molecules interspersed with proteins on the outer surface of the capsid. The surface proteins and lipids help recognize potential host cells and also provide targets for the immune response of animals combating a viral infection.

The structures of a large number of viruses have now been elucidated by electron microscopy and crystallography. The tobacco mosaic virus—the first to be seen, in 1939—is a rodlike particle about 300 nanometers long and 15 nanometers in diameter. The RNA of over 6000 nucleotide bases forms a single strand that fits into a groove in each of over 2000 identical protein molecules arranged with helical symmetry much like a coiled spring (Figure 14–24). Most plant viruses, like tobacco mosaic virus, are single-stranded RNA viruses. The few plant viruses with a double-stranded RNA genome all belong to the family *Reoviridae*. Rice gall dwarf disease is caused by a reovirus, specifically a phytoreovirus that is spread from plant to plant by leafhoppers.

The most common shape of a virus is an icosahedron, a 20-sided structure, in which the capsids are assembled from 180 or more protein molecules arranged in a symmetry similar to that of a geodesic dome (Figure 14–25). The size of most of the icosahedral plant viruses is about 30 nanometers in diameter. Viruses having this shape include tobacco ringspot, cucumber mosaic, cowpea chlorotic mottle (Figure 14–25a), and Tulare apple mosaic viruses (Figure 14–26a). Other plant viruses are bullet-shaped (Figure 14–26b) or appear as long, flexible rods or threads (Figure 14–26c).

Viruses Multiply by Taking Over the Genetic Machinery of the Host Cell

The transmission, or spread, of viruses from diseased to healthy plants most commonly involves insect vectors, such as aphids, leafhoppers, or whiteflies, with sucking mouthparts called stylets. Such insects that have fed on diseased plants carry viruses on their stylets and transmit the virus to a healthy plant when they feed upon it. Except for a few of the plant RNA viruses (see below), plant viruses do not actually infect their vectors, that is, the organisms that transmit them. Instead, plant viruses are simply carried in a very specific manner, and each type of plant virus is transmitted only by a particular

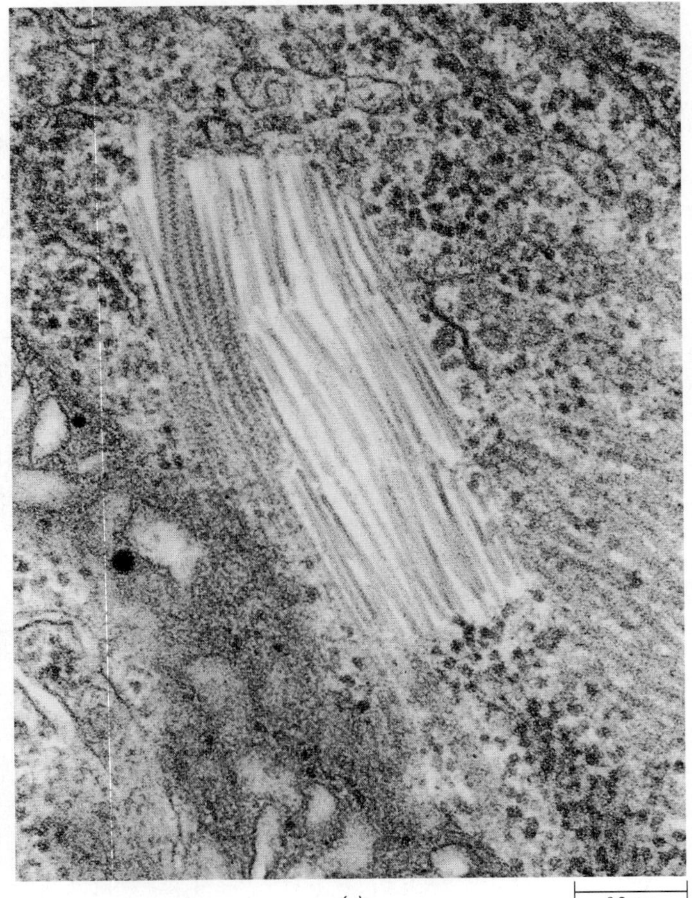

(a)

(b)

14–24
Tobacco mosaic virus (TMV). ***(a)*** *Electron micrograph showing virus particles in a mesophyll cell of a tobacco leaf.* ***(b)*** *A portion of the TMV particle, as determined by X-ray crystallography. The single-stranded RNA, shown here in red, fits into the grooves of the protein subunits that assemble into the helical arrangement of the capsid.*

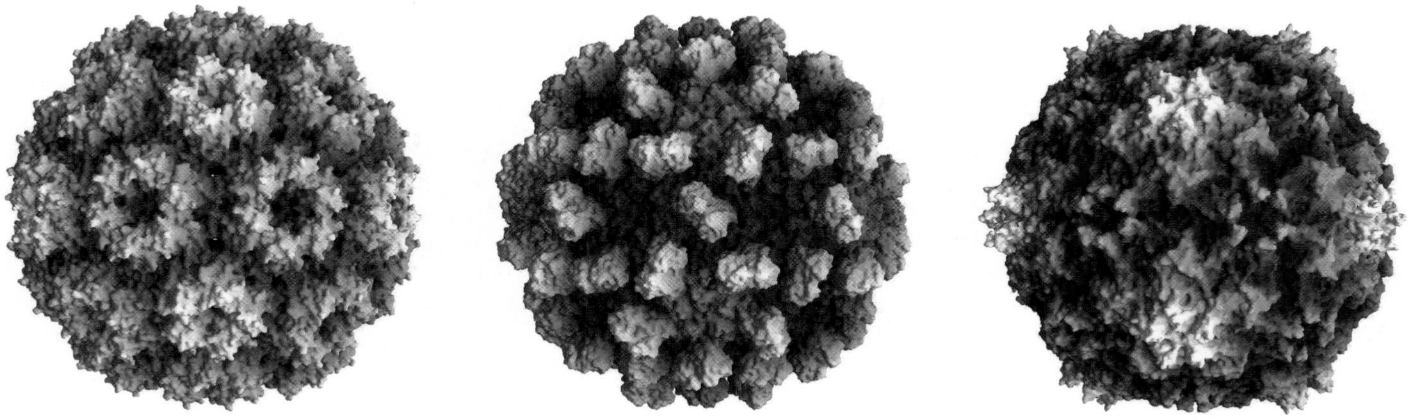

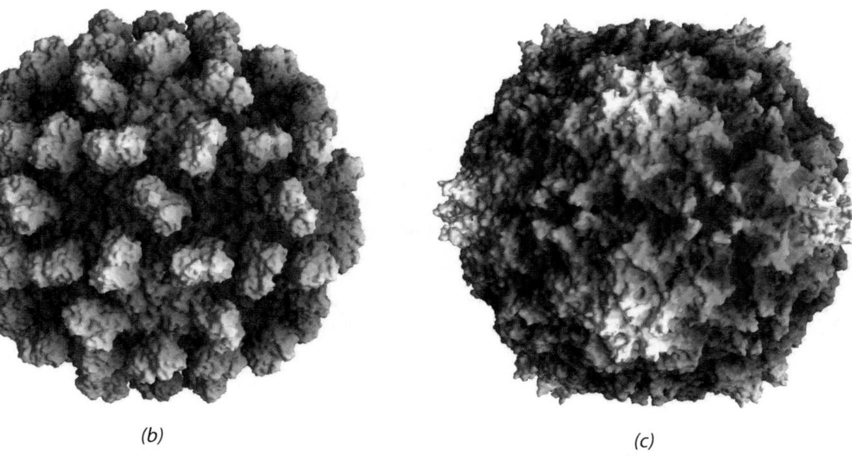

(a) (b) (c) 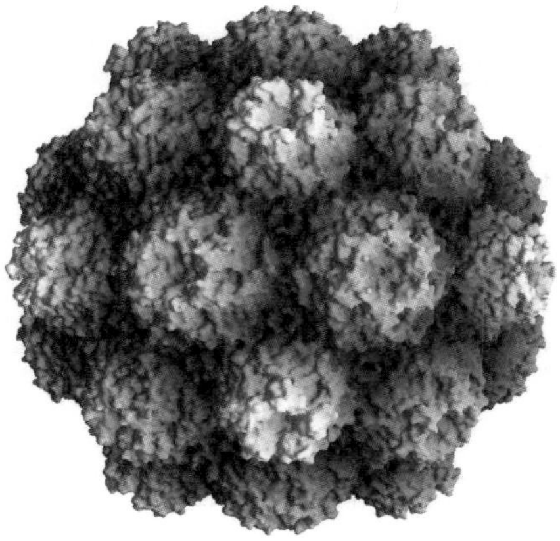 (d)

14–25

Plant viruses. Protein capsids, as determined by X-ray crystallography, for **(a)** *cowpea chlorotic mottle virus,* **(b)** *tomato bushy stunt virus,* **(c)** *southern bean mosaic virus, and* **(d)** *turnip yellow mosaic virus. All of these viruses are icosahedral; that is, their subunits are arranged in a 20-sided structure or a variation of it.*

14–26

Plant viruses; all of these are RNA viruses. **(a)** *A mixture of Tulare apple mosaic virus and tobacco mosaic virus. Tulare apple mosaic virus is icosahedral, whereas tobacco mosaic virus is rigid and elongated.* **(b)** *Particles of a rhabdovirus in a cell from a pepper plant* (Capsicum frutescens)*; this is a typical bullet-shaped virus. The rhabdovirus particles are seen here in the space between the inner and outer membranes of the nuclear envelope. As the virus migrates from this space through the inner nuclear membrane, it acquires a portion of that membrane, which then surrounds the particle.* **(c)** *Particles of a flexible, elongated virus in a cell from a Christmas cactus* (Zygocactus truncatus).

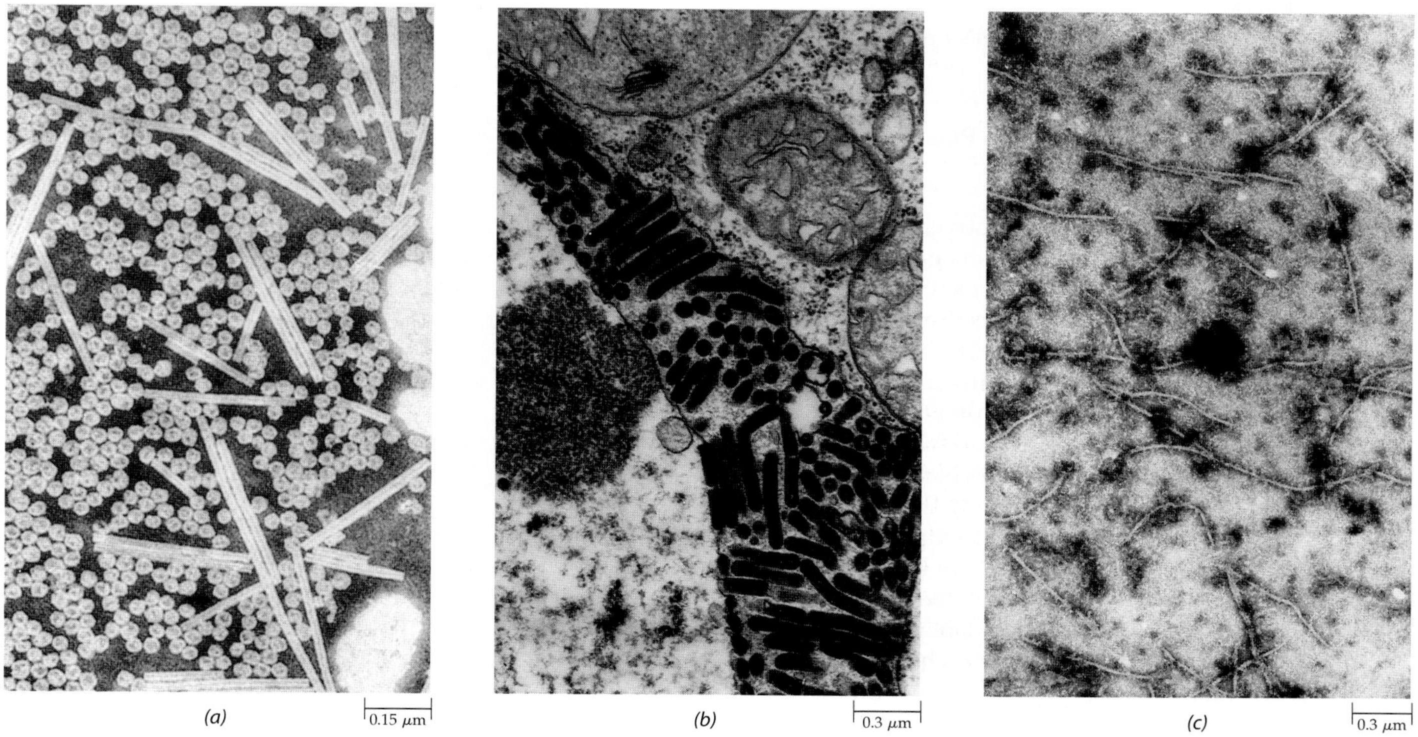

(a) 0.15 μm

(b) 0.3 μm

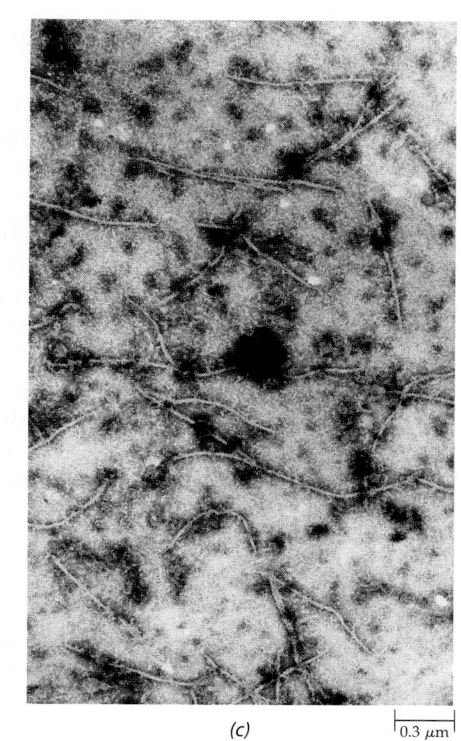

(c) 0.3 μm

vector or vector type. Other than by insect vectors, plant viruses can enter the plant only through wounds made mechanically or by transmission into an ovule via a pollen tube of an infected pollen grain.

Once inside a host cell, a virion sheds its capsid (and, if one is present, its envelope), freeing its nucleic acid. Within the cell, the RNA or DNA of the virus then multiplies by taking over the cell's genetic machinery, thus producing the nucleic acids and proteins necessary to assemble additional virus particles. In a single-stranded RNA virus, such as tobacco mosaic virus, the viral RNA directs the formation of a complementary strand of RNA, which then serves as the template for the production of new viral RNA molecules. The viral RNA, which is single-stranded, acts as messenger RNA, utilizing the ribosomes of its host cell and directing the synthesis of enzymes and of the protein subunits of the capsid. The new RNA strands and capsid protein subunits are assembled into complete virus particles within the host cell.

Three types of plant viruses, the geminiviruses, the badnaviruses, and the caulimoviruses, have DNA as their genetic material. Geminiviruses are small, spherical particles that almost always exist as connected pairs (Figure 14–27). Bean golden mosaic is a disease of bean plants caused by a geminivirus. It is spread from plant to plant by whiteflies and occurs mainly in tropical climates. Another such geminivirus causes maize (corn) streak, which is spread by leafhoppers and has the smallest known genome of any virus. Badnaviruses and caulimoviruses are larger viruses that replicate their DNA by means of an RNA intermediate, using a reverse transcription pathway very similar to that of the human immunodeficiency virus (HIV) and other retroviruses.

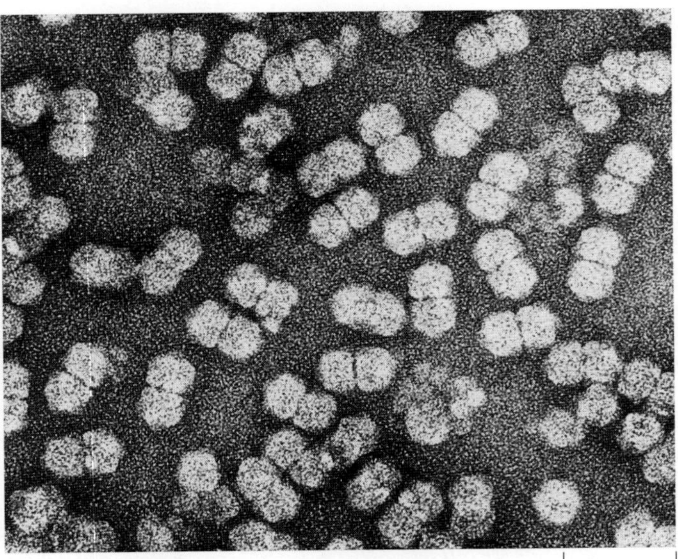

14–27
Purified geminivirus from the grass Digitaria, *negatively stained in 2 percent aqueous uranyl acetate. The geminiviruses, which have DNA as their genetic material, typically occur in pairs.*

Within the Plant, Viruses Move via Plasmodesmata, and Some Travel in the Phloem

Some viruses are confined to a relatively small area of the initial site of infection, whereas others move throughout the plant body and are said to be **systemic.** Short-distance movement from cell to cell occurs through plasmodesmata (Figure 14–28). Such movement is facilitated by viral-encoded proteins called **movement proteins,** which apparently bring about an increase in the size of the plasmodesmata and allow passage of the virus. Cell-to-cell movement via plasmodesmata is a slow process. In a leaf, for instance, the virus moves about 1 millimeter, or 8 to 10 parenchyma cells, per day.

Although some viruses appear to be restricted in their movement to parenchyma cells and their interconnecting plasmodesmata, a large number of viruses are rapidly transported in the conducting channels, or sieve tubes, of the phloem tissue. Once in the phloem, the virus moves systemically toward growing regions (for example, shoot tips and root tips) and regions of storage, such as rhizomes and tubers, where the virus reenters the parenchyma cells adjacent to the phloem. Viruses that depend on the phloem for successful establishment of infection are introduced by the vector directly into the phloem. Some phloem-dependent viruses, such as the beet yellows virus, seem to be limited to the phloem and a few adjacent parenchyma cells. Tobacco mosaic virus,

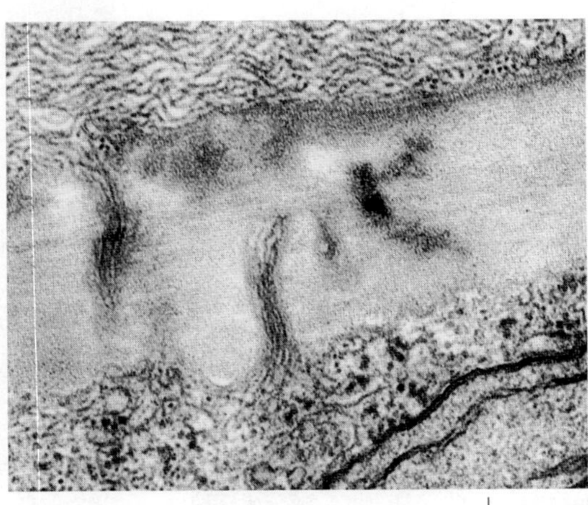

14–28
Beet yellows virus particles in plasmodesmata, moving from a sieve-tube member of the phloem (above) to its sister cell, a companion cell (below).

by contrast, invades all kinds of cells. Upon entering the sieve tubes, the viruses are transported to other parts of the plant along with the sugars and other substances present in the sieve-tube sap.

Viruses Cause a Variety of Plant Diseases

Well over 2000 kinds of plant diseases are known to be caused by the more than 600 different kinds of plant viruses identified. Viral diseases greatly reduce the productivity of many kinds of agricultural and horticultural crops. Worldwide losses due to viral diseases are estimated at about $15 billion annually.

Often the only symptom produced by virus infection is reduced growth rate, resulting in various degrees of dwarfing or stunting of the plant. The most obvious symptoms are usually those that appear on leaves, where the viruses interfere with chlorophyll production, thus affecting photosynthesis. Mosaics and ring spots are the most common symptoms produced by systemic viruses. In mosaic diseases, light green, yellow, or white areas—ranging in size from small flecks to large stripes—appear intermingled with the normal green of leaves and fruit (Figure 14–29). In ring spot diseases, chlorotic (yellow) or necrotic (dead tissue) rings appear on the leaves and sometimes also on the stems and fruits. Less common viral diseases include leaf roll (potato leaf roll), yellows (beet yellows), dwarf (barley yellow dwarf), canker (cherry black canker), and tumor (wound tumor) (Figure 14–30). The yellow blotches or borders on the leaves of some prized horticultural varieties may be caused by viruses, and the variegated appearance of some flowers is the result of viral infections that are passed on from generation to generation in vegetatively propagated plants (Figure 14–31).

14–29
Tobacco leaf infected with tobacco mosaic virus.

14–30
***(a)** Tumors produced by the wound tumor virus in sweet clover* (Melilotus alba). ***(b)** Wound tumor virus particles (arrows) visible in an electron micrograph of a cell of the host plant.* ***(c)** The virus is transmitted by the clover leafhopper* (Agallia constricta). *This electron micrograph shows a clover epidermal cell. Viruses are being produced in the honeycomblike area at the upper left. Individual viruses can be seen in the dark area below.*

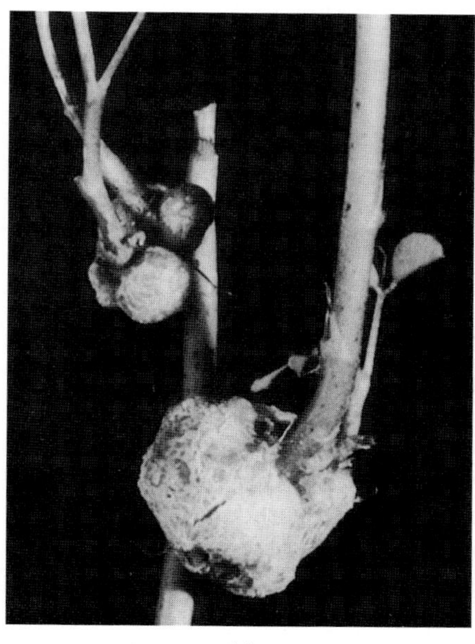

(a)

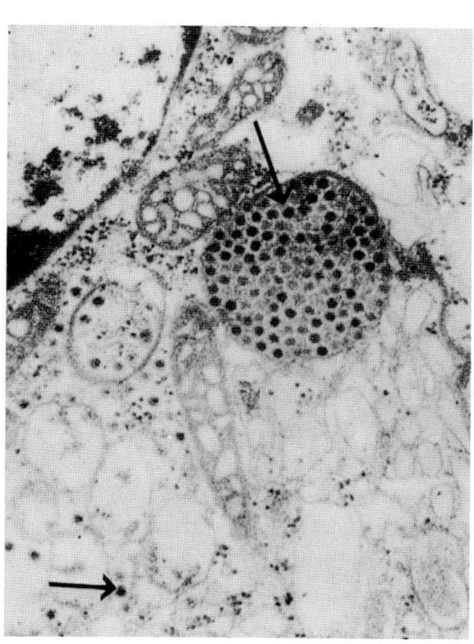

(b)

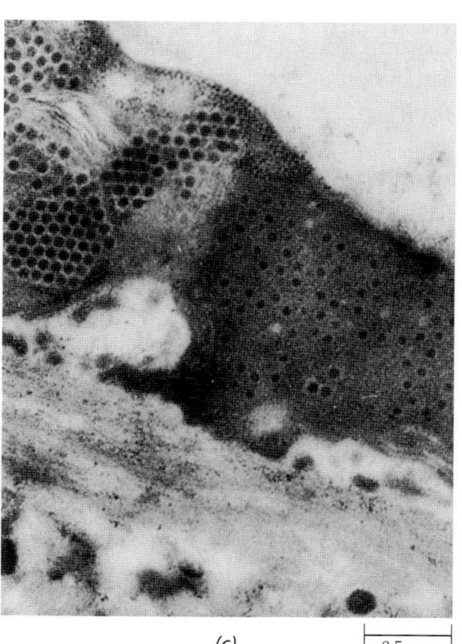

(c)

0.5 μm

14–31
Streaked flowers of Rembrandt tulips. The streaking is caused by a viral infection transmitted directly from plant to plant. The infection weakens the plant, and it subsequently dies. Current practice is to achieve streaking by selecting for transposon effects (page 193).

Viral Diseases Are Controlled in Various Ways Because viral diseases greatly reduce the productivity of many different kinds of crops worldwide, or decrease the desirability of the product, much effort is being expended to find effective ways to control these diseases. Chemicals sprayed onto infected plants are not effective, and vector control is often not practical and efficient. For the vegetatively propagated plants, including potatoes and many flower and fruit crops, meristem culture has been effective. The virus often does not invade the growing tip, so a new plant grown in tissue culture from the tip dissected away from the parent plant will often be virus-free. Such plants are then used as "mother" plants from which cuttings are produced to supply growers with virus-free plants. Eliminating seedborne virus is also effective in many crops, including peas, beans, barley, and lettuce. Samples of seeds are grown, and if the virus is present, the entire seed lot is discarded.

Another approach to the control of plant viruses has focused on the ability of a host plant to prevent the virus from replicating or from moving within the plant body, thus making the host plant resistant. As mentioned previously, viral-encoded movement proteins appear to modify plasmodesmata, allowing viruses to pass into adjacent cells. Some movement proteins display substantial host specificity and can only support significant cell-to-cell spread in certain hosts. Interestingly, in at least some cases, such misadapted movement proteins are competent for initial cell-to-cell spread in nonhost plants but lead to rapid induction of host resistance responses that block infection of further cells. Other types of plant resistance involve preventing the initial replication of the virus, blocking long-distance movement, or responding with a hypersensitive reaction that limits the virus by killing cells near the point of infection.

Resistance is known to be conditioned by specific host genes, and one such gene, the *N* gene from tobacco, has recently been cloned and characterized. Several other host resistance genes have been genetically mapped to chromosomal locations, and still others are widely used in developing varieties of crops that resist viruses. Viruses, like many other plant pathogens, often mutate and diversify to overcome these resistance genes. Plants can be genetically engineered, however, to resist virus infection by inserting any of several pieces of the viral nucleic acid into the host's genome. Some sequences work by completely blocking virus replication, whereas others greatly decrease the ability of the virus to move from cell to cell. Squash was the first crop to be marketed with genetically engineered resistance to viruses, but several others are expected to be used in cases where there is no other means of viral disease control.

Viroids: Other Infectious Particles

Viroids are the smallest known agents of infectious disease. They consist of small, circular, single-stranded molecules of RNA and lack capsids of any kind (Figure 14–32). Ranging in size from 246 base pairs (the coconut cadang-cadang viroid) to 375 base pairs (the citrus exocortis viroid), viroids are much smaller than the smallest viral genomes. Although the viroid RNA is a single-stranded circle, it can form a secondary structure that resembles a short double-stranded molecule with closed ends. The viroid RNA contains no protein-encoding genes, and hence it is totally dependent on the host for its replication. The viroid RNA molecule appears to be replicated in the nucleus of the host cell, where it apparently mimics DNA and allows the host cell RNA polymerase to replicate it. Viroids may cause their symptoms by interfering with gene regulation in the infected host cell.

The term "viroid" was first used by Theodor O. Diener of the U.S. Department of Agriculture in 1971, to describe the infectious agent causing the spindle tuber disease of potato. Potatoes infected by the potato spindle tuber viroid, or PSTVd (the d is added to the initials of viroids to distinguish them from abbreviations for virus names), are elongated (spindle-shaped) and gnarled.

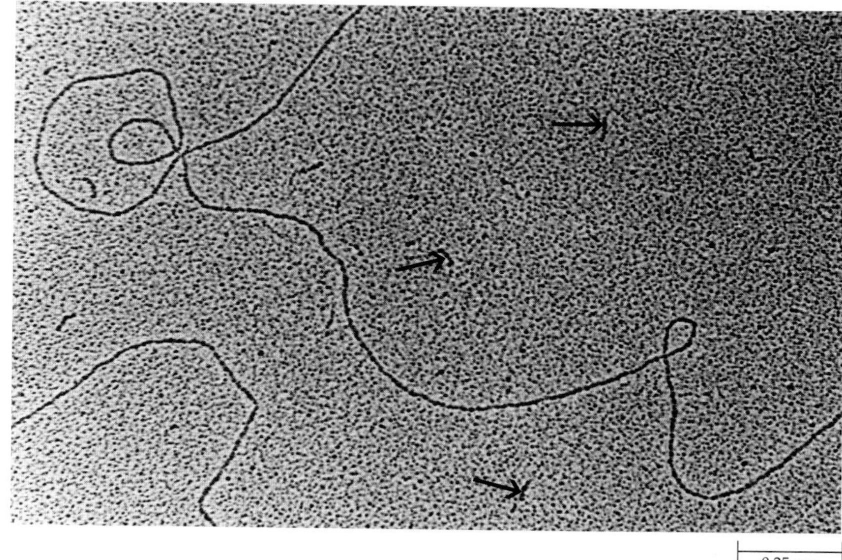

14–32
Electron micrograph of potato spindle tuber viroid (arrows) mixed with portions of the double-stranded DNA molecule of a bacteriophage. This micrograph illustrates the tremendous difference in size between a viroid and a virus.

They sometimes have deep crevices in their surfaces. Viroids have been identified as the cause of several economically important plant diseases. The coconut cadang-cadang viroid (CCCVd), for example, has resulted in the death of millions of coconut trees in the Philippines over the last half century, and the chrysanthemum stunt viroid (CSVd) nearly destroyed the chrysanthemum industry in the United States in the early 1950s.

The Origin of Viruses

Because of the simplicity of viral genomes and viral structure, some earlier investigators thought that viruses might represent the direct descendants of the self-replicating units from which the first cells evolved. Other investigators think that this line of descent is unlikely, for viruses exist only by virtue of their ability to utilize the genetic machinery of their host cells. They compete with the nucleic acids of these cells and take over the host genetic and metabolic activities in directing the formation of new viral particles. Therefore, viruses may have come into being after the evolution of cells in which the genetic code was already established.

Viruses may have originated as segments of host genomic material that escaped and acquired the ability to replicate independently in a cell and be transmitted to another cell. Acquisition of a protein that could protect the nucleic acid in the transmission process, and facilitate recognition of new host cells, was key to the evolution of viruses. Viruses can evolve with astonishing rapidity in response to strong selection pressures. It is likely that novel viruses are still evolving today from both bacteria and eukaryotes.

Summary

The *Bacteria* and *Archaea* Are the Two Prokaryotic Domains

The prokaryotes are the smallest and structurally simplest organisms. In evolutionary terms, they are also the oldest organisms on Earth, and they consist of two distinct lineages, the domains *Bacteria* and *Archaea.* Prokaryotes lack an organized nucleus and membrane-bounded cellular organelles. Most of their genetic material is incorporated into a single circular molecule of double-stranded DNA, which replicates before cell division. Often, additional small pieces of circular DNA, known as plasmids, are also present. Except for mycoplasmas, all prokaryotes have rigid cell walls. In the *Bacteria,* this wall is composed mainly of peptidoglycan. Gram-negative *Bacteria,* which have walls that do not retain the dye known as crystal violet, have an outer layer of lipopolysaccharides and proteins over the peptidoglycan layer. Many prokaryotes secrete slimy or gummy substances on the surface of their walls, forming a layer called the glycocalyx, or capsule. A wide variety of prokaryotes—both *Bacteria* and *Archaea*—contain granules of poly-β-hydroxybutyric acid and glycogen, which are food storage compounds.

Prokaryotic cells may be rod-shaped (bacilli), spherical (cocci), or spiral (spirilla). All prokaryotes are unicellular, but, if the cell wall does not divide completely following cell division, the daughter cells may adhere in groups, in filaments, or in solid masses. Many prokaryotes have flagella and, hence, are motile; the rotation of the flagella moves the cell through the medium. Lacking microtubules, the flagella of prokaryotes differ greatly from those of eukaryotes. Prokaryotes may also have fimbriae or pili.

Most prokaryotes reproduce by simple cell division, also called binary fission. Mutations, combined with the rapid generation time of prokaryotes, are responsible for their extraordinary adaptability. Further adaptability is provided by the genetic recombinations that take place as a result of conjugation, transformation, and transduction. Certain species of bacteria have the capacity to form endospores, dormant resting cells that allow the cell to survive unfavorable conditions.

Prokaryotes Exhibit Tremendous Metabolic Diversity

Although some are autotrophs, most prokaryotes are heterotrophs. The vast majority of heterotrophs are saprophytes and, together with the fungi, they are the recyclers of the biosphere. Some autotrophs—the photosynthetic autotrophs—obtain their energy from light. Other autotrophs obtain their energy from the reduction of inorganic compounds and are called chemosynthetic autotrophs. A number of genera play important roles in the cycling of nitrogen, sulfur, and carbon. Of all living organisms, only certain bacteria are capable of nitrogen fixation. Without bacteria, life on Earth as we know it would not be possible.

Some prokaryotes are aerobic, others are strict anaerobes, and still others are facultative anaerobes. Prokaryotes also vary with regard to the range of temperatures at which they grow, ranging from those that can grow at 0°C or lower (psychrophiles) to those that can grow at temperatures greater than 100°C (extreme thermophiles).

The *Bacteria* Include Pathogenic and Photosynthetic Organisms

Many bacteria are important pathogens in both plants and animals. One distinctive group of bacteria, the mycoplasmas (and mycoplasmalike organisms), which lack a cell wall and are very small, include a number of disease-causing organisms.

Photosynthetic bacteria can be divided into three major groups: the cyanobacteria, the prochlorophytes, and the purple and green bacteria. The cyanobacteria and prochlorophytes contain chlorophyll *a*, the same molecule that occurs in all photosynthetic eukaryotes. In addition, the prochlorophytes contain chlorophyll *b*, but lack phycobilins, which are present in cyanobacteria. Many genera of cyanobacteria can fix nitrogen. Clearly *Bacteria* seem to have been involved in the symbiotic origin of chloroplasts. Early in the history of eukaryotes, a symbiotic event similar to that occurring in the origin of chloroplasts seems to have given rise to mitochondria; purple nonsulfur bacteria apparently were the group involved in this event.

The *Archaea* Are Physiologically Diverse Organisms That Occupy a Wide Variety of Habitats

The *Archaea* can be divided into three large groups: extreme halophiles, methanogens, and extreme thermophiles. A fourth group is represented by a single genus, *Thermoplasma,* which lacks a cell wall. Once believed to occupy primarily hostile environments, the *Archaea* are now known to constitute a major component of the oceanic picoplankton.

Viruses Are Noncellular Structures Consisting of DNA or RNA Surrounded by a Coat of Protein

Viruses possess genomes that replicate within a living host by directing the genetic machinery of that cell to synthesize viral nucleic acids and proteins. Viruses contain either RNA or DNA—single-stranded or double-stranded—surrounded by an outer protein coat, or capsid, and sometimes also by a lipid-containing envelope. Viruses are comparable in size to large macromolecules and come in a variety of shapes. Most are spherical, with icosahedral symmetry, and a number of others are rod-shaped, with helical symmetry.

Viruses and Viroids Cause Diseases in Plants and Animals

Viruses are responsible for many diseases of humans and other animals, and also for more than 2000 known kinds of plant diseases. The transmission of viruses from diseased to healthy plants most commonly involves insect vectors. Once inside a host cell, the virus particle, or virion, sheds its capsid, freeing its nucleic acid. Most plant viruses are RNA viruses. In single-stranded RNA viruses, such as tobacco mosaic virus, the viral RNA directs the formation of a complementary strand of RNA, which then serves as a template for the production of new viral RNA molecules. Utilizing the ribosomes of the host cell, the viral RNA directs the synthesis of capsid proteins. The new RNA strands and capsid proteins then are assembled into complete virions within the host cell.

Short-distance movement of viruses from cell to cell within host plants occurs through plasmodesmata. Such movement is facilitated by viral-encoded proteins called movement proteins. A large number of plant viruses move systemically throughout the plant in the phloem.

Viroids, the smallest known agents of infection, consist of small, circular, single-stranded molecules of RNA. Unlike viruses, viroids lack protein coats. They are thought to interfere with gene regulation in infected host cells, where they occur mainly within the nucleus.

Selected Key Terms

aerobes p. 287
akinetes p. 290
anaerobes p. 287
bacilli p. 284
bacteriophage, or **phage** p. 297
binary fission p. 285
capsid p. 298
chemosynthetic autotrophs p. 287
cocci p. 284
conjugation p. 285
cyanobacteria p. 288
endospores p. 286
facultative anaerobes p. 287
fimbriae p. 284
glycocalyx p. 284
gram-negative p. 283
gram-positive p. 283
halophiles p. 295
heterocysts p. 290
heterotrophs p. 286
hormogonia p. 288
methanogens p. 295
movement proteins p. 300
mycoplasmalike organisms (MLOs) p. 292
mycoplasmas p. 292
nitrogen fixation p. 290
nucleoid p. 282
peptidoglycans p. 283
photosynthetic autotrophs p. 287
pili p. 284
plankton p. 289
plasmid p. 282
prochlorophytes p. 292
psychrophiles p. 287
saprophytes p. 286
spirilla p. 284
spiroplasmas p. 292
strict anaerobes p. 287
stromatolites p. 289
thermophiles p. 287
transduction p. 286
transformation p. 286
virion p. 297
viroids p. 302

Questions

1. Distinguish between the following: gram-positive bacterium/gram-negative bacterium; fimbria/pilus; endospore/akinete; virus/viroid.
2. Pear decline is so called because it causes a slow, progressive weakening and ultimate death of the pear tree. It is a systemic disease caused by an MLO, or phytoplasma. What is meant by a "systemic disease," and by what pathway do MLOs move through the tree?
3. What genetic factors contribute to the extraordinary adaptability of prokaryotes to a broad range of environmental conditions?
4. What are some of the ways by which plant viral diseases may be controlled?
5. One might argue that viruses should be considered living organisms. By what criteria might viruses be considered alive?

Chapter 15

Fungi

15–1
Mycelia of several different fungi growing under the bark of a fallen tree. Decomposition by fungi and bacteria makes it possible for the organic material incorporated into the bodies of organisms to be recycled in the ecosystem.

OVERVIEW

Fungi are literally everywhere on Earth. They absorb their food by secreting digestive enzymes into the immediate environment. These enzymes catalyze the breakdown of large food molecules into molecules small enough to be absorbed into the fungal cell. For this reason, fungi usually grow within or on top of their food supply.

This absorptive mode of nutrition is responsible for both the praises and curses earned by fungi. On the one hand, fungi are extremely beneficial because the digestive enzymes they secrete are responsible for decomposing dead plants and animals, making it possible to recycle chemical elements. However, fungi also secrete enzymes that digest materials valuable to us, causing mold on food, mildew on fabrics, dry rot on wood, and athlete's foot. Commercial use of fungi and their enzymes produces wine, beer, certain cheeses, and various antibiotics.

We first examine the groups that make up the kingdom Fungi *and then explore the important symbiotic relationships of fungi with other organisms.*

CHECKPOINTS

By the time you finish reading this chapter, you should be able to answer the following questions:

1. In what ways do the fungi differ from all other life forms? In other words, what are the distinctive characteristics of the *Fungi?*
2. From what type of organism is it thought that fungi evolved?
3. What are the distinguishing characteristics of the *Chytridiomycota, Zygomycota, Ascomycota,* and *Basidiomycota?*
4. What are the deuteromycetes, and what is their relationship to other groups of fungi?
5. What is a yeast, and what is the relationship of yeasts to the other groups of fungi?
6. What kinds of symbiotic relationships exist between fungi and other organisms?

The fungi are heterotrophic organisms that were once considered to be primitive or degenerate plants lacking chlorophyll. It is now clear, however, that the only characteristic fungi share with plants—other than those common to all eukaryotes—is a multicellular growth form. (A few fungi, including yeasts, are unicellular.) Recent molecular evidence strongly suggests that fungi are more closely related to animals than to plants. As we shall see, the fungi are a form of life so distinctive from all others that they have been assigned their own kingdom—the kingdom *Fungi.*

Over 70,000 species of fungi have been identified thus far, with some 1700 new species discovered each year. Conservative estimates for the total number of species exceed 1.5 million, placing the fungi second only to insects in this regard. The largest living organism on Earth today may be an individual of the tree root-rot fungus *Armillaria ostoyae,* which encompasses more than 600 hectares (1500 acres) of forest near Mount Adams in the state of Washington. This fungus is estimated to be 400 to 1000 years old. A close relative of *A. ostoyae, Armillaria gallica,* has been found occupying 15 hectares (35 acres) in northern Michigan. This "humongous fungus," as it has been dubbed, is estimated to be at least 1500 years old.

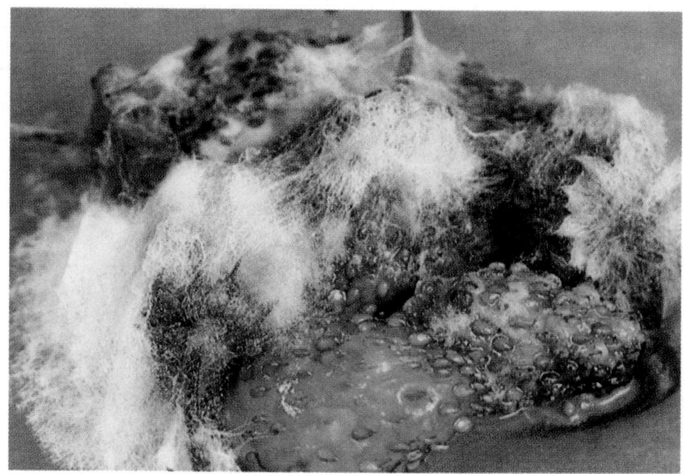

15–2
The common mold Rhizopus *growing on strawberries.*

The Importance of Fungi

Fungi Are Important Ecologically As Decomposers

The ecological impact of the fungi cannot be overestimated. Together with the heterotrophic bacteria, fungi are the principal decomposers of the biosphere (Figure 15–1). Decomposers are as necessary to the continued existence of the world as are food producers. Decomposition releases carbon dioxide into the atmosphere and returns nitrogenous compounds and other materials to the soil, where they can be used again—recycled—by plants and eventually by animals. Estimates are that, on average, the top 20 centimeters of fertile soil contains nearly 5 metric tons of fungi and bacteria per hectare (2.47 acres). Some 500 known species of fungi, representing a number of distinct groups, are marine, breaking down organic material in the sea just as their relatives do on land.

As decomposers, fungi often come into direct conflict with human interests. A fungus makes no distinction between a rotting tree that has fallen in the forest and a fence post; the fungus is just as likely to attack one as the other. Equipped with a powerful arsenal of enzymes that break down organic substances, including lignin and cellulose of wood, fungi are often nuisances and are sometimes highly destructive. Fungi attack cloth, paint, leather, waxes, jet fuel, petroleum, wood, paper, insulation on cables and wires, photographic film, and even the coating of the lenses of optical equipment—in fact, almost any conceivable substance. Although individual species of fungi are highly specific to particular substrates, as a group they attack virtually anything. Everywhere, they are the scourge of food producers, distributors, and sellers alike, for they grow on bread, fresh fruits (Figure 15–2), vegetables, meats, and other products. Fungi reduce the nutritional value, as well as the palatability, of such foodstuffs. In addition, some produce very toxic substances known as **mycotoxins** on certain plant materials.

Fungi Are Important Medically and Economically As Pests, Pathogens, and Producers of Certain Chemicals

The importance of fungi as commercial pests is enhanced by their ability to grow under a wide range of conditions. Some strains of *Cladosporium herbarum,* which attacks meat in cold storage, can grow at a temperature as low as −6°C. In contrast, one species of *Chaetomium* grows optimally at 50°C and survives even at 60°C.

Many fungi attack living organisms, rather than dead ones, and sometimes do so in surprising ways (see "Predaceous Fungi" on page 333). They are the most important causal agents of plant diseases. Well over 5000 species of fungi attack economically valuable crop and garden plants, as well as trees and many wild plants. Other fungi—over 150 species have been identified—cause serious diseases in domestic animals and humans.

Although fungal infections of humans are most common in tropical regions of the world, an alarming increase has occurred in the numbers of individuals infected with fungi in all regions of the world. This increase is due in part to the growing population of individuals with compromised immune systems, such as AIDS patients in hospitals. Nearly 40 percent of all deaths from hospital-acquired infections in the mid-1980s were found to be due not to bacteria or viruses but to fungi. Around 80 percent of AIDS deaths are due to pneumonia caused by *Pneumocystis carinii,* long thought to be a protozoan. Evidence now classifies *P. carinii* as a fungus, clearly an ascomycete. Another serious fungal pathogen of AIDS patients is *Candida,* which causes thrush and other infections of the mucous membranes.

The qualities that make fungi such important pests can also make them commercially valuable. Certain yeasts, such as *Saccharomyces cerevisiae,* are useful because they produce ethanol and carbon dioxide, which play a central role in baking, brewing, and winemaking. Other fungi provide the distinctive flavors and aromas of specific kinds of cheese. The commercial use of fungi in industry is growing, and many antibiotics—including penicillin, the first antibiotic to be used widely—are produced by fungi. Dozens of different kinds of fungi (mushrooms) are eaten regularly by humans, and some of them are cultivated commercially. The ability of fungi to break down substances is leading to investigations into the use of fungi in toxic waste cleanup programs. The white rot fungus *Phanerochaete chrysosporium,* which survives by digesting woody material, has been very effective in the degradation of toxic organic compounds.

A striking example of the potential value of compounds derived from fungi is cyclosporin, a "wonder drug" isolated from the soil-inhabiting fungus *Tolypocladium inflatum.* Cyclosporin suppresses the immune reactions that cause rejection of organ transplants, but without the undesirable side effects of other drugs used for this purpose. This remarkable drug became available in 1979, making it possible to resume organ transplants, which had essentially been abandoned. As a result of cyclosporin, successful organ transplants have today become almost commonplace.

Fungi Form Important Symbiotic Relationships

The kinds of relationships between fungi and other organisms are extremely diverse. For example, at least 80 percent of all vascular plants form mutually beneficial associations, called **mycorrhizae,** between their roots and fungi. These associations, which are discussed further beginning on page 340, play a critical role in plant nutrition. Lichens, many of which occupy extremely hostile habitats, are symbiotic associations between fungi and either algal or cyanobacterial cells (see page 334). Symbiotic relationships also exist between fungi and insects. In one such relationship, the fungi, which produce cellulase and other enzymes needed for digestion of plant material, are cultivated by ants in fungus "gardens." The ants supply the fungus with leaf cuttings and anal droppings, and the ants eat nothing but the fungus. Neither the fungus nor the ants can exist without the other. Other symbiotic relationships involve a great variety of fungi, known as **endophytes,** that live inside the leaves and stems of apparently healthy plants. Many of these fungi produce toxic secondary metabolites that appear to protect their hosts against pathogenic fungi and attack by insects and grazing mammals (see "From Pathogen to Symbiont: Fungal Endophytes" on page 335).

TABLE 15–1 **Major Characteristics of Fungal Phyla**

Phylum	Representatives	Nature of Hyphae	Method of Asexual Reproduction	Type of Sexual Spore	Common Plant Diseases
Chytridiomycota (790 species)	*Allomyces, Coelomomyces*	Aseptate, coenocytic	Zoospores	None	Brown spot of corn, crown wart of alfalfa, black wart of potato
Zygomycota (1060 species)	*Rhizopus* (common bread mold), *Glomus* (endomycorrhizal fungus)	Aseptate, coenocytic	Nonmotile spores	Zygospore (in zygosporangium)	Soft rot of various plant parts
Ascomycota (32,300 species)	*Neurospora,* powdery mildews, *Morchella* (edible morels), *Tuber* (truffles)	Septate	Budding, conidia (nonmotile spores), fragmentation	Ascospore	Powdery mildew, brown rot of stone fruits, chestnut blight, Dutch elm disease
Basidiomycota (22,244 species)	Mushrooms (*Amanita,* poisonous; *Agaricus,* edible), stinkhorns, puffballs, shelf fungi, rusts, smuts	Septate with dolipore	Budding, conidia (nonmotile spores, including urediniospores), fragmentation	Basidiospore	Black stem rust of wheat and other cereals, white pine blister rust, common corn smut, loose smut of oats, *Armillaria* root rot

15–3
Fungi. ***(a)*** *The chytrid* Polyphagus euglenae *parasitizing a* Euglena *cell. The cytoplasm of the rounded-up* Euglena *cell is degraded.* ***(b)*** *A flower fly* (Syrphus) *that has been killed by the fungus* Entomophthora muscae, *a zygomycete.* ***(c)*** *A common morel,* Morchella esculenta, *an ascomycete. The morels are among the most prized of the edible fungi.* ***(d)*** *A mushroom,* Hygrocybe aurantiosplendens, *a species of* Basidiomycetes. *A mushroom is made up of densely packed hyphae, collectively known as the mycelium.*

Today most mycologists recognize four phyla of fungi: *Chytridiomycota,* the chytrids; *Zygomycota,* zygomycetes; *Ascomycota,* ascomycetes; and *Basidiomycota, Basidiomycetes, Teliomycetes,* and *Ustomycetes* (Figure 15–3; Table 15–1). Reinclusion of the chytrids in the kingdom *Fungi* is very recent. About 15 to 20 years ago, the chytrids were included in the kingdom *Fungi,* but fungal systematists began to emphasize the absence of flagellated cells as a requirement for membership in the kingdom. Although chytrids were known to share many characteristics with the fungi, their formation of flagellated cells (zoospores) disqualified them as fungi. Instead, chytrids were placed in the kingdom *Protista.* Nevertheless, recent evidence obtained from comparisons of protein and nucleic acid sequences has tipped the scales in favor of placing the chytrids in the kingdom *Fungi.* We have therefore included the *Chytridiomycota* in the *Fungi,* but it should be noted that the systematics of the fungi is in a state of flux.

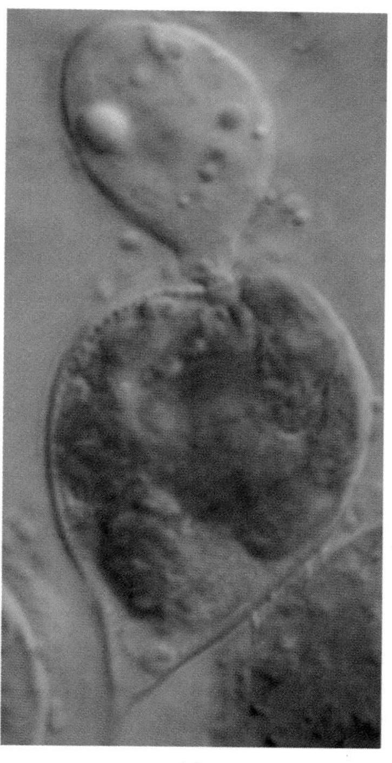

(a)

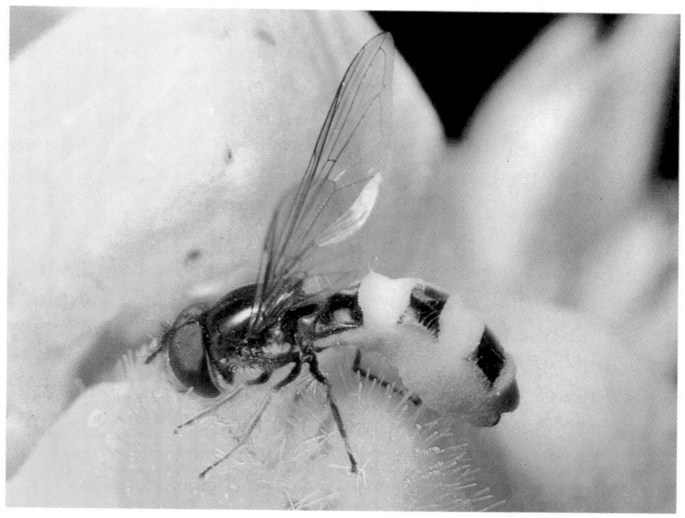

(b)

(c)

(d)

Biology and Characteristics of Fungi

Most Fungi Are Composed of Hyphae

Fungi are primarily terrestrial. Although some fungi are unicellular, most are filamentous, and structures such as mushrooms consist of a great many such filaments, packed tightly together (Figure 15–3c, d). Fungal filaments are known as **hyphae,** and a mass of hyphae is called a **mycelium** (Figure 15–1). Growth of hyphae occurs at their tips, but proteins are synthesized throughout the mycelium. The hyphae grow rapidly. An individual fungus may produce more than a kilometer of new hyphae within 24 hours. (The words "mycelium" and **mycology**—the study of fungi—are derived from the Greek word *mykēs,* meaning "fungus.")

The hyphae of most species of fungi are divided by partitions, or crosswalls, called **septa** (singular: septum). Such hyphae are said to be **septate.** In other species, septa typically occur only at the bases of reproductive structures (sporangia and gametangia) and in older, highly vacuolated portions of hyphae. Hyphae lacking septa are said to be **aseptate,** or **coenocytic,** which means "contained in a common cytoplasm" or multinucleate. In most fungi, the septa are perforated by a central pore so that the protoplasts of adjacent cells are essentially continuous from cell to cell. In members of the *Ascomycota,* the pores are usually unobstructed (Figure 15–4) and usually large enough to allow nuclei, which are quite small, to squeeze through. Such mycelia are therefore functionally coenocytic. The nuclei of fungal hyphae are haploid.

All fungi have cell walls. The cell walls of plants and many protists are built on a framework of cellulose microfibrils, interpenetrated by a matrix of noncellulosic molecules, such as hemicelluloses and pectic substances. In fungi, the cell wall is composed primarily of another polysaccharide—**chitin**—which is the same material found in the hard shells, or exoskeletons, of arthropods, such as insects, arachnids, and crustaceans. Chitin is more resistant to microbial degradation than is cellulose.

With their rapid growth and filamentous form, fungi have a relationship to their environment that is very different from that of any other group of organisms. The surface-to-volume ratio of fungi is very high, so they are in as intimate a contact with the environment as are the bacteria. Usually no somatic part of a fungus is more than a few micrometers from its external environment, being separated from it only by a thin cell wall and the plasma membrane. (The terms **somatic** and **soma,** from the Greek *sōma,* meaning "body," are equivalent to the term "vegetative" for plants.) With its extensive mycelium, a fungus can have a profound effect on its surroundings—for example, in binding soil particles together. Hyphae of individuals of the same species often fuse, thus increasing the intricacy of the network.

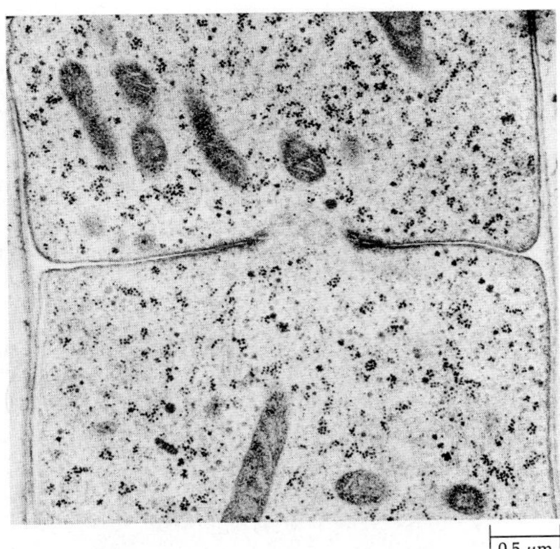

15–4
Electron micrograph of a septum between two cells in the ascomycete Gibberella acuminata. *The large globular structures are mitochondria, and the tiny dark granules are ribosomes. This specimen was thin-sectioned through the central pore region of a septum.*

Fungi Are Heterotrophic Absorbers

Because their cell walls are rigid, fungi are unable to engulf small microorganisms or other particles. Typically a fungus will secrete enzymes onto a food source and then absorb the smaller molecules that are released. Fungi absorb food mostly at or near the growing tips of their hyphae.

All fungi are heterotrophic. In obtaining their food, they function either as saprophytes (living on organic materials from dead organisms), as parasites, or as mutualistic symbionts (see page 334). Some fungi, mainly yeasts, obtain their energy by fermentation, producing ethyl alcohol from glucose. Glycogen is the primary storage polysaccharide in some fungi, as it is in animals and bacteria. Lipids serve an important storage function in other fungi.

Specialized hyphae, known as **rhizoids,** anchor some kinds of fungi to the substrate. Parasitic fungi often have similar specialized hyphae, called **haustoria** (singular: haustorium), which absorb nourishment directly from the cells of other organisms (Figure 15–5).

Fungi Have Unique Variations of Mitosis and Meiosis

One of the most characteristic features of the fungi involves nuclear division. In fungi, the processes of meio-

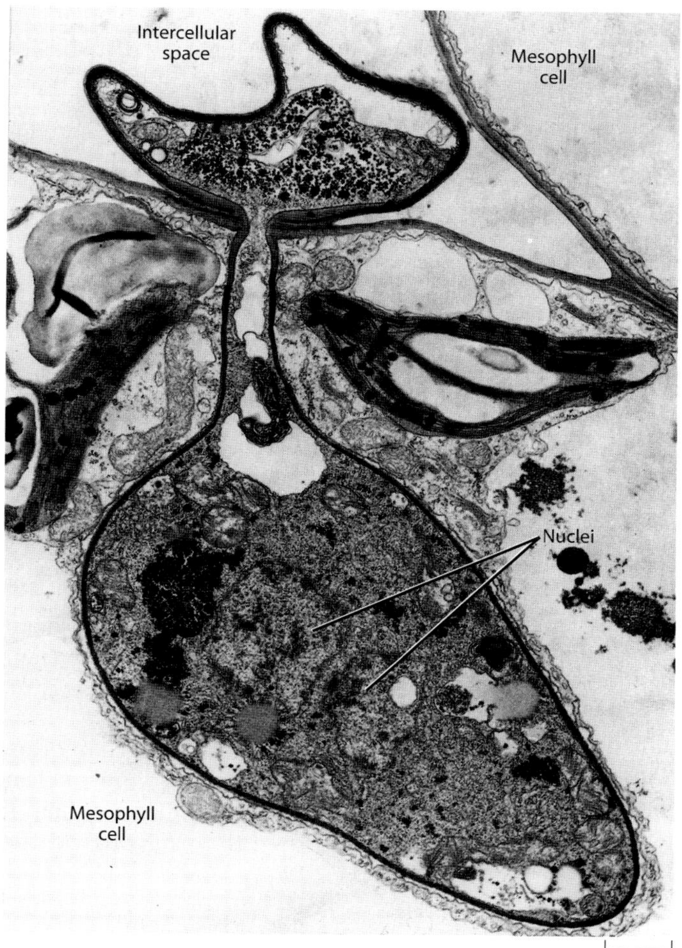

15–5
Electron micrograph of a haustorium of Melampsora lini, *a rust fungus, growing in a leaf cell of flax* (Linum usitatissimum). *In the intercellular space at the top of the micrograph is the haustorial mother cell. The narrow penetration hypha leads to the large, bulbous haustorium within the lower mesophyll cell.*

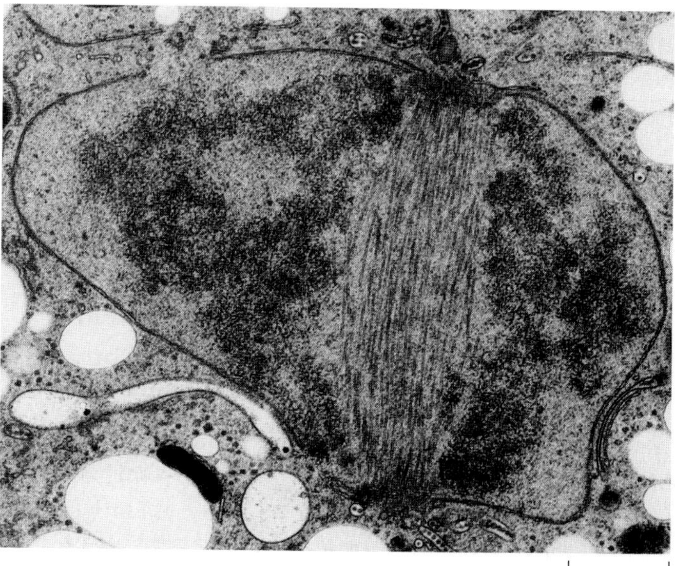

15–6
Electron micrograph of metaphase nucleus of Arthuriomyces peckianus, *a rust fungus, showing the spindle inside the nucleus and the two spindle pole bodies at either end of the spindle. Spindle pole bodies, which are microtubule organizing centers, are characteristic of the* Zygomycota, Ascomycota, *and* Basidiomycota.

sis and mitosis are different from those that occur in plants, animals, and many protists. In most fungi, the nuclear envelope does not disintegrate and re-form but is constricted near the midpoint between the two daughter nuclei. In others, it breaks down near the mid-region. In most fungi, the spindle forms within the nuclear envelope, but in some *Basidiomycota* it appears to form within the cytoplasm and move into the nucleus. Except for the chytrids, all fungi lack centrioles, but they form unique structures called **spindle pole bodies,** which appear at the spindle poles (Figure 15–6). Both spindle pole bodies and centrioles function as microtubule organizing centers during mitosis and meiosis.

Fungi Reproduce Both Asexually and Sexually

Fungi reproduce through the formation of spores that are produced either sexually or asexually. Except for the chytrids, nonmotile spores are the characteristic means of reproduction in fungi. Some spores are dry and very small. They can remain suspended in the air for long periods, thus being carried to great heights and for great distances. This property helps to explain the very wide distributions of many species of fungi. Other spores are slimy and stick to the bodies of insects and other arthropods, which may then spread them from place to place. The spores of some fungi are propelled ballistically into the air (see "Phototropism in a Fungus" on page 315). The bright colors and powdery textures of many types of molds are due to the spores. Some fungi never produce spores.

The most common method of asexual reproduction in fungi is by means of spores, which are produced either in **sporangia** (singular: sporangium) or from hyphal cells called **conidiogenous cells.** The spores produced by conidiogenous cells occur singly or in chains and are called **conidia** (singular: conidium). The sporangium is a saclike structure, the entire contents of which are converted into one or more—usually many—spores. Some fungi also reproduce asexually by fragmentation of their hyphae.

Sexual reproduction in fungi consists of three distinct phases: plasmogamy, karyogamy, and meiosis. The first two phases are phases of syngamy, or fertilization. **Plasmogamy** (the fusion of protoplasts) precedes **karyogamy** (the fusion of nuclei). In some species karyogamy follows plasmogamy almost immediately, whereas in others the two haploid nuclei do not fuse for some time, forming a **dikaryon** ("two nuclei"). Karyogamy may not take place for several months or even years. During that time the pairs of nuclei may divide in tandem, producing a dikaryotic mycelium. Eventually, the nuclei fuse to form a diploid nucleus, which sooner or later undergoes meiosis, reestablishing the haploid condition. Sexual reproduction in most fungi results in the formation of specialized spores such as zygospores, ascospores, and basidiospores.

It is important to emphasize that the diploid phase in the life cycle of a fungus is represented only by the zygote. Meiosis typically follows formation of the zygote; in other words, meiosis in fungi is zygotic (see Figure 9–3a). In addition, the gametes, when produced by fungi, are similar in appearance and size. Such gametes are referred to as **isogametes.** In most fungi, the nuclei act as the gametes. The general name of the gamete-producing structures is **gametangium** (plural: gametangia).

Evolution of the Fungi

As noted earlier, we include four phyla—*Chytridiomycota, Zygomycota, Ascomycota,* and *Basidiomycota*—in the kingdom *Fungi.* It now appears that the chytrids are the most primitive fungi and that the flagellated condition—represented by the flagellated zoospores—is a primitive character retained by the chytrids after they evolved from flagellated protists. Moreover, there is considerable molecular evidence that both animals and fungi diverged from a common ancestor, most likely a colonial protist resembling a choanoflagellate (Figure 15–7). If, in fact, the earliest fungi were aquatic flagellated organisms, the progenitors of the *Zygomycota, Ascomycota,* and *Basidiomycota* probably lost their flagellated stages fairly early in their evolutionary history. Although the *Fungi* appear to be a monophyletic lineage with a closer relationship to animals than to any other kingdom, relationships within the *Fungi* are far from certain.

Fungi have been around for a long time, but their history is poorly understood. The oldest fossils resembling fungi are represented by aseptate filaments from the Lower Cambrian, about 544 million years ago. They are believed to have been reef saprophytes. Fossil fungi morphologically similar to the chytrid *Allomyces* were found on the stems of *Aglaophyton major,* an Early Devonian plant more than 400 million years old. Highly branched hyphae were also found in cortical cells of *A. major.* Such fungi, which belong to the *Zygomycota,* form mycorrhizae, specifically **endomycorrhizae,** which penetrate cells (see page 341). They are one of the few symbiotic plant-fungus associations in the fossil record and are believed to have played a major role in the evolution of plants. The oldest known fossil *Ascomycota* are found in rocks of Silurian age (438 million years), and the oldest known *Basidiomycota* come from the Upper Devonian (380 million years ago).

Chytrids: Phylum *Chytridiomycota*

The *Chytridiomycota,* or chytrids, are a predominantly aquatic group, consisting of about 790 species. Soils from ditches and the banks of ponds and streams are also inhabited by chytrids, and some chytrids are even found in desert soils. Chytrids are varied not only in form, but also in the nature of their sexual interactions and in their life histories. The cell walls of chytrids contain chitin, and like other fungi, the chytrids store glycogen. Meiosis and mitosis in chytrids resemble these processes in other fungi in that they are intranuclear; that is, the nuclear envelope remains intact until late telophase, when it breaks in a median plane and then re-forms around the daughter nuclei.

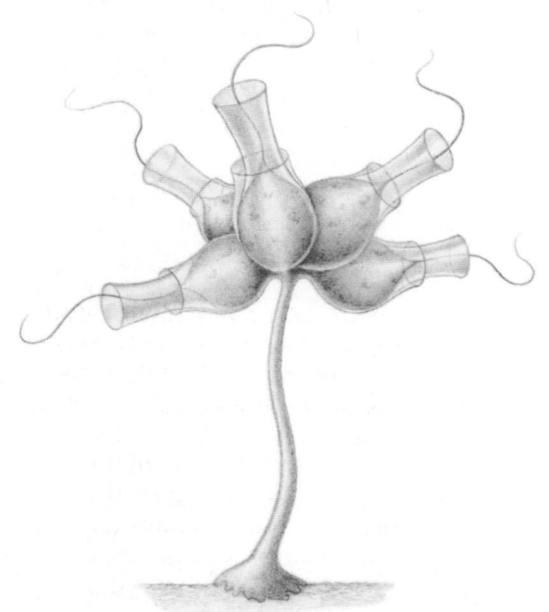

15–7
A choanoflagellate, a colonial protist that many zoologists believe is related to the ancestor of both animals and fungi.

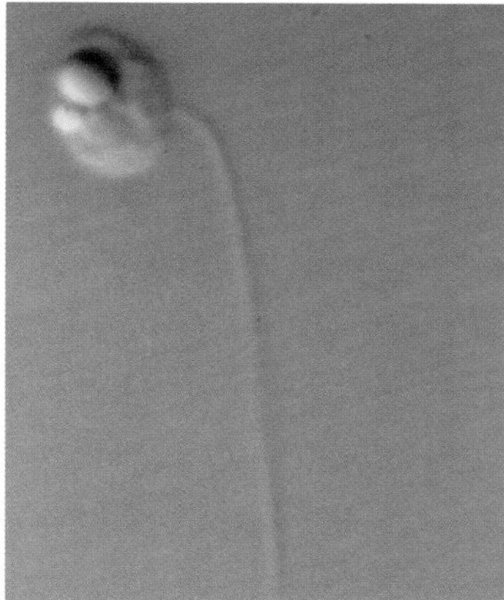

15–8
A uniflagellated zoospore of the chytrid Polyphagus euglenae. *The chytrids are distinguished from other fungi primarily by their characteristic motile zoospores and gametes.*

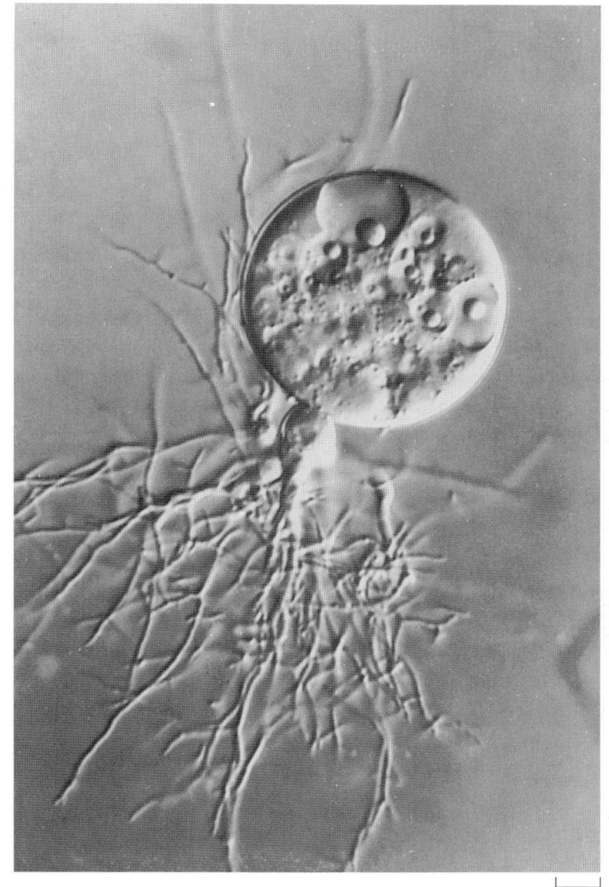

15–9
Chytridium confervae, *a common chytrid, as seen with the aid of differential interference contrast system optics. Note the slender rhizoids extending downward.*

Almost all chytrids are coenocytic, with few septa at maturity. They are distinguished from other fungi primarily by their characteristic motile cells (zoospores and gametes), most of which have a single, posterior, whiplash flagellum (Figure 15–8). Some chytrids are simple, unicellular organisms that do not develop a mycelium. In them, the whole organism is transformed into a reproductive structure at the appropriate time. Other chytrids have slender rhizoids that extend into the substrate and serve as an anchor (Figure 15–9). Some species are parasites of algae, protozoa, and aquatic oomycetes and of the spores, pollen grains, or other parts of plants. Some other chytrid species are saprophytic on such substrates as dead insects.

Several species of chytrids are plant pathogens, including *Physoderma maydis* and *Physoderma alfalfae,* which cause minor diseases known, respectively, as brown spot of corn and crown wart of alfalfa. *Synchytrium endobioticum* causes a disease of potatoes known as black wart disease, which is a serious problem in regions of Europe and Canada.

Chytrids exhibit a variety of modes of reproduction. Some species of *Allomyces,* for example, have an alternation of isomorphic generations like that shown in Figure 15–10, whereas in other species the alternating generations are heteromorphic—the haploid and diploid individuals do not resemble one another closely. Alternation of generations is characteristic of plants and of many algae but is otherwise found only in *Allomyces,* in one other closely related genus of chytrids, and in a very few heterotrophic protists not considered in this book. In terms of its life cycle, morphology, and physiology, *Allomyces* is the best-known chytrid.

Phylum *Zygomycota*

Most zygomycetes live on decaying plant and animal matter in the soil, while some are parasites of plants, insects, or small soil animals. Still others form symbiotic associations—endomycorrhizae—with plants, and a few occasionally cause severe infections in humans and domestic animals. There are approximately 1060 described species of zygomycetes. Most of them have coenocytic hyphae, within which the cytoplasm can often be seen streaming rapidly. Zygomycetes can usually be recognized by their profuse, rapidly growing hyphae, but some of them also exhibit a unicellular, yeastlike form of growth under certain conditions. Asexual reproduction by means of haploid spores pro-

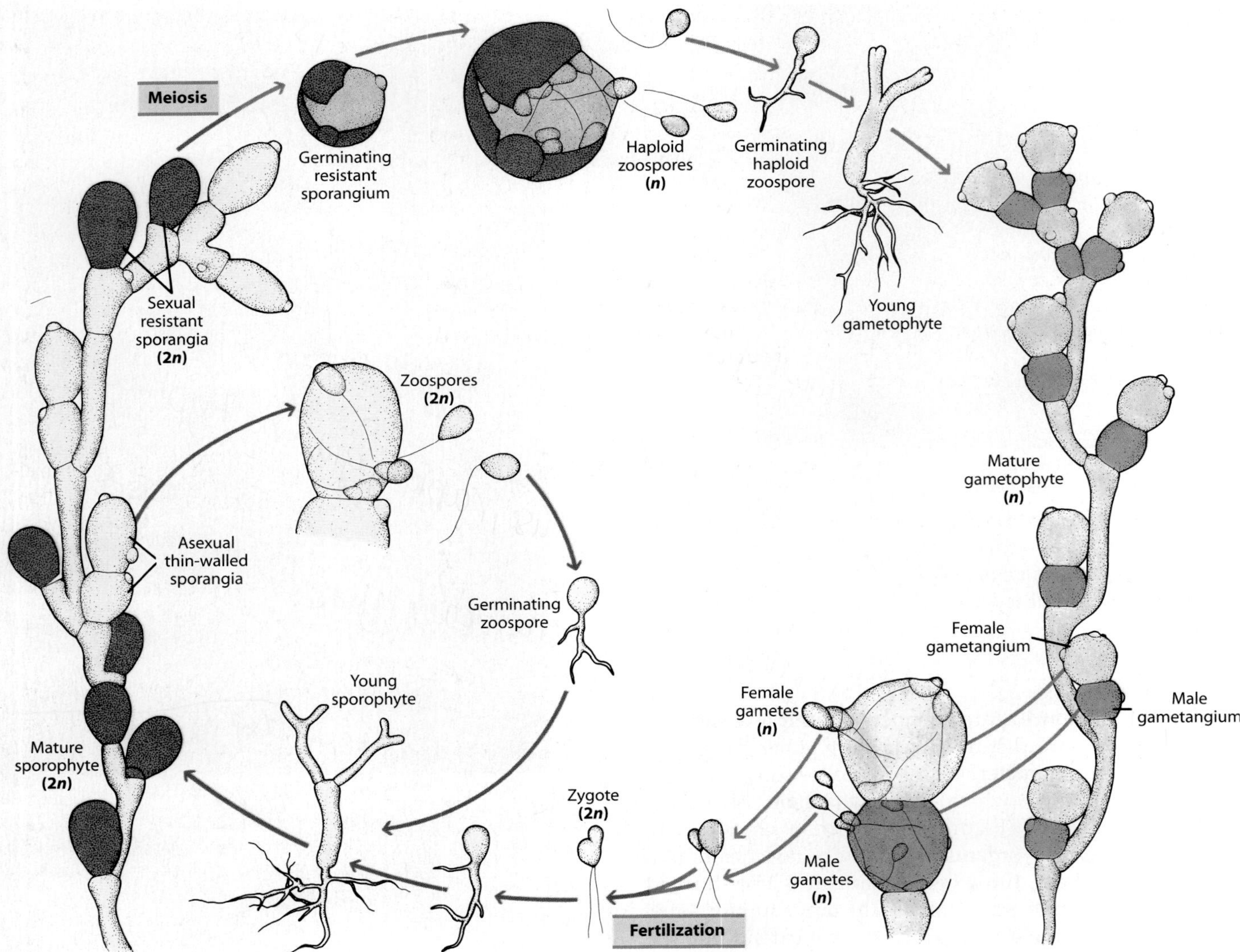

15–10

In the life cycle of the chytrid Allomyces arbusculus, *there is an alternation of isomorphic generations. The haploid and diploid individuals are indistinguishable until they begin to form reproductive organs. The haploid individuals (gametophytes) produce approximately equal numbers of colorless female gametangia and orange male gametangia (right). The male gametes, which are about half the size of the female gametes, are attracted by sirenin, a hormone produced by the female gametes. The zygote loses its flagella and germinates to produce a diploid individual. This sporophyte forms two kinds of sporangia. The first are asexual sporangia—colorless, thin-walled structures that release diploid zoospores—which in turn germinate and repeat the diploid generation. The second kind are sexual sporangia—thick-walled, reddish-brown structures that are able to withstand severe environmental conditions. After a period of dormancy, meiosis occurs in these sexual, resistant sporangia, resulting in the formation of haploid zoospores. These zoospores develop into gametophytes, which produce gametangia at maturity.*

Phototropism in a Fungus

Over the millennia, the fungi have evolved a variety of methods that ensure wide dispersal of their spores. One of the most ingenious is found in **Pilobolus,** ***a zygomycete that grows on dung. The sporangiophores of this fungus, which attain a height of 5 to 10 millimeters, are positively phototropic—that is, they grow toward the light. An expanded region of the sporangiophore located just below the sporangium (known appropriately as the subsporangial swelling) functions as a lens, focusing the sun's rays on a photoreceptive area at its base. Light focused elsewhere promotes maximum growth of the sporangiophore on the side away from the light, causing the sporangiophore to curve toward the light.***

The vacuole in the subsporangial swelling contains a high concentration of solutes, which results in water moving into it by osmosis. Eventually, the turgor pressure becomes so great that the swelling splits, shooting the sporangium in the direction of the light. The initial velocity may approach 50 kilometers per hour, and the sporangium may travel a distance greater than 2 meters. Considering that the sporangia are only about 80 micrometers in diameter, this is an enormous distance. This mechanism is adapted to shoot spores away from the dung—where animals do not feed—and into the grass where they can be eaten by herbivores and excreted in the fresh dung to continue the cycle anew.

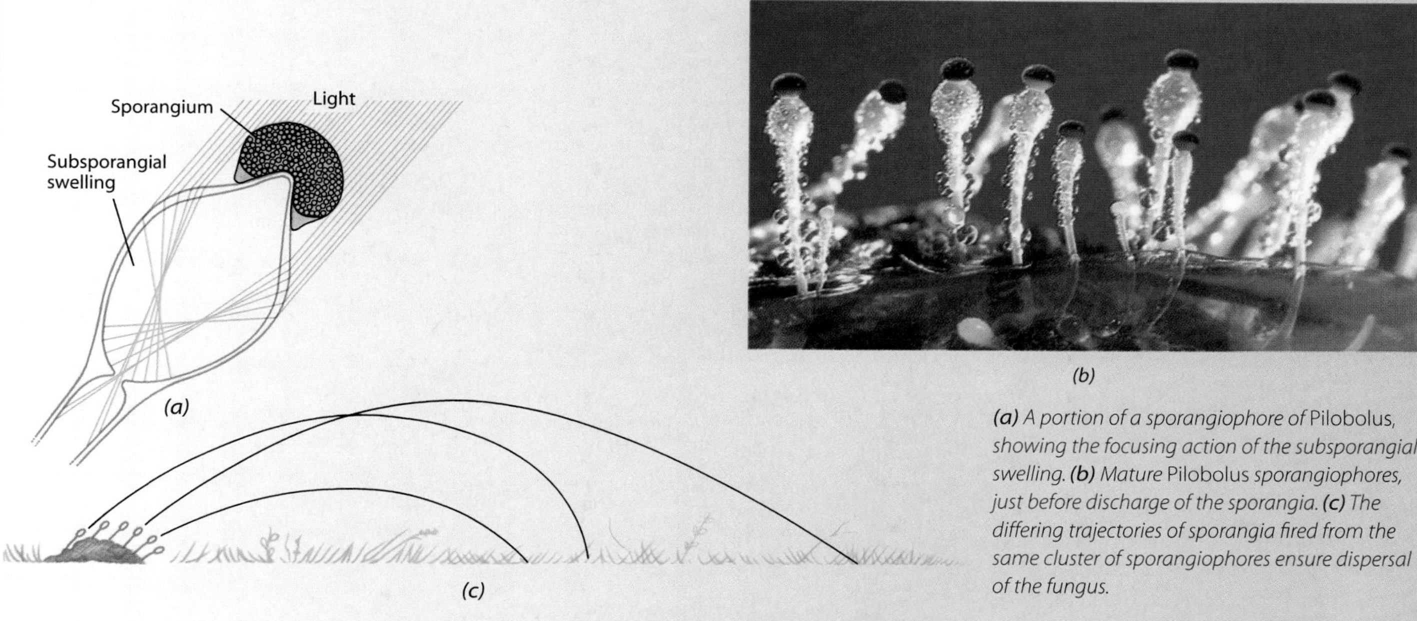

(a) A portion of a sporangiophore of Pilobolus, *showing the focusing action of the subsporangial swelling. (b) Mature* Pilobolus *sporangiophores, just before discharge of the sporangia. (c) The differing trajectories of sporangia fired from the same cluster of sporangiophores ensure dispersal of the fungus.*

duced in specialized sporangia borne on the hyphae is almost universal in zygomycetes.

One of the best-known and most familiar members of this phylum is *Rhizopus stolonifer,* a black mold that forms cottony masses on the surface of moist, carbohydrate-rich foods such as bread and similar substances that are exposed to air (Figure 15–2). This organism is also a serious pest of stored fruits and vegetables. The life cycle of *R. stolonifer* is illustrated in Figure 15–11. The mycelium of *Rhizopus* is composed of several distinct kinds of haploid hyphae. Most of the mycelium consists of rapidly growing, coenocytic hyphae that grow through the substrate, absorbing nutrients. From them, arching hyphae called **stolons** are formed. The stolons form rhizoids wherever their tips come into contact with the substrate. From each of these points, a sturdy, erect branch arises, which is called a **sporangiophore** ("sporangium bearer") because it produces a spherical sporangium at its apex. Each sporangium begins as a swelling, into which a number of nuclei flow. The sporangium is eventually isolated by the formation of a septum. The protoplasm within is cleaved, and a cell wall forms around each of the asexually produced nuclei to form spores. As the sporangium wall matures, it becomes black, giving the mold its characteristic color. With the breaking of the sporangium wall, the spores are liberated. Each spore can germinate to produce a new mycelium, completing the asexual cycle.

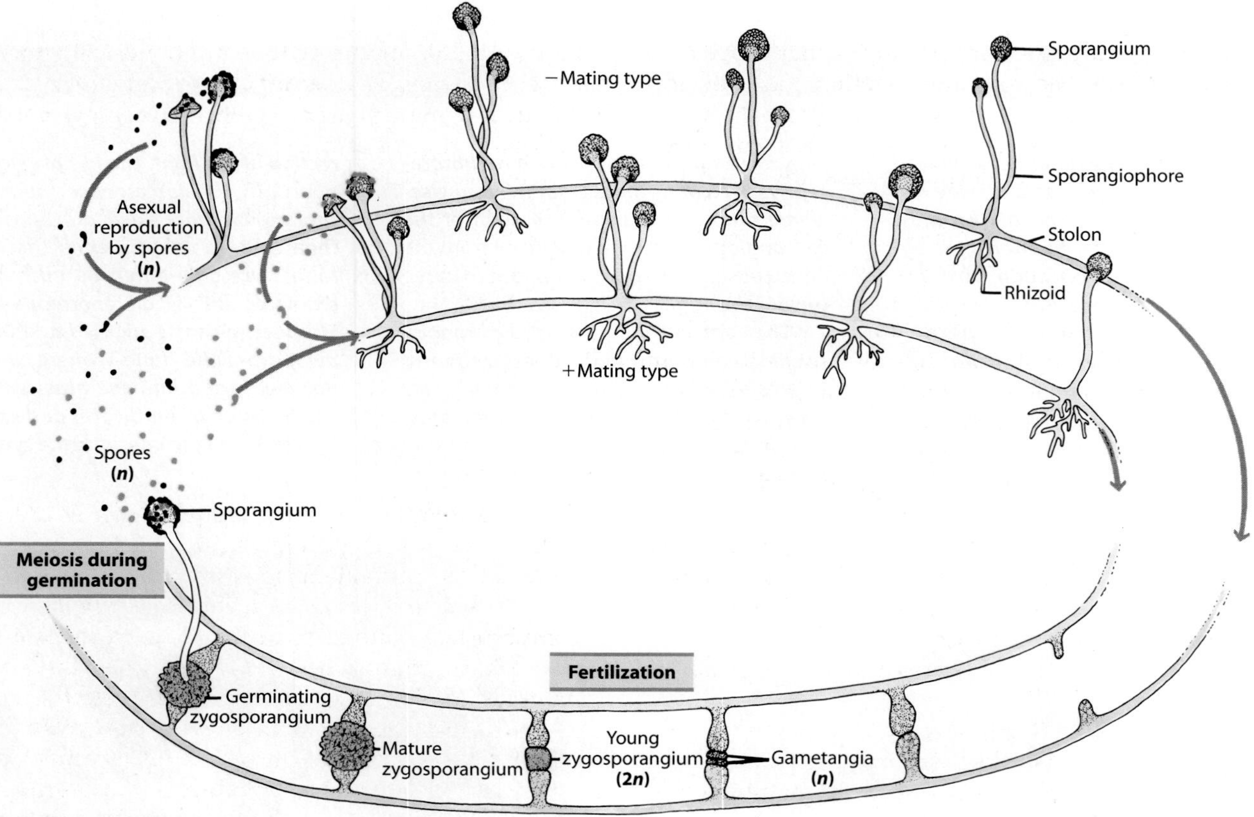

15–11
In Rhizopus stolonifer, *as in most other zygomycetes, asexual reproduction by means of haploid spores is the chief mode of reproduction. Less frequently, sexual reproduction occurs. The spores are formed in sporangia, whose black walls give the mold its characteristic color. In this common species, sexual reproduction involves genetically differentiated mating strains, which have traditionally been labeled + and − types. (Although the mating strains are morphologically indistinguishable from one another, they are shown here in two colors.) Sexual reproduction results in the formation of a resting spore called a zygospore, which develops within a zygosporangium. The zygosporangium in* Rhizopus *develops a thick, rough, black coat, and the zygospore remains dormant, often for several months.*

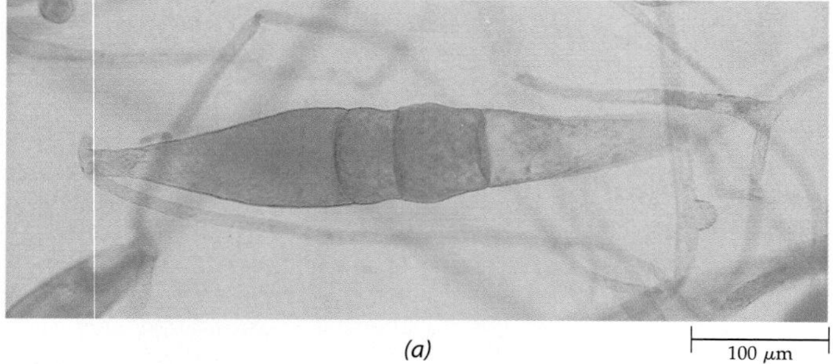

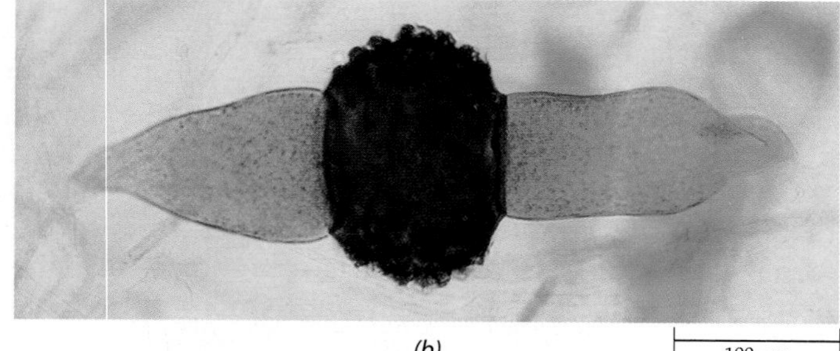

15–12
The zygomycete Rhizopus stolonifer, *a black mold.* ***(a)*** *Gametangia, the gamete-producing structures, are in the process of fusing to produce a zygospore.* ***(b)*** *The zygospore develops within a thick-walled zygosporangium.*

The phylum *Zygomycota* is named for its chief characteristic—the formation of sexually produced resting spores called **zygospores,** which develop within thick-walled structures called **zygosporangia** (Figure 15–12). The zygospores often remain dormant for long periods. Sexual reproduction in *R. stolonifer* requires the presence of two physiologically distinct mycelia, designated + and − strains. When two compatible individuals are in close proximity, hormones are produced that cause outgrowths of the hyphae to come together and develop into gametangia. Species such as *R. stolonifer* that require + and − strains for sexual reproduction are said to be **heterothallic,** whereas self-fertile species are called **homothallic.**

In any event, the gametangia become separated from the rest of the fungal body by the formation of septa (Figure 15–11). The walls between the two touching gametangia dissolve, and the two multinucleate protoplasts come together. Following plasmogamy (the fusion of the two multinucleate gametangia), the + and − nuclei pair, and a thick-walled zygosporangium is produced. Inside the zygosporangium, the paired + and − nuclei fuse (karyogamy) to form diploid nuclei, which develop into a single multinucleate zygospore. At the time of germination, the zygosporangium cracks open, and a sporangiophore emerges from the zygospore. Meiosis occurs at the time of germination so that the spores produced asexually within the new sporangium are haploid. When these spores germinate, the cycle begins again.

Only two genera of zygomycetes commonly cause disease in living plants and living plant tissue. One of them is *Rhizopus,* which causes soft rot of many flowers, fleshy fruits, seeds, bulbs, and corms. The other is *Choanephora,* which causes a soft rot of squash, pumpkin, okra, and pepper.

One of the most important groups of zygomycetes includes *Glomus* and related genera, which always grow in intimate association with the roots of plants, forming endomycorrhizae. Another group of zygomycetes that has great ecological significance, the order *Entomophthorales,* is parasitic on insects and other small animals (see Figure 15-3b). Species of this order, most of which reproduce by means of a terminal, asexual spore that is discharged at maturity, are being increasingly used in the biological control of insect pests of crops.

The trichomycetes, a third group of zygomycetes, have an intriguing relationship with arthropods. These fungi are found in aquatic insect larvae, millipedes, crayfish, and even crustaceans from undersea thermal vents. Trichomycetes are seldom seen because they occur in the gut of the host organism, where some are thought to provide vitamins to the animal.

Phylum *Ascomycota*

The ascomycetes, which comprise about 32,300 described species, include a number of familiar and economically important fungi. Most of the blue-green, red, and brown molds that cause food spoilage are ascomycetes. Ascomycetes are also the cause of a number of serious plant diseases, including the powdery mildews, which primarily attack leaves; brown rot of stone fruits (caused by *Monilinia fructicola*); chestnut blight (caused by the fungus *Cryphonectria parasitica,* which was accidentally introduced into North America from northern China); and Dutch elm disease (caused by *Ophiostoma ulmi* and *O. novo-ulmi,* which were introduced from northern Europe and the region of Romania and Ukraine, respectively). Many yeasts are also ascomycetes, as are the edible morels and truffles (Figure 15–13). Many new families and thousands of additional species of ascomycetes—some undoubtedly of great economic importance—await discovery and scientific description.

(a)

(b)

15–13
Ascomycetes. ***(a)*** *Eyelash cup,* Scutellinia scutellata. ***(b)*** *The highly prized, edible ascoma of a black truffle,* Tuber melanosporum. *In the truffles, this spore-bearing structure is produced below ground and remains closed, liberating its ascospores only when the ascoma decays or is broken open by digging animals. Truffles are mycorrhizal (see page 342), mainly on oaks and hazelnuts, and are searched for by specially trained dogs and pigs. The pigs used are sows because truffles emit a chemical that mimics the pheromone from the saliva of a hog. Recently, truffles have been cultivated commercially on a small scale by inoculating the roots of seedling host plants with their spores.*

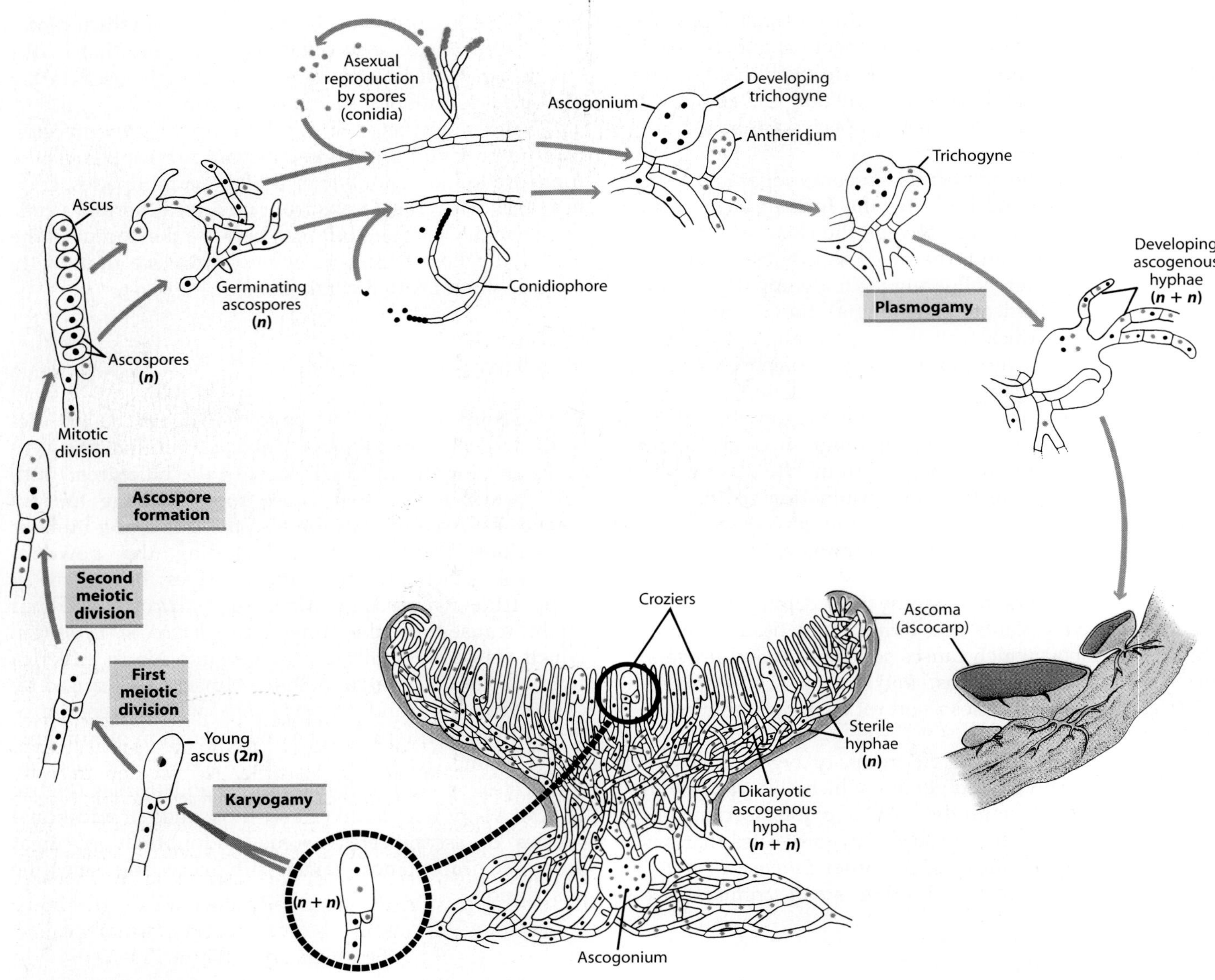

15–14
The typical life cycle of an ascomycete. Asexual reproduction occurs by way of specialized spores, known as conidia, which are usually uninucleate. Sexual reproduction involves the formation of asci and ascospores. Plasmogamy produces fused protoplasts with as yet unfused nuclei, designated n + n. *Karyogamy is followed immediately by meiosis in the ascus, resulting in the production of ascospores.*

Ascomycetes, with the exception of the unicellular yeasts, have filamentous growth forms. In general, their hyphae have perforated septa (Figure 15–4), which allow the cytoplasm and nuclei to move from one cell to the next. The hyphal cells of the vegetative mycelium are usually uninucleate. Some ascomycetes are homothallic, others heterothallic.

The life cycle of an ascomycete is diagrammed in Figure 15–14. In most species of this phylum, asexual reproduction is by the formation of usually multinucleate conidia. The conidia are formed from conidiogenous cells (Figure 15–15), which are borne at the tips of modified hyphae called **conidiophores** ("conidia bearers"). Unlike zygomycetes, which produce spores internally within a sporangium, ascomycetes produce their asexual spores externally as conidia.

Sexual reproduction in ascomycetes always involves the formation of an **ascus** (plural: asci), a saclike structure within which haploid **ascospores** are formed following meiosis. Because the ascus resembles a sac, the

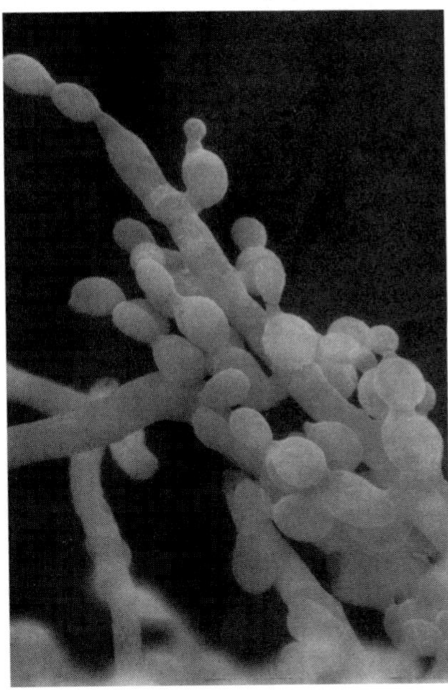

(a)

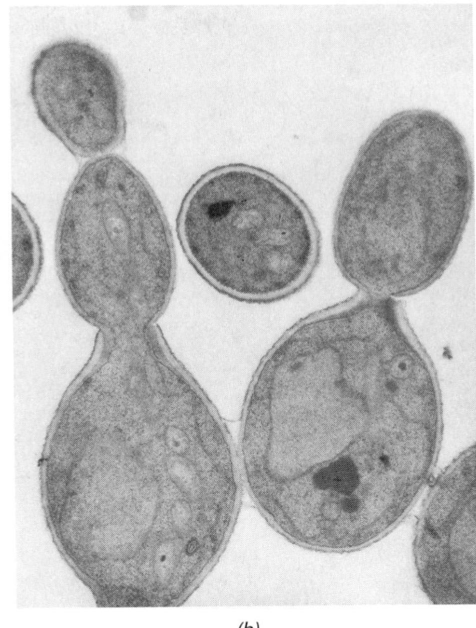

(b)

15–15

Conidia are the characteristic asexual spores of ascomycetes; they are usually uninucleate. These electron micrographs show stages in the formation of conidia in Nomuraea rileyi, *which infects the velvetbean caterpillar.* **(a)** *Scanning electron micrograph of conidia at various stages of development.* **(b)** *Transmission electron micrograph of conidia.*

ascomycetes commonly are referred to as the "sac fungi." Both asci and ascospores are unique structures that distinguish the ascomycetes from all other fungi (Figure 15–16). Ascus formation usually occurs within a complex structure composed of tightly interwoven hyphae—the **ascoma** (plural: ascomata), or ascocarp. Many ascomata are macroscopic. An ascoma may be open and more or less cup-shaped (an *apothecium;* Figure 15–13a), closed and spherical (a *cleistothecium;* Figure 15–16b), or spherical to flask-shaped with a small pore through which the ascospores escape (a *perithecium;* Figure 15–16c). The asci usually develop on the inner surface of the ascoma. The layer of asci is usually called the **hymenium,** or **hymenial layer** (Figure 15–17).

15–16

Asci and ascospores. **(a)** *An electron micrograph showing two asci of* Ascodesmis nigricans *in which ascospores are maturing.* **(b)** *Ascoma of* Erysiphe aggregata, *showing the enclosed asci and ascospores. This completely enclosed type of ascoma is called a cleistothecium.* **(c)** *An ascoma of* Coniochaeta, *showing the enclosed asci and ascospores. Note the small pore at the top. This sort of ascoma, with a small opening, is known as a perithecium.*

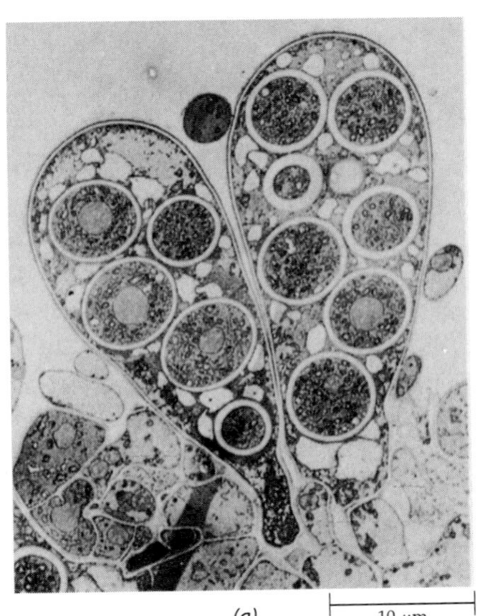

(a)

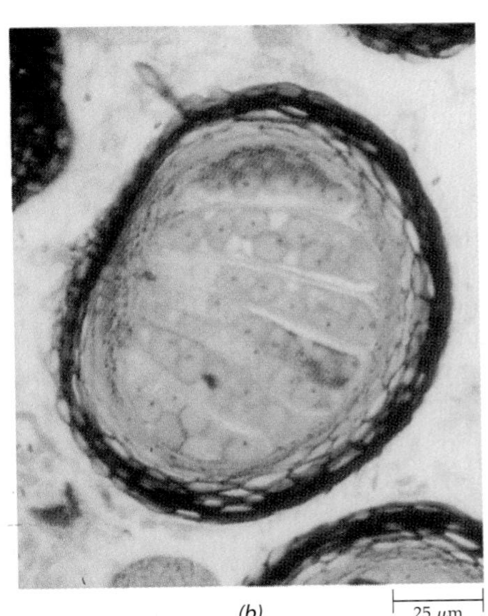

(b)

(c)

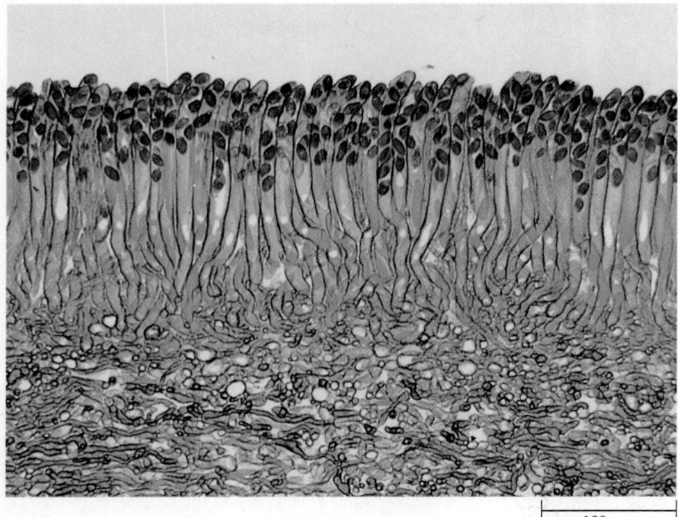

15–17
A stained thin section through the hymenial layer of a morel (Morchella), *showing asci with ascospores.*

15–18
Life cycle of a mushroom (phylum Basidiomycota, *a hymenomycete). Monokaryotic, primary mycelia are produced from basidiospores and give rise to dikaryotic, secondary mycelia, often following the fusion of different mating types, in which case the mycelia are heterokaryotic. Dikaryotic, tertiary mycelia form the basidioma, within which basidia form on the hymenia that line the gills, ultimately releasing up to billions of basidiospores.*

In the life cycle of an ascomycete (Figure 15–14, top left corner), the mycelium is initiated with the germination of an ascospore on a suitable substrate. Soon after, the mycelium begins to reproduce asexually by forming conidia. Many crops of conidia are produced during the growing season, and it is the conidia that are primarily responsible for propagating and disseminating the fungus.

Sexual reproduction, involving ascus formation, occurs on the same mycelium that produces conidia. The formation of multinucleate gametangia called **antheridia** (the male gametangia) and **ascogonia** (the female gametangia) precedes sexual reproduction. The male nuclei of the antheridium pass into the ascogonium via the **trichogyne,** which is an outgrowth of the ascogonium. Plasmogamy—the fusion of protoplasts—has now taken place. The male nuclei may then pair with the genetically different female nuclei within the common cytoplasm, but they *do not yet fuse with them.* **Ascogenous hyphae** now begin to grow out of the ascogonium. As they continue to develop, compatible pairs of nuclei migrate into them, and cell division occurs in such a way that the resultant cells are invariably **dikaryotic,** which means they contain two compatible haploid nuclei. (Monokaryotic cells contain only one nucleus.)

The asci form near the tips of the dikaryotic, ascogenous hyphae. Commonly, it is the apical cell of the dikaryotic hypha that grows to form a hook, or **crozier.** In this hooked cell, the two nuclei divide in such a way that their spindle fibers are parallel to each other. Two of the daughter nuclei are close to one another at the top of the hook; one of the remaining two daughter nuclei is near the tip, and the other is near the basal septum of the hook. Two septa are then formed; these divide the hook into three cells, of which the middle one becomes the ascus. It is in this middle cell that karyogamy occurs: the two nuclei fuse to form a diploid nucleus (zygote), the only diploid nucleus in the life cycle of the ascomycetes. Soon after karyogamy, the young ascus begins to elongate. The diploid nucleus then undergoes meiosis, which is generally followed by one mitotic division, producing an ascus with eight nuclei. These haploid nuclei are then cut off in segments of the cytoplasm to form ascospores. In most ascomycetes, the ascus becomes turgid at maturity and finally bursts, releasing its ascospores explosively into the air in the cup fungi and some of the perithecium-forming species. The ascospores are generally propelled about 2 centimeters from the ascus, but some species propel them as far as 30 centimeters. This initiates their airborne dispersal.

Phylum *Basidiomycota*

Phylum *Basidiomycota,* the last of the four phyla of fungi to be discussed, includes some of the most familiar fungi. Among the 22,300 distinct species of this phylum are the mushrooms, toadstools, stinkhorns, puffballs, and shelf fungi, as well as two important groups of plant pathogens, the rusts and smuts. Members of the *Basidiomycota* play a central role in the decomposition of plant litter, often constituting two-thirds of the living biomass (not including animals) in the soil.

A diagram of the life cycle of a mushroom will provide a convenient reference point as we proceed with our discussion (Figure 15–18). The *Basidiomycota* are distinguished from other fungi by their production of **basidiospores,** which are borne outside a club-shaped spore-producing structure called the **basidium** (plural: basidia) (Figure 15–19). In nature, most *Basidiomycota* reproduce primarily through the formation of basidiospores.

Germinating basidiospores and monokaryotic, primary mycelium (*n*)

Plasmogamy

(*n* + *n*)

Dikaryotic, secondary mycelium (*n* + *n*)

Young basidioma

Basidioma (basidiocarp)

Dikaryotic, tertiary mycelium (*n* + *n*)

Secondary mycelium (*n* + *n*)

Tertiary mycelium (*n* + *n*)

Secondary mycelium (*n* + *n*)

(*n* + *n*)

Karyogamy

Young basidium (2*n*)

First meiotic division

Second meiotic division

Basidiospore formation

Basidium

Basidiospores (*n*)

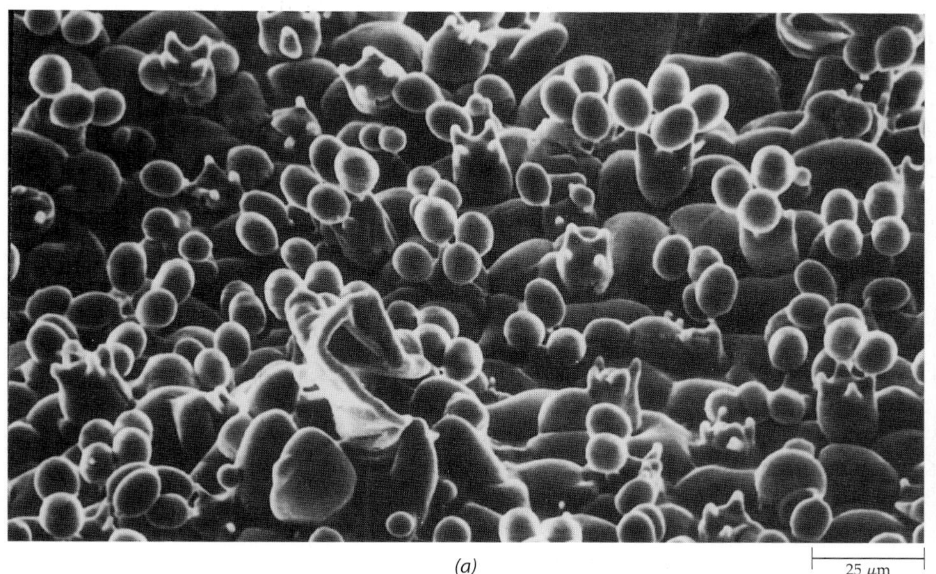

(a)

(b)

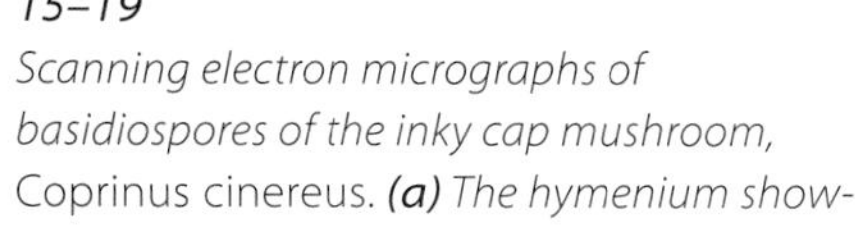

15–19

Scanning electron micrographs of basidiospores of the inky cap mushroom, Coprinus cinereus. ***(a)*** *The hymenium showing numerous basidia frozen at the time of basidiospore release.* ***(b)*** *The top of a basidium, with four basidiospores, each attached to a stalklike sterigma.*

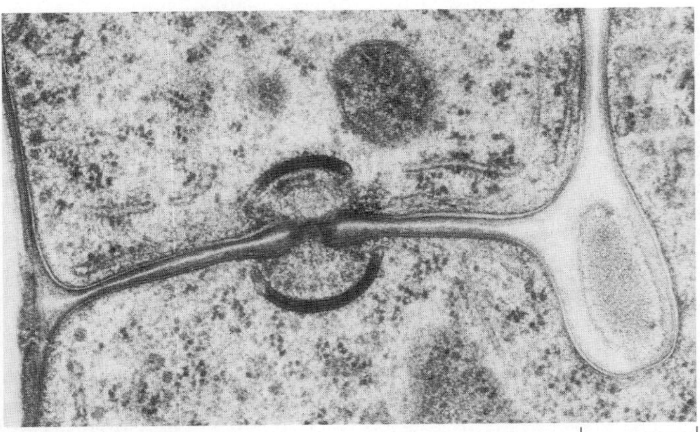

15–20
A dolipore septum, shown in Auricularia auricula, *a common wood-decay species of* Basidiomycetes. *Such septa are common in* Basidiomycetes. *Each dolipore septum is perforated by a pore. Parenthesomes are visible on either side of the dolipore.*

The mycelium of the *Basidiomycota* is always septate, but the septa are perforated. In many species, the pore of the septum has an inflated doughnut-like or barrel-shaped margin called a **dolipore.** Any fungus with dolipore septa belong to the *Basidiomycota.* On either side of the dolipore may be membranous caps called **parenthesomes** because in profile they resemble a pair of parentheses (Figure 15–20). Many *Basidiomycota,* including the rusts and smuts, have septa that resemble those of the ascomycetes.

In most species of *Basidiomycota,* the mycelium passes through two distinct phases—monokaryotic and dikaryotic—during the life cycle of the fungus. When it germinates, a basidiospore produces a mycelium that may be multinucleate initially. Septa are soon formed, however, and the mycelium is divided into **monokaryotic** (uninucleate) cells. This mycelium also is referred to as the **primary mycelium.** Commonly, the dikaryotic mycelium is produced by fusion of monokaryotic hyphae from different mating types (in which case it is heterokaryotic), resulting in formation of a **dikaryotic** (binucleate), or **secondary, mycelium,** since karyogamy does not immediately follow plasmogamy.

The apical cells of the dikaryotic mycelium usually divide by the formation of **clamp connections** (Figure 15–21). These clamp connections, which ensure the allocation of one nucleus of each type to the daughter cells, are found only in the *Basidiomycota,* although as many as 50 percent of the species may not form them.

The mycelium that forms the **basidiomata** (singular: basidioma)—fleshy, basidiospore-producing bodies, such as mushrooms and puffballs—also is dikaryotic. It is called the **tertiary mycelium.** The formation of the basidiomata may require light and low CO_2 levels, both of which signal to the mycelium that it is "outside" its substrate. As it forms the basidiomata, the tertiary mycelium becomes differentiated into specialized hyphae that play different functions within the basidiomata.

The *Basidiomycota* can be divided into three classes: *Basidiomycetes, Teliomycetes,* and *Ustomycetes.* The *Basidiomycetes* include all fungi that produce basidiomata, such as mushrooms, shelf fungi, and puffballs. Neither the *Teliomycetes* (the rusts) nor the *Ustomycetes* (the smuts) form basidiomata. Instead, these fungi produce their spores in masses called **sori** (singular: sorus). The basidiomata, which are characteristic of the *Basidiomycetes,* are analogous to the ascomata of the ascomycetes.

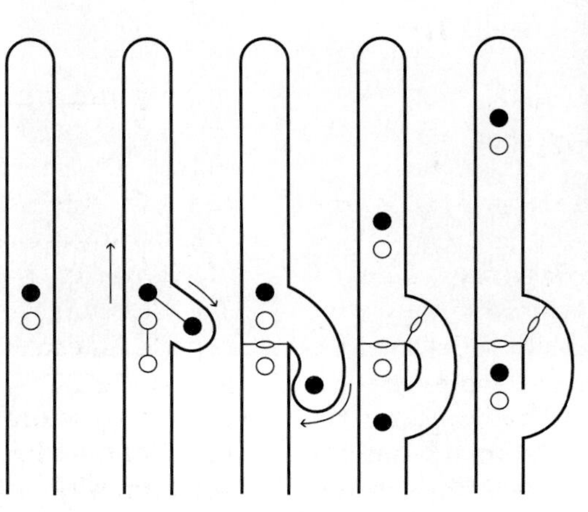

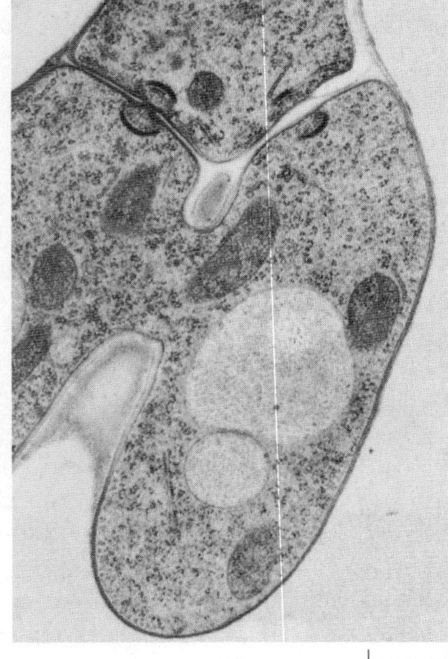

15–21
Clamp connections. ***(a)*** *In the* Basidiomycetes, *dikaryotic hyphae characteristically are distinguished by the formation of clamp connections during cell division in tips of hyphae. They presumably ensure the proper distribution of the two genetically distinct types of nuclei in the basidioma. Two septa form to divide the parent cell into two daughter cells.* ***(b)*** *Electron micrograph of a clamp connection and characteristic septa in a hypha of* Auricularia auricula.

15–22
Hymenomycetes. **(a)** *The fly agaric,* Amanita muscaria. *The mushrooms are at various stages of growth. Among the characteristics of this genus of mushrooms, many members of which are poisonous, are the scales on the cap, the ring on the stalk, and the cup, or volva, around its base.* **(b)** Polyporus arcularius, *a polypore. The polypores lack the gills found in most kinds of mushrooms. In* P. arcularius, *spores are shed through diamond-shaped pores.* **(c)** *A shelf fungus,* Ganoderma applanatum. *Shelf fungi are wood-rotting fungi.* **(d)** *An edible tooth fungus,* Hericium coralloides. *The hymenium, an outer spore-bearing layer of basidia, is borne on the surface of the downwardly directed teeth.*

Class *Basidiomycetes* Includes the Hymenomycetes and the Gasteromycetes

The class *Basidiomycetes* includes the edible and poisonous mushrooms, coral fungi, tooth fungi, and shelf or bracket fungi (Figure 15–22). These *Basidiomycetes* commonly are referred to as hymenomycetes, because they produce their basidiospores on a distinct fertile layer, the **hymenium,** which is exposed before the spores are mature (Figure 15–23). Another group of *Basidiomycetes,* the so-called gasteromycetes (literally the "stomach fungi"), include forms in which no distinct hymenium is visible at the time the basidiospores are released. Among the more familiar gasteromycetes are the stinkhorns, earthstars, false puffballs, bird's-nest fungi, and puffballs (see Figure 15–27). Most *Basidiomycetes* have club-shaped, aseptate (internally undivided) basidia, usually bearing four basidiospores, each on a minute projection called a **sterigma** (plural: sterigmata) (Figures 15–19b and 15–23). Others—the jelly fungi (Figure 15–24)—have septate (internally divided) basidia like those of the rusts and smuts.

The structure that one recognizes as a mushroom or toadstool is a basidioma (Figure 15–18). ("Mushroom" is sometimes popularly used to designate the edible forms of basidiomata, and "toadstool" is used to designate the

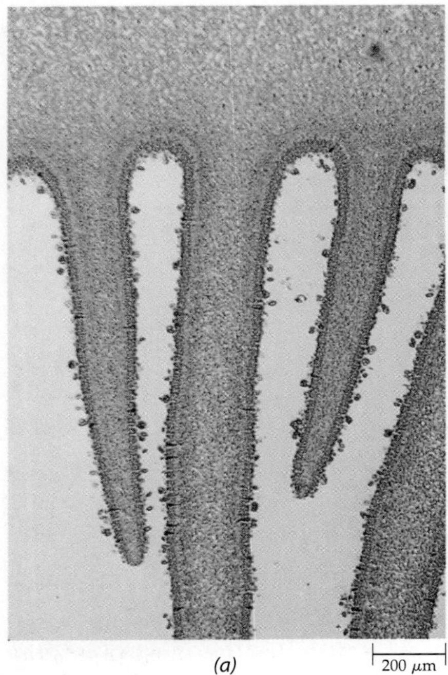

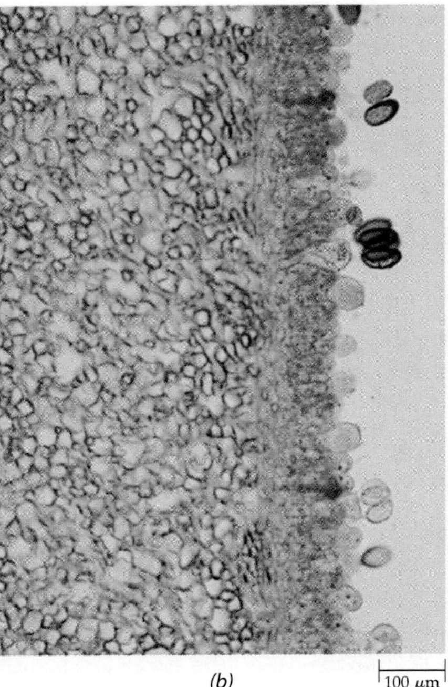

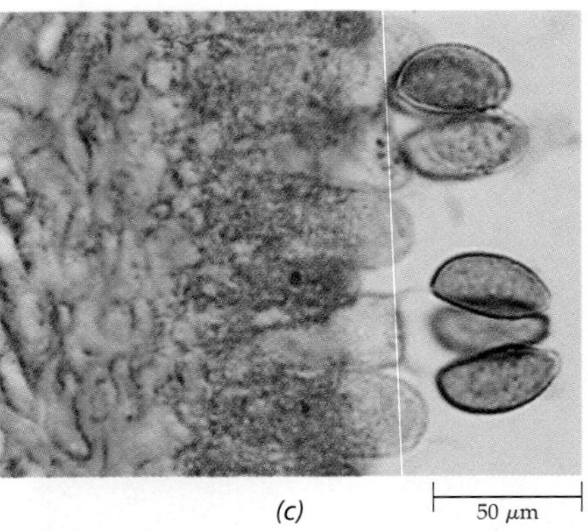

15–23
Stained sections through the gills of Coprinus, *a common mushroom, at progressively higher magnifications. The hymenial layer is stained darker in each of these preparations.* ***(a)*** *Outlines of some of the gills.* ***(b)*** *Developing basidia and basidiospores are shown in a section through the hymenial layer.* ***(c)*** *Nearly mature basidiospores attached to basidia by sterigmata.*

inedible ones, but mycologists do not recognize such a distinction and use only the term "mushroom." In this book, all such forms are referred to as mushrooms; this does not mean that they are all edible.) The mushroom generally consists of a **pileus,** or **cap,** that sits atop a **stipe,** or **stalk.** The masses of hyphae in the basidiomata usually form distinct layers. Early in its development—the "button" stage—the mushroom may be covered by a membranous tissue that ruptures as the mushroom enlarges. In some genera, remnants of this tissue are visible as patches on the upper surface of the cap and as a cup, or **volva,** at the base of the stipe (Figure 15–22a). In many hymenomycetes, the lower surface of the cap consists of radiating strips of tissue called **gills** (Figure 15–23), which are lined with the hymenium. In other members of the class, the hymenium is located elsewhere; for example, in the tooth fungi (Figure 15–22d) the hymenium covers downwardly directed pegs. In the shelf fungi and polypores (Figure 15–22b, c), the hymenium lines vertical tubes that open as pores.

15–24
A jelly fungus growing on a dead tree limb in the Amazon forest of Brazil. Jelly fungi form septate basidia. For this and other reasons, many mycologists no longer group the various jelly fungi with the hymenomycetes.

As mentioned previously, in hymenomycetes, basidia form in well-defined hymenia that are exposed before the basidiospores are mature. Each basidium develops from a terminal cell of a dikaryotic hypha. Soon after the young basidium enlarges, karyogamy occurs. This is followed almost immediately by meiosis of each diploid nucleus, resulting in the formation of four haploid nuclei (Figure 15–18). Each of the four nuclei then migrates into a sterigma, which enlarges at its tip to form a uninucleate, haploid basidiospore. At maturity, the basidiospores are discharged forcibly from the basidioma but depend on the wind for dispersal. The reproductive capacity of a single mushroom is tremendous, with billions of spores being produced by a single basidioma. This reproductive capacity is essential. Each species occupies a narrow niche in the environment, and the chance that a given spore will land on a substrate suitable for germination and growth is slim.

In relatively uniform habitats, such as lawns and fields, the mycelium from which mushrooms are produced spreads underground, growing downward and

15–25
A "fairy ring" formed by the mushroom Marasmius oreades. *Some fairy rings are estimated to be up to 500 years old. Because of the exhaustion of key nutrients, the grass immediately inside such a ring is often stunted and lighter green than the grass outside the ring.*

outward and forming a ring of mushrooms on the edge of the colony. This ring may grow as large as 30 meters in diameter. In an open area, the mycelium expands evenly in all directions, dying at the center and sending up basidiomata at the outer edges, where it grows most actively because this is the area in which the nutritive material in the soil is most abundant. As a consequence, the mushrooms appear in rings, and as the mycelium grows, the rings become larger. Such circles of mushrooms are known in European folk legends as "fairy rings" (Figure 15–25).

The best known hymenomycetes are the gill fungi, including *Agaricus campestris,* the common field mushroom. The closely related *Agaricus bisporus* is one of the few mushrooms cultivated commercially. It is now grown in more than 70 countries, and the value of the world crop exceeds $14 billion. Together with the Oriental shiitake mushroom, *Lentinula edodes, A. bisporus* makes up about 86 percent of the world's mushroom crop. Other mushrooms are also being cultivated, and some are gathered in large quantities in nature. An alarm has been sounded that mushrooms are declining both in total numbers of species and in the quantity of individual species in forests of Europe and the Pacific Northwest of the United States. If this trend continues, it could result in a dramatic decline in the health of the trees that are dependent on mycorrhizal fungi for nutrient uptake as well as disruption of the nutrient cycle in the ecosystem. The cause of the decline has not been identified, but pollutants such as nitrates are suspected.

The gill fungi also include many poisonous mushrooms. The genus *Amanita* includes the most poisonous of all mushrooms, as well as some that are edible. Just a few bites of the "destroying angel," *A. virosa,* can be fatal. Some *Basidiomycetes* contain chemicals that cause hallucinations in humans who eat them (Figure 15–26).

The gasteromycetes (Figure 15–27) are characterized by the fact that their basidiospores mature inside the basidiomata and are not discharged forcibly from them. Once considered a separate class, *Gasteromycetes,* this group is now known to be polyphyletic in nature. The basidiomata of the gasteromycetes possess a distinct outer covering, called the **peridium,** that varies from almost papery thin in some species to thick and rubbery or leathery in others. In some species the peridium opens naturally when the spores are mature; in others, it re-

(a)

(b)

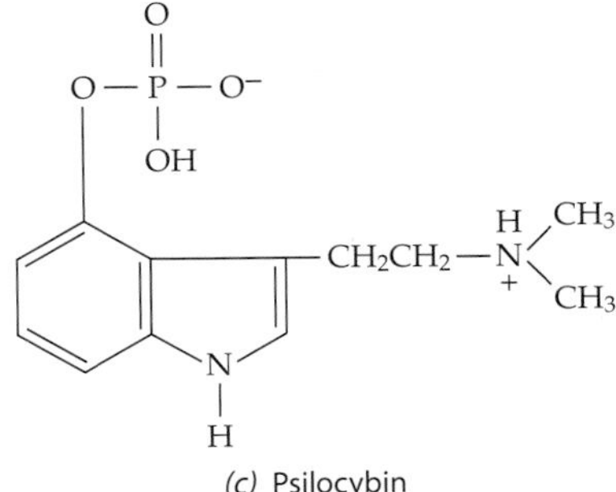

(c) Psilocybin

15–26
Mushrooms figure prominently in the religious ceremonies of several groups of Indians in southern Mexico and Central America. The Indians eat certain hymenomycetes for their hallucinogenic qualities. ***(a)*** *One of the most important of these mushrooms is* Psilocybe mexicana, *shown here growing in a pasture near Huautla de Jiménez, Oaxaca, Mexico.* ***(b)*** *The shaman María Sabina is shown eating* Psilocybe *in the course of a midnight religious ceremony.* ***(c)*** *Psilocybin, the chemical responsible for the colorful visions experienced by those who eat these "sacred" mushrooms, is a structural analogue of LSD and mescaline (see Figure 22–50a).*

(a)

(b)

(c)

(d)

15–27
Gasteromycetes. ***(a)*** *Puffballs,* Calostoma cinnabarina. *Raindrops cause the thin outer layer, or peridium, of the puffball to dimple, forcing out a puff of air, mixed with spores, through the opening.* ***(b)*** *The netted stinkhorn,* Dictyophora duplicata. *The basidiospores are released in a foul-smelling, sticky mass at the top of the fungus. Flies visit it for food and spread the spores, which adhere to their legs and bodies in great numbers.* ***(c)*** *White-egg bird's-nest fungus,* Crucibulum laeve. *The round structures ("eggs") of the basidiomata (the "nests") of these fungi contain the basidiospores, which are splashed out and dispersed by raindrops.* ***(d)*** *Earthstar,* Geastrum saccatum, *showing one fully opened individual and two others in earlier stages of development. The outer layers of the peridium fold back in this genus, raising the spore mass above the dead leaves.*

mains permanently closed, with the spores being liberated only after it has been ruptured through the action of an external agent.

Stinkhorns (Figure 15–27b) have a remarkable morphology. They develop underground as leathery, egg-shaped structures. At maturity, they differentiate into an elongating stalk and a pileus, or cap, bearing a **gleba,** which is the fertile portion of the basidioma. The gleba forms an unpleasant-smelling, sticky mass of spores that attracts flies and beetles, which disperse the spores.

The puffballs are familiar gasteromycetes. At maturity, the interior of a puffball dries up, and it releases a cloud of spores when struck (Figure 15–27a). Some giant puffballs may reach 1 meter in diameter and may produce several trillion basidiospores. The bird's-nest fungi (Figure 15–27c) begin their development like puffballs, but the disintegration of much of their internal structure leaves them looking like miniature bird's nests.

Class *Teliomycetes* Includes the Rusts

The class *Teliomycetes* consists of the fungi commonly referred to as rusts, approximately 7000 species of which have been described. Unlike the *Basidiomycetes,* the rusts do not form basidiomata. As mentioned previously, their spores occur in masses called sori (Figure 15–28). They do, however, form dikaryotic hyphae and basidia, which are septate like those of the jelly fungi (members of the class *Basidiomycetes*). As plant pathogens, the rusts are of tremendous economic importance, causing billions of dollars of damage to crops throughout the world each year. Among the more serious rust diseases are black stem rust of cereals, white pine blister rust, coffee rust, cedar-apple rust, and peanut rust.

The life cycles of many rusts are complex, and these pathogens are a constant challenge to the plant pathologist whose task it is to keep them under control. Until recently, the rusts were thought to be obligate parasites on vascular plants, but now several species are grown in artificial culture. Some smuts are also able to complete their development under laboratory conditions.

15–28
Orange sori of the blister rust (Kuehneola uredinis) *on a blackberry leaf photographed in San Mateo County, California.*

An example of a rust life cycle is provided by *Puccinia graminis,* the cause of black stem rust of wheat. Numerous strains of *P. graminis* exist, and, in addition to wheat, they parasitize other cereals such as barley, oats, and rye, and various species of wild grasses. *Puccinia graminis* is a continuous source of economic loss for wheat growers. In a single year, the losses in Minnesota, North Dakota, South Dakota, and the prairie provinces of Canada amounted to nearly 8 million metric tons. As early as A.D. 100, Pliny described wheat rust as "the greatest pest of the crops." Today plant pathologists combat black stem rust largely by breeding resistant wheat varieties, but mutation and recombination in the rust make any advantage short-lived.

Puccinia graminis is **heteroecious;** that is, it requires two different hosts to complete its life cycle (see Figure 15–29 on the following two pages). **Autoecious** parasites, in contrast, require only one host. *Puccinia graminis* can grow indefinitely on its grass host, but there it reproduces only asexually. In order for sexual reproduction to take place, the rust must spend part of its life cycle on barberry (*Berberis*) and part on a grass. One method of attempting to eliminate this rust has been to eradicate barberry bushes. For example, the crown colony of Massachusetts passed a law ordering "whoever . . . hath any barberry bushes growing in his or their land . . . shall cause the same to be extirpated or destroyed on or before the thirteenth day of June, A.D. 1760."

Infection of the barberry occurs in the spring (Figure 15–29, upper left), when uninucleate basidiospores infect the plant by forming haploid mycelia, which first develop **spermogonia,** primarily on the upper surfaces of the leaves. The form of *P. graminis* that grows on barberry consists of separate + and − strains, so that the basidiospores and the spermogonia derived from them are either + or −. Each spermogonium is a flask-shaped pustule lined by cells that form sticky, uninucleate cells called **spermatia.** The mouth of the spermogonium is surrounded by a brush of orange, stiff, unbranched, pointed hairs, the **periphyses,** which hold droplets of sugary, sweet-smelling nectar. The nectar, which is attractive to flies, contains the spermatia. Among the periphyses, branched **receptive hyphae** are also to be found. Flies visit the spermogonia and feed on the nectar. In moving from one spermogonium to another, they transfer spermatia. If a + spermatium of one spermogonium comes in contact with the − receptive hypha of another spermogonium, or vice versa, plasmogamy occurs and dikaryotic hyphae are produced. Aecial initials are produced from the dikaryotic hyphae that extend downward from the spermogonium. **Aecia** are then formed primarily on the lower surface of the leaf, where they produce chains of **aeciospores.** The dikaryotic aeciospores must then infect the wheat; they will not grow on barberry.

The first external manifestation of infection on the wheat is the appearance of rust-colored, linear streaks on the leaves and stems (the red stage). These streaks are **uredinia,** containing unicellular, dikaryotic **urediniospores.** Urediniospores are produced throughout the summer and reinfect the wheat; they are also the primary means by which the wheat rust has spread throughout the wheat-growing regions of the world. In late summer and early fall, the red-colored sori gradually darken and become **telia** with two-celled dikaryotic **teliospores** (the black stage). The teliospores are overwintering spores, which infect neither wheat nor barberries. Shortly after they are formed, karyogamy takes place, and the teliospores overwinter in the diploid state. Meiosis actually begins immediately but is arrested in prophase I. In early spring, prior to germination, meiosis is completed in the short, curved basidia that emerge from the two cells of the teliospore. Septa are formed between the resultant nuclei, which then migrate into the sterigmata and develop into basidiospores. Thus, the year-long cycle is completed.

In certain regions, the life cycle of wheat rust can be short-cut through the persistence of the uredinial state when actively growing plant tissues are available throughout the year. In the North American plains, urediniospores from winter wheat in the southwestern states and Mexico drift north to southern Manitoba. Later generations scatter westward to Alberta, and finally there is a southern drift at the end of summer, apparently moving along the eastern flank of the Rocky Mountains, and so back to the wintering grounds. Under such circumstances, wheat rust does not depend on barberry to persist, so that barberry eradication was not an effective tool for controlling wheat rust in this area. In contrast, the spread of urediniospores from south to north is prevented in Eurasia, where there are extensive east-west mountain ranges, and barberry *is* necessary for survival of the pathogen.

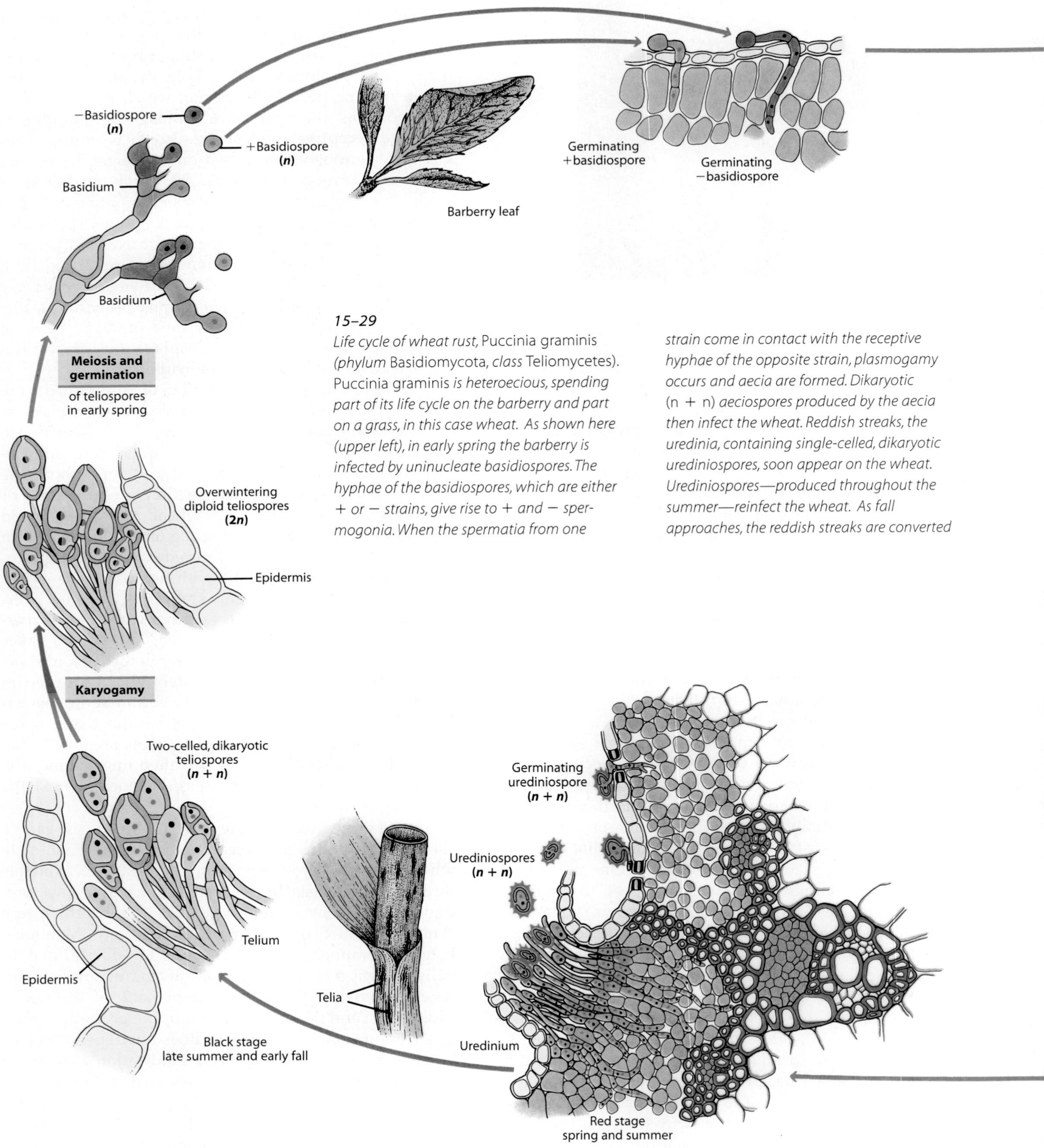

15–29
Life cycle of wheat rust, Puccinia graminis *(phylum* Basidiomycota, *class* Teliomycetes). Puccinia graminis *is heteroecious, spending part of its life cycle on the barberry and part on a grass, in this case wheat. As shown here (upper left), in early spring the barberry is infected by uninucleate basidiospores. The hyphae of the basidiospores, which are either + or − strains, give rise to + and − spermogonia. When the spermatia from one strain come in contact with the receptive hyphae of the opposite strain, plasmogamy occurs and aecia are formed. Dikaryotic* (n + n) *aeciospores produced by the aecia then infect the wheat. Reddish streaks, the uredinia, containing single-celled, dikaryotic urediniospores, soon appear on the wheat. Urediniospores—produced throughout the summer—reinfect the wheat. As fall approaches, the reddish streaks are converted*

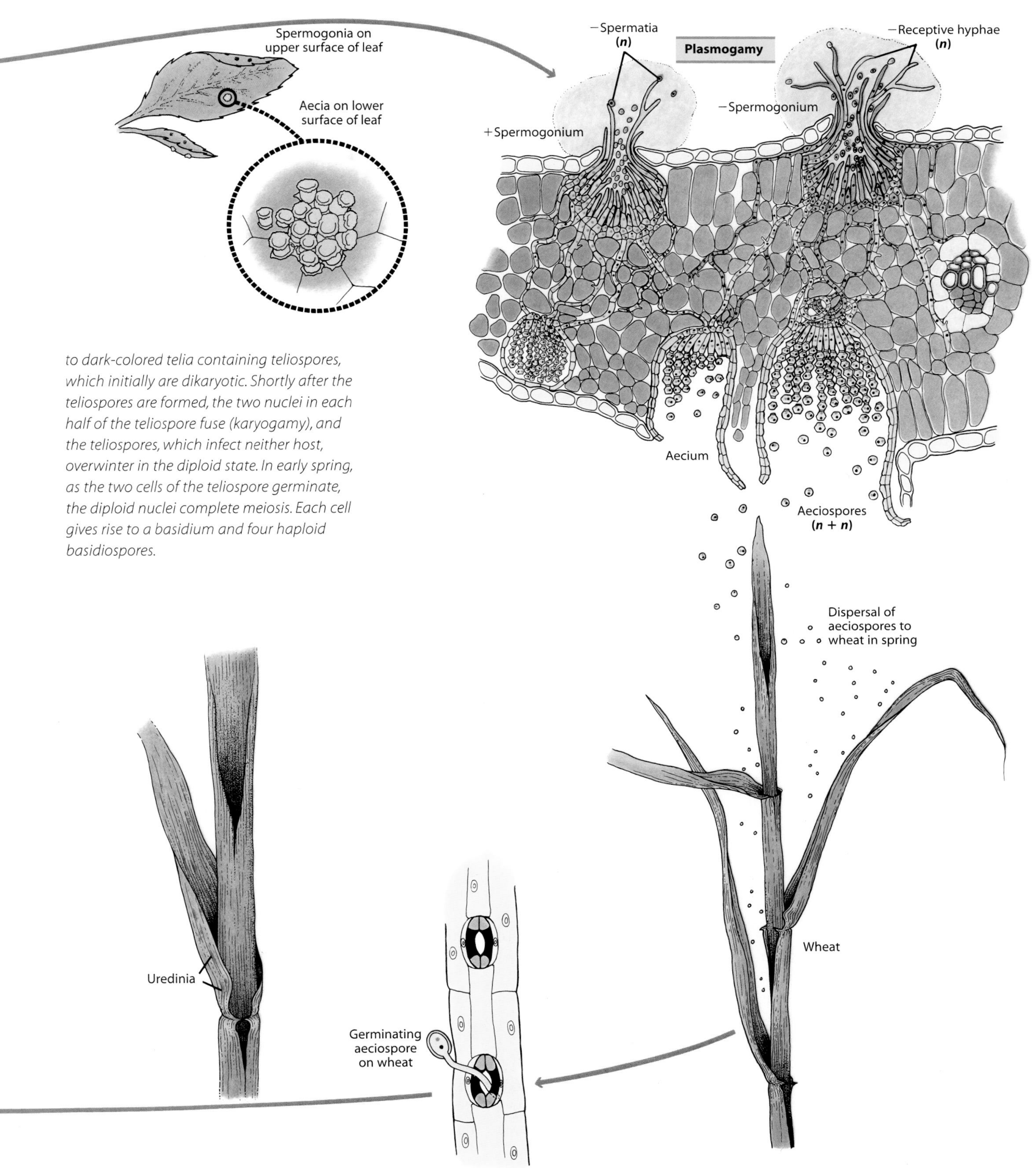

to dark-colored telia containing teliospores, which initially are dikaryotic. Shortly after the teliospores are formed, the two nuclei in each half of the teliospore fuse (karyogamy), and the teliospores, which infect neither host, overwinter in the diploid state. In early spring, as the two cells of the teliospore germinate, the diploid nuclei complete meiosis. Each cell gives rise to a basidium and four haploid basidiospores.

Class *Ustomycetes* Includes the Smuts

All members of the *Ustomycetes* are parasites of flowering plants and are commonly referred to as smuts. The name "smut" refers to the sooty or smutty appearance of the black, dusty masses of teliospores, which are the characteristic resting spores of the smut fungi. Approximately 1070 species of *Ustomycetes* have been described. Most smuts form septate basidia. Economically, the smuts are very important. They attack approximately 4000 species of flowering plants, including both food crops and ornamentals. Three of the better known smut fungi are *Ustilago maydis,* the cause of common corn smut (Figure 15–30); *Ustilago avenae,* which causes loose smut of oats; and *Tilletia tritici,* the cause of bunt or stinking smut of wheat.

The life cycle of a smut, which is autoecious (requiring only one host), is considerably simpler than that of *Puccinia graminis.* Take for example the *Ustilago maydis* life cycle. Infections by spores of *U. maydis* remain localized, producing sori or large tumors. The most conspicuous tumors, or galls, occur on the ear of corn (maize) where the kernels become much enlarged and unsightly due to the development within them of a massive mycelium. A dikaryotic mycelium eventually gives rise to the thick-walled teliospores, in which karyogamy and meiosis take place.

Upon germination, the teliospore gives rise to a four-celled basidium. Two + and two − haploid, uninucleate basidiospores are formed, one from each of the four cells of the basidium (*U. maydis,* like *P. graminis,* is heterothallic). The basidiospores may infect maize plants directly or give rise by budding to populations of uninucleate cells called **sporidia,** which can also infect maize plants. Upon germination the basidiospores or sporidia produce either a + or − mycelium. When mycelia of opposite strains come into contact, plasmogamy occurs, producing a dikaryotic ($n + n$) mycelium, most cells of which change into teliospores.

15–30
Corn smut is a familiar plant disease in which the fungus Ustilago maydis *produces black, dusty-looking masses of spores in ears of corn. When young and white, these are cooked and eaten in Mexico and Central America, where they are regarded as a delicacy.* Ustilago *is a smut, a member of the class* Teliomycetes *of the phylum* Basidiomycota.

Yeasts

A **yeast,** by definition, is simply a unicellular fungus that reproduces primarily by budding. Some fungi exhibit both unicellular and filamentous growth forms, shifting from one form to the other under changing environmental conditions. In many of these fungi the filamentous form is the phase in which the fungus spends most of its life cycle. Others exist primarily as yeasts, including the most familiar of yeasts, *Saccharomyces cerevisiae* (Figure 15–31a). Although *S. cerevisiae* has been known for some time to exhibit a filamentous phase (Figure 15–31b), it has only recently been discovered that the yeast-to-filament switch is triggered by a low level of available nitrogen. Generally, most laboratory cultures provide all of the nutrients essential for *S. cerevisiae* to continue its existence as a yeast. Apparently the filamentous phase is *S. cerevisiae*'s way of foraging for food.

The yeasts are not a formal taxonomic group. The yeast growth form is exhibited by a broad range of unrelated fungi encompassing the *Zygomycota, Ascomycota,* and *Basidiomycota.* There are at least 80 genera of yeasts, with approximately 600 known species. Most yeasts are ascomycetes, but at least a quarter of the genera are *Basidiomycota.*

A common form of reproduction exhibited by yeasts is **budding,** that is, the production of a small outgrowth, the bud, from the parent cell. Each yeast cell, therefore, can be regarded as a conidiogenous cell. Budding is, of course, an asexual method of reproduction. Some yeasts multiply asexually only in the haploid condition. Each haploid cell is capable of serving as a gamete, and at times two haploid cells may fuse to form a diploid cell, or zygote, which functions as an ascus (Figure 15–31c). (There is no dikaryotic phase in yeasts.)

Meiosis occurs within the ascus in ascomycomycetous yeasts. Usually four ascospores are produced per ascus, although in some species meiosis is followed by one or more mitotic divisions resulting in greater numbers of ascospores. In other yeasts, such as *S. cerevisiae,* meiosis is sometimes delayed and the zygote divides mitotically to form a population of diploid cells that reproduce

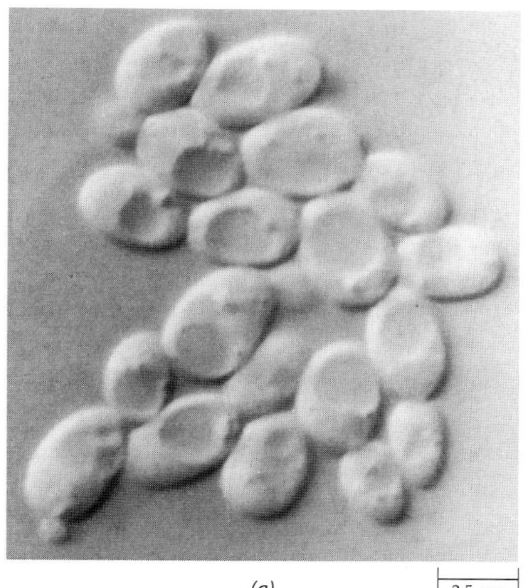

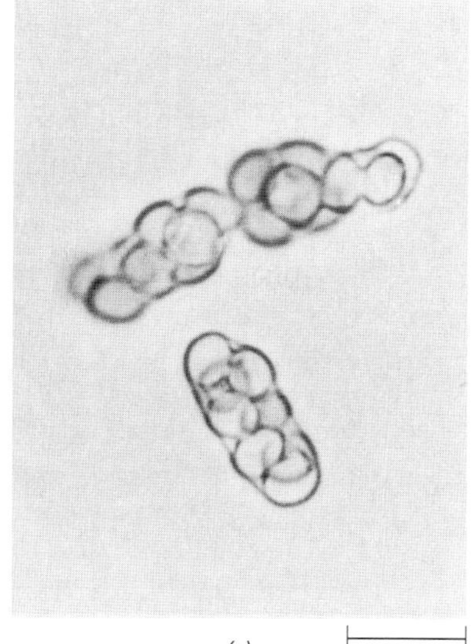

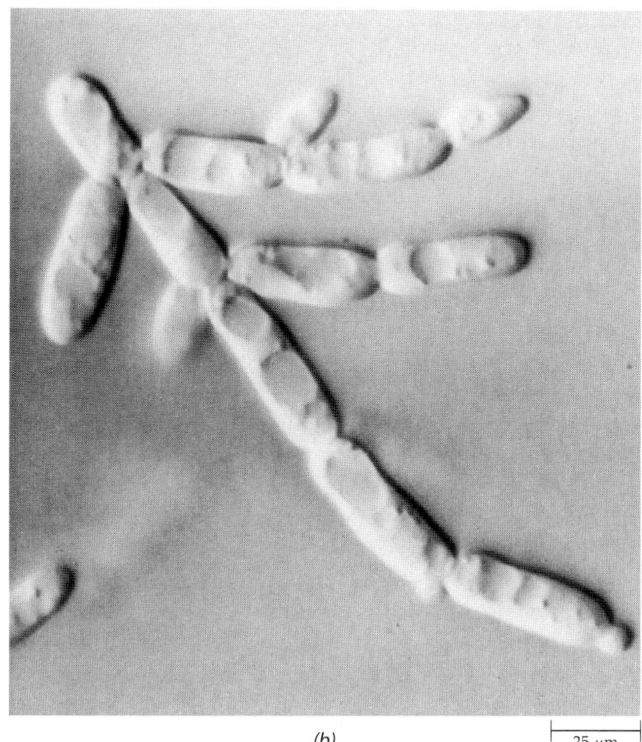

15–31
Yeasts. ***(a)*** *Single celled and* ***(b)*** *filamentous forms of bread yeast,* Saccharomyces cerevisiae. *Yeasts commonly reproduce by budding (lower left in each micrograph), an asexual process.* ***(c)*** *Asci, with eight ascospores each, of* Schizosaccharomyces octosporus.

asexually by budding. Thus, such yeasts have both haploid and diploid budding stages. Diploid cells may ultimately undergo meiosis and revert to the haploid condition. In still other yeasts, the ascospores fuse in pairs immediately after they are formed; the ascospore is the only haploid cell in the life cycle, which is predominantly diploid.

As mentioned earlier, yeasts are utilized by vintners as a source of ethanol, by bakers as a source of carbon dioxide, and by brewers as a source of both substances. Many domestically useful strains of yeast have been developed by selection and breeding, and the techniques of genetic engineering are now being used to improve these strains further through the addition of useful genes from other organisms. Some of the flavors of wine come directly from the grape, but most arise from the direct action of the yeast (Figure 15–32). Most of the yeasts important in the production of wine, cider, sake, and beer are strains of *S. cerevisiae,* although other species also play a role. Most lager beer, for example, is made using *Saccharomyces carlsbergensis. Saccharomyces cerevisiae* is now virtually the only species used in baking bread (Figure 15–31a). Some species of yeast are important as human pathogens, causing such diseases as thrush (caused by *Candida albicans*) and cryptococcosis, the latter caused by a species of *Basidiomycetes* (*Cryptococcus neoformans*) that has the growth habit of a yeast when it grows as a human pathogen.

15–32
In some areas of the world, yeasts that are present on grapes are still used to produce wine. When the grapes are crushed, the yeast mix with the juice and then break down the glucose in the juice to alcohol.

(a)

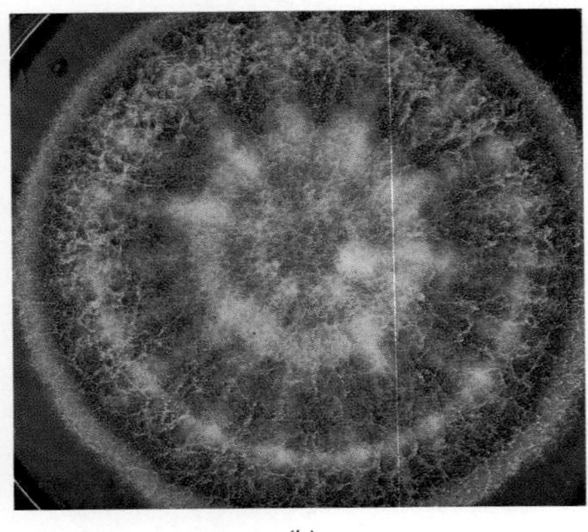

(b)

15–33
Penicillium *and* Aspergillus*—two of the common genera of deuteromycetes.* ***(a)*** *A culture of* Penicillium notatum, *the original penicillin-producing fungus, showing the distinctive colors produced during growth and spore development.* ***(b)*** *A culture of* Aspergillus fumigatus, *a fungus that causes respiratory disease in humans. Notice the concentric growth pattern produced by successive "pulses" of spore production.*

A number of yeasts, most notably *S. cerevisiae*, have become important laboratory organisms for genetic research. This yeast has become the organism of choice for studies of the metabolism, molecular genetics, and development of eukaryotic cells, and for chromosome studies. Haploid cells of *S. cerevisiae* have 16 chromosomes, and the DNA sequences of all 16 have been determined. *Saccharomyces cerevisiae* is thus the first eukaryote whose genome has been completely sequenced. In the future, such detailed knowledge, coupled with the ease of manipulating the genetics of yeasts growing on defined media, will doubtless greatly increase the importance of yeasts for scientific investigations and industrial applications.

Deuteromycetes

The deuteromycetes, or conidial fungi, are an artificial assemblage of about 15,000 distinct species of fungi for which only the asexually reproducing state is known or in which the sexual reproductive features are not used as the basis of classification (Figures 15–33 and 15–34). They have often been called "Fungi Imperfecti" because they were thought of as being second- (deutero-) class or "imperfect" members among sexually reproducing—"perfect"—organisms. Considering that many of the deuteromycetes are abundant, flourishing organisms, the term "imperfect" is somewhat misleading.

In some of the fungi classified as deuteromycetes, the sexual state of the life cycle has apparently been lost in the course of evolution. In others, the sexual state may simply not have been discovered yet, and in still others,

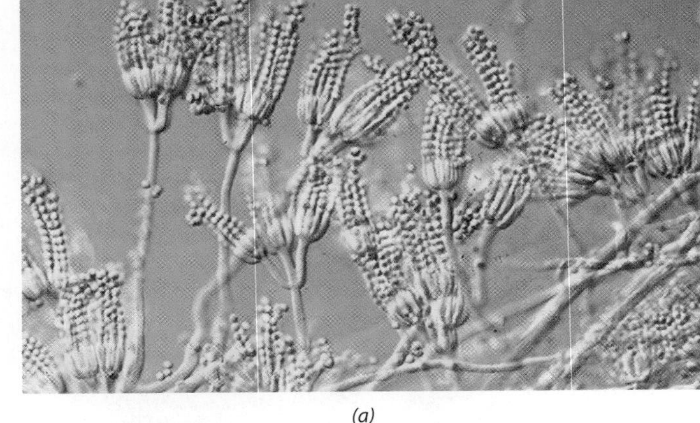

(a)

(b)

15–34
The conidiogenous cells and conidiophores—the specialized hyphae that bear the conidia—of deuteromycetes are used in their classification. ***(a)*** Penicillium *(brushlike) and* ***(b)*** Aspergillus *(tightly clumped and arising from the swollen top of the conidiophore). Note the long chains of small, dry conidia.*

Predaceous Fungi

Among the most highly specialized of the fungi are the predaceous fungi, which have developed a number of mechanisms for capturing small animals they use as food. Although microscopic fungi with such habits have been known for many years, recently it has been learned that a number of species of gilled fungi also attack and consume the small roundworms known as nematodes. The oyster mushroom, Pleurotus ostreatus, *for example, grows on decaying wood* (a, b). *Its hyphae secrete a substance that anesthetizes nematodes, after which the hyphae envelop and penetrate these tiny worms. The fungus apparently uses them primarily as a source of nitrogen, thus supplementing the low levels of nitrogen that are present in wood.*

Some of the microscopic deuteromycetes secrete on the surface of their hyphae a sticky substance in which passing protozoa, rotifers, small insects, or other animals become stuck (c). *More than 50 species of this group trap or snare nematodes. In the presence of these roundworms, the fungal hyphae produce loops that swell rapidly, closing the opening like a noose when a nematode rubs against its inner surface. Presumably the stimulation of the cell wall increases the amount of osmotically active material in the cell, causing water to enter the cells and increase their turgor pressure. The outer wall then splits, and a previously folded inner wall expands as the trap closes.*

(a)

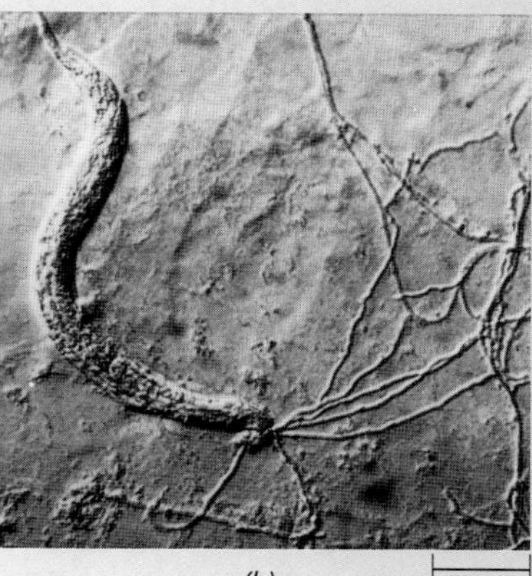

(b) 50 μm

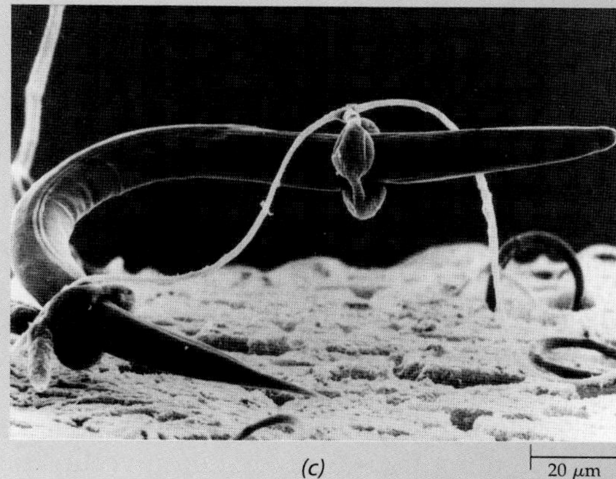

(c) 20 μm

(a) The oyster mushroom, Pleurotus ostreatus. *(b) Hyphae of the oyster mushroom, which produce a substance that anesthetizes, converging on the mouth of an immobilized nematode.* *(c) The predaceous deuteromycete* Arthrobotrys anchonia *has trapped a nematode. The trap consists of rings, each comprising three cells, which when triggered swell rapidly to about three times their original size in 0.1 second and strangle the nematode. Once the worm has been trapped, fungal hyphae grow into its body and digest it.*

it may not be used as the principal basis of classification because of the close resemblance between the conidium-producing features and those of other deuteromycetes. Thus for some members of the well-known deuteromycete genera *Aspergillus* and *Penicillium,* the sexual state is known, but the species are classified as members of these genera because of their overall resemblance to the other species. On the basis of their overall characteristics, most deuteromycetes are clearly ascomycetes, reproducing only by means of conidia. A few deuteromycetes are *Basidiomycetes* or zygomycetes, the former marked by the septa and clamp connections that are characteristic of the *Basidiomycetes.* New molecular techniques are now supplementing traditional morphological features in the evaluation of the relationship between particular deuteromycetes and their relatives that form sexual states. When the sexual and asexual states of a fungus are identified, the entire organism in all of its forms is given the name of the sexual state.

Many fungi exhibit the phenomenon of **heterokaryosis,** in which genetically different nuclei occur together in a common cytoplasm. The nuclei may differ from one another because of mutation or because of the fusion of genetically distinct hyphae. Since genetically different nuclei may occur in different proportions in different parts of a mycelium, these sectors may have different properties.

Among the deuteromycetes, as well as some other fungi, genetically distinct haploid nuclei occasionally fuse. Within the resulting diploid nuclei, the chromosomes may associate, recombination may follow, and genetically novel haploid nuclei may form. Restoration of the haploid condition does not involve meiosis. Instead, it results from a gradual loss of chromosomes, a process called *haploidization.* This genetic phenomenon, in which plasmogamy, karyogamy, and haploidization occur in sequence, is known as **parasexuality;** it was discovered in *Aspergillus,* a deuteromycete. Within the hyphae of this common fungus, there is one diploid nucleus, on average, for every 1000 haploid nuclei. Parasexual cycles may add considerably to genetic and evolutionary flexibility in fungi that lack a true sexual cycle.

Among the deuteromycetes are many organisms of great economic importance. For example, a number of the most important plant pathogens are deuteromycetes. Anthracnose diseases of plants, which cause lesions and blackening, are generally caused by deuteromycetes. In addition, an often fatal disease of dogwoods *(Cornus florida)* that was detected over a wide area of the eastern United States in the late 1980s is caused by the deuteromycete *Discula destructiva.*

Other members of the group produce effects that are valued by humans. For example, certain *Penicillium* species give some types of cheese the appearance, flavor, odor, and texture so highly prized by gourmets. Roquefort, Danish blue, Stilton, and Gorgonzola are all ripened by *Penicillium roqueforti.* Another species, *Penicillium camemberti,* gives Camembert and Brie cheeses their special qualities.

Soy paste (miso) is produced by fermenting soybeans with *Aspergillus oryzae,* and soy sauce by fermenting soybeans with a mixture of *A. oryzae* and *Aspergillus soyae,* as well as lactic acid bacteria. *Aspergillus oryzae* is also important for the initial steps in brewing sake, the traditional alcoholic beverage of Japan; *Saccharomyces cerevisiae* is important later in the process.

The ubiquitous soil deuteromycete *Trichoderma* has many commercial applications. For example, cellulose-degrading enzymes produced by *Trichoderma* are used by clothing manufacturers to give jeans a "stone-washed" look. The same enzymes are added to some household laundry detergents to help remove fabric nubs. *Trichoderma* is also used by farmers in the biological control of other fungi that attack crops and forest trees.

Many important antibiotics are produced by deuteromycetes. The first antibiotic was discovered by Sir Alexander Fleming, who noted in 1928 that a strain of *Penicillium* that had contaminated a culture of *Staphylococcus* growing on a nutrient agar plate had completely halted the growth of the bacteria. Ten years later, Howard Florey and his associates at Oxford University purified penicillin and later came to the United States to promote the large-scale production of the drug. Production of penicillin increased enormously with increasing demand during World War II. Penicillin is effective in curing a wide variety of bacterial diseases, including pneumonia, scarlet fever, syphilis, gonorrhea, diphtheria, and rheumatic fever.

Not all substances produced by deuteromycetes are useful to us. For example, the aflatoxins are potent causative agents for liver cancer in humans. These highly carcinogenic mycotoxins, which show their effects at concentrations as low as a few parts per billion, are secondary metabolites produced by certain strains of *Aspergillus flavus* and *Aspergillus parasiticus.* Both of these fungi frequently grow in stored food products, especially peanuts, maize, and wheat. In tropical countries, aflatoxins have been estimated to contaminate at least 25 percent of the food. Aflatoxins have been detected occasionally in maize harvested in the United States, although strong efforts have been made to detect and destroy contaminated maize.

One group of deuteromycetes, the dermatophytes (Gk. *derma,* "skin," + *phyton,* "plant"), are the cause of ringworm, athlete's foot, and other fungal skin diseases. Such diseases are especially prevalent in the tropics. The pathogenic stages of these fungi are asexual, but most of these organisms have now been correlated with species of ascomycetes. Nevertheless, they continue to be classified as deuteromycetes on the basis of their disease-causing forms. During World War II, more soldiers had to be sent back from the South Pacific because of skin infections than because of wounds received in battle.

Symbiotic Relationships of Fungi

Symbiosis—"living together"—is a close and long-term association between organisms of different species. Some symbiotic relationships, typically disease-causing, are **parasitic.** One species (the parasite) benefits from the association and the other (the host) is harmed. Although many fungi are parasites, other fungi are involved in symbiotic relationships that are **mutualistic**—that is, the association is beneficial to both organisms. Two of these mutualistic symbioses—lichens and mycorrhizae—have been and are of extraordinary importance in enabling photosynthetic organisms to become established in previously barren terrestrial environments.

A Lichen Consists of a Mycobiont and a Photobiont

A lichen is a mutualistic symbiotic association between a fungal partner and a population of unicellular or filamentous algal or cyanobacterial cells. The fungal component of the lichen is called the **mycobiont** (Gk. *mykēs,* "fungus," + *bios,* "life"), and the photosynthetic component is called the **photobiont** (Gk. *phōto-,* "light," + *bios,* "life"). The scientific name given to the lichen is the name of the fungus. About 98 percent of the lichen-forming fungi belong to the *Ascomycota,* the remainder to the *Basidiomycota.* Lichens are polyphyletic. Recent DNA evidence indicates that they have evolved independently on at least five occasions, and it is likely that they evolved independently many more times.

About 13,250 species of lichen-forming fungi have been described, representing almost half of all known ascomycetes. Some 40 genera of photobionts are found in combination with these ascomycetes. The most frequent

From Pathogen to Symbiont: Fungal Endophytes

The leaves and stems of plants are often riddled with fungal hyphae. Such fungi are called endophytes, a general term used for a plant or fungus growing within a plant. Although some fungal endophytes cause disease symptoms in the plants they inhabit, others produce no such effects. Instead, they protect host plants from their natural enemies: herbivores and, in some cases, pathogenic microbes.

In many species of grasses, endophytic fungi infect the flowers of the host and proliferate in the seeds. Eventually, a substantial mass develops throughout the stems and leaves of the mature grass plant, with the fungal hyphae growing between the host cells. A good example of such a relationship is provided by tall fescue, Festuca arundinacea, *a grass that covers more than 15,000 square kilometers (35 million acres) of lawns, fields, and pastures in the United States, especially in the East and Midwest. Tall fescue plants that are free of fungus provide good forage for livestock, but cattle feeding on infected plants become lethargic and stop grazing, often panting and drooling excessively. If the animals are not moved to other forage, they become feverish, gain weight slowly, produce little milk, have difficulty in conceiving, develop gangrene, and eventually die, if they have not been slaughtered by that time for their salvage value. In the 1970s, scientists at the University of Kentucky discovered that these symptoms were associated with the endophytic ascomycete* Sphacelia typhina. *In relationships with fungal endophytes, which are common in grasses, the host plant gains protection from its enemies (including, unfortunately for us, cattle), and the fungus obtains all of its nutritional needs from the grass. The deterrent effects on herbivores occur because the fungi produce alkaloids—bitter, nitrogen-rich compounds that are abundant in some plants. Alkaloids have physiological effects on humans and other animals (page 32).*

Some of the alkaloids produced by another fungus, Claviceps purpurea, *which infects rye* (Secale cereale) *and other grasses, are identical to those produced by* Sphacelia. *The rye-infecting fungus, which replaces infected grain with a hard mycelial mat, or sclerotium, causes a plant disease called ergot. The sclerotium—also called an ergot—contains lysergic acid amide (LDA), a precursor of lysergic acid diethylamide (LSD). LSD was first discovered in studies of the alkaloids of* C. purpurea. *Domestic animals and people who eat the infected grain develop a disease called ergotism, which is often accompanied by gangrene, nervous spasms, psychotic delusions, and convulsions. It occurred frequently during the Middle Ages, when it was known as St. Anthony's fire. It has been suggested that the widespread accusations of witchcraft in Salem Village (now Danvers) and other communities in Massachusetts and Connecticut in 1692, which led to a number of executions, may have resulted from an outbreak of convulsive ergotism. In one epidemic in Europe in the year 994, more than 40,000 people died. As recently as 1951, there was an outbreak of ergotism in a small French village, in which 30 people became temporarily insane, believing that they were pursued by demons or snakes. Five of the villagers died. Some of the ergot alkaloids have the property of enhancing muscle constriction and have been used for more than 400 years to hasten uterine contraction during childbirth. Other ergot alkaloids constrict the veins and are used in the treatment of migraine. Because of these useful medical properties, efforts are now under way to produce improved strains of* C. purpurea *and to grow them commercially.*

Sclerotia of Claviceps purpurea, *which causes the plant disease known as ergot, are seen here growing on stalks of rye* (Secale cereale).

are the green algae *Trebouxia, Pseudotrebouxia,* and *Trentepohlia,* and the cyanobacterium *Nostoc.* About 90 percent of all lichens have one of these four genera as their photobiont. A number of lichens incorporate two photobionts—a green alga and a cyanobacterium. Different species of the algal genus can serve as photobionts in a single lichen species. In addition, a single fungal species can form lichens with different algae or cyanobacteria.

Lichens are able to live in some of the harshest environments on Earth, and consequently they are extremely widespread (Figure 15–35). They occur from arid desert

(a)

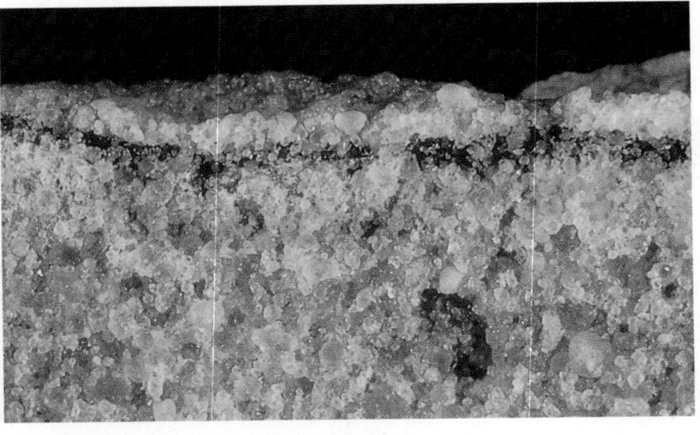

(b)

15–35
In this seemingly lifeless, dry region of Antarctica **(a)**, *lichens live just beneath the exposed surface of the sandstone.* **(b)** *In the fractured rock, the colored bands are distinct zones biologically. The white and black zones are formed by a lichen, while the lower green zone is produced by a nonlichenized unicellular green alga. The air temperatures in this part of Antarctica rise almost to freezing in the summer and probably fall to −60°C in winter.*

(a)

regions to the Arctic and grow on bare soil, tree trunks, sunbaked rocks, fence posts, and windswept alpine peaks all over the world (Figures 15–36 and 15–37). Some lichens are so tiny that they are almost invisible to the unaided eye; others, like the reindeer "mosses," may cover kilometers of land with ankle-deep growth. One species, *Verrucaria serpuloides*, is a permanently submerged marine lichen. Lichens are often the first colonists of newly exposed rocky areas. In Antarctica, there are more than 350 species of lichens (Figure 15–35) but only two species of vascular plants; seven species of lichens actually occur within 4 degrees of the South Pole! Although widely distributed, individual lichen species usually occupy fairly specific substrates, such as the surfaces or interiors of rocks, soil, leaves, and bark. Some lichens provide the substrate for other lichens and parasitic fungi that may be closely related to the lichen being parasitized.

(b)

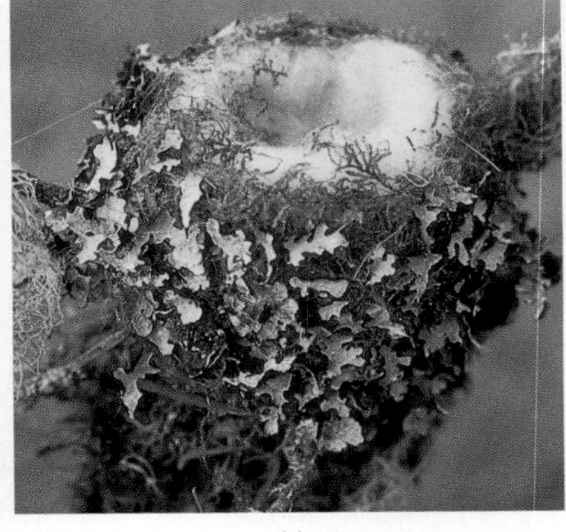

(c)

15–36
Crustose and foliose lichens. **(a)** *A crustose ("encrusting") lichen,* Caloplaca saxicola, *growing on a bare rock surface in central California.* **(b)** *Crustose and foliose ("leafy") lichens growing on the gravestone of Roland Thaxter (1858–1932), renowned mycologist from Harvard University.* **(c)** Parmelia perforata, *a foliose ("leafy") lichen, forms part of a hummingbird's nest on a dead tree branch in Mississippi.*

(a) (b) (c) (d)

15–37
Several fruticose ("shrubby") lichens. ***(a)*** *Goldeye lichen,* Teloschistes chrysophthalmus. ***(b)*** *British soldiers,* Cladonia cristatella, *are 1 to 2 centimeters tall.* ***(c)*** *Witch's hair* (Alectoria sarmentosa), *a hanging lichen that often occurs in masses on the limbs of trees. Although superficially similar to* Alectoria *and occupying a similar ecological niche, "Spanish moss," which is common throughout the southern United States, is not a lichen but a flowering plant—a member of the pineapple family* (Bromeliaceae). ***(d)*** Cladonia subtenuis, *commonly called "reindeer moss," is actually a lichen. Lichens of this group, which are abundant in the Arctic, concentrated radioactive substances following atmospheric nuclear tests and nuclear reactor accidents. Reindeer feeding on the lichens concentrated these radioactive substances still further and passed them on to the humans and other animals that consumed them or their products, especially milk and cheese.*

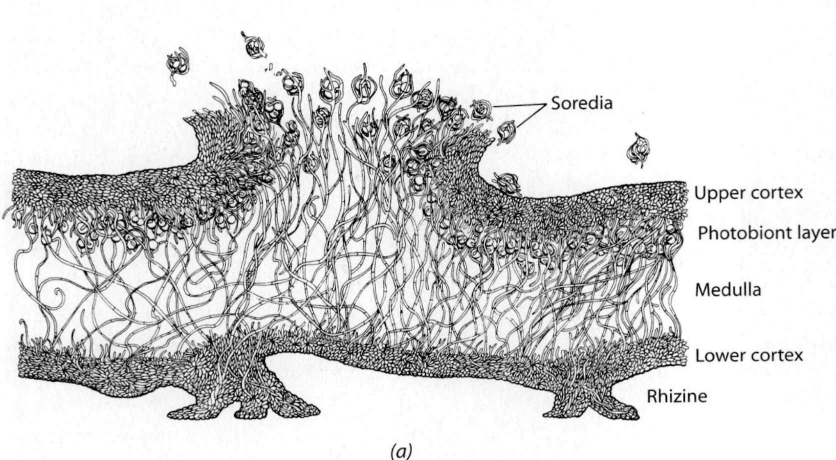

(a)

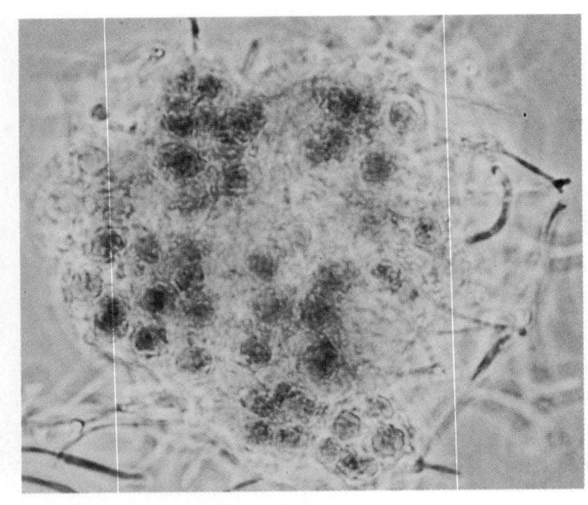

(b)

15–38

A stratified lichen. ***(a)*** *A cross section of the lichen* Lobaria verrucosa, *shown here releasing soredia consisting of hyphae wrapped around cyanobacteria. The simplest lichens consist of a crust of fungal hyphae entwining colonies of algae or cyanobacteria. More complex lichens, however, exhibit a definite growth form with a characteristic internal structure. The lichen shown here has four distinct layers: (1) the upper cortex, a protective surface of heavily gelatinized fungal hyphae; (2) the photobiont layer, which, in* Lobaria, *consists of cyanobacterial cells and loosely interwoven, thin-walled hyphae; (3) the medulla, consisting of loosely packed, weakly gelatinized hyphae, which makes up about two-thirds of the thickness of the thallus and appears to serve as a storage area; and (4) the lower cortex, which is thinner than the upper one and is covered with fine projections (rhizines) that attach the lichen to its substrate.* ***(b)*** *A soredium, which is composed of fungal hyphae and cells of the photobiont.*

In almost all cases, the fungus makes up most of the thallus and apparently plays the major role in determining the form of the lichen. There are two general types of lichen thalli. In one, the cells of the photobiont are more or less evenly distributed throughout the thallus; in the other, the cells of the photobiont form a distinct layer within the thallus (Figure 15–38). Three major growth forms are recognized among the latter, stratified lichens: **crustose,** which is flattened and adheres firmly to the substrate, having a "crusty" (encrusting) appearance; **foliose,** which is leaflike; and **fruticose,** which is erect and often branched and "bushy" (Figures 15–36 and 15–37).

The colors of lichens range from white to black, through shades of red, orange, brown, yellow, and green, and these organisms contain many unusual chemical compounds. Many lichens are used as sources of dyes; for example, the characteristic color of Harris tweed originally resulted from the treatment of the wool with a lichen dye. Many lichens have also been used as medicines, components of perfumes, or minor sources of food. Some species are being investigated for their ability to secrete anti-tumor compounds.

Lichens commonly reproduce by simple fragmentation, by the production of special powdery propagules known as **soredia** (Figure 15–38), or by small outgrowths known as **isidia.** Fragments, soredia, and isidia, which all contain both fungal hyphae and algae or cyanobacteria, act as small dispersal units to establish the lichen in new localities. The fungal component of the lichen produces ascospores, conidia, or basidiospores that are typical of its taxonomic group. If the fungus is an ascomycete, it may form ascomata, which are similar to those of other ascomycetes, except that in lichens the ascomata may endure and produce spores slowly but continuously over a number of years. Regardless, any of these spores may form new lichens when they germinate and come into contact with the appropriate green algae or cyanobacteria.

The Survival of Lichens Is Due to Their Ability to Dry Out Very Rapidly How can the lichens survive under environmental conditions so severe that they exclude any other form of life? At one time, it was thought that the secret of a lichen's success was that the fungal tissue protected the alga or cyanobacterium from drying out. Actually, one of the chief factors in their survival seems to be the fact that they dry out very rapidly. Lichens are frequently very desiccated in nature, with a water content ranging from only 2 to 10 percent of their dry weight. When a lichen dries out, photosynthesis ceases. In this state of "suspended animation," even blazing sunlight or great extremes of heat or cold can be endured by some species of lichens. Cessation of photosynthesis depends, in large part, on the fact that the upper cortex of the lichen becomes thicker and more opaque when dry, cutting off the passage of light energy. A wet lichen may be damaged or destroyed by light intensities or temperatures that do not harm a dry lichen.

The lichen reaches its maximum vitality, as judged by the rate of photosynthesis, after it has been soaked with water and has begun to dry. Its rate of photosynthesis reaches a peak when the water content is 65 to 90 percent of the maximum the organism can hold; below this level, if the lichen continues to lose water, the rate of photosynthesis decreases. In many environments, the water content of the lichen varies markedly in the course of a day, with most photosynthesis taking place only during a few hours, usually in the early morning after wetting by fog or dew. As a consequence, lichens have an extremely slow rate of growth, their radius increasing at rates ranging from 0.1 to 10 millimeters a year. Calculated on this basis, some mature lichens have been estimated to be as much as 4500 years old. They achieve their most luxuriant growth along seacoasts and on fog-shrouded mountains. The oldest known fossil lichens are found in the Early Devonian (400 million years old).

The Nature of the Relationship between Mycobiont and Photobiont Considerable debate has surrounded the nature of the relationship between mycobiont and photobiont, that is, whether the relationship is parasitic or mutualistic symbiosis. As noted by David L. Hawksworth, the issue is in reality one of scale. At the cellular level, individual photobiont cells can be considered parasitized by the mycobionts whose hyphae adhere closely to the phycobiont cell walls (Figure 15–39). These hyphae typically form haustoria or **appressoria,** which are specialized hyphae that penetrate photobiont cells by means of pegs. These structures are involved with the transfer of carbohydrates and nitrogen compounds (in the case of nitrogen-fixing cyanobacteria such as *Nostoc*) from the photobiont to the fungus. In addition, the mycobiont controls the division rate of its photobiont. The mycobiont in turn provides the photobiont with a suitable physical environment in which to grow and absorbs needed minerals from the air in the form of dust or from rain. At the whole lichen level, the association clearly is mutualistic as neither partner can flourish in niches in which they occur in nature without the other. Today mutualism generally is judged on the basis of the dual functional unit, and therefore lichen symbiosis is considered to be mutualistic.

Although at one time it was thought that separate dispersal of mycobiont and photobiont was uncommon, there now is evidence that independent dispersal and resynthesis of the lichen is common in nature in some species. Several lichenologists have developed methods for growing many photobionts and mycobionts separately in defined media and resynthesizing the lichen. The mechanism by which the germinating fungal spore recognizes the appropriate alga or cyanobacterium is not well understood. When grown together in culture, the fungus seems first to bring the photobiont partner under control and then to establish the characteristic form of the mature lichen (Figure 15–39).

Lichens Are Important Ecologically Lichens clearly play an important role in ecosystems. Mycobionts produce large numbers of secondary metabolites called **lichen acids,** which sometimes amount to 40 percent or more of the dry weight of a lichen. These metabolites are known to play a role in the biogeochemical weathering of rock

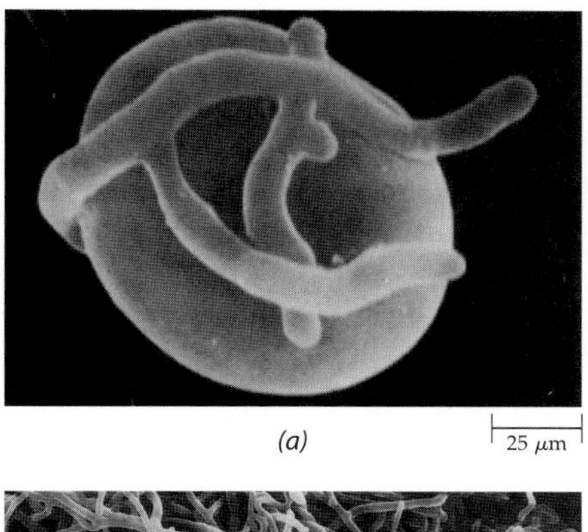

(a)

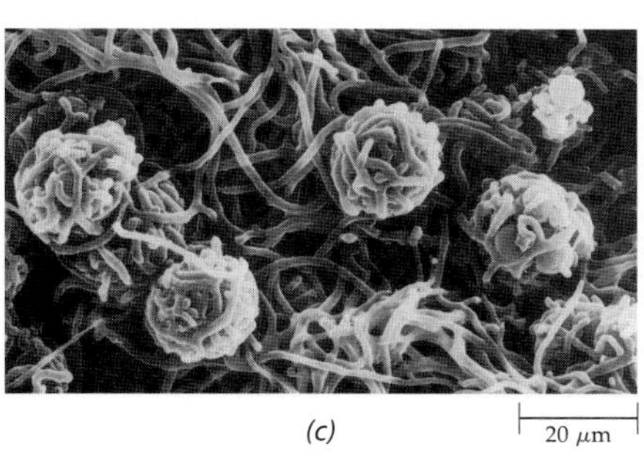

(c)

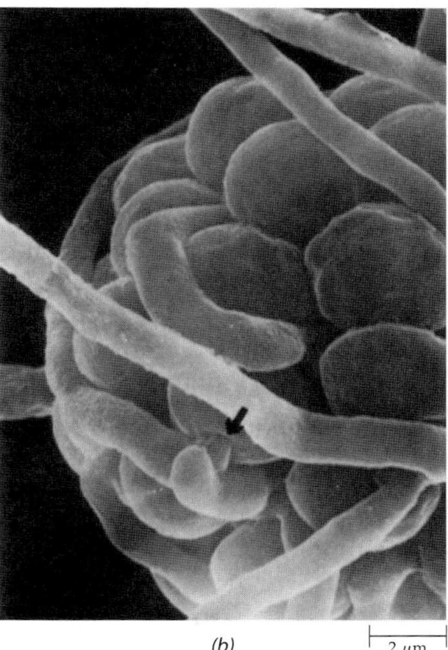

(b)

15–39
Scanning electron micrographs of early stages in the interaction between fungal and algal components of British soldier lichens (Cladonia cristatella) *in laboratory culture. The photosynthetic component in this lichen (Figure 15–37b) is* Trebouxia, *a green alga.* ***(a)*** *An algal cell surrounded by fungal hyphae.* ***(b)*** *Penetration of an algal cell by a fungal haustorium (arrow).* ***(c)*** *Mixed groups of fungal and algal components developing into the mature lichen.*

and soil formation. The lichens trap the newly formed soil, making it possible for a succession of plants to grow.

Lichens containing a cyanobacterium are of special importance because they contribute fixed nitrogen to the soil. Such lichens are an important factor in supplying nitrogen in many ecosystems, including the old-growth forests of the Pacific Northwest of the United States, some tropical forests, and certain desert and tundra sites.

Because they have no means of excreting the elements that they absorb, some lichens are particularly sensitive to toxic compounds. The toxins cause the limited amount of chlorophyll in their algal or cyanobacterial cells to deteriorate. Lichens are very sensitive indicators of the toxic components—particularly sulfur dioxide—of polluted air, and they are being used increasingly to monitor atmospheric pollutants, especially around cities. Since lichens containing cyanobacteria are especially sensitive to sulfur dioxide, air pollution can substantially limit nitrogen fixation in natural communities, causing long-range changes in soil fertility. Both the state of health of the lichens and their chemical composition are used to monitor the environment. For example, analysis of lichens can detect the distribution of heavy metals and other pollutants around industrial sites. Fortunately, many lichens have the ability to bind heavy metals outside their cells and thus escape damage themselves.

When nuclear tests were being conducted in the atmosphere, lichens were used to monitor the fallout. Now, lichens provide a useful way to monitor the contamination by radioactive substances following events such as the explosion at the Chernobyl nuclear power plant in 1986.

There are many interactions between lichens and other groups of organisms. Lichens are eaten by a number of vertebrate and invertebrate animals. They are an important winter food source for reindeer and caribou in the far-north regions of North America and Europe and are eaten by mites, insects, and slugs. Lichens are dispersed by birds that use lichens in their nests (Figure 15–36c). The nests of flying squirrels can contain up to 98 percent lichens. Some lichens have been found to have important functions as antibiotics.

Mycorrhizae Are Mutualistic Associations between Fungi and Roots

The most prevalent and possibly the most important mutualistic symbiosis in the plant kingdom is the **mycorrhiza,** which literally means "fungus root." Mycorrhizae are intimate and mutually beneficial symbiotic associations between fungi and roots, and they occur in the vast majority of vascular plants, both wild and cultivated. The few families of flowering plants that usually lack mycorrhizae include the mustard family *(Brassicaceae)* and the sedge family *(Cyperaceae).*

Mycorrhizal fungi benefit their host plants by increasing the plants' ability to capture water and essential elements (see Chapter 30), especially phosphorus. Increased absorption of zinc, manganese, and copper—three other essential nutrients—has likewise been demonstrated. For many forest trees, if seedlings are grown in a sterile nutrient solution and then transplanted to a grassland soil, they grow poorly, and many eventually die from malnutrition (Figure 15–40). Mycorrhizal fungi also provide protection against attack by pathogenic fungi and nematodes (small roundworms). In return for these benefits, the fungal partner receives from the host plant carbohydrates and vitamins essential for its growth.

15–40
Mycorrhizae and tree nutrition. Nine-month-old seedlings of white pine (Pinus strobus) *were raised for two months in a sterile nutrient solution and then transplanted to prairie soil. The seedlings on the left were transplanted directly. The seedlings on the right were grown for two weeks in forest soil containing fungi before being transplanted to the grassland soil.*

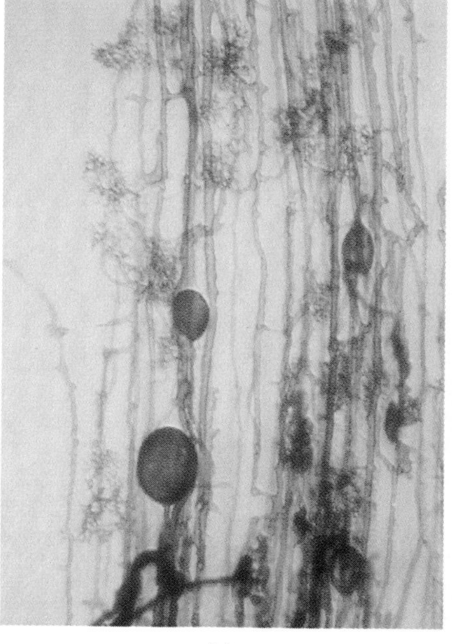
(a)

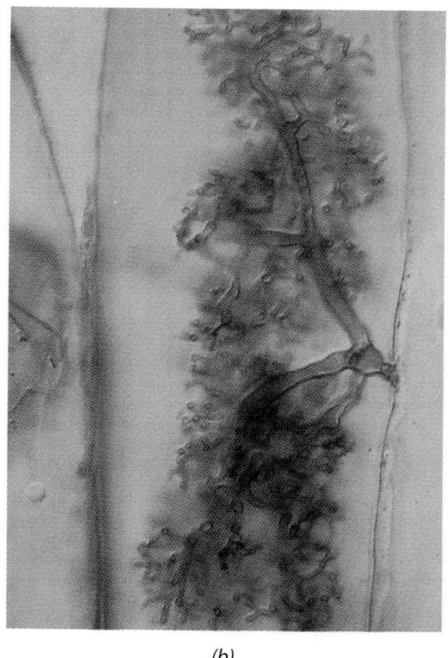
(b)

15–41
Endomycorrhizae. Glomus versiforme, *a zygomycete, is shown here growing in association with the roots of leeks,* Allium porrum. ***(a)*** *Arbuscules (highly branched structures) and vesicles (dark, oval structures). Arbuscules predominate in young infections, with vesicles becoming common later.* ***(b)*** *Arbuscules growing inside a leek root cell.*

Endomycorrhizae Penetrate Root Cells There are two major types of mycorrhizae: **endomycorrhizae,** which penetrate the root cells, and **ectomycorrhizae,** which surround the root cells. Of these two, endomycorrhizae are by far the more common, occurring in about 80 percent of all vascular plants. The fungal component is a zygomycete (order *Glomales*), with fewer than 200 species involved in such associations worldwide. Thus, endomycorrhizal relationships are not highly specific. The fungal hyphae penetrate the cortical cells of the plant root, where they form highly branched structures called **arbuscules** (Figures 15–41 and 15–42) and in some cases terminal swellings called **vesicles.** The arbuscules do not enter the protoplast but greatly invaginate the plasma membrane, increasing its surface area and presumably facilitating the bidirectional transfer of metabolites and nutrients by the two mycorrhizal partners. Most, or perhaps all, exchange between plant and fungus takes place at the arbuscules. Vesicles may also occur between host cells and are thought to function as storage compartments for the fungus. Such mycorrhizae are often called vesicular-arbuscular (or V/A) mycorrhizae. The hyphae extend out into the surrounding soil for several centimeters and thus greatly increase the potential for the absorption of water and the uptake of phosphates and other essential nutrients.

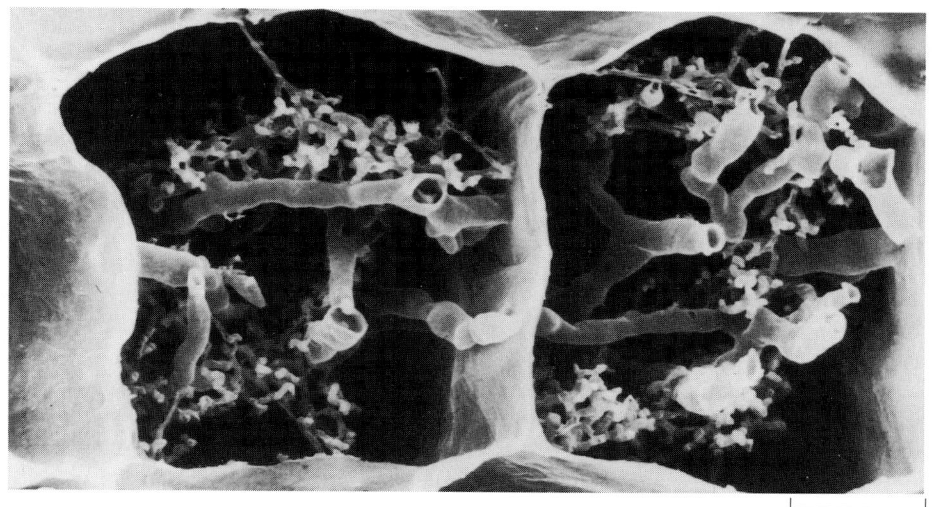

15–42
Endomycorrhizae. Scanning electron micrograph showing arbuscules of Glomus etunicatum, *a zygomycete, in cortical cells of a sugar maple* (Acer saccharum) *root. The protoplasts, which typically envelop the hyphae, are not apparent.*

15–43
Ectomycorrhizae. A section through the extensive ectomycorrhizae of a lodgepole pine (Pinus contorta) *seedling. The seedling extends about 4 centimeters above the soil surface.*

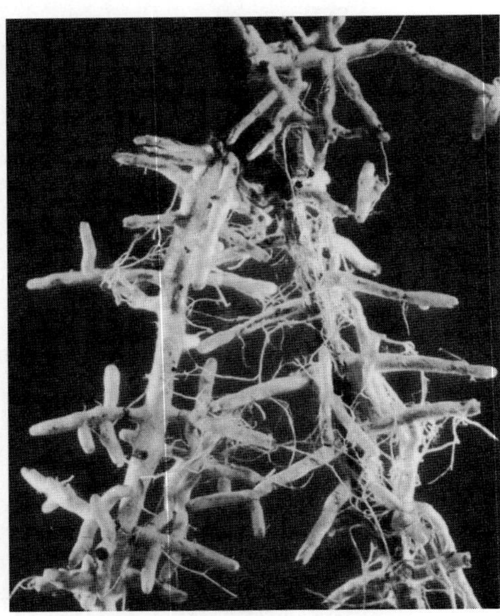

15–44
Ectomycorrhizae from a western hemlock (Tsuga heterophylla). *In such ectomycorrhizae, the fungus commonly forms a sheath of hyphae, called a fungal mantle, around the root. Hormones secreted by the fungus cause the root to branch. This growth pattern and the hyphal sheath impart a characteristic branched and swollen appearance to the ectomycorrhizae. The narrow mycelial strands extending from the mycorrhizae function as extensions of the root system.*

Ectomycorrhizae Surround but Do Not Penetrate Root Cells Ectomycorrhizae (Figures 15–43 and 15–44) are characteristic of certain groups of trees and shrubs that are found primarily in temperate regions. Among these groups are the beech family *(Fagaceae),* which includes the oaks; the willow, poplar, and cottonwood family *(Salicaceae);* the birches *(Betulaceae);* the pine family *(Pinaceae);* and certain groups of tropical trees that form dense stands of only one or a few species. The trees that grow at timberline (the upper altitudes and latitudes of tree growth) in different parts of the world—such as pines in the northern mountains, *Eucalyptus* in Australia, and southern beech (*Nothofagus;* see Figure 22–10)—almost always are ectomycorrhizal. The ectomycorrhizal association apparently makes the trees more resistant to the harsh, cold, dry conditions that occur at the limits of tree growth.

In ectomycorrhizae, the fungus surrounds but does not penetrate living cells in the roots. In the conifers, the hyphae grow between the cells of the root epidermis and cortex, forming a characteristic highly branched network, the **Hartig net** (Figure 15–45), which eventually surrounds many of the cortical and epidermal cells. In the roots of most angiosperms colonized by ectomycorrhizal fungi, the epidermal cells are triggered to enlarge primarily at right angles to the surface of the root, thickening the root rather than extending it, and the Hartig net is confined to this layer (Figure 15–45b). The Hartig net functions as the interface between the fungus and the plant. In addition to the Hartig net, ectomycorrhizae are characterized by a **mantle,** or sheath, of hyphae that covers the root surface. Mycelial strands extend from the mantle into the surrounding soil (Figure 15–44). Typically, root hairs do not develop on ectomycorrhizae, and the roots are short and often branched. Ectomycorrhizae are mostly formed with *Basidiomycetes,* including many genera of mushrooms, but some ectomycorrhizae involve associations with ascomycetes, including truffles *(Tuber)* and possibly the edible morels *(Morchella).* At least 5000 species of fungi are involved in ectomycorrhizal associations, often with a high degree of specificity.

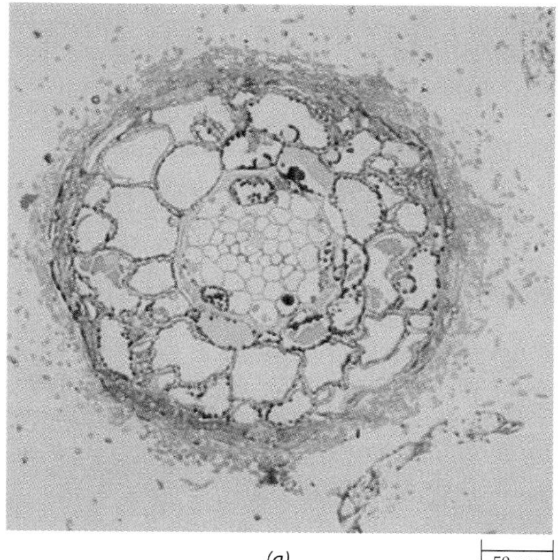

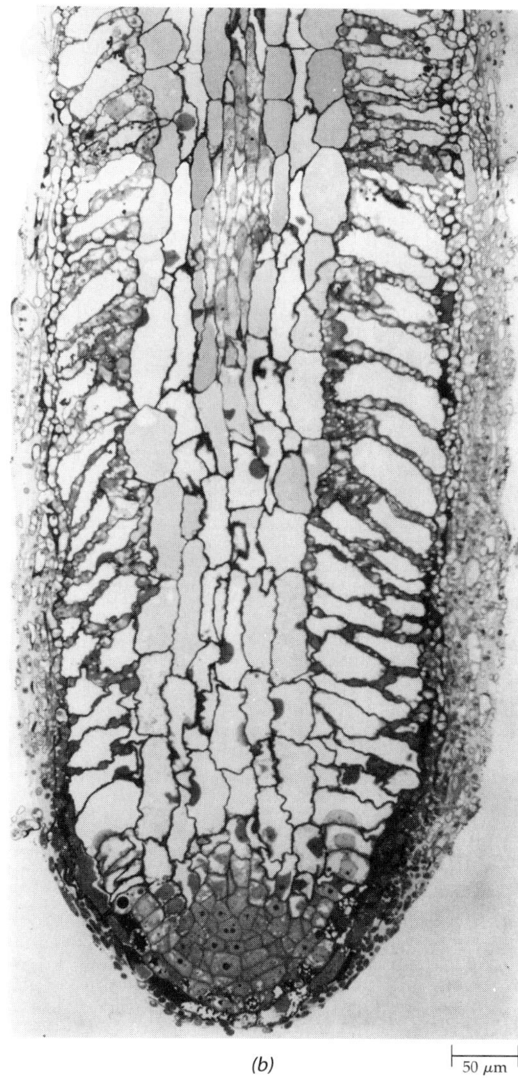

15–45
Ectomycorrhizae. ***(a)*** *Transverse section of an ectomycorrhiza of* Pinus. *The hyphae of the fungus form a mantle around the root and also penetrate between the epidermal and cortical cells, where they form the characteristic Hartig net.* ***(b)*** *Longitudinal section of ectomycorrhiza of birch* (Betula alleghaniensis). *The fungus ensheathes the root, forming a mantle around it. The Hartig net is confined to the layer of radially enlarged epidermal cells.*

Other Kinds of Mycorrhizae Are Found in the Heather and Orchid Families Two other kinds of mycorrhizae are those characteristic of the heather family *(Ericaceae)* and a few closely related groups, and those associated with the orchids *(Orchidaceae)*. In *Ericaceae,* the fungal hyphae form an extensive, loosely organized web over the root surface. Rather than extending the absorptive surface significantly, the principal role of the fungus is to release enzymes into the soil to break down certain compounds and make them available to the plant. The mycorrhizae of *Ericaceae* seem to function primarily in enhancing nitrogen, rather than phosphorus, uptake by the plants. This allows *Ericaceae* to colonize the kinds of infertile, acidic soils where they are especially frequent. Both ascomycetes and *Basidiomycetes* are involved in mycorrhizal associations with *Ericaceae.*

During their early stages of development, all species of orchids are heterotrophic and dependent upon the fungi for their nourishment. In this mycorrhizal association the fungi are internal and supply carbon to the host. The fungi in such associations are mostly *Basidiomycetes,* with more than 100 species involved.

Mycorrhizae Were Probably Important in the History of Vascular Plants A study of the fossils of early plants has revealed that endomycorrhizal associations were as frequent then as they are in their descendants (Figure 15–46). This finding led K. A. Pirozynski and D. W. Malloch to suggest that the evolution of mycorrhizal associations may have been a critical step in allowing colonization of the land by plants. Given the poorly developed soils that would have been available at the time of the first colonization, the role of mycorrhizal fungi (probably zygomycetes) may have been of crucial significance, particularly in facilitating the uptake of phosphorus and other nutrients. A similar relationship has been demonstrated among contemporary plants colonizing extremely nutrient-poor soils, such as slag

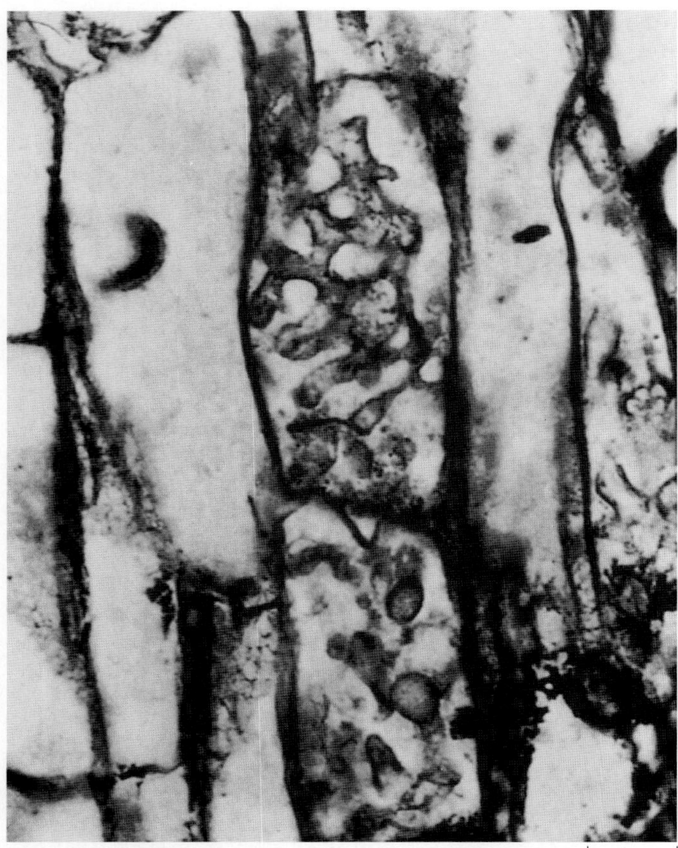

15–46
Silicified roots of a gymnosperm from the Triassic period of Antarctica, showing well-developed arbuscules. Endomycorrhizae were characteristic of the earliest plants for which fossils are known.

heaps: those individuals with endomycorrhizae have a much better chance of surviving. Thus it may have been not a single organism but rather a symbiotic association of organisms, comparable to a lichen, that initially invaded the land.

Summary

Fungi Are Ecologically and Economically Important

Fungi, together with the heterotrophic bacteria, are the principal decomposers of the biosphere, breaking down organic products and recycling carbon, nitrogen, and other components to the soil and air. As decomposers, they often come into direct conflict with human interests, attacking almost any conceivable substance. Most fungi are saprophytes; that is, they live on organic material from dead organisms. Many fungi attack living organisms, however, and cause diseases in plants, domestic animals, and humans. A number of fungi are economically important to humans as destroyers of foodstuffs and other organic materials. The kingdom *Fungi* also includes yeasts, cheese molds, edible mushrooms, and *Penicillium* and other producers of antibiotics.

Most Fungi Are Composed of Hyphae

Fungi are rapidly growing organisms that characteristically form filaments called hyphae, which may be septate or aseptate. In most fungi, the hyphae are highly branched, forming a mycelium. Parasitic fungi often have specialized hyphae (haustoria) by means of which they extract organic carbon and other substances from the living cells of other organisms.

Fungi Are Absorbers That Reproduce by Means of Spores

Fungi, which are almost all terrestrial, reproduce by means of spores, which are usually wind-dispersed. No motile cells are formed at any stage of the fungal life cycle, except by chytrids. The primary component of the fungal cell wall is chitin. Typically, a fungus secretes enzymes onto a food source and then absorbs the small molecules that are released. Glycogen is the primary storage polysaccharide.

The Diploid Phase in the Life Cycle of a Fungus Is Represented Only by the Zygote

The fungi are isogamous (gametes, when formed, are similar in size and shape), and meiosis in fungi is zygotic—that is, the zygote divides by meiosis to form four haploid cells.

There Are Four Phyla of Fungi

The kingdom *Fungi* includes four phyla: *Chytridiomycota, Zygomycota, Ascomycota,* and *Basidiomycota,* as well as an artificial assemblage known as the deuteromycetes. There is considerable evidence that animals and fungi diverged from a common ancestor, most likely a colonial protist resembling a choanoflagellate.

The *Chytridiomycota* Form Flagellated, Motile Cells

The *Chytridiomycota* are a predominantly aquatic group, and the only group of fungi with motile reproductive cells (zoospores and gametes). Most chytrids are coenocytic, with few septa at maturity. Some species are parasitic and others saprobic. Several are plant pathogens causing minor diseases such as brown spot of corn and crown wart of alfalfa.

The *Zygomycota* Form Zygospores in Zygosporangia

The zygomycetes (phylum *Zygomycota*) mostly have coenocytic mycelia. Their asexual spores are generally formed in sporangia—saclike structures in which the entire contents are converted to spores. The zygomycetes are so named because they form resting spores called zygospores during sexual reproduction. The zygospores develop within thick-walled structures called zygosporangia.

The *Ascomycota* Form Ascospores in Asci

Phylum *Ascomycota,* the members of which are commonly called ascomycetes, has some 32,300 named, distinct species, more than any other group of fungi. The distinguishing characteristic of ascomycetes is the ascus—a saclike structure in which the meiotic (sexual) spores, known as ascospores, are formed. In the ascomycete life cycle, the protoplasts of male and female gametangia fuse, and the female gametangia produce specialized hyphae that are dikaryotic (each compartment containing a pair of haploid nuclei). The ascus forms near the tip of a dikaryotic hypha. Ascospores are generally forcibly expelled. The asci are incorporated into complex spore-producing bodies called ascomata. Typically, asexual reproduction is by formation of usually multinucleate spores known as conidia.

The *Basidiomycota* Form Basidiospores Borne on Basidia

The phylum *Basidiomycota* includes many of the largest and most familiar fungi. They include mushrooms, puffballs, shelf fungi, stinkhorns, and others, as well as the rusts and smuts, which are important plant pathogens. The distinguishing characteristic of this phylum is the production of basidia. The basidium is produced at the tip of a dikaryotic hypha and is the structure in which meiosis occurs. Each basidium typically produces four basidiospores, and these are the principal means of reproduction in the *Basidiomycota.*

The Mushrooms, Rusts, and Smuts Are Representatives of the Three Classes of *Basidiomycota*

The *Basidiomycota* can be divided into three classes: *Basidiomycetes, Teliomycetes,* and *Ustomycetes.* In the *Basidiomycetes,* the basidia are incorporated into complex spore-producing bodies called basidiomata. In the hymenomycetes, which include the mushrooms and shelf fungi, the basidiospores are produced on a distinct fertile layer, the hymenium. This layer often lines the gills or tubes of the hymenomycetes and is exposed before the forcibly discharged spores are mature. In the gasteromycetes, which include the stinkhorns and puffballs, the basidiospores mature inside the basidiomata and are not discharged forcibly from them. The members of the classes *Teliomycetes* and *Ustomycetes,* the rusts and smuts, respectively, do not form basidiomata. Instead, rusts and smuts have septate basidia, as do the jelly fungi, which are members of the class *Basidiomycetes.* All *Basidiomycetes* other than the jelly fungi form aseptate basidia.

Yeasts Are Unicellular Fungi

Yeasts are unicellular fungi that typically reproduce by budding, an asexual method of reproduction. Some fungi exhibit both unicellular and filamentous growth forms, shifting from one form to the other under changing environmental conditions. Most yeasts are ascomycetes, which reproduce sexually through the production of ascospores.

Fungi with No Known Sexual State Are Placed in the Deuteromycetes

An artificial group of fungi—the deuteromycetes, also known as Fungi Imperfecti—includes many thousands of species with no known sexual cycle. Most of these deuteromycetes are allied with ascomycetes, but some are *Basidiomycetes* or zygomycetes.

Lichens Consist of a Mycobiont and a Photobiont

Lichens are mutualistic symbiotic partnerships between a fungal partner (the mycobiont) and a population of either green algae or cyanobacteria (the photobiont). About 98 percent of the fungal partners belong to the *Ascomycota,* the rest to the *Basidiomycota.* The fungi receive carbohydrates and nitrogen compounds from their photosynthetic partners and provide the photobiont with a suitable physical environment in which to grow. The ability of the lichen to survive under adverse environmental conditions is related to its ability to withstand desiccation and remain dormant when dry.

Mycorrhizae Are Mutualistic Associations between Fungi and Roots

Mycorrhizae—symbiotic associations between plant roots and fungi—characterize all but a few families of vascular plants. Endomycorrhizae, also called vesicular-arbuscular mycorrhizae, in which the fungal partners are zygomycetes, occur in about 80 percent of all kinds of vascular plants. In such associations, the fungus actually penetrates the cortical cells of the host, but does not enter the protoplasts. In the second major type of mycorrhizal association, ectomycorrhizae, the fungus does not penetrate the host cells but forms a sheath, or mantle, that surrounds the roots and also a network (the Hartig net) that grows around the cortical cells. Mostly *Basidiomycetes,* but also some ascomycetes, are involved in ectomycorrhizal associations. Mycorrhizal associations are important in obtaining phosphorus and other nutrients for the plant and in providing organic carbon for the fungus. Such associations were characteristic of the first plants to invade the land.

Selected Key Terms

aeciospores p. 327
aecium p. 327
antheridium p. 320
arbuscule p. 341
ascogonium p. 320
ascoma p. 319
ascospores p. 318
ascus p. 318
autoecious p. 327
basidioma p. 322
basidiospores p. 320
basidium p. 320
coenocytic p. 310
conidiophore p. 318
conidia p. 311
dolipore p. 322
endophytes p. 308
heteroecious p. 327
hymenium p. 319
mycobiont p. 334
mycology p. 310
mycorrhizae p. 340
mycotoxins p. 307
parasexuality p. 333
photobiont p. 334
septate p. 310
septum p. 310
soredia p. 338
sori p. 322
spermogonia p. 327
sporangiophore p. 315
stolons p. 315
telia p. 327
teliospores p. 327
uredinia p. 327
urediniospores p. 327
zygosporangium p. 316
zygospores p. 316

Questions

1. Distinguish between the following: hyphae/mycelium; somatic/vegetative; rhizoids/haustoria; plasmogamy/karyogamy; sporangium/gametangium; heterothallic/homothallic; dikaryotic/monokaryotic; parasitic/mutualistic; arbuscules/vesicles; endomycorrhizae/ectomycorrhizae.
2. "The fungi are of enormous importance both ecologically and economically." Elaborate upon this statement both in general terms and with specific reference to each of the major groups of fungi, including the yeasts, lichens, and mycorrhizal fungi.
3. How might one determine, on the basis of hyphal structure alone, whether a given fungus is a member of the *Zygomycota, Ascomycota,* or *Basidiomycota?*
4. What do zygospores, ascospores, and basidiospores have in common? Zoospores, conidia, aeciospores, and urediniospores?
5. Many fungi produce antibiotics. What do you think the function of the antibiotics might be for the fungi that produce them?
6. In the life cycle of a mushroom, three types of hyphae, or mycelia, can be recognized: primary, secondary, and tertiary. How do these three types of mycelia relate to one another, and how do they fit into the life cycle?
7. "Both the state of health of the lichens and their chemical composition are used to monitor the environment." Explain.
8. "The most prevalent and probably the most important mutualistic symbiosis in the plant kingdom is the mycorrhiza." Explain.

Chapter

Protista I: Euglenoids, Slime Molds, Cryptomonads, Red Algae, Dinoflagellates, and Haptophytes

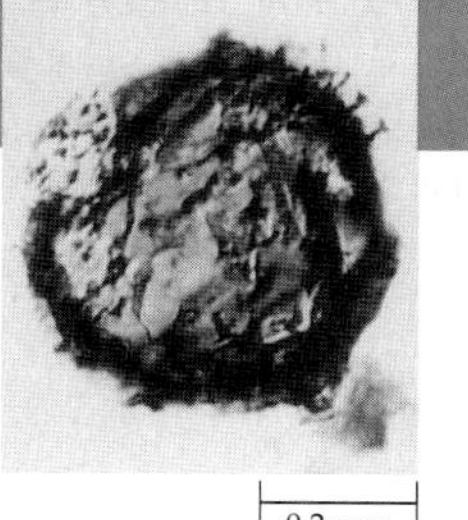

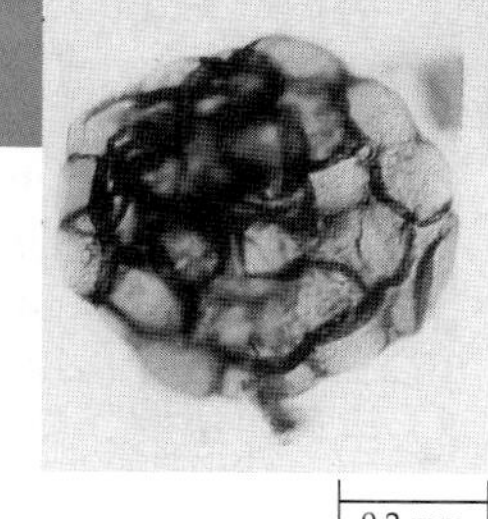

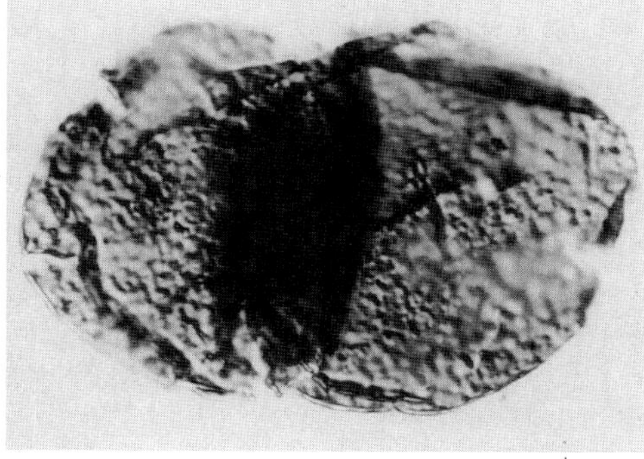

16–1
Three very different kinds of acritarchs that occurred together about 700 million years ago in the seas of what is now the Grand Canyon region of Arizona. Acritarchs, which first appeared in the fossil record about 1.5 billion years ago, were evidently eukaryotes, as indicated by their size and by the complexity of their cell walls. Acritarchs, which were very diverse, were probably planktonic autotrophs. Some closely resemble the resting zygotes of dinoflagellates (see page 365), whereas others resemble certain green algae.

OVERVIEW

In this chapter and the next we visit the kingdom Protista, *which encompasses a wide variety of organisms, including unicellular, colonial, and simple multicellular forms. This kingdom may be regarded as a "catchall" assemblage of organisms that simply do not fit well into any of the other kingdoms.*

The organisms considered in this chapter include two phyla of entirely heterotrophic organisms once treated as fungi and five phyla of algae. In general, algae are simple, photosynthetic organisms, although many are heterotrophic. Some, like the euglenoids, the cryptomonads, and the dinoflagellates, are largely unicellular. Others, like the haptophytes, are unicellular or colonial, and the red algae are mostly multicellular. The two phyla of heterotrophic organisms, on the other hand, exhibit strikingly different growth habits. The plasmodial slime molds grow mostly as large, multinucleate masses of protoplasm, whereas the cellular slime molds usually exist as unicellular amoeba-like cells that can aggregate to form a mass that resembles a multicellular organism.

As you will see, many of the organisms discussed in this chapter have tremendous ecological importance and provide clues concerning the evolution of eukaryotic cells and the origin of the other kingdoms of eukaryotic organisms.

CHECKPOINTS

By the time you finish reading this chapter, you should be able to answer the following questions:

1. What criteria are used to place an organism in the kingdom *Protista?* How is this kingdom different from the other four kingdoms of organisms?
2. How are algae of ecological importance?
3. In what ways are the cryptomonads and euglenoids similar? Why is it difficult to classify these organisms on the basis of the mechanisms by which they obtain their food?
4. What features distinguish the plasmodial from the cellular slime molds? Why are these organisms not considered to be algae?
5. What are the distinctive cellular characteristics of the red algae? In what ways is the life history of red algae unusual?
6. What are the distinctive features of the phylum *Haptophyta,* and how are haptophytes important in food webs?

With approximately 70 percent of its surface covered by water, Earth is known as the "water planet." This abundance of water provided the watery habitat in which life began, some 3.5 billion years ago, with the first appearance of prokaryotes. DNA evidence suggests that the first eukaryotes appeared between 2.5 and 1 billion years ago, at least a billion years after the first occurrence of prokaryotes.

Until recently, the oldest eukaryotic fossils known were the **acritarchs,** which first appeared in the fossil record some 1.5 billion years ago (Figure 16–1). Recently, however, eukaryotic fossils have been retrieved from sediments in China that are 1.7 billion years old. These fossils are said to resemble the modern aquatic eukaryotes known as brown algae. An even more ancient eukaryotic fossil, a 0.5-meter-long, thin fossil named *Grypania,* has been found in rocks that are 2.1 billion years old. Described as a photosynthetic eukaryote, *Grypania* may be the earliest known eukaryote. Although these interpretations are somewhat controversial, they suggest that the first eukaryotes are even more ancient than previously thought.

Today a diverse array of the descendants of these early eukaryotes—protists—populates the oceans and marine shorelines, as well as freshwater lakes, ponds, and streams. A few protists are able to live in terrestrial habitats, but their principal realm remains the water.

Biologists have grouped the protists together into the kingdom *Protista,* which contains eukaryotes that are not clearly assignable to the kingdoms *Fungi, Plantae,* or *Animalia* (see Figure 13–13). Most biologists would agree that fungi, plants, and animals are derived from ancient protists and that the study of modern protists sheds light on the origin of these important groups. In addition to the evolutionary importance of protists, some cause diseases of plants or animals, whereas others are of great ecological significance.

Protist groups covered in this book (in this and the following chapter) include photosynthetic organisms that function ecologically like plants—that is, as primary producers using light energy to manufacture their own food (see Chapter 32). Photosynthetic protists occur in the phyla made up of algae. Among the algae, the green algae are particularly significant because plants are derived from an ancestor that, if still alive, would be classified as a green alga. In addition to these autotrophs, we will describe some colorless, heterotrophic protists, including oomycetes. These heterotrophic protists are very closely related to the algae. We will also describe some protists, the mixotrophs, that display both photosynthesis and heterotrophy. The algae include autotrophic, heterotrophic, and mixotrophic eukaryotes. Also considered are two groups of protists—the cellular slime molds and the plasmodial slime molds—that are not algae. These two groups, together with the oomycetes, have traditionally been the concern of the *mycologist,* or student of fungi. Although these heterotrophic protists are not closely related to fungi, they continue to be described by terminology used for fungi. Similarities and differences among the protist groups discussed in this chapter are summarized in Table 16–1.

Protists exhibit an amazing array of body types that include amoeba-like cells; single cells—with or without cell walls—that may or may not have flagella; colonies consisting of aggregations of cells that may be flagellate or not; branched or unbranched filaments; one- or two-cell-thick sheets; tissue that resembles some tissues of plants and animals; and multinucleate masses of protoplasm with or without cell walls. Protists vary in size from the smallest known eukaryote—a tiny green alga—to 30-meter-long brown algae known as kelps. Both extremely small and very large sizes confer protection against being eaten by aquatic herbivores. Many protists reproduce sexually and have complex life histories, but some reproduce only by asexual means. All three types of sexual life cycles—zygotic, sporic, and gametic (see Figure 9–3)—occur among protists. It is common for different phases of the life history of a single protist species to be very different in size and appearance.

Studies of nuclear-encoded ribosomal RNA gene sequences have clarified the relationships of the various protist groups to one another. In this chapter, we will consider the groups believed to have appeared earliest: euglenoids, plasmodial and cellular slime molds, cryptomonads, red algae, dinoflagellates, and haptophyte algae. In the next chapter, we will discuss the protists that appeared somewhat more recently: water molds, diatoms, chrysophyte algae, brown algae, and the green algae.

Ecology of the Algae

In all bodies of water, minute photosynthetic cells and tiny animals occur as suspended **plankton** (Gk. *planktos,* "wandering"). The planktonic algae and the cyanobacteria, which together constitute the **phytoplankton,** are the beginning of the food chain for the heterotrophic organisms that live in the ocean and in bodies of fresh water. The heterotrophic plankton—**zooplankton**—consists mainly of tiny crustaceans, the larvae of many different phyla of animals, and many heterotrophic protists and bacteria.

In the sea, most small and some large fish, as well as most of the great whales, feed on the plankton, and still larger fish feed on the smaller ones. In this way, the "great meadow of the sea," as the phytoplankton is sometimes called, can be likened to the meadows of the land, serving as the source of nourishment for het-

TABLE 16–1 Comparative Summary of Characteristics of *Protista* I

Phylum	Number of Species	Photosynthetic Pigments	Carbohydrate Food Reserve	Flagella	Cell Wall Component	Habitat
Euglenophyta (euglenoids)	900	Most have none, or chlorophylls *a* and *b;* carotenoids	Paramylon	Usually 2; often unequal, with 1 forward and 1 behind or reduced to stub; extend apically; various hairs	No cell wall; have a flexible or rigid pellicle of proteinaceous strips beneath the plasma membrane	Mostly freshwater, some marine
Myxomycota (plasmodial slime molds)	700	None	Glycogen	Usually 2; apical; unequal; in reproductive cells only, whiplash	None on plasmodium	Terrestrial
Dictyosteliomycota (dictyostelids, or cellular slime molds)	50	None	Glycogen	None (amoeboid)	Cellulose	Terrestrial
Cryptophyta (cryptomonads)	200	None, or chlorophylls *a* and *c;* phycobilins; carotenoids	Starch	2; unequal; subapical; tinsel (1 with one row, 1 with two rows of hairs)	No cell wall; have a stiff layer of proteinaceous plates beneath plasma membrane	Marine and freshwater; cold waters
Rhodophyta (red algae)	4000–6000	Chlorophyll *a;* phycobilins; carotenoids	Floridean starch	None	Cellulose microfibrils embedded in matrix (usually galactans); deposits of calcium carbonate in many	Predominantly marine, about 100 freshwater species; many tropical species
Dinophyta (dinoflagellates)	2000–4000	None in many, or chlorophylls *a* and *c;* carotenoids, mainly peridinin	Starch	None (except in gametes) or 2, dissimilar; lateral (1 transverse, 1 longitudinal); both have hairs	Have a layer of vesicles beneath the plasma membrane with or without cellulose plates	Mostly marine, many freshwater; some in symbiotic relationships
Haptophyta (haptophyte algae)	300	Chlorophylls *a* and *c;* carotenoids, especially fucoxanthin	Chrysolaminarin	None or 2; equal or unequal; most have no hairs; most have haptonema	Scales of cellulose; scales of calcified organic material in some	Great majority marine, a few freshwater

erotrophic organisms. Floating or swimming single-celled or colonial chrysophytes, diatoms, green algae, and dinoflagellates are the most important organisms at the base of freshwater food chains. In marine waters, unicellular or colonial haptophytes, dinoflagellates, and diatoms are the most important eukaryotic members of the marine phytoplankton and therefore essential to the support of marine animal life (Figure 16–2).

16–2
Marine phytoplankton. The organisms shown here are dinoflagellates and filamentous and unicellular diatoms.

Marine phytoplankton is being increasingly used as "fodder" to support the growth of shrimp, shellfish, and other marketable seafood products. Seaweeds can be grown in aquatic farming operations to produce edible and industrially useful products (see "Algae and Human Affairs" on page 380). Both types of applications of algae are examples of *mariculture,* by which marine organisms are cultivated in their natural environment, and are analogous to terrestrial agricultural systems.

In both marine and fresh waters relatively free of serious human disturbance, phytoplankton populations are usually held in check by seasonal climate changes, nutrient limitation, and predation. When humans pollute aquatic systems, however, some algae may be released from these constraints, and their populations grow to undesirable, "bloom" proportions. In the ocean, some of these blooms are known as "red tides" or "brown tides" because the water becomes colored by large numbers of algal cells containing red or brown accessory pigments. Algal blooms correlate with the release of large quantities of toxic compounds into the water. These toxins, which may have evolved as a defense against protist or animal predators, can cause human illness and massive die-offs of fishes, birds, or aquatic mammals (see "Red Tides/Toxic Blooms" on page 364). In recent years the frequency of toxic marine blooms has increased globally, although only a few dozen species of phytoplankton are toxic. Some ecologists link this increase with the worldwide decline in water quality caused by increased human populations.

Algae, which are abundant on this watery planet, are important in carbon cycling (see "The Carbon Cycle" on page 150). They are able to transform carbon dioxide (CO_2)—a so-called "greenhouse gas" that contributes to global warming—into carbohydrates by photosynthesis and into calcium carbonate by calcification. Large amounts of organic carbon and calcium carbonate are incorporated into algae and have been transported to the ocean bottom. Today, the marine phytoplankton absorbs about one-half of all the CO_2 that results from human activities such as the burning of fossil fuels. The extent to which such carbon is transported to the ocean depths, and thus does not contribute to global temperature increases, is a matter of controversy.

The mechanism by which some phytoplanktonic organisms reduce the amount of CO_2 in the atmosphere is by favoring the formation of calcium carbonate as they fix CO_2 during photosynthesis. The calcium carbonate is deposited as tiny scales covering the phytoplankton. The CO_2 removed from the water by this calcification process and by photosynthesis is replaced by atmospheric CO_2, creating a suction effect, also known as "CO_2 drawdown." Phytoplankton covered with calcium carbonate, upon settling to the ocean floor, gave rise to much of the famous White Cliffs of Dover and to the economically important North Sea oil deposits. Several kinds of red, green, and brown seaweeds can also become encrusted with calcium carbonate. The effect of calcification of these seaweeds on the global carbon cycle is not as well understood as the effect of phytoplankton calcification.

Some marine phytoplankton, particularly haptophytes and dinoflagellates, produce significant amounts of a sulfur-containing organic compound that aids in regulating osmotic pressure within their cells. A volatile compound derived from it is excreted by the cells and subsequently is converted to sulfur oxides in the atmosphere, which increase cloud cover and, hence, reflectiveness. Sulfur oxides, which are also generated by burning fossil fuels, contribute to acid rain and have a cooling effect on the climate. Scientists who try to predict future climates must take such cooling effects into account, along with the climate-warming effects of the greenhouse gases.

In this chapter we will consider the following, largely unrelated protist phyla: *Euglenophyta,* the euglenoids; *Myxomycota* and *Dictyosteliomycota,* the plasmodial and cellular slime molds; *Cryptophyta,* the cryptomonads; *Rhodophyta,* the red algae; *Dinophyta,* the dinoflagellates; and *Haptophyta,* the haptophytes. All of these groups of organisms and those considered in the chapter that follows are of great ecological importance.

Euglenoids: Phylum *Euglenophyta*

Flagellates—protists bearing flagella—known as euglenoids make up the phylum *Euglenophyta.* There are 900 known species of euglenoids. Molecular evidence suggests that the earliest euglenoids were phagocytes (parti-

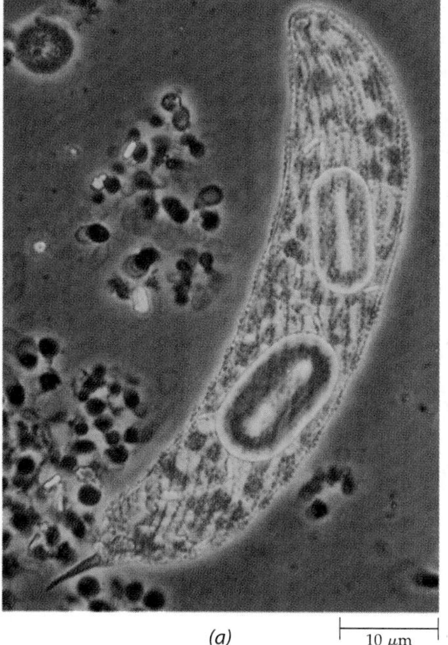

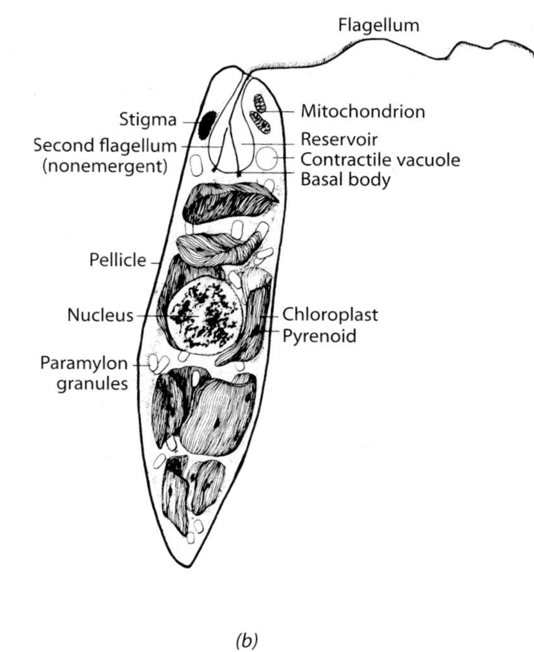

(a) (b)

16–3
Euglena. *(a) Electron micrograph showing two large storage bodies of paramylon and the helically arranged proteinaceous strips of the pellicle. A red eyespot, or stigma, is visible at the upper end of the cell. (b) The structure of* Euglena, *as interpreted from electron micrographs.*

cle eaters). About one-third of the genera, including the common *Euglena,* contain chloroplasts. The similarities between the chloroplasts of the euglenoids and those of the green algae—both possess chlorophylls *a* and *b* together with several carotenoids—suggest that the euglenoid chloroplasts were derived from endosymbiotic green algae. About two-thirds of the genera are colorless heterotrophs that rely upon particle feeding or absorption of dissolved organic compounds. These feeding habits, along with a more general requirement for vitamins, explain why many of the euglenoids occur in fresh waters that are rich in organic particles and compounds.

The structure of *Euglena* (Figure 16–3) illustrates many of the features typical of euglenoids in general. With one exception (the colonial genus *Colacium*), euglenoids are unicellular. *Euglena,* like most euglenoids, does not have a cell wall or any other rigid structure covering the plasma membrane. The genus *Trachelomonas* (Figure 16–4), however, has a wall-like covering made of iron and manganese minerals. The plasma membrane of euglenoids is supported by an array of helically arranged proteinaceous strips, which are in the cytoplasm, immediately beneath the plasma membrane. These strips form a structure called the **pellicle,** which may be flexible or rigid. *Euglena*'s flexible pellicle allows the cell to change its shape, facilitating movement in muddy habitats where flagellar swimming is difficult. Swimming *Euglena* cells have a single, long flagellum that emerges from the base of an anterior depression called the **reservoir,** and a second, nonemergent flagellum (Figure 16–3). A swelling at the base of the emergent flagellum, together with a nearby, distinctive red **eyespot,** or **stigma,** in the cytoplasm, constitute a light-sensing system in euglenoids.

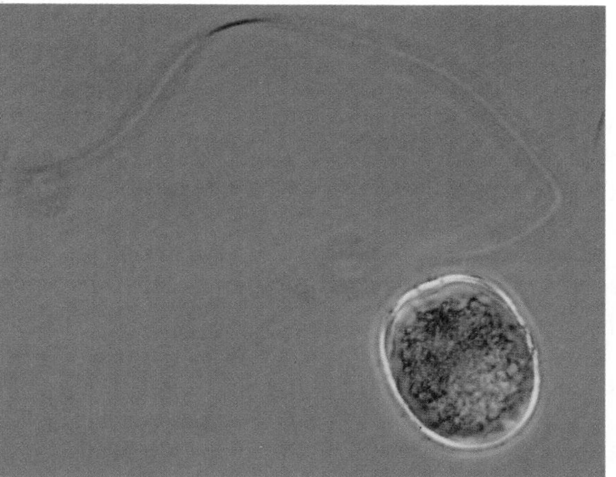

16–4
Trachelomonas, *a widespread euglenoid. Its green-pigmented cell is encased in a rigid, often ornamented wall-like covering called a lorica, which may vary from clear to black-brown in color. A single flagellum protrudes through a pore at the apex of the lorica. Reproduction is accomplished by mitosis.*

A **contractile vacuole** collects excess water from all parts of the euglenoid cell. The water is discharged from the cell via the reservoir. After each discharge, a new contractile vacuole is formed by coalescence of small vesicles. Contractile vacuoles are commonly observed in freshwater protists, which need to eliminate the excess water that accumulates as a result of osmosis. If the water is not removed, the cells will burst.

In contrast to the chloroplasts of green algae, euglenoid plastids do not store starch. Instead, granules of a unique polysaccharide known as **paramylon** form in the cytoplasm. Euglenoid plastids often resemble plastids of many green and other algae in their possession of a protein-rich region called a **pyrenoid,** which is the site of Rubisco and some other enzymes involved in photosynthesis (page 140).

Euglenoids reproduce by mitosis and lengthwise cytokinesis, continuing to swim while they divide. The nuclear envelope remains intact during mitosis as it does in many other protists and in fungi. This suggests that an intact mitotic nuclear envelope is a primitive feature and that nuclear envelopes that break down during mitosis, as they do in plants, animals, and various protists, are a derived characteristic. Sexual reproduction and meiosis do not seem to occur in euglenoids, suggesting that these processes had not yet evolved when this group diverged from the main lineage of protists.

16–5
Plasmodium of a plasmodial slime mold, Physarum, *growing on a tree trunk.*

Plasmodial Slime Molds: Phylum *Myxomycota*

The plasmodial slime molds, or myxomycetes, are a group of about 700 species that seems to have no direct relationship to the cellular slime molds, the fungi, or any other group. Although referred to as molds, as are certain fungi, molecular evidence demonstrates that neither the plasmodial slime molds nor cellular slime molds are closely related to the fungi. When conditions are appropriate, the plasmodial slime molds exist as thin, streaming multinucleate masses of protoplasm that creep along in amoeboid fashion. Lacking a cell wall, this "naked" mass of protoplasm is called a **plasmodium.** As these plasmodia travel, they engulf and digest bacteria, yeast cells, fungal spores, and small particles of decaying plant and animal matter. Plasmodia can also be cultured on media that do not contain particulate matter, suggesting that plasmodia can also obtain food by absorption of dissolved organic compounds.

The plasmodium may grow to weigh as much as 20 to 30 grams and, because slime molds are spread out thinly, this amount can cover an area of up to several square meters (Figure 16–5). The plasmodium, with its many nuclei, is not divided by cell walls. As it grows, the nuclei divide repeatedly and synchronously; that is, all the nuclei in a plasmodium divide at the same time. Centrioles are present, and mitosis is similar to that of plants and higher animals, although the chromosomes are very small.

Typically, the moving plasmodium is fan-shaped, with flowing protoplasmic tubules that are thicker at the base of the fan and spread out, branch, and become thinner toward their outer ends. The tubules are composed of slightly solidified protoplasm through which more liquefied protoplasm flows rapidly. The foremost edge of the plasmodium consists of a thin film of gel separated from the substrate by only a plasma membrane and a slime sheath.

Plasmodial slime molds have a life history that involves sexual reproduction and therefore may be among the earliest divergent protists to have acquired sex. Plasmodial growth continues as long as an adequate food supply and moisture are available. Generally, when either of these is in short supply, the plasmodium migrates away from the feeding area. At such times, plasmodia may be found crossing roads or lawns, climbing trees, or in other unlikely places. In many species, when the plasmodium stops moving, it divides into a large number of small mounds. The mounds are similar in size and volume, and so their formation is probably controlled by chemical effects within the plasmodium. The life cycle of a typical plasmodial slime mold is summarized in Figure 16–6. Each mound produces a sporangium, usually borne at the tip of a stalk. The mature sporangium is often extremely ornate (Figure 16–7). The protoplasm of the young sporangium contains many nuclei, which increase in number by mitosis. Progressively, the protoplasm is cleaved into a large number of spores, each containing a single diploid nucleus. Meiosis then occurs, giving rise to four haploid nuclei per spore. Three of the four nuclei disintegrate, however, leaving each spore with a single haploid nucleus. In some members of this group, discrete sporangia are not produced, and the entire plasmodium may develop either into a **plasmodiocarp** (Figure 16–7c), which retains the former shape of the plasmodium, or into an **aethalium** (Figure 16–7d), in which the plasmodium forms a large mound that is essentially a single large sporangium.

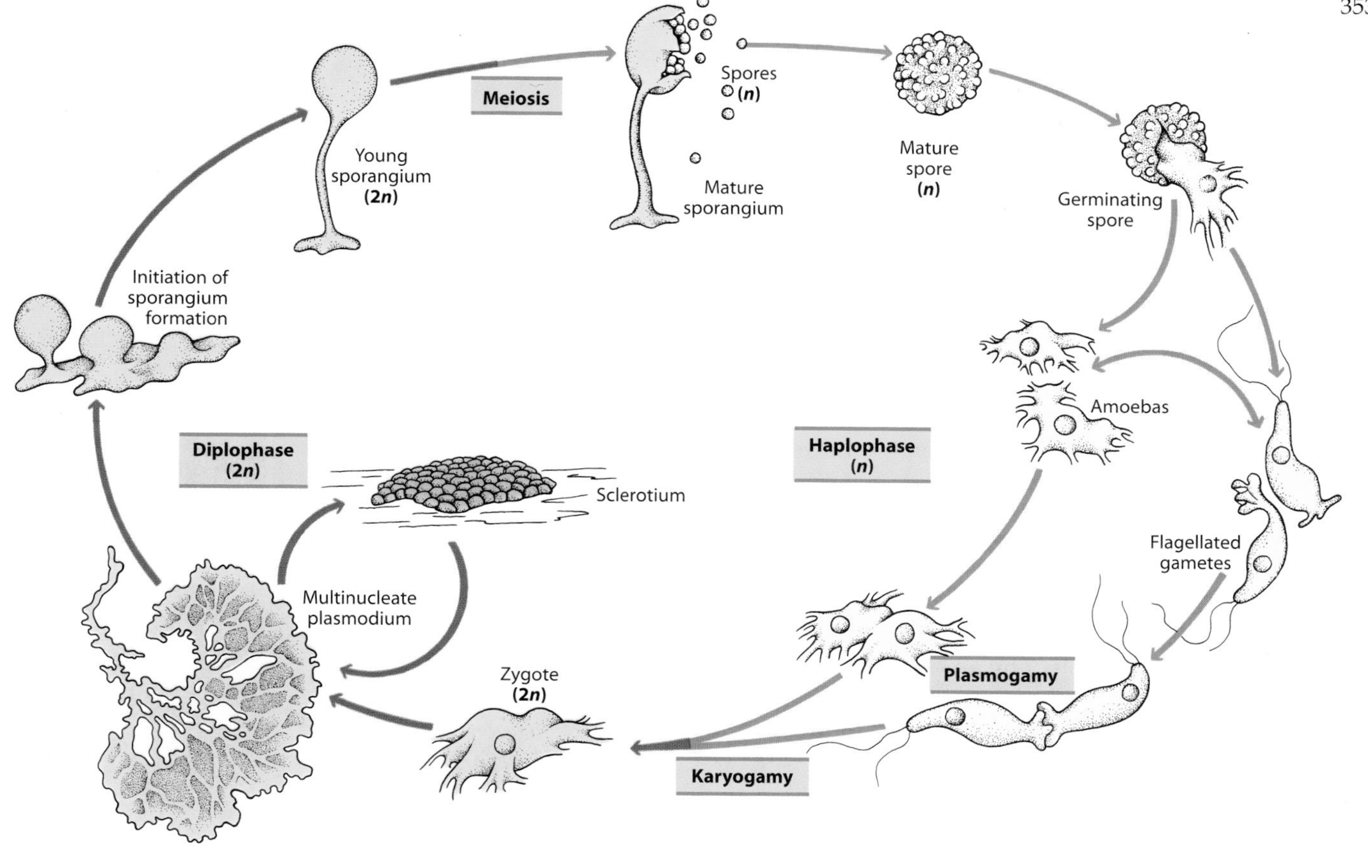

16–6
Life cycle of a typical myxomycete. Sexual reproduction in the plasmodial slime molds consists of three distinct phases: plasmogamy, karyogamy, and meiosis. Plasmogamy is the union of two protoplasts, which may be amoebas or flagellated gametes derived from germinated spores; it brings two haploid nuclei together in the same cell. Karyogamy is the fusion of these two nuclei, resulting in the formation of a diploid zygote and initiation of the so-called diplophase of the life cycle. The plasmodium is a multinucleate, free-flowing mass of protoplasm that can pass through a silk cloth or a piece of filter paper and remain unchanged. Plasmodia often form a hardened stage, the sclerotium, in nature and are able to survive dry periods well in this condition. In any event, the active plasmodium may ultimately form sporangia. Within the developing sporangia, meiosis restores the haploid condition and thus initiates the haplophase of the life cycle.

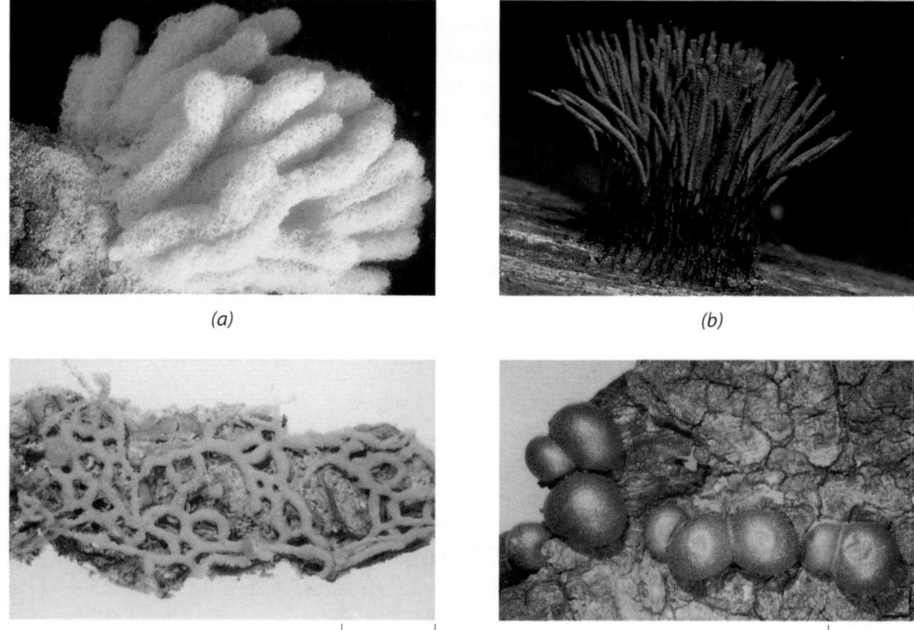

16–7
*Spore-producing structures in plasmodial slime molds. **(a)** Sporangium of* Arcyria nutans. ***(b)*** *Sporangia of* Stemonitis splendens. ***(c)*** *Plasmodiocarp of* Hemitrichia serpula. ***(d)*** *Aethalia of* Lycogala *growing on bark.*

When habitats dry out, plasmodia can quickly form an encysted stage, the **sclerotium.** Sclerotia can be seen easily on firewood piles because these organisms are often brightly colored in shades of yellow and orange. They are of great importance for the survival of plasmodial slime molds, especially in fast-drying habitats such as dead cacti or soil in deserts, where these organisms are abundant.

The spores of myxomycetes are also resistant to environmental extremes and can be very long-lived. Some have germinated after being kept in the laboratory for more than 60 years. Thus, spore formation in this group seems to make possible not only genetic recombination but also survival under adverse conditions.

Under favorable conditions, the spores split open and the protoplast slips out (Figure 16–6). The protoplast may remain amoeboid, or it may develop one to four whiplash flagella. The amoeboid and flagellated stages are readily interconvertible. The amoebas feed by the ingestion of bacteria and organic material and multiply by mitosis and cell cleavage. If the food supply is used up or conditions are otherwise unfavorable, an amoeba may cease moving about, become round, and secrete a thin wall to form a **microcyst.** These microcysts can remain viable for a year or more, resuming activity when favorable conditions return.

After a period of growth, plasmodia appear in the amoeba population. Their appearance is governed by a number of factors, including cell age, environment, density of amoebas, and cyclic adenosine monophosphate (cyclic AMP, or cAMP). These factors play roles similar to those in the cellular slime mold *Dictyostelium discoideum,* discussed in the following section. One method of plasmodium formation is by the fusion of gametes. The gametes are usually genetically different from one another, in which case they are ultimately derived from different haploid spores. These gametes are simply some of the amoebas or flagellated cells that now have a new role. In some species and strains, however, the plasmodium is known to form directly from a single amoeba. Such plasmodia are usually haploid, like the amoebas from which they arose.

Cellular Slime Molds: Phylum *Dictyosteliomycota*

The cellular slime molds, or dictyostelids—a group of about 50 species in four genera—are probably more closely related to the amoebas (phylum *Rhizopoda,* not considered in this book) than to any other group. They are common inhabitants of most litter-rich soils, where they usually exist as free-living amoeba-like cells, or **myxamoebas.** These myxamoebas feed on bacteria by phagocytosis (Figure 16–8). Unlike fungi, with which

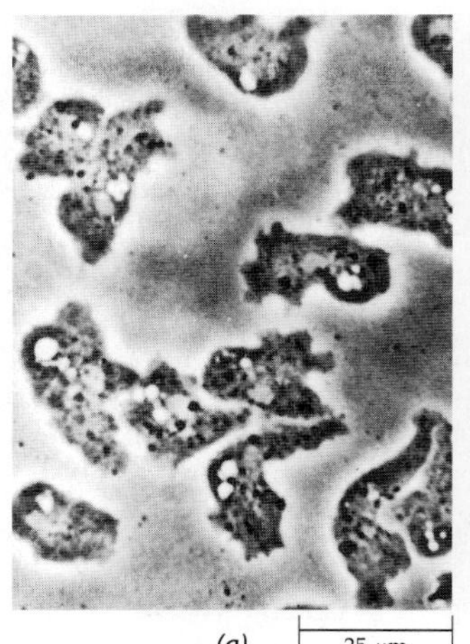

(a)

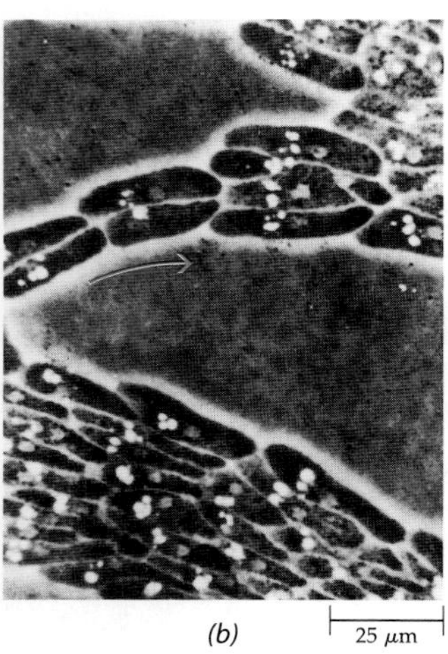

(b)

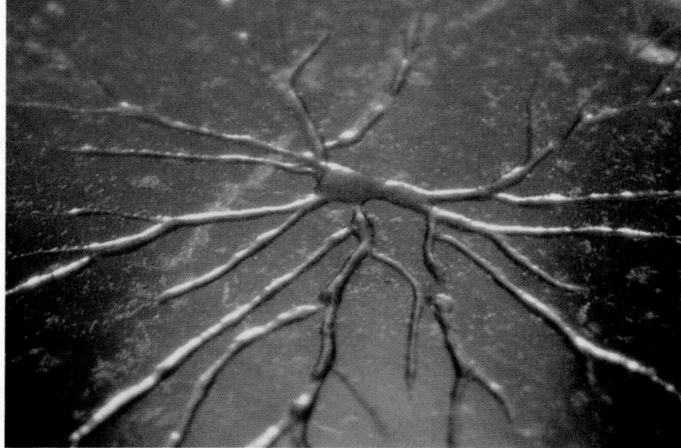

(c)

16–8
Life cycle of the cellular slime mold Dictyostelium discoideum. ***(a)*** *The feeding stage of the myxamoebas. The light gray area in the center of each cell is the nucleus, and the white areas are contractile vacuoles.* ***(b, c)*** *Myxamoebas aggregating. The direction in which the stream is moving is indicated by an arrow.* ***(d)*** *Migrating pseudoplasmodium, formed of many myxamoebas. Each sluglike mass deposits a thick slime sheath, which collapses behind it.* ***(e–g)*** *At the end of the migration, the pseudoplasmodium gathers itself into a mound and begins to rise vertically, differentiating into a stalk and a mass of spores, as seen in these scanning electron micrographs.*

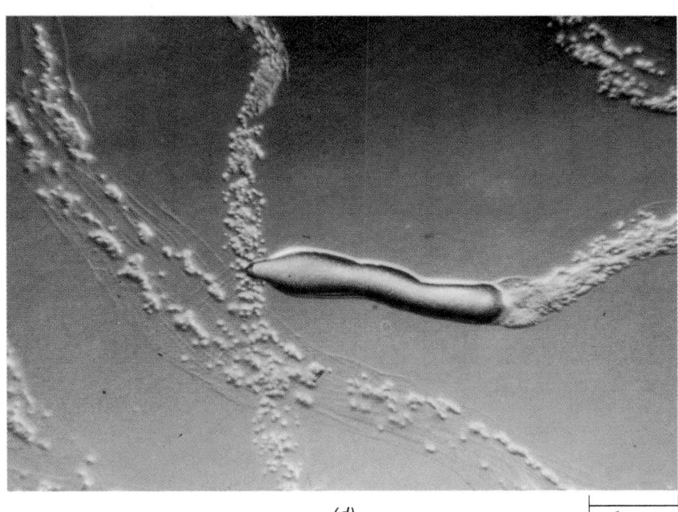

(d)

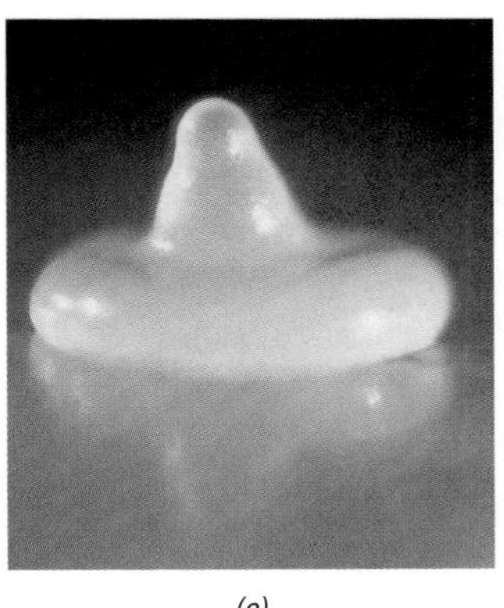

(e)

(f)

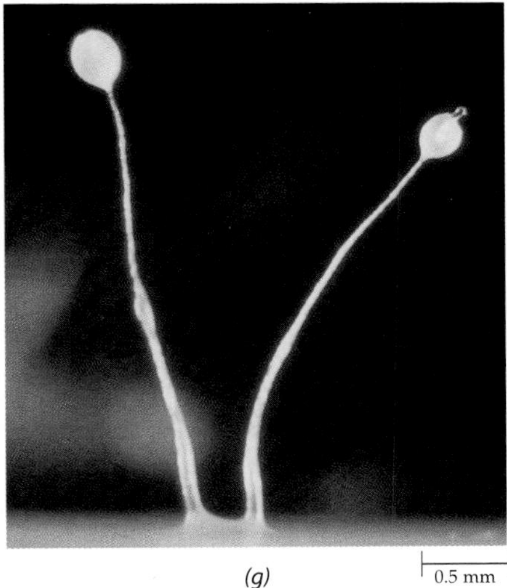

(g)

they were once grouped, the cellular slime molds have cellulose-rich cell walls during part of their life cycle and undergo the type of mitosis found in plants and animals, in which the nuclear envelope breaks down. In addition, again unlike the fungi, cellular slime molds have centrioles.

Dictyostelium discoideum has become an important model system for laboratory study of eukaryotic gene expression and developmental processes. One such process is programmed cell death, or **apoptosis,** normal cell death in which the cell activates a death program and kills itself. *Dictyostelium* reproduces by cell division and shows little morphological differentiation until it exhausts the available supply of bacteria. In response to starvation, the cellular slime molds produce asexual spores, but not in a simple way. Individual cells first aggregate to form a motile sluglike mass. The uninucleate, haploid myxamoebas retain their individuality in the mass, which usually contains 10,000 to 125,000 individuals. This mass, called a **pseudoplasmodium** (Figure 16–8d), or **slug,** migrates to a new place before differentiating and releasing spores. Thus, it avoids releasing new spores into habitats depleted of bacteria.

The myxamoebas aggregate by **chemotaxis,** migrating toward a source of cAMP, which is secreted by the starved myxamoebas. The cAMP diffuses away from the cells, establishing a concentration gradient along which the surrounding cells move toward the cells that are secreting the cAMP. The secreting cells, in turn, are stimulated to emit a new "pulse" of cAMP after a period of about five minutes. At least three waves of cells are recruited in this fashion. Binding of the cAMP signal to receptors in the plasma membrane triggers massive rearrangement of actin filaments, enabling the myxamoeba to crawl toward the source of cAMP. As the cells accumulate at the aggregation center, their plasma membranes become sticky; this causes them to adhere to each other and results in formation of a pseudoplasmodium.

The eventual developmental fate of an individual cell is determined at an early stage by its position in the aggregation, which appears to be controlled by the stage of the cell cycle the individual myxamoeba was in when aggregation began. Cells that divide between about 1.5 hours before and 40 minutes after the onset of starvation enter the aggregation last. When migration ceases, cells in the forward, or anterior, end of the slug become the stalk cells of the developing spore-bearing body. The stalk cells become coated with cellulose, providing rigidity to the stalk, and then die by apoptosis. Meanwhile, the posterior cells of the pseudoplasmodium move to the top of the stalk and become dormant spores. Eventually, the spores are dispersed. If spores fall on a warm, damp surface, they germinate. Each spore releases a single myxamoeba, and the cycle is repeated.

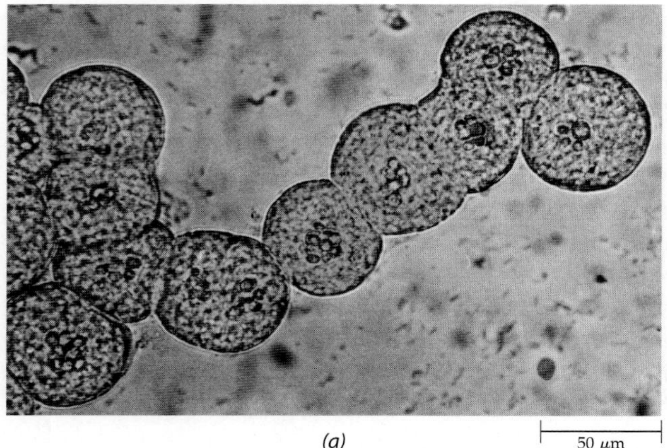

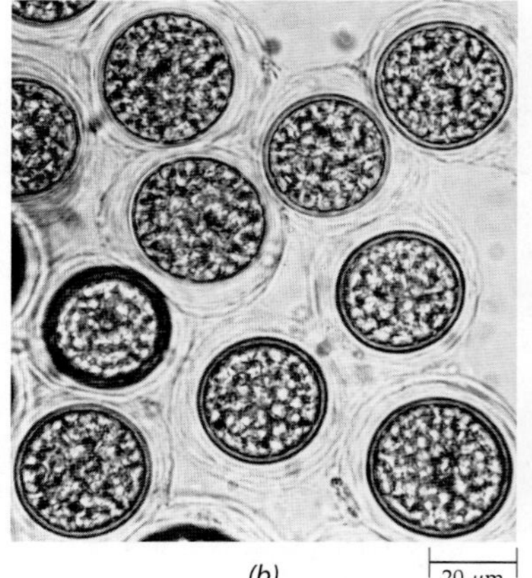

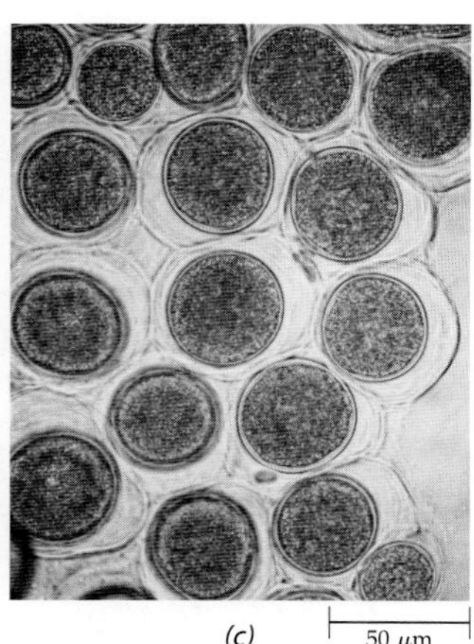

16–9
The process of macrocyst formation in Dictyostelium mucoroides. ***(a)*** *Each zygote, or giant cell, is beginning to engulf the myxamoebas around it.* ***(b)*** *Giant cells have engulfed all of the myxamoebas, and each has laid down a cellulose wall.* ***(c)*** *Mature macrocysts; at this stage they appear homogeneous.*

Reproduction involving asexual spores is common in the cellular slime molds. Sexual reproduction also occurs frequently and results in the formation of walled zygotes called **macrocysts.** The macrocysts are formed by aggregations of myxamoebas that are smaller than those involved in the formation of slugs. In addition, these aggregations are rounded in outline, rather than being elongate (Figure 16–9). During the formation of a macrocyst, two haploid myxamoebas first fuse, forming a single large myxamoeba, the zygote, which becomes actively phagocytic. The zygote continues to feed voraciously until all of the surrounding myxamoebas have been engulfed, becoming a giant cell in the process. At this stage, a thick cell wall, rich in cellulose, is laid down around the giant cell, and a mature macrocyst is formed. Within the macrocyst, the zygote—the only diploid cell of the life cycle—undergoes meiosis and several mitotic divisions before germination and the release of numerous, haploid myxamoebas.

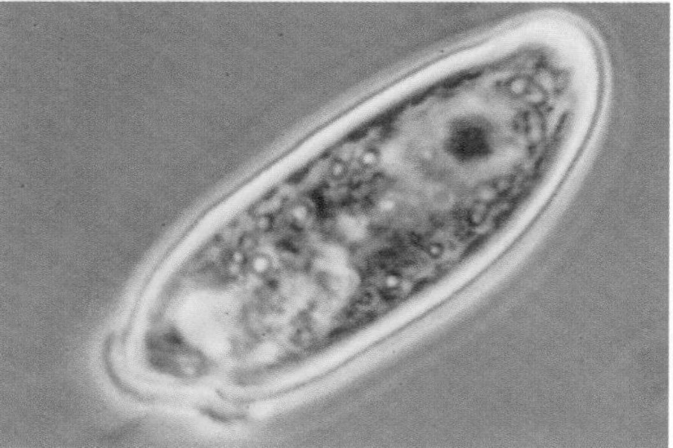

16–10
Cryptomonas, *a cryptomonad. Each of these unicellular algae has two, slightly unequal flagella (not apparent here) of about the same length as the cell. Both flagella emerge from an anterior depression of the cell.*

Cryptomonads: Phylum *Cryptophyta*

The cryptomonads are chocolate-brown, olive, blue-green, or red, fast-growing, single-celled flagellates that occur in marine and fresh waters (Figure 16–10). There are 200 known species of cryptomonads (Gk. *kryptos,* "hidden," + *monos,* "alone"). They are aptly named because their small size (3 to 50 micrometers) often makes them inconspicuous. They occur primarily in cold or subsurface waters, are not easily preserved, and are readily eaten by aquatic herbivores. Cryptomonads are particularly rich in polyunsaturated fatty acids, which are essential to the growth and development of zooplankton. Cryptomonads are ecologically important phytoplanktonic algae because of their edibility, and because they are frequently the dominant algae in lakes and coastal waters when diatom and dinoflagellate populations subside on a seasonal basis.

Cryptomonads are like euglenoids not only in requiring certain vitamins but also in having both pigmented photosynthetic members and colorless phagocytic members that consume particles such as bacteria. Cryptomonads provide some of the best evidence that colorless eukaryotic hosts can obtain chloroplasts from endosymbiotic eukaryotes. The chloroplasts of cryptomonads and certain other algae have four bounding membranes. Evidence indicates that the cryptomonads arose through the fusion of two different eukaryotic cells, one het-

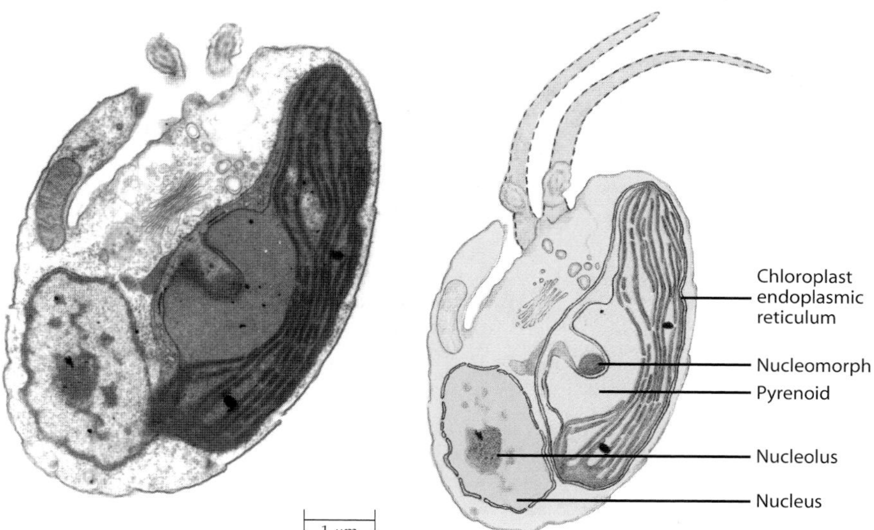

16–11
Electron micrograph of a cryptomonad, showing the nucleus and the nucleomorph, which is wedged in a cleft in the chloroplast. The nucleomorph is considered to be a vestigial nucleus belonging to the endosymbiont—a red algal cell—engulfed by a heterotrophic host. The outer of the four membranes surrounding the chloroplast is the chloroplast endoplasmic reticulum. The connection between the nuclear membrane and the chloroplast membranes is not visible in the plane of this section.

erotrophic and the other photosynthetic, establishing a **secondary endosymbiosis.** In addition to chlorophylls *a* and *c* and carotenoids, some cryptomonad chloroplasts contain a phycobilin, either phycocyanin or phycoerythrin. These water-soluble accessory pigments are otherwise known only in cyanobacteria and red algae, providing evidence for the origin of the cryptomonad chloroplast.

The outer of the four membranes surrounding the cryptomonad chloroplast is continuous with the nuclear envelope and is called the **chloroplast endoplasmic reticulum** (Figure 16–11). The space between the second and third chloroplast membranes contains starch grains and the remains of a reduced nucleus, complete with three linear chromosomes and a nucleolus with typical eukaryotic RNA. The reduced nucleus, called a **nucleomorph** (meaning "looks like a nucleus"), is interpreted as the remains of the nucleus of a red algal cell that was ingested and enslaved for its photosynthetic capabilities by a heterotrophic host. The endosymbiont resembles the chloroplasts of other algae in that most of its genes have been transferred to the host's nucleus so that it is no longer capable of independent existence.

Red Algae: Phylum *Rhodophyta*

Red algae are particularly abundant in tropical and warm waters, although many are found in the cooler regions of the world. There are 4000 to 6000 known species in 680 or so genera, only a very few of which are unicellular—such as *Cyanidium*, one of the only organisms capable of growth in acidic hot springs—or microscopic filaments. The vast majority of red algae are more structurally complex, macroscopic seaweeds. Fewer than 100 different species of red algae occur in fresh water (Figure 16–12), but in the sea the number of species is greater

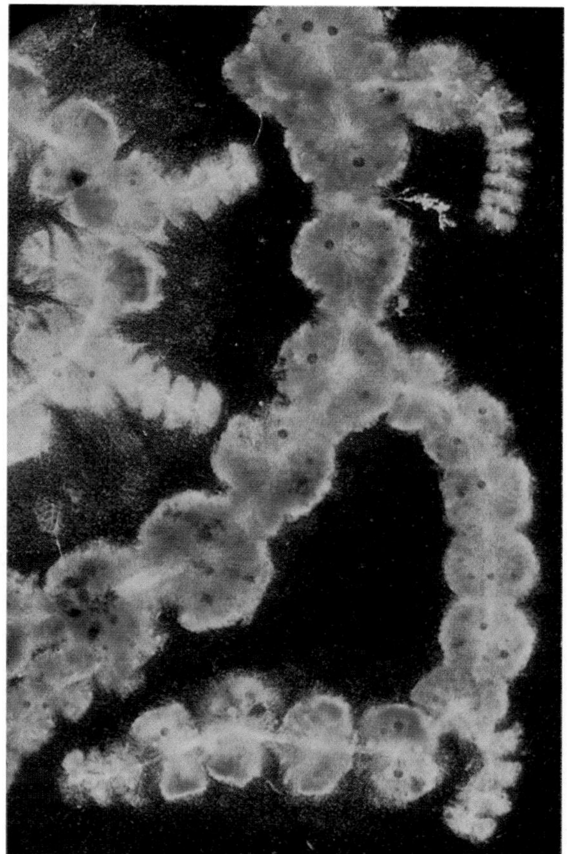

16–12
The simple, filamentous, freshwater red alga Batrachospermum moniliforme. *The soft, gelatinous, branched axes of this freshwater red alga are most frequently found in cold streams, ponds, and lakes, where they occur throughout the world.*

(a) (b) (c) (d)

16–13
Marine red algae. ***(a)*** *In* Bonnemaisonia hamifera, *the basically filamentous structure of the red algae is clearly evident. The branched filaments of this red alga are hooked, enabling it to cling to other seaweeds.* ***(b)*** *Jointed coralline algae in a tidal pool in California.* ***(c)*** *The reef-stabilizing, crustose coralline red alga* Porolithon craspedium. ***(d)*** *Irish moss* (Chondrus crispus), *an important source of carrageenan.*

than that of other types of seaweeds combined (Figure 16–13). Red algae usually grow attached to rocks or to other algae, but there are a few floating forms.

The chloroplasts of red algae contain phycobilins, which mask the color of chlorophyll *a* and give red algae their distinctive color. These pigments are particularly well suited to the absorption of the green and blue-green light that penetrates into deep water, where red algae are well represented. Biochemically and structurally, the chloroplasts of red algae closely resemble the cyanobacteria from which they were almost certainly derived, directly following endosymbiosis. Some red algae have lost most or all of their pigments and grow as parasites upon other red algae. Chloroplasts of a few primitive red algae have starch-forming pyrenoids, but pyrenoids appear to have been lost prior to the origin of the more complex forms.

Cells of Red Algae Have Some Unique Features

Red algae are unique among algal phyla in that they have neither centrioles nor flagellated cells. Whether they lost centrioles and flagella at some ancient time or their ancestors never had these structures is not known. In place of centrioles, which occur in many other eukaryotes, the red algae have microtubule organizing centers called **polar rings.** The main food reserves of the red algae are granules of **floridean starch,** which are stored in the cytoplasm. Floridean starch, a unique molecule that resembles the amylopectin portion of starch, is actually more like glycogen than starch (page 20).

The cell walls of most red algae include a rigid, inner component consisting of microfibrils of cellulose or of another polysaccharide and an outer mucilaginous layer, usually a sulfated polymer of galactose, such as agar or carrageenan (see "Algae and Human Affairs" on page 380). It is this mucilaginous layer that gives red algae their characteristic flexible, slippery texture. Continuous production and sloughing of the mucilage helps red algae rid themselves of other organisms that might colonize their surfaces and reduce their exposure to sunlight.

In addition, certain red algae deposit calcium carbonate in their cell walls. The function of algal calcification is uncertain. One hypothesis is that calcification helps algae to obtain carbon dioxide from the water for photo-

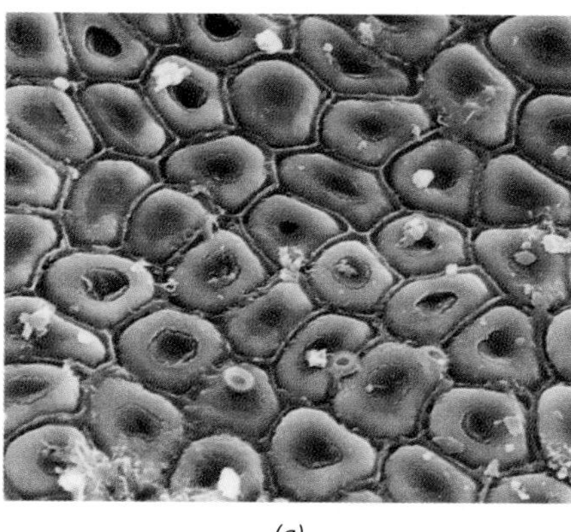

(a)

(b)

16–14
Coralline algae. ***(a)*** *Scanning electron micrograph of an unidentified species of crustose red alga from a depth of 268 meters on a seamount in the Bahamas, nearly 100 meters below the lowest limits established for any other photosynthesizing organism. At this depth, the light intensity was estimated as 0.0005 percent of its value at the ocean surface. The alga formed patches about a meter across, covering about 10 percent of the rock surface. When tested in the laboratory, this alga was found to be about 100 times more efficient than its shallow-water relatives in capturing and using light energy.* ***(b)*** *Purple crustose coralline algae from the same seamount.*

synthesis. Many of the calcified red algae are especially tough and stony and constitute the family *Corallinaceae*, the **coralline algae.** Calcification explains the occurrence of possible fossil coralline algae that are more than 700 million years old. Coralline algae are common throughout the oceans of the world, growing on stable surfaces that receive enough light, including 268-meter-deep seabed rocks (Figure 16–14). Other habitats include tidal pool rocks, on which jointed coralline red algae grow (Figure 16–13b), and the surf-pounded shoreward surfaces of coral reefs, where crust-forming (crustose) coralline algae (Figure 16–13c) help to stabilize reef structure. Large areas of diverse coral reefs around the world owe their survival, in part, to the architectural strength conferred by coralline algae. In recent years, a bright orange bacterium that causes a lethal disease of coralline algae has been spreading throughout the South Pacific, endangering thousands of kilometers of reef.

Many red algae produce unusual toxic terpenoids (page 33) that may assist in deterring herbivores. Some red algal terpenoids have anti-tumor activity and are currently being tested for possible use as anti-cancer drugs.

A few genera, such as *Porphyra*, have cells densely packed together into one- or two-layered sheets (Figure 16–15). However, most red algae are composed of filaments that are often densely interwoven and held to-

16–15
The red alga, Porphyra nereocystis, *the life history of which includes both bladelike and filamentous phases.* ***(a)*** *The bladelike phase, which is a gametophyte, produces both female gametangia, called carpogonia (reddish cell aggregates on left), and male gametangia, called spermatangia (cell aggregates on right).* ***(b)*** *After fertilization, the diploid carpospores give rise to a system of branched filaments. Meiosis occurs during germination of the spores (conchospores) produced by the filamentous phase. These haploid cells then grow into the bladelike gametophyte.*

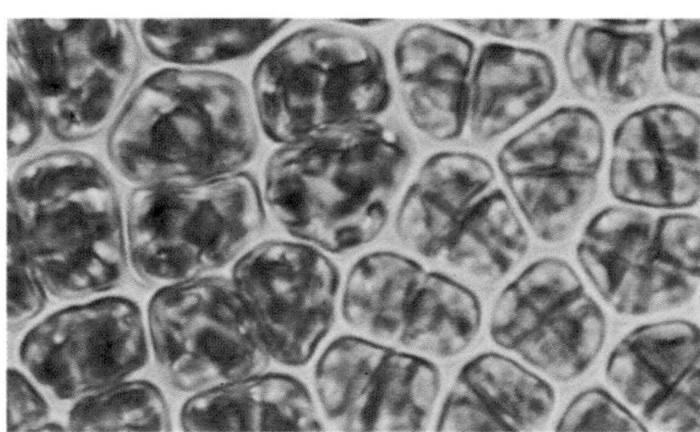

(a)

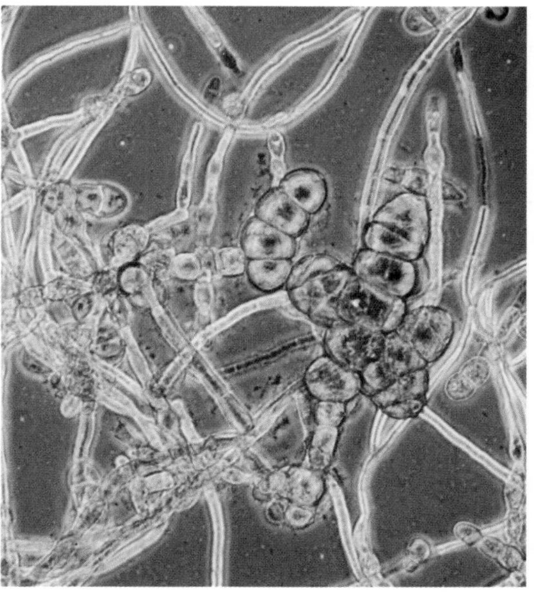

(b)

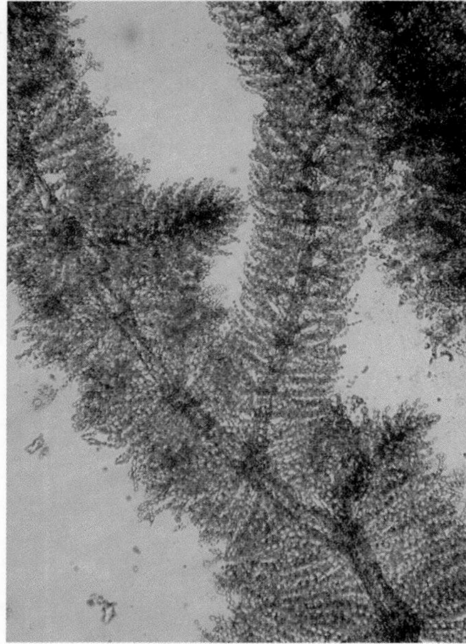

16–16
Batrachospermum sirodotia, *showing the whorls of lateral branches.*

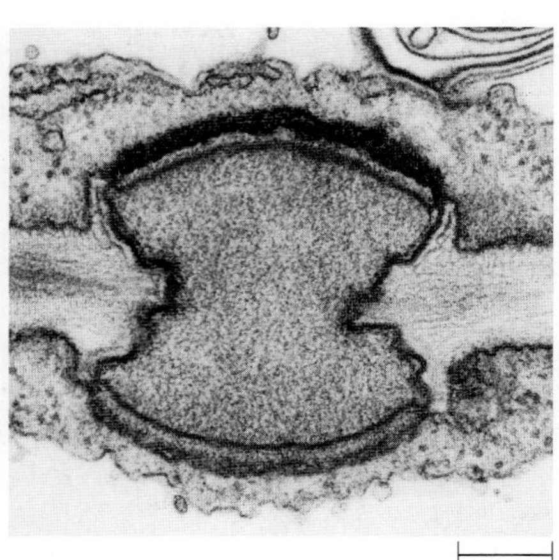

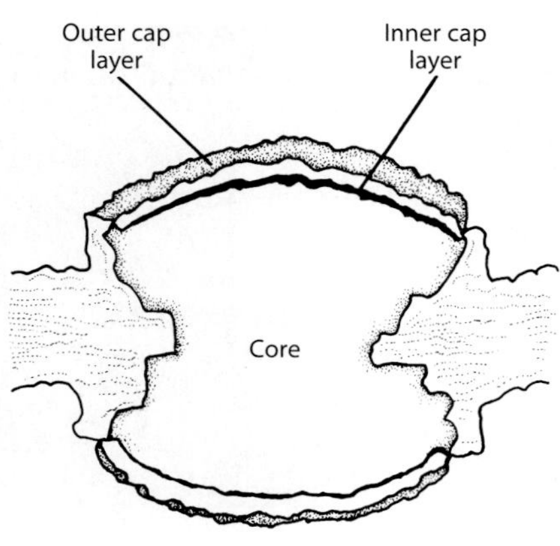

16–17
A pit connection in the red alga Palmaria. *Pit connections are distinct, lens-shaped plugs that form between the cells of red algae as the cells divide. These connections are also formed frequently between the cells of adjacent filaments that come into contact with each other, linking together the bodies of red algae. Pit-connection cores are protein, and their outer cap layers are, at least in part, polysaccharide.*

gether by the mucilaginous layer, which has a rather firm consistency. Growth in filamentous red algae is initiated by a single, dome-shaped apical cell that cuts off segments sequentially to form an axis. This axis, in turn, forms whorls of lateral branches (Figure 16–16). In most red algae, the cells are interconnected by primary **pit connections** (Figure 16–17) that develop at cytokinesis. Many red algae are multiaxial—that is, made up of many coherent filaments forming a three-dimensional body. In such forms, the filaments are interconnected by the formation of secondary pit connections. These pits form between cells of different filaments when the filaments come into contact with one another.

Red Algae Have Complicated Life Histories

Many red algae reproduce asexually by discharging spores, called **monospores,** into the water. If conditions are suitable, monospores may attach to a substrate. By repeated mitoses, a new seaweed similar to the monospore-producing parent is produced. Sexual reproduction also occurs widely among the multicellular red algae and can involve very complex life histories.

The simplest type of sexual life history in red algae involves an alternation of generations between two separate, multicellular forms of the same species—a haploid, gamete-producing **gametophyte** and a diploid, spore-producing **sporophyte.** The gametophyte produces **spermatangia** (singular: spermatangium), structures that generate and release nonmotile **spermatia** (singular: spermatium), or male gametes, which are carried to female gametes by water currents. The female gamete, or egg, is the lower, nucleus-containing portion of a structure known as the **carpogonium** (plural: carpogonia), which is borne on the same gametophyte as the spermatangia and remains attached to it. The carpogonium develops a protuberance, the **trichogyne,** for reception of spermatia. When a spermatium comes in contact with a trichogyne, the two cells fuse. The male nucleus then migrates down the trichogyne to the female nucleus and fuses with it. The resultant diploid zygote then produces a few diploid **carpospores,** which are released from the parent gametophyte into the water. If they survive, the carpospores attach to a surface and grow into the sporophyte, which produces haploid spores by sporic meiosis. If these haploid spores survive, they in turn attach to a surface and grow into gametophytes, thus completing the life cycle.

Experts believe that the red algae acquired an alternation of two multicellular generations early in their evolutionary history as an adaptive response to their lack of flagellated male gametes. Such nonflagellated gametes cannot swim toward female gametes, as do the flagellated male gametes of some other protists, animals, and some plants. Hence, fertilization may be more a matter of chance, with the consequence that zygote formation may be relatively rare. Alternation of generations is regarded as an adaptation that increases the number and genetic diversity of the progeny resulting from each individual fertilization event, or zygote. This is because a multicellular sporophyte can produce many more spores—and more diverse haploid spores—than could a single, meiotic zygote nucleus. Alternation of two multicellular generations also occurs in several other protist groups, such as green and brown seaweeds (discussed in Chapter 17), and in bryophytes and vascular plants (discussed in Chapters 18 to 21), where it may have similar ecological and genetic benefits.

A further evolutionary advance has occurred in most red algae. Rather than immediately producing spores, the zygote nucleus divides repeatedly by mitosis, generating a third multicellular life-cycle phase, the diploid **carposporophyte** generation. The carposporophyte generation remains attached to its parental gametophyte and probably receives organic nutrients from it. These nutrients help to support rapid proliferation of cells by mitosis. When the carposporophyte reaches its mature size, mitosis occurs in apical cells, giving rise to carpospores. The carpospores are released into the water, settle onto a substrate, and grow into separate diploid sporophytes.

In many red algae, a mitotically produced copy of the diploid zygote nucleus is transferred to another cell of the gametophyte. This cell, known as an **auxiliary cell,** serves as a host and nutritional source for repeated mitoses by the adopted nucleus. Proliferation of diploid filaments from the auxiliary cell generates a carposporophyte and carpospores. In many forms, multiple copies of the diploid zygote nucleus are carried by the growth of long tubular cells throughout the algal body and are deposited into many additional auxiliary cells. Each diploid nucleus then produces many carposporophytes, which release very large numbers of carpospores into the water. In one case, each zygote nucleus is known to result in the release of some 4500 carpospores. Each carpospore is capable of growing into a usually free-living, multicellular diploid generation, called the **tetrasporophyte.** Meiosis occurs in specialized cells of the tetrasporophyte, called **tetrasporangia.** Each of the tetraspores produced can germinate into a new gametophyte, if conditions are favorable. *Polysiphonia* provides an example of this kind of life cycle (Figure 16–18).

The life history of most red algae thus consists of three phases: (1) a haploid gametophyte; (2) a diploid phase, called a carposporophyte; and (3) another diploid phase, called a tetrasporophyte. The red algal carposporophyte generation is regarded as an additional way to increase the genetic products of sexual reproduction when fertilization rates are low. Alternation of generations involving three multicellular generations is unique to the red algae. The ability to produce many carposporophytes with the resultant larger numbers of carpospores and potentially huge numbers of tetraspores, all from a single zygote, has helped the red algae to conquer the sexual disability imposed by their lack of flagella.

In most red algae, the gametophyte and tetrasporophyte generations resemble one another closely and are therefore said to be isomorphic, as in *Polysiphonia.* Coralline algae also have isomorphic life cycles. However, an increasing number of heteromorphic life cycles are also being discovered. In these species, the tetrasporophytes either are microscopic and filamentous or consist of a thin crust that is tightly attached to a rock substrate. *Phycologists*—scientists who study algae—speculate that differences in appearance have selective advantages in responding to seasonal changes or other environmental variations. The development of techniques for cultivating algae in the laboratory has led to the discovery that what appear to be distinct species are in some cases actually alternating generations of the same species.

The most important discovery of this kind resulted from British phycologist Kathleen Drew Baker's laboratory cultivation research. She discovered that the inconspicuous red filaments of *Conchocelis,* which grows in seashells, are the diploid phase of the edible, blade-forming seaweed *Porphyra* (the haploid phase). Her work, published in the journal *Nature* in 1949, has formed the basis for the modern, billion-dollar-per-year nori production industry in Japan, Korea, and China (see the essay on page 380). In gratitude, a memorial park was established in Drew Baker's honor in Kumamoto Prefecture, Japan, where she is annually revered with a ceremony dedicated to "the mother of the sea."

Dinoflagellates: Phylum *Dinophyta*

Molecular systematic data indicate that the dinoflagellates are closely related to ciliate protozoa such as *Paramecium* and *Vorticella* (see Figure 13–12). They are also closely related to the sporozoa, which include malarial and other parasites of animals and humans.

Most dinoflagellates are unicellular biflagellates (Figure 16–19). Some 2000 to 4000 species are known, many of them abundant and highly productive members of the marine phytoplankton; others occur in fresh water. Dinoflagellates are unique in that their flagella beat within two grooves. One groove encircles the body like a belt, and the second groove is perpendicular to the first. The beating of the flagella in their respective grooves causes the dinoflagellate to spin like a top as it moves. The encircling flagellum is ribbon-like. There are also numerous nonmotile dinoflagellates, but they typically produce reproductive cells having flagella in grooves, from which their relationship to other dinoflagellates is deduced.

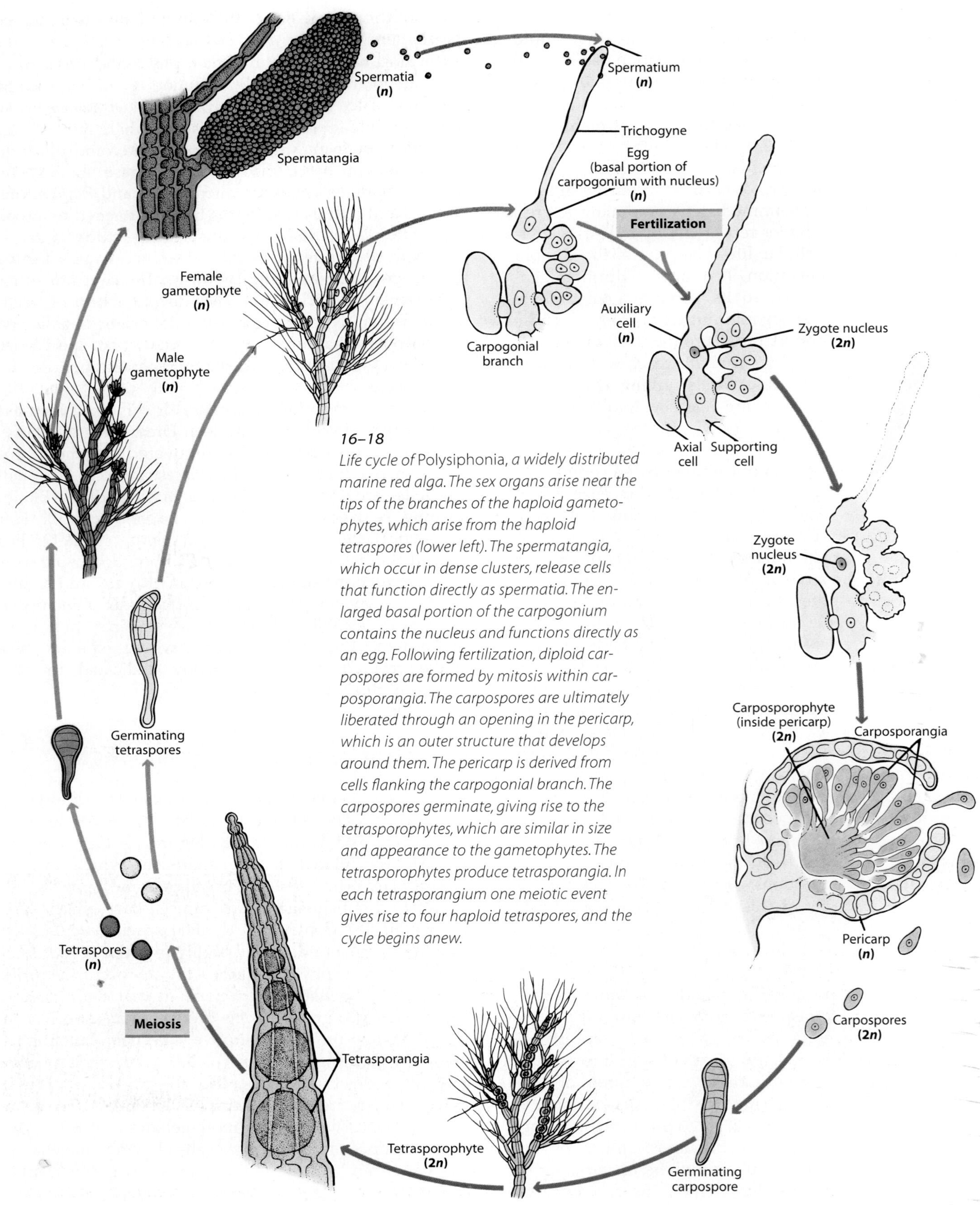

16–18

Life cycle of Polysiphonia, *a widely distributed marine red alga. The sex organs arise near the tips of the branches of the haploid gametophytes, which arise from the haploid tetraspores (lower left). The spermatangia, which occur in dense clusters, release cells that function directly as spermatia. The enlarged basal portion of the carpogonium contains the nucleus and functions directly as an egg. Following fertilization, diploid carpospores are formed by mitosis within carposporangia. The carpospores are ultimately liberated through an opening in the pericarp, which is an outer structure that develops around them. The pericarp is derived from cells flanking the carpogonial branch. The carpospores germinate, giving rise to the tetrasporophytes, which are similar in size and appearance to the gametophytes. The tetrasporophytes produce tetrasporangia. In each tetrasporangium one meiotic event gives rise to four haploid tetraspores, and the cycle begins anew.*

Dinoflagellates are unusual, though not unique, in having permanently condensed chromosomes. This feature, along with some unusual aspects of mitosis, was once thought to indicate that dinoflagellates were quite primitive. Current opinion is that dinoflagellates are a highly derived protist group. The chief method of reproduction of dinoflagellates is by longitudinal cell division, with each daughter cell receiving one of the flagella and a portion of the wall, or theca. Each daughter cell then reconstructs the missing parts in a very intricate sequence.

Many of the dinoflagellates are bizarre in appearance, with stiff cellulose plates forming the **theca,** which often looks like a strange helmet or part of an ancient coat of armor (Figures 16–19 and 16–20). The cellulose plates of the wall are within vesicles just inside the outermost membrane of the cell. Dinoflagellates in the open ocean often have large, elaborate sail-like thecal plates that aid in flotation. Other dinoflagellates have very thin or no cellulose plates and therefore do not appear to have a theca.

16–19
The "armor" of some dinoflagellates consists of cellulose plates in vesicles inside the plasma membrane. The vesicles of genera that appear to be naked may or may not contain cellulose plates.

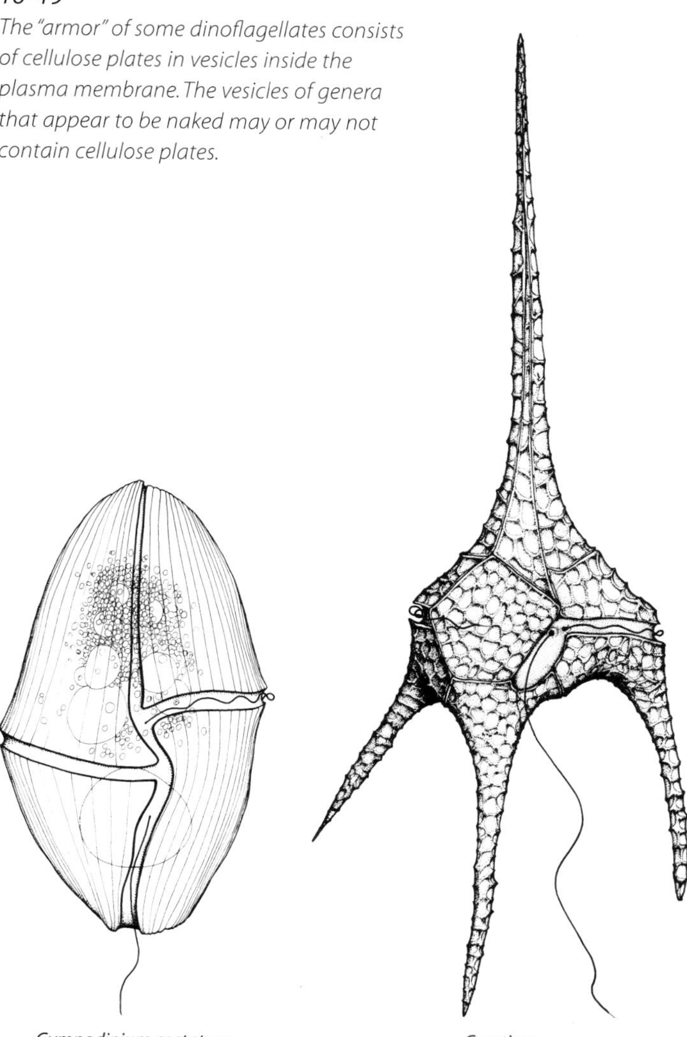

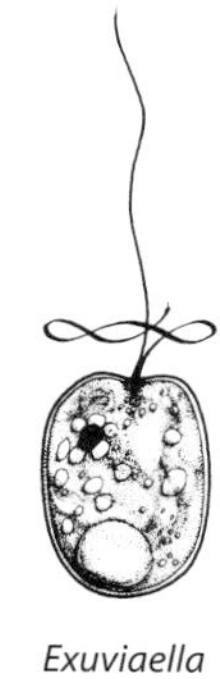

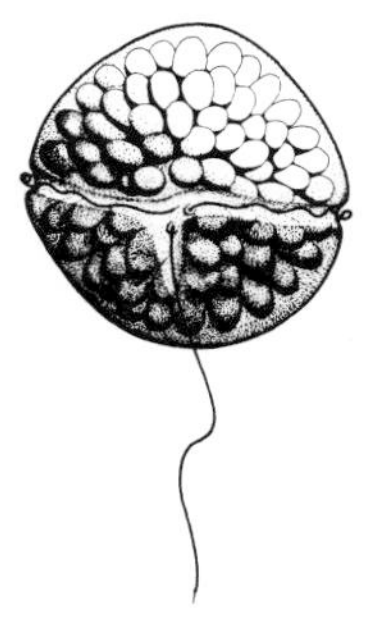

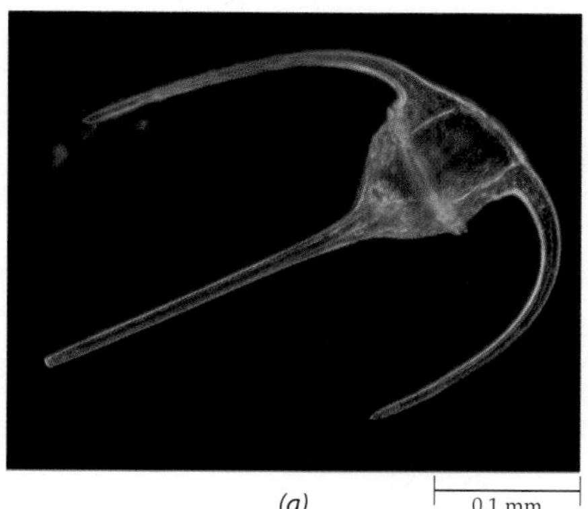

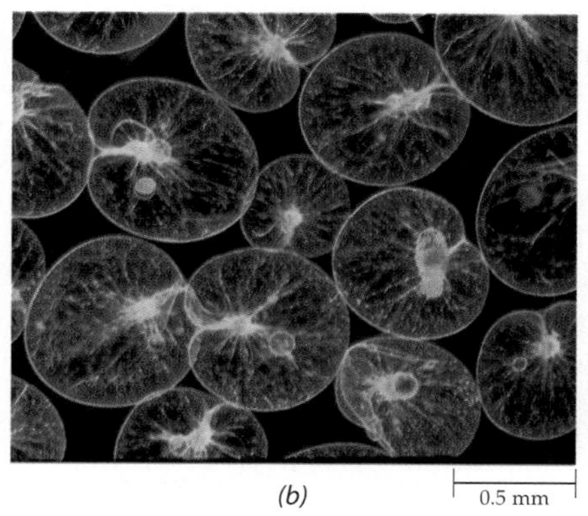

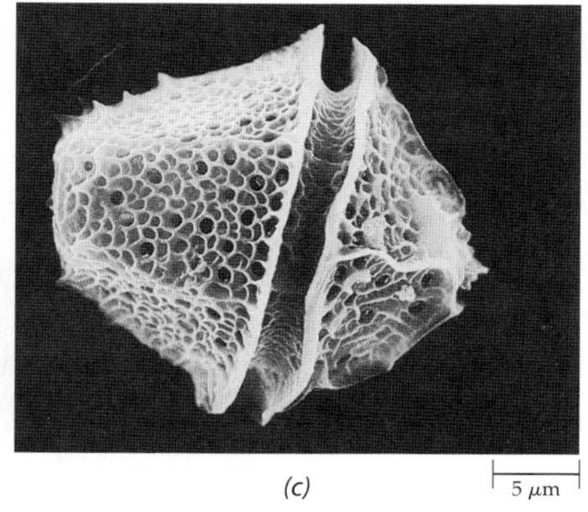

16–20
Dinoflagellates. ***(a)*** Ceratium tripos, *an armored dinoflagellate.* ***(b)*** Noctiluca scintillans, *a bioluminescent marine dinoflagellate.* ***(c)*** Gonyaulax polyedra, *the dinoflagellate responsible for the spectacular red tides along the coast of southern California.*

Red Tides/Toxic Blooms

In late August 1987, the west coast of Florida was ravaged by a major red tide incident—one of dozens that are known to have occurred there over the past 150 years. Hundreds of thousands of dead fish littered the beaches, and millions of tourist dollars were lost. The organism responsible for most red tides and their associated effects in the Gulf of Mexico is the dinoflagellate Gymnodinium breve. *Following a period of rapid dinoflagellate reproduction, a toxic bloom formed in which* G. breve *was so abundant that the sea was colored reddish brown. Environmental factors that favor such blooms include warm surface temperatures, high nutrient content in the water, low salinity (which often occurs during rainy periods), and calm seas. Thus rainy weather followed by sunny weather in the summer months is often associated with red tide outbreaks.*

Two months later, G. breve *invaded estuaries along the North Carolina coast, areas where it had never before been observed. The dinoflagellates became so abundant that the waters turned yellowish, devastating the tourist industry in this region. At least 41 cases of respiratory, gastrointestinal, or neurological illness were reported in people who swam in these waters—all associated with the toxins produced by the dinoflagellate bloom.* Gymnodinium *is thought to have traveled north from Florida in the Gulf Stream; with it came schools of a fish called menhaden, which consumed large amounts of the dinoflagellates. Bottle-nosed dolphins ate the menhaden, and fully half of the dolphin population in the western Atlantic died. The fish had not been injured by the toxic dinoflagellates in their guts, but the dolphins, eating the fish whole, had been poisoned. In their weakened condition, dolphins proved easy prey for bacterial and viral diseases.*

Red tides have been around for a long time. Records of their presence are found in the Old Testament *and Homer's* Iliad. *There is considerable concern, however, that their incidence is increasing and spreading around the globe. In 1990, there were twice as many outbreaks as in 1970. Ecologists are not sure whether this increase is merely an upturn in a natural cycle or the beginning of a serious global epidemic. The increased frequency is correlated with increased nutrients in coastal waters originating from runoff from human development, the heavy use of fertilizers, and livestock farms.*

The most recently identified toxin-producing dinoflagellate is Pfiesteria piscicida *(see page 366), which has been responsible for massive fish kills in the Neuse and Pamlico Rivers in North Carolina. The toxins produced by* Pfiesteria *are powerful enough to tear bleeding sores in fish. Special care must be taken by people working with the toxins, which have been reported to cause nausea, vomiting, headaches, burning eyes, memory lapses, breathing difficulty, mood swings, and impaired speech.* Pfiesteria *is the first known toxin-producing dinoflagellate with multiple stages (at least 24), including flagellated, amoeboid, and cyst stages with transition forms between each of these (see Figure 16–22).*

Other dinoflagellates are responsible for the formation of red tides in different areas. Thus Gonyaulax tamarensis *is the organism involved along the northeastern Atlantic coast, from the Canadian Maritime provinces to southern New England, while* Gymnodinium catenella *causes red tides at times along the Pacific Coast from Alaska to California,* Ptichodiscus brevis *in the Gulf of Mexico, and* Protogonyaulax tamarensis *in the North Sea off the coast of Northumberland in the United Kingdom. Over 40 marine species of dinoflagellates have been identified as producing toxins that kill birds and mammals, render shellfish toxic, or produce a widespread tropical fish-poisoning disease called ciguatera. The poisons produced by some dinoflagellates, such as* G. catenella, *are extraordinarily powerful nerve toxins. The chemical nature and biological activity of most of these toxins are relatively well known.*

When shellfish, such as mussels and clams, ingest toxic dinoflagellates or other organisms, they accumulate and concentrate the toxins. Depending on the species of toxic organisms the shellfish have consumed, they themselves may become dangerously toxic to the people who eat them. Along the Atlantic Coast, fisheries are commonly closed in summer, and people regularly suffer from poisoning after consuming mussels, oysters, scallops, or clams taken from certain regions. Research with Pfiesteria piscicida *has revealed that the dinoflagellates can attack humans not only indirectly, through shellfish, but directly.*

(a)

(b)

(a) Fish killed by Pfiesteria. *(b)* Gymnodinium breve, *the unarmored dinoflagellate responsible for the outbreaks of red tide along the west coast of Florida. The curved, transverse flagellum lies in a groove encircling the organism. Its longitudinal flagellum, only a portion of which is visible, extends from the middle of the organism toward the lower left. The apical groove at the top is an identifying characteristic of* Gymnodinium.

Many Dinoflagellates Ingest Solid Food Particles or Absorb Dissolved Organic Compounds

About half of all dinoflagellates lack a photosynthetic apparatus and hence obtain their nutrition either by ingesting solid food particles or by absorbing dissolved organic compounds. Even many pigmented—photosynthetic—and heavily armored dinoflagellates can feed in these ways. Some feeding dinoflagellates extrude a tubular process known as a peduncle, which can suction organic materials into the cell. The peduncle is retracted into the cell when feeding is finished.

Most pigmented dinoflagellates typically contain chlorophylls *a* and *c*, which are generally masked by carotenoid pigments, including **peridinin,** which is similar to fucoxanthin, an accessory pigment typical of chrysophytes (see Chapter 17). The presence of peridinin supports the hypothesis that the chloroplasts of many dinoflagellates were derived from ingested chrysophytes by endosymbiosis, as described in Chapter 13. Other dinoflagellates have green or blue-green plastids obtained from ingested green algae or cryptomonads. The carbohydrate food reserve in dinoflagellates is starch, which is stored in the cytoplasm.

Pigmented dinoflagellates occur as symbionts in many other kinds of organisms, including sponges, jellyfish, sea anemones, tunicates, corals, octopuses and squids, gastropods, turbellarians, and certain protists. In the giant clams of the family *Tridachnidae,* the dorsal surface of the inner lobes of the mantle may appear chocolate-brown as a result of the presence of symbiotic dinoflagellates. When they are symbionts, the dinoflagellates lack armored plates and appear as golden spherical cells called **zooxanthellae** (Figure 16–21).

Zooxanthellae are primarily responsible for the photosynthetic productivity that makes possible the growth of coral reefs in tropical waters, which are notoriously nutrient-poor. Coral tissues may contain as many as 30,000 symbiotic dinoflagellates per cubic millimeter, primarily in cells that line the gut of the coral polyps. Amino acids produced by the polyps stimulate the dinoflagellates to produce glycerol instead of starch. The glycerol is used directly for coral respiration. Because the dinoflagellates require light for photosynthesis, the corals containing them grow mainly in ocean waters less than 60 meters deep. Many of the variations in the shapes of coral are related to the light-gathering properties of different geometric arrangements. This relationship is somewhat similar to the ways in which various branching patterns of trees serve to maximally expose their leaves to sunlight.

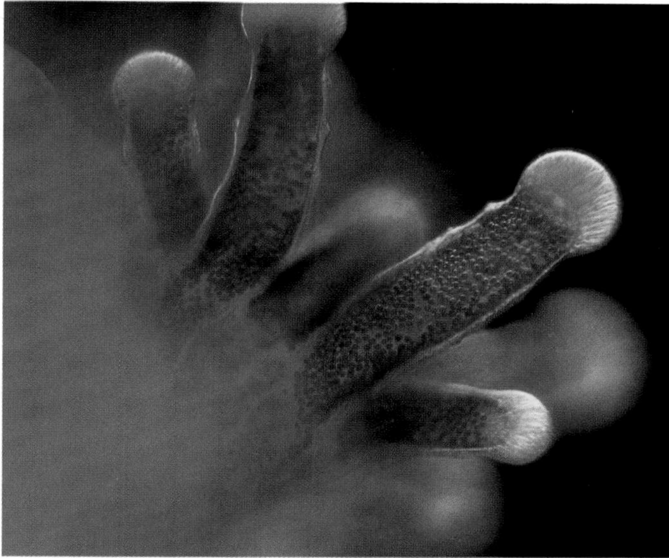

16–21
Zooxanthellae, the symbiotic form of dinoflagellates, shown here in a tentacle of a coral animal. These symbionts are responsible for much of the productivity of coral reefs.

During Periods of Low Nutrient Levels, Dinoflagellates Form Resting Cysts

Under conditions that do not allow continued population growth, such as low nutrient levels, dinoflagellates may produce nonmotile resting cysts that drift to the lake or ocean bottom, where they remain viable for years. Ocean currents may transport these benthic (meaning "bottom of the water") cysts to other locations. When conditions become favorable, the cysts may germinate, reviving the population of swimming cells. Cyst production, movement, and germination explain many aspects of the ecology and geography of toxic, dinoflagellate blooms. They explain why blooms do not necessarily occur at the same site every year, and why blooms are associated with nutrient pollution of the ocean with sewage and agricultural runoff. In addition, they explain why blooms appear to move from one location to another from year to year.

Sexual reproduction has been found in a number of species of dinoflagellates. Dinoflagellate zygotes form distinctive thick, chemically inert, ornamented cell walls that resemble some ancient fossil acritarchs (Figure 16–1). The life histories of dinoflagellates may be very complex, involving multiple forms, some of which resemble amoebas. As is the case for certain red algae, by growing the organisms in laboratory culture, it is possible to show that morphologically different protists may be components of the life history of a single species. Understanding life histories is important in understanding the roles of phytoplankton, including dinoflagellates, in food webs and in toxic bloom formation.

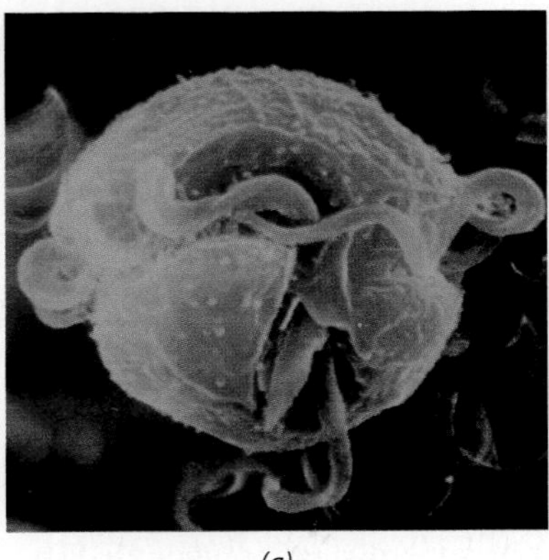
(a)

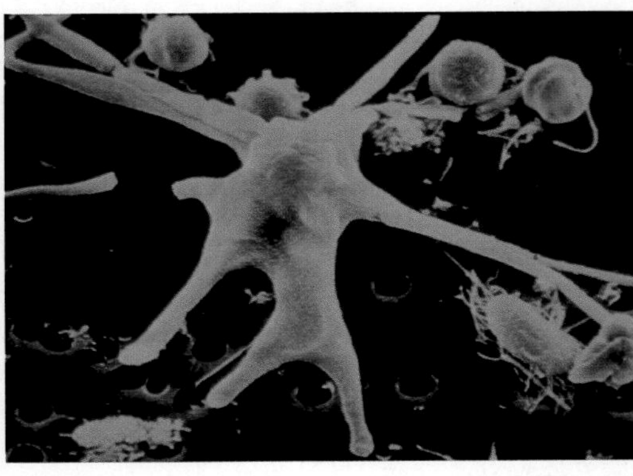
(b)

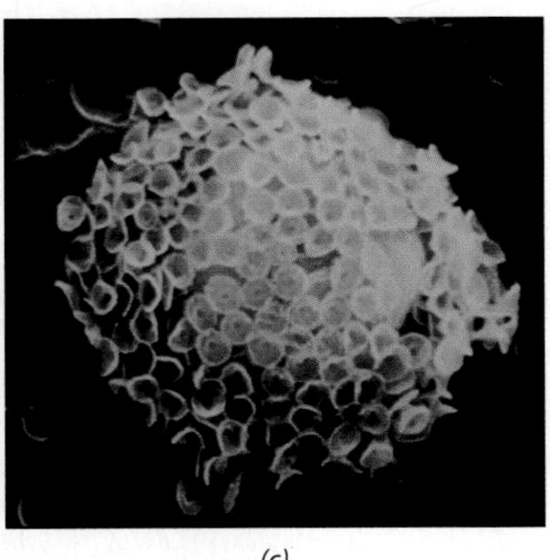
(c)

16–22
Three of the known stages in the complex life cycle of Pfiesteria piscicida, *a dinoflagellate.* ***(a)*** *A biflagellated cell showing flagella and the peduncle, which is used in feeding.* ***(b)*** *One of the amoeboid stages, which dominate the life cycle.* ***(c)*** *An amoeboid cyst stage. The biflagellated cyst can transform into an amoeboid stage within minutes.* Pfiesteria piscicida *has been termed an "ambush predator" because it releases a toxin that kills fish and then consumes the sloughed tissue of these fish.*

Many Dinoflagellates Produce Toxic or Bioluminescent Compounds

About 20 percent of all known dinoflagellate species produce one or more highly toxic compounds that are economically and ecologically significant (see essay on page 364). Dinoflagellate toxins may provide protection from predation, but at least one recently discovered dinoflagellate, *Pfiesteria piscicida,* uses its deadly toxin in a "hit and run" feeding strategy (Figure 16–22). The presence of fish, such as the menhaden, stimulates benthic cysts to germinate into swimming *Pfiesteria* cells that produce a toxin that paralyzes the fishes' respiratory systems, causing death by suffocation. As the fish decay, the dinoflagellates extend their peduncles and feed frenetically upon bits of fish flesh. When the food has been consumed, the phantom dinoflagellates rapidly return to the more obscure benthic cyst stage, sometimes within a period of just two hours. As a result, the cause of massive fish kills can be difficult to ascertain unless water samples are taken at the very beginning of the killing event. Ocean nutrient pollution is related to fish kills in that excess phosphorus in the water stimulates at least one phase in the life cycle of *Pfiesteria.*

Marine dinoflagellates are also famous for their bioluminescent capabilities (Figure 16–20b). These are responsible for the attractive sparkling of ocean waters that is commonly observed at night as boats or swimmers agitate the water. When dinoflagellate cells are disturbed, a series of well-understood biochemical events results in a reaction involving luciferin and the enzyme luciferase that, as in fireflies and other organisms, creates a brief flash of light (see Chapter 28). Bioluminescence is thought to serve as protection against predators, such as copepods, small crustaceans that are the most numerous members of the zooplankton. One hypothesis is that the dinoflagellates' light flashes directly disrupt feeding by startling the predators. Another hypothesis suggests a more indirect process—copepods that have fed upon luminescent dinoflagellates become more visible to the fish that feed upon them.

Haptophytes: Phylum *Haptophyta*

The phylum *Haptophyta* consists of a diverse array of primarily marine phytoplankton, though a few freshwater and terrestrial forms are known. The phylum consists of unicellular flagellates, colonial flagellates, and nonmotile single cells and colonies. There are about 300 known species in 80 genera, but new species are being discovered continuously. Haptophyte species diversity is highest in the tropics.

The most distinctive feature of the haptophyte algae is the **haptonema** (Gk. *haptein,* "to fasten"; relates to sense of touch). The haptonema is a threadlike structure that extends from the cell along with two flagella of equal length (Figure 16–23). It is structurally distinct from a flagellum. Although microtubules are present in the haptonema, they do not have the 9-plus-2 arrangement that is typical of eukaryotic flagella and cilia. The haptonema can bend and coil but cannot beat like a flagellum. In some cases, it allows the haptophyte cell to catch prey food particles, functioning somewhat like a fishing rod. In other cases, it seems to help the cells sense and avoid obstacles.

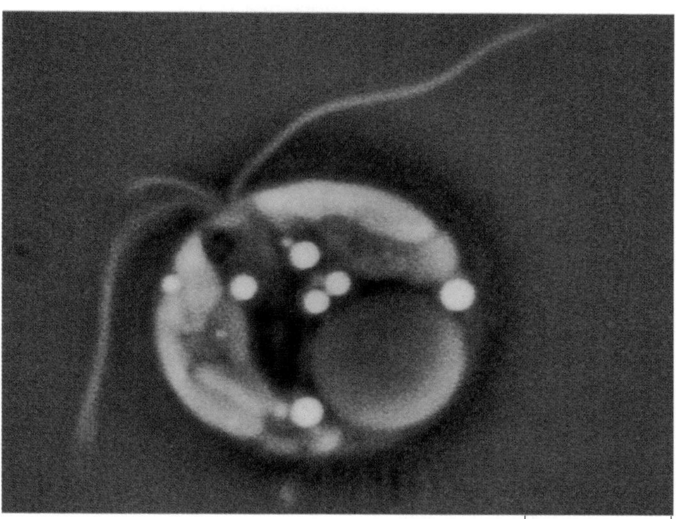

16–23
The haptophyte Prymnesium parvum. *Most haptophytes have two smooth and nearly equal flagella. Many also possess a haptonema (the smaller appendage shown here), which can bend or coil but cannot beat as do the flagella.*

Another characteristic of haptophyte algae is the presence of small, flat scales on the outer surface of the cell (Figure 16–24). These scales are composed of organic material or calcified organic material. The calcified scales are known as **coccoliths,** and the 12 or more families of organisms that are decorated with coccoliths are known as the coccolithophorids. Coccoliths are of two types. Those produced inside the cell—in Golgi vesicles—are transported to the cell's exterior. The other type of coccolith is generated outside the cell. Coccoliths form the basis for a continuous fossil record back to their first appearance in the Late Triassic, some 230 million years ago.

Most haptophytes are photosynthetic, with chlorophyll *a* and some variation of chlorophyll *c*. At least one nonphotosynthetic representative is known. Some have the accessory pigment fucoxanthin, in common with chrysophytes and diatoms (see Chapter 17). Other haptophytes lack fucoxanthin.

In common with the cryptomonads, the plastids of haptophytes are surrounded by a chloroplast endoplasmic reticulum, which is continuous with the nuclear envelope. As in cryptomonads, the chloroplast endoplasmic reticulum is evidence that the plastids were acquired by secondary endosymbiosis. Sexual reproduction and alternation of heteromorphic generations occur in haptophytes, but the chromosome levels and life histories of many forms are as yet unknown.

Marine haptophytes are significant components of food webs, serving both as producers and, even though most are autotrophic, as consumers. As consumers, they graze on small particles such as cyanobacteria or absorb dissolved organic carbon. They are an important means by which organic carbon and two-thirds of the oceans' calcium carbonate are transported to the deep ocean. In addition, they are important producers of sulfur oxides connected with acid rain (page 350). The gelatinous, colonial stage of *Phaeocystis* dominates the phytoplankton of the marginal ice zone in polar regions and contributes some 10 percent of the atmospheric sulfur compounds that are generated by phytoplankton. In addition, its gelatin contributes significant organic carbon to the water. In all of the oceans, especially at mid-latitudes, *Emiliania huxleyi* can form blooms covering thousands of square kilometers of ocean. The two haptophyte genera *Chrysochromulina* and *Prymnesium* are notorious for forming marine toxic blooms that kill fishes and other marine life.

In the next chapter, we will turn our attention to the remaining protist phyla traditionally studied by botanists. Special emphasis is placed on the green algae, which include representatives considered to be more closely related than any other living organisms to the ancestors of bryophytes and vascular plants.

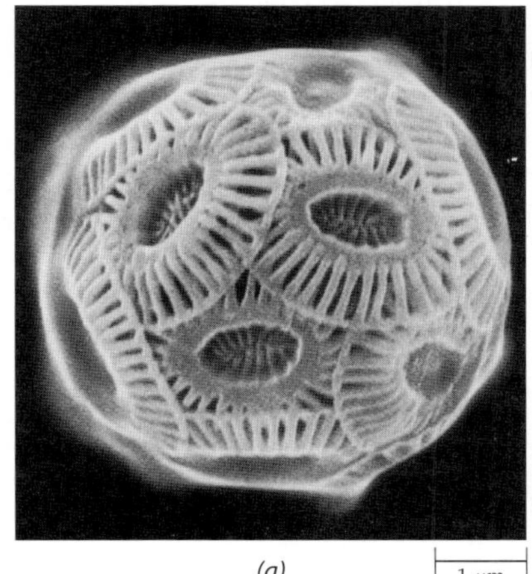

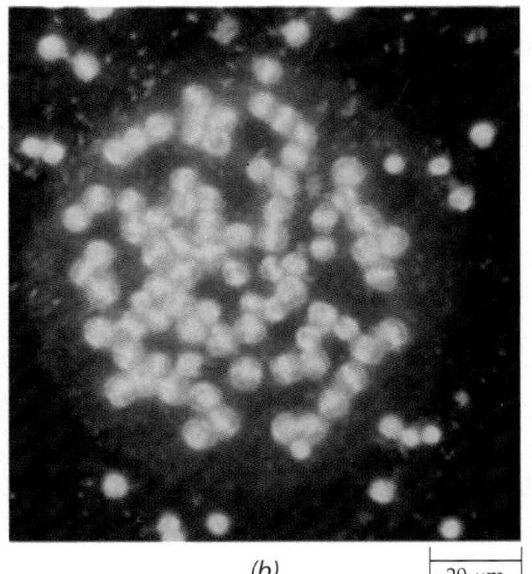

16–24
Haptophyte algae. ***(a)*** Emiliania huxleyi, *a coccolithophorid. This is the most widespread and abundant of the estimated 300 species of this group of extremely minute algae. The platelike scales covering cells of the coccolithophorids consist of calcium carbonate.* ***(b)*** *A fluorescence micrograph of a young colony of* Phaeocystis, *stained with acridine orange. The cells are embedded in a polysaccharide mucilage. In temperate oceans, such as the North Sea, massive* Phaeocystis *blooms clog fishing nets and wash up onto beaches, producing meter-thick mounds of foam.*

Summary

The Kingdom *Protista* Includes a Variety of Autotrophic and Heterotrophic Organisms

Protists are eukaryotic organisms that are not included in the fungal, plant, or animal kingdoms. Those considered in this book include both photosynthetic (autotrophic) organisms, or algae, and heterotrophic organisms once treated as fungi. The latter include the *Myxomycota* and *Dictyosteliomycota,* as well as the *Oomycota,* which are considered in the next chapter.

The algae include photosynthetic protists as well as colorless relatives. Algae are important components of aquatic food webs and, together with the cyanobacteria, constitute the phytoplankton. The algae also play significant roles in global carbon and sulfur cycling.

The Plasmodium of the *Myxomycota* Is a "Naked" Mass of Protoplasm

The plasmodial slime molds, phylum *Myxomycota,* may exist as streaming, multinucleate masses of protoplasm called plasmodia, which are usually diploid. These plasmodia typically form sporangia in which diploid spores are formed. Meiosis occurs within each of the spores, and three of the resulting nuclei disintegrate, leaving one haploid nucleus in each spore. Under favorable conditions, the spores split open, producing amoebas, which may become flagellated. These amoebas or flagellated cells may function as gametes. Plasmodium formation often, but not always, follows fusion of gametes.

The Pseudoplasmodium of the *Dictyosteliomycota* Is an Aggregation of Myxamoebas

The cellular slime molds, phylum *Dictyosteliomycota,* are a group of amoeba-like organisms that aggregate together at one stage of their life cycle to form pseudoplasmodia, or slugs. Cyclic AMP plays a major role in the aggregation of individual myxamoebas into these slugs, which undergo complex patterns of differentiation and programmed cell death that have led to their being used as models in this field. Flagellated cells are not known. Reproduction involving asexual spores is common. Sexual reproduction involves the formation of walled zygotes called macrocysts. In contrast to fungi, but like many other protists, the plasmodial and cellular slime molds ingest particulate food.

Algae Obtain Nutrients in a Variety of Ways

Although many algae are capable of photosynthesis, uptake and utilization of dissolved organic compounds is also common. Particle feeding occurs in at least some members of the euglenoids, cryptomonads, dinoflagellates, and haptophytes. Phagocytosis is regarded as a method by which the ancestors of the pigmented members of these algal groups probably acquired their chloroplasts.

The Euglenoids, Phylum *Euglenophyta,* Occur Primarily in Fresh Water and Are Unicellular

About one-third of the euglenoids are photosynthetic and have chloroplasts containing chlorophylls *a* and *b* and several carotenoids. The plastids do not store starch; rather, granules of a unique polysaccharide called paramylon are stored in the cytoplasm. Euglenoids lack a cell wall but have a series of helically arranged proteinaceous strips, called a pellicle, which lies just beneath the plasma membrane. The cells contain a contractile vacuole and bear flagella. Sexual reproduction is unknown in euglenoids.

Cryptomonads, Phylum *Cryptophyta,* and Dinoflagellates, Phylum *Dinophyta,* Typically Occur in Both Freshwater and Marine Habitats

The cryptomonads are single-celled flagellates, which appear to have arisen through the fusion of two different eukaryotic cells, one heterotrophic and the other photosynthetic. In addition to chlorophylls *a* and *c* and carotenoids, some cryptomonad chloroplasts contain either phycocyanin or phycoerythrin, water-soluble pigments otherwise known only in cyanobacteria and red algae.

The dinoflagellates are unicellular biflagellates. Many are bizarre in appearance, with stiff cellulose plates forming a wall, or theca. Other dinoflagellates have very thin or no cellulose plates at all. About half of the dinoflagellates are photosynthetic, containing chlorophylls *a* and *c,* which are generally masked by carotenoid pigments, including peridinin. Some marine dinoflagellates produce toxic compounds and/or harmful red tides. Other marine dinoflagellates serve as food-producing endosymbionts within the cells of reef-forming corals.

Red Algae, Phylum *Rhodophyta,* Are a Large Group Particularly Common in Warmer Marine Waters

Red algae usually grow attached to a substrate, and some grow at great depths (down to 268 meters). The chloroplasts contain phycobilins, which mask the color of chlorophyll *a* and give the red algae their distinctive color. The chloroplasts of the red algae are biochemically and structurally very similar to the cyanobacteria from which they very likely were derived. Red algae are notable for their complex life cycles, which commonly involve alternation of three distinct generations—gametophyte, carposporophyte, and tetrasporophyte. Red algae are also sources of useful, valuable carbohydrates such as agar and carrageenan.

The Haptophytes, Phylum *Haptophyta,* Are Primarily Marine Phytoplankton

The most distinctive feature of the haptophytes is the haptonema, a threadlike structure that seems to help the cell to sense and avoid obstacles or to catch prey food particles. Most haptophytes are photosynthetic, with chlorophylls *a* and *c*. They are ecologically important in both the carbon and sulfur cycles on a worldwide basis.

Selected Key Terms

acritarchs p. 348
aethalium p. 352
apoptosis p. 355
carposporophyte p. 361
chloroplast endoplasmic reticulum p. 357
contractile vacuole p. 351
coralline algae p. 359
eyespot, or **stigma** p. 351
floridean starch p. 358
gametophyte p. 360
haptonema p. 366
monospores p. 360
myxamoebas p. 354
nucleomorph p. 357
pellicle p. 351
phytoplankton p. 348
pit connections p. 360
plasmodiocarp p. 352
plasmodium p. 352
pseudoplasmodium p. 355
pyrenoid p. 352
sclerotium p. 354
secondary endosymbiosis p. 357
sporophyte p. 360
tetrasporophyte p. 361
theca p. 363
zooplankton p. 348
zooxanthellae p. 365

Questions

1. Expound upon the phytoplankton as the "great meadow of the sea."
2. By means of a simple, labeled diagram, explain the structure-function relationships of *Euglena.*
3. "When the going gets tough, the tough get going." Describe how each of the following organisms adapt to tough times, such as periods of inadequate nutrient or moisture levels: plasmodial slime molds, cellular slime molds, dinoflagellates.
4. Of what advantage is the diploid carposporophyte generation to the red algae?
5. What do such organisms as *Gymnodinium breve, Pfiesteria piscicida,* and *Gonyaulax tamarensis* have in common?

Chapter 17

Protista II: Heterokonts and Green Algae

17–1
Multicellular photosynthetic organisms anchored themselves to rocky shores early in the course of their evolution. These kelp, seen at low tide on the rocks at Botanical Beach on Vancouver Island, British Columbia, are brown algae (Phaeophyta)*, a group in which multicellularity evolved independently of other groups of organisms.*

OVERVIEW

Algae are crucially important components of aquatic ecosystems, producing oxygen and serving as food for aquatic animals. In this chapter, we continue our study of the algae. We first examine the heterokonts, which consist of four phyla whose members have two flagella of different length and structure. The heterokonts include three mostly photosynthetic phyla—algae of various types—and one entirely heterotrophic phylum, the oomycetes.

We then turn our attention to the green algae—phylum Chlorophyta*—whose motile members have flagella of similar length and structure. In this phylum you will find organisms with almost every type of body plan, including unicellular, colonial, filamentous, and flat, tissuelike forms. Chlorophytes are important evolutionarily because some ancestral members are thought to have given rise to the bryophytes and vascular plants, setting the stage for the story continued in the following chapters.*

CHECKPOINTS

By the time you finish reading this chapter, you should be able to answer the following questions:

1. What do all phyla of heterokonts have in common?
2. How do the oomycetes differ from the other heterokonts, and what are some important plant diseases caused by oomycetes?
3. What is unique about the cell walls of diatoms? What effect does this feature have on the diatom life cycle, generation after generation?
4. What are the basic characteristics of the brown algae?
5. What characteristics of the green algae have led botanists to conclude that the green algae are the protist group from which plants evolved?
6. How does the mode of cell division in the *Chlorophyceae* differ from that of the other classes of green algae?

The open sea, the shore, and the land are the three life zones that make up our biosphere. Of these zones, the sea and the shore are more ancient. Here algae play a role comparable to the role played by plants in the far younger terrestrial world. Often, algae are also dominant in freshwater habitats—ponds, streams, and lakes—where they may be the most important contributors to the productivity of these ecosystems. Everywhere they grow, algae play an ecological role comparable to that of the plants in land habitats.

Along rocky shores can be found larger, more complex seaweeds, such as the red algae described in the previous chapter and members of the brown and green algae. At low tide, it is easy to observe fairly distinct banding patterns or layers that reflect the positions of seaweed species in relation to their ability to survive exposure (Figure 17–1). Seaweeds of this intertidal zone are subjected twice a day to large fluctuations of humidity, temperature, salinity, and light, in addition to the pounding action of the surf and forceful, abrasive water motions. Polar seaweeds must endure months of darkness under the sea ice. Seaweeds are also prey to a host of herbivores as well as microbial pathogens. Their complex biochemistries, structures, and life histories reflect adaptation to these physical and biological challenges.

Anchored offshore beyond the zone of waves, massive brown kelps form forests that provide shelter for a rich diversity of fishes and invertebrate animals, some of which are valued human foods. Many large carnivores, including sea otters and tuna, find food and refuge in kelp beds, such as those occurring off the coast of California. The kelps themselves, along with some red algal species, are harvested by humans for extraction of industrial products (see "Algae and Human Affairs" on page 380).

In this chapter we continue our study of the protists by examining the green and brown algae, as well as three groups of protists that are closely related to the brown algae. Green algae include the smallest known eukaryotes, which are unicellular or colonial members of the phytoplankton. They also include attached filaments, macroscopic sheets, and tissuelike and coenocytic (multinucleate) forms. The brown algae have more complex structures than the green algae. Generally, they are attached to a substrate, and they range from microscopic branched filaments to 50-meter-long giant kelps. Chrysophytes and diatoms include many unicellular and colonial members of marine and freshwater phytoplankton. Oomycetes, formerly thought to be related to the fungi, are now known to be more closely related to the chrysophytes, diatoms, and brown algae. Oomycetes are microscopic, heterotrophic residents of both aquatic and terrestrial habitats. Similarities and differences among these protist groups are summarized in Table 17–1.

The Heterokonts

Electron microscopists have long suspected that oomycetes, chrysophytes, diatoms, brown algae, and certain other groups (not discussed in this book) are closely related, based on the common occurrence of similar flagella. These organisms are known as **heterokonts** (meaning "different flagella") because their flagella differ in length and ornamentation. The flagella occur in pairs, with one flagellum long and ornamented with distinctive hairs (tinsel) and the other shorter and smooth (whiplash), as shown in Figure 3–28. Molecular sequence analyses have confirmed the suspicion, once based only on these unique flagella, that oomycetes, chrysophytes, diatoms, and brown algae are indeed closely related.

Molecular studies have further revealed that: (1) the fungus-like heterokonts (including oomycetes) diverged relatively early; (2) the pigmented heterokonts are derived from a single, common ancestor; and (3) there was an early separation of the pigmented forms into two clades. One of these pigmented lineages consists of the diatoms, the other includes the rest of the pigmented heterokonts.

Oomycetes: Phylum *Oomycota*

The phylum *Oomycota,* with about 700 species, is a distinctive group, commonly called oomycetes. In common with dinoflagellates and many green algae, the cell walls of these organisms are composed largely of cellulose or cellulose-like polymers. The oomycetes range from unicellular to highly branched, coenocytic, and filamentous forms. The latter somewhat resemble the hyphae that are characteristic of the fungi, and for this reason, oomycetes have been grouped with fungi in the past.

Most species of oomycetes can reproduce both sexually and asexually. Asexual reproduction among the oomycetes is by means of motile zoospores, which have two flagella that characterize the heterokonts—one tinsel, which is distinctive, and one whiplash. Sexual reproduction is oogamous, meaning that the female gamete is a relatively large, nonflagellated egg, and the male gamete is noticeably smaller and flagellate (Figure 17–2).

TABLE 17–1 Comparative Summary of Characteristics of *Protista* II

Phylum	Number of Species	Photosynthetic Pigments	Carbohydrate Food Reserve	Flagella	Cell Wall Component	Habitat
Oomycota (including water molds)	700	None	Glycogen	2; in zoospores and male gametes only; apical or lateral; 1 tinsel (two rows of hairs) forward, 1 whiplash behind	Cellulose or cellulose-like	Marine, freshwater, and terrestrial (need water)
Bacillariophyta (diatoms)	100,000	None, or chlorophylls *a* and *c*; carotenoids, mainly fucoxanthin	Chrysolaminarin	None or 1; only in male gametes of centric type; apical; tinsel (two rows of hairs)	Silica	Marine and freshwater
Chrysophyta (chrysophytes)	1000	None, or chlorophylls *a* and *c*; carotenoids, mainly fucoxanthin	Chrysolaminarin	None or 2; apical; tinsel (two rows of hairs) forward, whiplash behind	Wall-less or silica scales or cellulosic fibrils	Predominantly freshwater, a few marine
Phaeophyta (brown algae)	1500	Chlorophylls *a* and *c*; carotenoids, mainly fucoxanthin	Laminarin, mannitol (transported)	2; only in reproductive cells; lateral; tinsel (two rows of hairs) forward, whiplash behind	Cellulose embedded in matrix of mucilaginous algin; plasmodesmata in some	Almost all marine; mostly temperate and polar, flourish in cold ocean waters
Chlorophyta (green algae)	17,000	Chlorophylls *a* and *b*; carotenoids	Starch	None or 2 (or more); apical or subapical; equal or unequal; whiplash (some with hairs)	Glycoproteins, noncellulose polysaccharides or cellulose; plasmodesmata in some	Mostly aquatic, freshwater or marine; many in symbiotic relationships

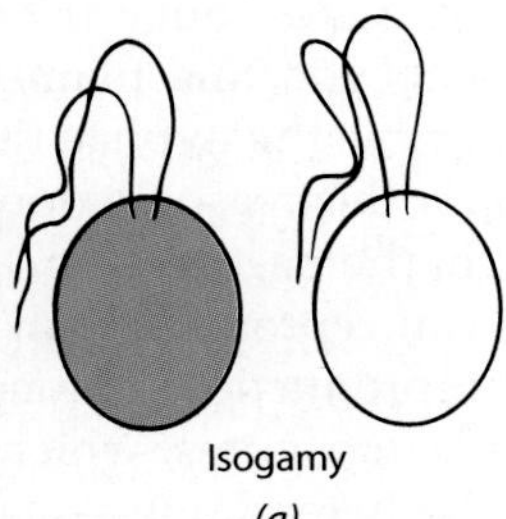

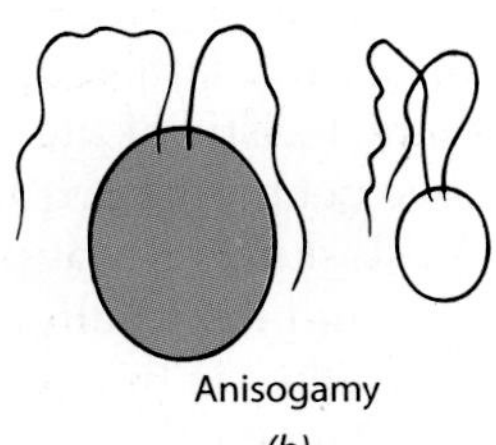

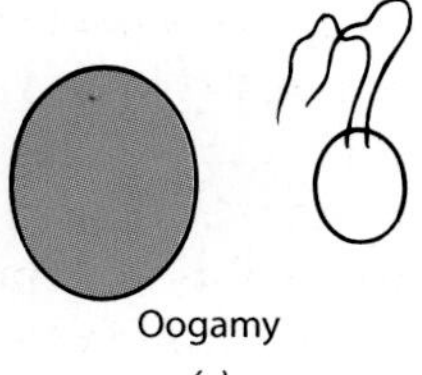

17–2
Types of sexual reproduction, based on gamete form. ***(a)*** *Isogamy—the gametes are equal in size and shape.* ***(b)*** *Anisogamy—one gamete, conventionally termed male, is smaller than the other.* ***(c)*** *Oogamy—one gamete, usually the larger, is nonmotile and female.*

Oogamy is also characteristic of some of the brown and green algae covered in this chapter and of bryophytes and vascular plants, as well as the red algae described in Chapter 16.

In the oomycetes, one to many eggs are produced in a structure called an **oogonium** (plural: oogonia), and an **antheridium** (plural: antheridia) contains numerous male nuclei (Figure 17–3). Fertilization results in the formation of a thick-walled zygote, the **oospore,** for which this phylum is named. The oospore serves as a resting stage that can tolerate stressful conditions. When conditions improve, the oospore germinates. Other heterokonts, as well as many green algae that live in similar habitats, also produce resistant resting stages that result directly from sexual reproduction.

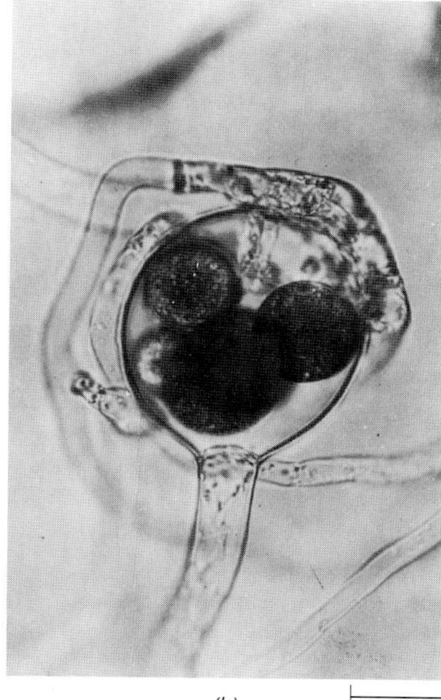

17–3
Achlya ambisexualis, *an oomycete that reproduces both asexually and sexually.* ***(a)*** *Empty sporangium with zoospores encysted about its opening, a distinctive feature of* Achlya. ***(b)*** *Sex organs, showing fertilization tubes extending from the antheridium through the wall of the oogonium to the eggs. Fertilization results in the formation of thick-walled zygotes known as oospores.*

The Water Molds Are Aquatic Oomycetes

One large group of the phylum *Oomycota* is aquatic. The members of this group—the so-called water molds—are abundant in fresh water and are easy to isolate from it. Most of them are saprophytic, living on the remains of dead plants and animals, but a few are parasitic, including species that cause diseases of fishes and fish eggs.

In some water molds, such as *Saprolegnia* (Figure 17–4), sexual reproduction can occur with male and female sex organs borne on the same individual; in other words, these water molds are **homothallic.** Other water molds, such as some species of *Achlya* (Figure 17–3), are **heterothallic**—male and female sex organs are borne by different individuals, or if they are borne on one individual, that individual is genetically incapable of fertilizing itself. Both *Saprolegnia* and *Achlya* can reproduce sexually and asexually.

Some Terrestrial Oomycetes Are Important Plant Pathogens

Another group of oomycetes is primarily terrestrial, although the organisms still form motile zoospores when liquid water is available. Among this group—the order *Peronosporales*—are several forms that are economically important. One of these is *Plasmopara viticola,* which causes downy mildew in grapes. Downy mildew was accidentally introduced into France in the late 1870s on grape stock from the United States that had been imported because of its resistance to other diseases. The mildew soon threatened the entire French wine industry. It was eventually brought under control by a combination of good fortune and skillful observation. French vineyard owners in the vicinity of Medoc customarily put a distasteful mixture of copper sulfate and lime on vines growing along the roadside to discourage passersby from picking the grapes. A professor from the University of Bordeaux who was studying the problem of downy mildew noticed that these plants were free from symptoms of the disease. After conferring with the vineyard owners, the professor prepared his own mixture of chemicals—the Bordeaux mixture—which was made generally available in 1882. The Bordeaux mixture was the first chemical used in the control of a plant disease.

Another economically important member of this group is the genus *Phytophthora* (meaning "plant destroyer"). *Phytophthora,* with about 35 species, is a particularly important plant pathogen that causes widespread destruction of many crops, including cacao, pineapples, tomatoes, rubber, papayas, onions, strawberries, apples, soybeans, tobacco, and citrus. A widespread member of this genus, *Phytophthora cinnamomi,* which occurs in soil, has killed or rendered unproductive millions of avocado trees in southern California and elsewhere. It has also destroyed tens of thousands of hectares of valuable eucalyptus timberland in Australia. The zoospores of *P. cinnamomi* are attracted to the plants they infect by chemicals exuded by the roots. This oomycete also produces resistant spores that can survive for up to six years in moist soil. Extensive breeding efforts are now under way with avocados and other susceptible crops to produce strains resistant to this oomycete.

The best-known species of *Phytophthora,* however, is *Phytophthora infestans* (Figure 17–5), the cause of the late blight of potatoes, which produced the great potato famine of 1846–1847 in Ireland. The population of Ireland, which had risen from 4.5 million people in 1800 to about 8.5 million in 1845, fell to 6.5 million people in 1851 as a result of this famine. About 800,000 people starved to death, and great numbers of people emigrated, many of them to the United States. Virtually the entire Irish potato crop was wiped out in a single week in the summer of 1846. This was a disaster for the Irish peasants, who depended almost entirely on potatoes for their food. Each adult consumed between 4 and 6 kilograms (nearly 9 pounds and more) of potatoes every day—the amount required to supply sufficient quantities of protein to keep alive. Even today, *Phytophthora infestans* is still a serious pest of the potato crop.

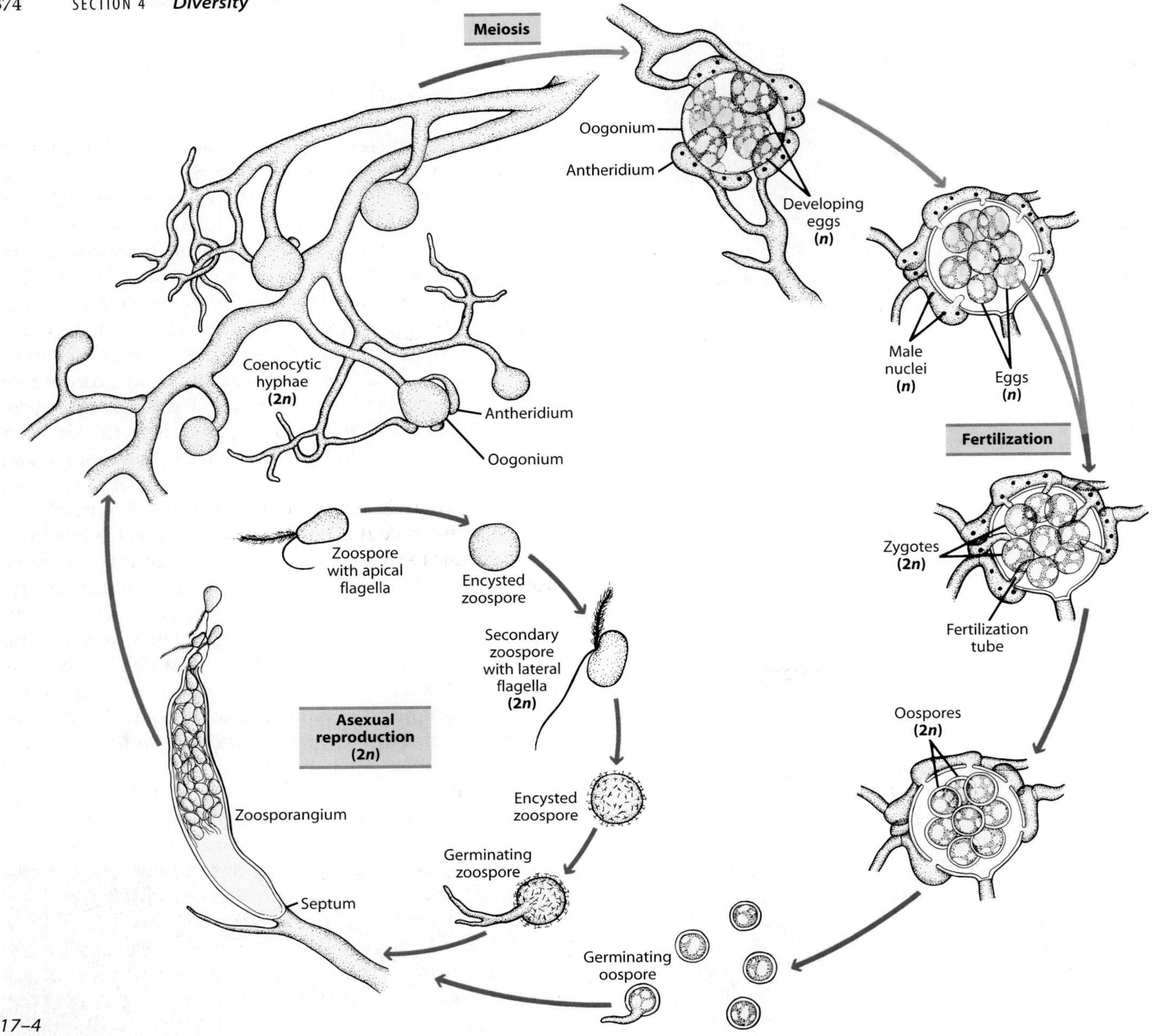

17–4
Life cycle of Saprolegnia, *an oomycete. The mycelium of this water mold is diploid. Reproduction is mainly asexual (lower left). Biflagellated zoospores released from a sporangium, known as a zoosporangium because it produces zoospores, swim for a while and then encyst. Each eventually gives rise to a secondary zoospore, which also encysts and then germinates to produce a new mycelium.*

During sexual reproduction, oogonia and antheridia are formed, in this species, on the same hypha (upper left). Meiosis occurs within these structures. The oogonia are enlarged cells in which a number of spherical eggs are produced. The antheridia develop from the tips of other filaments of the same individual and produce numerous male nuclei. In mating, the antheridia grow toward the oogonia and develop fertilization tubes, which penetrate the oogonia, as seen in Achlya *in Figure 17–3b.*

Male nuclei travel down the fertilization tubes to the female nuclei and fuse with them. Following each nuclear fusion, a thick-walled zygote—the oospore—is produced. On germination, the oospore develops into a hypha, which eventually produces a zoosporangium, beginning the cycle anew.

Also deserving of mention is the genus *Pythium,* the members of which are soil-inhabiting organisms found all over the world. *Pythium* species are the most important causes of **damping-off** diseases of seedlings. They attack a wide variety of economically important crops and can even be a very serious problem in turf grasses used for golf courses and football fields. Some *Pythium* species attack and rot seeds in the field and may destroy seedlings either before they emerge from the soil (preemergence damping-off) or afterward (postemergence damping-off). Postemergence damping-off, in which the young seedling rots near the soil line and then falls over, is a special problem in greenhouses, where large numbers of seedlings are grown in dense stands. Many home gardeners encounter this problem in the spring when renewing their flower beds with recently purchased annuals.

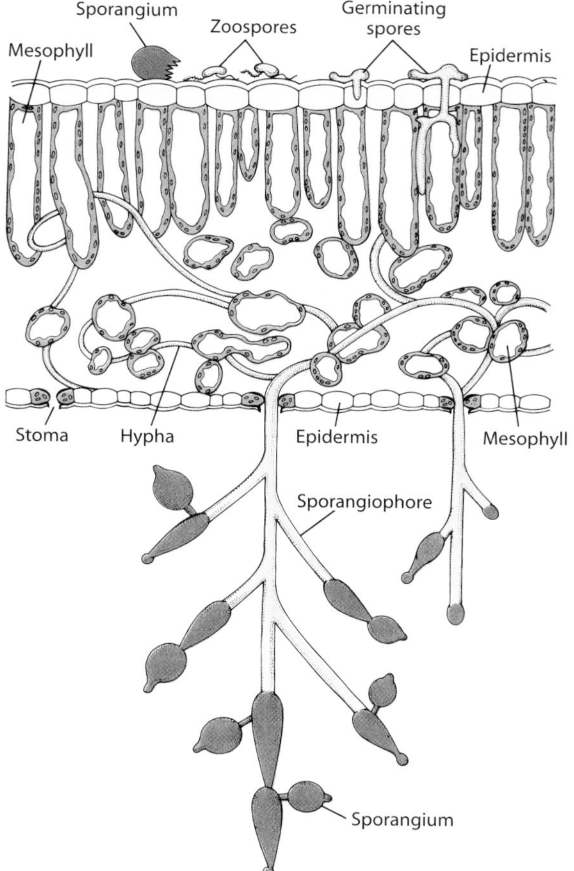

17–5
Phytophthora infestans, *the cause of the late blight of potatoes. The cells of the potato leaf are shown in green. In the presence of water and low temperatures, either of two events can occur. Zoospores can be released from the sporangia and swim to the germination site (as shown here), or the sporangia can germinate directly through a germ tube.*

Diatoms: Phylum *Bacillariophyta*

The diatoms are unicellular or colonial organisms that are exceedingly important components of the phytoplankton (Figure 17–6). It has been estimated that marine planktonic diatoms account for as much as 25 percent of the total primary production on Earth. Diatoms, especially very small forms, account for the greatest biomass and species diversity of phytoplankton in polar waters. Diatoms are a primary source of food for aquatic animals in both marine and freshwater habitats. Species such as *Thalassiosira pseudonana* are commonly used as food in marine cultures, or mariculture, of economically valuable bivalves, such as oysters. Diatoms provide essential carbohydrates, fatty acids, sterols, and vitamins to the animals.

It is estimated that there are 250 genera and 100,000 living species of diatoms, and many phycologists believe that the number may be a great deal higher. There are also thousands of extinct species, known from the remains of silica-containing cell walls. Diatoms first appeared about 250 million years ago. They first became abundant in the fossil record some 100 million years ago, during the Cretaceous period. Many of the fossil species are identical to those still living today, which indicates an unusual persistence through geologic time.

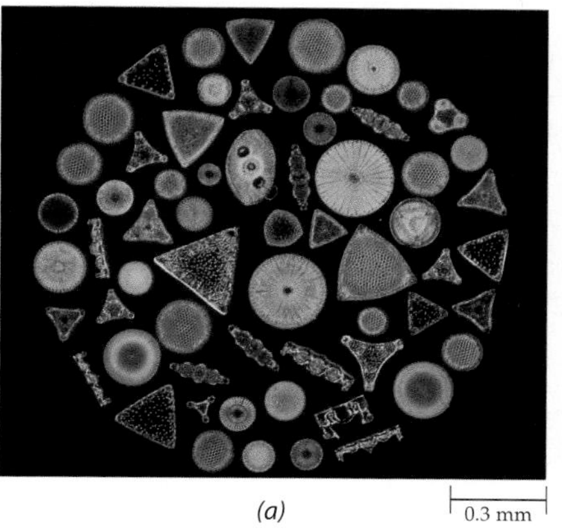
(a) 0.3 mm

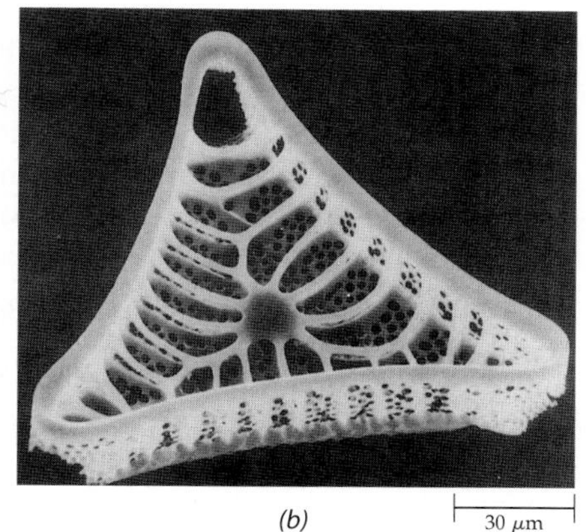
(b) 30 μm

(c)

(d) 5 μm

17–6
Diatoms. ***(a)*** *A selected array of marine diatoms, as seen with a light microscope.* ***(b)*** *Scanning electron micrograph of one half of an* Entogonia *frustule.* ***(c)*** Licmophora flagellata, *a stalked pennate diatom, as seen with a light microscope.* ***(d)*** *Scanning electron micrograph of* Cyclotella meneghiniana, *a centric diatom that occurs in brackish water.*

There are often tremendous numbers of individuals in small areas. For example, over 30 to 50 million individuals of the freshwater genus *Achnanthes* may occur on 1 square centimeter of a submerged rock in the streams of North America. Large numbers of species likewise may occur together. In two small samples of mud from the ocean near Beaufort, North Carolina, for example, 369 species of diatoms were identified. Most species of diatoms occur in plankton, but some are bottom dwellers or grow on other algae or on submerged plants.

The Walls of Diatoms Consist of Two Halves

Diatoms differ from *Chrysophyta* in lacking flagella, except for some male gametes, and in the unique structure of their two-part cell walls. Known as **frustules,** the walls, which are made of polymerized, opaline silica ($SiO_2 \cdot nH_2O$), consist of two overlapping halves. The two halves fit together like a laboratory Petri dish. Electron microscopy has shown that fine tracings on the diatom frustules are actually composed of a large number of minute, intricately shaped depressions, pores, or passageways, some of which connect the living protoplasm within the frustule to the outside environment (Figure 17–6b, d). Species can be distinguished by differences in frustule ornamentation. In most cases, both halves of the frustule have exactly the same ornamentation, but in some cases, the ornamentation may differ.

On the basis of symmetry, two types of diatoms are recognized: the **pennate** diatoms, which are bilaterally symmetrical (Figure 17–6c), and the **centric** diatoms, which are radially symmetrical (Figure 17–6d). Centric diatoms, which have a larger surface-to-volume ratio than pennate ones and consequently float more easily, are more abundant than pennate in large lakes and marine habitats.

Reproduction in Diatoms Is Mainly Asexual, Occurring by Cell Division

When cell division takes place, each daughter cell receives one half of the frustule of its parental cell and constructs a new half (Figure 17–7). As a consequence, one of the two new cells is typically somewhat smaller than the parental cell, and after a long series of cell divisions, the size of the diatoms in the resulting population will often have declined. In some diatom populations, when the size decreases to a critical level, sexual reproduction occurs. The cells that develop from division of the zygote typically regain maximum size for the species. In some other cases, sexual reproduction is triggered by changes in the physical environment.

The sexual life history of diatoms is gametic, like that of animals and certain brown and green seaweeds described later in this chapter. When sexual reproduction occurs in the centric diatoms, it is oogamous. The male gametes, which may have a single, tinsel flagellum, are the only flagellated cells found in diatoms at any stage of their life cycle. In the pennate diatoms, sexual reproduction is isogamous, and both male and female gametes are nonflagellate. Both types of sexual reproduction result in empty frustules that readily sediment (meaning "to sink down"). Researchers have observed that mass sexual reproduction of marine diatoms can result in the formation of layers of silica in southern ocean sediments.

Unfavorable conditions, such as low levels of mineral nutrients, can cause marine coastal or benthic diatoms to form resting stages. The resting cells have heavy frustules, which allow them to sink readily to the bottom if

17–7

Generalized life cycle in a centric diatom. Reproduction in the diatoms is mainly asexual, occurring by cell division. The cell walls, or frustules, in all diatoms are composed of two overlapping portions, one of which encloses the other. When cell division takes place, each daughter cell (bottom left) receives one half of the parental frustule (bottom right) and constructs a new frustule half. The existing half is always the larger part of the silica wall, with the new half fitting inside it. Thus one daughter cell of each new pair tends to be smaller than the parental cell from which it is derived.

In some species, the frustules are expandable and are enlarged by the growing protoplasm within them. In other species, however, the frustules are more rigid. Thus, in a population, the average cell size decreases through successive cell divisions. When the individuals of these species have decreased in size to about 30 percent of the maximum diameter, sexual reproduction may occur (top). Certain cells function as male gametangia, and each produces sperm through meiosis. Other cells function as female gametangia. In these female gametangia, two or three of the four products of meiosis are nonfunctional, so that one or two eggs are produced per cell. This is an example of gametic meiosis (see Figure 9–3b). After fertilization, the resulting auxospore, or zygote, expands to the full size characteristic of the species. The walls formed by the auxospore are often different from those of the asexually reproducing cells of the same species. Once the auxospore is mature, it divides and produces new frustule halves with intricate markings typical of the asexually reproducing cells.

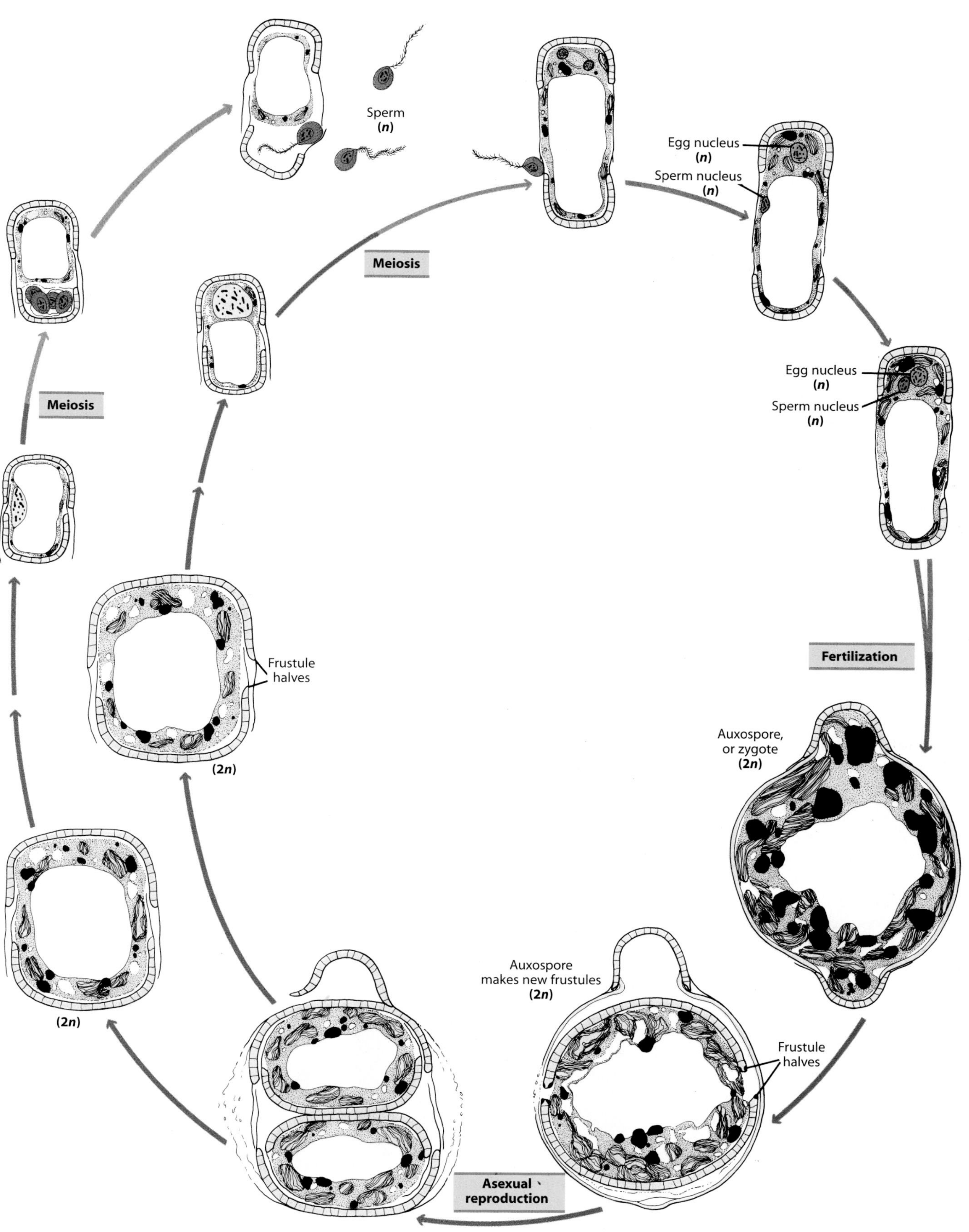
Sperm
(n)
Egg nucleus
(n)
Sperm nucleus
(n)
Meiosis
Meiosis
Egg nucleus
(n)
Sperm nucleus
(n)
Frustule
halves
Fertilization
(2n)
Auxospore,
or zygote
(2n)
Auxospore
makes new frustules
(2n)
Frustule
halves
(2n)
Asexual
reproduction

they are not already there. These cells will germinate when nutrient conditions improve. Diatoms are frequently most abundant in the spring and fall, when upwelling in the oceans or wind-driven turnover of stratified lakes occurs. These processes resuspend sufficient silica for diatom growth. When the silica becomes depleted, diatom blooms give way to dominance by other phytoplankton that do not require silica. Diatoms may bloom beneath the winter ice cover of lakes because herbivorous animals do not actively feed during the cold season.

The silica frustules of diatoms have accumulated in ocean sediments over millions of years, forming the fine, crumbly substance known as diatomaceous earth. This substance is used as an abrasive in silver polish and as a filtering and insulating material. In the Santa Maria, California, oil fields there is a subterranean deposit of diatomaceous earth that is 900 meters thick, and near Lompoc, California, more than 270,000 metric tons of diatomaceous earth are quarried annually for industrial use.

The most conspicuous features within the protoplast of diatoms are the brownish plastids that contain chlorophylls *a* and *c* as well as **fucoxanthin,** a golden-brown carotenoid. There are usually two large plastids in the cells of pennate diatoms, whereas centric diatoms have numerous discoid plastids. Diatom reserve storage materials include lipids and the water-soluble polysaccharide **chrysolaminarin,** which is stored in vacuoles. Chrysolaminarin is similar to the laminarin found in brown algae.

Although most species of diatoms are autotrophic, some are heterotrophic, absorbing dissolved organic carbon. These heterotrophic species are primarily pennate diatoms that live on the bottom of the sea in relatively shallow habitats. A few diatoms are obligate heterotrophs. They lack chlorophyll and thus are not capable of producing their own food through photosynthesis. On the other hand, some diatoms, lacking their characteristic frustules, live symbiotically in large marine protozoa (order *Foraminifera*) and provide organic carbon to their hosts. Certain diatoms are associated with production of the neurotoxin domoic acid, which causes amnesiac shellfish poisoning in humans.

Chrysophytes: Phylum *Chrysophyta*

Chrysophytes are primarily unicellular or colonial organisms that are abundant in fresh and salt water throughout the world (Figure 17–8). There are a few plasmodial, filamentous, and tissuelike forms and about 1000 known species. Some chrysophytes are colorless, whereas others have chlorophylls *a* and *c*, the color of which is largely masked by an abundance of fucoxanthin. The golden color of this pigment gives rise to the name "chrysophyte" (Gk. *chrysos*, "gold," + *phyton*, "plant"). An individual pigmented cell usually contains one or two large chloroplasts. As in the diatoms, the carbohydrate food reserve is chrysolaminarin. It is stored in a vacuole usually found in the posterior of the cell.

Several chrysophytes are known to ingest bacteria and other organic particles. Individual cells of the freshwater, pigmented, motile colony *Dinobryon* can each consume about 36 bacteria per hour. *Dinobryon* and its chrysophyte relatives are the major consumers of bacteria in some of the cooler lakes of North America. *Poterioochromonas* can ingest motile algal cells that are two to three times larger in diameter than itself. Its cell volume can expand as much as 30 times to accommo-

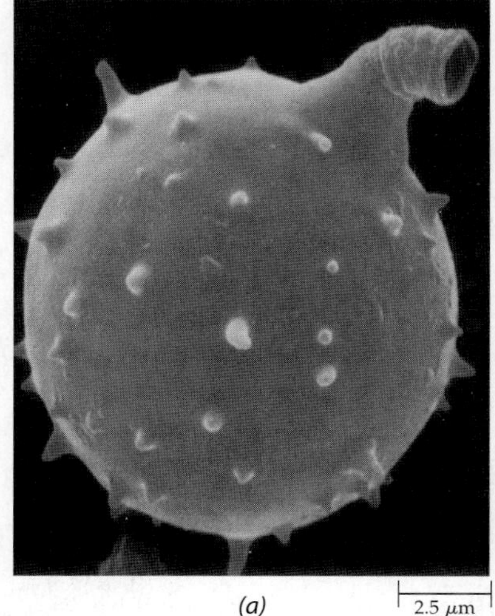

17–8
Representative chrysophytes. ***(a)*** *Scanning electron micrograph of a mature resting cyst of* Dinobryon cylindricum. *The hooked, cylindrical collar contains a pore through which the amoeboid cell emerges when it is ready to germinate. Surface spines represent ornamentation.* ***(b)*** *Scanning electron micrograph of silica scales and flagella of* Synura *cells in a colony. The scales are arranged in very regular overlap patterns in spiral rows.*

date the food it has eaten. *Uroglena americana,* another relative, seems to require prey food as a source of an essential phospholipid. Such forms can be dominant components of the phytoplankton in temperate lakes of low nutrient content (oligotrophic). The ability to consume particulate food provides an advantage under such low-nutrient conditions.

Some chrysophytes have walls made of interwoven cellulosic fibrils that may be impregnated with minerals. Others are wall-less, and some look very much like amoebas with plastids. The members of one group, known as the synurophytes after the motile, colonial genus *Synura* (Figure 17–8b), are covered with overlapping, ornamented silica scales. These are produced inside the cells in vesicles and are then transported to the outside of the cell. The presence of a scaly enclosure prevents these algae from feeding on particles. In cold, acidic lakes, their scales may persist in the sediments and provide useful information about past habitats.

Reproduction in most chrysophytes is asexual and, for some forms, involves zoospore formation. Sexual reproduction is also known for some species. Characteristic resting cysts are commonly formed at the end of the growing season. Sometimes, but not always, cysts are the result of sexual reproduction. In some groups the resting cysts contain silica, which, like silica-containing scales, can sediment and form valuable records of past ecological conditions.

The marine chrysophytes *Heterosigma* and *Aureococcus* have generated toxic "brown tides" that have caused millions of dollars worth of damage to the shellfish and salmon fisheries. Some freshwater chrysophytes can also form blooms and are blamed for unpleasant tastes and odors that arise from their excretion of organic compounds into drinking water.

Brown Algae: Phylum *Phaeophyta*

The brown algae, an almost entirely marine group, include the most conspicuous seaweeds of temperate, boreal (or northern), and polar waters. Although there are only about 1500 species, the brown algae dominate rocky shores throughout the cooler regions of the world (Figure 17–9). Most people have observed seashore rocks covered with rockweeds, the common name for members of the brown algal order *Fucales.* The larger brown algae of the order *Laminariales,* a number of which form extensive beds offshore, are called kelps. In clear water, brown algae flourish from low-tide level to a depth of 20 to 30 meters. On gently sloping shores, they may extend 5 to 10 kilometers from the coastline.

(a)

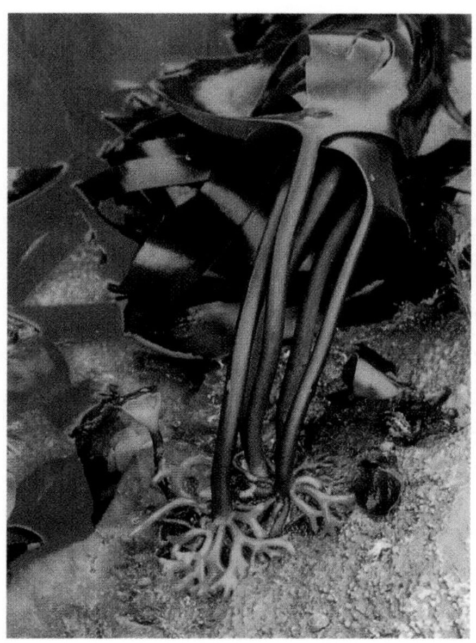

(b)

(c)

17–9

Brown algae. ***(a)*** *Bull kelp* (Durvillea antarctica) *exposed at low tide off a rocky coast in New Zealand.* ***(b)*** *Detail of the kelp* Laminaria, *showing holdfasts, stipes, and the bases of several fronds.* ***(c)*** *Rockweed* (Fucus vesiculosus) *densely covers many rocky shores that are exposed at low tide. When submerged, the air-filled bladders on the blades carry them up toward the light. Photosynthetic rates of frequently exposed marine algae are one to seven times as great in air as in water, whereas the rates are higher in water for those rarely exposed. This difference accounts in part for the vertical distribution of seaweeds in intertidal areas.*

Algae and Human Affairs

People of various parts of the world, especially in the Far East, eat both red and brown algae. Kelps ("kombu") are eaten regularly as vegetables in China and Japan. They are sometimes cultivated but are mainly harvested from natural populations. Porphyra *("nori"), a red alga, is eaten by many inhabitants of the north Pacific Basin and has been cultivated in Japan, Korea, and China for centuries (page 359). Various other red algae are eaten on the islands of the Pacific and also on the shores of the North Atlantic. Seaweeds are generally not of high nutritive value as a source of carbohydrates because humans, like most other animals, lack the enzymes necessary to break down most of the materials in cell walls, such as cellulose and the protein-rich intercellular matrix. Seaweeds do, however, provide necessary salts, as well as a number of important vitamins and trace elements, and so are valuable supplementary foods. Some green algae, such as* Ulva*, or sea lettuce, are also eaten as greens.*

In many northern temperate regions, kelp is harvested for its ash, which is rich in sodium and potassium salts and is therefore valued for industrial processes. Kelp is also harvested regionally and used directly for fertilizer.

Alginates, which are a group of substances derived from kelps, such as Macrocystis*, are widely used as thickening agents and colloid stabilizers in the food, textile, cosmetic, pharmaceutical, paper, and welding industries. Off the west coast of the United States,* Macrocystis *kelp beds can be harvested several times a year by cropping them just below the water surface.*

One of the most useful, direct commercial applications of any alga is the preparation of agar, which is made from a mucilaginous material extracted from the cell walls of a number of genera of red algae. Agar is used to make the capsules that contain vitamins and drugs, as well as for dental-impression material, as a base for cosmetics, and as a culture medium for bacteria and other microorganisms. Purified agarose is the gel often used in electrophoresis in biochemical experimentation. Agar is also employed as an anti-drying agent in bakery goods, in the preparation of rapid-setting jellies and desserts, and as a temporary preservative for meat and fish in tropical regions. Agar is produced in many parts of the world, but Japan is the principal manufacturer. A similar algal colloid called carrageenan is used in preference to agar for the stabilization of emulsions, such as paints, cosmetics, and dairy products. In the Philippines, the red alga Eucheuma *is cultivated commercially as a source of carrageenan.*

(a)

(b)

(c)

(a) A forest of giant kelp (Macrocystis pyrifera) *growing off the coast of California. (b) Harvesting the seaweed* Nudaria *by hand from submerged ropes in Japan. (c) A kelp harvester operating in the nearshore waters of California. Cutting racks at the rear of the ship are lowered 3 meters below the water's surface, and the ship moves backward, cutting through the kelp canopy. The harvested kelp is moved via conveyor belts to a collecting bin on board the ship.*

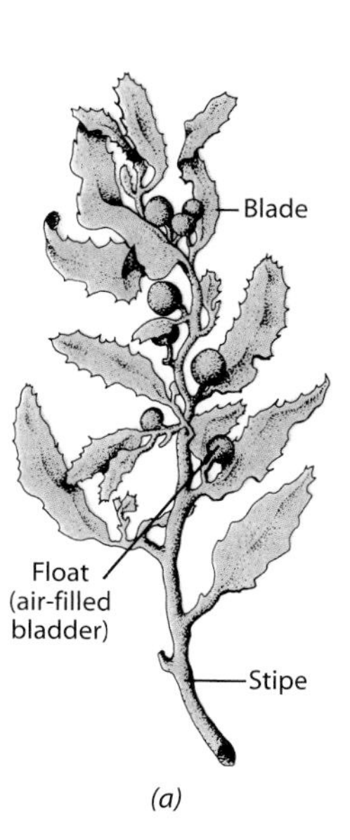

(a)

(b)

17–10
(a) *The brown alga* Sargassum *has a complex pattern of organization.* Sargassum, *like* Fucus, *is a member of the order* Fucales *and has a life cycle like that shown in Figure 17–14.* ***(b)*** *Two species of this genus, which lack sexual reproduction, form the great free-floating masses of the Sargasso Sea.*

Even in the tropics, where the brown algae are less common, there are immense floating masses of *Sargassum* (Figure 17–10) in such areas as the Sargasso Sea (so named because of the abundance of *Sargassum*) in the Atlantic Ocean northeast of the Caribbean islands. *Sargassum muticum* and some other brown algae can form nuisance growths when introduced into non-native areas and can seriously interfere with mariculture operations. *Sargassum* can also outcompete and replace members of the *Laminariales* and *Fucales,* which are regarded as critical or keystone species in their communities.

The Basic Form of a Brown Alga Is a Thallus

Though they are a monophyletic group, brown algae range in size from microscopic forms to the largest of all seaweeds, the kelps, which are as much as 60 meters long and weigh more than 300 kilograms. The basic form of a brown alga is a **thallus** (plural: thalli)—a simple, relatively undifferentiated vegetative body. The thalli range in complexity from simple branched filaments (Figure 17–11), to aggregations of branched filaments that are called pseudoparenchyma because they look tissuelike, to authentic tissues (Figure 17–12). As in some green algae and plants, adjacent cells are typically linked by plasmodesmata. Unlike plant plasmodesmata, those of brown algae do not seem to have desmotubules connecting the endoplasmic reticulum of adjacent cells (see page 87).

In the past, brown algae have been classified into orders on the basis of thallus structure. However, recent molecular and cellular studies have revealed that thallus organization is not a good indicator of brown algal relationships. Closely related species of brown algae may have quite different thallus organization, and unrelated genera may appear structurally similar. The expression of similar morphologies by unrelated species is a response to similar selective pressures. This phenomenon, which also occurs in the green algae and the angiosperms, is known as convergent evolution (see page 266).

The Pigment Fucoxanthin Gives the Brown Algae Their Characteristic Color

Brown algal cells typically contain numerous disk-shaped, golden-brown plastids that are similar both biochemically and structurally to the plastids of chrysophytes and diatoms, with which they probably had a common origin. In addition to chlorophylls *a* and *c*, the chloroplasts of brown algae also contain various carotenoids, including an abundance of the xanthophyll fucoxanthin, which gives the members of this phylum their characteristic dark-brown or olive-green color. The reserve storage material in brown algae is the carbohydrate laminarin, which is stored in vacuoles. Molecular analyses suggest that there are two major lineages of brown algae: those with starch-producing pyrenoids in

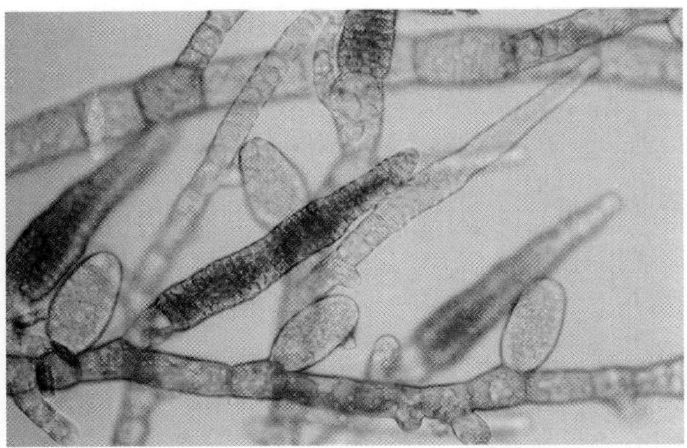

17–11
Ectocarpus, *a brown alga that has simple branched filaments. This micrograph of* E. siliculosus *shows unilocular sporangia (the short, rounded, light-colored structures) and plurilocular sporangia (the longer, dark-colored structures), which are borne on sporophytes. Meiosis, resulting in haploid zoospores, takes place within the unilocular sporangia. Diploid zoospores are formed in the plurilocular sporangia.* Ectocarpus *occurs in shallow water and estuaries throughout the world, from cold Arctic and Antarctic waters to the tropics.*

their plastids, including *Ectocarpus* (Figure 17–11), and those without pyrenoids, such as *Laminaria* and its relatives. These two groups also consistently differ in the structure of their sperm. Oogamy and a heteromorphic type of life cycle occur only in the group that lacks pyrenoids.

The Kelps Have the Most Highly Differentiated Bodies among the Algae

Large kelps, such as *Laminaria,* are differentiated into regions known as the holdfast, stipe, and blade, with a meristematic region located between the blade and stipe (Figure 17–9b). The pattern of growth resulting from this type of meristematic activity is particularly important in the commercial use of *Macrocystis,* which is harvested along the California coast. When the older blades are harvested at the surface by kelp-cutting boats, *Macrocystis* is able to regenerate new blades. Giant kelps, such as *Macrocystis* and *Nereocystis,* may be more than 60 meters long. They grow very rapidly so that a considerable amount of material is available for harvest. One of the most important products derived from the kelps is a mucilaginous intercellular material called **algin,** which is important as a stabilizer and emulsifier for some foods and for paint and as a coating for paper. Algin, together with cellulose in inner cell wall layers, provides the flexibility and toughness that allow seaweeds to withstand mechanical stress imposed by waves and currents. Algin, which also helps to reduce drying when the seaweeds are exposed by low tides, increases buoyancy and helps to slough off organisms that attempt to colonize the algal blades.

The internal structure of kelps is complex. Some of them have, in the center of the stipe, elongated cells that are modified for food conduction. These cells resemble the food-conducting cells in the phloem of vascular plants, including the presence of sieve plates (Figure 17–12b). The cells are able to conduct food material rapidly—at rates as high as 60 centimeters per hour—from the blades at the water surface to the poorly illuminated stipe and holdfast regions far below. Lateral translocation from the outer photosynthetic layers to the inner cells takes place in many relatively thick kelps. **Mannitol** is the primary carbohydrate that is translocated, along with amino acids.

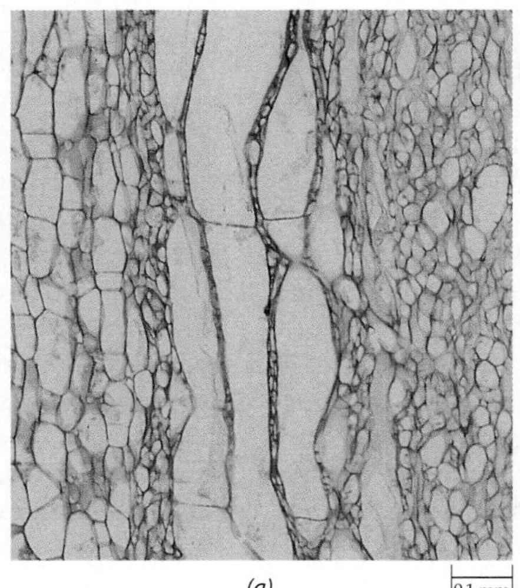

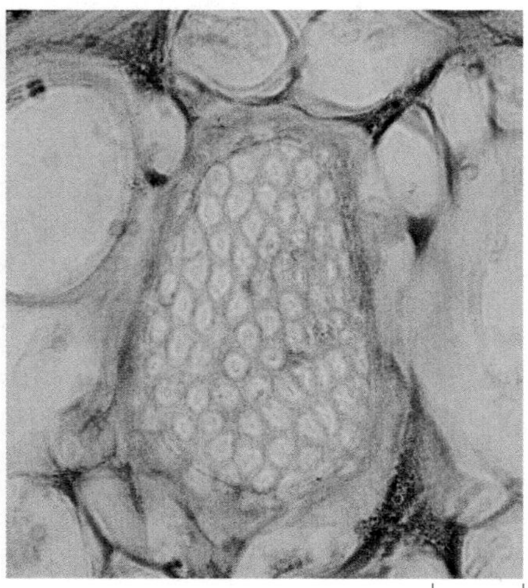

17–12
Some brown algae, such as the giant kelp Macrocystis integrifolia, *have evolved sieve tubes comparable to those found in food-conducting tissue of vascular plants.* ***(a)*** *Longitudinal section of part of a stipe, with sieve tubes, the relatively wide elements in the middle of the micrograph. The sieve-tube members are joined end-on-end by the sieve plates, which appear as narrow cross walls here.* ***(b)*** *Cross section showing a sieve plate.*

The rockweed *Fucus* (Figure 17–9c) is a dichotomously branching brown alga that has air bladders near the ends of its blades. The pattern of differentiation of *Fucus* otherwise resembles that of the kelps. *Sargassum* (Figure 17–10) is related to *Fucus*. Some species of *Sargassum* remain attached, whereas in others the individuals form floating masses in which the holdfasts have been lost. Both forms occur within certain species. *Fucus* and *Sargassum* and some other brown algae grow by means of repeated divisions from a single apical cell and not from a meristem located within the body, as is characteristic of the kelps.

The Life Cycles of Most Brown Algae Involve Sporic Meiosis

The life cycles of most brown algae involve an alternation of generations and, therefore, sporic meiosis (see Figure 9–3c). The gametophytes of the more primitive brown algae, such as *Ectocarpus,* produce multicellular reproductive structures called **plurilocular gametangia.** They may function as male or female gametangia or produce flagellated haploid spores that give rise to new gametophytes. The diploid sporophytes produce both **plurilocular** and **unilocular sporangia** (Figure 17–11). The plurilocular sporangia form diploid zoospores that produce new sporophytes. Meiosis takes place within the unilocular sporangia, producing haploid zoospores that germinate to produce gametophytes. Unilocular sporangia, along with algin and plasmodesmata, are defining features of the brown algae.

In *Ectocarpus,* the gametophyte and sporophyte are similar in size and appearance (isomorphic). Many of the larger brown algae, including the kelps, undergo an alternation of heteromorphic generations—a large sporophyte and a microscopic gametophyte, as in the common kelp *Laminaria* (Figure 17–13). In *Laminaria,* the unilocular sporangia are produced on the surface of the mature blades. Half of the zoospores that the sporangia produce have the potential to grow into male gametophytes and half into female gametophytes. According to one hypothesis, the plurilocular gametangia borne on these gametophytes have become modified during the course of their evolution into one-celled antheridia and one-celled oogonia. Each antheridium releases a single sperm, and each oogonium contains a single egg. The fertilized egg in *Laminaria* remains attached to the female gametophyte and develops into a new sporophyte. In several genera of brown algae, the female gametes attract the male gametes by organic compounds.

Fucus and its close relatives have a gametic life cycle (see Figure 9–3b and Figure 17–14), as do the diatoms and certain green seaweeds. Understanding the evolutionary pressures that stimulated the origin of gametic life cycles in these protists may illuminate the early appearance of the gametic life cycle in our own metazoan (Gk. *meta,* "between," + *zōion,* "animal") lineage. *Fucus* and other brown seaweeds may contain large amounts of phenolic compounds that discourage herbivores. Other brown seaweeds tend to produce terpenes for the same purpose (page 33). In some cases, these compounds also have anti-microbial or anti-tumor activities, prompting investigations of their potential use in human medicine.

Green Algae: Phylum *Chlorophyta*

The green algae, including at least 17,000 species, are diverse in structure and life history. Although most green algae are aquatic, they are found in a wide variety of habitats, including on the surface of snow (Figure 17–15), on tree trunks, in the soil, and in symbiotic associations with lichens, freshwater protozoa, sponges, and coelenterates. Some green algae—such as species of the unicellular genera *Chlamydomonas* and *Chloromonas,* found growing on the surface of snow, and *Trentepohlia,* a filamentous alga that grows on rocks and tree trunks or branches—produce large amounts of carotenoids that function as a shield against intense light. These accessory pigments give the algae an orange, red, or rust color. Most aquatic green algae are found in fresh water, but a number of groups are marine. Many green algae are microscopic, although some of the marine species are large. *Codium magnum* of Mexico, for example, sometimes attains a breadth of 25 centimeters and a length of more than 8 meters.

The *Chlorophyta* resemble plants in several important characteristics. They contain chlorophylls *a* and *b,* and they store starch, their food reserve, inside plastids. Green algae and plants are the only two groups to do so. Some, but not all, green algae are like plants in having firm cell walls composed of cellulose, hemicelluloses, and pectic substances. In addition, the microscopic structure of the flagellated reproductive cells in some green algae resembles that of plant sperm. Many other biochemical details, such as the production of phytochrome (see Chapter 29) in at least two genera of green algae, also indicate that there is a very close relationship between the two groups. For these reasons, green algae are believed to be the protist group from which plants evolved.

Molecular and cellular studies of green algae have revealed that at least two multicellular lineages have evolved separately from unicellular flagellates. Each lineage includes a diverse mixture of morphological types. Unicellular forms, colonies, unbranched filaments, branched filaments, and other forms may be found.

Traditional classification systems group green algae according to their outward structure. Unicellular flagel-

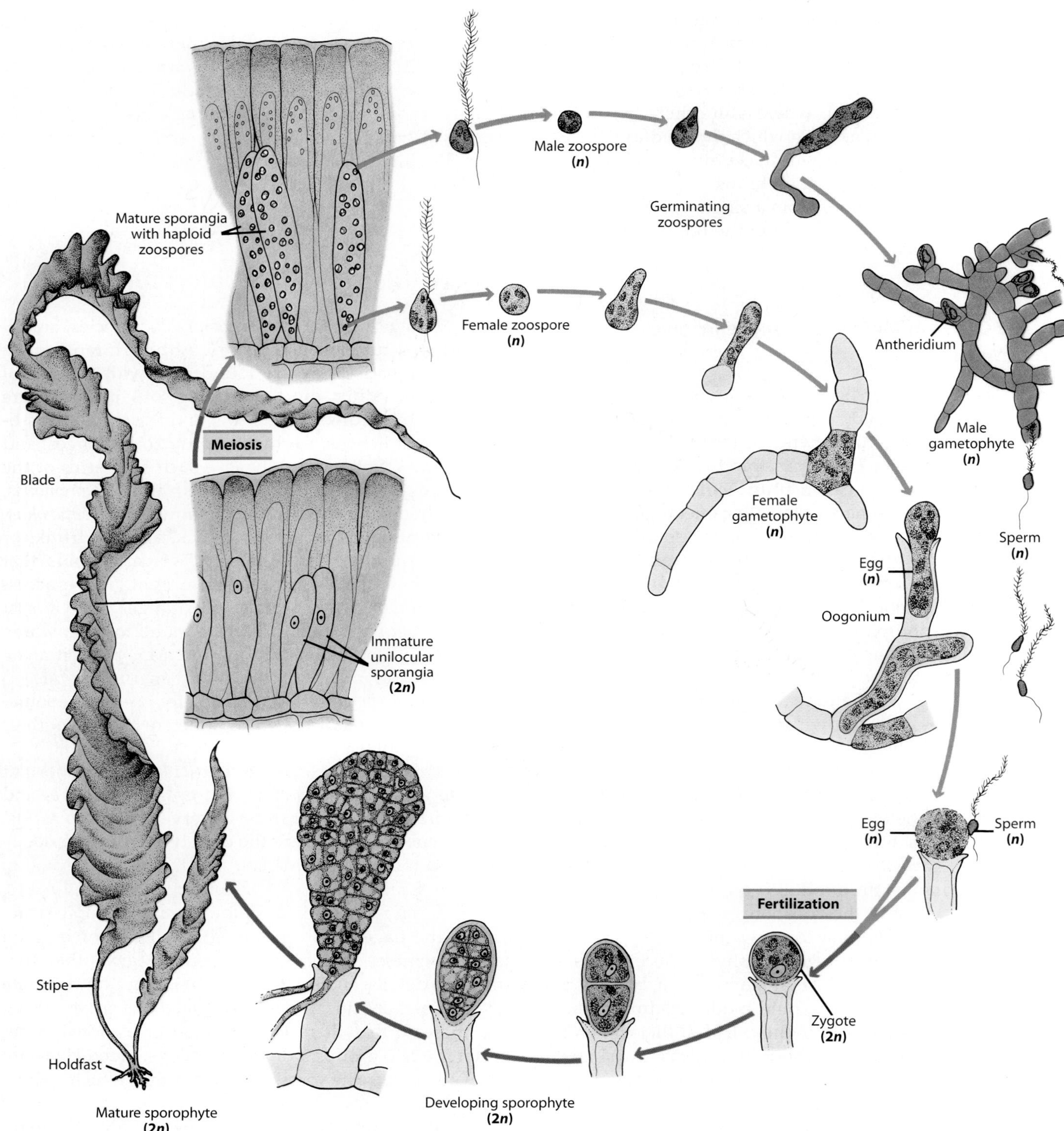

17–13
Life cycle of the kelp Laminaria, *an example of sporic meiosis. Like many of the brown algae,* Laminaria *has an alternation of heteromorphic generations in which the sporophyte is conspicuous. Motile haploid zoospores are produced in the sporangia following meiosis (upper left). From these zoospores grow the microscopic, filamentous gametophytes, which in turn produce the motile sperm and nonmotile eggs. In some other brown algae, the sporophyte and gametophyte are often similar; they have an alternation of isomorphic generations.*

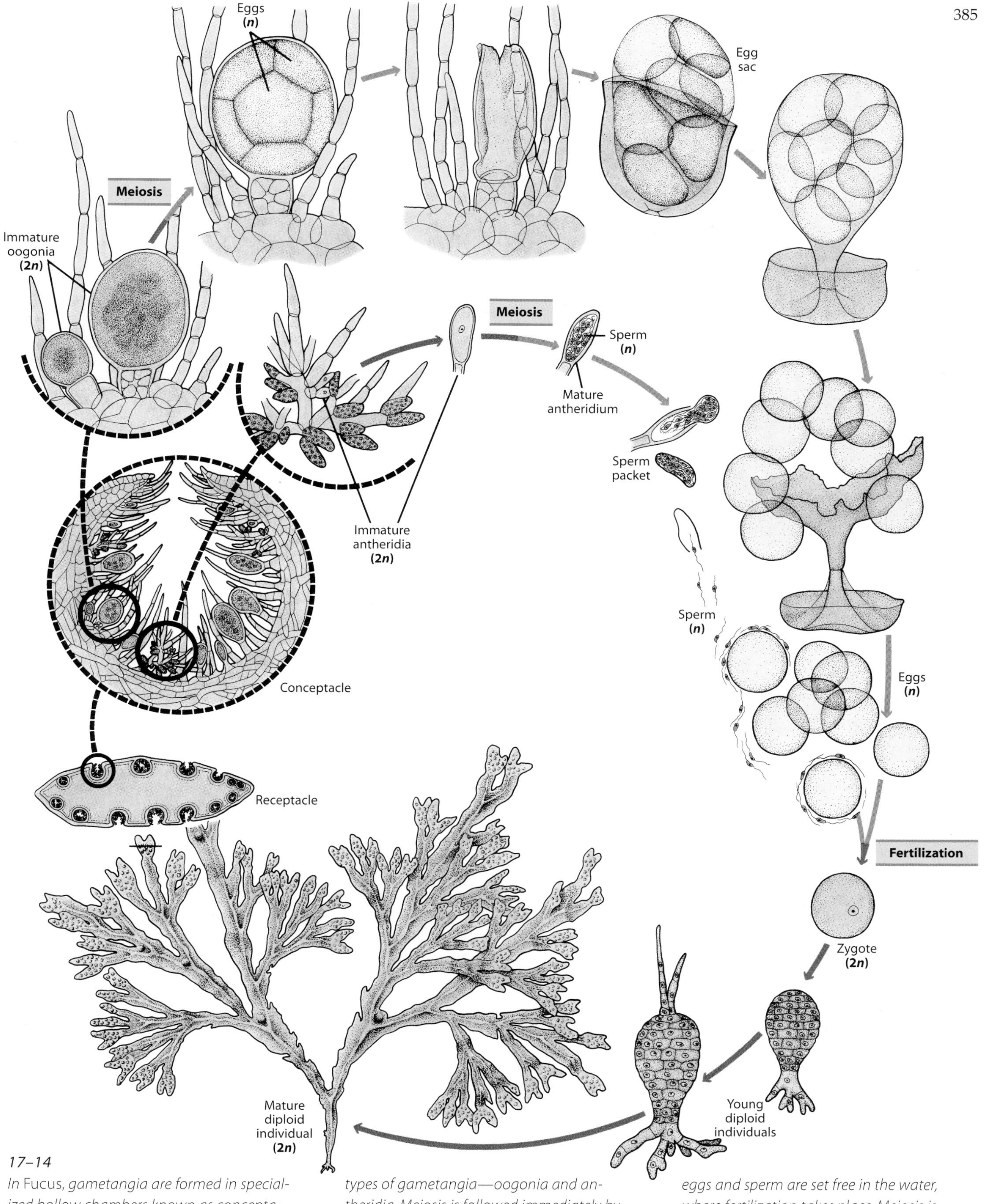

17–14

In Fucus, *gametangia are formed in specialized hollow chambers known as conceptacles, which are found in fertile areas called receptacles at the tips of the branches of diploid individuals (lower left). There are two types of gametangia—oogonia and antheridia. Meiosis is followed immediately by mitosis to give rise to 8 eggs per oogonium and 64 sperm per antheridium. Eventually the eggs and sperm are set free in the water, where fertilization takes place. Meiosis is gametic, and the zygote grows directly into the new diploid individual.*

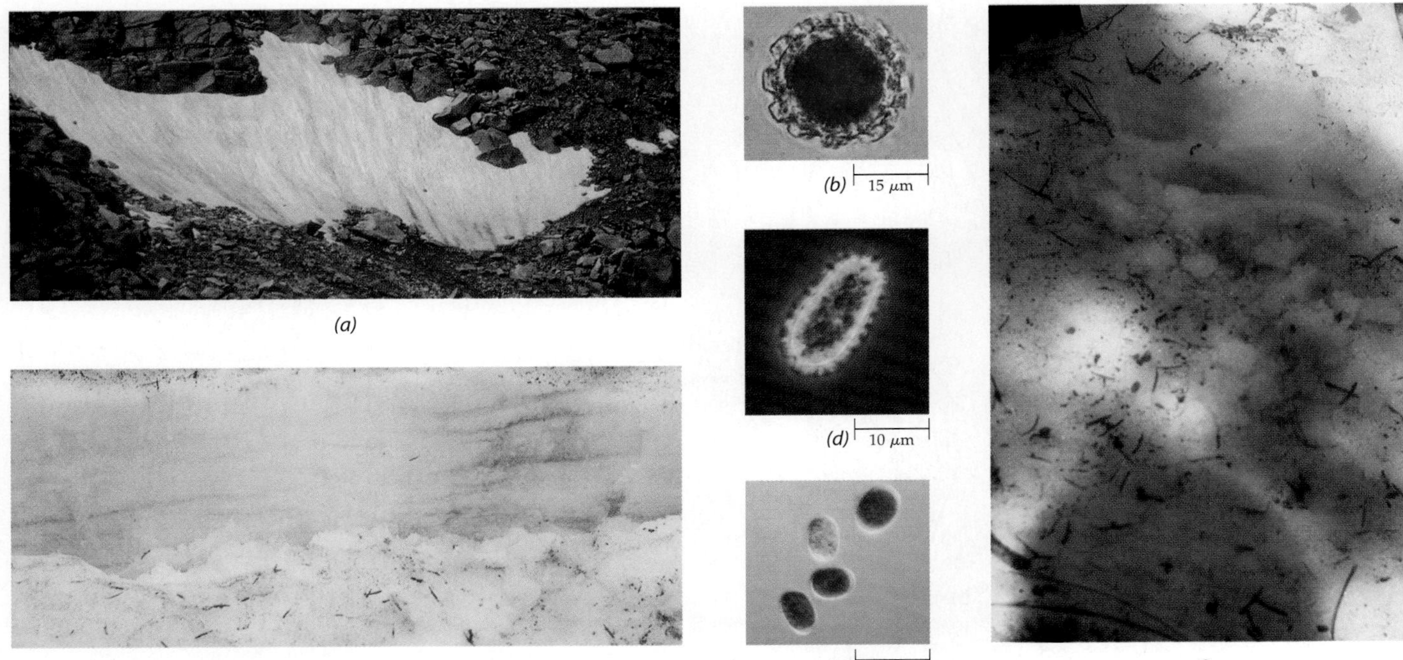

17–15
Snow algae. Snow algae are unique because of their tolerance to temperature extremes, acidity, high levels of irradiation, and minimal nutrients for growth. ***(a)*** *In many parts of the world, as in alpine areas from northern Mexico to Alaska, the presence of large numbers of snow algae produces "red snow" during the summer. This photograph was taken near Beartooth Pass in Montana.* ***(b)*** *Dormant zygote of the snow alga* Chlamydomonas nivalis. *The red color results from carotenoids that serve to protect the chlorophyll in the zygote.*

(c) *Green snow occurs just below the surface, usually near tree canopies in alpine forests. It is widespread, occurring as far south as Arizona and as far north as Alaska and Quebec. This photograph was taken at Cayuse Pass, Mt. Rainier National Park, Washington.* ***(d)*** *Resting zygote of the alga* Chloromonas brevispina, *found in green snow.*

(e) *Three brilliant orange resting zygotes of* Chloromonas granulosa, *the alga responsible for orange snow. The single yellow zygote will change to orange as it matures.* ***(f)*** *Orange snow, associated with trees in forested alpine regions, receives more irradiation than green snow and less than red snow. Orange snow, which is not as common as the other two kinds of colored snow, occurs from Arizona and New Mexico to Alaska. This photograph was taken at Bill Williams Mountain in Arizona.*

lates are grouped together, filamentous types are grouped together, and so forth. As we observed earlier in the brown algae, related green algae cannot always be recognized by their outward structure. However, evidence of a relationship is revealed by ultrastructural studies of mitosis, cytokinesis, and reproductive cells, as well as molecular similarities. This new information has resulted in a new systematic alignment of the green algae into several classes, three of which—the *Chlorophyceae*, the *Ulvophyceae*, and the *Charophyceae*—are discussed in this book.

Cell Division and Motile Cell Differences Exist among Classes of Green Algae

The members of the largest class of green algae, the freshwater *Chlorophyceae*, have a unique mode of cytokinesis involving a **phycoplast** (Figure 17–16). In these algae, the daughter nuclei move toward one another as the nonpersistent mitotic spindle collapses, and a new system of microtubules, the phycoplast, develops parallel to the plane of cell division. Presumably, the role of the phycoplast is to ensure that the **cleavage furrow,** resulting from infolding of the plasma membrane, will pass between the two daughter nuclei. The nuclear envelope persists throughout mitosis. In motile cells of the *Chlorophyceae*, there is a cross-shaped pattern of four narrow bands of microtubules known as **flagellar roots,** which are associated with the basal bodies (centrioles) of the flagella (Figure 17–17).

In other green algal classes, spindles may remain present throughout cytokinesis until they are disrupted, either by furrowing or by growth of a cell plate. A cell plate originates in the central region of the cell and grows outward to its margins. Some of the members of the class *Charophyceae* produce a new cytokinetic micro-

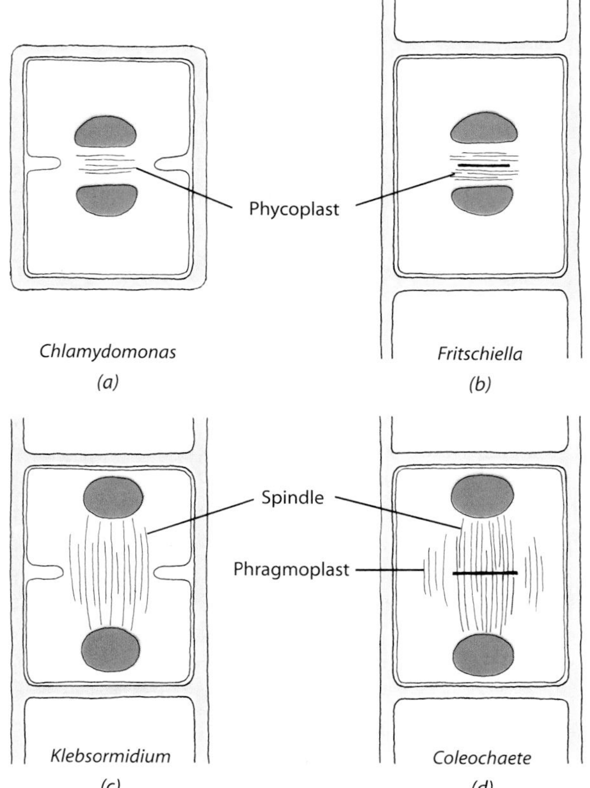

17–16
Cytokinesis in two classes of the phylum Chlorophyta. ***(a), (b)*** *In the class* Chlorophyceae, *the mitotic spindle is nonpersistent and the daughter nuclei, which are relatively near one another, are separated by a phycoplast. Cytokinesis in* Chlamydomonas *is by furrowing, while cytokinesis in* Fritschiella *occurs by cell plate formation.* ***(c)*** *In simpler members of the class* Charophyceae, *such as* Klebsormidium, *the mitotic spindle is persistent and the daughter nuclei are relatively far apart. Cytokinesis occurs by furrowing.* ***(d)*** *Advanced charophytes like* Coleochaete *and* Chara *have a plantlike phragmoplast, and cytokinesis occurs by cell plate formation as in plants.* Ulvophyceae, *like* Charophyceae, *also have a persistent spindle, but do not have a phragmoplast or a cell plate.*

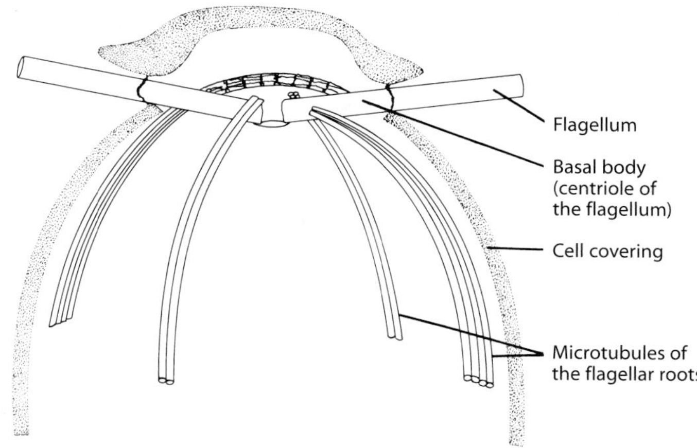

17–17
Diagram of the cross-shaped arrangement of four narrow bands of microtubules known as flagellar roots. These flagellar roots are associated with the flagellar basal bodies (centrioles) and are characteristic of green algae of the class Chlorophyceae.

tubular system, the **phragmoplast,** which is nearly identical to that present in plants. The microtubules in a phragmoplast are oriented perpendicular to the plane of cell division (Figure 17–16).

The flagellated cells of the *Charophyceae* are distinct from those of the other classes in that they have an asymmetrical flagellar root system of microtubules (Figure 17–18). One role of the flagellar root system is to provide the means by which a flagellum is anchored in place. Quite often, an entity known as a multilayered structure is associated with one of the flagellar roots. The type of flagellar root is often an important taxonomic character. The flagellar root system of the *Charophyceae,* with its multilayered structure, is very similar to that found in the sperm of bryophytes and some vascular plants. For these and other reasons, including biochemical and molecular similarities, *Charophyceae* is the group

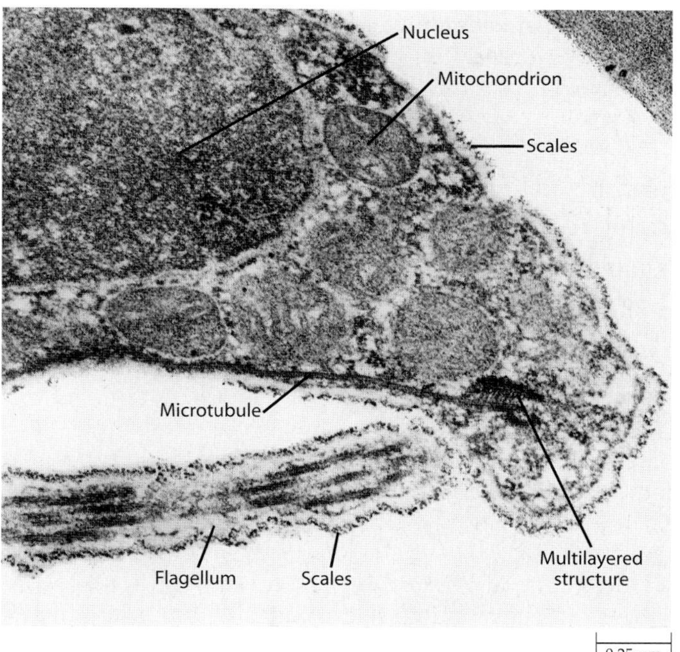

17–18
Electron micrograph of the anterior portion of a motile sperm of the green alga Coleochaete *(class* Charophyceae*). Shown here is the multilayered structure, which is associated with the flagellar root system at the base of the flagellum. The multilayered structure is also characteristic of the sperm of bryophytes and some vascular plants, and it is one of the features linking these plants with* Charophyceae, *their ancestors. As seen here, a layer of microtubules extends from the multilayered structure down into the posterior end of the cell and serves as a cytoskeleton for these wall-less cells. The flagellar and plasma membranes are covered by a layer of small scales.*

of living green algae considered to be closest to the ancestry of bryophytes and vascular plants.

Class *Chlorophyceae,* the Chlorophytes, Consists of Mainly Freshwater Species

The class *Chlorophyceae* includes flagellated and nonflagellated unicellular algae, motile and nonmotile colonial algae, filamentous algae, and algae consisting of flat sheets of cells. The members of this class live mainly in fresh water, although a few unicellular, planktonic species occur in coastal marine waters. A few *Chlorophyceae* are essentially terrestrial, occurring in habitats such as on snow (Figure 17–15), in soil, or on wood.

Chlamydomonas* Is an Example of a Motile Unicellular *Chlorophyceae The common freshwater green alga *Chlamydomonas* (Figure 17–19), a unicellular form with two equal flagella, has been widely used as a model system for molecular studies of the genes regulating photosynthesis and other cell processes. Molecular analyses have revealed that *Chlamydomonas* is a polyphyletic group; that is, it consists of several distinct lineages, all of which coincidentally are unicellular with two equal flagella.

The chloroplast of *Chlamydomonas,* in having a red photosensitive eyespot, or stigma, that aids in the detection of light, is similar to that of many other green flagellates and of zoospores of multicellular green algae. The chloroplast also contains a pyrenoid, which is typically surrounded by a shell of starch. Similar pyrenoids occur in many other green algal species. The uninucleate protoplast is surrounded by a thin glycoproteinaceous (carbohydrate-protein) cell wall, inside which is the plasma membrane. There is no cellulose in the cell wall of *Chlamydomonas.* At the anterior end of the cell, there are two contractile vacuoles, which collect excess water and ultimately discharge it from the cell.

Chlamydomonas reproduces both sexually and asexually. During asexual reproduction, the haploid nucleus usually divides by mitosis to produce up to 16 daughter cells within the parent cell wall. Each cell then secretes a wall around itself and develops flagella. The cells secrete an enzyme that breaks down the parental wall, and the daughter cells can then escape, although fully formed daughter cells are often retained for some time within the parent cell wall.

Sexual reproduction in *Chlamydomonas* involves the fusion of individuals belonging to different mating types (Figure 17–20). The vegetative cells are induced to form gametes by nitrogen starvation. The gametes, which re-

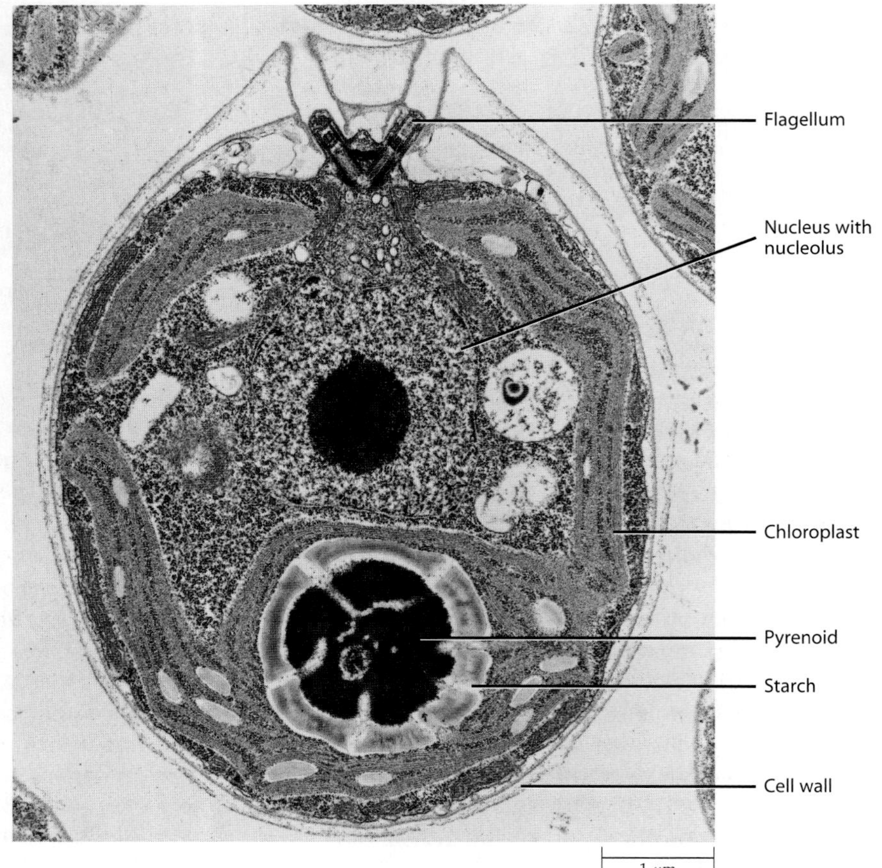

17–19
Chlamydomonas, *a unicellular green alga. The stigma was not preserved in this cell, and only the bases of the flagella can be seen in this electron micrograph. The nucleus of this uninucleate flagellate contains a prominent nucleolus. A shell of starch surrounds the pyrenoid, which is located within the chloroplast.*

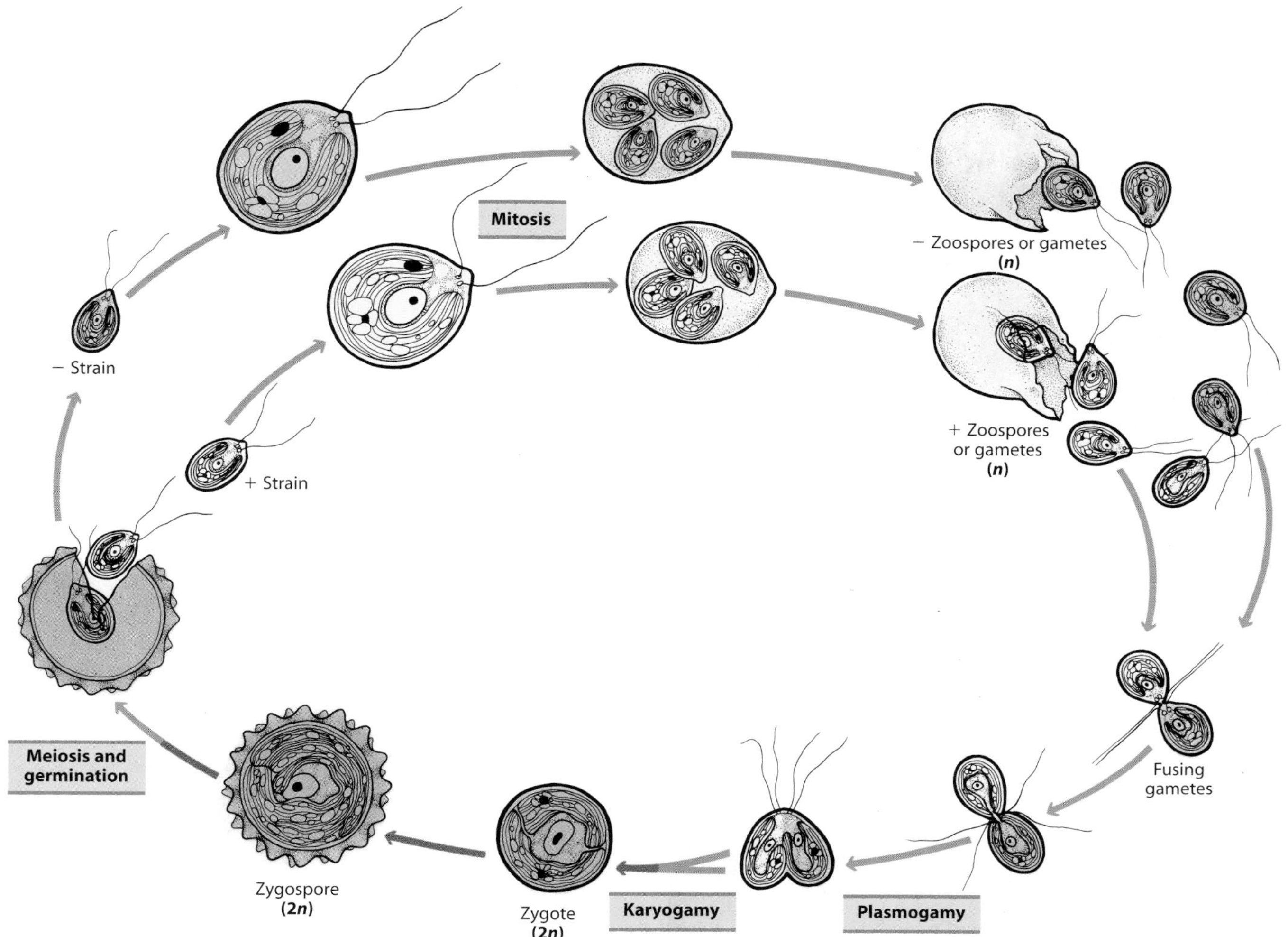

17–20
Life cycle of Chlamydomonas. *Sexual reproduction occurs when gametes of different mating types come together, cohering at first by their flagellar membranes and then by a slender protoplasmic thread—the conjugation tube (lower right). The protoplasts of the two cells fuse completely (plasmogamy), followed by the union of their nuclei (karyogamy). A thick wall is then formed around the diploid zygote, known at this point as a zygospore. After a period of dormancy, meiosis occurs, followed by germination, and four haploid cells emerge. Asexual reproduction of the haploid individuals by cell division is the most frequent mode of reproduction.*

semble the vegetative cells, first become aggregated in clumps. Within these clumps, pairs are formed that stick together, first by their flagellar membranes and later by a slender protoplasmic thread—the conjugation tube—that connects them at the base of their flagella. As soon as this protoplasmic connection is formed, the flagella become free, and one or both pairs of flagella propel the partially fused gametes through the water. The protoplasts of the two gametes fuse completely (plasmogamy), followed by fusion of their nuclei (karyogamy), forming the zygote. Soon the four flagella shorten and eventually disappear, and a thick cell wall forms around the diploid zygote. This thick-walled, resistant zygote, or zygospore, then undergoes a period of dormancy. Meiosis occurs at the end of the dormant period, resulting in the production of four haploid cells, each of which develops two flagella and a cell wall. These cells can either divide asexually or mate with a cell of another mating strain to produce a new zygote. Thus, *Chlamydomonas* exhibits zygotic meiosis (see Figure 9–3a), and the haploid phase is the dominant phase in its life cycle.

Volvox Is the Most Spectacular of the Motile Colonial _Chlorophyceae_ The class *Chlorophyceae* also includes colonies of various shapes and sizes that are aggregates of flagellated cells resembling *Chlamydomonas.* Examples of such colonies are *Gonium, Pandorina, Eudorina,* and *Volvox* (Figures 17–21 and 17–22). Like *Chlamydomonas, Volvox* has been the subject of much laboratory investigation. With the exception of their zygotes, *Chlamydomonas* and *Volvox* are both haploid. Therefore, mutant

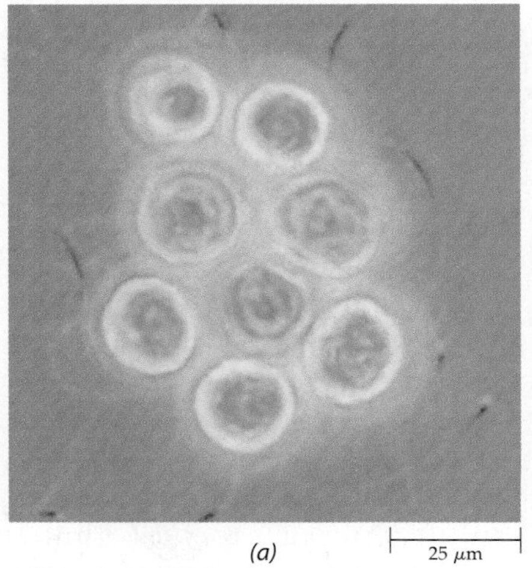

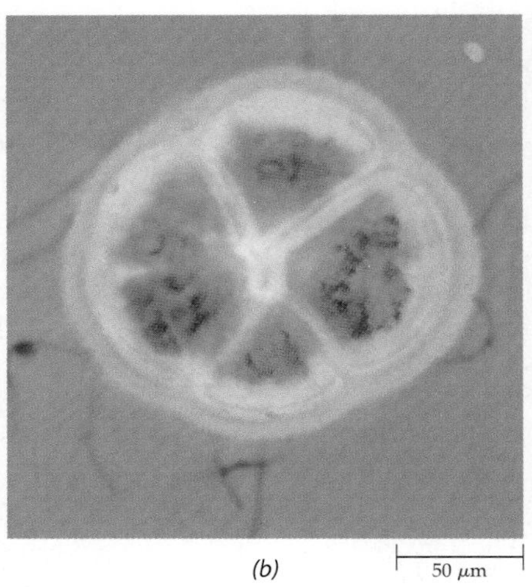

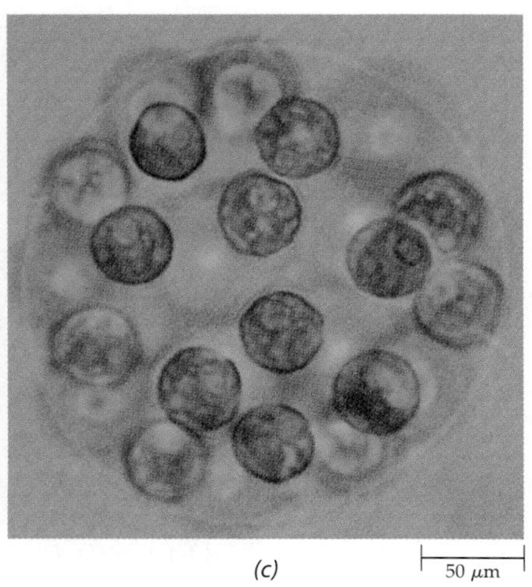

17–21
Several colonial Chlorophyceae. ***(a)*** Gonium, ***(b)*** Pandorina, *and* ***(c)*** Eudorina. *In these algae, cells similar to those of* Chlamydomonas *adhere in a gelatinous matrix to form multicellular colonies propelled by the beating of the flagella of the individual cells. Varying degrees of cellular specialization are found in different genera.*

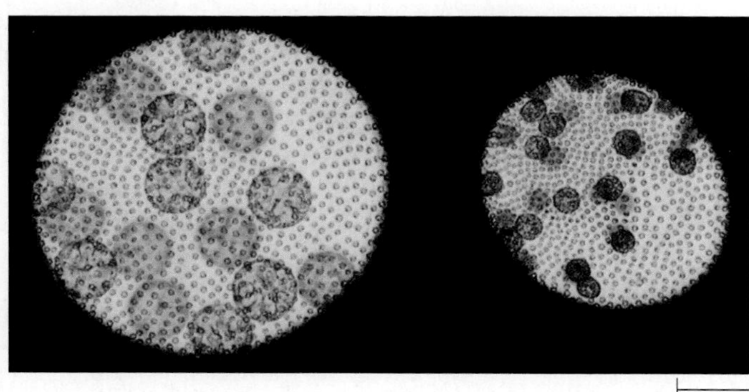

17–22
Two spheroids (individuals) of Volvox carteri. *The approximately 2000 small cells that dot the periphery of the spheroids are the* Chlamydomonas*-like somatic cells. In* V. carteri, *somatic cells are not interconnected in mature colonies as they are in some members of the genus. A few cells are capable of becoming reproductive, either sexually or asexually. The spheroid at the left is asexual. Mitotic division has already occurred, producing 16 juvenile spheroids, each of which will eventually digest a passageway out of the parent colony and swim away.*

The spheroid at the right is sexual. When heat-shocked, a female spheroid produces egg-bearing sexual female individuals and a male spheroid produces sperm-laden sexual male individuals. The zygotes produced by fertilization are heat- and desiccation-resistant.

genes are not masked by dominant alleles, and mutations affecting development can be readily detected. Hundreds of mutant strains with specific, inheritable developmental defects have been isolated and are being studied in order to learn how specific genes regulate photosynthesis, sexual reproduction, cellular differentiation, and other cell processes.

Volvox consists of a hollow sphere, the spheroid, made up of a single layer of 500 to 60,000 vegetative, biflagellated cells that serve primarily a photosynthetic function, and a small number of larger, nonflagellated reproductive cells. The specialized reproductive cells undergo repeated mitoses to form many-celled, juvenile spheroids, which "hatch" from the parental spheroid by releasing an enzyme that dissolves the transparent parental matrix. Interestingly, as the spheroids first develop, all of the flagella face the hollow center, so the colony must turn inside out before it can become motile.

Sexual reproduction in *Volvox* is always oogamous (Figure 17–2). In all species that have been studied, sexual reproduction is synchronized within the population of colonies by a sexual inducer molecule, a glycoprotein with a molecular weight of about 30,000. This inducer molecule is produced by a spheroid that has itself become sexual by some other, as yet poorly understood, mechanism. One male colony of *V. carteri* may produce enough inducer to induce sexual reproduction in over half a billion other colonies.

The Class *Chlorophyceae* Also Includes Nonmotile Unicellular Members One such member is *Chlorococcum,* which is very commonly found in the microbial flora of soils (Figure 17–23). There is a very large number of uni-

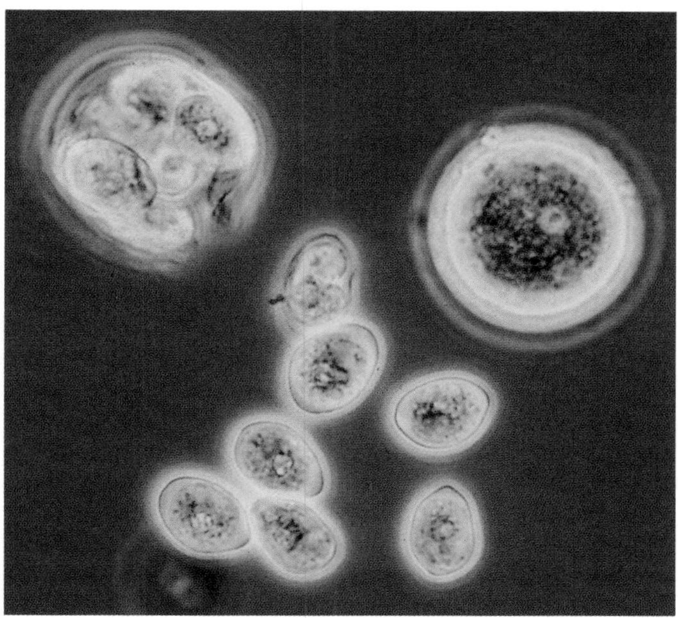

17–23
Chlorococcum echinozygotum. *At the upper left is a cell filled with asexual zoospores, which have formed mitotically within the cell. The smaller cells in the middle are biflagellate zoospores (the flagella are not visible). A nonmotile vegetative cell is at the upper right.*

cellular soil algal genera that superficially resemble *Chlorococcum* but that can be distinguished on the basis of cellular, reproductive, and molecular features. *Chlorococcum* and its relatives reproduce asexually by producing biflagellated zoospores, which are released from the parental cell. Sexual reproduction is accomplished by the release of flagellated gametes, which fuse in pairs to form zygotes. Meiosis is zygotic, as it is in all members of the *Chlorophyceae*.

Some *Chlorophyceae* Are Nonmotile Colonies Nonmotile colonial members of the *Chlorophyceae* include *Hydrodictyon*, the "water net" (Figure 17–24). Under favorable conditions, it forms massive surface blooms in ponds, lakes, and gentle streams. Each colony consists of many large, cylindrical cells arranged in the form of a large, lacy, hollow cylinder. Initially uninucleate, each cell eventually becomes multinucleate. At maturity, each cell contains a large central vacuole and peripheral cytoplasm containing the nuclei and a large reticulate (resembling a net) chloroplast with numerous pyrenoids. *Hydrodictyon* reproduces asexually through the formation of large numbers of uninucleate, biflagellated zoospores in each cell of the net. The zoospores are not released from parental cells but, rather amazingly, group themselves into geometric arrays of four to nine (most typically six) within the cylindrical parent cell. Zoospores then lose their flagella and form the component cells of daughter mini-nets. These are eventually released from the parent cell and grow into large mature nets by dramatic cell enlargement. In view of this mode of reproduction, it is easy to see how *Hydrodictyon* can form such conspicuous blooms in nature. Sexual reproduction in *Hydrodictyon* is isogamous, and meiosis is zygotic, as in all sexually reproducing *Chlorophyceae*.

17–24
(a) A stained portion of the "water net," Hydrodictyon, *a colonial member of the* Chlorophyceae. *(b) A higher magnification of a portion of* Hydrodictyon reticulatum.

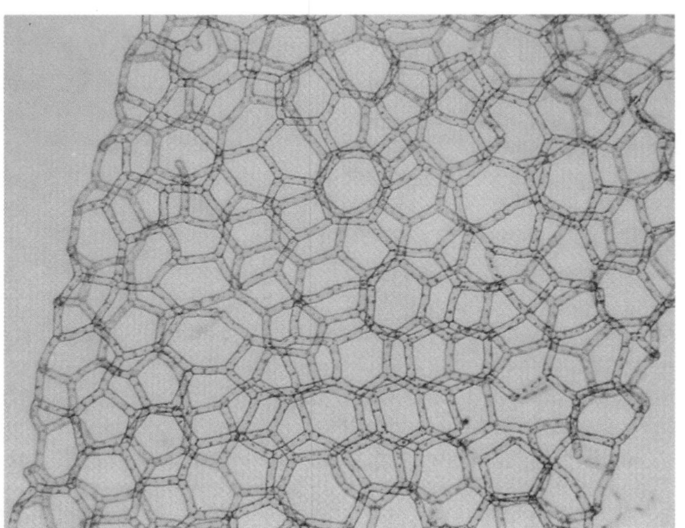

(a)

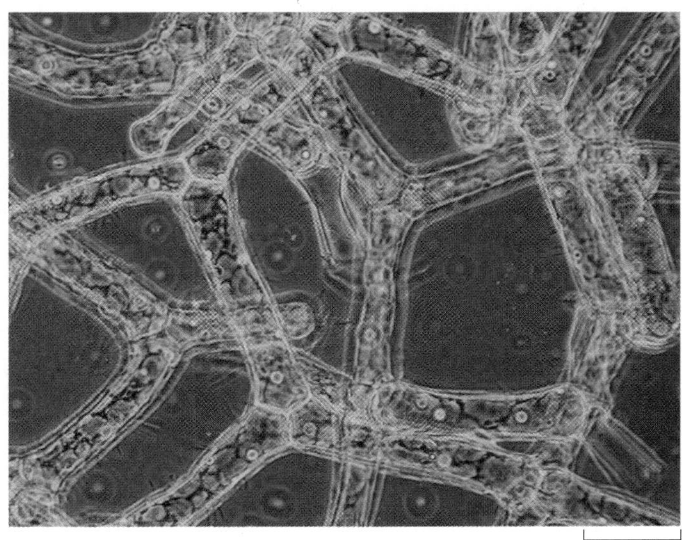

(b)

There Are Also Filamentous and Parenchymatous *Chlorophyceae* *Oedogonium* is an example of an unbranched filamentous member of the *Chlorophyceae.* Filaments begin their development attached to underwater substrates by a holdfast, but massive growths may later break away to form noticeable floating blooms in lakes. The mode of cell division in *Oedogonium* results in the formation of characteristic "caps" or annular scars with each cell division (Figure 17–25). Thus, these scars reflect the number of divisions that have occurred in a given cell.

Asexual reproduction in *Oedogonium* takes place by means of zoospore formation, with a single zoospore produced per cell. Each zoospore has a crown of about 120 flagella. Sexual reproduction is oogamous (Figure 17–26). Each antheridium produces two multiflagellated sperm, and each oogonium produces a single egg. Like the brown algae, *Oedogonium* uses water-soluble chemical attractants to bring sperm to eggs. Meiosis in *Oedogonium* is zygotic, as in all *Chlorophyceae.*

The branched filamentous and parenchymatous, or tissuelike, *Chlorophyceae* include algae that have the most complex structures found in the class. Their cells can be specialized with respect to particular functions or positions in the algal body and, like plant cells, are sometimes connected by plasmodesmata. *Stigeoclonium* consists of branched filaments (Figure 17–27), and *Fritschiella* is composed of subterranean rhizoids—a parenchymatous prostrate system near the soil surface—and two kinds of erect branches (Figure 17–28). *Fritschiella* grows on damp surfaces, such as tree trunks, moist walls, and leaf surfaces.

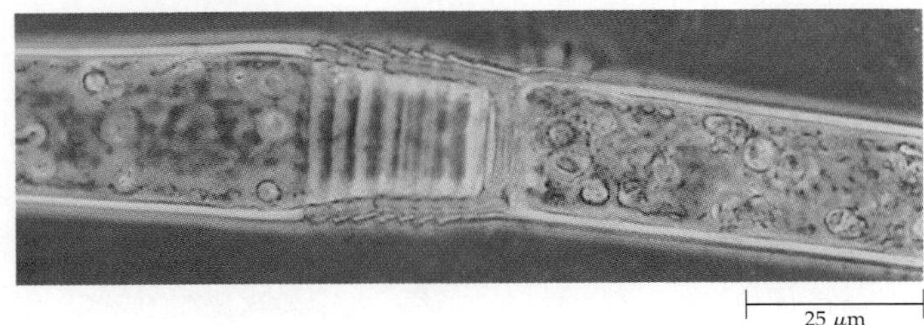

17–25
Oedogonium, *an unbranched, filamentous member of the* Chlorophyceae. *A section of vegetative filament showing annular scars.*

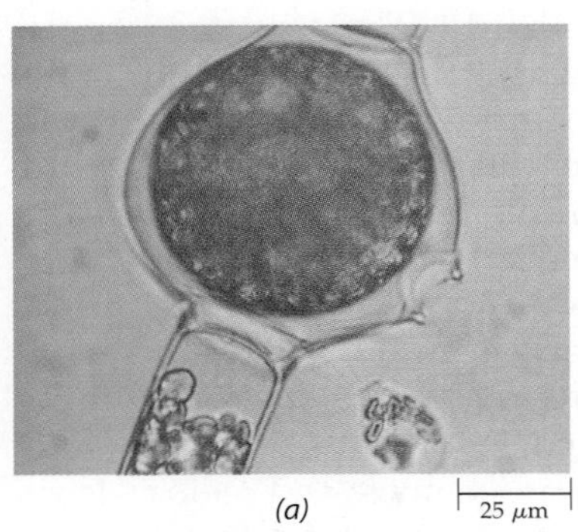

(a)

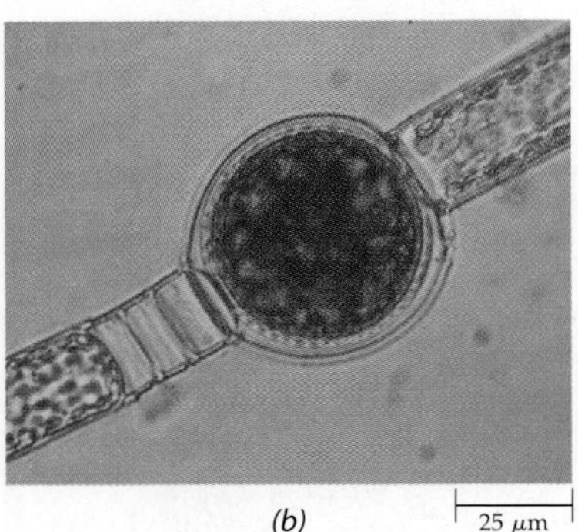

(b)

17–26
Sexual reproduction in Oedogonium *is oogamous. Each oogonium produces a single egg, whereas each antheridium produces two multiflagellated sperm.* **(a)** *An oogonium of* Oedogonium cardiacum, *with a large, round egg and, at the lower right, a sperm cell.* **(b)** *An oogonium of* Oedogonium foveolatum, *with a young zygote and, to the left, a chain of empty antheridia.*

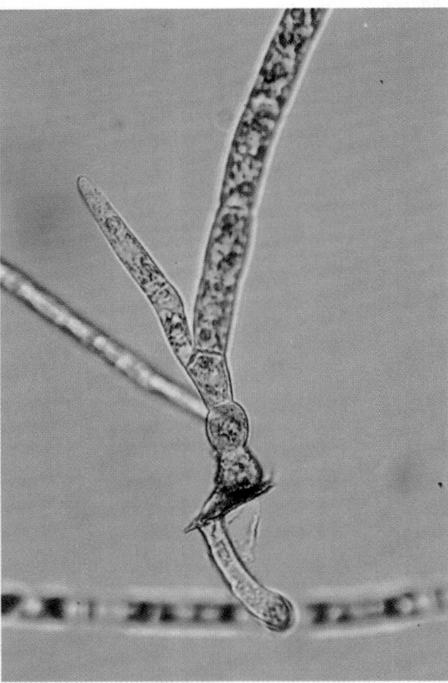

17–27
Stigeoclonium, *a branched, filamentous member of the* Chlorophyceae *that was formerly considered to belong to the genus* Ulothrix (Ulvophyceae). *Branching has just begun in this young individual, with a dark holdfast at its base and a rhizoid extending downward from just above the holdfast.*

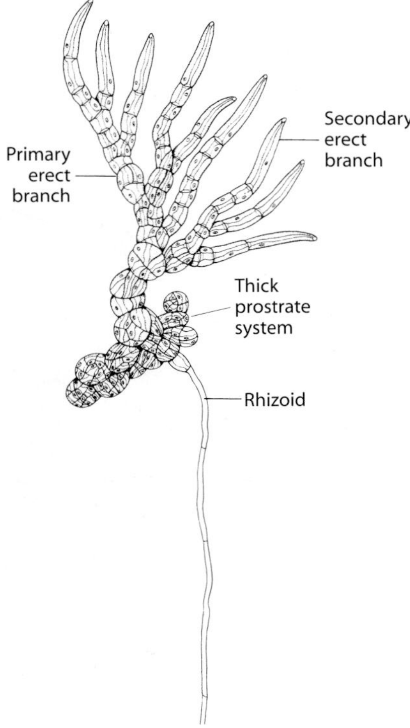

17–28
Fritschiella, *a terrestrial member of the* Chlorophyceae. *In its adaptation to a terrestrial habitat,* Fritschiella *has independently evolved some of the features that are characteristic of plants.*

Class *Ulvophyceae*, the Ulvophytes, Consists of Mainly Marine Species

The *Ulvophyceae* are primarily marine, but a few important representatives occur in fresh water, probably having moved there from the marine habitat at some time in the past. Ulvophytes may be filamentous or composed of flat sheets of cells, or they may be macroscopic and multinucleate. Ulvophytes have a closed mitosis in which the nuclear envelope persists; the spindle is persistent through cytokinesis.

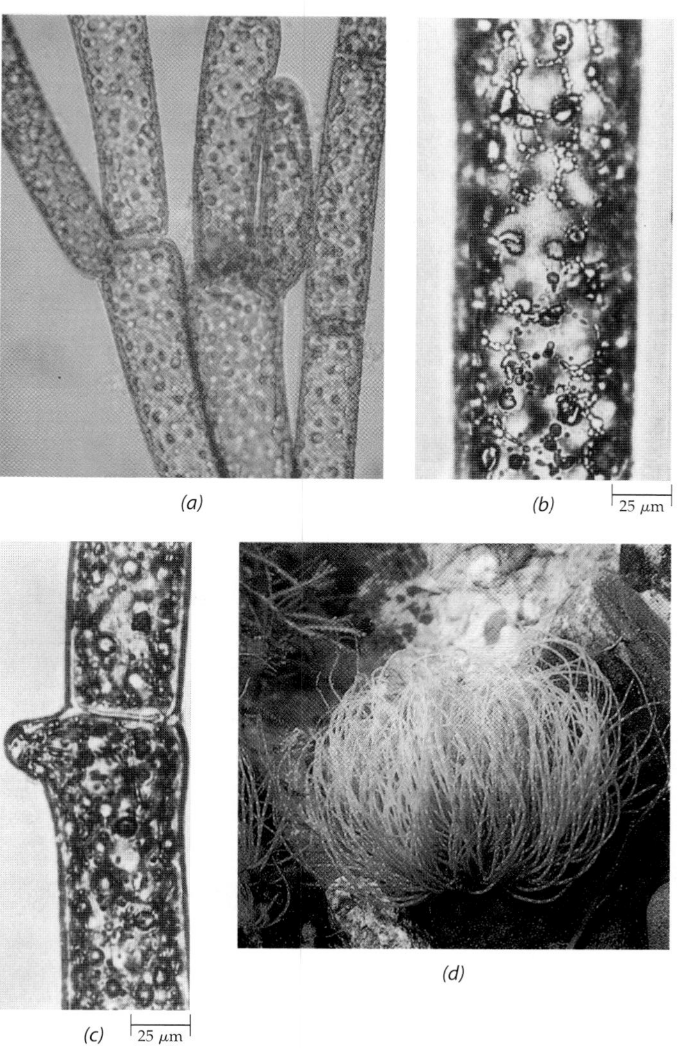

17–29
Cladophora, *a member of the class* Ulvophyceae, *is widespread in marine and freshwater habitats. The marine species, like most* Ulvophyceae, *have an alternation of generations, while the freshwater species do not.* ***(a)*** *Branched filaments of* Cladophora. ***(b)*** *Part of an individual cell, showing the netlike chloroplast.* ***(c)*** *Commencement of branching at the apical end of a cell.* ***(d)*** *An individual of* Cladophora *growing in a sluggish stream in California.*

The flagellated cells of ulvophytes are scaly or naked like those of *Charophyceae,* but in contrast to motile charophytes, ulvophytes are nearly radially symmetrical and have apical, forward-directed flagella like those of the chlorophytes. The flagellated cells of *Ulvophyceae* may have two, four, or many flagella, like *Chlorophyceae.* Those of *Charophyceae* are always biflagellate. Ulvophytes are the only green algae that have an alternation of generations with sporic meiosis or a diploid, dominant life history involving gametic meiosis.

One evolutionary line of *Ulvophyceae* consists of filamentous algae with large, multinucleate septate cells. One of the genera of this group, *Cladophora* (Figure 17–29), is widespread in both salt water and fresh waters, sometimes forming nuisance blooms in the latter habitat. Its filaments commonly grow in dense mats, which are either free-floating or attached to rocks and vegetation. The filaments elongate and branch near the ends. Each cell contains many nuclei and a single, peripheral, netlike chloroplast with many starch-forming pyrenoids. Marine species of *Cladophora* have an alternation of isomorphic generations. Most of the freshwater species, however, do not have an alternation of generations, apparently having lost this characteristic during the course of their transition from marine to fresh waters.

17–30
Sea lettuce, Ulva, *a common member of the class* Ulvophyceae *that grows on rocks, pilings, and similar places in shallow seas worldwide.*

A second kind of growth habit among the *Ulvophyceae* is that of *Ulva,* commonly known as sea lettuce (Figure 17–30). This familiar alga is common along temperate seashores throughout the world. Individuals of *Ulva* consist of a glistening, flat thallus that is two cells thick and a meter or more long in exceptionally large individuals. The thallus is anchored to the substrate by a holdfast produced by extensions of the cells at its base. Each cell of the thallus contains a single nucleus and chloroplast. *Ulva* is anisogamous (Figure 17–2) and has an alternation of isomorphic generations like that of many other *Ulvophyceae* (Figure 17–31).

The siphonous marine algae, characterized by very large, branched, coenocytic cells that are rarely septate, constitute additional evolutionary lines in the *Ulvophyceae* (Figure 17–32). These algae, which are very diverse, develop as a result of repeated nuclear division without the formation of cell walls. Cell walls are produced only in the reproductive phases of the siphonous green algae.

Codium, mentioned on page 383, is a member of this group. It is a spongy mass of densely intertwined coenocytic filaments (Figure 17–32a). Strains of *Codium fragile* may be quite weedy, and nuisance growths of this seaweed are spreading in quiet waters of the temperate zone. *Ventricaria* (also known as *Valonia*), which is common in tropical waters, has been widely used in studies of cell walls and in physiological experiments requiring large amounts of cell sap. *Ventricaria* appears to be unicellular but is actually a large, multinucleate vesicle at-

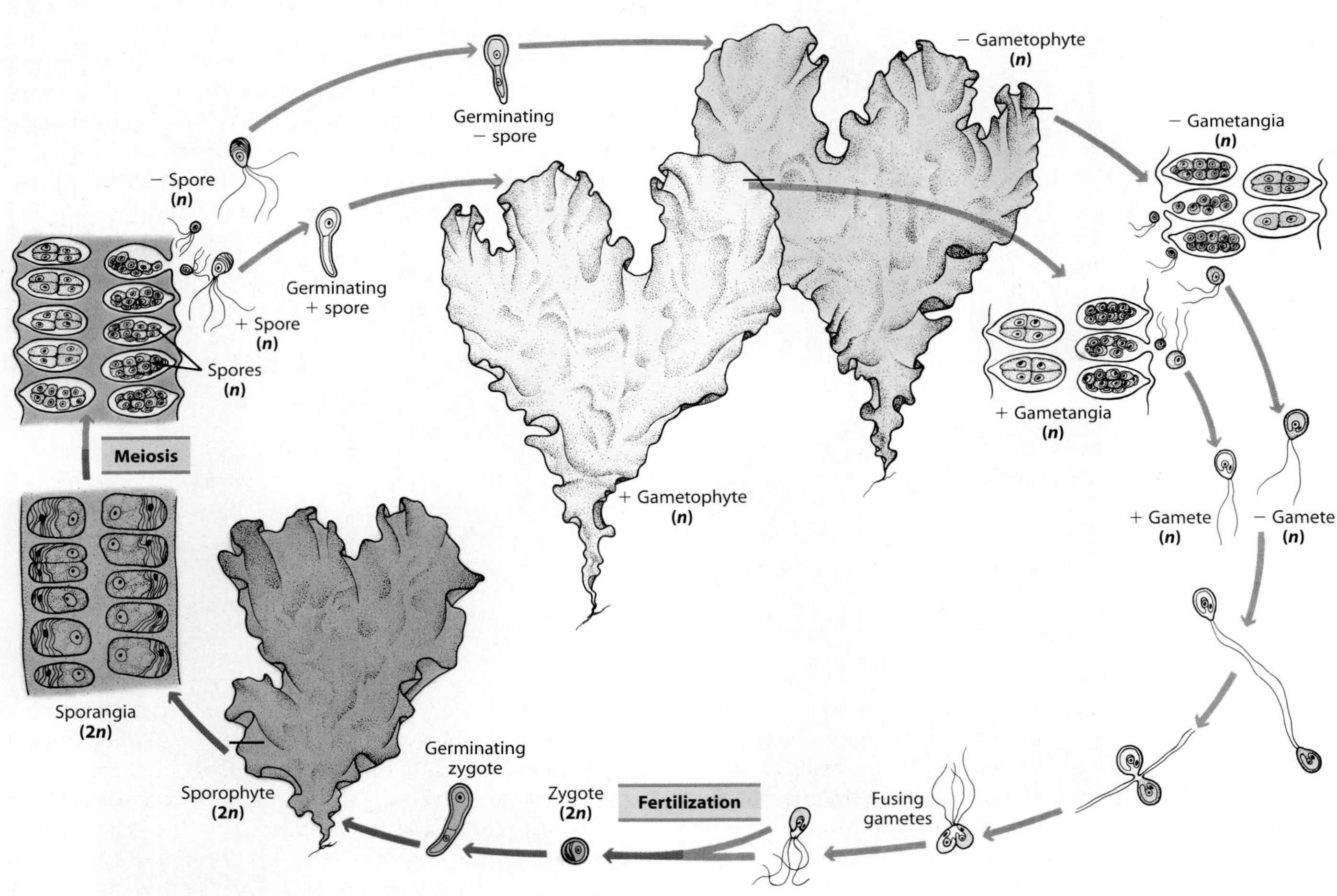

17–31
In the sea lettuce, Ulva, *we can see the reproductive pattern known as an alternation of generations, in which one generation produces spores (left), the other gametes (right). The haploid* (n) *gametophyte produces haploid isogametes, and the gametes fuse to form a diploid* (2n) *zygote. A sporophyte, a multicellular body in which all the cells are diploid, develops from the zygote. The sporophyte produces haploid spores by meiosis. The haploid spores develop into haploid gametophytes, and the cycle begins again.*

17–32
Four genera of siphonous green algae of the class Ulvophyceae. ***(a)*** *A species of* Codium, *abundant along the Atlantic coast.* ***(b)*** Ventricaria, *common in tropical waters; individuals are often about the size of a hen's egg.* ***(c)*** Acetabularia, *the "mermaid's wine glass," a mushroom-shaped siphonous green alga. The siphonous green alga in the background is* Dasycladus. *This photograph was taken in the Bahamas.* ***(d)*** Halimeda, *a siphonous green alga that is often dominant in reefs in warmer waters throughout the world. This alga produces distasteful compounds that retard grazing by fishes and other marine herbivores.*

tached to the substrate by several rhizoids (Figure 17–32b). It grows to the size of a hen's egg. Another well-known siphonous green alga is *Acetabularia* (Figure 17–32c), which has been widely used in experiments on the genetic basis of differentiation. The siphonous green algae are primarily diploid. The gametes are the only haploid cells in the life cycle.

The chloroplasts of some siphonous green algae, including *Codium,* are harvested by sea slugs (nudibranchs), which are marine mollusks that lack shells (Figure 17–33). The sea slugs eat the algae, and the algal chloroplasts persist in the cells that line the animal's respiratory chamber. In the presence of light, these chloroplasts carry on photosynthesis so efficiently that individuals of the nudibranch *Placobranchus ocellatus* are reported to evolve more oxygen than they consume.

Halimeda and related genera of siphonous green algae are notable for their calcified cell walls. When these algae die and disintegrate, they play a major role in the generation of the white carbonate sand that is so characteristic of tropical waters. A number of genera of algae, including *Halimeda* (Figure 17–32d), contain secondary metabolites that significantly reduce feeding by herbivorous fishes. The fronds of *Halimeda* expand and grow rapidly at night, building up toxic compounds that deter herbivores, which are active mainly during the day.

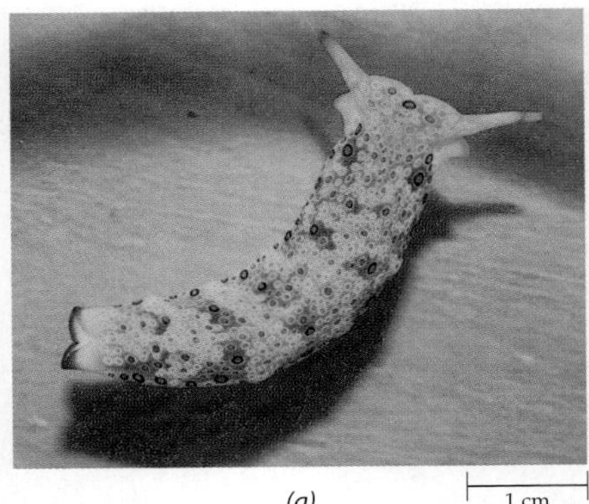

(a)

(b)

17–33
Placobranchus, *a sea slug. The tissues of this animal contain chloroplasts, which it obtains by eating certain green algae.* **(a)** *Ordinarily, flaplike structures called parapodia are folded over the animal's back, hiding the chloroplast-containing tissues.* **(b)** *When the parapodia are spread apart, however, the deep green tissues become visible. The chloroplasts carry on photosynthesis so efficiently that, within a 24-hour cycle of light and darkness, some of the animals produce more oxygen than they consume.*

Within an hour after sunrise, the fronds turn from white to green as chloroplasts from the lower portions of the thallus migrate upward and begin to photosynthesize rapidly.

Class *Charophyceae,* the Charophytes, Includes Members That Closely Resemble Plants

Charophyceae consist of unicellular, colonial, filamentous, and parenchymatous genera. Their relationship with one another, and to plants, is revealed by many fundamental structural, biochemical, and genetic similarities. These include the presence of asymmetrical flagellated cells, some of which have distinctive multilayered structures (Figure 17–18). Other similarities are breakdown of the nuclear envelope at mitosis, persistent spindles or phragmoplasts at cytokinesis, the presence of phytochrome, and other molecular features.

Spirogyra (Figure 17–34) is a well-known genus of unbranched, filamentous *Charophyceae* that often forms frothy or slimy floating masses in bodies of fresh water. Each filament is surrounded by a watery sheath. The name *Spirogyra* refers to the helical arrangement of the one or more ribbon-like chloroplasts with numerous pyrenoids found within each uninucleate cell. Asexual reproduction occurs in *Spirogyra* by cell division and fragmentation. There are no flagellated cells at any stage of the life cycle, but, as noted earlier, flagellated reproductive cells do occur in other genera of *Charophyceae.* During sexual reproduction in *Spirogyra,* a conjugation tube forms between two filaments. The contents of the two cells that are joined by the tube serve as isogametes. Fertilization may occur in the tube, or one of the gametes may migrate into the other filament, where fertilization then takes place. The zygotes become surrounded by thick walls that contain **sporopollenin.** This protective substance is the most resistant biopolymer known. Sporopollenin enables the zygotes to survive harsh con-

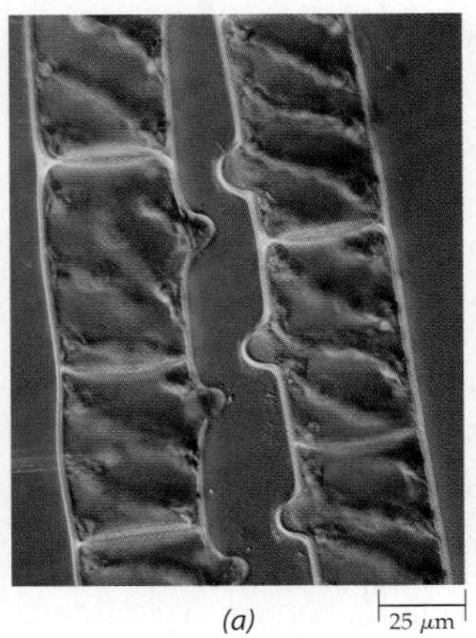

(a)

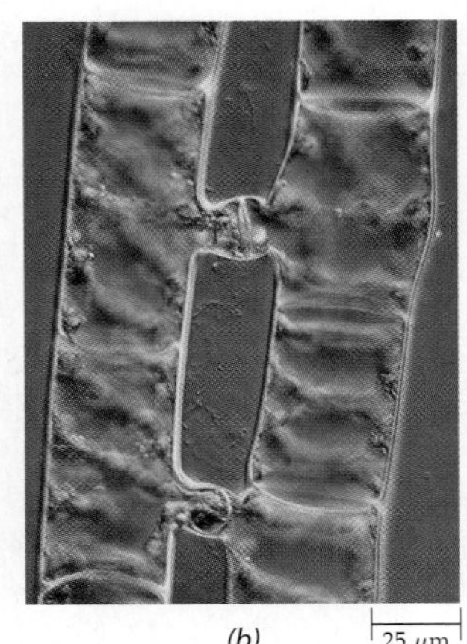

(b)

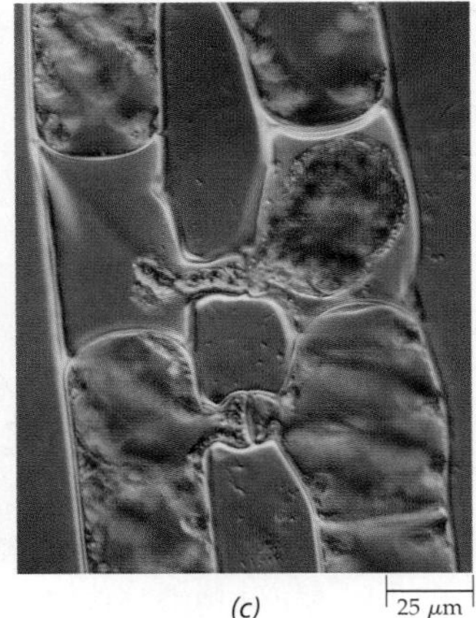

(c)

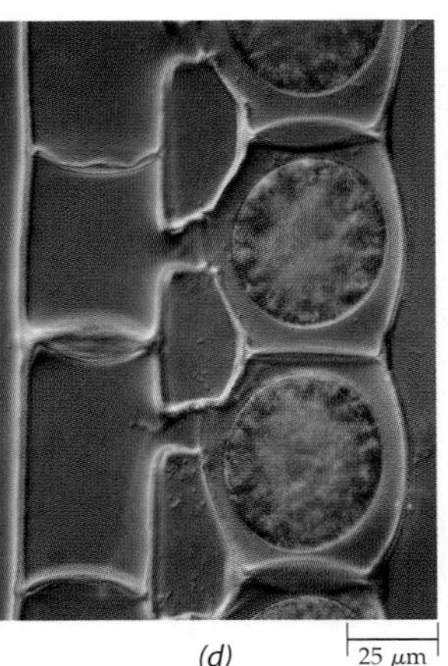

(d)

17–34
Sexual reproduction in Spirogyra. **(a), (b)** *The formation of conjugation tubes between the cells of adjacent filaments.* **(c)** *The contents of the cells of the − strain (on the left) pass through these tubes into the cells of the + strain.* **(d)** *Fertilization occurs within these cells. The resulting zygote develops a thick, resistant cell wall and is termed a zygospore. The vegetative filaments of* Spirogyra *are haploid, and meiosis occurs during germination of the zygospores, as it does in all* Charophyceae.

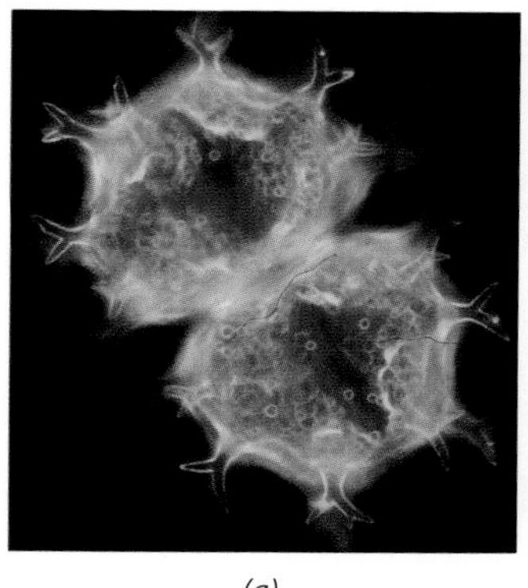

(a)

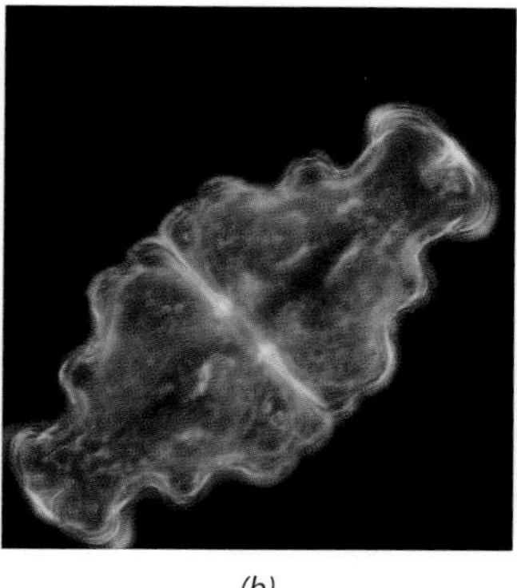

(b)

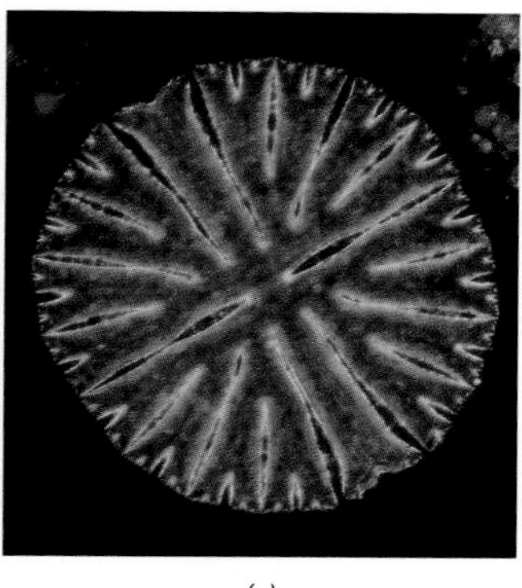

(c)

(d)

17–35

Desmids. The desmids are a group of thousands of species of unicellular, freshwater Charophyceae. *Many desmids are constricted, which gives them a very characteristic appearance.* ***(a)*** Xanthidium armatum. ***(b)*** Euastrum affine. ***(c)*** Micrasterias radiosa. ***(d)*** *Cell division in* Micrasterias thomasiana. *The smaller half of each daughter individual will grow to the size of the larger half, forming a full-sized desmid.*

ditions for long periods of time before germinating when conditions improve. Meiosis is zygotic, as in all members of the *Charophyceae*.

The desmids are a large group of freshwater green algae related to *Spirogyra*. Like *Spirogyra*, they lack flagellated cells. Some desmids are filamentous, but most are unicellular. Most desmid cells consist of two sections, or semi-cells, joined by a narrow constriction—the isthmus—between them (Figure 17–35). Desmid cell division and sexual reproduction are very similar to those of *Spirogyra*. There are thousands of desmid species. They are most abundant and diverse in peat bogs and ponds that are poor in mineral nutrients. Some are associated with distinctive and possibly symbiotic bacteria, which live in the mucilaginous sheaths.

The *Coleochaetales* and *Charales* Have Plantlike Characteristics Two orders of charophycean green algae, the *Coleochaetales* and the *Charales*, resemble plants more closely than do other charophytes in the details of their cell division and sexual reproduction. These orders have a plantlike microtubular phragmoplast operating during cytokinesis. Like plants, they are oogamous, and their sperm are ultrastructurally similar to those of bryophytes. Plants were probably derived from an extinct member of the *Charophyceae* that in many respects resembled members of the *Coleochaetales* and *Charales*.

The order *Coleochaetales* includes both branched filamentous genera and discoid (disk-shaped) genera that grow by division of apical or peripheral cells (Figure 17–36a). *Coleochaete*, which grows on the surface of submerged rocks or freshwater plants, has uninucleate vegetative cells that each contain one large chloroplast with an embedded pyrenoid. Very similar chloroplasts and pyrenoids occur in some of the hornworts, a group of bryophytes that is discussed in Chapter 18. *Coleochaete*, like a number of charophytes, reproduces asexually by zoospores that are formed singly within cells. Sexual reproduction is oogamous. The zygotes, which remain attached to the parental thallus, stimulate the growth of a layer of cells that covers the zygotes (Figure 17–36b). In at least one species, these parental cells have wall in-

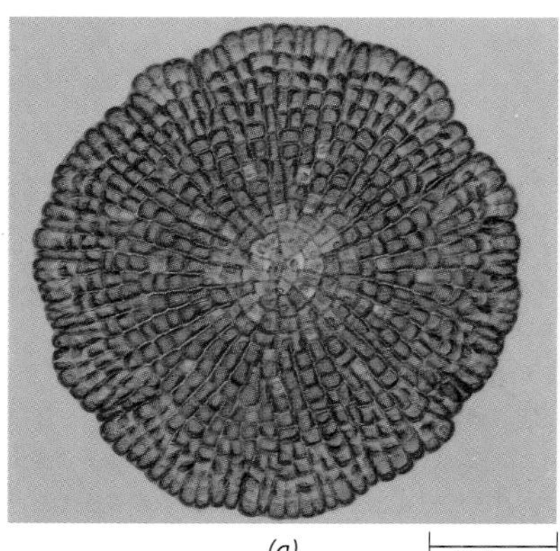

(a) 0.25 μm

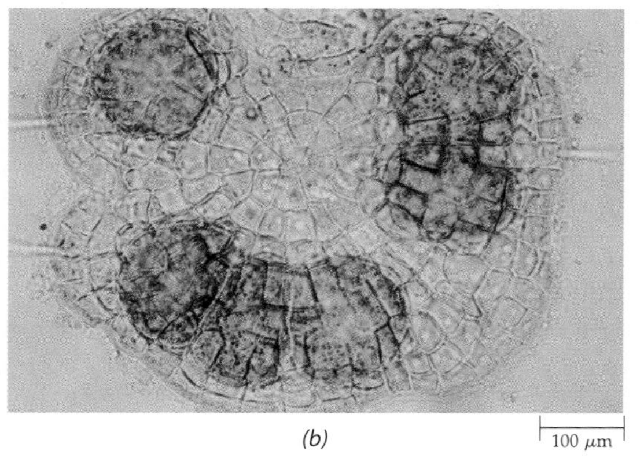

(b) 100 μm

17–36

(a) Coleochaete *grows on rocks and on stems of aquatic flowering plants in shallow lake waters.* ***(b)*** *Individuals of this species of* Coleochaete *consist of a parenchymatous disk, which is generally one cell thick. The large cells are zygotes, which are protected by a cellular covering. The hair cells extending from the disk are ensheathed at the base;* Coleochaete *means "sheathed hair." These hairs are thought to discourage aquatic animals from feeding on the alga.*

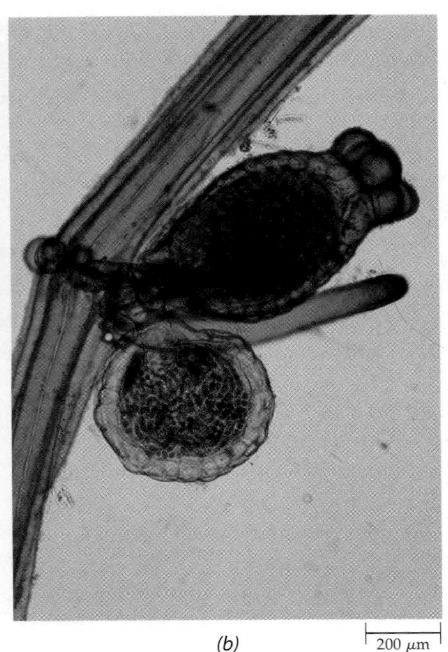

17–37
(a) Chara, *a stonewort (class* Charophyceae*) that grows in shallow waters of temperate lakes.* *(b)* Chara *showing gametangia. The top structure is an oogonium, and that below is an antheridium.*

(a) (b)

growths similar to those occurring at the gametophyte-sporophyte junction of bryophytes (see Chapter 18) and other plants. These specialized cells, called transfer cells, are believed to function in nutrient transport between the gametophyte and sporophyte. The wall ingrowths suggest how the distinctive life cycle of plants and the sporophyte generation might have evolved in the long-extinct ancestors of modern plants.

The order *Charales* includes about 250 living species of distinctive green algae found primarily in fresh water or sometimes in brackish water. Modern forms such as *Chara* (Figure 17–37) are commonly known as the stoneworts, because some of them have heavily calcified cell walls. Calcification of the distinctive reproductive structures of ancient relatives has resulted in a good fossil record extending back to Late Silurian times (about 410 million years ago).

The *Charales*, like *Coleochaete* and plants, exhibit apical growth. In addition, the thallus of the *Charales* is differentiated into nodal and internodal regions. The tissue organization in the nodal regions resembles the parenchyma of plants, as does the pattern of plasmodesmatal connections. From the nodal regions arise whorls of branches. In some species, files of cells grow over the central, filamentous axis, creating a thicker, stronger thallus. The sperm of *Charales* are produced in multicellular antheridia that are more complex than those found in any other group of protists. Their eggs are borne in oogonia enclosed by several long, tubular, twisted cells. These cells are in a position analogous to the distinctive female gametangia of seedless plants, and they may serve similar functions. Sperm are the only flagellated cells in the *Charales* life cycle, and they very closely resemble bryophyte sperm. Zygotes are believed to germinate by meiosis, though this has been difficult to study because zygotes are enclosed in very tough walls that include sporopollenin. Sporopollenin is also a component of the walls of plant spores and pollen and is responsible for the widespread occurrence of these cells in the fossil record.

Summary

Heterokonts Have One Long Tinsel Flagellum and One Short Whiplash Flagellum

Brown algae (phylum *Phaeophyta*) are closely related to a diverse array of microscopic protists, including oomycetes (phylum *Oomycota*), diatoms (phylum *Bacillariophyta*), and chrysophytes (phylum *Chrysophyta*). Collectively, these organisms are known as heterokonts because characteristically they have two flagella that differ in length and ornamentation.

Phylum *Oomycota* Includes Both Aquatic and Terrestrial Heterotrophs That Form Thick-Walled Oospores

The oomycetes range from unicellular forms to highly branched coenocytic filamentous forms. The cell walls are composed largely of cellulose or cellulose-like polymers. Asexual reproduction is by means of zoospores. Sexual reproduction involves a large, immobile egg and a small, motile male gamete. Three of the partly terrestrial members of this phylum are *Phytophthora*, a very important causative agent of plant diseases, including late blight of Irish potatoes; *Plasmopara viticola*, which causes downy mildew of grapes; and *Pythium*, which causes damping-off diseases of seedlings.

Diatoms Belong to the Phylum *Bacillariophyta*

The diatoms are unicellular or colonial organisms that occur in fresh as well as in marine waters. They are important components of the phytoplankton and are unique for their two-part siliceous cell walls.

Chrysophytes Belong to Phylum *Chrysophyta*

Chrysophytes are colorless or gold-pigmented, unicellular or colonial organisms that are abundant in fresh waters and in the oceans of the world. Some chrysophytes feed on bacteria and other organic particles, whereas others cannot feed because they are coated by overlapping silica scales.

Brown Algae Belong to Phylum *Phaeophyta*

The brown algae include the largest and most complex of marine algae and are the most conspicuous seaweeds of temperate, boreal, and polar waters. In many types, the vegetative body is well differentiated into holdfast, stipe, and blade. Some have food-conducting tissues that approach, in their complexity, those of vascular plants. Giant brown kelps form ecologically important marine forests, and these algae are the source of industrially important polysaccharides. The life cycles of most brown algae involve an alternation of generations.

Chlorophyceae, Ulvophyceae, and *Charophyceae* Are Classes of the Phylum *Chlorophyta*

The green algae include a wide diversity of freshwater and terrestrial forms, in addition to seaweeds. Several classes of green algae have been defined on the basis of cell division, reproductive cell structure, and other features. The *Ulvophyceae* are primarily marine, and alternation of generations occurs in some forms. The *Chlorophyceae* are primarily freshwater species, and two members of this group, *Chlamydomonas* and *Volvox,* are important laboratory model systems. The *Chlorophyceae* have a unique mode of cytokinesis—the mitotic spindle collapses at telophase, and a phycoplast develops parallel to the plane of cell division. The freshwater *Charophyceae* are the green algae that are most closely related to bryophytes and vascular plants. The modern genus *Coleochaete* and the order *Charales* have cell division, reproductive, and other features that link them particularly closely to the ancestry of plants.

Selected Key Terms

algin p. 382
antheridium p. 372
centric p. 376
chrysolaminarin p. 378
cleavage furrow p. 386
damping-off p. 374
flagellar roots p. 386
frustules p. 376
fucoxanthin p. 378
heterokonts p. 371
heterothallic p. 373
homothallic p. 373
mannitol p. 382
oogonium p. 372
oospore p. 372
pennate p. 376
phragmoplast p. 387
phycoplast p. 386
plurilocular gametangia p. 383
plurilocular sporangia p. 383
sporopollenin p. 396
thallus p. 381
unilocular sporangia p. 383

Questions

1. Distinguish between each of the following: oogonium/antheridium; homothallic/heterothallic; pennate/centric; phycoplast/phragmoplast.
2. Identify the plant diseases caused by each of the following oomycetes: *Plasmopara viticola, Phytophthora infestans,* and *Pythium* sp.
3. The diatoms may be characterized as "the algae that live in glass houses." Explain.
4. What pigments do the diatoms, chrysophytes, and brown algae have in common? Which of these pigments is responsible for the color of these algae?
5. The kelps have the most highly differentiated bodies among the algae. How so?
6. *Fucus* has a life cycle that is, in some ways, similar to our own. Explain.
7. Distinguish among the three classes of *Chlorophyta*: *Chlorophyceae, Ulvophyceae,* and *Charophyceae.*
8. What features of the modern genus *Coleochaete* and the order *Charales* link them closely to the ancestry of plants?

Chapter 18

Bryophytes

18–1
Dense growth of the moss Fissidens *on limestone rocks in a waterfall. This photograph was taken in a nature reserve just west of Yalta on the Crimean Peninsula in Ukraine.*

OVERVIEW

With the bryophytes—liverworts, hornworts, and mosses—we see the important evolutionary move from the water to the land. Such a move involved solving a variety of problems—the most crucial of which was how to avoid drying out. The gametes of bryophytes are enclosed in multicellular protective structures—an antheridium encloses the sperm and an archegonium encloses the egg. But a vestige of their aquatic algal ancestors persists in that sperm must still swim through water to reach the egg. As you will see, fertilization takes place within the archegonium, and the resulting embryo is nourished and protected by the maternal plant. The growing embryo (the young sporophyte) never becomes independent of the parent (the gametophyte), and in fact one of the characteristics of the bryophytes is that the free-living gametophyte is almost always larger than the sporophyte. Meiosis quickly occurs in the sporophyte, and the resulting spores are dispersed into the environment, where they germinate and give rise to new gametophytes.

CHECKPOINTS

By the time you finish reading this chapter, you should be able to answer the following questions:

1. What are the general characteristics of the bryophytes? In other words, what does it take to be a bryophyte?
2. What are the three phyla of bryophytes? How are they similar to and different from one another?
3. How does sexual reproduction occur in bryophytes? What are the principal parts of the resulting sporophyte of most bryophytes?
4. How can the three types of liverworts be distinguished from one another?
5. What are the distinguishing features of the hornworts?
6. What are the distinguishing features of each of the three classes of mosses?

Bryophytes—liverworts, hornworts, and mosses—are small "leafy" or flat plants that most often grow in moist locations in temperate and tropical forests or along the edges of streams and wetlands (Figure 18–1). Bryophytes are not confined, however, to such habitats. Many species of mosses are found in relatively dry deserts, and several can form extensive masses on dry, exposed rocks that can become very hot (Figure 18–2).

18–2
A dry-land moss, Tortula obtusissima, *that lives on and around limestone in the central plateau of Mexico. Having no roots, the plants obtain their moisture directly from the external environment in the form of dew or rain. They can recover physiologically from complete dryness in less than five minutes.*

Mosses sometimes dominate the terrain to the exclusion of other plants over large areas north of the Arctic Circle. Mosses are also the dominant plants on rocky slopes above timberline in the mountains, and a significant number of mosses are able to withstand the long periods of severe cold on the Antarctic continent (Figure 18–3). A few bryophytes are aquatic, and some are even found on rocks splashed by ocean waves, although none are truly marine.

Bryophytes contribute significantly to plant biodiversity and are also important in some parts of the world for the large amounts of carbon they store, thereby playing a significant role in the global carbon cycle (page 150). Increasing evidence indicates that the first plants were much like modern, or extant, bryophytes, and even today, together with lichens, bryophytes are important initial colonizers of bare rock and soil surfaces. Like the lichens, some bryophytes are remarkably sensitive to air pollution, and they are often absent or represented by only a few species in highly polluted areas.

The Relationships of Bryophytes to Other Groups

In many respects, bryophytes are transitional between the charophycean green algae, or charophytes (see Chapter 17), and the vascular plants, which are discussed in Chapters 19 through 22. In the last chapter, we considered some of the features shared by charophytes and plants. Both contain chloroplasts with well-devel-

(a)

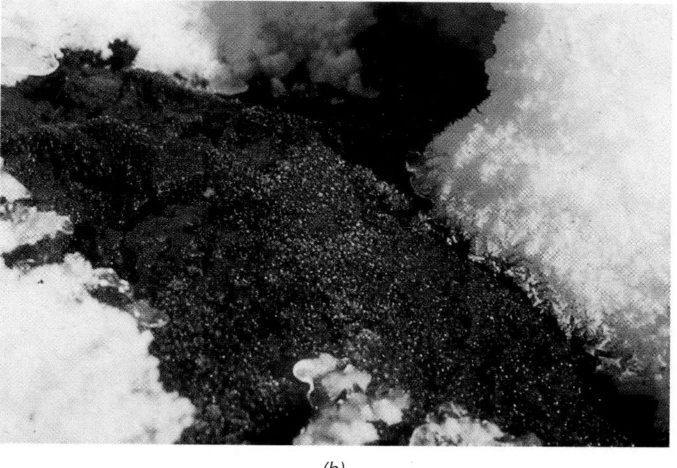

(b)

18–3
***(a)** At about 3000 meters elevation on Mount Melbourne, Antarctica, the daily temperatures in summer mainly range from −10° to −30°C. In this incredibly harsh environment, botanists from New Zealand discovered patches of a moss of the genus* Campylopus ***(b)**, growing in the bare areas visible in the photograph, where volcanic activity produces temperatures that may reach 30°C. The growth of* Campylopus *in this locality demonstrates the remarkable dispersal powers of mosses as well as their ability to survive in harsh habitats.*

oped grana, and both have motile cells that are asymmetrical, with flagella that extend from the side rather than the end of the cell. During the cell cycle, both charophytes and plants exhibit breakdown of the nuclear envelope at mitosis and persistent spindles or phragmoplasts during division of the cytoplasm (cytokinesis). In addition, you may recall that, among the charophytes, the *Charales* and *Coleochaetales* appear to be more closely related to the plants than any others. In *Coleochaete*, the zygotes are retained within the parental thallus and, in at least one species of *Coleochaete*, the cells covering the zygotes develop wall ingrowths. These covering cells apparently function as transfer cells involved with the transport of sugars to the zygotes. Like *Coleochaete*, all plants are oogamous, that is, they have an egg cell that is fertilized by sperm.

Bryophytes and vascular plants share a number of characters that distinguish them from the charophytes. These include: (1) the presence of male and female gametangia, called **antheridia** and **archegonia,** respectively, with a protective layer called a sterile jacket layer; (2) retention of both the zygote and the developing multicellular embryo, or young sporophyte, within the archegonium or the female gametophyte; (3) the presence of a multicellular diploid sporophyte, which results in an increased number of meioses and an amplification of the number of spores that can be produced following each fertilization event; (4) multicellular sporangia consisting of a sterile jacket layer and internal spore-producing **(sporogenous)** tissue; and (5) spores with walls containing **sporopollenin,** which resists decay and drying. Charophytes lack all of these shared bryophyte and vascular plant characters, which are correlated with the existence of plants on land.

Living bryophytes lack the water- and food-conducting (vascular) tissues called xylem and phloem, respectively, which are present in vascular plants. Although some bryophytes have specialized conducting tissues, the cell walls of the bryophyte water-conducting cells are not lignified, as are those of the vascular plants. Also, there are differences in the life cycles of bryophytes and vascular plants, both of which exhibit alternating heteromorphic gametophytic and sporophytic generations. In the bryophytes, the gametophyte is dominant and free-living and the sporophyte is small, permanently attached to and nutritionally dependent upon its parental gametophyte. By contrast, the sporophyte of vascular plants is the dominant generation and is also ultimately free-living. In addition, the bryophyte sporophyte is unbranched and bears only a single sporangium, whereas the sporophytes of extant vascular plants are branched and bear many more sporangia. Vascular plant sporophytes therefore produce a great many more spores than do the sporophytes of bryophytes.

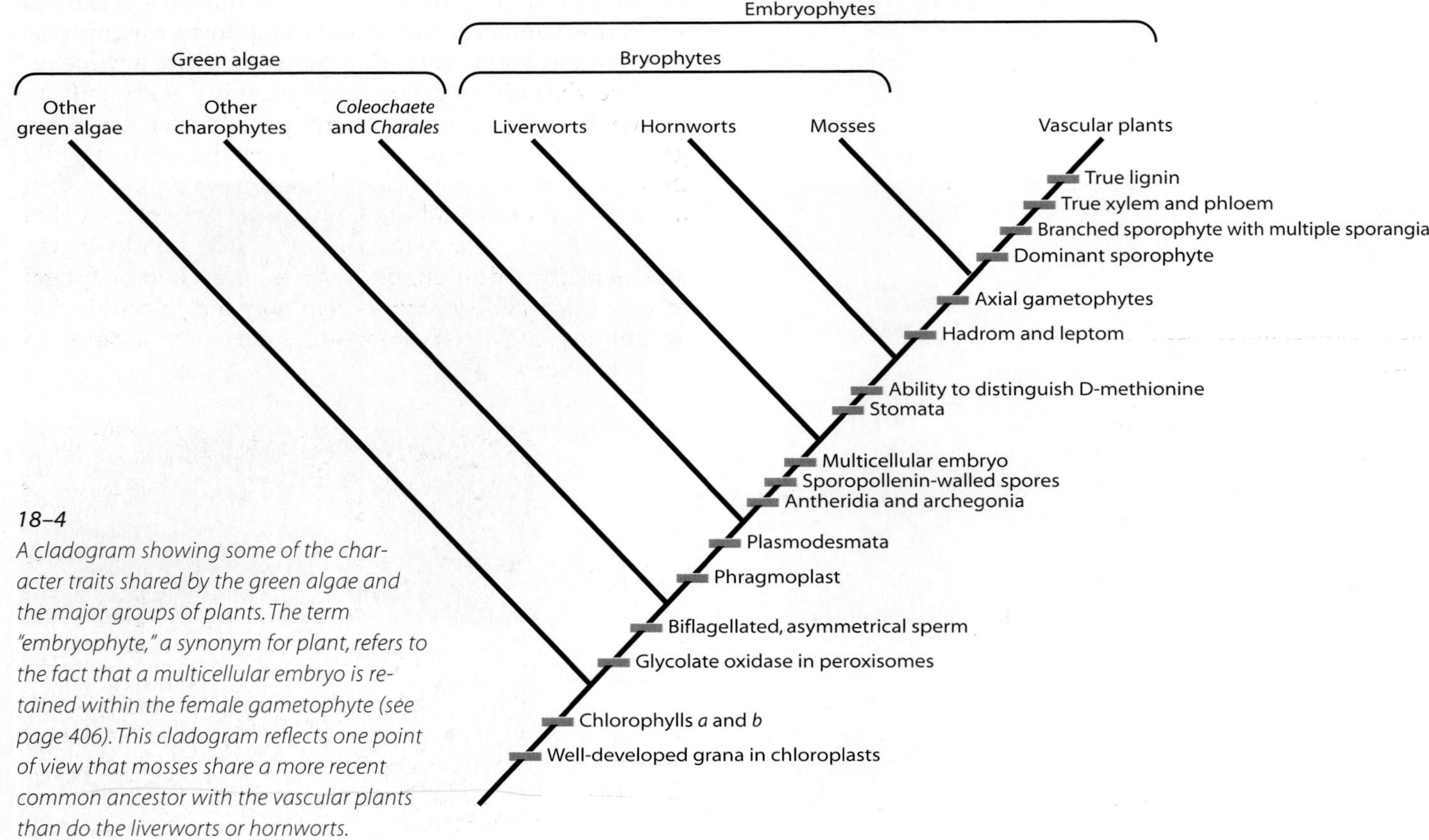

18–4
A cladogram showing some of the character traits shared by the green algae and the major groups of plants. The term "embryophyte," a synonym for plant, refers to the fact that a multicellular embryo is retained within the female gametophyte (see page 406). This cladogram reflects one point of view that mosses share a more recent common ancestor with the vascular plants than do the liverworts or hornworts.

In the past, the relationship of bryophytes to vascular plants was controversial. Today, however, nucleotide sequence data and recent fossil discoveries, together with classical morphological characters and newly described ultrastructural features, reveal that bryophytes include the earliest of the extant plant groups to have diverged within a monophyletic plant lineage (Figure 18–4). Modern bryophytes can therefore provide important insights into the nature of the earliest plants and the process by which vascular plants evolved. A comparison of the structure and reproduction of extant bryophytes with those of ancient fossils and modern vascular plants can show how various features of vascular plants may have evolved.

Comparative Structure and Reproduction of Bryophytes

Some bryophytes, namely the hornworts and certain liverworts, are described as "thalloid" because their gametophytes, which are generally flat and dichotomously branched (forking repeatedly into two equal branches), are **thalli** (singular: thallus). Thalli are undifferentiated bodies, or bodies not differentiated into roots, leaves, and stems. Such thalli are often relatively thin, which may facilitate the uptake of both water and CO_2. Some bryophyte gametophytes have specialized adaptations on their upper surface for increasing CO_2 permeability while at the same time reducing water loss. The surface pores of the liverwort *Marchantia* are one such example (Figure 18–5). The gametophytes of some liverworts (leafy liverworts) and the mosses are said to be differentiated into "leaves" and "stems," but it could be argued that these are not true leaves and stems because they occur in the gametophytic generation and do not contain xylem and phloem. However, the thalli of certain liverworts and mosses do contain centrally located strands of cells that appear to have conducting functions. Such cells may be similar to ancient evolutionary precursors of phloem and lignified vascular tissues. Inasmuch as the terms "leaf" and "stem" are commonly used when referring to the leaflike and stemlike structures of the gametophytes of leafy liverworts and mosses, this practice will be followed in this book.

Surface layers reminiscent of the cuticles commonly found on surfaces of the true leaves and stems of vascular plants also occur on the surfaces of some bryophytes. The cuticle of sporophytes is closely correlated with the presence of stomata, which function primarily in the regulation of gas exchange. The pores seen in some bryophyte gametophytes, such as those of *Marchantia*, are considered to be analogous to stomata (Figure 18–5). The biochemistry and evolution of the bryophyte cuticle are poorly understood, however, primarily because bryophyte cuticles are more difficult to remove for chemical analysis than are cuticles of vascular plants.

The gametophytes of both thalloid and leafy bryophytes are generally attached to the substrate, such as soil, by **rhizoids** (Figure 18–5). The rhizoids of mosses are multicellular, each consisting of a linear row of cells, whereas those of liverworts and hornworts are unicellular. The rhizoids of bryophytes generally serve only to anchor the plants, since absorption of water and inorganic ions commonly occurs directly and rapidly throughout the gametophyte. Mosses, in particular, often have special hairs and other structural adaptations that aid in external water transport and absorption by leaves and stems. In addition, bryophytes often harbor fungal or cyanobacterial symbionts that may aid in acquisition of mineral nutrients. Rootlike organs are lacking in the bryophytes.

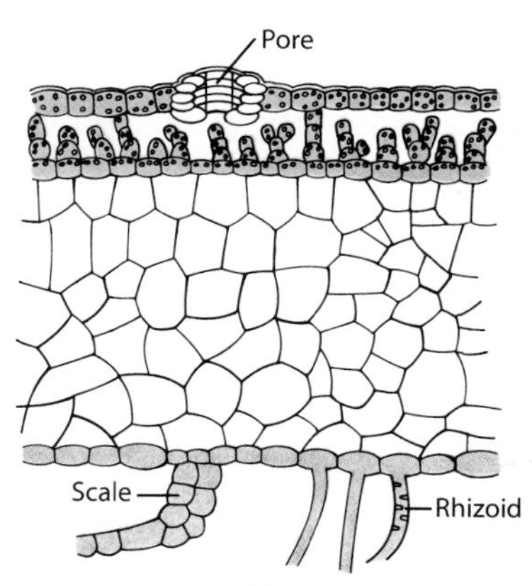

(a)

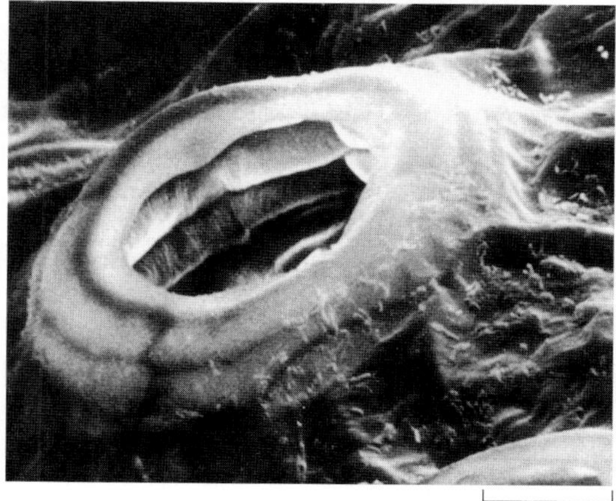

(b)

18–5
Surface pores of Marchantia, *a thalloid liverwort.* ***(a)*** *Transverse section of the gametophyte of* Marchantia. *Numerous chloroplast-bearing cells are evident in the upper layers, and there are several layers of colorless cells below them, as well as rhizoids that anchor the plant body to the substrate. Pores permit the exchange of gases in the air-filled chambers that honeycomb the upper photosynthetic layer. The specialized cells that surround each pore are usually arranged in four or five superimposed circular tiers of four cells each, and the whole structure is barrel-shaped. Under dry conditions, the cells of the lowermost tier, which usually protrude into the chamber, become juxtaposed and retard water loss, whereas under moist conditions they separate. Thus the pores serve a function similar to that of the stomata of vascular plants.* ***(b)*** *A scanning electron micrograph of a pore on the dorsal surface of a gametophyte of* Marchantia.

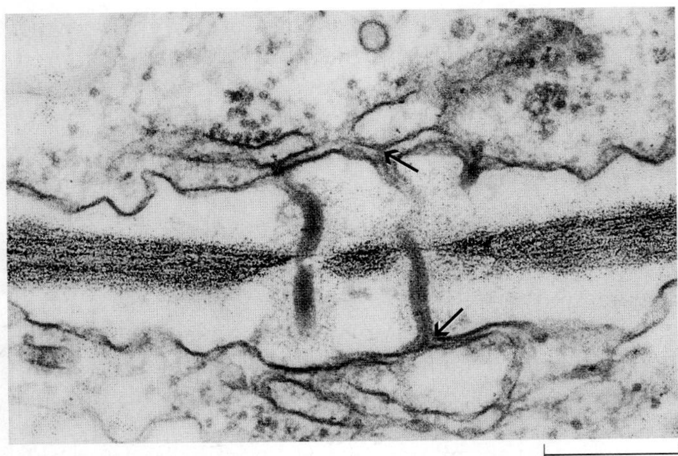

18–6

Longitudinal view of plasmodesmata in the liverwort Monoclea gottschei. *Note (arrows) that the desmotubule in the plasmodesma on the right is continuous with the endoplasmic reticulum in the cytosol.*

The cells of bryophyte tissues are interconnected by plasmodesmata. Bryophyte plasmodesmata are similar to those of vascular plants in possessing an internal component known as the desmotubule (Figure 18–6). The desmotubule is derived from a segment of tubular endoplasmic reticulum that becomes entrapped in developing cell plates during cytokinesis (see Figure 8–11). Certain charophytes also possess plasmodesmata, but the degree to which they are similar to those of bryophytes and represent the ancient forerunners of plant plasmodesmata is still under study (page 398). The cells of most bryophytes resemble those of vascular plants in having many small, disk-shaped plastids. All of the cells of some hornwort species, and the apical and/or reproductive cells of many bryophytes, by contrast, have only a single large plastid per cell. This characteristic is believed to be an evolutionary holdover from ancestral green algae, which, like modern *Coleochaete*, probably contained only a single large plastid per cell. During cell division, the cells of bryophytes and vascular plants produce preprophase bands consisting of microtubules that specify the position of the future cell wall. Such bands are lacking in the charophytes.

Sperm Are the Only Flagellated Cells Produced by Bryophytes, and They Require Water to Swim to the Egg

Many bryophytes can reproduce asexually by fragmentation (vegetative propagation), whereby small fragments, or pieces, of tissue produce an entire gametophyte. Another widespread means of asexual reproduction in both liverworts and mosses is the production of **gemmae**—multicellular bodies that give rise to new gametophytes (see Figure 18–15). Unlike some charophytes, which can generate flagellated zoospores for asexual reproduction, sperm are the only flagellated cells produced by bryophytes. Loss of ability to produce zoospores, which are likely to be less useful on land than in the water, is probably correlated with the absence of centrioles from the spindles of bryophytes and other plants (page 158). Mitosis in certain liverworts and hornworts shows features that are intermediate between those of charophytes and vascular plants, suggesting evolutionary stages leading to the absence of centrioles in plant mitosis.

Sexual reproduction in bryophytes involves production of antheridia and archegonia, often on separate male or female gametophytes. In some species, sex is known to be controlled by the distribution at meiosis of distinctive sex chromosomes. In fact, sex chromosomes in plants were first discovered in bryophytes. The spherical or elongated antheridium is commonly stalked and consists of a sterile jacket layer, one cell thick, that surrounds numerous **spermatogenous cells** (Figure 18–7a). The "jacket" layer of cells is said to be "sterile" because it cannot produce sperm. Each spermatogenous cell forms a single biflagellated sperm that must swim through water to reach the egg located inside an archegonium. Liquid water is therefore required for fertilization in bryophytes.

The archegonia of bryophytes are flask-shaped, with a long neck and a swollen basal portion, the **venter,** which encloses a single egg (Figure 18–7b). The outer layer of cells of the neck and venter forms the sterile protective layer of the archegonium. The central cells of the neck, the **neck canal cells,** disintegrate when the egg is mature, resulting in a fluid-filled tube through which the sperm swim to the egg. During this period, chemicals are released that attract sperm. After fertilization, the zygote remains within the archegonium where it is nourished by sugars, amino acids, and probably other substances provided by the maternal gametophyte. This form of nutrition is known as **matrotrophy** ("food derived from the mother"). Thus supplied, the zygote undergoes repeated mitotic divisions, generating the multicellular embryo (Figure 18–8), which eventually develops into the mature sporophyte (Figure 18–9). There are no plasmodesmatal connections between cells of the two adjacent generations. Nutrient transport is thus apoplastic—that is, nutrients move along the cell walls. This transport is facilitated by a **placenta** that occurs at the interface between the sporophyte and parental gametophyte (Figure 18–10) and is therefore analogous to the placenta of mammals. The bryophyte placenta is composed of transfer cells with an extensive labyrinth of highly branched cell wall ingrowths that vastly increase the surface area of the plasma membrane across which active nutrient transport takes place. Similar transfer cells occur at the gametophyte-sporophyte interface of vascular plants (for example, *Arabidopsis* and soybean) and at the haploid-diploid junction of *Coleochaete* (page 397). The occurrence of pla-

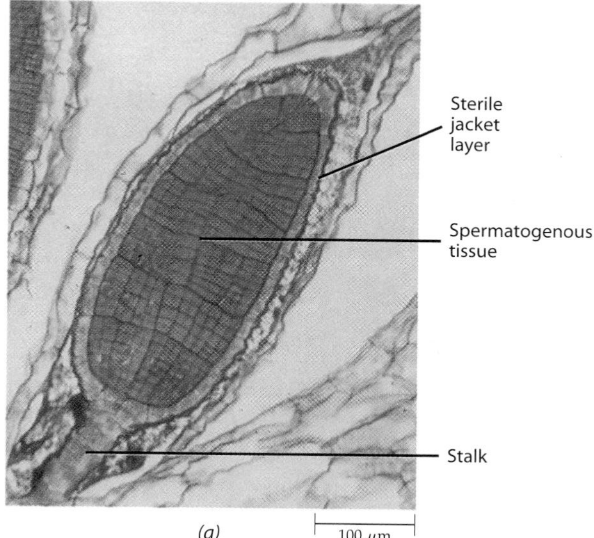

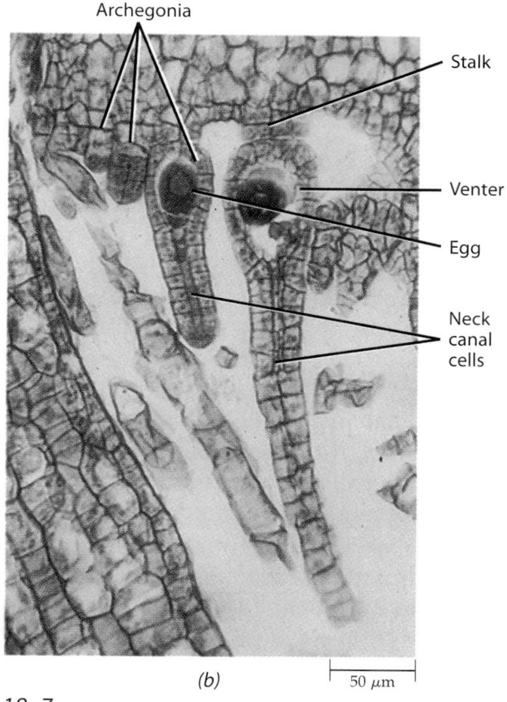

18–7
Gametangia of Marchantia, *a liverwort.* ***(a)*** *A developing antheridium, consisting of a stalk and sterile—that is, non-sperm-forming—jacket layer enclosing spermatogenous tissue. The spermatogenous tissue develops into spermatogenous cells, each of which forms a single sperm propelled by two flagella.*

(b) *Several archegonia at different stages of development. An egg is contained in the venter, a swollen portion at the base of each flask-shaped archegonium. When the egg is mature, the neck canal cells disintegrate, creating a fluid-filled tube through which the biflagellated sperm swim to the egg in response to chemical attractants. In* Marchantia, *the archegonia and antheridia are borne on different gametophytes.*

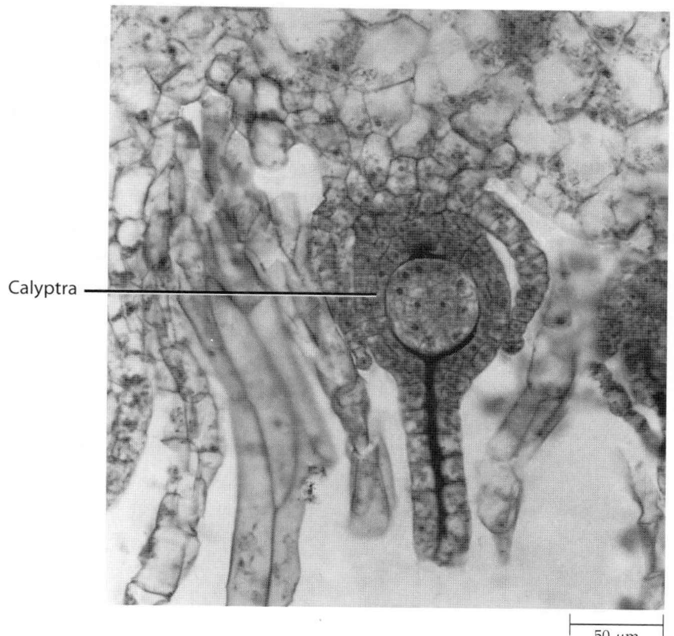

18–8
An early stage in development of the embryo, or young sporophyte, of Marchantia. *Here the young sporophyte is nothing more than an undifferentiated spherical mass of cells within the enlarged venter, or calyptra.*

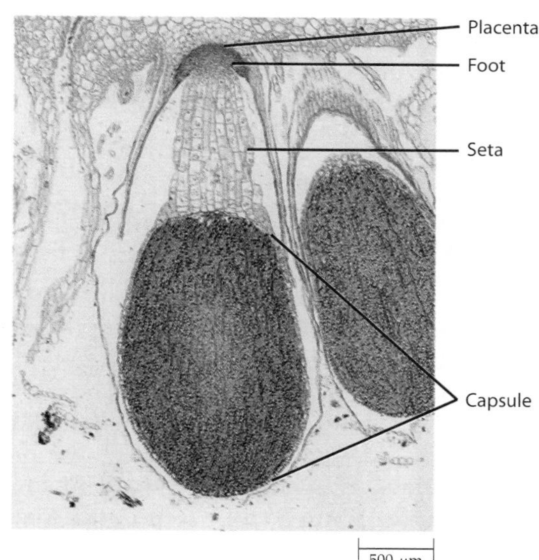

18–9
A nearly mature sporophyte of Marchantia. *At this stage of development the foot, seta, and capsule, or sporangium, of the sporophyte are distinct. The placenta occurs at the interface between the foot and gametophyte and consists of transfer cells of both sporophyte and gametophyte.*

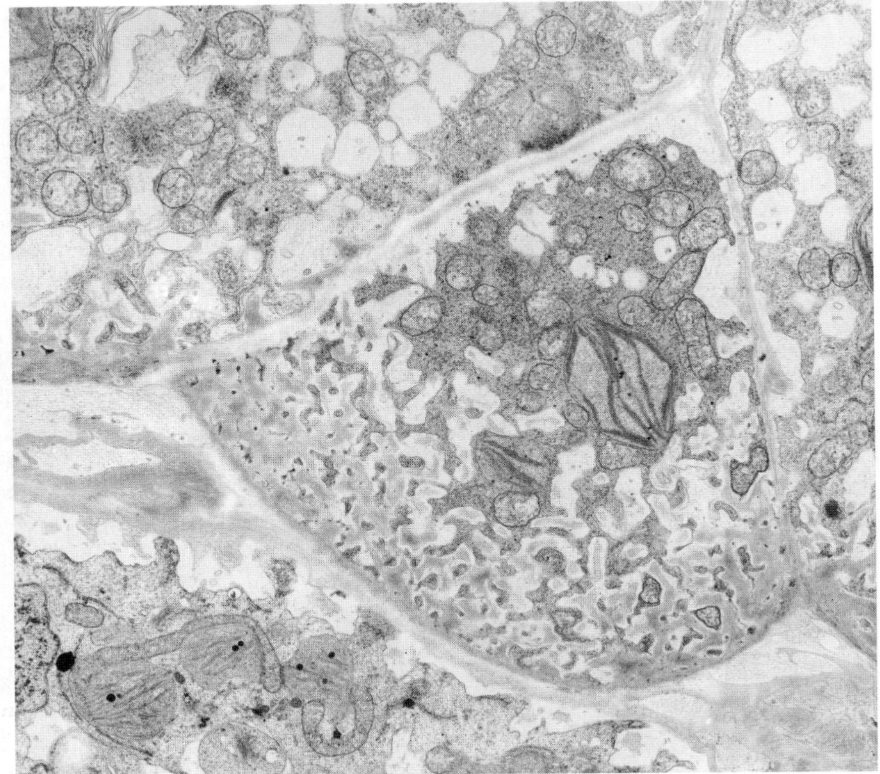

18–10
The gametophyte-sporophyte junction (placenta) in the liverwort Carrpos monocarpos. *Extensive wall ingrowths develop in the single cell layer of transfer cells in the sporophyte (upper three cells). There are several layers of transfer cells in the gametophyte (lower left corner), but their wall ingrowths are not as highly branched as those of the sporophyte layer. Numerous chloroplasts and mitochondria are present in the placental cells of both generations.*

cental cells in *Coleochaete* suggests that matrotrophy had already evolved in the charophyte ancestors of plants.

As the bryophyte embryo develops, the venter undergoes cell division, keeping pace with the growth of the young sporophyte. The enlarged venter of the archegonium is called a **calyptra.** At maturity the sporophyte of most bryophytes consists of a **foot,** which remains embedded in the archegonium, a **seta,** or stalk, and a **capsule,** or **sporangium** (Figure 18–9). The transfer cells at the junction between the foot and archegonium constitute the placenta.

The Term "Embryophytes" Is an Appropriate Synonym for Plants

The occurrence of a multicellular, matrotrophic embryo in all groups of plants, from bryophytes through angiosperms, is the basis for the term **embryophytes** as a synonym for plants (Figure 18–4). The advantage of matrotrophy and the plant placenta is that they fuel the production of a many-celled diploid sporophyte, each cell of which is genetically equivalent to the fertilized egg. These cells can be used to produce many genetically diverse haploid spores upon meiosis in the sporangium. This condition may have provided a significant advantage to early plants as they began to occupy the land. Production of greater numbers of spores per fertilization event may also have helped compensate for low fertilization rates when water became scarce. The sporophytic generation of plants is thought to have evolved from a zygote, such as those produced by charophytes, in which meiosis had been delayed until after at least a few mitotic divisions had occurred. The more mitotic divisions that occur between fertilization and meiosis, the larger the sporophyte that can be formed and the greater the number of spores that can be produced. Throughout the evolutionary history of plants, there has been a tendency for sporophytes to become increasingly larger in relation to the size of the gametophyte generation.

The sporophyte epidermis of hornworts and mosses typically contains stomata—each bordered by two guard cells—that resemble the stomata of vascular plants. The stomata apparently have similar functions. One such function is to aid in the uptake of CO_2 by the sporophyte for photosynthesis. Another function is to generate a flow of water and nutrients between the sporophyte and gametophyte induced by the loss of water vapor through the stomata, and, of course, when closed, the stomata retard water loss. Liverwort sporophytes, which are typically smaller and more ephemeral than those of mosses and hornworts, lack stomata. The epidermal cell walls of the moss and liverwort sporophytes are impregnated with decay-resistant phenolic materials that may protect developing spores. Those of the hornwort sporophyte are covered with a protective cuticle.

The Sporopollenin Walls of Bryophyte Spores Have Survival Value

Bryophyte spores, like those of all other plants, are encased in a substantial wall impregnated with the most decay- and chemical-resistant biopolymer known, sporopollenin. The sporopollenin walls enable bryophyte spores to survive dispersal through the air from one moist site to another. The spores of charophytes, which are typically dispersed in water, are not enclosed by a sporopollenin wall. Charophyte zygotes, however, are lined with sporopollenin and can therefore tolerate exposure and microbial attack, remaining viable for long periods. The sporopollenin-walled spores of plants are thought to have originated from charophyte zygotes by change in the timing of sporopollenin deposition.

Bryophyte spores germinate to form juvenile developmental stages, which in mosses are called **protonemata** (singular: protonema; from the Greek, *prōtos*, "first," and *nēma*, "thread"). From the protonemata develop gametophytes and gametangia. Protonemata, which are characteristic of all mosses, are also found in some liverworts, but not in hornworts.

The bryophytes are grouped into three phyla: *Hepatophyta* (the liverworts), *Anthocerophyta* (the hornworts), and *Bryophyta* (the mosses). In addition, there is an enigmatic bryophyte known as *Takakia* (Figure 18–11). In the past, *Takakia* was classified as a liverwort, but now it is considered to be a very divergent moss. The relative order of evolutionary divergence of these four groups is controversial. There is some disagreement as to which group represents the earliest divergent bryophytes and which is most closely related to the vascular plants. With the availability of more molecular sequence information and integration of information from the fossil record, however, it seems most likely that the liverworts diverged first and mosses are more closely related to vascular plants.

18–11
Takakia ceratophylla, *a bryophyte that for many years was known only from its unusual gametophytes. On the left is a female gametophyte with an attached immature sporophyte; on the right is a male gametophyte with orange antheridia.*

Liverworts: Phylum *Hepatophyta*

Liverworts, or hepatics, are a group of about 6000 species of plants that are generally small and inconspicuous, although they may form relatively large masses in favorable habitats, such as moist, shaded soil or rocks, tree trunks, or branches. A few kinds of liverworts grow in water. The name "liverwort" dates from the ninth century, when it was thought, because of the liver-shaped outline of the gametophyte in some genera, that these plants might be useful in treating diseases of the liver. According to the medieval "Doctrine of Signatures," the outward appearance of a body signaled the possession of special properties. The Anglo-Saxon ending *-wort* (originally *wyrt*) means "herb"; it appears as a part of many plant names in the English language.

Most liverwort gametophytes develop directly from spores, but some genera first form a protonema-like filament of cells, from which the mature gametophyte develops. Gametophytes continue to grow from an apical meristem. There are three major types of liverworts that can be differentiated on the basis of structure and grouped into two clades (page 267). One clade consists of the complex thalloid liverworts, which have internal tissue differentiation. The other clade contains the leafy liverworts and the simple thalloid types, which consist of ribbons of relatively undifferentiated tissue.

Complex Thalloid Liverworts Include *Riccia, Ricciocarpus,* and *Marchantia*

Thalloid liverworts can be found on moist, shaded banks and in other suitable habitats, such as flowerpots in a cool greenhouse. The thallus, which is about 30 cells thick at the midrib and approximately 10 cells thick in the thinner portions, is sharply differentiated into a thin, chlorophyll-rich upper (dorsal) portion and a thicker, colorless lower (ventral) portion (Figure 18–5a). The lower surface bears rhizoids, as well as rows of scales. The upper surface is often divided into raised regions, each with a large pore that leads to an underlying air chamber (Figure 18–5b).

The sporophyte structure of *Riccia* and *Ricciocarpus* is among the simplest seen in liverworts (Figure 18–12). *Ricciocarpus,* which grows in water or on damp soil, is bisexual—that is, both sex organs arise on the same plant. Some species of *Riccia* are aquatic, although most are terrestrial. *Riccia* gametophytes may be either unisexual or bisexual. In both *Riccia* and *Ricciocarpus,* the sporophytes are deeply embedded within the dichotomously branched gametophytes and consist of little more than a sporangium. No special mechanism for spore dispersal occurs in these sporophytes. When the portion of the gametophyte containing mature sporophytes dies and decays, the spores are liberated.

(a)

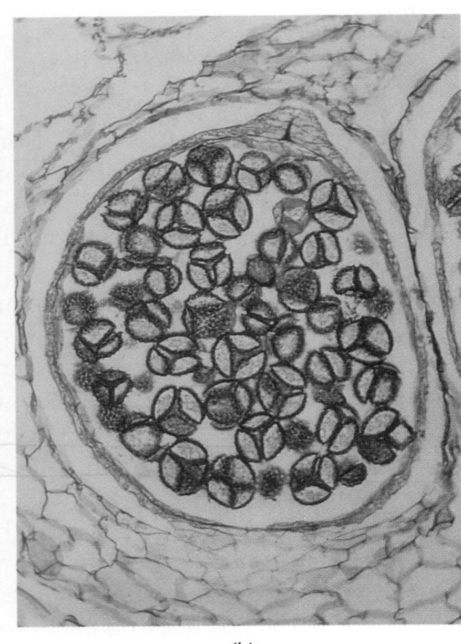

(b)

18–12
Riccia, *one of the simplest of liverworts.* **(a)** *The system of branching of* Riccia *gametophytes is dichotomous, that is, the main and subsequent axes fork into two branches.* **(b)** *The sporophyte, which is embedded within the gametophyte, consists solely of a spherical capsule.*

One of the most familiar of liverworts is *Marchantia,* a widespread genus that grows on moist soil and rocks (see Figure 13-15b). Its dichotomously branched gametophytes are larger than those of *Riccia* and *Ricciocarpus.* Unlike the latter two genera, in which the sex organs are distributed along the dorsal surface of the thallus, *Marchantia* has its gametangia borne on specialized structures called **gametophores.**

The gametophytes of *Marchantia* are unisexual, and the male and female gametophytes can be readily distinguished by their distinctive gametophores. The antheridia are borne on disk-headed gametophores called **antheridiophores,** whereas the archegonia are borne on umbrella-headed gametophores called **archegoniophores** (Figure 18–13). In *Marchantia,* the sporophyte generation consists of a foot, a short seta, and a capsule (Figure 18–9). In addition to spores, the mature sporangium contains elongate cells called **elaters,** which have helically arranged hygroscopic (moisture-absorbing) wall thickenings (Figure 18–14). The walls of elaters are sensitive to slight changes in humidity, and, after the capsule dehisces (dries out and opens) into a number of petal-like segments, the elaters undergo a twisting action that helps disperse the spores.

Fragmentation is the principal means of asexual reproduction in liverworts, but another widespread mechanism is the production of gemmae. In *Marchantia,* the gemmae are produced in special cuplike structures—called **gemma cups**—located on the dorsal (upper) surface of the gametophyte (Figure 18–15). The gemmae are dispersed primarily by splashes of rain.

The life cycle of *Marchantia* is illustrated in Figure 18–17 (pages 410 and 411).

(a)

(b)

18–13
Gametophytes of Marchantia. *The antheridia* **(a)** *and archegonia* **(b)** *are elevated on stalks—the antheridiophores and archegoniophores, respectively—above the thallus.*

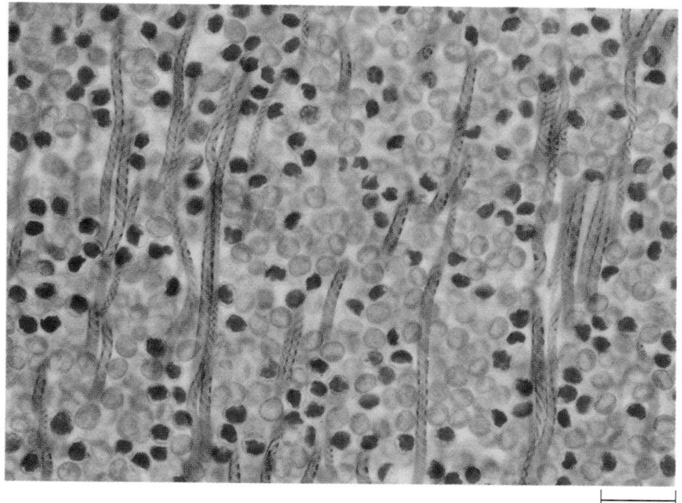

18–14
Mature spores (red spheres) and elaters (green strands) from a capsule of Marchantia.

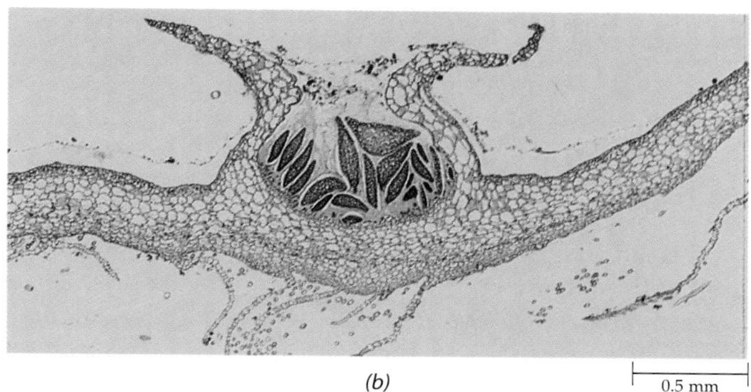

18–15
Gemma cups. ***(a)*** *Gametophytes of* Marchantia, *with gemma cups containing gemmae. The gemmae appear as more or less spherical pieces of tissue. The gemmae are splashed out by the rain and may then grow into new gametophytes, each genetically identical to the parent plant from which it was derived by mitosis.* ***(b)*** *Longitudinal section of a gemma cup. The gemmae are the dark, more or less lens-shaped structures.*

Leafy Liverworts Have a Distinctive Leaf Structure and/or Arrangement

The leafy liverworts are a very diverse group that includes more than 4000 of the 6000 species of the phylum *Hepatophyta* (Figure 18–16). The leafy liverworts are especially abundant in the tropics and subtropics, in regions of heavy rainfall or high humidity, where they grow on the leaves and the bark of trees, as well as on other plant surfaces (Figure 18–18, page 412). There are probably many tropical species that have not yet been described. Leafy liverworts are also well represented in temperate regions. The plants are usually well branched and form small mats.

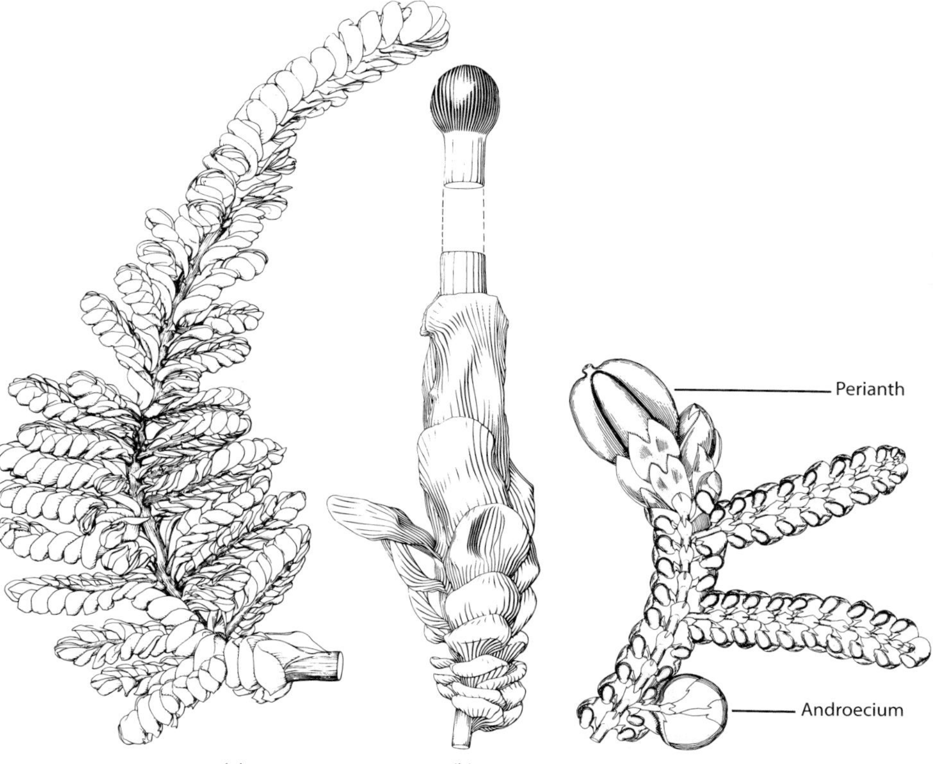

18–16
Leafy liverworts. ***(a)*** Clasmatocolea puccionana, *showing the characteristic arrangement of the leaves.* ***(b)*** *The end of a branch of* Clasmatocolea humilis. *The capsule and the long stalk of the sporophyte are visible.* ***(c)*** *A portion of a branch of* Frullania, *showing the characteristic arrangement of its leaves. The antheridia are contained within the androecium. The archegonium and developing sporophyte are contained within the perianth.*

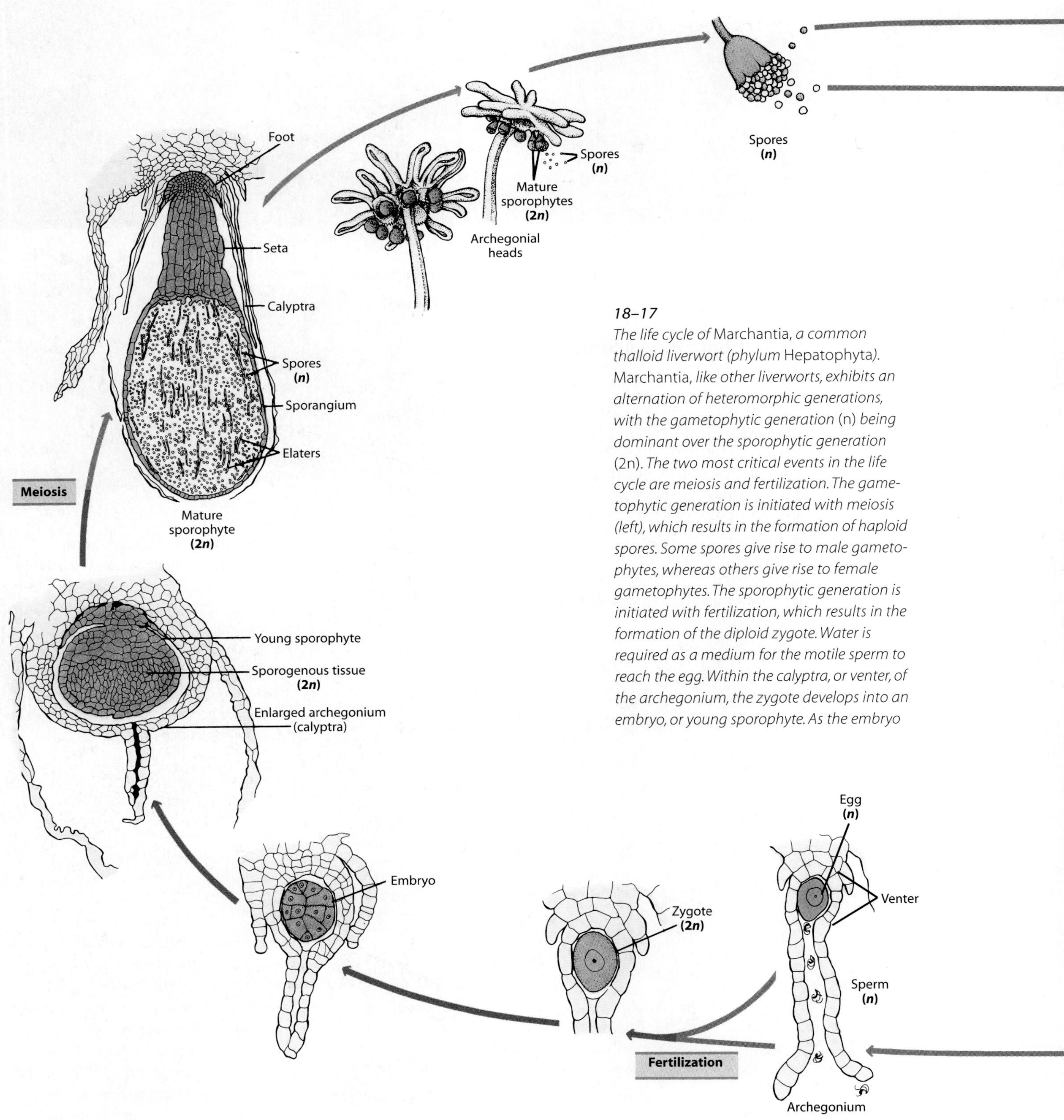

18–17

The life cycle of Marchantia, *a common thalloid liverwort (phylum* Hepatophyta*).* Marchantia, *like other liverworts, exhibits an alternation of heteromorphic generations, with the gametophytic generation* (n) *being dominant over the sporophytic generation* (2n)*. The two most critical events in the life cycle are meiosis and fertilization. The gametophytic generation is initiated with meiosis (left), which results in the formation of haploid spores. Some spores give rise to male gametophytes, whereas others give rise to female gametophytes. The sporophytic generation is initiated with fertilization, which results in the formation of the diploid zygote. Water is required as a medium for the motile sperm to reach the egg. Within the calyptra, or venter, of the archegonium, the zygote develops into an embryo, or young sporophyte. As the embryo*

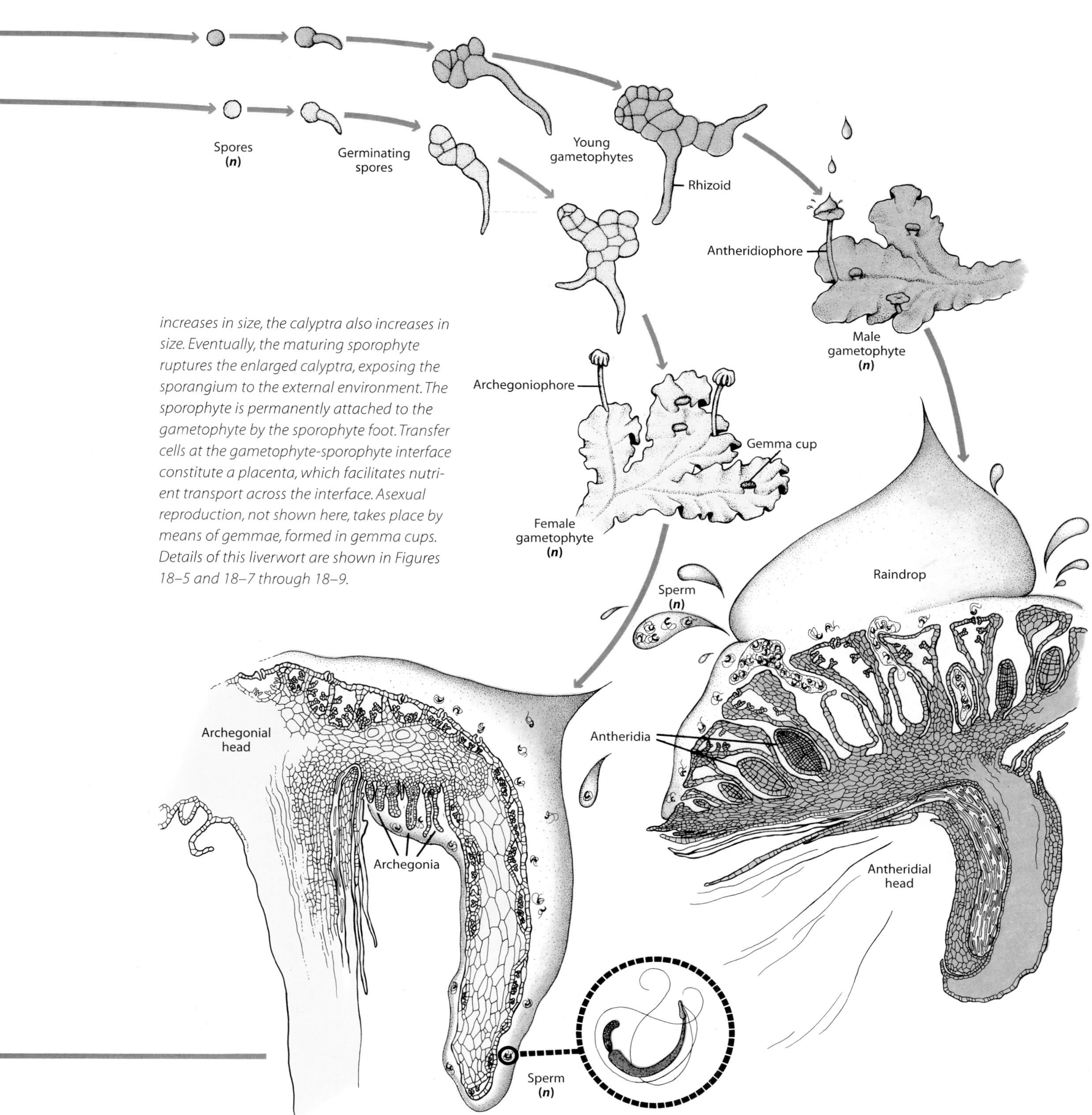

increases in size, the calyptra also increases in size. Eventually, the maturing sporophyte ruptures the enlarged calyptra, exposing the sporangium to the external environment. The sporophyte is permanently attached to the gametophyte by the sporophyte foot. Transfer cells at the gametophyte-sporophyte interface constitute a placenta, which facilitates nutrient transport across the interface. Asexual reproduction, not shown here, takes place by means of gemmae, formed in gemma cups. Details of this liverwort are shown in Figures 18–5 and 18–7 through 18–9.

Liverwort leaves, like moss leaves, generally consist of only a single layer of undifferentiated cells. One way to distinguish liverworts from mosses is that moss leaves are usually equal in size and spirally arranged around the stem, whereas many liverworts have two rows of equal-sized leaves and a third row of smaller leaves along the lower surface of the gametophyte. Most moss leaves splay outward from the stem in three dimensions, but a few have leaves flattened into one plane, as do many liverworts. In addition, moss leaves sometimes have a thickened "midrib," but liverwort leaves lack this structure. Moss leaves are most often entire, in contrast to liverwort leaves, which can be highly lobed or dissected. In *Frullania,* a common liverwort that grows on bark, the leaves consist of a large dorsal lobe and a small, helmet-shaped ventral lobe (Figure 18–16c).

In the leafy liverworts, the antheridia generally occur on a short side branch with modified leaves known as the **androecium.** The developing sporophyte, as well as the archegonium from which it develops, is characteristically surrounded by a tubular sheath known as the **perianth** (Figure 18–16c).

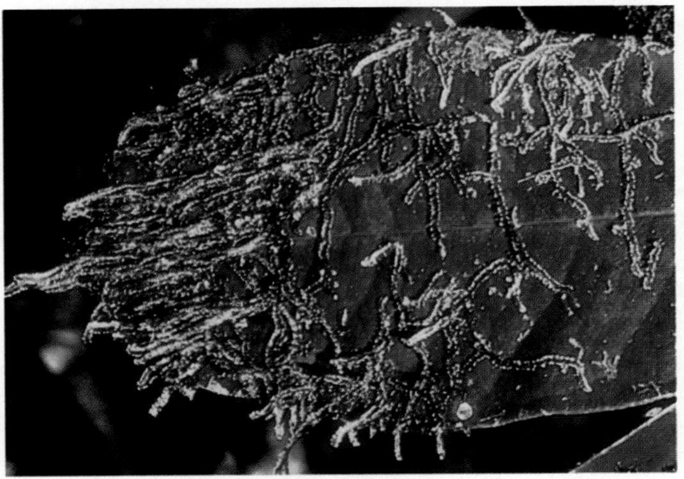

18–18
A leafy liverwort growing on the leaf of an evergreen tree in the rainforest of the Amazon Basin, near Manaus, Brazil.

Hornworts: Phylum *Anthocerophyta*

Hornworts constitute a small phylum of about 100 species. The members of the genus *Anthoceros* are the most familiar of the six genera. The gametophytes of hornworts (Figure 18–19a) superficially resemble those of the thallose liverworts, but there are many features that indicate a relatively distant relationship. For example, the cells of most species usually have a single large chloroplast with a pyrenoid, as in the green alga *Coleochaete.* Some hornwort species have cells containing many small chloroplasts lacking pyrenoids, as do most plant cells, but even in these hornworts the apical cell contains a single plastid, reflecting the ancestral condition (page 404).

Hornwort gametophytes are often rosettelike, and their dichotomous branching is often not apparent. They are usually about 1 to 2 centimeters across. *Anthoceros* has extensive internal cavities that are often inhabited by cyanobacteria of the genus *Nostoc* that fix nitrogen and supply it to their host plants.

The gametophytes of some species of *Anthoceros* are unisexual, whereas others are bisexual. The antheridia and archegonia are sunken on the dorsal surface of the gametophyte, with the antheridia clustered in chambers. Numerous sporophytes may develop on the same gametophyte.

The sporophyte of *Anthoceros,* which is an upright elongated structure, consists of a foot and a long, cylindrical capsule, or sporangium (Figures 18–19 and 18–20). A unique aspect of hornwort sporophytes is that early in their development, a meristem, or zone of actively dividing cells, develops between the foot and the sporangium. This basal meristem remains active as long as conditions are favorable for growth. As a result, the sporophyte continues to elongate for a prolonged period of time. It is green, having several layers of photosynthetic cells. It is also covered with a cuticle and has stomata (Figure 18–19c). The presence of stomata on the sporophytes of hornworts and mosses is regarded as evidence of an important evolutionary link to the vascular plants. Maturation of the spores, and, ultimately, dehiscence of the sporangium, begins near its tip and extends toward the base as the spores mature (Figure 18–19b). Among the spores are sterile, elongate, often multicellular structures that resemble the elaters of liverworts. The dehiscing sporangium splits longitudinally into ribbonlike halves.

Mosses: Phylum *Bryophyta*

Many groups of organisms contain members that are commonly called "mosses"—reindeer "mosses" are lichens, scale "mosses" are leafy liverworts, while club "mosses" and Spanish "moss" belong to different groups of vascular plants. Sea "moss" and Irish "moss" are algae. The genuine mosses, however, are members of the phylum *Bryophyta,* which consists of three classes: *Sphagnidae* (the peat mosses), *Andreaeidae* (the granite mosses), and *Bryidae* (often referred to as the "true mosses"). These groups are very distinct from one another, differing in many important features. Molecular and other information suggests that the peat mosses and granite mosses diverged earlier from the main line of moss evolution. The class *Bryidae* contains the vast majority of moss species. There are at least 9500 species of mosses, with new forms being discovered constantly, especially in the tropics.

(a)

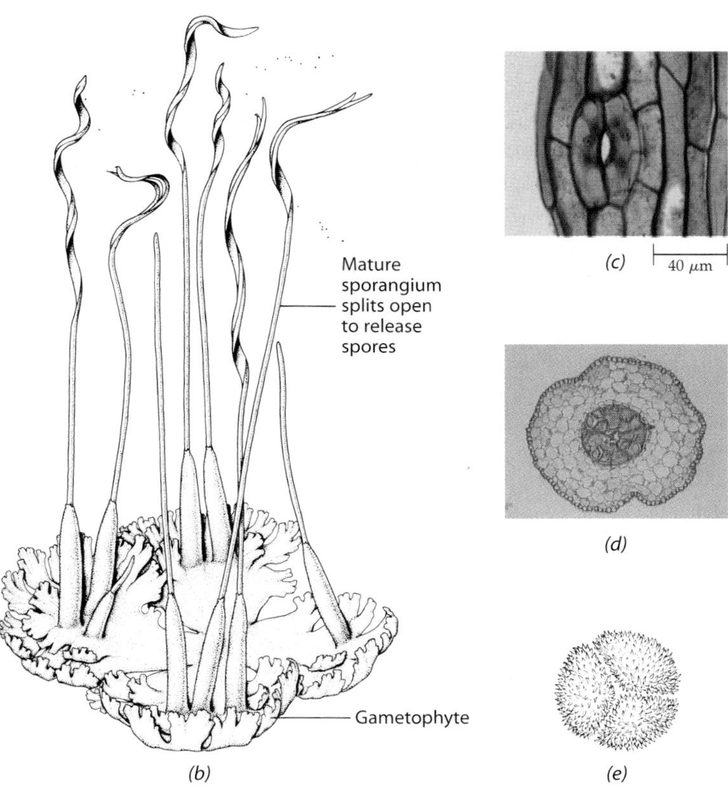

18–19

Anthoceros, *a hornwort.* ***(a)*** *A gametophyte with attached (elongate) sporophytes.* ***(b)*** *When mature, the sporangium splits, and the spores are released.* ***(c)*** *A stoma; stomata are abundant on the sporophytes of the hornworts, which are green and photosynthetic.* ***(d)*** *Developing spores, visible in the center of this cross section of a sporangium, and* ***(e)*** *mature spores still held in a tetrad, a group of four spores—three of which are visible here—formed from a spore mother cell by meiosis.*

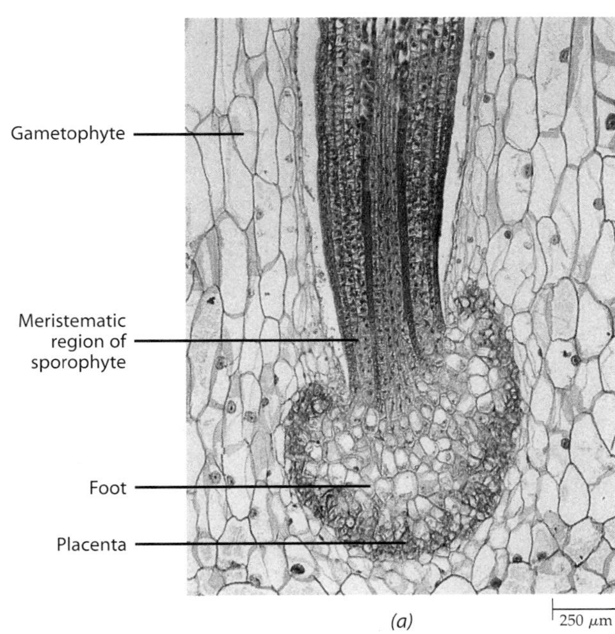

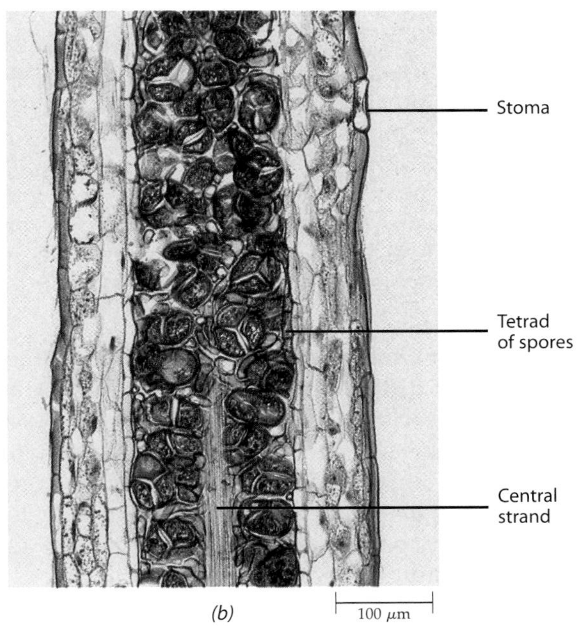

18–20

Anthoceros. ***(a)*** *Longitudinal section of the lower portion of a sporophyte, showing its foot embedded in the tissue of the gametophyte.* ***(b)*** *Longitudinal section of a portion of a sporangium, showing tetrads of spores with elater-like structures among them. The central strand of tissue in the lower part of the sporangium consists of tissue that may function in conduction.*

Peat Mosses Belong to the Class *Sphagnidae*

The class *Sphagnidae* contains primarily one living genus, *Sphagnum*. Distinctive features of its gametophytes and sporophytes (Figure 18–21), as well as comparative DNA sequences, indicate that *Sphagnum* diverged early from the main line of moss evolution. The time of its first appearance is not known, but the fossil order *Protosphagnales*, consisting of several genera of Permian age (about 290 million years ago; see inside front cover), is clearly very closely related to modern *Sphagnum*. There are at least 150 species of living *Sphagnum*. Although more than 300 species have been described, many of these are thought to be polymorphic forms whose structure differs somewhat in different environments. *Sphagnum* is distributed worldwide, in wet areas such as the extensive bog regions of the Northern Hemisphere, and is commercially and ecologically valuable.

Sexual reproduction in *Sphagnum* involves formation of antheridia and archegonia at the ends of special branches located at the tips of the moss gametophyte. Fertilization takes place in late winter, and four months later mature spores are discharged from sporangia.

Among mosses, the sporophytes of *Sphagnum* (Figure 18–21a) are quite distinctive. The red to blackish-brown capsules are nearly spherical and are raised on a stalk, the **pseudopodium,** which is part of the gametophyte and may be up to 3 millimeters long. The sporophyte has a very short seta, or stalk. Spore discharge in *Sphagnum* is spectacular (Figure 18–21c). At the top of the capsule is a lidlike **operculum,** separated from the rest of the capsule by a circular groove. As the capsule matures and dries, its internal tissues shrink and the internal air pressure builds to several bars, similar to pressures found in the tires of trailer trucks. The operculum is eventually blown off with an audible click, and the escaping gas carries a cloud of spores out of the capsule.

Asexual reproduction by fragmentation is very common. Young branches and stem pieces that break off from the gametophyte and injured leaves can regenerate new gametophytes. As a result, *Sphagnum* then forms large, densely packed clumps.

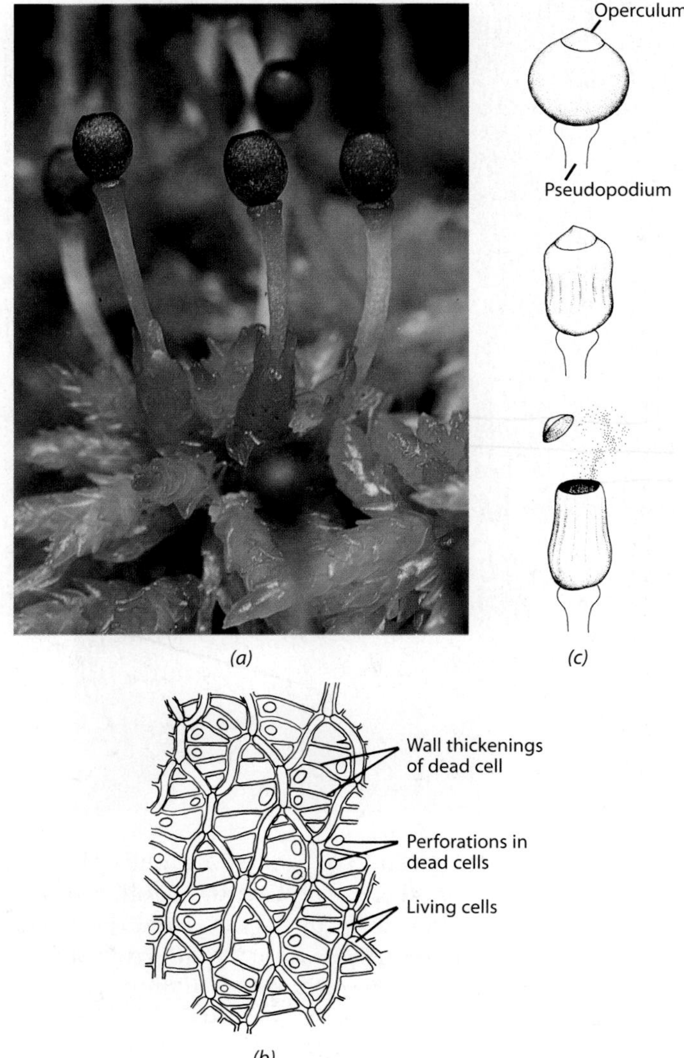

18–21
A peat moss, Sphagnum. ***(a)*** *A gametophyte, with many attached sporophytes. Some of the capsules, such as the two at the front, have already discharged their spores.* ***(b)*** *Structure of a leaf. Large, dead cells are surrounded by smaller living cells, rich in chloroplasts.* ***(c)*** *Dehiscence of a capsule. As the capsule dries, it contracts, changing from a spherical to a cylindrical shape. This change in shape causes compression of trapped gas within the capsule. When the compressed gas reaches a pressure of about 5 bars, the pressure inside the capsule explosively blows off the operculum, releasing a cloud of spores.*

Three Features Distinguish the *Sphagnidae* from Other Mosses The most distinctive differences between the class *Sphagnidae* and other mosses are its unusual protonema, the peculiar morphology of its gametophyte, and its explosive operculum mechanism. The protonema—the first stage of development of the gametophyte—of the *Sphagnidae* does not consist of an extensive array of multicellular, branched filaments as in most other mosses. Instead, each protonema consists of a plate of cells, one layer thick, that grows by a marginal meristem, most of whose cells can divide in one of only two possible directions. In these respects, the protonema of *Sphagnum* is remarkably similar to the disk-shaped thalli of *Coleochaete* (page 397). The erect gametophyte arises from a budlike structure that grows from one of the marginal cells (Figure 18–22). This structure contains an apical meristem that divides in three directions, forming leaf and stem tissues.

The stems of *Sphagnum* gametophytes bear clusters of branches, often five at a node, which are more densely tufted near the tips of the stem, resulting in a moplike head. Both branches and stems bear leaves, but stem leaves often have little or no chlorophyll, whereas most

18–22
Young leafy gametophyte of Sphagnum.

branch leaves are green. The leaves are one cell layer thick and have a distinctive cell pattern unlike that of any other moss. Each leaf consists of large dead cells with circular wall thickenings, surrounded by narrow, green or occasionally red-pigmented, living cells containing several discoid plastids (Figure 18–21b). The walls of dead leaf cells, and the outermost stem cells, are perforated so that they readily become filled with water. As a result, the water-holding capacity of the peat mosses is up to 20 times their dry weight. By comparison, cotton absorbs only four to six times its dry weight. Both living and dead cell walls of *Sphagnum* are impregnated with decay-resistant phenolic compounds and have antiseptic properties. In addition, the peat mosses contribute to the acidity of their own environment by releasing H^+ ions; in the center of the bogs the pH is often less than 4—very unusual for a natural environment.

Because of their superior absorptive and antiseptic qualities, *Sphagnum* mosses have been used as diaper material by native peoples and, in Europe from the 1880s to World War I, as dressings for wounds and boils. *Sphagnum* is still very widely used in horticulture as a packing material for plant roots, as a planting medium, and as a soil additive. Gardeners mix peat moss with soil to increase the water-holding capacity of the soil and make it more acidic. The harvesting and processing of *Sphagnum* from peat bogs for these purposes is a multimillion dollar industry and is of ecological concern because it can result in severe degradation of some wetlands. Efforts are under way to develop techniques for regenerating peatlands because of their ecological importance.

The Ecology of *Sphagnum* Is of Worldwide Importance

Sphagnum-dominated peatlands occupy more than 1 percent of the Earth's surface, an enormous area equal to about one-half that of the United States. *Sphagnum* is thus one of the most abundant plants in the world. Peatlands are of particular importance in the global carbon cycle because peat stores very large amounts (about 400 gigatons, or 400 billion metric tons, on a global basis) of organic carbon that is not readily decayed to CO_2 by microorganisms. Peat is formed from the accumulation and compression of the mosses themselves, as well as the sedges, reeds, grasses, and other plants that grow among them. In Ireland and some other northern regions, dried peat is burned and used widely as industrial fuel, as well as for domestic heating. Ecologists are concerned that global warming brought about by increasing amounts of CO_2 and other gases in the atmosphere—due in large part to human activities—might result in oxidation of peatland carbon. This could further increase CO_2 levels and global temperatures.

(a)

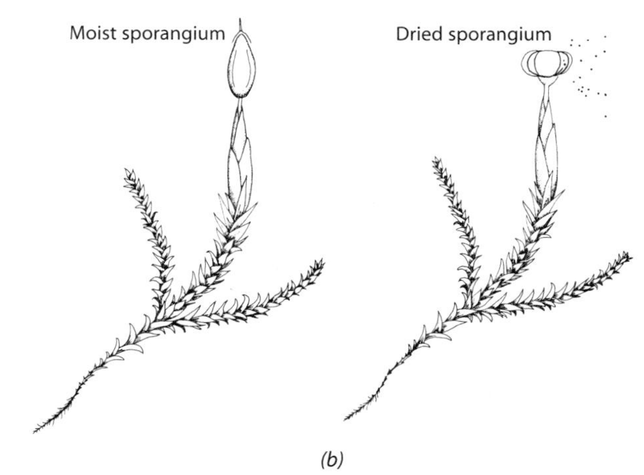

(b)

18–23
(a) Andreaea rothii *growing on granite rock in Devon, England.* *(b)* *The* Andreaea *sporangium (or capsule) contracts and splits as it dries out, allowing the spores to fall out under dry conditions.*

Granite Mosses Belong to the Class *Andreaeidae*

The genus *Andreaea* consists of about 100 species of small, blackish-green or dark reddish-brown tufted rock mosses (Figure 18–23), which in their own way are as peculiar as *Sphagnum*. *Andreaea* occurs in mountainous or Arctic regions, often on granite rocks—hence its common name. A second genus, *Andreaeobryum*, is restricted to northwestern Canada and adjacent Alaska and grows primarily on calcareous (calcium-containing) rocks. In this group, the protonema is unusual in having two or more rows of cells, rather than one row as in most mosses. The rhizoids are unusual in consisting of two rows of cells. The minute capsules are marked by four vertical lines of weaker cells along which the capsule splits. The capsule remains intact above and below these dehiscence lines. The resulting four valves are very sensitive to the humidity of the surrounding air, opening when it is dry—the spores can be carried far by the wind under such circumstances—and closing when it is moist. This mechanism of spore discharge, by means of slits in the capsule, is different from that of any other moss (Figure 18–23). Recent evidence indicates that *Takakia* (page 407) is closely allied to *Andreaea* and *Andreaeobryum*.

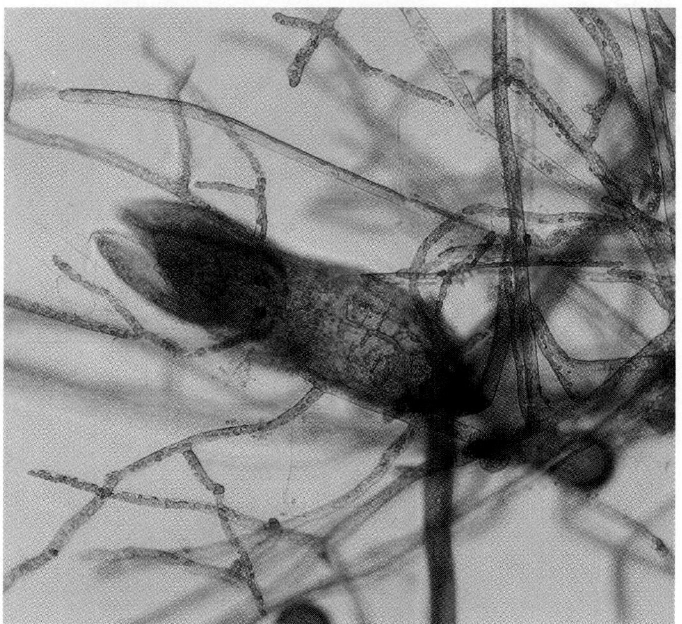

18–24
A "true moss." Protonema of a moss with a budlike structure, from which the leafy gametophyte will develop. Protonemata are the first stage of the gametophyte generation of mosses and some liverworts. They often resemble filamentous green algae.

"True Mosses" Belong to the Class *Bryidae*

The class *Bryidae* contains most of the species of mosses. In this group of mosses—the "true mosses"—the branching filaments of the protonemata are composed of a single row of cells and resemble filamentous green algae (Figure 18–24). They can usually be distinguished from the green algae, however, by their slanted cross walls. Leafy gametophytes develop from minute budlike structures on the protonemata. In a few genera of mosses the protonema is persistent and assumes the major photosynthetic role, whereas the leafy shoots of the gametophyte are minute.

Many Mosses Have Tissues Specialized for Water and Food Conduction Moss gametophytes, which exhibit varying degrees of complexity, can be as small as 0.5 millimeter or 50 centimeters or more long. All have multicellular rhizoids, and the leaves are normally only one cell layer thick except at the midrib (which is lacking in some genera). In many mosses, the stems of the gametophytes and sporophytes have a central strand of conducting tissue called **hadrom.** The water-conducting cells are known as **hydroids** (Figure 18–25). Hydroids are elongate cells with inclined end walls that are thin and highly permeable to water, making them the preferred pathways for water and solutes. Hydroids resemble the water-conducting tracheary elements of vascular plants because both lack a living protoplast at maturity (see Chapter 24). Unlike tracheary elements, however, hydroids lack specialized, lignin-containing wall thickenings. In some moss genera, food-conducting cells, also known as **leptoids,** surround the strand of hydroids (Figure 18–25). The food-conducting tissue is called **leptom.** The leptoids are elongate cells that have some structural and developmental similarities to the food-conducting sieve elements of seedless vascular plants (see page 427). At maturity, both cell types have inclined end walls with small pores and living protoplasts with degenerate nuclei. The conducting cells of mosses—the hydroids and leptoids—apparently are similar to those of certain fossil plants known as **protracheophytes,** which may represent an intermediate stage in the evolution of vascular plants, or **tracheophytes** (see page 434).

Sexual Reproduction in Mosses Is Similar to That of Other Bryophytes The sexual cycle of mosses is similar to that of liverworts and hornworts in that it involves production of male and female gametangia (Figure 18–26), an unbranched matrotrophic (maternally nourished) sporophyte, and specialized spore dispersal processes (see the moss life cycle in Figure 18–27, pages 418 and 419).

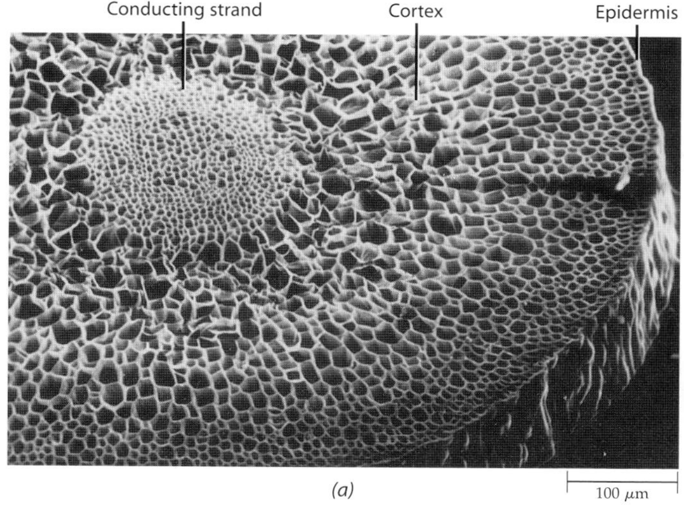

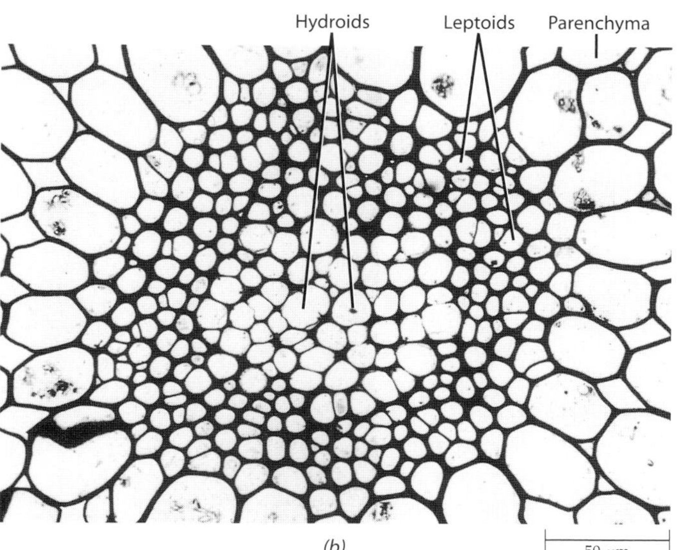

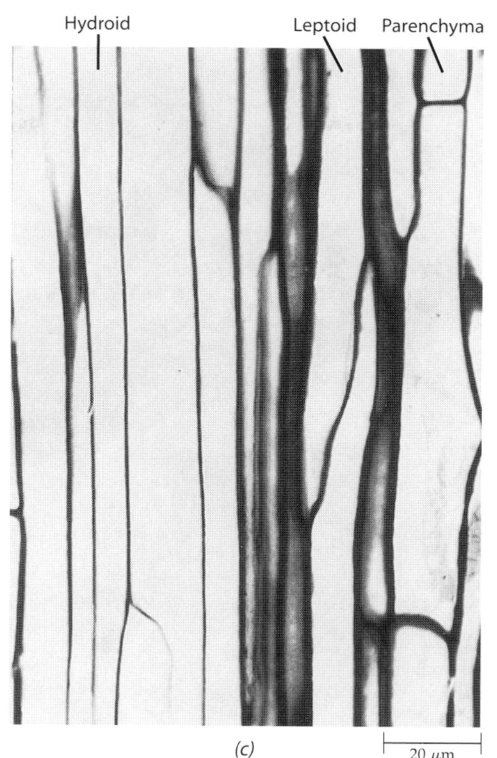

18–25

Conducting strands in the seta, or stalk, of a sporophyte of the moss Dawsonia superba. ***(a)*** *General organization of the seta as seen in transverse section with the scanning electron microscope.* ***(b)*** *Transverse section showing the central column of water-conducting hydroids surrounded by a sheath of food-conducting leptoids and the parenchyma of the cortex.* ***(c)*** *Longitudinal section of a portion of the central strand, showing (from left to right) hydroids, leptoids, and parenchyma.*

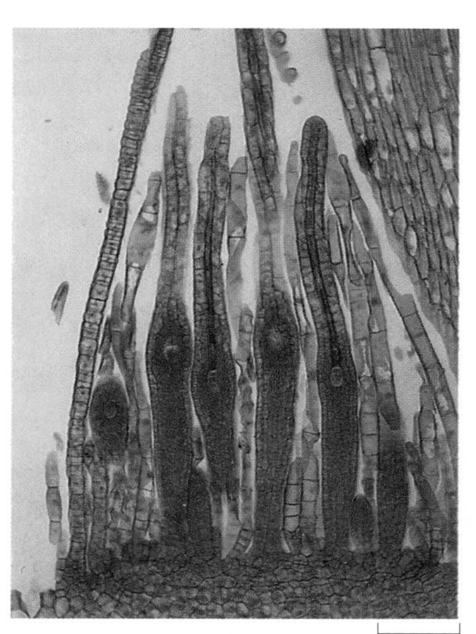

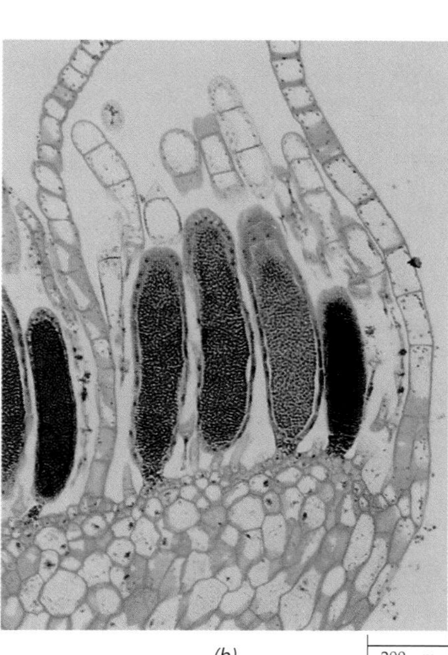

18–26

Gametangia of Mnium, *a unisexual moss.* ***(a)*** *Longitudinal section through an archegonial head showing the pink-stained archegonia surrounded by sterile structures called paraphyses.* ***(b)*** *Longitudinal section through an antheridial head showing antheridia surrounded by paraphyses.*

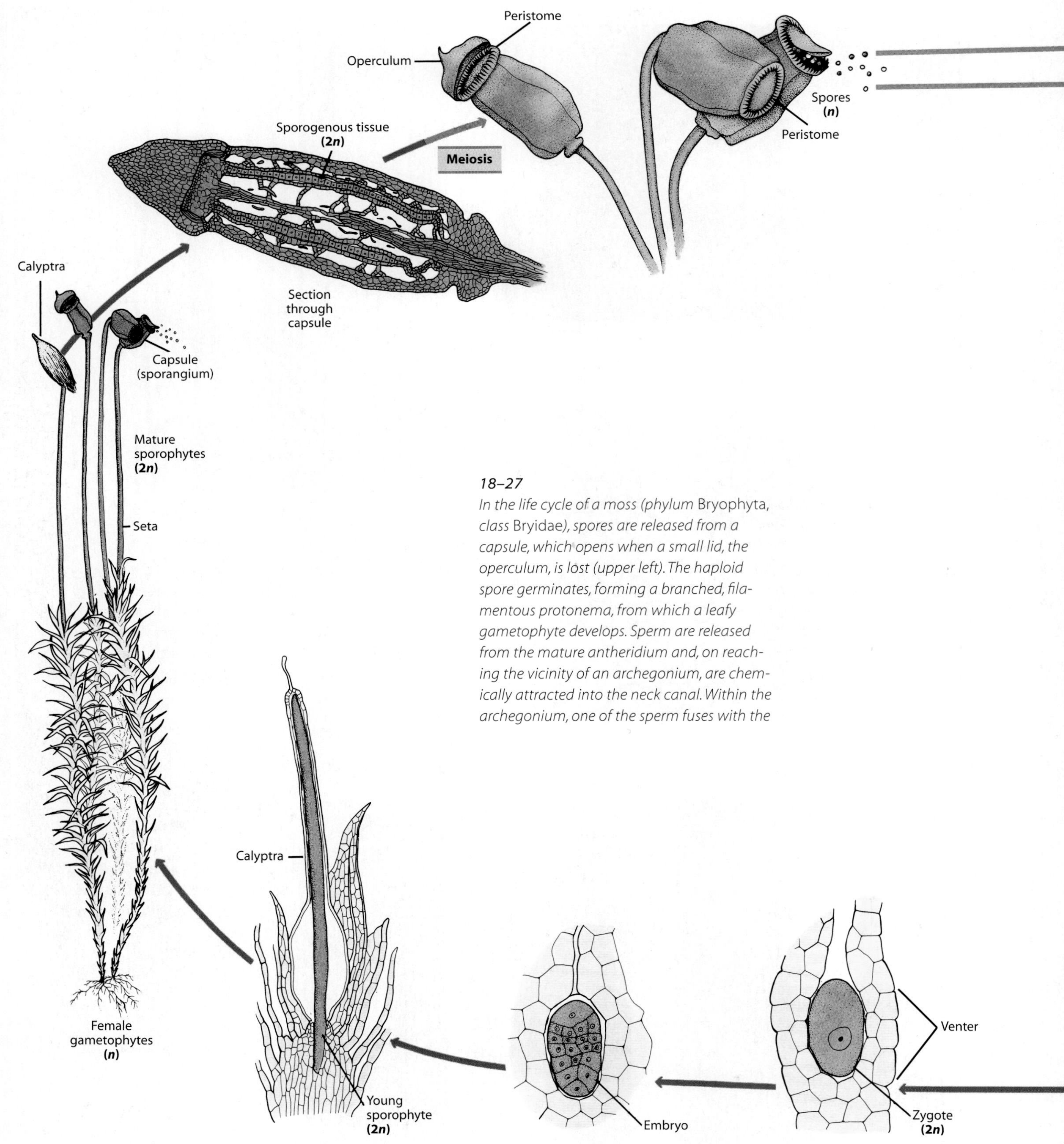

18–27
In the life cycle of a moss (phylum Bryophyta, *class* Bryidae*), spores are released from a capsule, which opens when a small lid, the operculum, is lost (upper left). The haploid spore germinates, forming a branched, filamentous protonema, from which a leafy gametophyte develops. Sperm are released from the mature antheridium and, on reaching the vicinity of an archegonium, are chemically attracted into the neck canal. Within the archegonium, one of the sperm fuses with the*

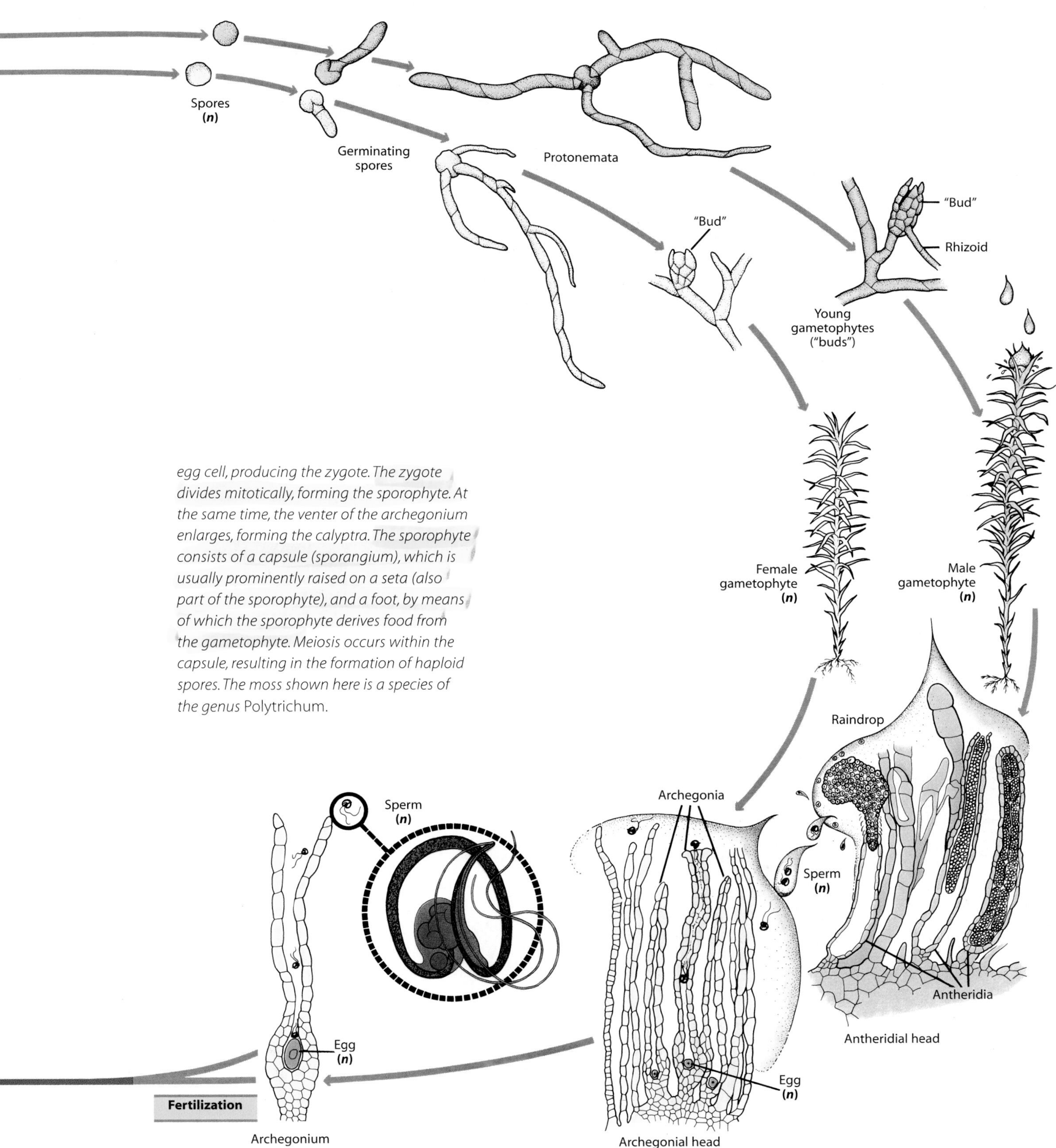

egg cell, producing the zygote. The zygote divides mitotically, forming the sporophyte. At the same time, the venter of the archegonium enlarges, forming the calyptra. The sporophyte consists of a capsule (sporangium), which is usually prominently raised on a seta (also part of the sporophyte), and a foot, by means of which the sporophyte derives food from the gametophyte. Meiosis occurs within the capsule, resulting in the formation of haploid spores. The moss shown here is a species of the genus Polytrichum.

Gametangia may be produced by mature leafy gametophytes, either at the tip of the main axis or on a lateral branch. In some genera, the gametophytes are unisexual (Figure 18–26), but in other genera both archegonia and antheridia are produced by the same plant. Antheridia are often clustered within leafy structures called splash cups (Figure 18–28). The sperm from several to many antheridia are discharged into a drop of water within each cup and are then dispersed as a result of the action of raindrops falling into the cup. Insects may also carry drops of water, rich in sperm, from plant to plant.

Moss sporophytes, like those of hornworts and liverworts, are borne on the gametophytes, which supply the sporophytes with nutrients. A short foot at the base of the stalklike seta is embedded in the tissue of the gametophyte, and cells of both foot and adjacent gametophyte function as transfer cells in the placenta. In the moss *Polytrichum*, simple sugars have been shown to move across the junction between generations. The capsules, or sporangia, usually take 6 to 18 months to reach maturity in temperate species and are generally elevated on a seta into the air, thus facilitating spore dispersal. Some mosses produce brightly colored sporangia that attract insects. Setae may reach 15 to 20 centimeters in length in a few species, but may be very short or entirely absent in other species. The setae of many moss sporophytes contain a central strand of hydroids, which in some genera is surrounded by leptoids (Figure 18–25). Stomata are normally present on the epidermis of moss sporophytes. Some moss stomata, however, are bordered by only a single, doughnut-shaped guard cell (Figure 18–29).

18–28
Leafy male gametophytes of the moss Polytrichum piliferum, *showing the mature antheridia clustered together in cup-shaped heads. The sperm are discharged into drops of water held within these leafy heads and are then splashed out of them by raindrops, sometimes reaching the vicinity of archegonia on other gametophytes (Figure 18–27).*

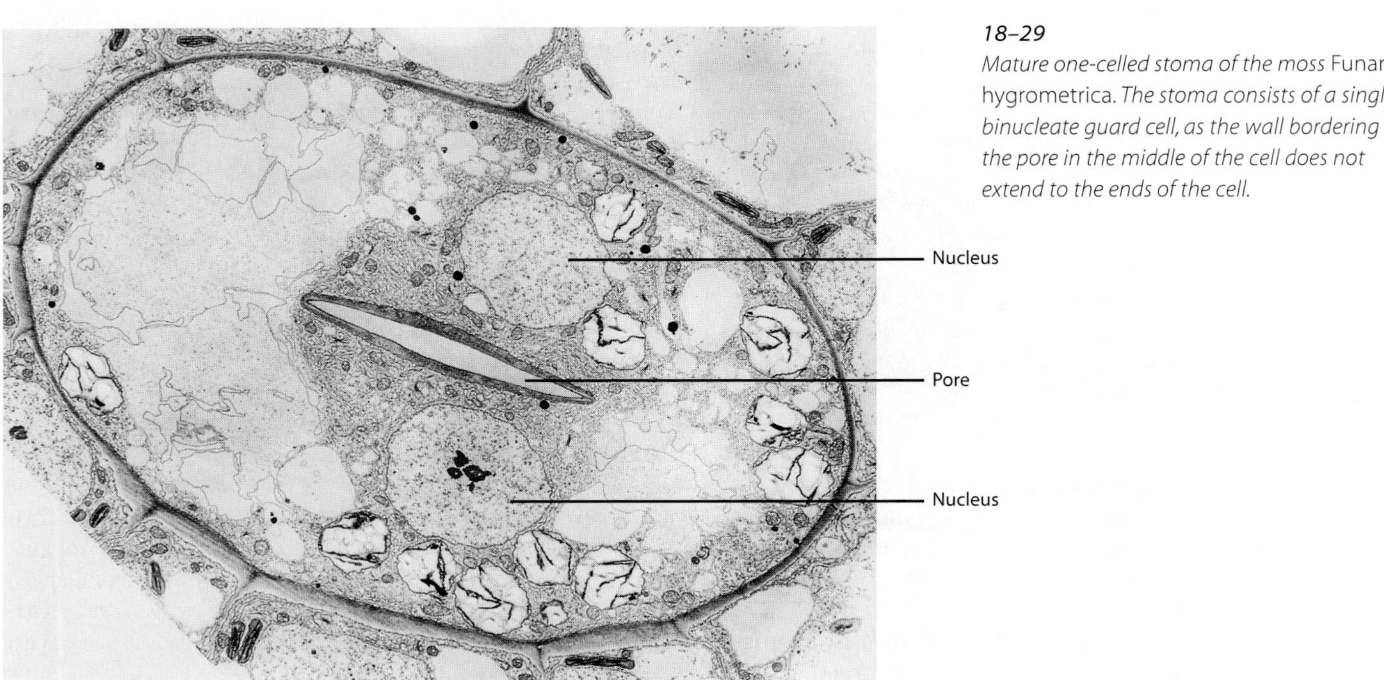

18–29
Mature one-celled stoma of the moss Funaria hygrometrica. *The stoma consists of a single, binucleate guard cell, as the wall bordering the pore in the middle of the cell does not extend to the ends of the cell.*

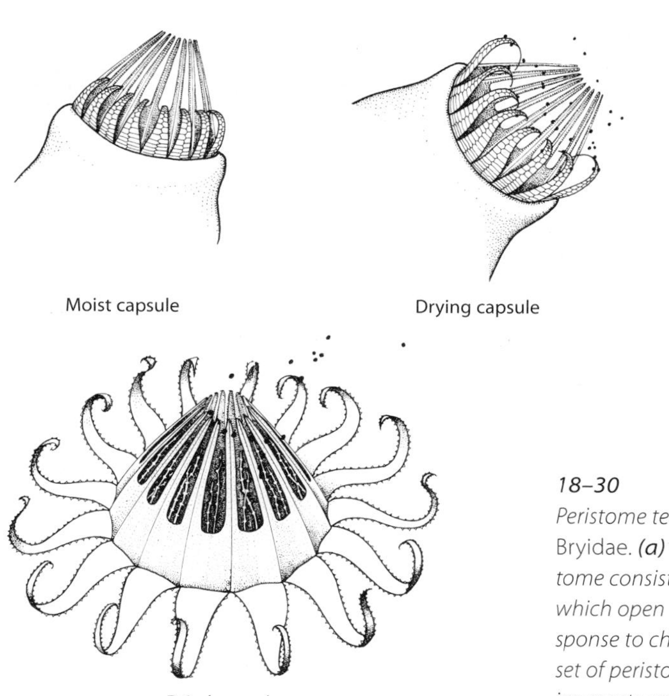

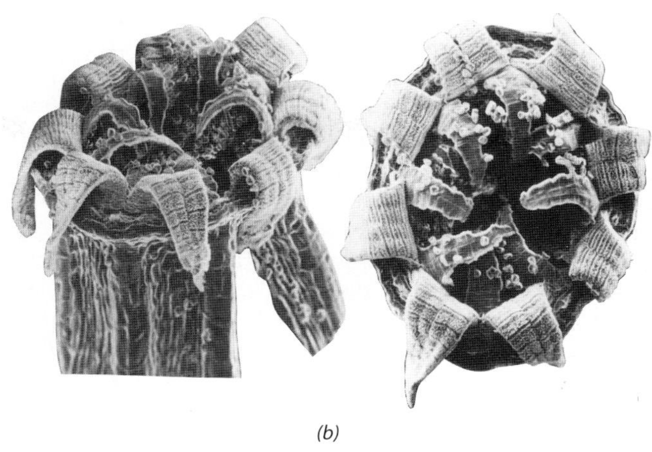

18–30
Peristome teeth in mosses of the class Bryidae. ***(a)*** Brachythecium *has a peristome consisting of two rings of teeth, which open to release the spores in response to changes in moisture. The outer set of peristome teeth interlocks with the inner set under damp conditions. As the capsule dries out, the outer teeth pull away, allowing the dispersal of spores by the wind.* ***(b)*** *Scanning electron micrographs of the peristome teeth of two capsules of* Orthotrichum, *showing the inner teeth curved inward and the outer teeth curved outward in dry conditions.*

Generally, the cells of the young and maturing sporophyte contain chloroplasts and carry out photosynthesis. When a moss sporophyte is mature, however, it gradually loses its ability to photosynthesize and turns yellow, then orange, and finally brown. The calyptra, derived from the archegonium, is commonly lifted upward with the capsule as the seta elongates. Prior to spore dispersal, the protective calyptra falls off, and the operculum of the capsule bursts off, revealing a ring of teeth—the **peristome**—surrounding the opening (Figure 18–30). The teeth of the peristome are formed by the splitting, along a zone of weakness, of a cellular layer near the end of the capsule. In most mosses, the teeth uncurl slowly when the air is relatively dry and curl up again when it is moist. The movements of the teeth expose the spores, which are gradually released. A capsule sheds up to 50 million haploid spores, each of which is capable of giving rise to a new gametophyte. The peristome is a characteristic of the class *Bryidae* and is lacking in the other two classes of mosses. Distinctive features of the peristomes of different moss groups are used in classifying and identifying mosses.

Asexual reproduction usually occurs by fragmentation as virtually any portion of the moss gametophyte is capable of regeneration. However, some mosses produce specialized asexual reproductive structures.

Mosses Exhibit "Cushiony" or "Feathery" Growth Patterns

Two patterns of growth are common among *Bryidae* (Figure 18–31). In the "cushiony" mosses, the gametophytes are erect and little branched, usually bearing terminal sporophytes. In the second pattern, the gametophytes are highly branched and "feathery," the plants are creeping, and the sporophytes are borne laterally. This second type of growth pattern is commonly found in those mosses that hang in masses from the branches of trees in rainforests and tropical cloud forests. Organisms such as these, which grow upon other organisms but are not parasitic on them, are called **epiphytes.** Trees provide a diverse array of microhabitats that are occupied by moss and other bryophyte species. Among these microhabitats are tree bases and buttress roots, fissures and ridges of bark, irregular surfaces of twigs, depressions at the bases of branches, and the surfaces of leaves.

A number of moss genera and species are highly endemic, that is, restricted to very limited geographic areas. Many of the endemic mosses occur as epiphytes in high-altitude temperate and tropical cloud forests, where the bryophyte biodiversity is poorly known. Bryophytes also have important but poorly catalogued interactions with a variety of invertebrates, some of which live, breed, and feed preferentially in mosses. Some experts are concerned that growth of human populations may drastically alter natural environments, leading to extensive loss of bryophyte species and associated animals before many of the organisms are even described.

(a)

(b)

18–31
The two common growth forms found in the gametophytes of different genera of mosses in the class Bryidae. ***(a)*** *"Cushiony" form, in which the gametophytes are erect and have few branches, shown here in* Polytrichum juniperinum. *Sporophytes can be seen rising above the gametophytes. Each consists of a spore capsule atop a long, slender seta.* ***(b)*** *"Feathery" form, with matted, creeping gametophytes, shown here in* Thuidium abietinum.

Summary TABLE Comparative Summary of Characteristics of Bryophyte Phyla

PHYLUM	NUMBER OF SPECIES	GENERAL CHARACTERISTICS OF GAMETOPHYTE	GENERAL CHARACTERISTICS OF SPOROPHYTE	HABITATS
Hepatophyta (liverworts)	6000	Dominant and free-living generation; both thalloid and leafy genera; pores in some thalloid types; unicellular rhizoids; most cells have numerous chloroplasts; many produce gemmae; protonema stage in some; growth from apical meristem	Small and nutritionally dependent on gametophyte; unbranched; consists of little more than sporangium in some genera, and of foot, short seta, and sporangium in others; phenolic materials in epidermal cell walls; lacks stomata	Mostly moist temperate and tropical; a few aquatic; often as epiphytes
Anthocerophyta (hornworts)	100	Dominant and free-living generation; thalloid; unicellular rhizoids; most have single chloroplast per cell	Small and nutritionally dependent on gametophyte; unbranched; consists of foot and long, cylindrical sporangium, with a meristem between foot and sporangium; cuticle; stomata; no specialized conducting tissues	Moist temperate and tropical
Bryophyta (mosses)	9500	Dominant and free-living generation; leafy; multicellular rhizoids; most cells have numerous chloroplasts; many produce gemmae; protonema stage that grows by marginal meristem followed by further growth from an apical meristem in *Sphagnum*; growth by apical meristem only in *Bryidae;* some species have leptoids and nonlignified hydroids	Small and nutritionally dependent on gametophyte; unbranched; consists of foot, long seta, and sporangium in *Bryidae;* phenolic materials in epidermal cell walls; stomata; some species have leptoids and nonlignified hydroids	Mostly moist temperate and tropical; some Arctic and Antarctic; many in dry habitats; a few aquatic

Summary

Plants Most Likely Evolved from a Charophyte Ancestor

Plants, collectively known as embryophytes, seem to have been derived from a charophycean green alga. The two groups share many unique features, including a phragmoplast and cell plate at cytokinesis. Molecular and other evidence strongly suggests that plants are descended from a single common ancestor and that bryophytes include the earliest living plants to have diverged from the main line of plant evolution. The earliest plants were probably similar in several respects to modern bryophytes. The characteristics they share are thalli constructed of tissues produced by an apical meristem, a life history involving alternation of heteromorphic generations, protective walled gametangia, matrotrophic embryos, and sporopollenin-walled spores.

The Bryophytes Are the Liverworts, Hornworts, and Mosses

The bryophytes consist of three phyla of structurally rather simple, small plants. Their gametophytes are always nutritionally independent of the sporophytes, whereas the sporophytes are permanently attached to the gametophytes and nutritionally dependent upon them for at least some time early in embryo development. Thus, the gametophyte is the dominant generation. The male sex organs, antheridia, and female sex organs, archegonia, both have protective jacket layers. Each archegonium contains a single egg, whereas each antheridium produces numerous sperm. The biflagellated sperm are free-swimming and require water to reach the egg. The sporophyte is typically differentiated into a foot, a seta, and a capsule, or sporangium. Maturing hornwort and moss sporophytes are green and become less nutritionally dependent on their gametophytes than those of liverworts, which usually remain completely dependent on their gametophyte.

The Bryophytes Differ from One Another in the Presence or Absence of Stomata and Conducting Tissues and in Types of Meristems

Liverworts (phylum *Hepatophyta*) differ from mosses and hornworts in lacking stomata. Hornworts (phylum *Anthocerophyta*) have a unique basal meristem and lack specialized conducting tissue. Mosses (phylum *Bryophyta*) have, at least in some groups, both specialized conducting tissue and stomata that resemble those of vascular plants. The conducting tissue of mosses, when present, consists of hydroids, water-conducting cells, and leptoids, food-conducting cells.

Bryophytes Are Important Ecologically

Bryophytes are particularly abundant and diverse in temperate rainforests and tropical cloud forests. The moss *Sphagnum* occupies more than 1 percent of the Earth's surface, is economically valuable, and plays an essential role in the global carbon cycle.

Selected Key Terms

antheridia p. 402
archegonia p. 402
calyptra p. 406
capsule, or **sporangium** p. 406
elaters p. 408
embryophytes p. 406
foot p. 406
gametophores p. 408
gemma cups p. 408
gemmae p. 404
hadrom p. 416
hydroids p. 416
leptoids p. 416
leptom p. 416
matrotrophy p. 404
neck canal cells p. 404
operculum p. 414
peristome p. 421
placenta p. 404
protonema/protonemata p. 407
rhizoids p. 403
seta p. 406
spermatogenous cells p. 404
sporopollenin p. 402
tracheophytes p. 416
venter p. 404

Questions

1. By means of a simple, labeled diagram, outline a generalized life cycle of a bryophyte. Explain why it is referred to as an alternation of heteromorphic generations.
2. What evidence is there in support of a charophyte ancestry for plants?
3. Bryophytes and vascular plants share a number of characters that distinguish them from charophytes and that adapt them for existence on land. What are those characters?
4. In your opinion, which of the bryophytes has the most highly developed sporophyte? Which has the most highly developed gametophyte? In each case, give the reasons for your answer.
5. Judging from the cladogram in Figure 18–4, what group of bryophytes is most closely related to vascular plants? What characters shared by vascular plants are lacking in bryophytes?
6. Describe the structural modifications related to water absorption in *Sphagnum*. Why is *Sphagnum* of such great ecological importance?

Chapter 19

Seedless Vascular Plants

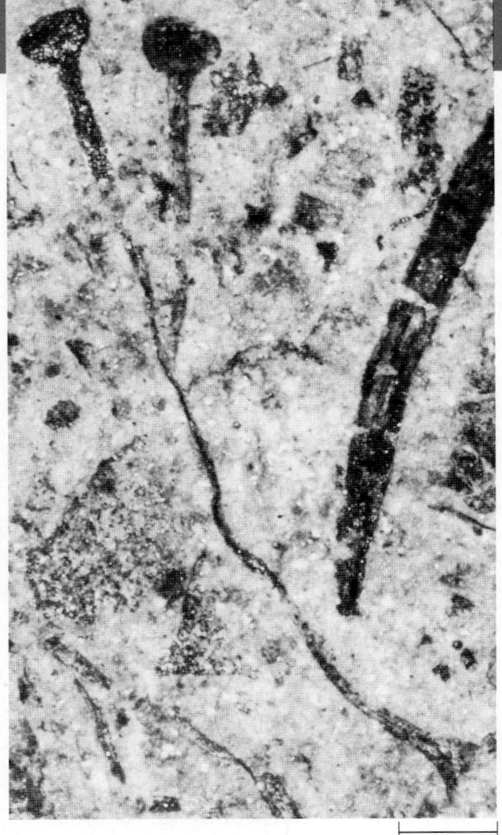

19–1
Cooksonia, *the oldest known vascular plant, consisted of dichotomously branching axes. This fossil, found in New York State, is from the Late Silurian (414–408 million years ago). Its leafless aerial stems ranged up to about 6.5 centimeters long and terminated in sporangia, or spore-producing structures. These small plants probably lived in moist environments such as mud flats.*

OVERVIEW

In this chapter you will learn not only about the ferns, but also about their closest relatives, often referred to as the "fern allies." All have two characteristics in common: they lack seeds but have vascular tissues, the xylem and phloem, that transport water, minerals, and sugars throughout the plant. Because of their vascular tissues, vascular plants can grow taller than the bryophytes discussed in the previous chapter. And as we will see in the following three chapters, the vascular plants became the dominant plants on Earth.

From the fossil record, we can piece together some of the major steps in the evolution of this group. The earliest vascular plants consisted only of stems, but from this humble beginning we can trace both the evolution of roots and the evolution of leaves from simple scalelike structures to large, complex fronds. Associated with these adaptations came modifications in the types of spores produced, the arrangement of vascular tissues, and the size of the gametophyte—all of which are key events in the story that will be continued in subsequent chapters.

CHECKPOINTS

By the time you finish reading this chapter, you should be able to answer the following questions:

1. What "pivotal steps" in the early history of plant evolution contributed to the success of vascular plants in their occupation of the land?
2. What explanations are there for the evolutionary origin of microphylls and megaphylls? Which groups of seedless vascular plants have microphylls? Which have megaphylls?
3. What is meant by homospory and heterospory? What are the relative sizes of the gametophytes produced by homosporous and heterosporous plants?
4. What are the characteristics of each of the following phyla of seedless vascular plants: *Rhyniophyta, Zosterophyllophyta, Lycophyta, Trimerophytophyta, Psilotophyta, Sphenophyta,* and *Pterophyta?* Which of these are exclusively fossil phyla?
5. In terms of their structure and method of development, how do eusporangia differ from leptosporangia? Which ferns are eusporangiate? Which are leptosporangiate?

Plants, like all living organisms, had aquatic ancestors. The story of plant evolution is therefore inseparably linked with their progressive occupation of the land and their increasing independence from water for reproduction. In this chapter, we shall first discuss the general features of vascular plant evolution—features linked with life on land—and then describe the features of the seedless vascular plants, those with the most generalized features. This chapter will tell the story of the club mosses, horsetails, ferns, and other groups of vascular plants that retain some of the simpler features characteristic of the earliest members of this great group of plants.

Evolution of Vascular Plants

In the previous chapter, we noted that the bryophytes and vascular plants share a number of important characters, and that together these two groups of plants—both of which have multicellular embryos—form a monophyletic lineage, the embryophytes. You will recall that it has been hypothesized that the ultimate origin of this lineage can be traced to a *Coleochaete*-like organism. Both bryophytes and vascular plants have a basically similar life cycle—an alternation of heteromorphic generations—in which the gametophyte differs from the sporophyte. An important characteristic of bryophytes, however, is the dominant gametophyte, whereas in vascular plants the sporophyte dominates (Figure 19–1). Thus, the occupation of the land by the bryophytes was undertaken with emphasis on the gamete-producing generation, which requires water to enable its motile sperm to swim to the eggs. This requirement for water undoubtedly accounts for the small size and ground-hugging form of most bryophyte gametophytes.

Relatively early in the history of plants, the evolution of efficient fluid-conducting systems, consisting of xylem and phloem, solved the problem of water and food transport throughout the plant body—a serious problem for any large organism growing on land. The ability to synthesize lignin, which is incorporated into the cell walls of supporting and water-conducting cells, was also a pivotal step in the evolution of plants. It has been suggested that the earliest plants could stand erect only by means of turgor pressure, which limited not only the environments in which they could live but also the stature of such plants. Lignin adds rigidity to the walls, making it possible for the vascularized sporophyte—the dominant generation of vascular plants—to reach great heights. Vascular plants are also characterized by the capacity to branch profusely through the activity of apical meristems located at the tips of stems and branches. In the bryophytes, on the other hand, increase in the length of the sporophyte is subapical; that is, it occurs below the tip of the stem. In addition, each bryophyte sporophyte is unbranched and produces a single sporangium. By contrast, the branched sporophytes of vascular plants produce multiple sporangia. Picture a pine tree—a single individual—with its numerous branches and many cones, each of which contains multiple sporangia, and below it a carpet of moss gametophytes—many individuals—each bearing a single unbranched sporophyte tipped by a single sporangium.

The belowground and aboveground parts of the sporophytes of early vascular plants differed little from one another structurally, but ultimately the ancient vascular plants gave rise to more specialized plants with a more highly differentiated plant body. These plants consisted of roots, which function in anchorage and absorption of water and minerals, and of stems and leaves, which provide a system well suited to the demands of life on land—namely, the acquisition of energy from sunlight, of carbon dioxide from the atmosphere, and of water. Meanwhile, the gametophytic generation underwent a progressive reduction in size and gradually became more protected and nutritionally dependent upon the sporophyte. Finally, seeds evolved in one evolutionary line. **Seeds** are structures that provide the embryonic sporophyte with nutrients and that also help protect it from the rigors of life on land—thus providing a means of withstanding unfavorable environmental conditions. Obviously, the seedless vascular plants lack seeds. Moreover, the gametophytes of most seedless vascular plants, like those of the bryophytes, are free-living, and water is required in the environment for their motile sperm to swim to the eggs.

Because of their adaptations for existence on land, the vascular plants have been ecologically successful and are the dominant plants in terrestrial habitats. They were already numerous and diverse by the Devonian period (408–362 million years ago; see inside of front cover) (Figure 19–2). There are nine phyla with living representatives. In addition, there are several phyla that consist entirely of extinct vascular plants. In this chapter we shall describe some of the characteristic features of vascular plants and discuss seven phyla of seedless vascular plants, three of which are extinct. In Chapters 20 through 22 we shall discuss the seed plants, which include five phyla with living representatives.

Organization of the Vascular Plant Body

The sporophytes of early vascular plants were dichotomously branched (evenly forked) axes that lacked roots and leaves. With evolutionary specialization, morphological and physiological differences arose between various parts of the plant body, bringing about the differentiation of roots, stems, and leaves—the organs of

19–2
By the Early Devonian, some 408 to 387 million years ago, small leafless plants with simple vascular systems were growing upright on land. It is thought that their pioneering ancestors were bryophyte-like plants, seen here near the water at the center, that invaded land sometime in the Ordovician (510 to 439 million years ago). The vascular colonizers shown are, from left to center, very tiny Cooksonia *with rounded sporangia,* Zosterophyllum *with clustered sporangia, and* Aglaophyton *with solitary, elongated sporangia. During the Middle Devonian (387 to 374 million years ago), larger plants with more complex features became established. Seen here on the right are, from back to front,* Psilophyton, *a robust trimerophyte with plentiful sterile and fertile branchlets, and two lycophytes with simple microphyllous leaves,* Drepanophycus *and* Protolepidodendron.

the plant (Figure 19–3). Collectively, the roots make up the **root system,** which anchors the plant and absorbs water and minerals from the soil. The stems and leaves together make up the **shoot system,** with the stems raising the specialized photosynthetic organs—the leaves—toward the sun. The vascular system conducts water and minerals to the leaves and the products of photosynthesis away from the leaves to other parts of the plant.

The different kinds of cells of the plant body are organized into tissues, and the tissues are organized into still larger units called tissue systems. Three tissue systems, dermal, vascular, and ground, which occur in all organs of the plant, are continuous from organ to organ and reveal the basic unity of the plant body. The **dermal tissue system** makes up the outer, protective covering of the plant. The **vascular tissue system** comprises the conductive tissues—**xylem** and **phloem**—and is embedded in the **ground tissue system** (Figure 19–3). The principal differences in the structures of root, stem, and leaf lie in the relative distribution of the vascular and ground tissue systems, as will be discussed in Section 5.

Primary Growth Involves the Extension of Roots and Stems, and Secondary Growth Increases Their Thickness

Primary growth may be defined as the growth that occurs relatively close to the tips of roots and stems. It is initiated by the apical meristems and is primarily in-

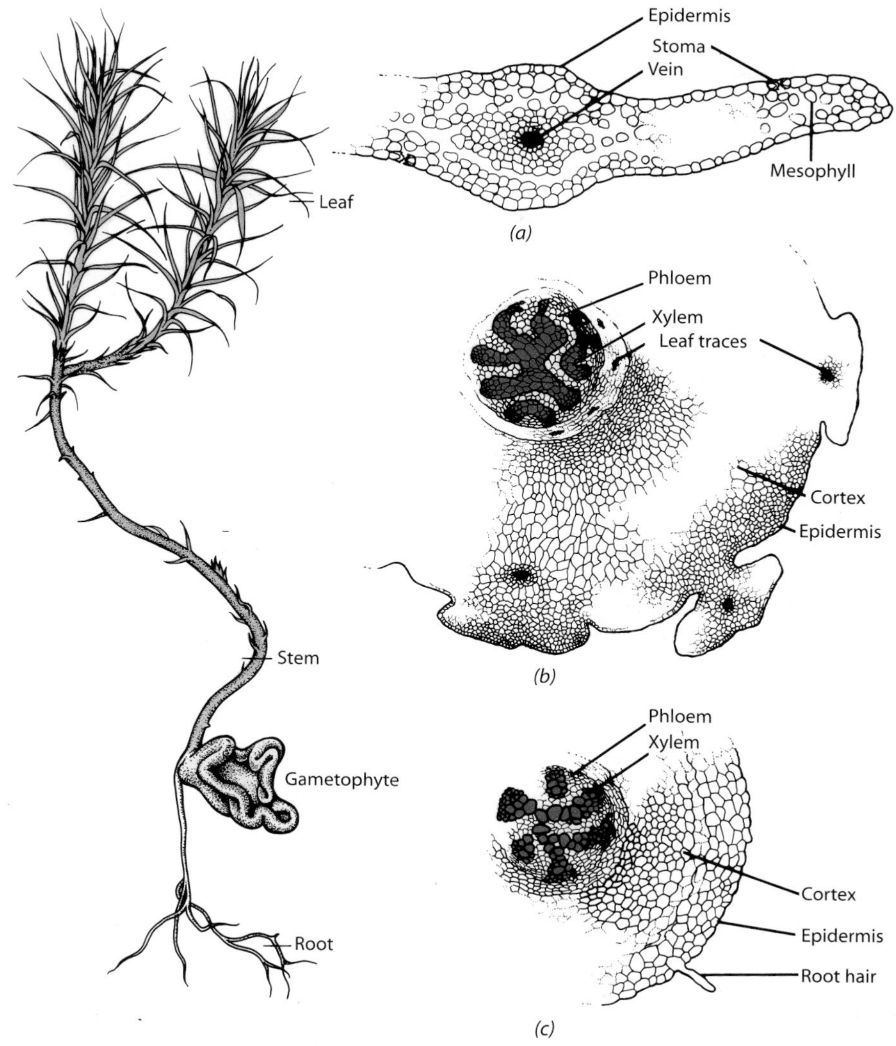

19–3
A diagram of a young sporophyte of the club moss Lycopodium lagopus, *which is still attached to its subterranean gametophyte. The dermal, vascular, and ground tissues are shown in transverse sections of* **(a)** *leaf,* **(b)** *stem, and* **(c)** *root. In all three organs, the dermal tissue system is represented by the epidermis, and the vascular tissue system, consisting of xylem and phloem, is embedded in the ground tissue system. The ground tissue in the leaf—in* Lycopodium, *a microphyll—is represented by the mesophyll, and in the stem and root by the cortex, which surrounds a solid strand of vascular tissue, or protostele. The leaf is specialized for photosynthesis, the stem for support of the leaves and for conduction, and the root for absorption and anchorage.*

volved with extension of the plant body—often the vertical growth of a plant. The tissues arising during primary growth are known as **primary tissues,** and the part of the plant body composed of these tissues is called the **primary plant body.** Ancient vascular plants, and many contemporary ones as well, consist entirely of primary tissues.

In addition to primary growth, many plants undergo additional growth that thickens the stem and root; such growth is termed **secondary growth.** It results from the activity of lateral meristems, one of which, the **vascular cambium,** produces **secondary vascular tissues** known as secondary xylem and secondary phloem (see Figure 27–6). The production of secondary vascular tissues is commonly supplemented by the activity of a second lateral meristem, the **cork cambium,** which forms a **periderm,** composed mostly of cork tissue. The periderm replaces the epidermis as the dermal tissue system of the plant. The secondary vascular tissues and periderm make up the **secondary plant body.** Secondary growth appeared in the Middle Devonian period, about 380 million years ago, in several unrelated groups of vascular plants.

Tracheary Elements—Tracheids and Vessel Elements—Are the Conducting Cells of the Xylem

Sieve elements, the conducting cells of the phloem, have soft walls and often collapse after they die, so they are rarely well preserved as fossils. In contrast, **tracheary elements,** the conducting cells of the xylem, have distinctive, lignified wall thickenings (Figure 19–4) and are frequently well preserved in the fossil record. Because of their various wall patterns, the tracheary elements provide valuable clues to the interrelationships of the different groups of vascular plants.

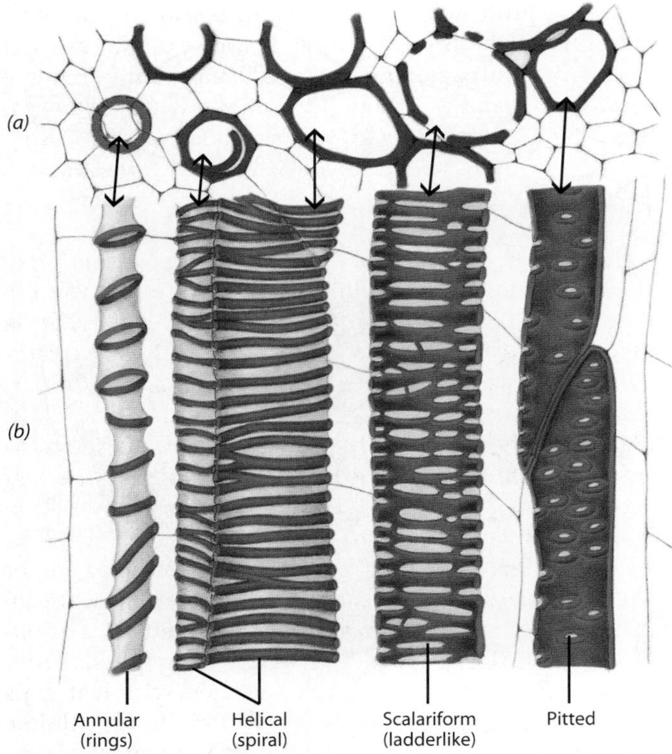

19–4
Tracheary elements, the conducting cells of the xylem. A portion of the stem of Dutchman's pipe (Aristolochia) *in* ***(a)*** *transverse and* ***(b)*** *longitudinal views, showing some of the distinctive types of wall thickenings exhibited by tracheary elements. Here, the wall thickenings vary, left to right, from those elements formed earliest in the development of the plant to those formed more recently.*

In fossil vascular plants from the Silurian and Devonian periods, the tracheary elements are elongate cells with long, tapering ends. Such tracheary elements, called **tracheids,** were the first type of water-conducting cell to evolve; they are the only type of water-conducting cell in most vascular plants other than angiosperms and a peculiar group of gymnosperms, the *Gnetophyta* (see page 490). Tracheids not only provide channels for the passage of water and minerals, but in many modern plants they also provide support for stems. Water-conducting cells are rigid, mostly because of the lignin in their walls. This rigidity made it possible for plants to evolve an upright habit and eventually for some of them to become trees.

Tracheids are more primitive (less specialized) than **vessel elements**—the principal water-conducting cells in angiosperms. Vessel elements apparently evolved independently from tracheids in several groups of vascular plants. This is an excellent example of convergent evolution—the independent development of similar structures by unrelated or only distantly related organisms (see the essay on page 266).

Vascular Tissues Are Located in the Vascular Cylinders, or Steles, of Roots and Stems

The primary vascular tissues—primary xylem and primary phloem—and, in some vascular plants, a central column of ground tissue known as the **pith** make up the central cylinder, or **stele,** of the stem and root in the primary plant body. Several types of steles are recognized, among them the protostele, the siphonostele, and the eustele (Figure 19–5).

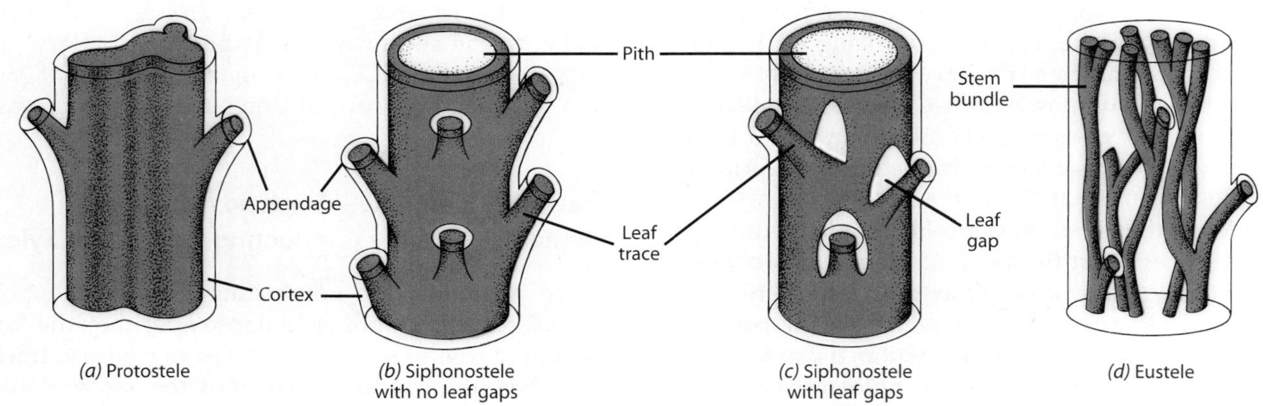

19–5
Steles. ***(a)*** *A protostele, with diverging traces of appendages (leaves or leaf precursors), the evolutionary precursors of leaves.* ***(b)*** *A siphonostele with no leaf gaps; the vascular traces leading to the leaves simply diverge from the solid cylinder. This sort of siphonostele is found in* Selaginella, *among other plants.* ***(c)*** *A siphonostele with leaf gaps, commonly found in ferns.* ***(d)*** *A eustele, found in almost all seed plants. Siphonosteles and eusteles appear to have evolved independently from protosteles.*

The **protostele**—the simplest and most ancient type of stele—consists of a solid cylinder of vascular tissue in which the phloem either surrounds the xylem or is interspersed within it (Figures 19–3 and 19–5a). It is found in the extinct groups of seedless vascular plants discussed below, as well as in the *Psilotophyta* and *Lycophyta* (composed primarily of club mosses) and in the juvenile stems of some other living groups. In addition, it is the type of stele found in most roots.

The **siphonostele**—the type of stele found in the stems of most species of seedless vascular plants—is characterized by a central pith surrounded by the vascular tissue (Figure 19–5b). The phloem may form only outside the cylinder of xylem or on both sides of it. In the siphonosteles of ferns, the departure from the stem of the vascular strands leading to the leaves—the **leaf traces**—generally is marked by gaps—**leaf gaps**—in the siphonostele (as in Figure 19–5c). These leaf gaps are filled with parenchyma cells just like those that occur within and outside the vascular tissue of the siphonostele. Although the leaf traces in seed plants are associated with parenchymatous areas reminiscent of leaf gaps, these areas are generally not considered to be homologous to leaf gaps. Therefore, we will refer to these areas in seed plants as **leaf trace gaps.**

If the primary vascular cylinder consists of a system of discrete strands around a pith, as it does in almost all seed plants, the stele is called a **eustele** (Figure 19–5d). Comparative studies of living and fossil vascular plants have suggested that the eustele of seed plants evolved directly from a protostele. Eusteles appeared first among the progymnosperms, a group of spore-bearing plants that are discussed in Chapter 20 (pages 470–472). Siphonosteles evidently evolved independently from protosteles. This relationship indicates that none of the groups of seedless vascular plants with living representatives gave rise to any living seed plants.

Roots and Leaves Evolved in Different Ways

Although the fossil record reveals little information on the origins of roots as we know them today, they must have evolved from the lower, often subterranean, portions of the axis of ancient vascular plants. For the most part, roots are relatively simple structures that seem to have retained many of the ancient structural characteristics no longer present in the stems of modern plants.

Leaves are the principal lateral appendages of the stem. Regardless of their ultimate size or structure, they arise as protuberances (leaf primordia) from the apical meristem of the shoot. From an evolutionary perspective, there are two fundamentally distinct kinds of leaves—microphylls and megaphylls.

Microphylls are usually relatively small leaves that contain only a single strand of vascular tissue (Figure 19–6a). Microphylls are typically associated with stems possessing protosteles and are characteristic of the lycophytes. The leaf traces leading to microphylls are not associated with gaps, and there is usually only a single vein in each leaf. Even though the name *microphyll* means "small leaf," some species of *Isoetes* have fairly long leaves (see Figure 19–19). In fact, certain Carboniferous and Permian lycophytes had microphylls a meter or more in length.

According to different theories, microphylls may have evolved as superficial lateral outgrowths of the stem (Figure 19–7a) or from the sterilization of sporangia in lycophyte ancestors. According to one theory, microphylls began as small, scalelike or spinelike outgrowths, called enations, devoid of vascular tissue. Gradually, rudimentary leaf traces developed, which initially extended only to the base of the enation. Finally, the leaf traces extended into the enation, resulting in formation of the primitive microphyll.

Most **megaphylls,** as the name implies, are larger than most microphylls. With few exceptions, they are associated with stems that have either siphonosteles or eusteles. The leaf traces leading to the megaphylls from siphonosteles and eusteles are associated with leaf gaps and leaf trace gaps, respectively (Figure 19–6b). Unlike the microphylls, the blade, or **lamina,** of most megaphylls has a complex system of branching veins.

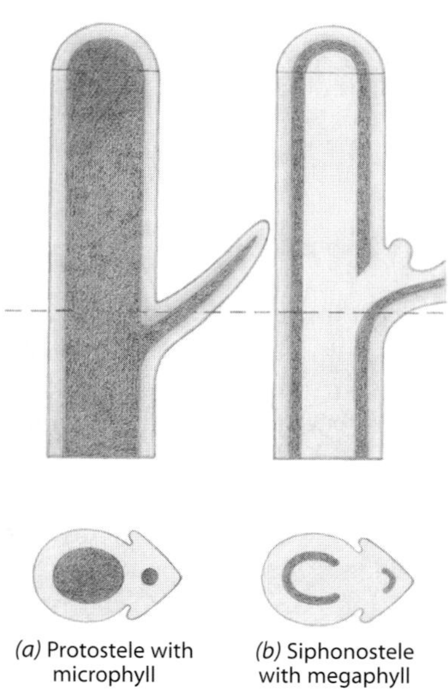

19–6
*Microphylls and megaphylls. Longitudinal and transverse sections through **(a)** a stem with a protostele and a microphyll and **(b)** a stem with a siphonostele and a megaphyll, emphasizing the nodes, or regions where the leaves are attached. Note the presence of pith and a leaf gap in the stem with a siphonostele and their absence in the stem with a protostele. Microphylls are characteristic of lycophytes, while megaphylls are found in all other vascular plants.*

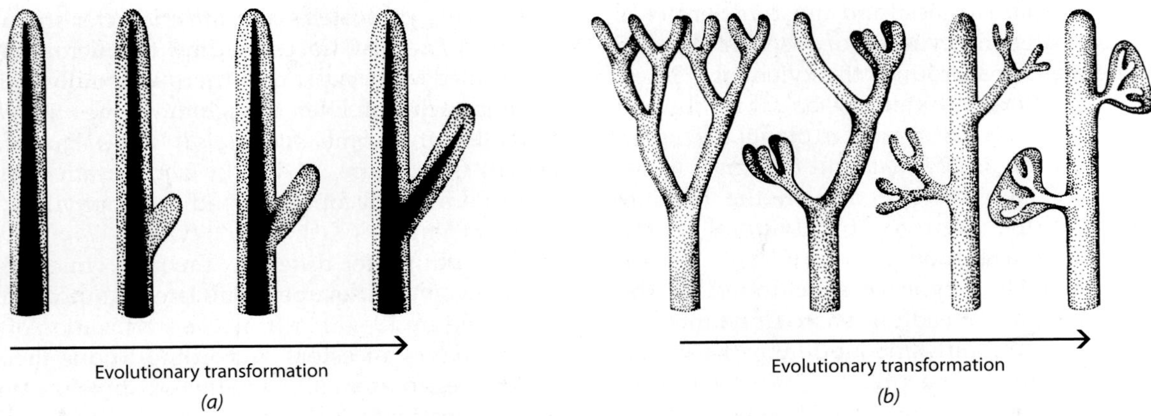

19–7
Evolution of microphylls and megaphylls. ***(a)*** *According to one widely accepted theory, microphylls evolved as outgrowths, called enations, of the main axis of the plant.* ***(b)*** *Megaphylls evolved by fusion of branch systems.*

It seems likely that megaphylls evolved from entire branch systems by a series of steps similar to that shown in Figure 19–7b. The earliest plants had a leafless, dichotomously branching axis, without distinction between axis and megaphylls. Unequal branching resulted in more aggressive branches "overtopping" the weaker ones. The subordinated, overtopped lateral branches represented the beginning of leaves, and the more aggressive portions became stemlike axes. This was followed by flattening out, or "planation," of the lateral branches. The final step was fusion, or "webbing," of the separate lateral branches to form a primitive lamina.

Reproductive Systems

As mentioned previously, all vascular plants are oogamous—that is, they have large nonmotile eggs and small sperm that swim or are conveyed to the egg. In addition, all vascular plants have an alternation of heteromorphic generations, in which the sporophyte, the dominant phase of the life cycle, is larger and structurally much more complex than the gametophyte (Figure 19–8). Oogamy is clearly favored in plants, since only one of the kinds of gametes must navigate across a hostile environment outside the plant.

Homosporous Plants Produce Only One Kind of Spore, whereas Heterosporous Plants Produce Two Types

Early vascular plants produced only one kind of spore as a result of meiosis; such vascular plants are said to be **homosporous.** Among living vascular plants, homospory is found in the psilotophytes, sphenophytes (horsetails), some of the lycophytes, and almost all ferns. Upon germination, such spores have the potential to produce bisexual gametophytes—that is, gametophytes that bear both antheridia and archegonia. Studies have revealed, however, that the gametophytes of diploid species of homosporous ferns are functionally unisexual. For example, if a sperm from a bisexual gametophyte were to fertilize an egg from that same gametophyte, the resulting sporophyte would be homozygous for all loci. Genetic studies indicate, however, that the sporophytes of most ferns are heterozygous, so they cannot have resulted from self-fertilization. Even when bisexual gametophytes of diploid species were isolated and "forced" to interbreed, they failed to produce new sporophytes.

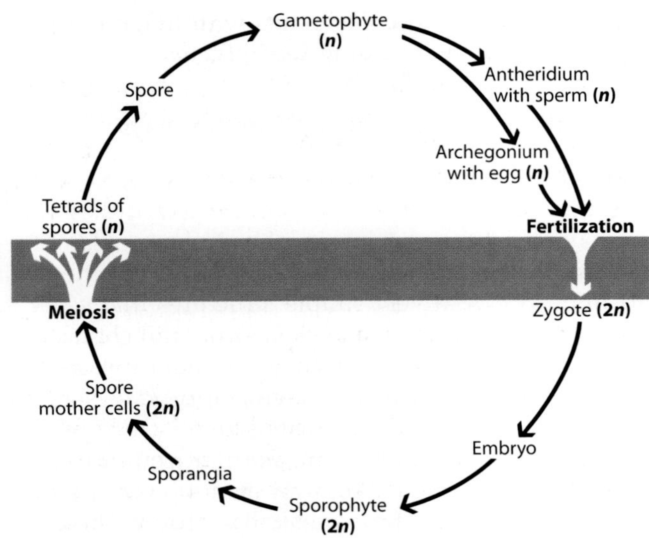

19–8
Generalized life cycle of a vascular plant, in which the sporophyte is the dominant phase of the life cycle.

Thus, rather than fertilizing their own eggs, the sperm produced by bisexual gametophytes usually fertilize the eggs of neighboring, genetically different gametophytes. Moreover, in most natural populations, although any one gametophyte may produce both antheridia and archegonia, both sex organs are not mature at the same time.

Heterospory—the production of two types of spores in two different kinds of sporangia—is found in some of the lycophytes, in a few ferns, and in all seed plants. Heterospory arose many times in unrelated groups during the evolution of vascular plants. It was common as early as the Devonian period, with the earliest record from about 370 million years ago. The two types of spores are called **microspores** and **megaspores,** and they are produced in **microsporangia** and **megasporangia,** respectively. The two types of spores are differentiated on the basis of function and not necessarily relative size. Microspores give rise to male gametophytes (microgametophytes), and megaspores give rise to female gametophytes (megagametophytes). Both of these types of unisexual gametophytes are much reduced in size compared with the gametophytes of homosporous vascular plants. Another difference is that in heterosporous plants, the gametophytes develop within the spore wall (endosporic development), while in homosporous plants the gametophytes develop outside the spore wall (exosporic development).

Over Evolutionary Time, the Gametophytes of Vascular Plants Have Become Smaller and Simpler

The relatively large gametophytes of homosporous plants are independent of the sporophyte for their nutrition, although the subterranean gametophytes of some species—such as those of *Psilotum* (see Figure 19–22) and several genera of club mosses *(Lycopodiaceae)*—are heterotrophic, depending on endomycorrhizal fungi for their nutrients. Other genera of club mosses, like most ferns and the horsetails, have free-living, photosynthetic gametophytes. In contrast, the gametophytes of many heterosporous vascular plants, and especially those of the seed plants, are dependent on the sporophyte for their nutrition.

While the initial stages of plant evolution from *Coleochaete*-like ancestors involved elaboration and modification of both the gametophyte and sporophyte, within vascular plants the evolution of the gametophyte is characterized by an overall trend toward reduction in size and complexity, and angiosperm gametophytes are the most reduced of all (see page 503). The mature megagametophyte of angiosperms commonly consists of only seven cells, one of them an egg cell. When mature, the microgametophyte contains only three cells, and two of them are sperm. Archegonia and antheridia, which are found in all seedless vascular plants, have apparently been lost in the lineage leading to angiosperms. All but a few gymnosperms, which will be discussed in the next chapter, produce archegonia but lack antheridia. In the seedless vascular plants, the motile sperm swim through water to the archegonium. These plants must therefore grow in habitats where water is at least occasionally plentiful. In angiosperms and in most gymnosperms, entire immature microgametophytes **(pollen grains)** are carried to the vicinity of the megagametophyte. This transfer of the pollen grains is called **pollination.** Germination of the pollen grains produces special structures called **pollen tubes,** through which motile sperm (in cycads and *Ginkgo*) swim to or nonmotile sperm (in conifers and *Gnetales*) are transferred to the egg to achieve fertilization.

The Phyla of Seedless Vascular Plants

Several groups of seedless vascular plants flourished during the Devonian, of which the three most important have been recognized as the *Rhyniophyta, Zosterophyllophyta,* and *Trimerophytophyta.* All three groups were extinct by about the end of the Devonian, 360 million years ago. All three phyla consisted of seedless plants that were relatively simple in structure. A fourth phylum of seedless vascular plants, *Progymnospermophyta,* or progymnosperms, will be discussed in Chapter 20 because the members of this group may have been ancestral to the seed plants, both gymnosperms and angiosperms (Figure 19–9). In addition to these extinct phyla, we shall discuss in this chapter the *Psilotophyta, Lycophyta, Sphenophyta,* and *Pterophyta,* the four phyla of seedless vascular plants that have living representatives.

The overall pattern of diversification of plants may be interpreted in terms of the successive rise to dominance of four major plant groups that largely replaced the groups that were dominant earlier. In each instance, numerous species evolved in the groups that were rising to dominance. The major groups involved are as follows:

(1) Early vascular plants—characterized by relatively small stature and a simple and presumably primitive morphology. These plants included the rhyniophytes, zosterophyllophytes, and trimerophytes (Figure 19–10), which were dominant from the mid-Silurian to the mid-Devonian, from about 425 to 370 million years ago (Figure 19–2).

(2) Ferns, lycophytes, sphenophytes, progymnosperms. These more complex groups were dominant from the Late Devonian period through the Carboniferous period (see Figures 19–11 and 20–1), from about 375 to about 290 million years ago (see "Coal Age Plants" on pages 456 and 457).

(3) Seed plants arose starting in the Late Devonian period, at least 380 million years ago, and evolved many new lines by the Permian period. Gymnosperms dominated land floras throughout most of the Mesozoic era until about 100 million years ago.

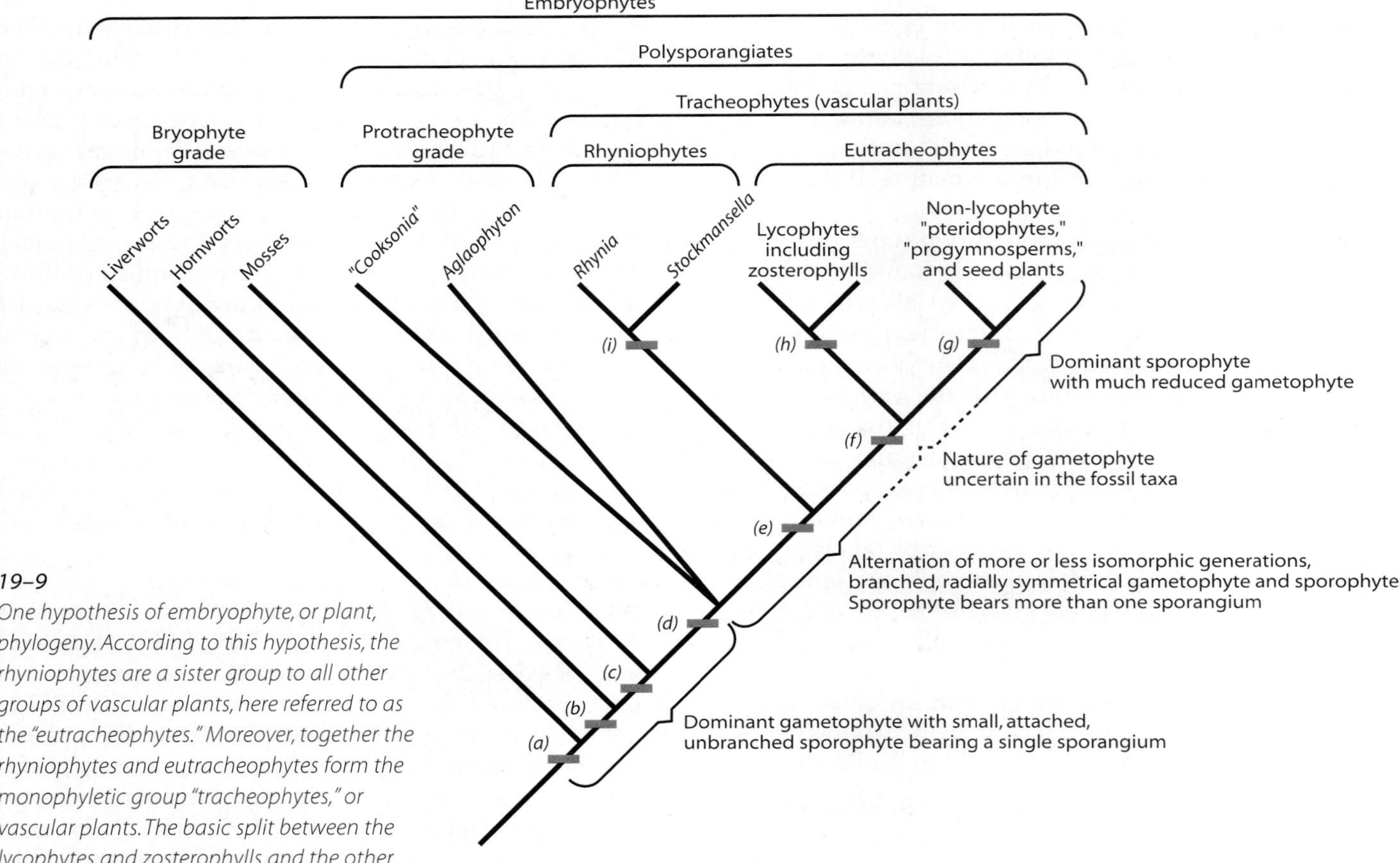

19–9
One hypothesis of embryophyte, or plant, phylogeny. According to this hypothesis, the rhyniophytes are a sister group to all other groups of vascular plants, here referred to as the "eutracheophytes." Moreover, together the rhyniophytes and eutracheophytes form the monophyletic group "tracheophytes," or vascular plants. The basic split between the lycophytes and zosterophylls and the other vascular plants is supported by both morphological and molecular data.

Characters are as follows: **(a)** *Embryophytes: multicellular embryos; antheridia and archegonia; cuticle; preprophase band microtubules.* **(b)** *Hornwort-moss plus tracheophyte clade: ability to distinguish* D*-methionine; stomata.* **(c)** *Moss plus polysporangiate clade: axial gametophyte; terminal gametangia; persistent and internally differentiated sporophyte.* **(d)** *Polysporangiates: branched sporophyte with multiple sporangia; dominant sporophyte; alternation of more or less isomorphic generations.* **(e)** *Tracheophytes: tracheids with internal annular or helical wall thickenings.* **(f)** *Eutracheophytes: distinctive tracheids found in zosterophylls and lycophytes and comparable to the earliest-formed elements in some modern seedless vascular plants.* **(g)** *Non-lycophyte clade: pitted water-conducting cells; longitudinal dehiscence, or opening, of sporangia; sporangia arranged in terminal clusters.* **(h)** *Lycophyte clade: kidney-shaped sporangia; sporangia lateral on short stalks.* **(i)** *Rhyniophytes: separation or isolation layer at base of sporangium; sporangium attached directly to main axis or terminal on short branch.*

(4) Flowering plants, which appeared in the fossil record about 130 million years ago. This phylum became abundant in most parts of the world within 30 to 40 million years and has remained dominant ever since.

Phylum *Rhyniophyta*

The earliest known vascular plants that we understand in detail belong to the phylum *Rhyniophyta*, a group that dates back to the mid-Silurian, at least 425 million years ago. The group became extinct in the mid-Devonian (about 380 million years ago). Earlier vascular plants were probably similar; their remains go back at least another 15 million years. Rhyniophytes were seedless plants, consisting of simple, dichotomously branching axes, or stems, with terminal sporangia. Their plant bodies were not differentiated into roots, stems, and leaves, and they were homosporous. The name of the phylum comes from the good representation of these primitive plants as fossils preserved in chert near the village of Rhynie, in Scotland.

Among the first rhyniophytes to be described is *Rhynia gwynne-vaughanii*. Probably a marsh plant, it consisted of an upright, dichotomously branched aerial system attached to a dichotomously branched rhizome (underground stem) system with rhizoids. Among the

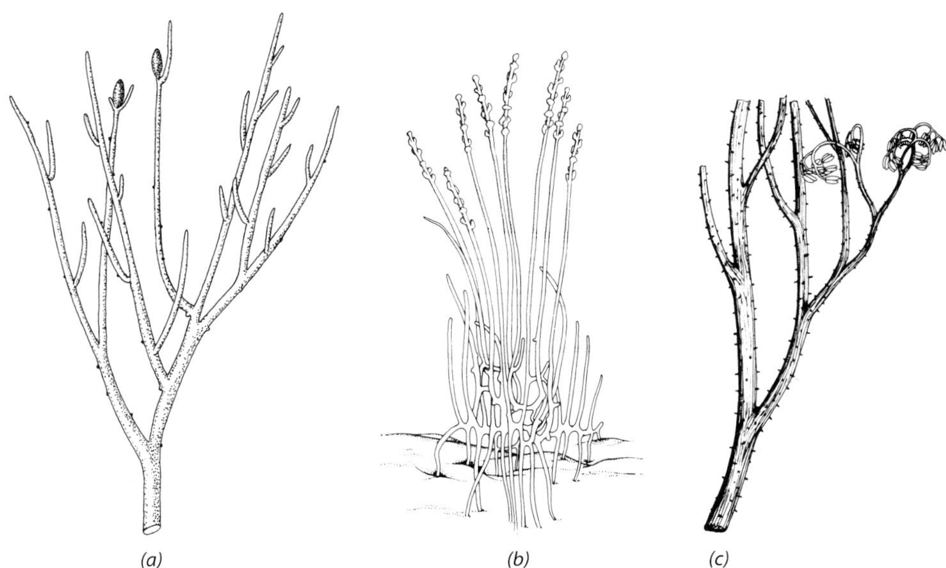

19–10
Early vascular plants. ***(a)*** Rhynia gwynne-vaughanii *is a rhyniophyte, one of the simplest of the known vascular plants. The axis was leafless and dichotomously branched, with numerous lateral branches. The sporangia were terminal on some of the upright main branches and usually overtopped by development of lateral branches.* ***(b)*** *In* Zosterophyllum *and the other zosterophyllophytes, the generally kidney-shaped sporangia were borne laterally in a helix or in two rows on the stems. Sporangia split along definite slits that formed around the outer margin. The zosterophyllophytes were larger than the rhyniophytes, but like the latter, they were mostly dichotomously branched plants that were either naked, spiny, or toothed.* ***(c)*** *The trimerophytes were larger plants with more complex branching that was generally differentiated into a strong central axis with smaller side branches. The side branches were dichotomously branched and often had terminal masses of paired sporangia that tapered at both ends. The best-known genera included in this group are* Psilophyton *and* Trimerophyton. *A reconstruction of* Psilophyton princeps *is shown here. Individuals of* R. gwynne-vaughanii *ranged up to about 18 centimeters tall, while some of the trimerophytes were up to 1 meter or more in height. See also Figure 19–2.*

distinctive features of *R. gwynne-vaughanii* are the numerous lateral branches that arose from the dichotomized axes (Figure 19–10a) and the short branches on which the sporangia were often borne. The aerial branch system, which was about 18 centimeters in height, was covered with a cuticle and bore stomata. Lacking leaves, the aerial axes served as the photosynthetic organs.

The internal structure of *R. gwynne-vaughanii* was similar to that of many of today's vascular plants. A

19–11
Reconstruction of a Carboniferous period swamp forest. See also Figure 20–1.

single layer of superficial cells—the epidermis—surrounded the photosynthetic tissue of the cortex, and the center of the axis consisted of a solid strand of xylem surrounded by one or two layers of phloemlike cells. The tracheids were different from those of most vascular plants, and although they had internal thickenings, they share some features with the water-conducting cells of mosses.

Probably the best known plant found in the Rhynie chert is the plant originally called *Rhynia major* (Figure 19–12). A more robust plant than *R. gwynne-vaughanii*, reaching heights of about 50 centimeters, it consisted of an extensive dichotomously branched rhizome system with a limited number of upright stems that branched dichotomously. All axes terminated in sporangia. For over 60 years, *R. major* was considered to be a vascular plant. Then it was shown that the cells forming the central strand of conducting tissue lack the wall thickenings typical of tracheids. Rather than being tracheids, these cells are more like the hydroids of modern mosses. For this reason, this fossil plant has been removed from the genus *Rhynia* and assigned to a new genus, *Aglaophyton*. *Aglaophyton major*, with its branched axes and multiple sporangia, may represent an intermediate stage—known as a **protracheophyte**—in the evolution of vascular plants and should probably not be retained in the phylum *Rhyniophyta*.

Cooksonia, a rhyniophyte that is believed to have inhabited mud flats, has the distinction of being the oldest known vascular plant (see Figures 1–7 and 19–1). Specimens of *Cooksonia* have been found in Wales, Scotland, Bohemia, Canada, and the United States. *Cooksonia* is the smallest and simplest vascular plant known from the fossil record. Its slender, leafless aerial stems ranged up to about 6.5 centimeters long and terminated in globose sporangia. Tracheids have been identified in the central region of the axes of *Cooksonia pertoni* from the Lower Devonian. Plants similar in form to *C. pertoni* occur in the Silurian, but it is questionable whether they contained vascular tissues. The genus *Cooksonia* may contain early simple fossil plants of diverse relationships. Some of these plants may be associated with the protracheophyte *Aglaophyton*, but others are almost certainly true vascular plants. *Cooksonia* had become extinct by the Lower Devonian period, about 390 million years ago.

Evidence from the Rhynie chert and the Lower Devonian of Germany indicates that the gametophytes of plants such as *Aglaophyton* and, by implication, *Rhynia* and *Cooksonia* (among others) were relatively large, branched structures. Some of these gametophytes apparently had water-conducting cells, cuticles, and stomata. Hence, some of these plants had an alternation of isomorphic generations, in which sporophyte and gametophyte were basically similar except for their sporangia and gametangia, respectively.

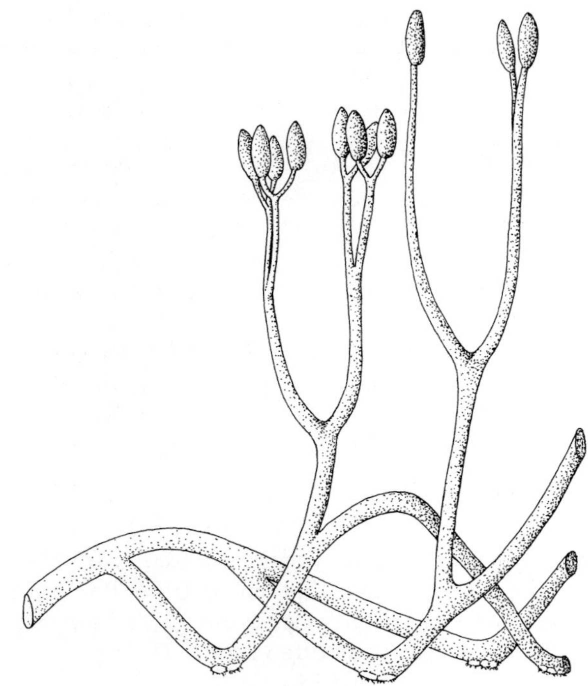

19–12
Aglaophyton major, *previously known as* Rhynia major *when it was considered to be a vascular plant. Its central strand of conducting tissue lacks tracheids but contains cells similar to the hydroids of mosses.* Aglaophyton major *may be an intermediate stage, known as a protracheophyte, in the evolution of vascular plants.*

Phylum *Zosterophyllophyta*

The fossils of a second phylum of extinct seedless vascular plants—the *Zosterophyllophyta*—have been found in strata from the Early to Late Devonian period, from about 408 to 370 million years ago. Like the rhyniophytes, the zosterophyllophytes, or zosterophylls, were leafless and dichotomously branched. The aerial stems were covered with a cuticle, but only the upper ones contained stomata, indicating that the lower branches may have been embedded in mud. In *Zosterophyllum*, it has been suggested that the lower branches frequently produced lateral branches that forked into two axes, one that grew upward, the other downward (Figure 19–10b). The downward-growing branches may have functioned like a root, permitting the plant to spread outwardly from the center by providing support. The zosterophylls are so named because of their general resemblance to the modern seagrass *Zostera*, marine angiosperms that superficially resemble grasses.

Unlike those of the rhyniophytes, the globose or kidney-shaped sporangia of the zosterophylls were borne laterally on short stalks. These plants were homosporous. The internal structure of the zosterophylls was essentially similar to that of the rhyniophytes, except that in the zosterophylls the first xylem cells to mature were located around the periphery of the xylem strand and the last to mature were located in the center. This process, known as centripetal differentiation, is the opposite of the centrifugal differentiation found in the rhyniophytes.

The zosterophylls were almost certainly the ancestors of the lycophytes. The sporangia of zosterophylls and early lycophytes are very similar and in both groups were also borne laterally. The xylem in both phyla also differentiated centripetally.

Phylum *Lycophyta*

The 10 to 15 living genera and approximately 1000 living species of *Lycophyta* are the representatives of an evolutionary line that extends back to the Devonian period. The progenitors of the lycophytes were almost certainly early zosterophylls (Figure 19–9). There are a number of orders of lycophytes, and at least three of the extinct ones included small to large trees. The three orders of living lycophytes, however, consist entirely of herbs; each order includes a single family. All lycophytes, living and fossil, possess microphylls, and this type of leaf, which shows relatively little diversity in form, is highly characteristic of the phylum. Tree lycophytes were among the dominant plants of the coal-forming forests of the Carboniferous period (see pages 456 and 457). Most lines of woody lycophytes—those lycophytes that exhibited secondary growth—became extinct before the end of the Paleozoic era, 248 million years ago.

The Club Mosses Belong to the Family *Lycopodiaceae*

Perhaps the most familiar living lycophytes are the club mosses, family *Lycopodiaceae* (see Figure 13-15c). All but two genera of living lycophytes belong to this family, most members of which were formerly grouped in the collective genus *"Lycopodium."* Seven of these genera are represented in the United States and Canada, but most of the estimated 400 species in the family are tropical. The taxonomic boundaries of the predominantly tropical genera of this family are poorly understood, and as many as 15 genera may ultimately be recognized. *Lycopodiaceae* extend from Arctic regions to the tropics, but they rarely form conspicuous elements in any plant community. Most tropical species, many of which belong to the genus *Phlegmariurus,* are epiphytes and thus rarely seen, but several of the temperate species form mats that may be evident on forest floors. Because they are evergreen, they are most noticeable in winter.

The sporophyte of most genera of *Lycopodiaceae* consists of a branching rhizome from which aerial branches and roots arise (Figure 19–14, pages 436 and 437). Both stem and root are protostelic (Figure 19–13). The microphylls of *Lycopodiaceae* are usually spirally arranged, but

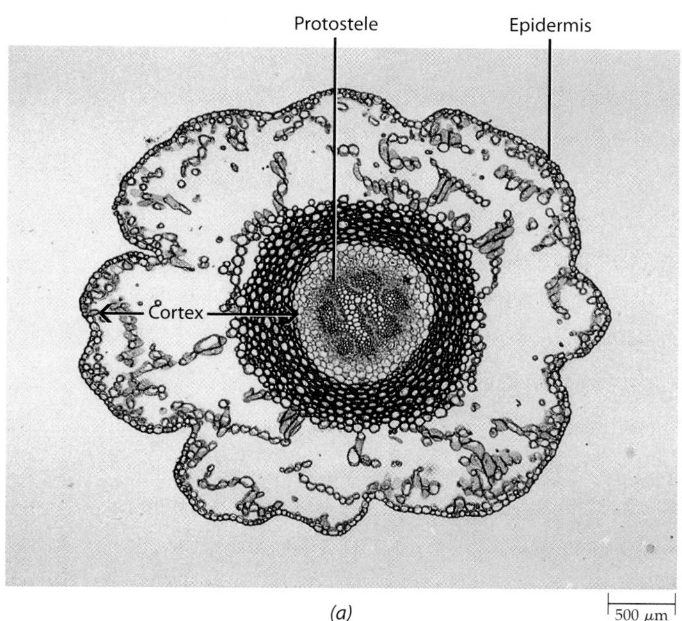

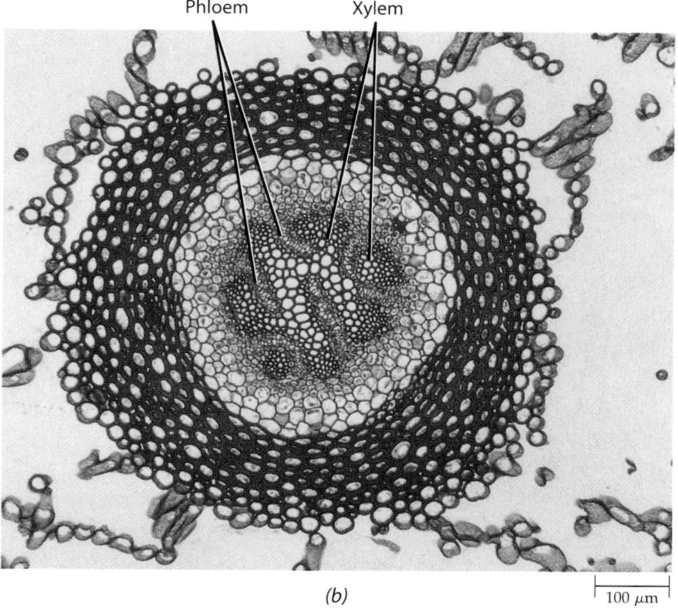

19–13
Both the stem and root of Lycopodiaceae *are protostelic.* ***(a)*** *Transverse section of* Diphasiastrum complanatum *stem, showing mature tissues. Note the large air spaces in the cortex, which surrounds the central protostele.* ***(b)*** *Detail of protostele of* D. complanatum, *showing xylem and phloem. See also Figure 19–3.*

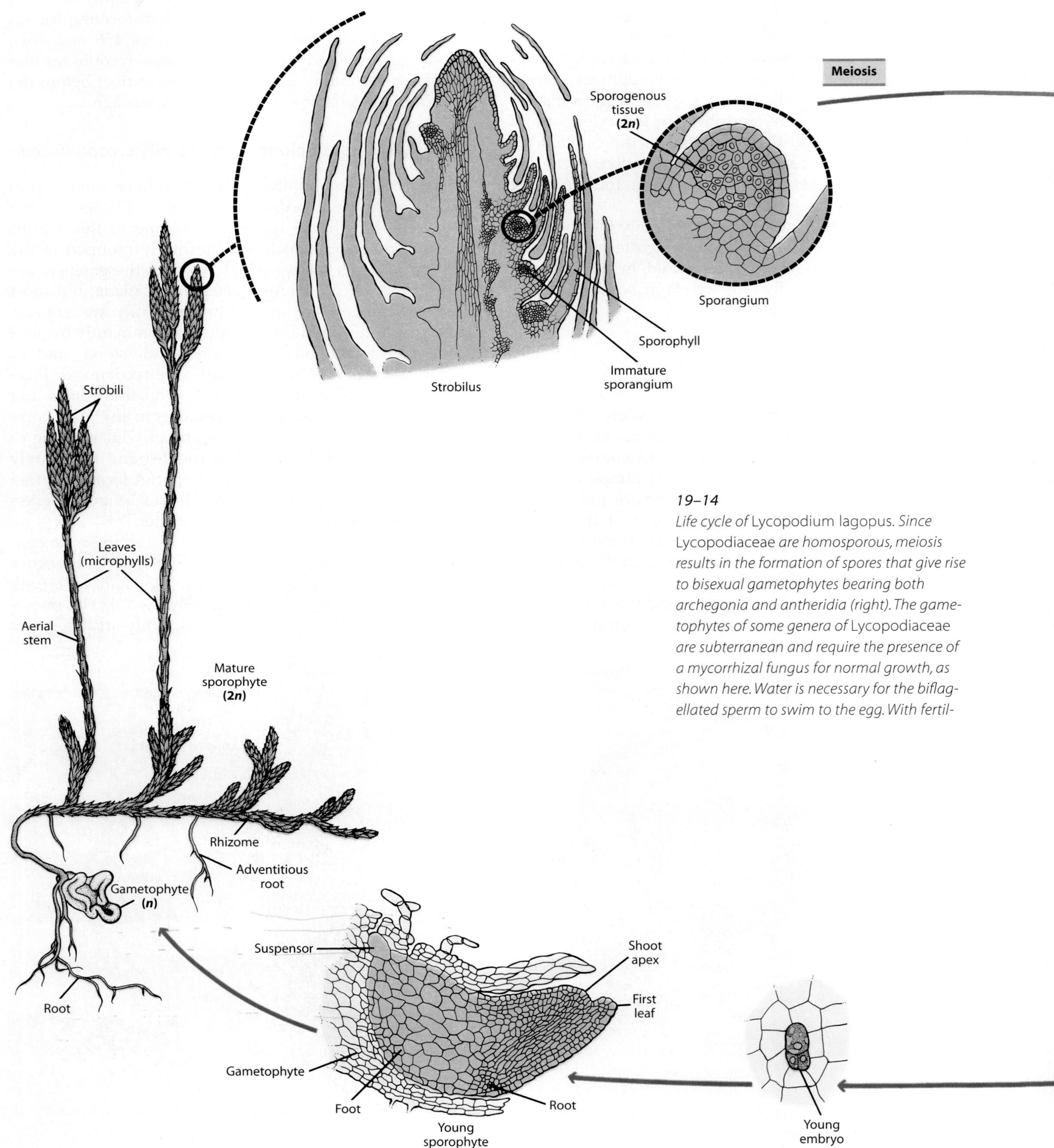

19–14
Life cycle of Lycopodium lagopus. *Since* Lycopodiaceae *are homosporous, meiosis results in the formation of spores that give rise to bisexual gametophytes bearing both archegonia and antheridia (right). The gametophytes of some genera of* Lycopodiaceae *are subterranean and require the presence of a mycorrhizal fungus for normal growth, as shown here. Water is necessary for the biflagellated sperm to swim to the egg. With fertil-*

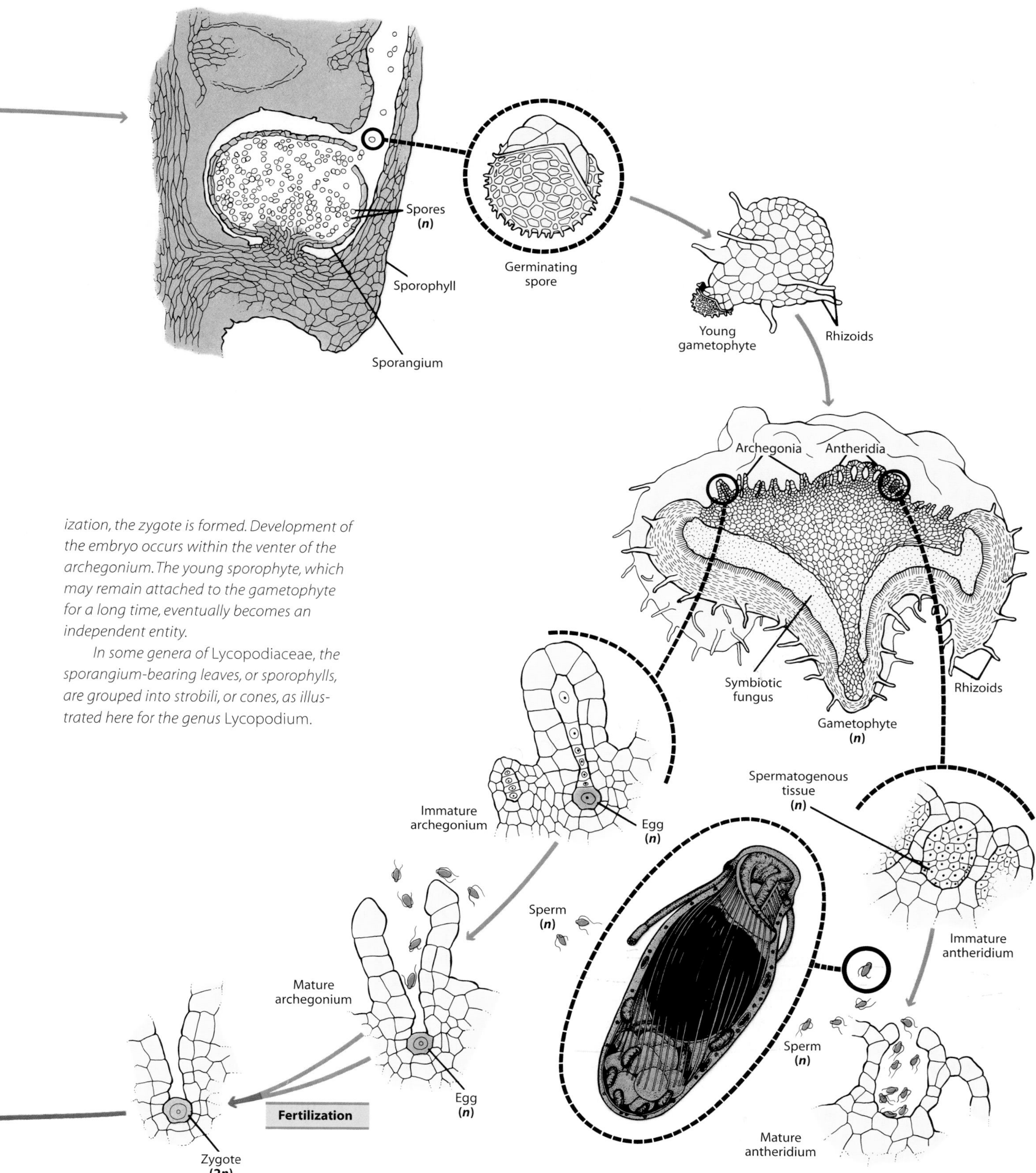

ization, the zygote is formed. Development of the embryo occurs within the venter of the archegonium. The young sporophyte, which may remain attached to the gametophyte for a long time, eventually becomes an independent entity.

In some genera of Lycopodiaceae, *the sporangium-bearing leaves, or sporophylls, are grouped into strobili, or cones, as illustrated here for the genus* Lycopodium.

(a)

(b)

19–15
Sporophylls and strobili. ***(a)*** Huperzia lucidula *is a representative of those genera of* Lycopodiaceae *that lack differentiated strobili. The sporangia (small yellow structures along stem) are borne in the axils of fertile microphylls known as sporophylls. Areas of fertile sporophylls alternate with regions of sterile microphylls.* ***(b)*** *The terminal branches in* Lycopodium lagopus *are terminated by sporophylls grouped into strobili.*

they appear opposite or whorled in some members of the group. *Lycopodiaceae* are homosporous; the sporangia occur singly on the upper surface of fertile microphylls called **sporophylls,** which are modified leaves or leaflike organs that bear sporangia. In *Huperzia* (Figure 19–15a) and *Phlegmariurus,* the sporophylls are similar to ordinary microphylls and are interspersed among the sterile microphylls. In the other genera of *Lycopodiaceae* found in the United States and Canada, including *Diphasiastrum* (see Figure 13–15c) and *Lycopodium,* nonphotosynthetic sporophylls are grouped into **strobili,** or cones, at the ends of the aerial branches (Figure 19–15b).

Upon germination, the spores of *Lycopodiaceae* give rise to bisexual gametophytes that, depending on the genus, are either green, irregularly lobed masses (*Lycopodiella, Palhinhaea,* and *Pseudolycopodiella,* among the genera that occur in the United States and Canada) or subterranean, nonphotosynthetic, mycorrhizal structures (*Diphasiastrum, Huperzia, Lycopodium,* and *Phlegmariurus,* among the genera represented in the United States and Canada). The development and maturation of archegonia and antheridia in a gametophyte of *Lycopodiaceae* may require from 6 to 15 years, and their gametophytes may even produce a series of sporophytes in successive archegonia as they continue to grow. Despite the fact that the gametophytes are bisexual, it is known that, in at least some species of lycophytes, self-fertilization rates are very low. The gametophytes of these species predominantly cross-fertilize.

Water is required for fertilization in *Lycopodiaceae.* The biflagellated sperm swim through water to the archegonium. Following fertilization, the zygote develops into an embryo, which grows within the venter of the archegonium. The young sporophyte may remain attached to the gametophyte for a long time, but eventually it becomes independent. The life cycle of *Lycopodium lagopus,* a representative of those *Lycopodiaceae* that have a subterranean, mycorrhizal gametophyte and form a strobilus, is illustrated in Figure 19–14.

Among the genera of *Lycopodiaceae* found in the United States and Canada, *Huperzia* (Figure 19–15a), the fir mosses, consists of 7 species; *Lycopodium* (Figure 19–14), the tree club mosses, of 5; *Diphasiastrum* (Figure 13–15c), the club mosses and running pines, of 11; and *Lycopodiella* of 6. These genera, and the others now recognized in *Lycopodiaceae,* differ from one another in various technical characteristics, including the arrangement of the sporophylls, the presence of rhizomes and organization of the vegetative body, the nature of the gametophyte, and basic chromosome numbers.

(a) (b) (c) (d)

19–16
Representative Selaginella. ***(a)*** Selaginella lepidophylla, *the resurrection plant, a plant that becomes completely dried out when water is not available but quickly revives following a rain. This plant was growing in Big Bend National Park, in Texas.* ***(b)*** Selaginella rupestris, *the rock spikemoss, with strobili.* ***(c)*** Selaginella kraussiana, *a prostrate, creeping plant. Adventitious roots can be seen arising from the stems.* ***(d)*** Selaginella willdenovii, *from the Old World tropics. Shade-loving, it climbs to 7 meters and has peacock-blue leaves with a metallic sheen. Note the clearly evident rhizomes.*

The Resurrection Plant Belongs to the Family *Selaginellaceae*

Among the living genera of lycophytes, *Selaginella,* the only genus of the family *Selaginellaceae,* has the most species, about 700. Most of them are tropical in distribution. Many grow in moist places, although a few inhabit deserts, becoming dormant during the driest part of the year. Among the latter is the so-called resurrection plant, *Selaginella lepidophylla,* which occurs in Texas, New Mexico, and Mexico (Figure 19–16a).

Basically, the herbaceous sporophyte of *Selaginella* is similar to that of some *Lycopodiaceae* in that it bears microphylls and its sporophylls are arranged in strobili (Figure 19–16b). Unlike *Lycopodiaceae,* however, *Selaginella* has a small, scalelike outgrowth, called a **ligule,** near the base of the upper surface of each microphyll and sporophyll (Figure 19–17). The stem and root are protostelic (Figure 19–18).

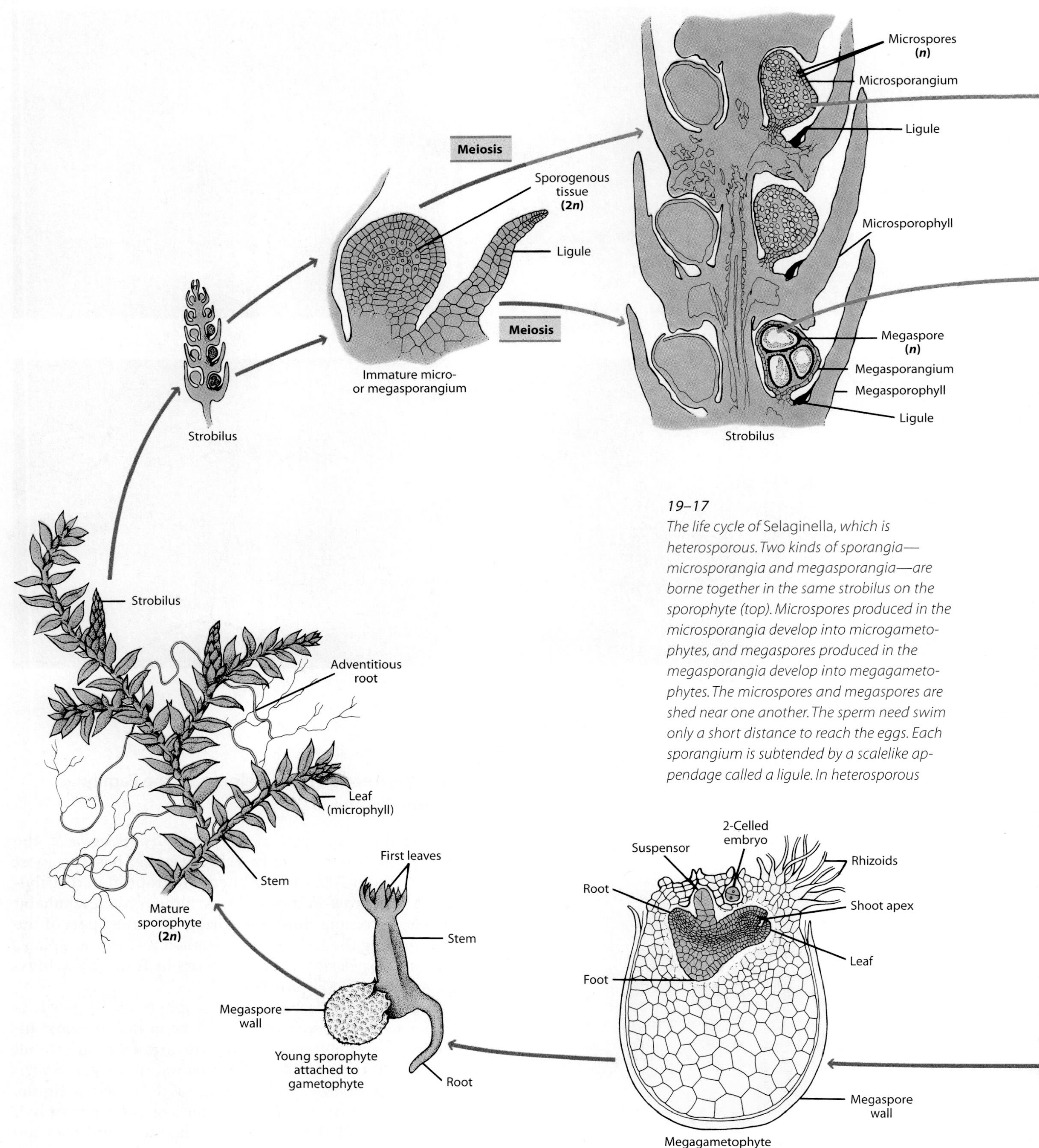

19–17
The life cycle of Selaginella, *which is heterosporous. Two kinds of sporangia—microsporangia and megasporangia—are borne together in the same strobilus on the sporophyte (top). Microspores produced in the microsporangia develop into microgametophytes, and megaspores produced in the megasporangia develop into megagametophytes. The microspores and megaspores are shed near one another. The sperm need swim only a short distance to reach the eggs. Each sporangium is subtended by a scalelike appendage called a ligule. In heterosporous*

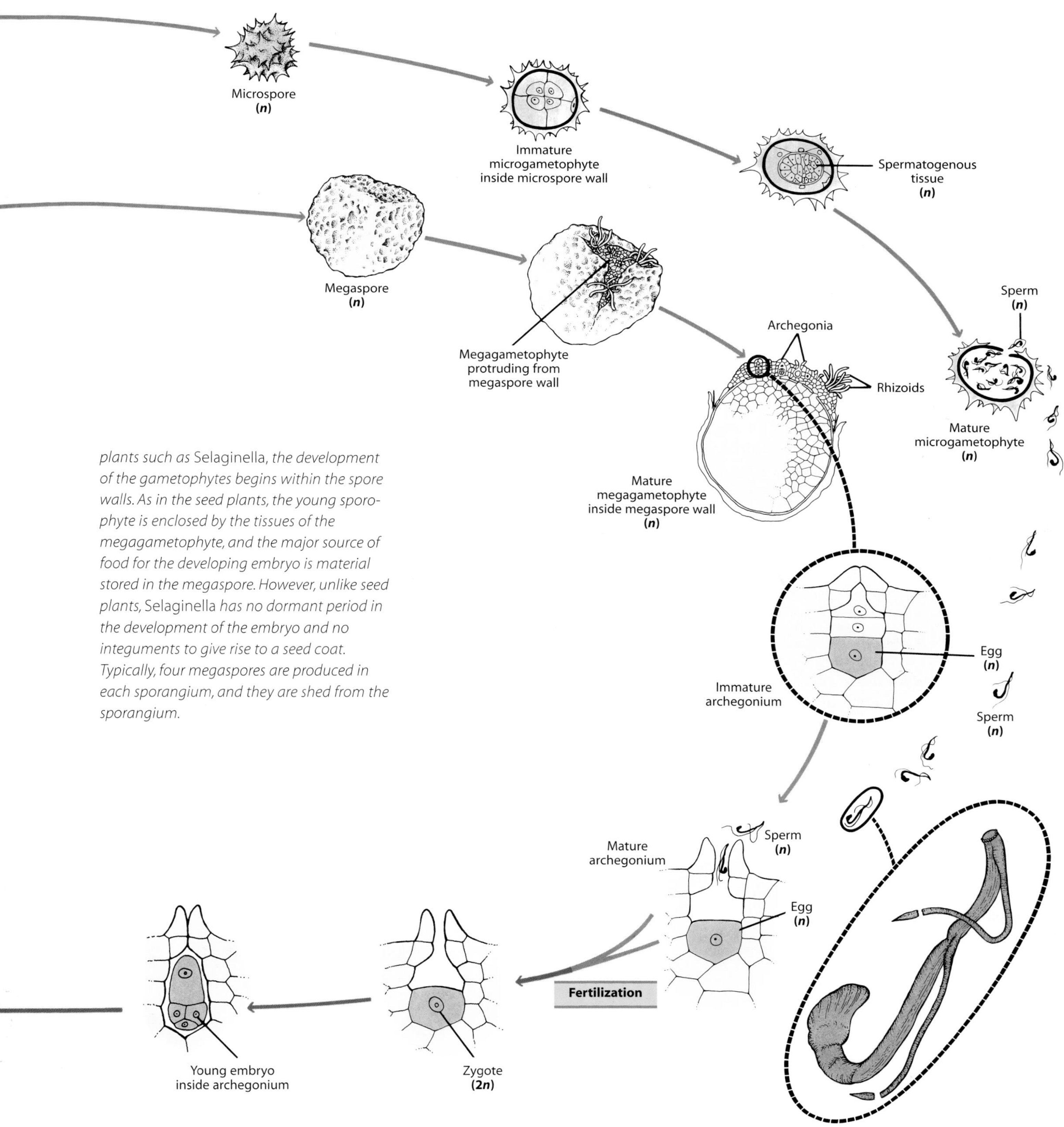

plants such as Selaginella, *the development of the gametophytes begins within the spore walls. As in the seed plants, the young sporophyte is enclosed by the tissues of the megagametophyte, and the major source of food for the developing embryo is material stored in the megaspore. However, unlike seed plants,* Selaginella *has no dormant period in the development of the embryo and no integuments to give rise to a seed coat. Typically, four megaspores are produced in each sporangium, and they are shed from the sporangium.*

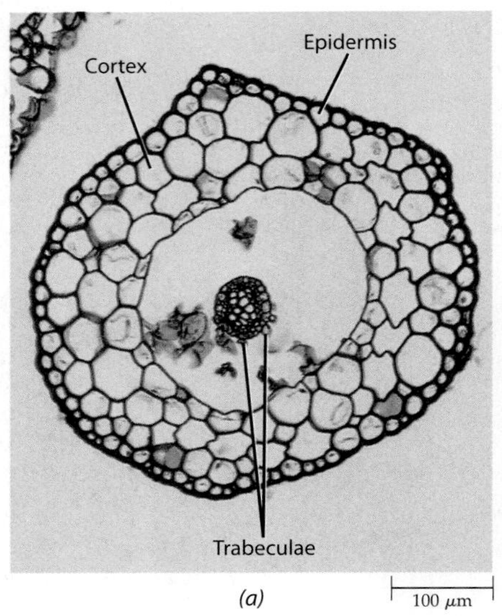

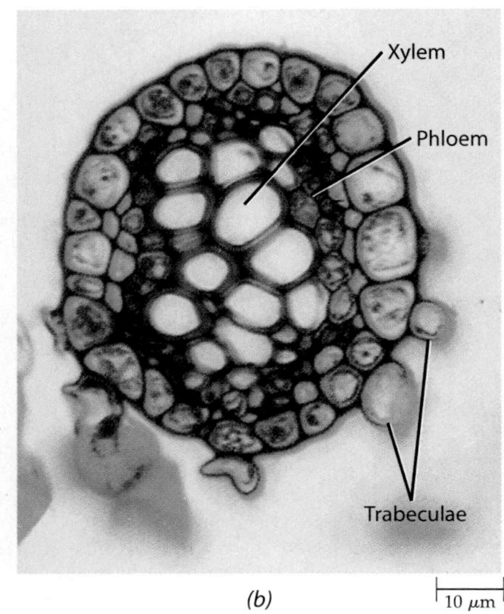

19–18
Selaginella *protostele.* ***(a)*** *Transverse section of stem, showing mature tissues. The protostele is suspended in the middle of the hollow stem by elongate cortical cells (endodermal cells), called trabeculae. Only a portion of each trabecula can be seen here.* ***(b)*** *Detail of protostele.*

Whereas *Lycopodiaceae* are homosporous, *Selaginella* is heterosporous, with unisexual gametophytes. Each sporophyll bears a single sporangium on its upper surface. Megasporangia are borne by **megasporophylls,** and microsporangia are borne by **microsporophylls.** Both kinds of sporangia occur in the same strobilus.

The male gametophytes (microgametophytes) in *Selaginella* develop within the microspore, and they lack chlorophyll. At maturity the male gametophyte consists of a single prothallial, or vegetative, cell and an antheridium, which gives rise to many biflagellated sperm. The microspore wall must rupture in order for the sperm to be liberated.

During development of the female gametophyte (megagametophyte), the megaspore wall ruptures, and the gametophyte protrudes through the rupture to the outside. This is the portion of the female gametophyte in which the archegonia develop. It has been reported that the female gametophytes sometimes develop chloroplasts, although most *Selaginella* gametophytes derive their nutrition from food stored within the megaspores.

Water is required in order for the sperm to swim to the archegonia and fertilize the eggs. Commonly, fertilization occurs after the gametophytes have been shed from the strobilus. During development of the embryos in both *Lycopodiaceae* and *Selaginella,* a structure called a **suspensor** is formed. Although inactive in *Lycopodiaceae* and some species of *Selaginella,* in other species of *Selaginella* the suspensor serves to thrust the developing embryo deep within the nutrient-rich tissue of the female gametophyte. Gradually, the developing sporophyte emerges from the gametophyte and becomes independent.

The life cycle of *Selaginella* is illustrated in Figure 19–17.

The Quillworts Belong to the Family *Isoetaceae*

The only genus of the family *Isoetaceae* is *Isoetes,* the quillworts. *Isoetes* is the nearest living relative of the ancient tree lycophytes of the Carboniferous (see page 456). Plants of *Isoetes* may be aquatic, or they may grow in pools that become dry at certain seasons. The sporophyte of *Isoetes* consists of a short, fleshy underground stem (corm) bearing quill-like microphylls on its upper surface and roots on its lower surface (Figure 19–19). In *Isoetes,* each leaf is a potential sporophyll.

19–19
Isoetes storkii. *View of the sporophyte showing quill-like leaves (microphylls), stem, and roots.* Isoetes *is the last living representative of the group that included the extinct tree lycophytes of the Carboniferous coal swamps.*

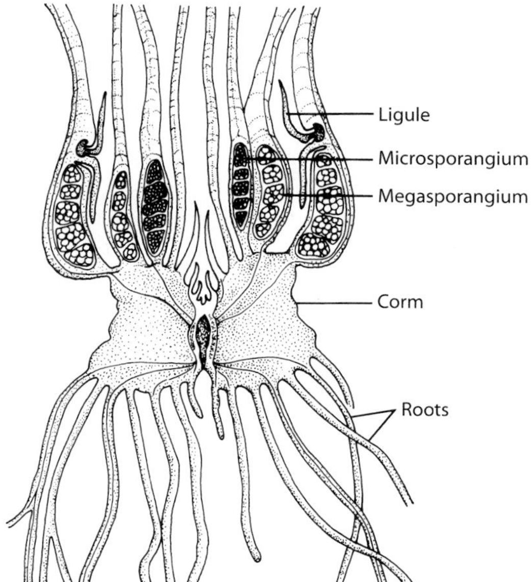

19–20
Diagram of a vertical section of an Isoetes *plant. Leaves are borne on the upper surface, and roots on the lower surface, of a short, fleshy underground stem, or corm. Some leaves (megasporophylls) bear megasporangia, and others (microsporophylls) bear microsporangia. The microsporophylls are located nearer the center of the plant.*

Like *Selaginella, Isoetes* is heterosporous. The megasporangia are borne at the base of megasporophylls, and the microsporangia are borne at the base of microsporophylls, similar to the megasporophylls but located nearer the center of the plant (Figure 19–20). A ligule is present just above the sporangium of each sporophyll.

One of the distinctive features of *Isoetes* is the presence of a specialized cambium that adds secondary tissues to the corm. Externally the cambium produces only parenchyma tissue, whereas internally it produces a peculiar vascular tissue consisting of sieve elements, parenchyma cells, and tracheids in varying proportions.

Some species of *Isoetes* (assigned to another genus, *Stylites,* by some) from high elevations in the tropics have the unique characteristic of obtaining their carbon for photosynthesis from the sediment in which they grow rather than from the atmosphere. The leaves of these plants lack stomata, have a thick cuticle, and carry on essentially no gas exchange with the atmosphere. Like at least some of the other species of *Isoetes* in which the plants dry out for part of the year, these species have CAM photosynthesis (page 147).

Phylum *Trimerophytophyta*

The phylum *Trimerophytophyta,* which probably evolved directly from the rhyniophytes, most likely contains plants of diverse evolutionary relationships and seems to represent the ancestral stock of the ferns, the progymnosperms, and perhaps the horsetails as well. The trimerophytes, which were larger and more complex plants than the rhyniophytes or zosterophyllophytes (Figure 19–10c), first appeared in the Early Devonian period about 395 million years ago and had become extinct by the end of the mid-Devonian, about 20 million years later—a relatively short period of existence.

Although generally larger and evolutionarily more specialized than the rhyniophytes, trimerophytes still lacked leaves. Branching, however, was more complex, with the main axis forming lateral branch systems that dichotomized several times. The trimerophytes, like the rhyniophytes and zosterophyllophytes, were homosporous. Some of their smaller branches terminated in elongate sporangia, while others were entirely vegetative. Besides their more complex branching pattern, the trimerophytes had a more massive vascular strand than the rhyniophytes. Together with a broad band of thick-walled cells in the cortex, the large vascular strand probably was capable of supporting fairly large plants, over a meter in height. As in the rhyniophytes, the xylem of the trimerophytes differentiated centrifugally. The name of the phylum comes from the Greek words *tri-, meros,* and *phyton,* meaning "three-parted plant," because of the organization of the secondary branches into three rows in the genus *Trimerophyton.*

Phylum *Psilotophyta*

The phylum *Psilotophyta* includes two living genera, *Psilotum* and *Tmesipteris. Psilotum,* the whisk fern, is tropical and subtropical in distribution. In the United States, it occurs in Alabama, Arizona, Florida, Hawaii, Louisiana, North Carolina, and Texas, as well as Puerto Rico, and it is a common greenhouse weed. *Tmesipteris* is restricted in distribution to Australia, New Caledonia, New Zealand, and other regions of the South Pacific. Although both genera are very simple plants that resemble the rhyniophytes in some aspects of their basic structure, it seems likely that they may reflect an early-diverging lineage related to living ferns. Their simple structure appears to have resulted from reduction from more complex ancestors.

Psilotum is unique among living vascular plants in that it lacks both roots and leaves. The sporophyte consists of a dichotomously branching aerial portion with small scalelike outgrowths and a branching underground portion, or a system of rhizomes with many rhi-

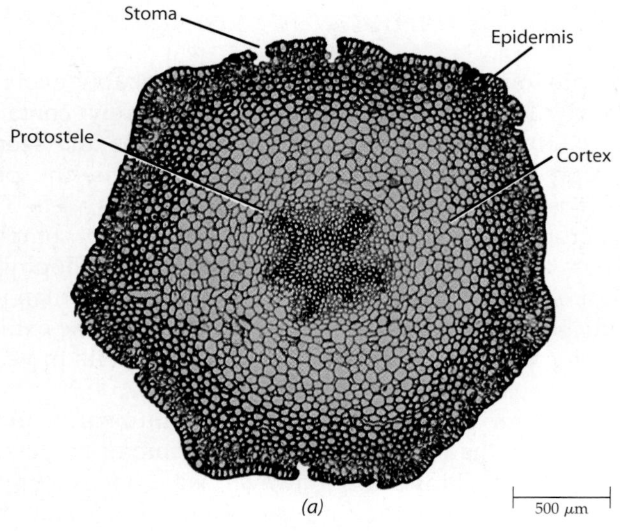

(a)

Phloem

Xylem

(b)

100 μm

19–21
Psilotum nudum. *(a) Transverse section of stem, showing mature tissues.* ***(b)*** *Detail of protostele, showing xylem and phloem.*

zoids (see the mature sporophyte at the left in Figure 19–24). A symbiotic fungus—an endomycorrhizal zygomycete (page 341)—is present in the outer cortical cells of the rhizomes. *Psilotum* has a protostele (Figure 19–21).

Psilotum is homosporous, the spores being produced in sporangia that are generally aggregated in groups of three on the ends of short, lateral branches. Upon germination, the spores give rise to bisexual gametophytes, which resemble portions of the rhizome (Figure 19–22).

19–22
(a) The subterranean gametophyte of Psilotum nudum. *The gametophytes of psilotophytes are bisexual; that is, they bear both antheridia and archegonia.* ***(b)*** Psilotum nudum, *known locally as the moa plant, growing on a 1955 lava flow on the island of Hawaii. The yellow sporangia are clearly evident.*

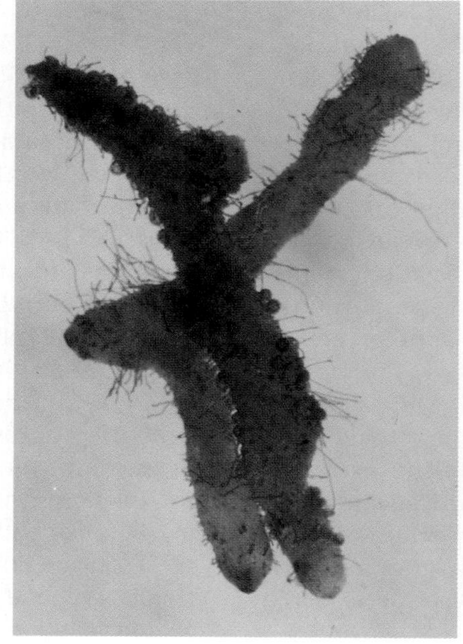

(a)

(b)

(a)

(b)

19–23
(a) Tmesipteris parva *growing on the trunk of the tree fern* Cyathea australis *in New South Wales, Australia.* *(b)* Tmesipteris lanceolata, *in New Caledonia, an island of the southwest Pacific.*

Like the rhizome, the subterranean gametophyte contains a symbiotic fungus. In addition, some gametophytes contain vascular tissue. The sperm of *Psilotum* are multiflagellated and require water to swim to the egg. Initially the sporophyte is attached to the gametophyte by a foot, a structure that absorbs nutrients from the gametophyte. Eventually the sporophyte becomes detached from the foot, which remains embedded in the gametophyte. The life cycle of *Psilotum* is illustrated in Figure 19–24 (pages 446 and 447).

Tmesipteris grows as an epiphyte on tree ferns and other plants (Figure 19–23) and in rock crevices. The leaflike appendages of *Tmesipteris* are larger than the scalelike outgrowths of *Psilotum,* but in other respects *Tmesipteris* is essentially similar to *Psilotum.*

Phylum *Sphenophyta*

Like the *Lycophyta,* the *Sphenophyta* extend back to the Devonian period. The sphenophytes reached their maximum abundance and diversity later in the Paleozoic era, about 300 million years ago. During the Late Devonian and Carboniferous periods, they were represented by the calamites (see page 456), a group of trees that reached 18 meters or more in height, with a trunk that could be more than 45 centimeters thick. Today the *Sphenophyta* are represented by a single herbaceous genus, *Equisetum* (Figure 19–25), which consists of 15 species. Since *Equisetum* is essentially identical to *Equisetites,* a plant that appeared about 300 million years ago, in the Carboniferous period, *Equisetum* may be the oldest surviving genus of plants on Earth.

The species of *Equisetum* are known as the "horsetails"; they are widespread in moist or damp places, by streams, and along the edge of woods. Horsetails are easily recognized because of their conspicuously jointed stems and rough texture. The small, scalelike leaves, or microphylls, are whorled at the nodes. When present, the branches arise laterally at the nodes and alternate with the leaves. The internodes (the portions of the stems between successive nodes) are ribbed, and the ribs are tough and strengthened with siliceous deposits in the epidermal cells. Horsetails have been used to scour pots and pans, particularly in colonial and frontier times, and have thus earned the name "scouring rushes." The roots are adventitious, arising at the nodes of the rhizomes, which are important in vegetative propagation.

The aerial stems of *Equisetum* arise from branching underground rhizomes, and, although the plants may die back during unfavorable seasons, the rhizomes are perennial. The aerial stem is complex anatomically (Figure 19–26). At maturity, its internodes contain a hollow pith surrounded by a ring of smaller canals called **carinal canals.** Each of these smaller canals is associated with a strand of xylem and phloem.

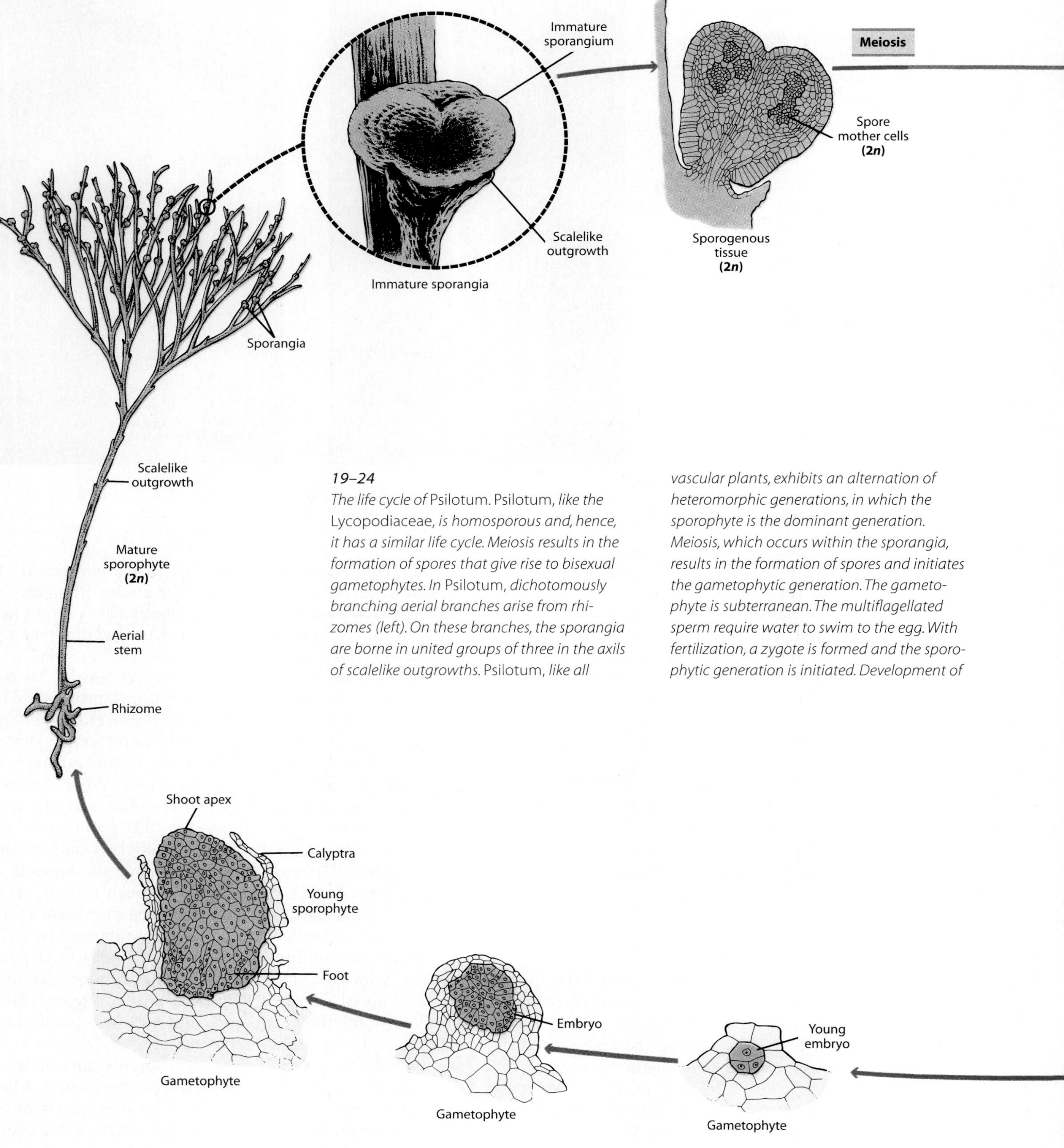

19–24
The life cycle of Psilotum. Psilotum, *like the* Lycopodiaceae, *is homosporous and, hence, it has a similar life cycle. Meiosis results in the formation of spores that give rise to bisexual gametophytes. In* Psilotum, *dichotomously branching aerial branches arise from rhizomes (left). On these branches, the sporangia are borne in united groups of three in the axils of scalelike outgrowths.* Psilotum, *like all vascular plants, exhibits an alternation of heteromorphic generations, in which the sporophyte is the dominant generation. Meiosis, which occurs within the sporangia, results in the formation of spores and initiates the gametophytic generation. The gametophyte is subterranean. The multiflagellated sperm require water to swim to the egg. With fertilization, a zygote is formed and the sporophytic generation is initiated. Development of*

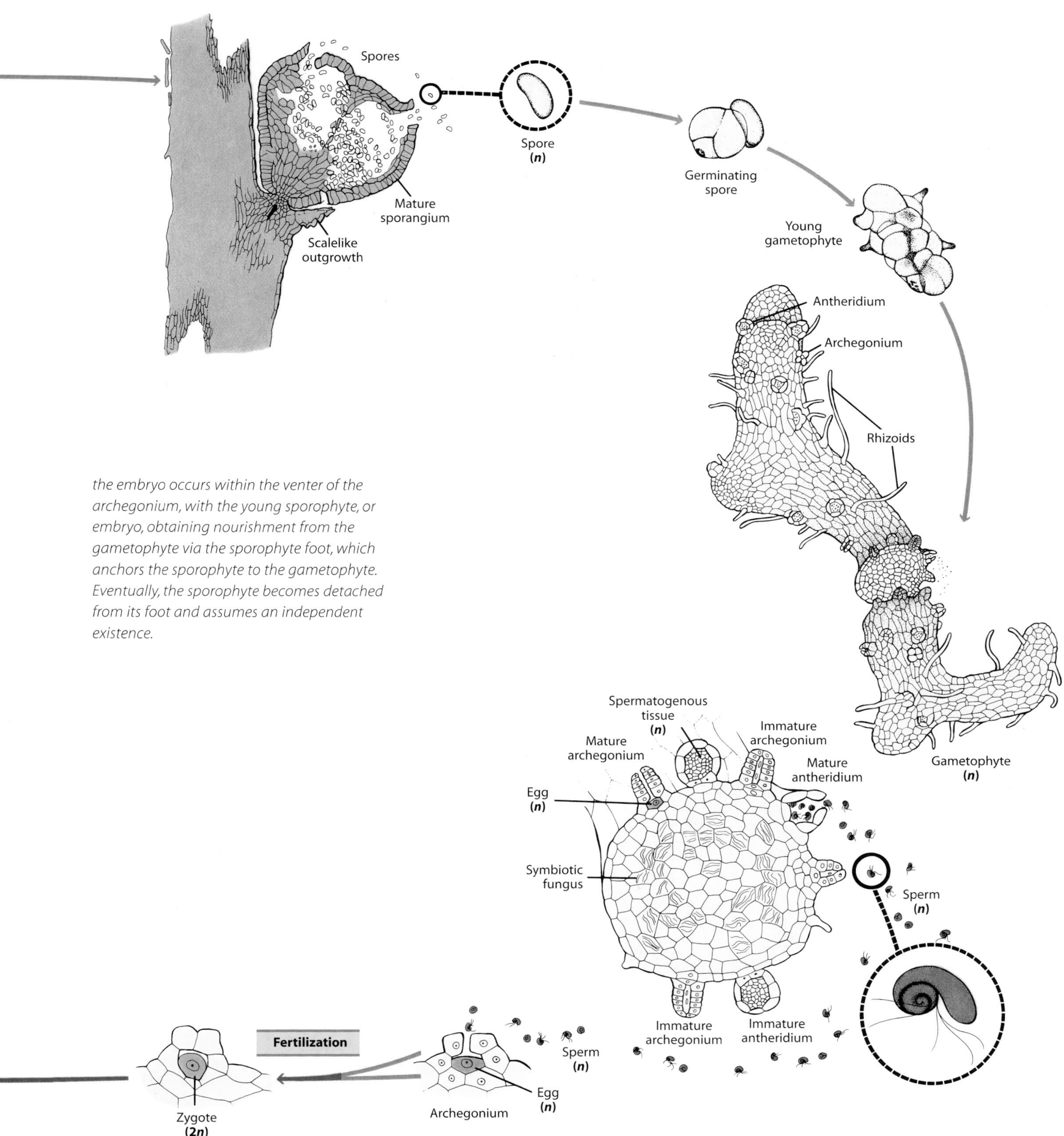

the embryo occurs within the venter of the archegonium, with the young sporophyte, or embryo, obtaining nourishment from the gametophyte via the sporophyte foot, which anchors the sporophyte to the gametophyte. Eventually, the sporophyte becomes detached from its foot and assumes an independent existence.

19–25
Equisetum. ***(a)*** *A species of* Equisetum *in which there are separate fertile and vegetative shoots. The fertile shoots essentially lack chlorophyll and are very different in appearance from the vegetative shoots. Each fertile shoot has a terminal strobilus. Notice the whorls of scale-like leaves at each node.* ***(b)*** *Branching vegetative shoots of* Equisetum arvense.

Equisetum is homosporous. Sporangia are borne in groups of five to 10 along the margins of small umbrella-like structures known as **sporangiophores** (sporangia-bearing branches), which are clustered into strobili at the apex of the stem (Figures 19–25a and 19–29). The fertile stems of some species do not contain much chlorophyll. In these species, the fertile stems are sharply distinct from the vegetative stems, often appearing before the latter early in the spring (Figure 19–25). In other species of *Equisetum,* the strobili are borne at the tips of otherwise vegetative stems (see Figure 13–15d). When the spores are mature, the sporangia contract and split along their inner surface, releasing numerous spores. Elaters—thickened bands that arise from the outer layer of the

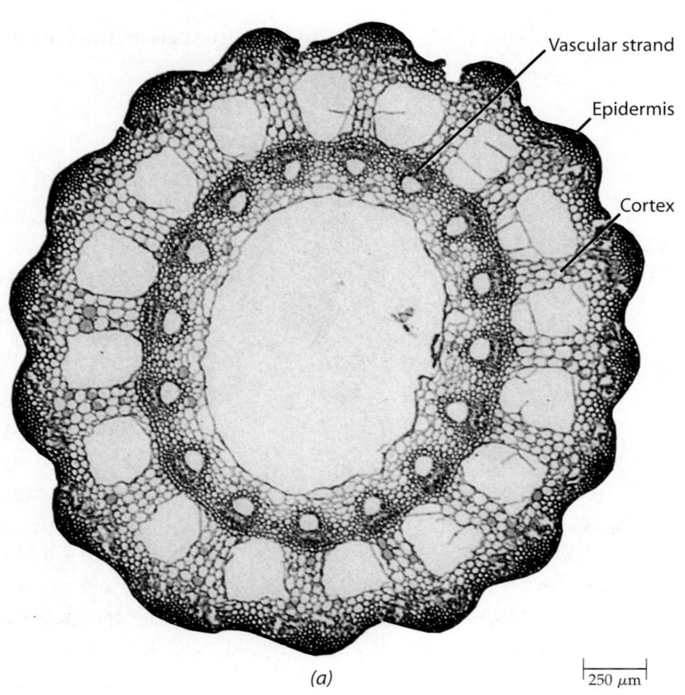

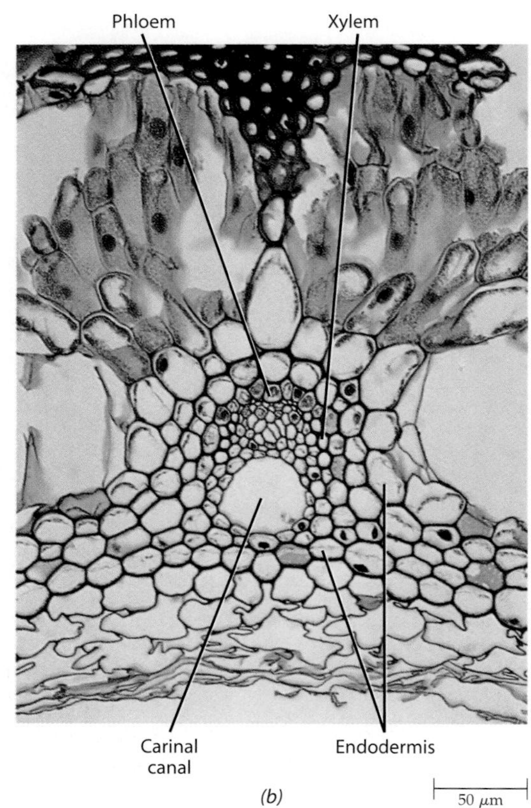

19–26
Stem anatomy of Equisetum. ***(a)*** *Transverse section of stem, showing mature tissues.* ***(b)*** *Detail of vascular strand, showing xylem and phloem.*

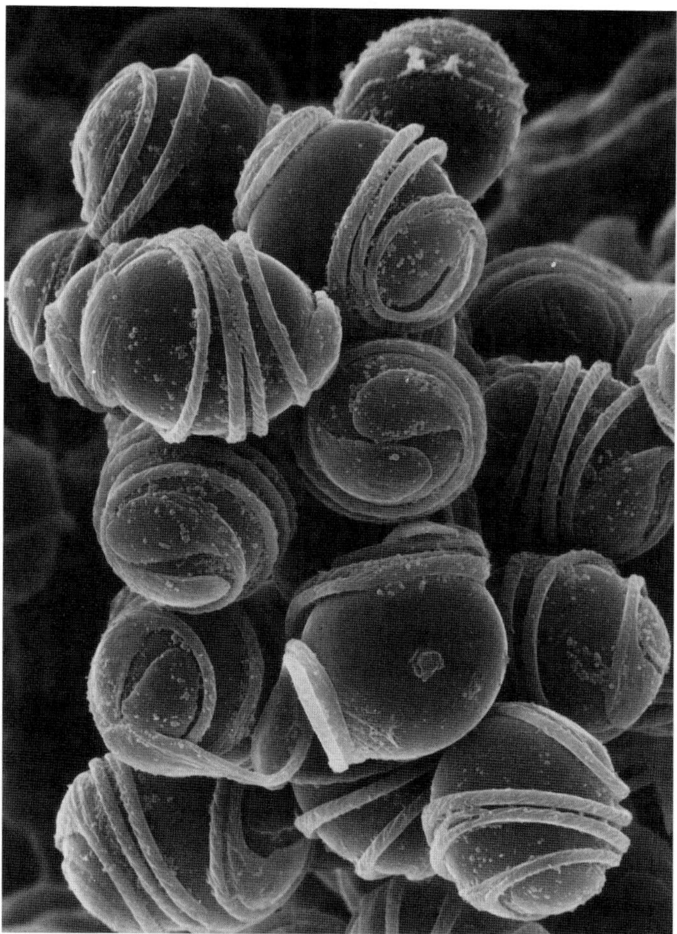

19–27
Spores of the horsetail Equisetum arvense, *as seen in a scanning electron micrograph. Shown here are moist spores wrapped tightly by thickened bands, known as elaters, which are attached to the spore walls. As the spores dry, the elaters uncoil, helping to disperse the spores from the sporangium.*

spore wall—coil when moist and uncoil when dry, thus playing a role in spore dispersal (Figures 19–27 and 19–29). These are quite distinct from the elaters that aid in spore dispersal in *Marchantia.* There, the elaters are elongated cells with helically arranged wall thickenings (see Figure 18–14).

The gametophytes of *Equisetum* are green and free-living and range in diameter from a few millimeters to 1 centimeter or even 3 to 3.5 centimeters in some species. Gametophytes become established mainly on mud that has recently been flooded and is rich in nutrients. The gametophytes, which reach sexual maturity in three to five weeks, are either bisexual or male (Figure 19–28). In bisexual gametophytes, the archegonia develop before

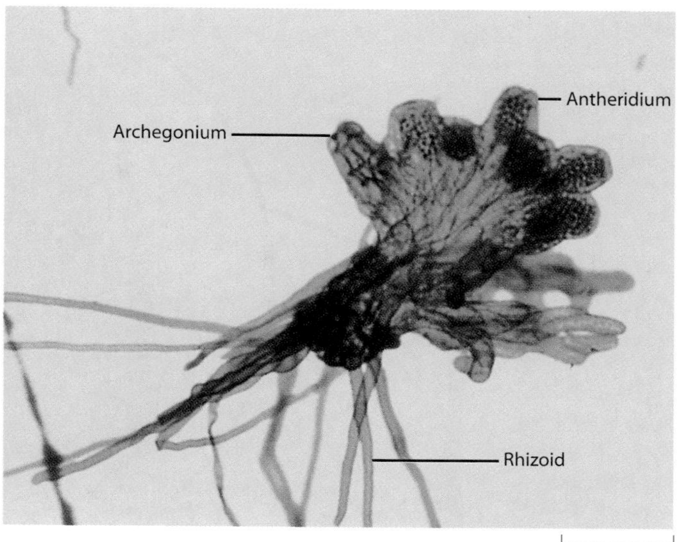

19–28
Bisexual gametophyte of Equisetum, *showing male and female gametangia, antheridia and archegonia, respectively. Rhizoids can be seen extending from the lower surface of the gametophyte.*

the antheridia; this developmental pattern increases the probability of cross-fertilization. The sperm are multiflagellated and require water to swim to the eggs. The eggs of several archegonia on a single gametophyte may be fertilized and develop into embryos, or young sporophytes.

The life cycle of *Equisetum* is illustrated in Figure 19–29 (pages 450 and 451).

Phylum *Pterophyta*

Ferns have been relatively abundant in the fossil record from the Carboniferous period to the present (see pages 456 and 457, and Figure 20–1). Today, ferns number about 11,000 species; they are the largest group of plants other than the flowering plants and are the most diverse in both form and habit (Figure 19–30).

The diversity of ferns is greatest in the tropics, where about three-fourths of the species are found. Here, not only are there many species of ferns, but ferns are abundant in many plant communities. Only about 380 species of ferns occur in the United States and Canada, whereas about 1000 occur in the small tropical country of Costa Rica in Central America. Approximately a third of all species of tropical ferns grow upon the trunks or branches of trees as epiphytes (Figure 19–30).

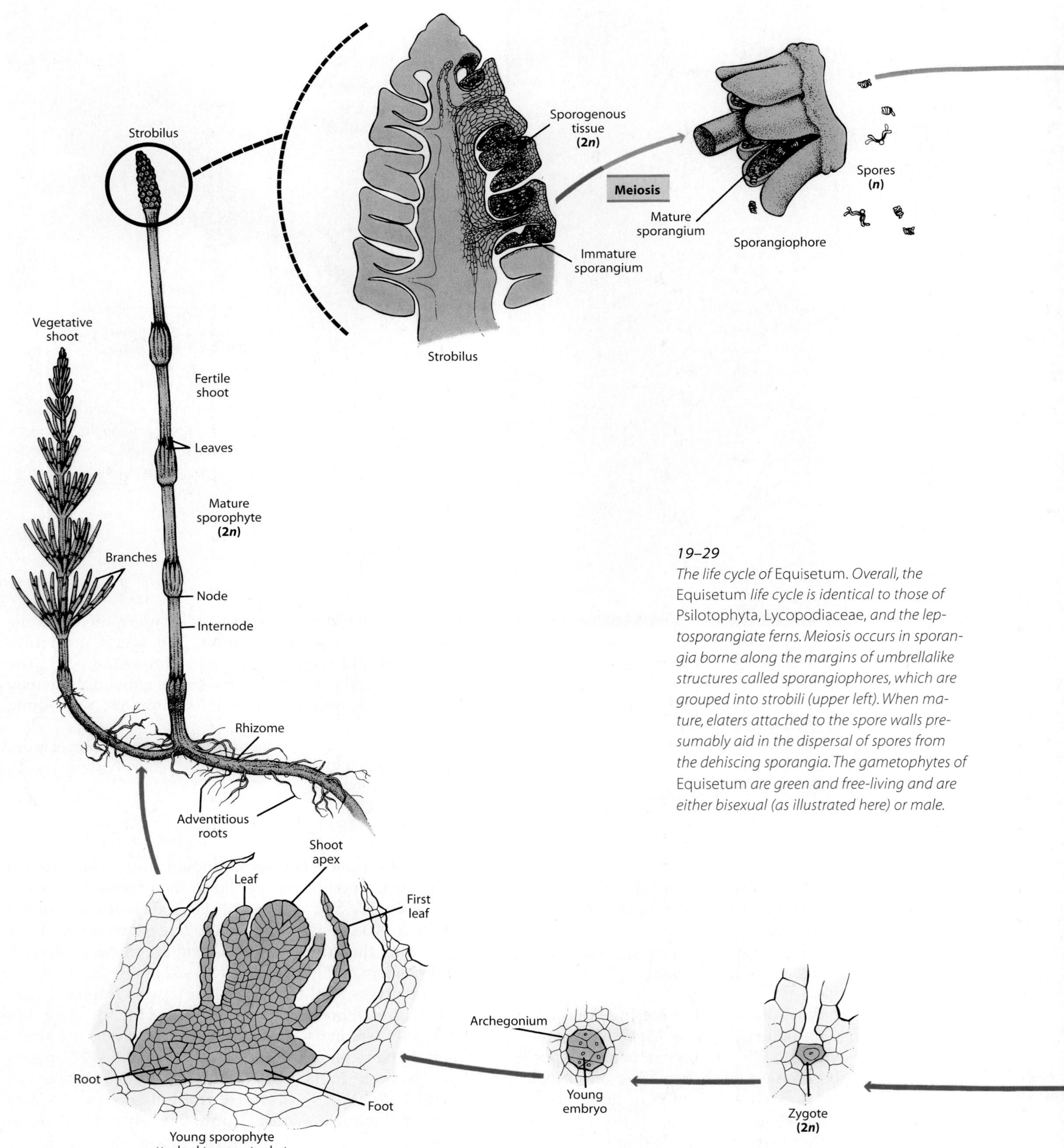

19–29

The life cycle of Equisetum. *Overall, the* Equisetum *life cycle is identical to those of* Psilotophyta, Lycopodiaceae, *and the leptosporangiate ferns. Meiosis occurs in sporangia borne along the margins of umbrellalike structures called sporangiophores, which are grouped into strobili (upper left). When mature, elaters attached to the spore walls presumably aid in the dispersal of spores from the dehiscing sporangia. The gametophytes of* Equisetum *are green and free-living and are either bisexual (as illustrated here) or male.*

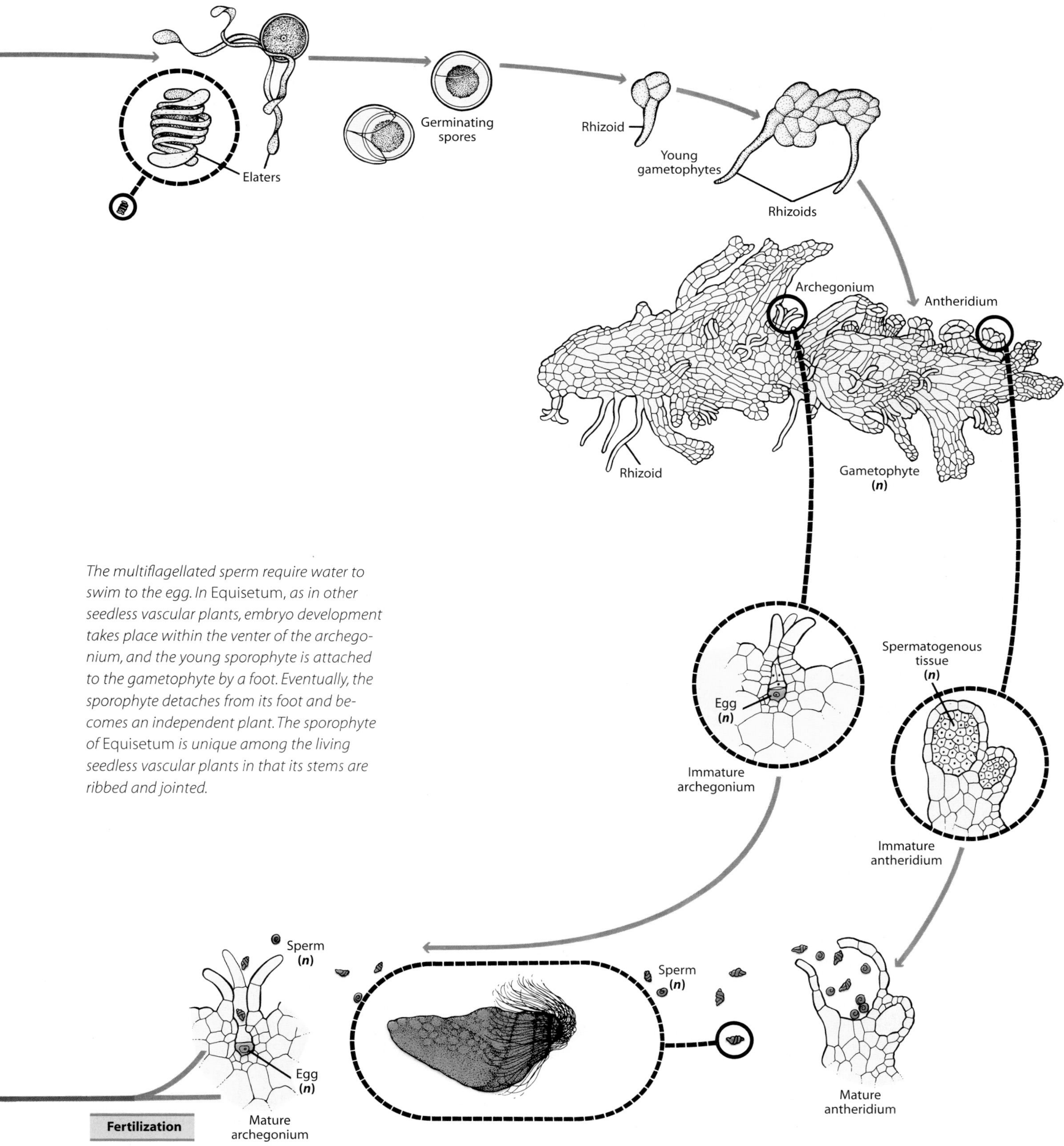

The multiflagellated sperm require water to swim to the egg. In Equisetum, *as in other seedless vascular plants, embryo development takes place within the venter of the archegonium, and the young sporophyte is attached to the gametophyte by a foot. Eventually, the sporophyte detaches from its foot and becomes an independent plant. The sporophyte of* Equisetum *is unique among the living seedless vascular plants in that its stems are ribbed and jointed.*

(a) (b) (c) (d) (e) (f) (g)

19–30
The diversity of ferns, as illustrated by a few genera of the largest order of ferns, Filicales. *(a)* Lindsaea, *Volcán Barba, Costa Rica. (b) A tree fern,* Cyathea, *at Monteverde, Costa Rica. (c)* Plagiogyria, *with distinct fertile and vegetative leaves, Volcán Poás, Costa Rica. (d)* Elaphoglossum, *with thick, undivided leaves, near Cuzco, Peru. (e)* Asplenium septentrionale, *a small fern that occurs all around the Northern Hemisphere, growing on metal-rich soil near a lead-silver mine in Wales. (f)* Pleopeltis polypodioides, *growing as an epiphyte on a juniper trunk in Arkansas. (g)* Hymenophyllum *species, one of the filmy ferns, so-called because of their delicate leaves. Filmy ferns occur as epiphytes primarily in tropical rainforests or wet temperate regions.*

Some ferns are very small and have undivided leaves. *Lygodium,* a climbing fern, has leaves with a long, twining rachis (an extension of the leaf stalk, or petiole) that may be up to 30 meters or more in length. Some tree ferns (Figure 19–30b), such as those of the genus *Cyathea,* have been recorded to reach heights of more than 24 meters and to have leaves 5 meters or more in length. Although the trunks of such tree ferns may be 30 centimeters or more thick, their tissues are entirely primary in origin. Most of this thickness is the fibrous root mantle; the true stem is only four to six centimeters in diameter. The herbaceous genus *Botrychium* (see Figure 19–32a) is the only living fern known to form a vascular cambium.

In terms of the structure and method of development of their sporangia, ferns may be classified as either **eusporangiate** or **leptosporangiate** (Figure 19–31). The distinction between these two types of sporangia is important for understanding relationships among vascular plants. In a **eusporangium,** the parent cells, or initials, are located at the surface of the tissue from which the sporangium is produced (Figure 19–31a). These initials divide by the formation of walls parallel to the surface, resulting in the formation of an inner and an outer series of cells. The outer cell layer, by further divisions in both planes, builds up the several-layered wall of the spo-

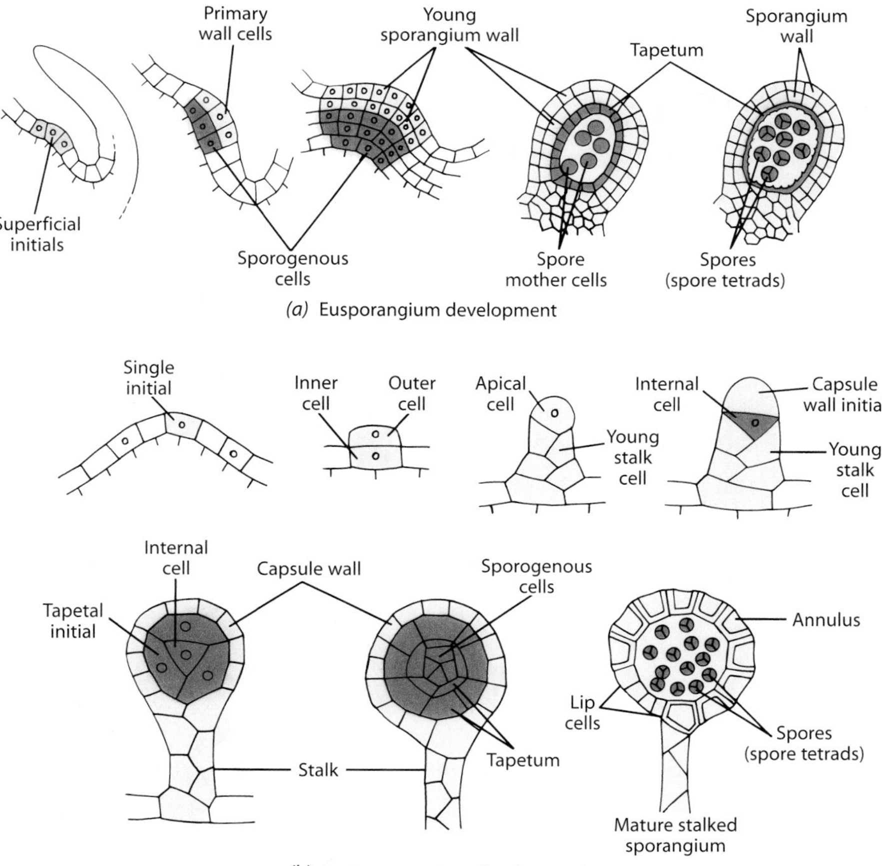

19–31
Development and structure of the two principal types of fern sporangia. (a) The eusporangium originates from a series of superficial parent cells, or initials. Each eusporangium develops a wall two or more layers thick (although at maturity the inner wall layers may be crushed) and a high number of spores. (b) The leptosporangium originates from a single initial cell, which first produces a stalk and then a capsule. Each leptosporangium gives rise to a relatively small number of spores.

rangium. The inner layer gives rise to a mass of irregularly oriented cells from which the spore mother cells ultimately arise. In many eusporangia, the inner wall layers are stretched and compressed during the course of development, so that the walls may apparently consist of a single layer of cells at maturity. Eusporangia, which are larger than leptosporangia and contain many more spores, are characteristic of all vascular plants—including those we have considered thus far—except for the leptosporangiate ferns.

In contrast to the multicellular origin of eusporangia, **leptosporangia** arise from a single superficial initial cell, which divides transversely or obliquely (Figure 19–31b). The inner of the two cells produced by this division may either contribute cells that produce a large part of the sporangial stalk or remain inactive and play no role in the further development of the sporangium, which is the more common condition. By a precise pattern of divisions, the outer cell ultimately gives rise to an elaborate, stalked sporangium, with a globose capsule having a wall that is one cell layer thick. Within this wall is a nutritive structure two cell layers thick called the **tapetum.** The inner mass of the leptosporangium eventually differentiates into spore mother cells, which undergo meiosis to produce four spores each. After it nourishes the young dividing cells within the sporangium, the tapetum is deposited around the spores, creating ridges, spines, and other types of surface features that are often characteristic for individual families and genera. The spores are exposed following the development of a crack in the so-called *lip cells* of the sporangium. The sporangia are stalked, and each contains a special layer of unevenly thick-walled cells called an **annulus.** As the sporangium dries out, contraction of the annulus causes tearing in the middle of the capsule. The sudden explosion and snapping back of the annulus to its original position then result in a catapultlike discharge of the spores. In eusporangia, the stalks are more massive and, while there may be preformed lines of dehiscence, there is no annulus and no catapultlike discharge of spores.

Most living ferns are homosporous; heterospory is restricted to two orders of living water ferns (see Figure 19–39), which will be discussed further below. A few extinct ferns also were heterosporous.

We shall now consider as examples three very different kinds of ferns: (1) the orders *Ophioglossales* and *Marattiales,* as examples of eusporangiate ferns; (2) the *Filicales,* or homosporous leptosporangiate ferns; and (3) the water ferns, orders *Marsileales* and *Salviniales,* the heterosporous leptosporangiate ferns.

The Orders *Ophioglossales* and *Marattiales* Are Eusporangiate Ferns

Of the three genera of the order *Ophioglossales, Botrychium,* the grape ferns (Figure 19–32a), and *Ophioglossum,* the adder's tongues (Figure 19–32b), are widespread in the north temperate region. In both of these genera, a single leaf typically is produced each year from the stem. Each leaf consists of two parts: (1) a vegetative portion, or blade, which is deeply dissected in *Botrychium* and undivided in most species of *Ophioglossum,* and (2) a fertile segment. In *Botrychium,* the fertile segment is dissected in the same way as the vegetative portion and bears two rows of eusporangia on the outermost segments. In *Ophioglossum,* the fertile portion is undivided and bears two rows of sunken eusporangia.

(a)

(b)

19–32

Representatives of the two genera of Ophioglossales *that occur in North America.* ***(a)*** Botrychium parallelum. *In the genus* Botrychium, *the lower, vegetative portion of the leaf is divided. This is the only fern genus to form a vascular cambium.* ***(b)*** *In* Ophioglossum, *the lower portion of the leaf is undivided. In both genera, the erect, fertile, upper part of the leaf is sharply distinct from the vegetative portion.*

The gametophytes of *Botrychium* and *Ophioglossum* are subterranean, tuberous, elongate structures with numerous rhizoids; they have endophytic fungi and resemble the gametophytes of *Psilotophyta*. In *Botrychium*, the gametophytes usually possess a dorsal ridge in which the antheridia are embedded, with the archegonia generally located along the sides of the ridge. In the nature of their gametophytes, the structure of their leaves, and several other anatomical details, the *Ophioglossales* are sharply distinct from other living ferns and are clearly an early diverging and distinct group. Unfortunately, the group has no well-established fossil record before about 50 million years ago. One member of the *Ophioglossales*, *Ophioglossum reticulatum*, has the highest chromosome number known in any living organism, with a diploid complement of about 1260 chromosomes.

The only other order of ferns that has eusporangia, the tropical *Marattiales*, is an ancient group with a fossil record that extends back to the Carboniferous period. The members of this order resemble more familiar groups of ferns more closely than they do *Ophioglossales*. *Psaronius*, the extinct tree fern illustrated in Figure 20–1, was a member of this order. The six living genera of *Marattiales* include about 200 species.

Filicales Is an Order of Homosporous Leptosporangiate Ferns

Nearly all familiar ferns are members of the larger order *Filicales*, with at least 10,500 species. About 35 families and 320 genera are recognized in the order. *Filicales* differ from *Ophioglossales* and *Marattiales* in being leptosporangiate, and from the water ferns, which we shall discuss next, in being homosporous. All ferns other than *Ophioglossales* and *Marattiales*, in fact, are leptosporangiate, and very few have the subterranean gametophytes with endophytic fungi that are characteristic of the *Ophioglossales* and *Marattiales*. Clearly, leptosporangia and the other distinctive features of most ferns are specialized characters since they occur nowhere else among the vascular plants, including *Ophioglossales* and *Marattiales*, which share more features in common with other groups of ancient plants.

Most garden and woodland ferns of temperate regions have siphonostelic rhizomes (Figure 19–33) that produce new sets of leaves each year. The fern embryo produces a true root, but this soon withers, and the rest of the roots are adventitious—that is, they arise from the rhizomes near the bases of the leaves. The leaves, or **fronds,** are megaphylls and represent the most conspicuous part of the sporophyte. Their high surface-to-volume ratio allows them to capture sunlight much more effectively than the microphylls of the lycophytes. The ferns are the only seedless vascular plants to possess well-developed megaphylls. Commonly, the fronds are compound; that is, the lamina is divided into leaflets, or **pinnae,** which are attached to the **rachis,** an extension of the leaf stalk, or petiole. In nearly all ferns, the young leaves are coiled (circinate); they are commonly referred to as "fiddleheads" (Figure 19–34). This type of leaf de-

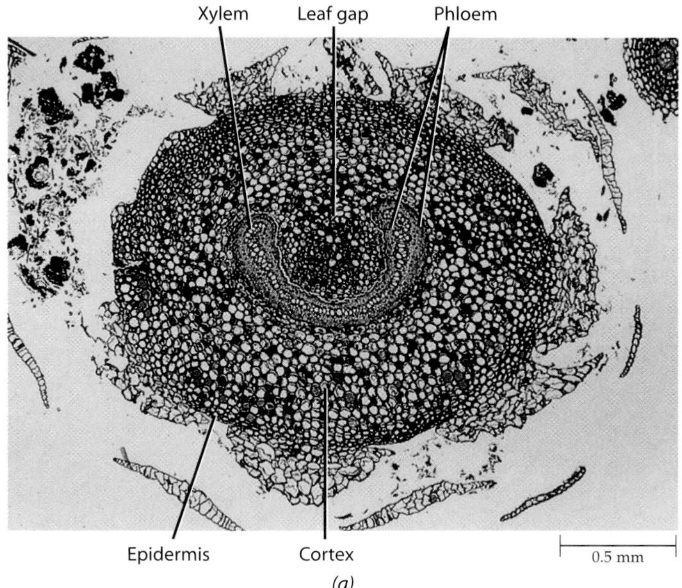

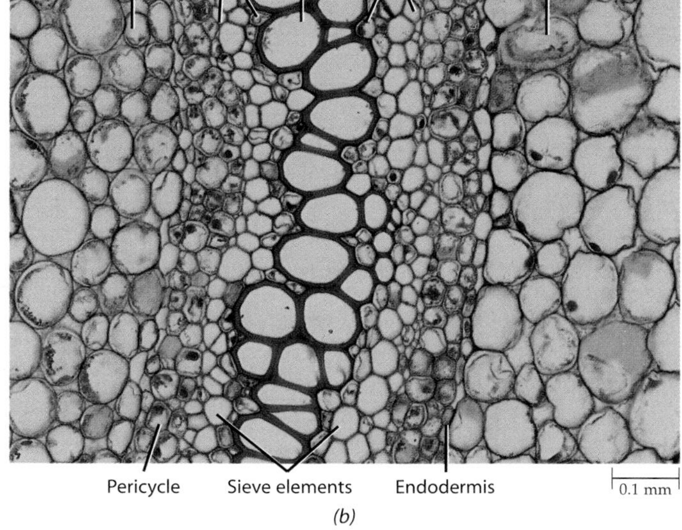

19–33

The anatomy of fern rhizomes. ***(a)*** Adiantum, *or maidenhair fern. Transverse section of a rhizome, showing the siphonostele. Note the wide leaf gap.* ***(b)*** *Transverse section of part of the vascular region of a rhizome of the tree fern* Dicksonia. *The phloem is composed mainly of sieve elements; the xylem is composed entirely of tracheids.*

Coal Age Plants

The amount of carbon dioxide used in photosynthesis is about 100 billion metric tons annually, about a tenth of the total carbon dioxide present in the atmosphere. The amount of carbon dioxide returned as a result of oxidation of these living materials is about the same, differing only by 1 part in 10,000. This very slight imbalance is caused by the burying of organisms in sediment or mud under conditions in which oxygen is excluded and decay is only partial. This accumulation of partially decayed plant material is known as peat (page 415). The peat may eventually become covered with sedimentary rock and thus placed under pressure. Depending on time, temperature, and other factors, peat may become compressed into soft or hard coal, one of the so-called fossil fuels.

During certain periods in the Earth's history, the rate of fossil-fuel formation was greater than at other times. One such time was the Carboniferous period, which extended from about 362 to 290 million years ago (see Figures 19–11 and 20–1). The lands were low, covered by shallow seas or swamps, and, in what are now temperate regions of Europe and North America, conditions were favorable for year-round growth. These regions were tropical to subtropical, with the equator then arcing across the Appalachians, over northern Europe, and through Ukraine. Five groups of plants dominated the swamplands, and three of them were seedless vascular plants—lycophytes, sphenophytes (calamites), and ferns. The other two were gymnospermous types of seed plants—seed ferns (Pteridospermales) *and cordaites* (Cordaitales).

Lycophyte Trees

For most of the "Age of Coal" in the late Carboniferous period (Pennsylvanian), lycophyte trees dominated the coal-forming swamps. Most of these plants grew to heights of 10 to 35 meters and were sparsely branched (a). *After the plant attained most of its total height, the trunk branched dichotomously. Successive branching produced progressively smaller branches until, finally, the tissues of the branch tips lost their ability to grow further. The branches bore long microphylls. The tree lycophytes were largely supported by a massive periderm surrounding a relatively small amount of xylem.*

Like Selaginella *and* Isoetes, *lycophyte trees were heterosporous, and their sporophylls were aggregated into cones. Some of these trees produced structures analogous to seeds.*

As the swamplands began to dry up and the climate in Euramerica began to change toward the end of the Carboniferous period, the lycophyte trees vanished almost overnight, geologically speaking. The only remaining living relative of the group is the genus Isoetes. *Herbaceous lycophytes essentially similar to* Lycopodium *and* Selaginella *existed in the Carboniferous period, and representatives of some of them have survived to the present; there are 10 to 15 living genera.*

Calamites

Calamites, or giant horsetails, were plants of treelike proportions, reaching heights of 18 meters or more (see Figure 20–1). Like the plant body of Equisetum, *that of the calamites consisted of a branched aerial portion and an underground rhizome system. In addition, the leaves and branches were whorled at the nodes. Even the stems were remarkably similar to those of* Equisetum, *except for the presence of secondary xylem in the calamites, which accounted for much of the great diameter of the stems (trunks up to one-third of a meter in diameter). The similarities between calamites and living* Equisetum *are so strong that they are now regarded as belonging to the same order.*

The fertile appendages, or sporangio-

(a)

(a) Scientists believe that once a lycophyte tree was stabilized by its shallow, forking, rootlike axes, it grew rapidly skyward. These underground stigmarian axes produced spirally arranged rootlets, seen here as slender projections emerging from the forest floor. From left to right, a young leafy form, a pole-like juvenile, and a giant adult of 35 meters. *(b)* One of the most interesting gymnosperm groups is the seed ferns, a large, artificial group of primitive seed-bearing plants that appeared in the Late Devonian and flourished for about 125 million years. Fossils of these bizarre plants are common in rocks of Carboniferous age and have been well known to paleobotanists for a century or more. Their vegetative parts are so fernlike that for many years

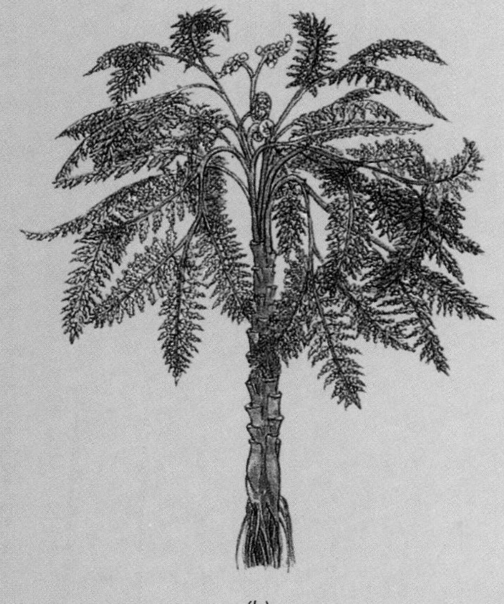

(b)

(c)

they were grouped with the ferns. This drawing is a reconstruction of the Carboniferous seed fern Medullosa noei. *The plant was about 5 meters tall.* **(c)** *Tip of young branch of the primitive conifer-like* Cordaites, *with long, straplike leaves.*

phores, of the calamites were aggregated into cones. Although most were homosporous, a few giant horsetails were heterosporous. Like most of the lycophyte trees, the giant horsetails declined in importance toward the end of the Paleozoic but persisted in much reduced form through the Mesozoic and Tertiary. Today they are the horsetails and are represented by only one genus, Equisetum.

Ferns

Many of the ferns represented in the fossil record are recognizable as members of today's primitive fern families. The "Age of Ferns" in the late Carboniferous period was dominated by tree ferns such as Psaronius, *one of the* Marattiales—*a eusporangiate group. Up to 8 meters tall,* Psaronius *had a stele that expanded toward the apex; the stele was covered below with adventitious roots, which played the key role in supporting the plant. The stem of* Psaronius *ended in an aggregate of large, pinnately compound fronds (see Figure 20–1).*

Seed Plants

The two remaining plant groups that dominated the tropical lowlands of Euramerica were the seed ferns and the cordaites. The fossil plants that are usually grouped as seed ferns are probably of diverse evolutionary relationships. The remains of seed ferns are common fossils in rocks of Carboniferous age (b). *Their large, pinnately compound fronds were so fernlike that these plants were long regarded as ferns. Then in 1905, F. W. Oliver and D. H. Scott demonstrated that they bore seeds and so were gymnosperms. Many species were small, shrubby, or scrambling plants.*

Other probable seed ferns were tall, woody trees. The fronds of seed ferns were borne at the top of their stem, or trunk, with microsporangia and seeds borne on them. The seed ferns survived into the Mesozoic era.

The cordaites were widely distributed during the Carboniferous period both in swamps and in drier environments. Although some members of the order were shrubs, many were tall (15 to 30 meters), highly branched trees that perhaps formed extensive forests. Their long (up to 1 meter), straplike leaves were spirally arranged at the tips of the youngest branches (c). *The center of the stem was occupied by a large pith, and a vascular cambium gave rise to a complete cylinder of secondary xylem. The root system, located at the base of the plant, also contained secondary xylem. The plants bore pollen-bearing cones and seed-bearing, conelike structures on separate branches. The cordaites persisted into the Permian period (286 to 248 million years ago), the drier and cooler period that followed the Carboniferous, but were apparently extinct by the beginning of the Mesozoic.*

In Conclusion

The dominant tropical coal-swamp plants of the Carboniferous period in Euramerica—the lycophyte trees—became extinct during the Late Paleozoic, a time of increasing tropical drought. Only the herbaceous relatives of the tree lycophytes and horsetails of the Carboniferous period continued to flourish and exist today, as do several groups of ferns that appeared in the Carboniferous period. Both the seed ferns and the cordaites eventually disappeared. Only one group of Carboniferous gymnosperms, the conifers (not a dominant group at the time), survived and went on to produce new types during the Permian period. The living conifers are discussed in detail in Chapter 20.

19–34
"Fiddleheads" of the ostrich fern (Matteuccia struthiopteris). *Ostrich fern fiddleheads are gathered commercially in upper New England and New Brunswick; they are marketed fresh, canned, and frozen. These fiddleheads, which taste somewhat like crisp asparagus, should be picked when they are less than 15 centimeters long. The fiddleheads of many ferns are considered toxic. The ostrich fern is the most widely planted fern around house foundations in the eastern United States and Canada.*

(a)

(b)

(c)

(d)

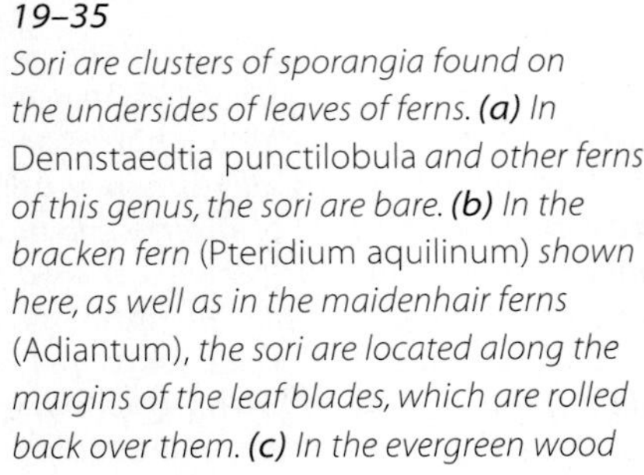

19–35
Sori are clusters of sporangia found on the undersides of leaves of ferns. ***(a)*** *In* Dennstaedtia punctilobula *and other ferns of this genus, the sori are bare.* ***(b)*** *In the bracken fern* (Pteridium aquilinum) *shown here, as well as in the maidenhair ferns* (Adiantum), *the sori are located along the margins of the leaf blades, which are rolled back over them.* ***(c)*** *In the evergreen wood fern* (Dryopteris marginalis), *the sori, which are also located near the margins of the leaf blades, are completely covered by kidney-shaped indusia.* ***(d)*** *In* Onoclea sensibilis, *the sori are enfolded by globular lobes of the pinna (leaflet) and therefore not visible. After overwintering, the lobes separate slightly, and the spores are released early in the spring, often over the snow.*

velopment is known as **circinate vernation.** Uncoiling of the fiddlehead results from more rapid growth on the inner than on the outer surface of the leaf early in development and is mediated by the hormone auxin (page 671), produced by the young pinnae on the inner side of the fiddlehead. This type of vernation protects the delicate embryonic leaf tip during development. Both fiddleheads and rhizomes are usually clothed with either hairs or scales, both of which are epidermal outgrowths; the characteristics of these structures are important in fern classification.

The sporangia of *Filicales,* all of which are homosporous, occur on the margins or lower surfaces of the leaves, on specially modified leaves, or on separate stalks. The sporangia commonly occur in clusters called **sori** (singular: sorus) (Figures 19–35 and 19–36), which may appear as yellow, orange, brownish, or blackish lines, dots, or broad patches on the lower surface of a frond. In many genera, the young sori are covered by specialized outgrowths of the leaf, the **indusia** (singular: indusium), which may shrivel when the sporangia are ripe and ready to shed their spores. The shape of the sorus, its position, and the presence or absence of an indusium are important characteristics in the taxonomy of the *Filicales*.

The spores of ferns in the *Filicales* give rise to free-living, bisexual gametophytes, which are often found in moist places, such as the sides of pots in greenhouses. The gametophyte typically develops rapidly into a flat, heart-shaped, usually membranous structure, the **prothallus,** that has numerous rhizoids on its central lower surface. Both antheridia and archegonia develop on the

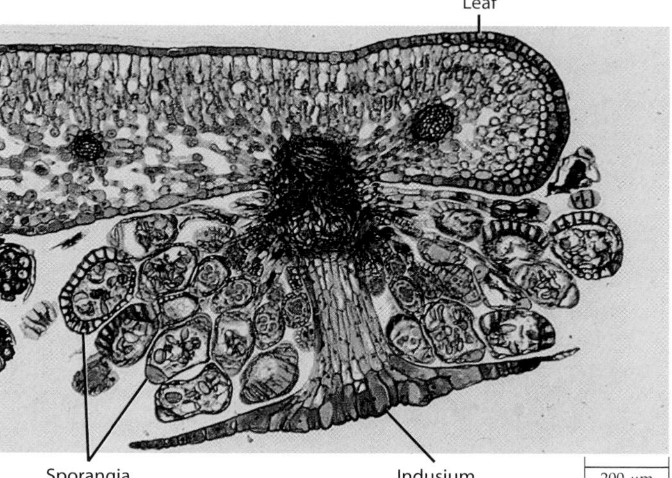

19–36
Cyrtomium falcatum, a homosporous fern. Transverse section of a leaf, showing a sorus on the lower surface. The sporangia are in different stages of development and are protected by an umbrellalike indusium.

ventral surface of the prothallus. The antheridia occur more typically among the rhizoids, while the archegonia are usually formed near the notch, an indentation at the anterior end of the gametophyte. The order of appearance of these gametangia is controlled genetically and can be mediated by special chemicals produced by the gametophytes. The timing of the appearance of the gametangia can influence whether the breeding system is primarily inbreeding or outcrossing. Water is required for the multiflagellated sperm to swim to the eggs.

Early in its development, the embryo, or young sporophyte, receives nutrients from the gametophyte through a foot. Development is rapid, and the sporophyte soon becomes an independent plant, at which time the gametophyte disintegrates.

The life cycle of one of the *Filicales* is shown in Figure 19–38 (pages 460 and 461).

Typically, the sporophyte is the perennial stage in ferns, and the small, thalloid gametophyte is short-lived. Remarkably, the strap-shaped or filamentous gametophytes of some species of ferns, including three genera with six tropical species found in the southern Appalachians, persist indefinitely without ever producing sporophytes. In addition, these species have never yet been induced to produce sporophytes in the laboratory (Figure 19–37). They reproduce by vegetative out-

(a)

(b)

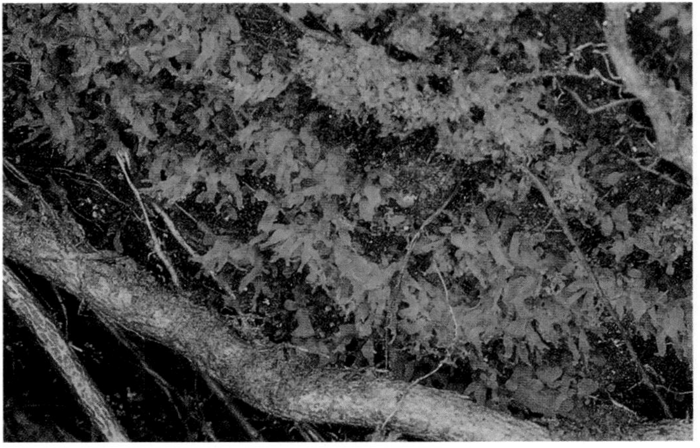

(c)

19–37
In some ferns from widely scattered parts of the world, the gametophytes reproduce asexually and persist; sporophytes are not formed, either in the field or in the laboratory. These photographs show two of the three fern genera known to exhibit this habit in the eastern United States. ***(a)*** *Typical habitat of persistent gametophytes of* Vittaria *and* Trichomanes*, Ash Cave, Hocking County, Ohio.* ***(b)*** Trichomanes *gametophytes, Lancaster County, Pennsylvania.* ***(c)*** Vittaria *gametophytes, Franklin County, Alabama.*

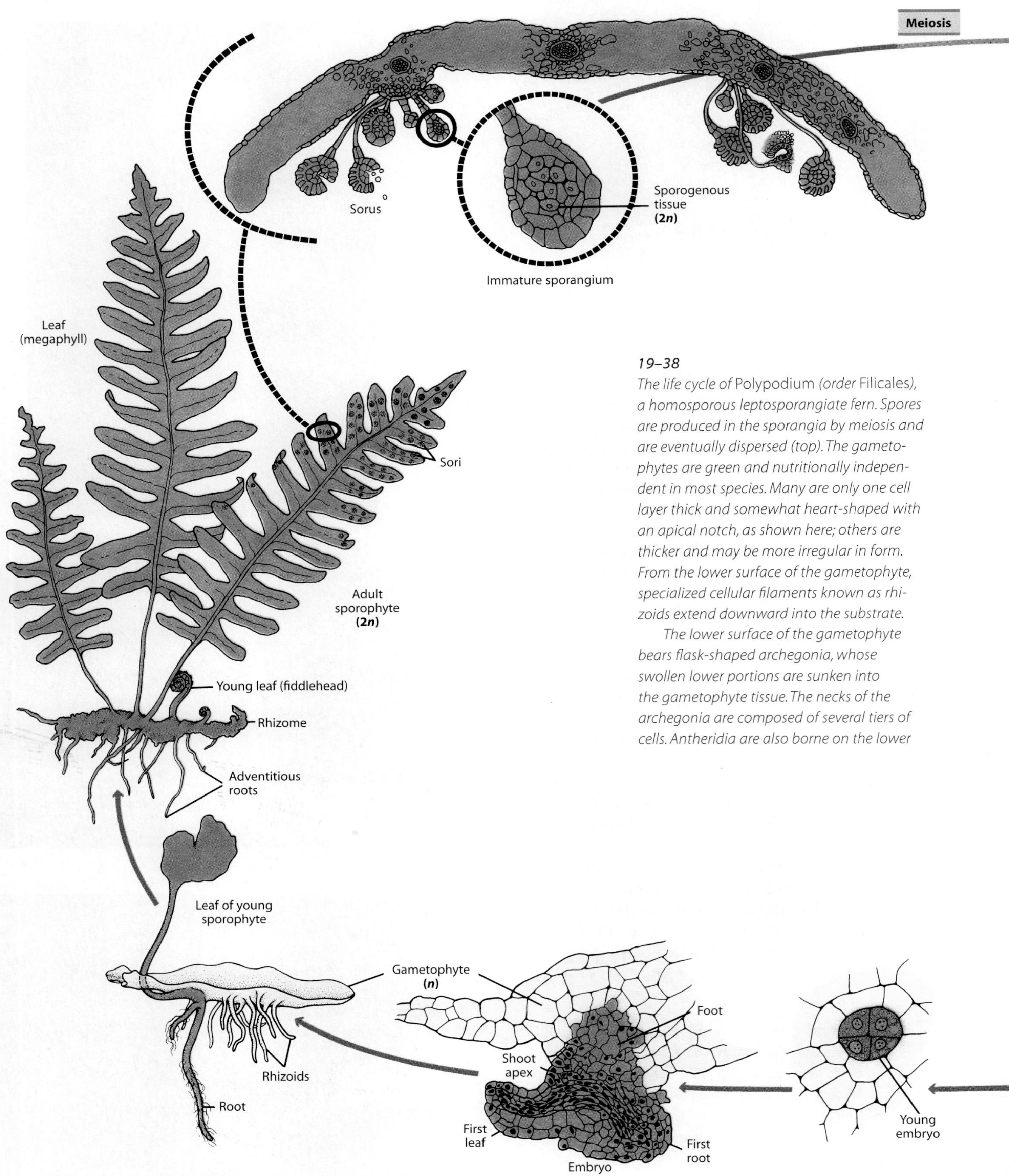

19–38

The life cycle of Polypodium *(order* Filicales*), a homosporous leptosporangiate fern. Spores are produced in the sporangia by meiosis and are eventually dispersed (top). The gametophytes are green and nutritionally independent in most species. Many are only one cell layer thick and somewhat heart-shaped with an apical notch, as shown here; others are thicker and may be more irregular in form. From the lower surface of the gametophyte, specialized cellular filaments known as rhizoids extend downward into the substrate.*

The lower surface of the gametophyte bears flask-shaped archegonia, whose swollen lower portions are sunken into the gametophyte tissue. The necks of the archegonia are composed of several tiers of cells. Antheridia are also borne on the lower

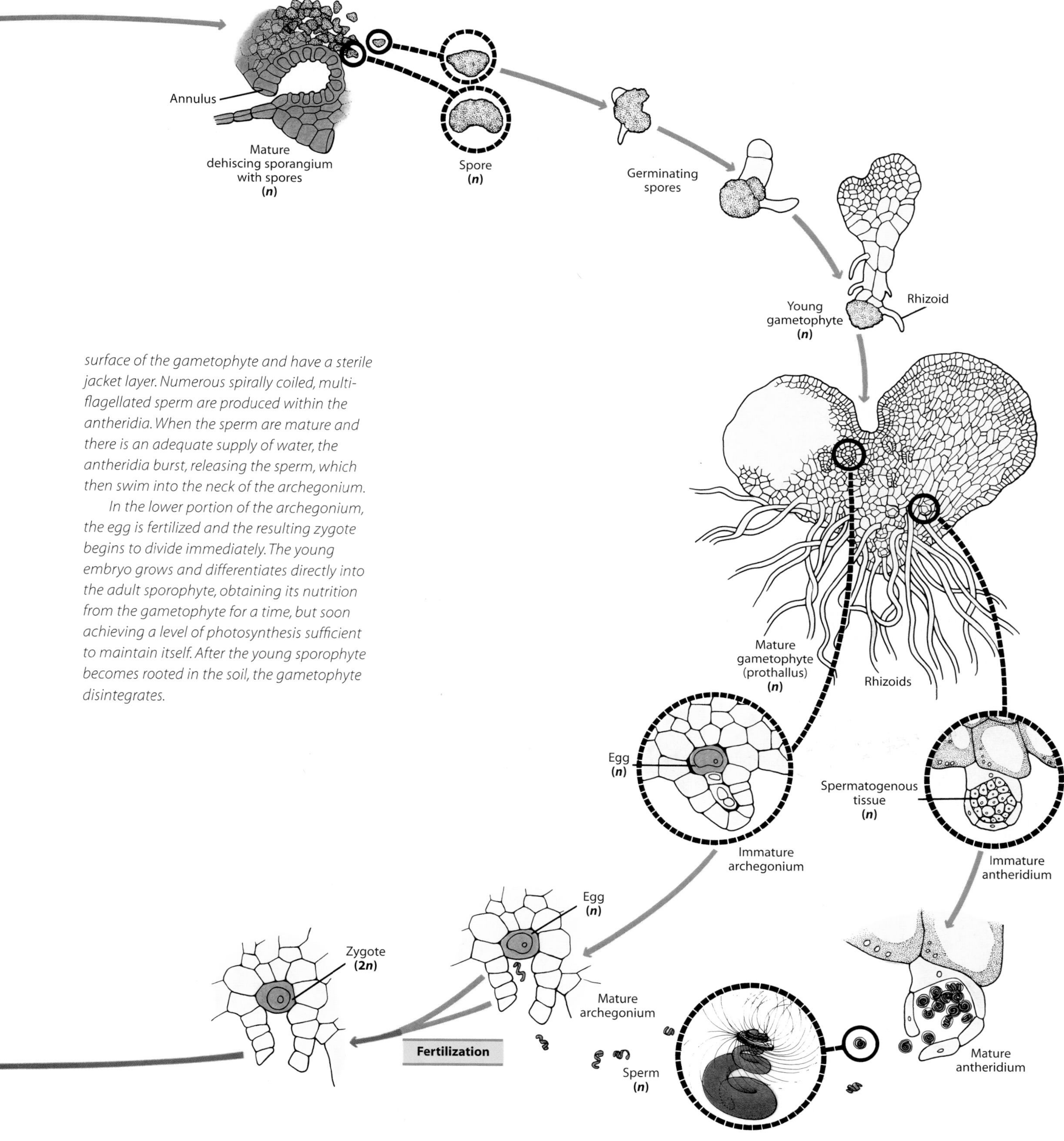

surface of the gametophyte and have a sterile jacket layer. Numerous spirally coiled, multi-flagellated sperm are produced within the antheridia. When the sperm are mature and there is an adequate supply of water, the antheridia burst, releasing the sperm, which then swim into the neck of the archegonium.

In the lower portion of the archegonium, the egg is fertilized and the resulting zygote begins to divide immediately. The young embryo grows and differentiates directly into the adult sporophyte, obtaining its nutrition from the gametophyte for a time, but soon achieving a level of photosynthesis sufficient to maintain itself. After the young sporophyte becomes rooted in the soil, the gametophyte disintegrates.

growths called gemmae that fall off and are blown away to found new colonies. These ferns appear to be distinct from the other, sporophyte-producing species of their respective genera, as judged by differences in their enzymes, and probably should be treated as distinct species. Such situations are common in mosses and are being discovered in ferns much more widely than was previously expected. Populations of perennial, free-living gametophytes of *Trichomanes speciosum,* recently discovered in the Elbsandsteingebirge (a mountain range shared by Germany and the Czech Republic), are estimated to be over 1000 years old. The possibility exists that they are relics of former populations that included both sporophytes and gametophytes. The extinction of the sporophytes possibly occurred as a result of climatic changes during the glacial intervals of the last 2 million years.

The Water Ferns—Orders *Marsileales* and *Salviniales*—Are Heterosporous Leptosporangiate Ferns

The water ferns constitute two orders, *Marsileales* and *Salviniales.* Although they are structurally very different from each other, recent evidence from molecular analyses indicates that the two orders were derived from a common terrestrial ancestor. All water ferns are heterosporous, and they are the only living heterosporous ferns. There are five genera of water ferns. The slender rhizomes of the three genera of *Marsileales,* including *Marsilea* (which has about 50 species), grow in mud, on damp soil, or often with the leaves floating on the surface of water (Figure 19–39a). The leaves of *Marsilea* resemble those of a four-leaf clover. Drought-resistant, bean-shaped reproductive structures called **sporocarps,** which may remain viable even after 100 years of dry storage, germinate when placed in water to produce chains of sori, each bearing series of megasporangia and microsporangia (Figure 19–39b). The extremely specialized gametophytes and the heterospory of *Marsileales* are the primary reasons that we recognize these ferns as a distinct order.

The two genera of *Salviniales, Azolla* (see Figure 30–11) and *Salvinia* (Figure 19–39c), are small plants that float on the surface of water. Both genera produce their sporangia in sporocarps that are quite different in structure from those of *Marsileales.* In *Azolla,* the tiny, crowded, bilobed leaves are borne on slender stems. A pouch that forms on the upper, photosynthetic lobe of each leaf is inhabited by colonies of the cyanobacterium *Anabaena azollae.* The lower, smaller lobe of each leaf is often nearly colorless. Because of the nitrogen-fixing abilities of the *Anabaena, Azolla* is important in maintaining the fertility of rice paddies and of certain natural

(a)

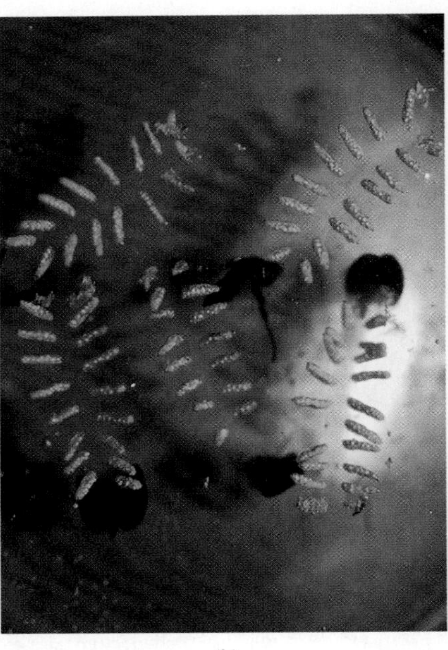

(b)

(c)

19–39
Water ferns. The two very distinct orders of water ferns are the only living heterosporous ferns. ***(a)*** Marsilea polycarpa, *with its leaves floating on the surface of the water, photographed in Venezuela.* ***(b)*** Marsilea, *showing the germination of a sporocarp, with chains of sori. Each sorus contains a series of megasporangia and microsporangia.* ***(c)*** Salvinia, *with two floating leaves and one feathery dissected, submerged leaf at each node. These two genera are representatives of the orders* Marsileales *and* Salviniales, *respectively.*

ecosystems. The undivided leaves of *Salvinia,* which are up to 2 centimeters long, are borne in whorls of three on the floating rhizome. One of the three leaves hangs down below the surface of the water and is highly dissected, resembling a mass of whitish roots. These "roots," however, bear sporangia, which reveals that they are actually leaves. The two upper leaves, which float on the water, are covered by hairs that protect their surface from getting wet, and the leaves float back to the surface if they are temporarily submerged.

Summary

Vascular plants are characterized by the possession of the vascular tissues xylem and phloem, and they exhibit an alternation of heteromorphic generations in which the sporophyte is large and complex and is the nutritionally independent phase.

The Primary Vascular Tissues Are Arranged in Steles of Three Basic Types

The plant bodies of many vascular plants consist entirely of primary tissues. Today, secondary growth is confined largely to the seed plants, although it occurred in several unrelated fossil groups of seedless vascular plants. The primary vascular tissues and associated ground tissues exhibit three basic arrangements: (1) the protostele, which consists of a solid core of vascular tissue; (2) the siphonostele, which contains a pith surrounded by vascular tissue; and (3) the eustele, which consists of a system of strands surrounding a pith, with the strands separated from one another by ground tissue.

Roots and Leaves Evolved in Different Ways

Roots evolved from the underground portions of the ancient plant body. Leaves originated in more than one way. Microphylls, single-veined leaves whose leaf traces

Summary TABLE A Comparison of Some of the Main Features of the Seedless Vascular Plants

PHYLUM	DICHOTOMOUSLY BRANCHED?	DIFFERENTIATED INTO ROOTS, STEMS, AND LEAVES?	HOMOSPOROUS OR HETEROSPOROUS	TYPE OF LEAVES	TYPE OF STELE	SPORANGIA	MISCELLANEOUS CHARACTERISTICS
Rhyniophyta (rhyniophytes)	Often	Stem only	Homosporous	None	Protostele	Terminal	Exclusively fossils; likely ancestors of trimerophytes
Zosterophyllophyta (zosterophyllophytes)	Often	Stem only	Many homosporous; some heterosporous	None	Protostele	Lateral	Exclusively fossils; closely related to lycophytes
Lycophyta (lycophytes)	Some are more or less dichotomous	Yes	*Lycopodiaceae* homosporous; *Selaginellaceae* and *Isoetaceae* heterosporous	Microphyll	Most with protostele or modified protostele	On or in the axils of sporophylls	Members of the *Selaginellaceae* and *Isoetaceae* have ligules; many extinct representatives
Trimerophytophyta (trimerophytes)	Most are not	Stem only	Homosporous	None	Protostele	Terminal on ultimate dichotomies	Exclusively fossils; likely ancestors of ferns, progymnosperms, and perhaps horsetails
Psilotophyta (psilotophytes)	Yes	Stem only	Homosporous	None	Protostele	Lateral	Resemble rhyniophytes in aspects of structure; *Psilotum* and *Tmesipteris* only modern genera
Sphenophyta (sphenophytes)	No	Yes	Homosporous; some fossils heterosporous	Microphyll through reduction	Eustele-like siphonostele	On sporangiophores in strobili	Represented today by single genus, *Equisetum,* the horsetails
Pterophyta (ferns)	No	Yes	All homosporous except for *Marsileales* and *Salviniales,* which are heterosporous	Megaphyll	Protostele in some; siphonostele or more complex types in others	On sporophylls; some clustered in sori	*Ophioglossales* and *Marattiales* eusporangiate; *Filicales, Marsileales,* and *Salviniales* leptosporangiate

are not associated with leaf gaps, evolved either as superficial lateral outgrowths of the stem or from sterile sporangia. They are associated with protosteles and are characteristic of the lycophytes. Megaphylls, leaves with complex venation, evolved from branch systems. They are associated with siphonosteles and eusteles. In the siphonosteles of ferns, the leaf traces are associated with leaf gaps.

Vascular Plants May Be Either Homosporous or Heterosporous

Homosporous vascular plants produce only one type of spore, which has the potential to give rise to a bisexual gametophyte. Heterosporous plants produce microspores and megaspores, which germinate and give rise to male gametophytes and female gametophytes, respectively. The gametophytes of heterosporous plants are much reduced in size, compared with those of homosporous plants. In the history of the vascular plants, heterospory has evolved several times. There has been a long, continuous evolutionary trend toward a reduction in the size and complexity of the gametophyte, which culminated in the angiosperms. Seedless vascular plants have archegonia and antheridia. All but a few gymnosperms have archegonia, but both archegonia and antheridia have been lost in all angiosperms.

Seedless Vascular Plants Exhibit an Alternation of Heteromorphic Generations

The life cycles of the seedless vascular plants all represent modifications of an essentially similar alternation of heteromorphic generations in which the sporophyte is dominant and free-living. The gametophytes of the homosporous species are independent of the sporophyte for their nutrition. Although potentially bisexual, producing both antheridia and archegonia, these gametophytes are functionally unisexual. The gametophytes of heterosporous species are unisexual, much reduced in size, and, except for the few genera of heterosporous ferns, dependent on stored food derived from the sporophyte for their nutrition. All of the seedless vascular plants have motile sperm, and the presence of water is necessary for the sperm to swim to the eggs.

The Oldest Fossils of Vascular Plants Belong to the Phylum *Rhyniophyta*

Vascular plants go back at least 430 million years; the earliest ones about which we have many structural details belong to the phylum *Rhyniophyta,* the oldest fossils of which are from the mid-Silurian period, about 425 million years ago. Some fossils, once considered rhyniophytes, have conducting cells similar to bryophyte hydroids rather than to tracheids. These plants, which had branched axes and multiple sporangia, may represent an intermediate stage in the evolution of vascular plants. They are called protracheophytes. The plant bodies of the rhyniophytes and other contemporary plants were simple, dichotomously branching axes lacking roots and leaves. With evolutionary specialization, morphological and physiological differences arose between various parts of the plant body, bringing about the differentiation of root, stem, and leaf.

The Living Seedless Vascular Plants Are Classified in Four Phyla

The phyla of seedless vascular plants are the *Lycophyta* (including *Lycopodium, Selaginella,* and *Isoetes*), the *Psilotophyta (Psilotum* and *Tmesipteris),* the *Sphenophyta (Equisetum),* and the *Pterophyta* (ferns). Most are homosporous, but heterospory is exhibited by *Selaginella, Isoetes,* and the water ferns *(Salviniales* and *Marsileales).*

The *Lycophyta* are believed to have evolved from the *Zosterophyllophyta,* a phylum of entirely extinct vascular plants. The *Trimerophytophyta,* another phylum of entirely extinct vascular plants, apparently represent the ancestral stock of the ferns, the progymnosperms, and perhaps the sphenophytes as well.

Psilotophytes differ from other living vascular plants in their lack of leaves (with the possible exception of *Tmesipteris*) and roots. Lycophytes are characterized by microphylls. The members of the other phyla have megaphylls.

Two orders of ferns *(Ophioglossales* and *Marattiales)* possess eusporangia, like those of other seedless vascular plants. In eusporangia, the walls are several cell layers thick, and a number of cells participate in the initial stages of sporangial development. Other ferns--*Filicales* and the two orders of water ferns—form leptosporangia, specialized structures in which the wall consists of a single layer of cells that develop from a single initial cell.

Two of the four phyla of seedless vascular plants that contain living representatives, the lycophytes and sphenophytes, extend back to the Devonian period. Among the seedless vascular plants, only the ferns, which first appear in the fossil record in the Carboniferous period, are represented by a large number of living species, about 11,000.

Five groups of vascular plants dominated the swamplands of the Carboniferous period ("Age of Coal"), and three of them were seedless vascular plants—lycophytes, sphenophytes, and ferns. The other two were gymnosperms—the seed ferns and the cordaites.

Selected Key Terms

annulus p. 454
eusporangium p. 453
eustele p. 429
frond p. 455
heterosporous p. 431
homosporous p. 430
indusium p. 458
leaf gap p. 429
leaf trace gap p. 429
leptosporangium p. 454
megaphyll p. 429
megasporangium p. 431
megaspores p. 431
microphyll p. 429
microsporangium p. 431
microspores p. 431
phloem p. 426
prothallus p. 458
protostele p. 429
protracheophyte p. 434
seeds p. 425
sieve elements p. 427
siphonostele p. 429
sorus p. 458
sporangiophore p. 448
sporocarp p. 462
sporophyll p. 438
tapetum p. 454
tracheary elements p. 427
tracheids p. 428
vessel elements p. 428
xylem p. 426

Questions

1. What basic structural features do the *Rhyniophyta, Zosterophyllophyta,* and *Trimerophytophyta* have in common?
2. The vessel elements and heterospory present in several unrelated groups of vascular plants represent excellent examples of convergent evolution. Explain.
3. With the use of simple, labeled diagrams, describe the structure of the three basic types of steles.
4. Compare the life cycle of a moss with that of a homosporous leptosporangiate fern.
5. What is coal? How was it formed? What plants were involved in its formation?
6. The bryophytes often are referred to as the "amphibians of the plant kingdom," but that characterization might also be applied to the seedless vascular plants. Can you explain why?
7. What is the probable evolutionary origin of the *Lycophyta?* What evidence is there for this view?

20–1

Tropical swamps of the Late Carboniferous were dominated by several genera of giant lycophyte trees that, when mature, formed a forest canopy of airy, diffusely branched crowns, seen here in the background. These trees had massive trunks (left foreground and elsewhere) that were stabilized in the swampy mire by long stigmarian axes, from which numerous rootlets, possibly photosynthetic, extended. Note also some low-growing lycophytes with cones, in the center foreground. Flooded and boggy areas of the forest floor favored other lycophyte types, such as branchless Chaloneria *(center and right foreground), and horsetail types, such as the shrub* Sphenophyllum *(right bottom corner) and the Christmas-tree-shaped* Diplocalamites *(far right midground).*

Less wet, or slightly elevated ground, as seen here at the left, fostered mixed vegetation, including early conifers, ground-cover ferns, tall tree ferns, and seed ferns. Among the seed plants were, from front to back, Cordaixylon, *a shrubby conifer relative with strap-shaped leaves;* Callistophyton, *a scrambling seed fern growing at the base of the largest lycophyte tree; and* Psaronius, *an early tree fern (left background). Elsewhere, robust plants like the upright umbrella-shaped seed fern* Medullosa *are seen occupying sunnier, more disturbed sites opened up by previous channel flooding (right midground and background).*

Chapter 20
Gymnosperms

OVERVIEW

In this chapter we will examine the gymnosperms, a series of evolutionary lines of seed-bearing plants that include such common trees as pine, spruce, fir, hemlock, and cedar. Conifers, such as these trees, are a major source of lumber and paper pulp. They also include the tallest and oldest of trees.

We begin our journey through the gymnosperms with a discussion of the evolutionary origin of the ovule, the structure that develops into a seed. A marvelous innovation, the seed—consisting of a seed coat, an embryo, and stored food—replaced the spore as the unit of dispersal, and provided seed plants with an enormous advantage over the seedless vascular plants. As the name implies (the Greek word gymnos *means "naked," and* sperma *means "seed"), the seeds of gymnosperms lack the protection of an enveloping structure such as the fruit wall that encloses the seeds of angiosperms, or flowering plants.*

Another major advantage that both gymnosperms and angiosperms have over all other plants is their independence from water as a medium of transport of the sperm to the egg. In gymnosperms, the partly developed male gametophyte—the pollen grain—is transferred bodily to the vicinity of the female gametophyte within the ovule, after which it produces a pollen tube. Although not originally a sperm conveyor, the pollen tube eventually evolved into a conveyor of nonmotile sperm to the eggs of the female gametophyte within the ovule.

CHECKPOINTS

By the time you finish reading this chapter, you should be able to answer the following questions:

1. What is a seed, and why was the evolution of the seed such an important innovation for plants?
2. From which group of plants is it hypothesized that seed plants evolved? Why?
3. How do the mechanisms by which sperm reach the eggs differ in gymnosperms and seedless vascular plants?
4. What are the distinguishing features of the four phyla of living gymnosperms?
5. In what ways do gnetophytes resemble angiosperms?

One of the most dramatic innovations to arise during the evolution of the vascular plants was the seed. Seeds are one of the principal factors responsible for the dominance of seed plants in today's flora—a dominance that has become progressively greater over a period of several hundred million years. The reason is simple: the seed has great survival value. The protection that a seed affords the enclosed embryo, and the stored food that is available to that embryo at the critical stages of germination and establishment, give seed plants a great selective advantage over their free-sporing relatives and ancestors, that is, over plants that shed their spores.

Evolution of the Seed

All seed plants are heterosporous, producing megaspores and microspores that give rise respectively to megagametophytes and microgametophytes—but that trait is not unique to seed plants. As we have discussed in Chapter 19, some seedless vascular plants are also heterosporous. The production of seeds is, however, a particularly extreme form of heterospory that has been modified to form an ovule, the structure that develops into the seed. Indeed, a **seed** is simply a mature ovule containing an embryo. The immature ovule consists of a megasporangium surrounded by one or two additional layers of tissue, the **integuments** (Figure 20–2).

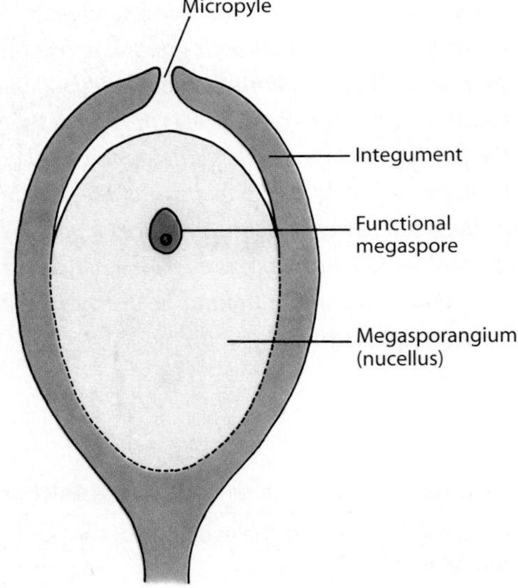

20–2
Longitudinal section of an ovule, which consists of a megasporangium (nucellus) enveloped by an integument with an opening, the micropyle, at its apical end. A single functional megaspore is retained within the megasporangium (not shed) and will give rise to a megagametophyte, which is retained within the megasporangium. Following fertilization, the ovule matures into a seed, which becomes the unit of dispersal.

Several events led to the evolution of an ovule, including: (1) retention of the megaspores within the megasporangium, which is fleshy and called the **nucellus** in seed plants—in other words, the megasporangium no longer releases the spores; (2) reduction in the number of megaspore mother cells in each megasporangium to one; (3) survival of only one of the four megaspores produced by the spore mother cell, leaving a single functional megaspore in the megasporangium; (4) formation of a highly reduced megagametophyte inside the single functional megaspore—that is, formation of an endosporic (within the wall) megagametophyte that is no longer free living—which is retained within the megasporangium; (5) development of the embryo, or young sporophyte, within the megagametophyte retained within the megasporangium; (6) formation of an integument that completely envelops the megasporangium except for an opening at its apex called the **micropyle;** and (7) modification of the apex of the megasporangium to receive microspores or pollen grains. Related to these events is a basic shift in the unit of dispersal from the megaspore to the seed, the integumented megasporangium containing the mature embryo.

The Fossil Record Provides Clues to Ovule Evolution

The exact order in which these events occurred is unknown because of the incompleteness of the fossil record. They occurred fairly early in the history of vascular plants, however, because the oldest ovules or seeds are from the Late Devonian (about 365 million years ago). One of these early seed plants is *Elkinsia polymorpha* (Figure 20–3). The ovule of *Elkinsia* consisted of a nucellus and four or five integumentary lobes with little or no fusion between lobes, which led some paleobotanists to name these structures "preovules." The integumentary lobes curved inward at their tips, forming a ring around the apex of the nucellus. The apex of the nucellus was modified into a barrel-like structure for the reception of pollen. The ovules were surrounded by dichotomously branched, sterile structures called cupules.

A slightly younger Late Devonian seed plant is *Archaeosperma arnoldii* (Figure 20–4). Only the apical portion of the integument of *Archaeosperma* was divided into lobes, which formed a rudimentary micropyle. The integuments of ovules apparently evolved through the gradual fusion of integumentary lobes until the only opening left was the micropyle (Figure 20–5).

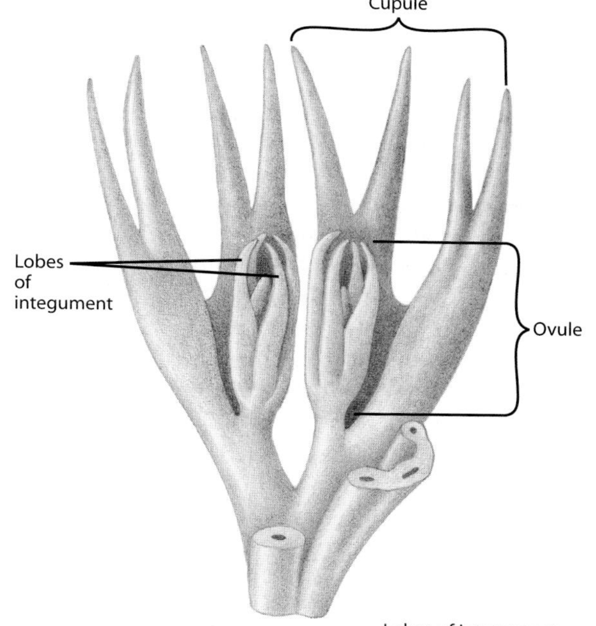

20–3

Reconstruction of a fertile branch of the Late Devonian plant Elkinsia polymorpha, *showing its ovules. Each ovule was overtopped by a dichotomously branched, sterile structure called a cupule. Note the more or less free lobes of the integument.*

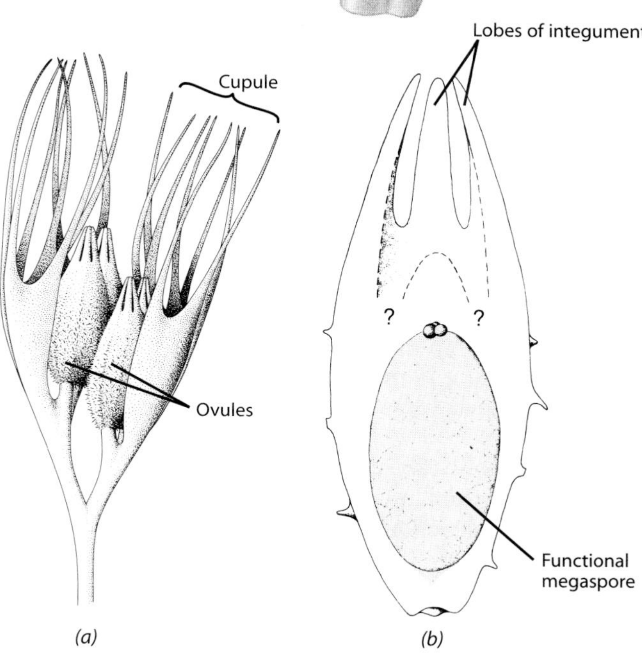

20–4

(a) *Reconstruction of a fertile branch of the Late Devonian plant* Archaeosperma arnoldii, *showing four ovules. The cupules, which partly enclose the ovules, are arranged in pairs, and each cupule contains two flask-shaped ovules about 4 millimeters long. The apex of each integument was lobed.* ***(b)*** *Diagram of the ovule, showing the position of a megaspore tetrad. The three aborted megaspores are found at the top of the functional megaspore. The lobes of the integument form a rudimentary micropyle. The question marks indicate the presumed position of the nucellus.* ***(c)*** *A megaspore. This fossil, from Pennsylvania, is the oldest known seedlike structure—about 360 million years in age.*

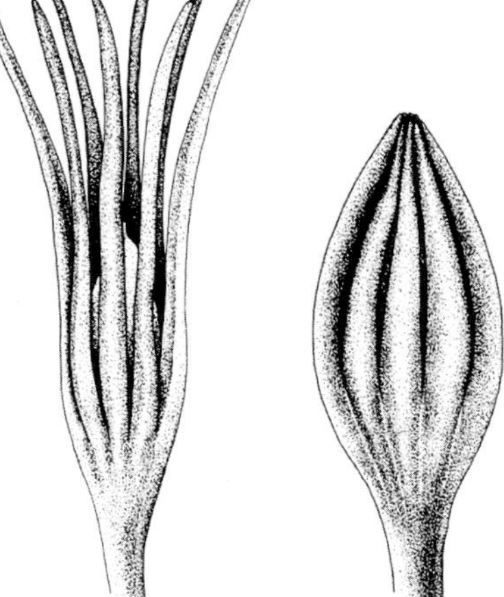

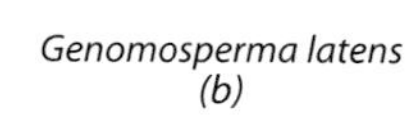

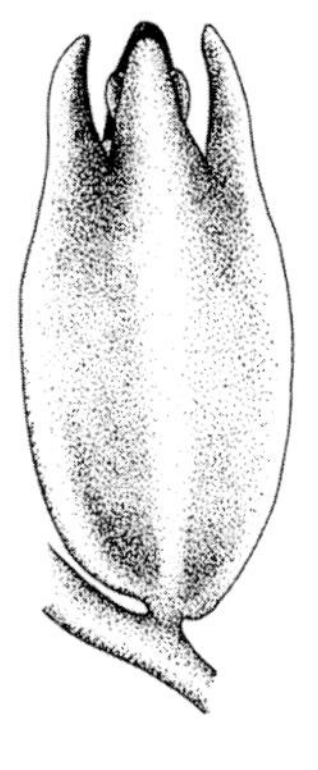

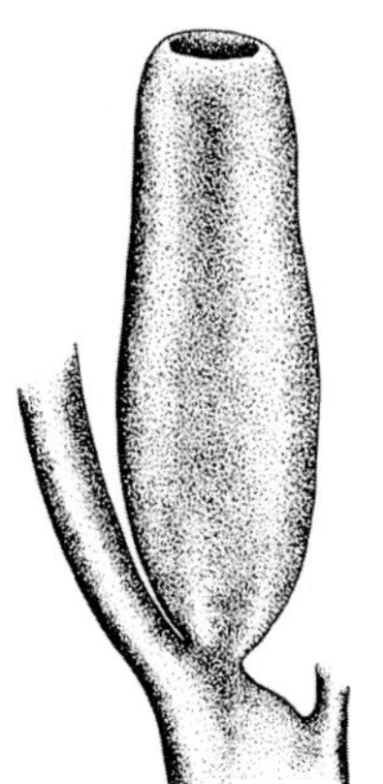

20–5

Seedlike structures in a number of Paleozoic plants, showing some of the potential stages in the evolution of the integument. ***(a)*** *In* Genomosperma kidstonii *(the Greek word* genomein *means "to become," and* sperma *means "seed"), eight fingerlike projections arise at the base of the megasporangium and are separated for their entire length.* ***(b)*** *In* Genomosperma latens *the integumentary lobes are fused from the base of the megasporangium for about a third of their length.* ***(c)*** *In* Eurystoma angulare *fusion is almost complete, while* ***(d)*** *in* Stamnostoma huttonense *it is complete, with only the micropyle remaining open at the top.*

A Seed Consists of an Embryo, Stored Food, and a Seed Coat

In modern seed plants the ovule consists of a nucellus enveloped by one or two integuments with a micropyle. When the ovules of most gymnosperms are ready for fertilization, the nucellus contains a megagametophyte composed of nutritive tissue and archegonia. After fertilization, the integuments develop into a **seed coat,** and a seed is formed. In most modern seed plants, an embryo develops within the seed before dispersal. In addition, all seeds contain stored food.

There Are Five Phyla of Seed Plants with Living Representatives

Seed plants arose starting in the Late Devonian period, at least 365 million years ago. During the next 50 million years, a wide array of seed-bearing plants evolved, many of which are grouped together as the so-called seed ferns, while others are recognized as cordaites and conifers (see "Coal Age Plants" on pages 456 and 457).

The seed plants, all of which typically possess megaphylls (generally large leaves with several to many veins, but modified to needles or scales in some groups), include five phyla with living representatives: the *Cycadophyta,* the *Ginkgophyta,* the *Coniferophyta,* the *Gnetophyta,* and the *Anthophyta*. The phylum *Anthophyta* comprises the angiosperms, or flowering plants, and the remaining four phyla are often grouped together as the gymnosperms.

The oldest fossil angiosperms are only about 135 million years old, from the early part of the Cretaceous period. Although the angiosperms—overwhelmingly the most successful vascular plants at the present time—actually may be somewhat older than our current understanding of the fossil record indicates, they are still relative newcomers in the broad picture of vascular plant evolution. Because they are such a large and important group, the angiosperms are considered in detail in Chapters 21 and 22.

The gymnosperms do not constitute an evolutionary line that is equivalent to the angiosperms. Instead, the gymnosperms represent a series of evolutionary lines of seed-bearing plants that lack the distinctive characteristics of the angiosperms. Although there are only about 720 species of living gymnosperms—compared with some 235,000 species of angiosperms—individual gymnosperm species are often dominant over wide areas.

Before beginning our discussion of seed plants, we shall briefly examine one more group of seedless vascular plants—the progymnosperms. They are discussed here, rather than in Chapter 19, because they are the likely progenitors of seed plants.

Progymnosperms

In the Late Paleozoic era, there existed a group of plants called the progymnosperms (phylum *Progymnospermophyta*), which had characteristics intermediate between those of the seedless vascular trimerophytes and those of the seed plants. Although the progymnosperms reproduced by means of freely dispersed spores, they produced secondary xylem (wood) remarkably similar to that of living conifers (Figure 20–6). The progymnosperms were unique among woody Devonian plants in that they also produced secondary phloem. Both the progymnosperms and Paleozoic ferns probably evolved from the more ancient trimerophytes (see Figure 19–10c), from which they differed primarily by having more elaborate and more highly differentiated branch systems and correspondingly more complex vascular systems.

In the progymnosperms, the most important evolutionary advance over both the trimerophytes and the ferns is the presence of a *bifacial vascular cambium*—that is, one that produces both secondary xylem and secondary phloem. Vascular cambia of this type are characteristic of seed plants and apparently evolved first in the progymnosperms.

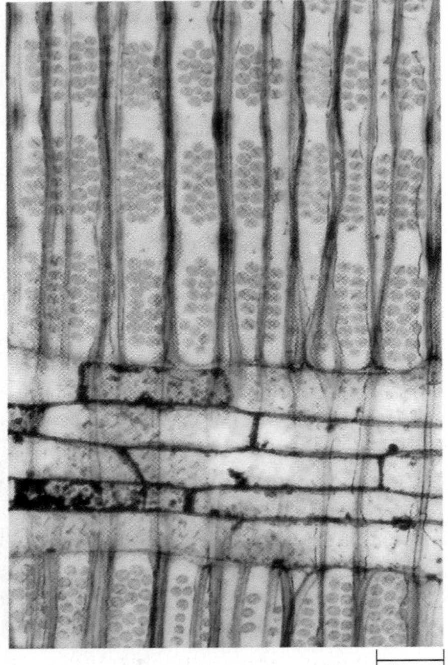

20–6
Radial view of the secondary xylem, or wood, of the progymnosperm Callixylon newberryi. *This fossil wood, with its regular series of pitted tracheids, is remarkably similar to the wood of certain conifers.*

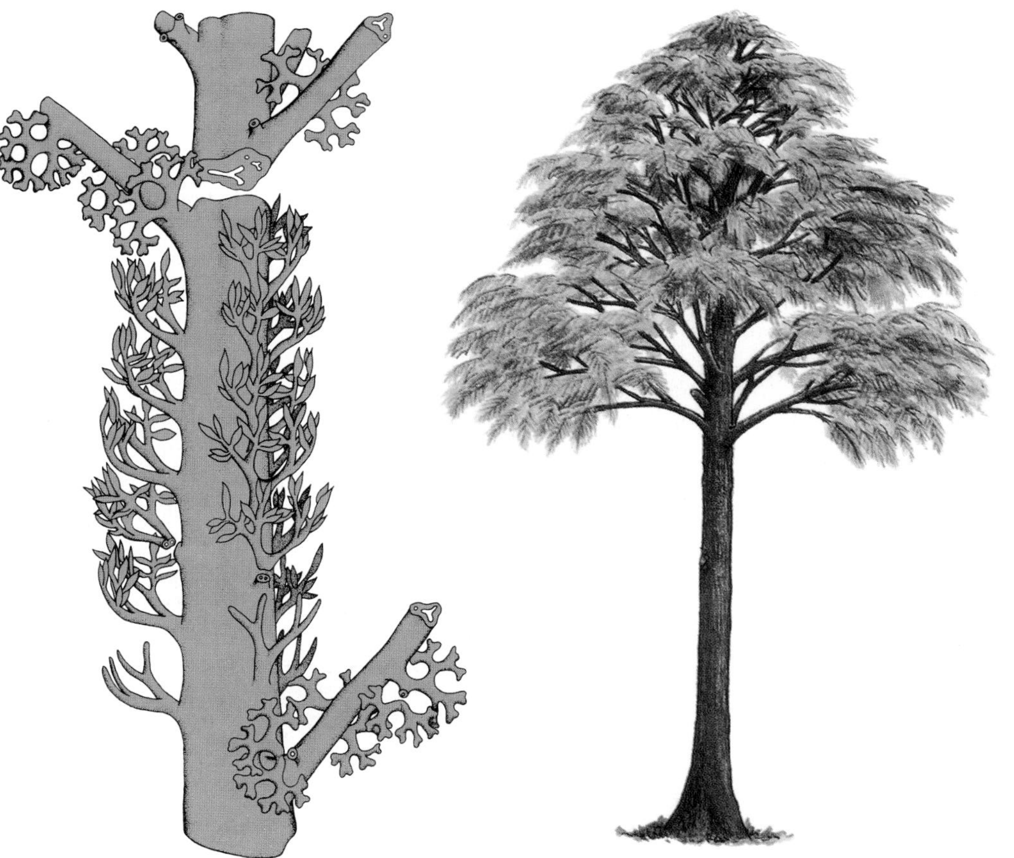

20–7
Reconstruction of a portion of the branch system of Triloboxylon ashlandicum, *an* Aneurophyton*-type progymnosperm. The main axis bears vegetative branches at the top and bottom and fertile organs with sporangia in between.*

20–8
Reconstruction of the progymnosperm Archaeopteris, *which is common in the fossil record of eastern North America. Specimens of* Archaeopteris *attained heights of 20 meters or more, and some of them seem to have formed forests.*

20–9
Reconstruction of a frondlike lateral branch system of the progymnosperm Archaeopteris macilenta. *Fertile leaves can be seen bearing maturing sporangia (seen here as brown) on centrally located primary branches.*

One kind of progymnosperm, the *Aneurophyton* type, which occurred in the Devonian period approximately 362 to 380 million years ago, was characterized by complex three-dimensional branching (Figure 20–7) and had a solid cylinder of vascular tissue, or protostele. In many respects these progymnosperms were similar to the more complex trimerophytes, but they also resemble some of the early seed ferns, which has led some paleobotanists to suggest that the branch systems of the *Aneurophyton*-type progymnosperms may have been the precursors of the fernlike leaves of early seed ferns.

A second major kind of progymnosperm, the *Archaeopteris* type, also appeared in the Devonian period, about 370 million years ago, and extended into the Mississippian period, about 340 million years ago (Figure 20–8). In this group, the lateral branch systems were flattened in one plane and bore laminar structures considered to be leaves (Figure 20–9). A eustele—that is, an arrangement of vascular tissues in discrete strands around a pith—apparently evolved in this group of progymnosperms and is a strong similarity linking this group with the living seed plants (page 429). The larger branches of *Archaeopteris*-type progymnosperms had a pith. Although most progymnosperms were homosporous, some species of *Archaeopteris* were heterosporous.

Fossil logs of *Archaeopteris*, called *Callixylon*, may be up to a meter or more in diameter and 10 meters long, indicating that at least some species of this group were large trees. They appear to have formed extensive forests in some regions. As the reconstruction in Figure 20–8 suggests, individuals of *Archaeopteris* may have resembled conifers in their branching patterns.

Evidence accumulated during the past several decades strongly indicates that the seed plants evolved from the progymnosperms following the appearance of the seed in what now seems to have been the common

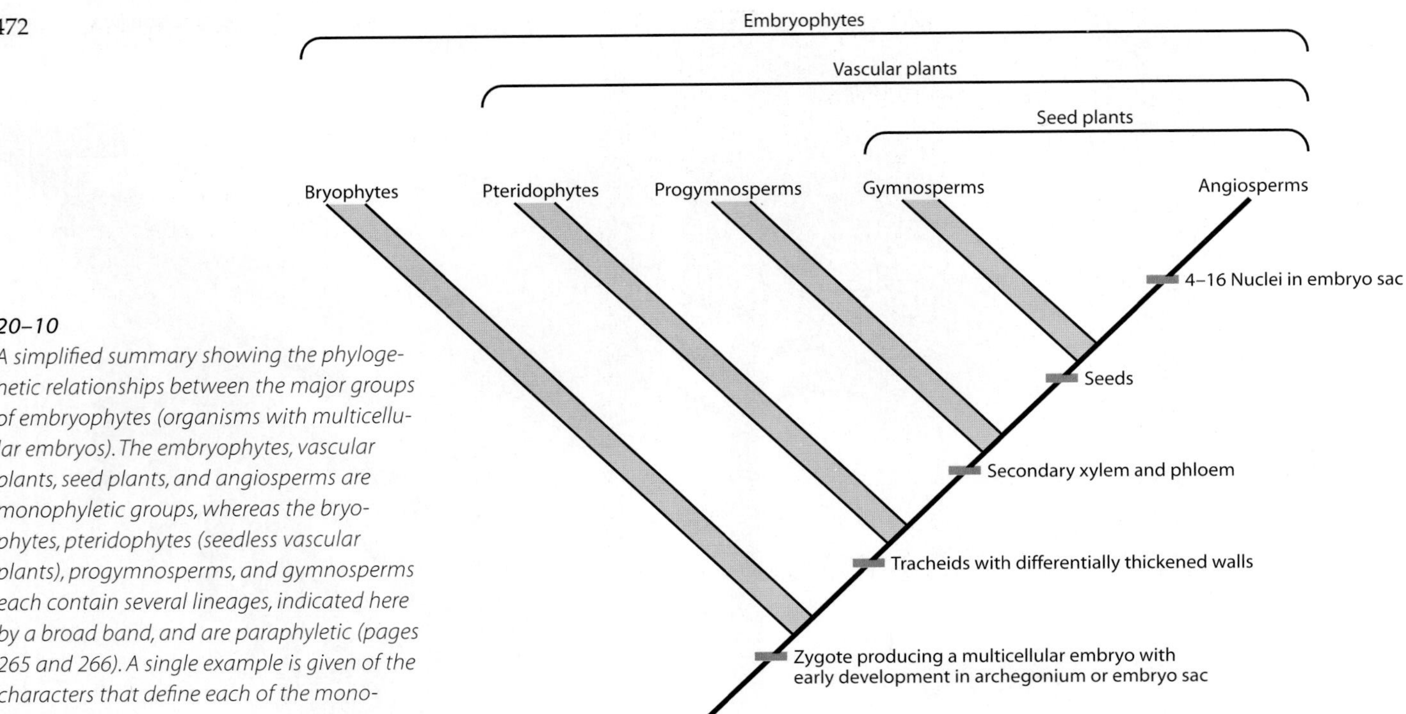

20–10
A simplified summary showing the phylogenetic relationships between the major groups of embryophytes (organisms with multicellular embryos). The embryophytes, vascular plants, seed plants, and angiosperms are monophyletic groups, whereas the bryophytes, pteridophytes (seedless vascular plants), progymnosperms, and gymnosperms each contain several lineages, indicated here by a broad band, and are paraphyletic (pages 265 and 266). A single example is given of the characters that define each of the monophyletic groups.

ancestor of all seed plants (Figure 20–10). However, many problems still remain to be solved in developing a more detailed understanding of the early evolution of seed plants.

Extinct Gymnosperms

Two groups of extinct gymnosperms—the seed ferns (phylum *Pteridospermophyta*) and the *Cordaitales,* primitive coniferlike plants—were discussed and illustrated in Chapter 19 (pages 456 and 457). The seed ferns, or pteridosperms, are a very diverse and highly unnatural group that range in age from the Devonian to the Jurassic. They range in form from Late Devonian slender, branched plants with ovules, such as *Elkinsia* and *Archaeosperma,* to Carboniferous plants, such as *Medullosa,* that had the appearance of tree ferns (see Figure 20–1). Several groups of extinct Mesozoic plants are also sometimes included in the seed ferns. Exactly how these different groups of seed ferns relate to the living gymnosperms remains uncertain.

Another group of extinct gymnosperms—the cycadeoids, or *Bennettitales*—consisted of plants with palmlike leaves, somewhat resembling the living cycads (Figure 20–11a; see page 486). The *Bennettitales* are an enigmatic group of Mesozoic gymnosperms that disappeared from the fossil record during the Cretaceous. Some paleobotanists believe that the *Bennettitales* may have been members of the same evolutionary line as angiosperms. Currently, it is still uncertain exactly where the *Bennettitales* fit in phylogenetically. The *Bennettitales* lived at the same time as the extinct cycads, and both groups produced very similar leaves through most of the Mesozoic. Reproductively, however, *Bennettitales* were distinct from the cycads in several respects, including the presence of flowerlike reproductive structures that were bisexual in some species (Figure 20–11b).

Living Gymnosperms

There are four phyla of gymnosperms with living representatives: *Cycadophyta* (cycads), *Ginkgophyta* (maidenhair tree, or ginkgo), *Coniferophyta* (conifers), and *Gnetophyta* (gnetophytes). The name *gymnosperm,* which literally means "naked seed," points to one of the principal characteristics of the plants belonging to these four phyla—namely, that their ovules and seeds are exposed on the surface of sporophylls and analogous structures.

With few exceptions, the female gametophyte of gymnosperms produces several archegonia. As a result, more than one egg may be fertilized, and several embryos may begin to develop within a single ovule—a phenomenon known as **polyembryony.** In most cases, however, only one embryo survives and therefore relatively few fully developed seeds contain more than one embryo.

In the seedless vascular plants, water is required for the motile, flagellated sperm to reach and fertilize the eggs. In the gymnosperms, however, water is not required as a medium of transport of the sperm to the eggs. Instead, the partly developed male gametophyte, the **pollen grain,** is transferred bodily (usually passively, by the wind) to the vicinity of a female gametophyte within an ovule. This process is called **pollination.** After pollination, the endosporic male gametophyte produces a tubular outgrowth, the **pollen tube.** The male gameto-

(a)

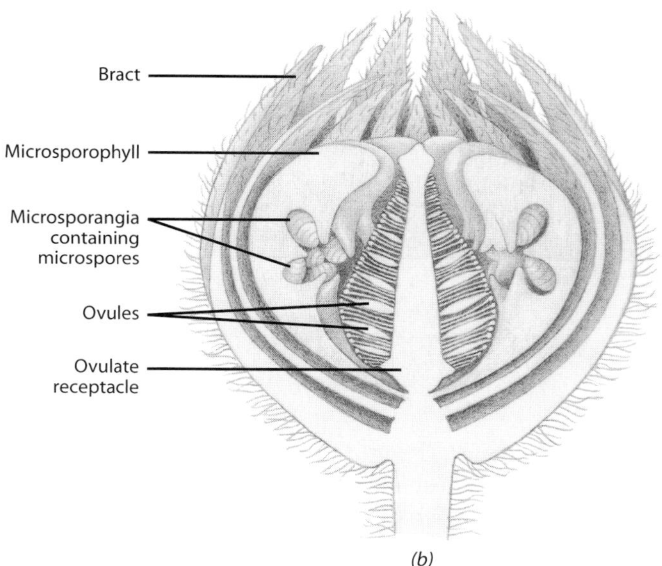

(b)

20–11
Bennettitales. ***(a)*** *Reconstruction of* Wielandiella, *an extinct gymnosperm from the Triassic.* Wielandiella *exhibits a forked branching pattern. A single strobilus, or cone, is borne at each fork.* ***(b)*** *A diagrammatic reconstruction of the bisporangiate, or bisexual, strobilus of* Williamsoniella coronata *from the Jurassic. The strobilus consists of a central ovulate receptacle surrounded by a whorl of microsporophylls bearing microsporangia containing microspores, which develop into male gametophytes (pollen grains). Hairy bracts enclose the reproductive parts.*

phytes of gymnosperms and other seed plants do not form antheridia.

In the conifers and gnetophytes, the sperm are nonmotile, and the pollen tubes convey them directly to the archegonia. In the cycads and *Ginkgo*, fertilization is transitional between the condition found in ferns and other seedless plants, in which free-swimming sperm occur, and the condition found in other seed plants, which have nonmotile sperm. The male gametophytes of cycads and *Ginkgo* produce a pollen tube, but this does not penetrate the archegonium (Figure 20–12). Instead, it is haustorial and may grow for several months in the tissue of the nucellus where it appears to play a role in absorbing nutrients. Eventually, the pollen grain bursts in the vicinity of the archegonium, releasing multiflagellated, swimming sperm cells (see Figure 20–39). The sperm then swim to an archegonium, and one of them fertilizes the egg.

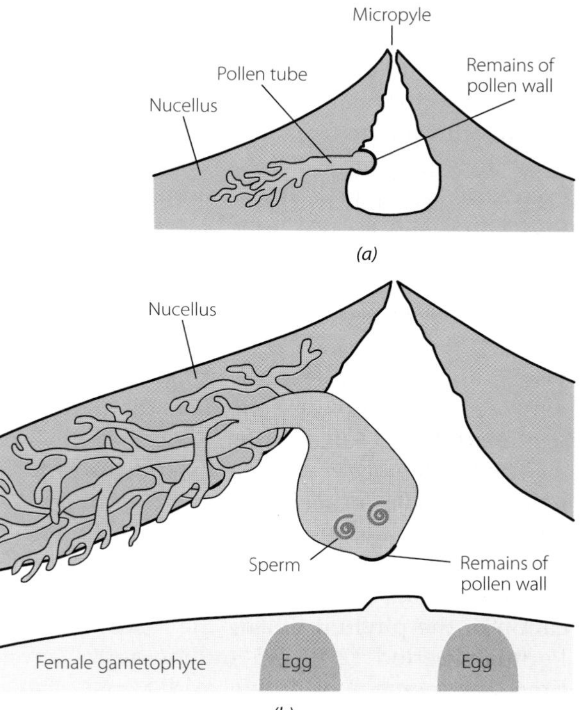

20–12
Development of the male gametophyte of Ginkgo biloba. ***(a)*** *Early in its development the pollen tube grows by tip growth and begins to form what will become a highly branched haustorial structure. The pollen tube in* Ginkgo *grows intercellularly in the nucellus.* ***(b)*** *Late in development, the basal end of the pollen tube enlarges into a saclike structure that contains the two multiflagellated sperm. Subsequently, the basal end of the pollen tube ruptures, releasing the two sperm, which then swim to the eggs contained in the archegonia of the female gametophyte.*

In conifers, gnetophytes, and angiosperms the pollen tube conveys the sperm to the egg cell. With this innovation, seed plants were no longer dependent on the presence of free water to ensure fertilization—a necessity for all seedless plants. The presence of haustorial pollen tubes in *Ginkgo* and cycads suggests that, originally, the pollen tube evolved to absorb nutrients for the production of sperm by the male gametophyte during its growth within the ovule. From this perspective, the conveyance of nonmotile sperm by a pollen tube that grows directly to an egg can be seen as a later evolutionary modification of a structure initially developed for another purpose.

Phylum *Coniferophyta*

By far the most numerous, most widespread, and most ecologically important of the gymnosperm phyla living today are the *Coniferophyta,* which comprise some 50 genera with about 550 species. The tallest vascular plant, the redwood *(Sequoia sempervirens)* of coastal California and southwestern Oregon, is a conifer. Redwood trees attain heights of up to 117 meters and trunk diameters in excess of 11 meters. The conifers, which also include pines, firs, and spruces, are of great commercial value. Their stately forests are one of the most important natural resources in vast regions of the north temperate zone (see "Jobs versus Owls" in Chapter 33). During the Early Tertiary period, some genera were more widespread than they are now, and a diverse conifer flora was present across huge expanses on all of the northern continents.

The history of the conifers extends back at least to the Late Carboniferous period, some 300 million years ago. The leaves of modern conifers have many drought-resistant features, which may bring ecological advantages in certain habitats and may also be related to the diversification of the phylum during the relatively dry and cold Permian period (290–245 million years ago). At that time, increasing worldwide aridity may have favored structural adaptations such as those of conifer leaves.

Pines Are Conifers with a Unique Leaf Arrangement

The pines (genus *Pinus*), which include perhaps the most familiar of all gymnosperms (Figure 20–13), dominate broad stretches of North America and Eurasia and are widely cultivated even in the Southern Hemisphere. There are about 90 species of pines, all of which are characterized by an arrangement of the leaves that is unique among living conifers. In pine seedlings, the needlelike leaves are spirally arranged and borne singly on the stems (Figure 20–14). After a year or two of growth, a pine begins to produce its leaves in bundles, or fascicles, each of which contains a specific number of leaves (needles)—from one to eight, depending on the species (Figure 20–15). These fascicles, wrapped at the base by a series of short, scalelike leaves, are actually short shoots in which the activity of the apical meristem is suspended. Thus, a fascicle of needles in a pine is morphologically a **determinate** (restricted in growth) branch. Under unusual circumstances, the apical meristem within a fascicle of needles in a pine may be reactivated and grow into a new shoot with **indeterminate** growth, or sometimes may even produce roots and grow into an entire pine tree (Figure 20–16).

20–13
Longleaf pines, Pinus palustris, *growing in North Carolina.*

The leaves of pines, like those of many other conifers, are impressively suited for growth under conditions where water may be scarce or difficult to obtain (Figure 20–17). A thick cuticle covers the epidermis, beneath which are one or more layers of compactly arranged, thick-walled cells—the hypodermis. The stomata are sunken below the surface of the leaf. The mesophyll, or ground tissue of the leaf, consists of parenchyma cells with conspicuous wall ridges that project into the cells, increasing their surface area. Commonly, the mesophyll is penetrated by two or more resin ducts. One vascular bundle, or two bundles side by side, are found in the center of the leaf. The vascular bundles, made up of xylem and phloem, are surrounded by transfusion tis-

(a)

(b)

20–14
(a) *Seedlings of longleaf pine,* Pinus palustris, *in Georgia, showing the long leaves in fascicles, or bundles, of three.* ***(b)*** *A seedling of pinyon pine,* Pinus edulis, *showing spirally arranged juvenile leaves and a young taproot system. The mature leaves of this species are borne in fascicles of two needles each.*

20–15
Bristlecone pine, Pinus longaeva, *in the White Mountains of California. Branch showing fascicles of five needles each, a mature ovulate cone on the right, and a young ovulate cone on the left. The individual needles of this species may remain functional for up to 45 years. It is also the longest-lived tree (see Figure 27–28).*

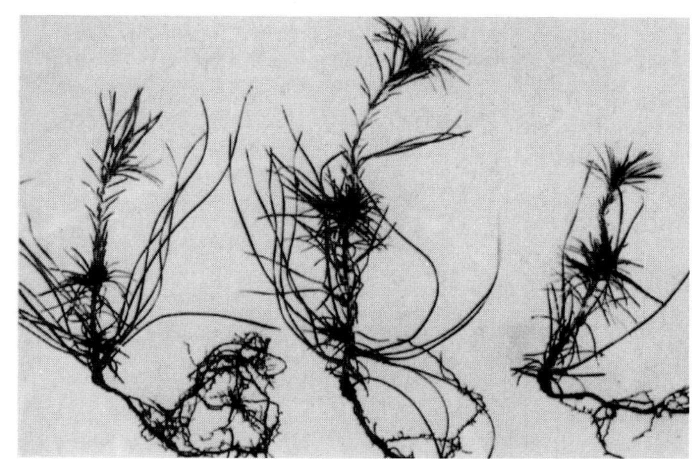

20–16
One-year-old Monterey pines (Pinus radiata) *grown from rooted fascicles of needles. This experiment demonstrates that a fascicle of pine needles is actually a short shoot in which the activity of the apical meristem has been suspended but can be regenerated.*

sue, composed of living parenchyma cells and short, nonliving tracheids. The transfusion tissue is believed to conduct materials between the mesophyll and the vascular bundles. A single layer of cells known as the endodermis surrounds the transfusion tissue, separating the transfusion tissue from the mesophyll.

Most pine species retain their needles for two to four years, and the overall photosynthetic balance of a given plant depends on the health of several years' crops of needles. In bristlecone pine *(Pinus longaeva)*, the longest-lived tree (Figure 20–15), the needles are retained for up to 45 years and remain photosynthetically active the en-

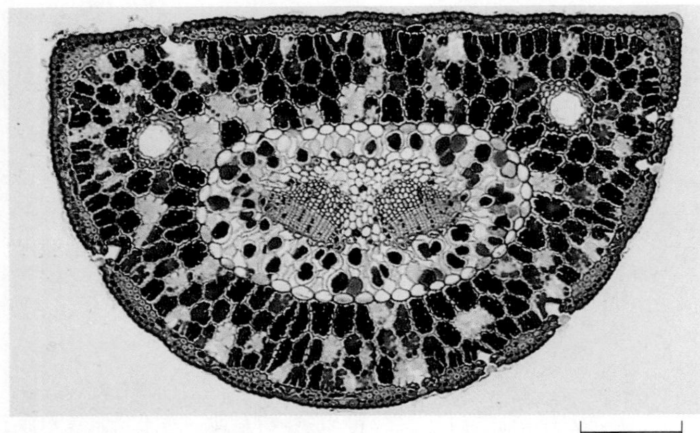

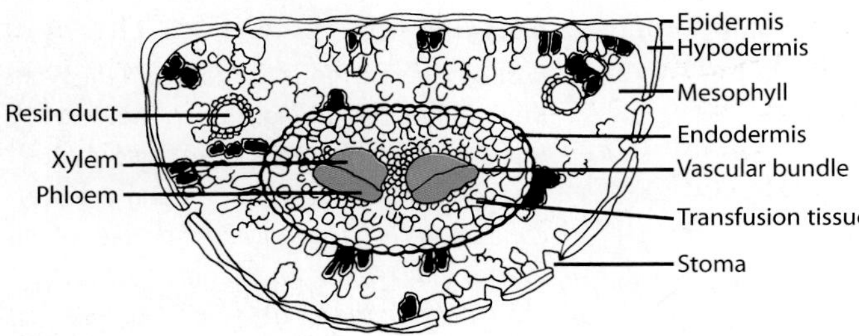

20–17
Pinus. *Transverse section of needle, showing mature tissues.*

tire time. Because the leaves of pines and other evergreens function for more than one season, they are exposed to possible damage by drought, freezing, or air pollution for much longer than the leaves of deciduous plants, which are replaced each year.

In the stems of pines and other conifers, secondary growth begins early and leads to the internal formation of substantial amounts of secondary xylem, or wood (Figure 20–18). Secondary xylem is produced toward the inside of the vascular cambium, and secondary phloem is produced toward the outside. The xylem of conifers consists primarily of tracheids, while the phloem consists of sieve cells, which are the typical food-conducting cells of gymnosperms (see page 574). Both kinds of tissues are traversed radially by narrow rays. With the initiation of secondary growth, the epidermis is eventually replaced with a periderm, which has its origin in the outer layer of cortical cells. As secondary growth continues, subsequent periderms are produced by active cell division deeper in the bark.

The Pine Life Cycle Extends over a Period of Two Years
As you read this account of reproduction in pines, you may find it useful to refer, from time to time, to the pine life cycle (Figure 20–22) on pages 478 and 479.

The microsporangia and megasporangia in pines and most other conifers are borne in separate cones, or strobili, on the same tree. Ordinarily, the microsporangiate (pollen-producing) cones are borne on the lower branches of the tree and the megasporangiate, or *ovulate*, cones are

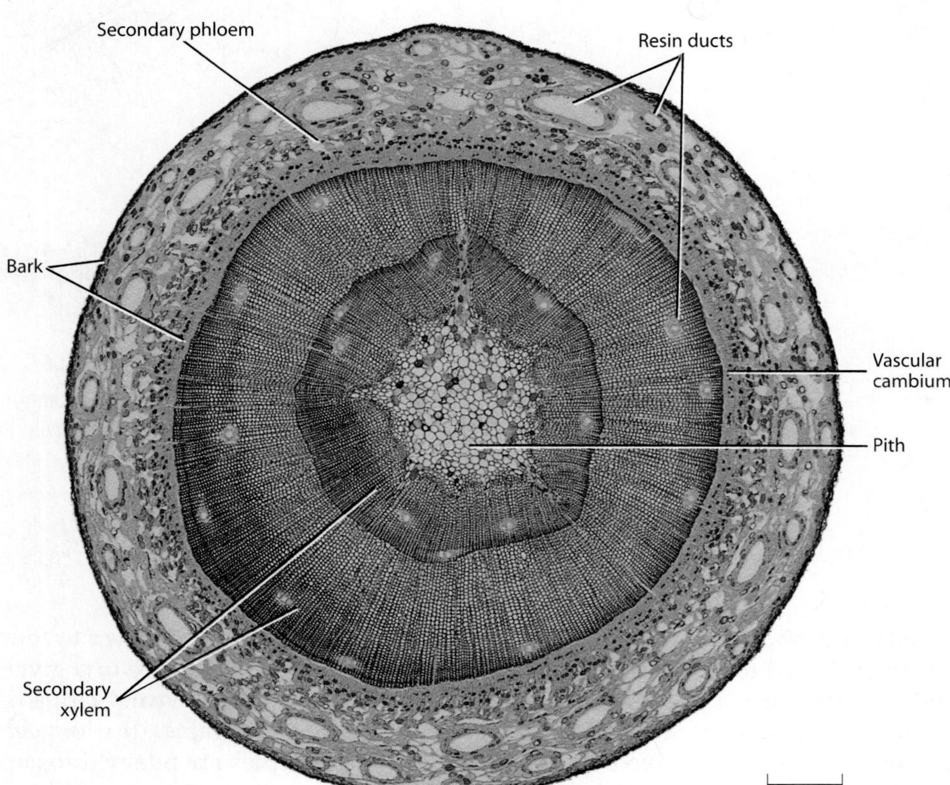

20–18
Pinus. *Cross section of stem, showing secondary xylem and secondary phloem separated from one another by vascular cambium. All of the tissues outside the vascular cambium, including the phloem, make up the bark.*

20–19
Monterey pine, Pinus radiata. *Microsporangiate cones shedding pollen, which is blown about by the wind. Some of the pollen reaches the vicinity of the ovules in the ovulate cones and then germinates, producing pollen tubes and eventually bringing about fertilization.*

borne on the upper branches. In some pines, they are borne on the same branch, with the ovulate cones closer to the tip. Since the pollen is not normally blown straight upward, the ovulate cones are usually pollinated by pollen from another tree, thus enhancing outcrossing.

Microsporangiate cones in the pines are relatively small, usually 1 to 2 centimeters long (Figure 20–19). The microsporophylls (Figure 20–20) are spirally arranged and more or less membranous. Each bears two microsporangia on its lower surface. A young microsporangium contains many **microsporocytes,** or **microspore mother cells.** In early spring the microsporocytes undergo meiosis and each produces four haploid microspores. Each microspore develops into a winged pollen grain, consisting of two **prothallial cells,** a **generative cell,** and a **tube cell** (Figure 20–21). This four-celled pollen grain is the immature male gametophyte. It is at this stage that the pollen grains are shed in enormous quantities; some are carried by the wind to the ovulate cones.

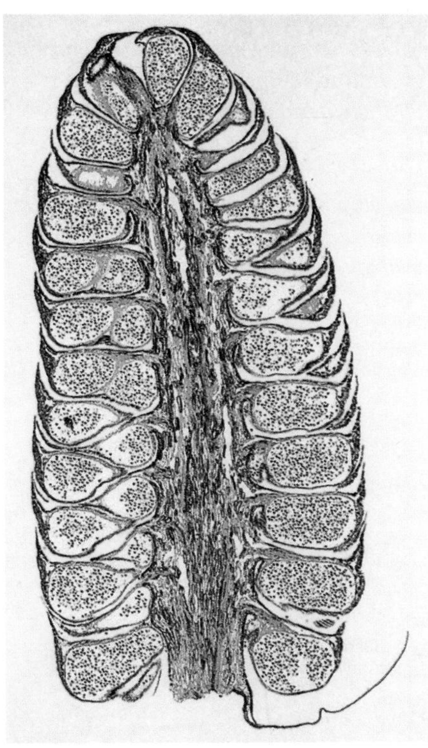

20–20
Pinus. *Longitudinal view of a microsporangiate (pollen-producing) cone, showing microsporophylls and microsporangia containing mature pollen grains.*

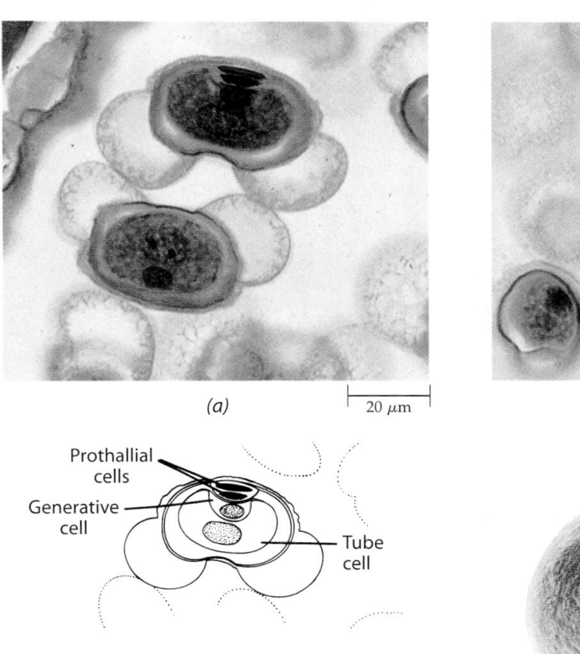

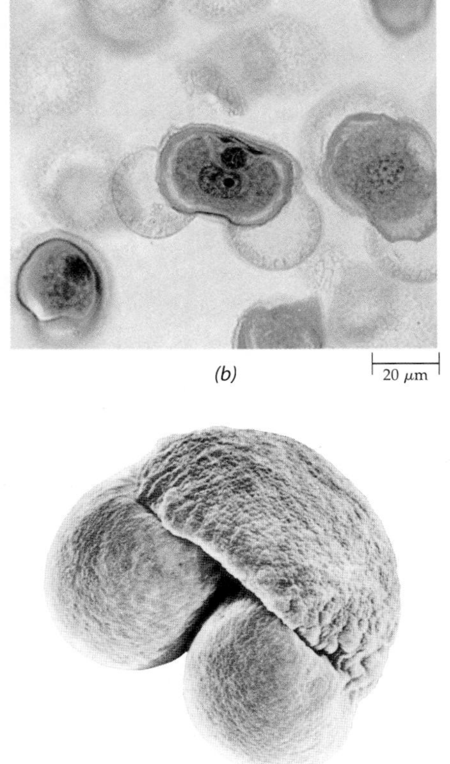

20–21
Pinus. *(a) Pollen grains with enclosed immature male gametophytes. Each gametophyte consists of two prothallial cells, a relatively small generative cell, and a relatively large tube cell. (b) A somewhat older pollen grain. Here the prothallial cells, which have no apparent function, have degenerated. (c) Scanning electron micrograph of a pine pollen grain, with its two bladder-shaped wings. When the pollen grain germinates, the pollen tube emerges from the lower end of the grain between the wings.*

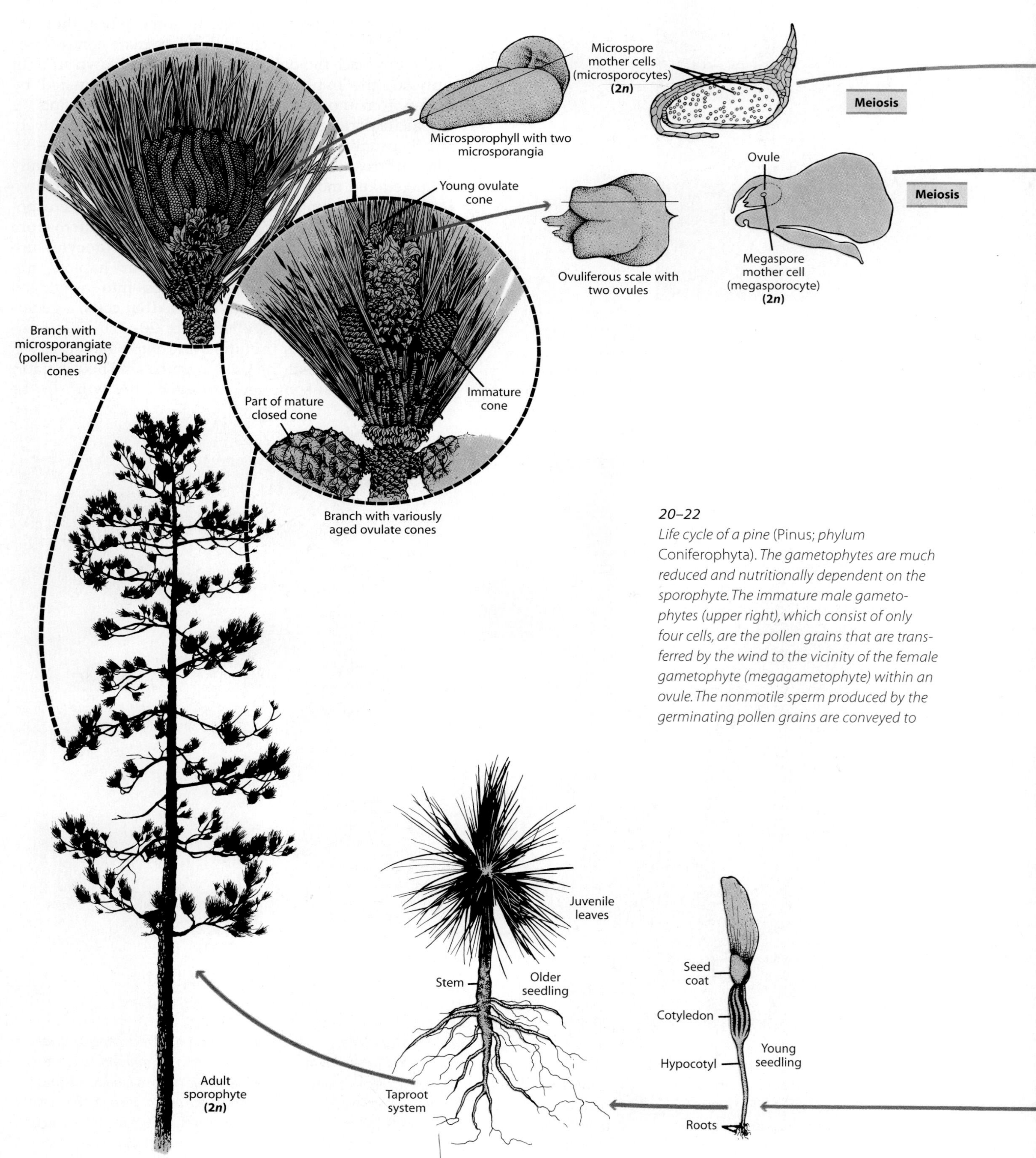

20–22
Life cycle of a pine (Pinus; *phylum* Coniferophyta). *The gametophytes are much reduced and nutritionally dependent on the sporophyte. The immature male gametophytes (upper right), which consist of only four cells, are the pollen grains that are transferred by the wind to the vicinity of the female gametophyte (megagametophyte) within an ovule. The nonmotile sperm produced by the germinating pollen grains are conveyed to*

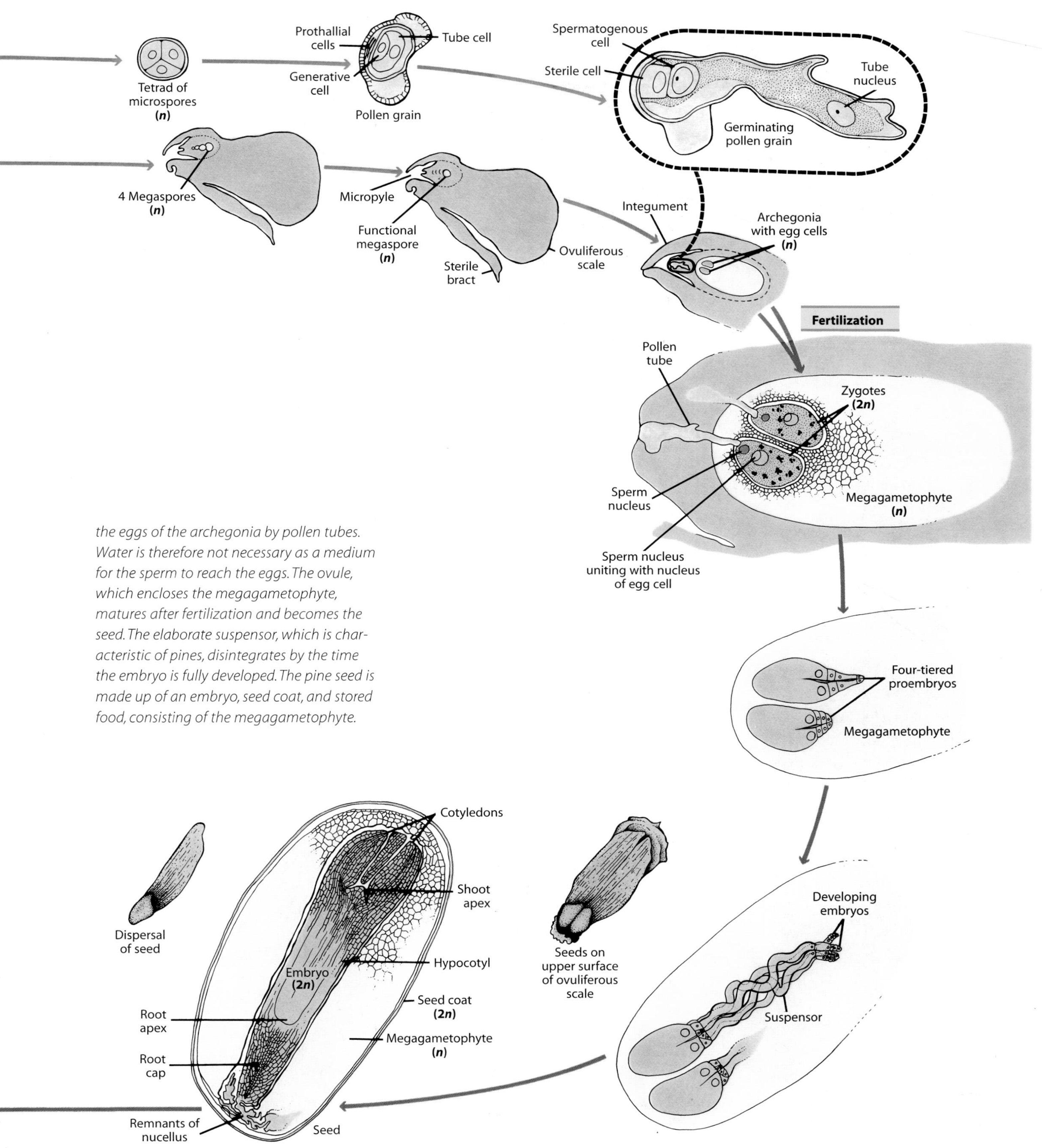

the eggs of the archegonia by pollen tubes. Water is therefore not necessary as a medium for the sperm to reach the eggs. The ovule, which encloses the megagametophyte, matures after fertilization and becomes the seed. The elaborate suspensor, which is characteristic of pines, disintegrates by the time the embryo is fully developed. The pine seed is made up of an embryo, seed coat, and stored food, consisting of the megagametophyte.

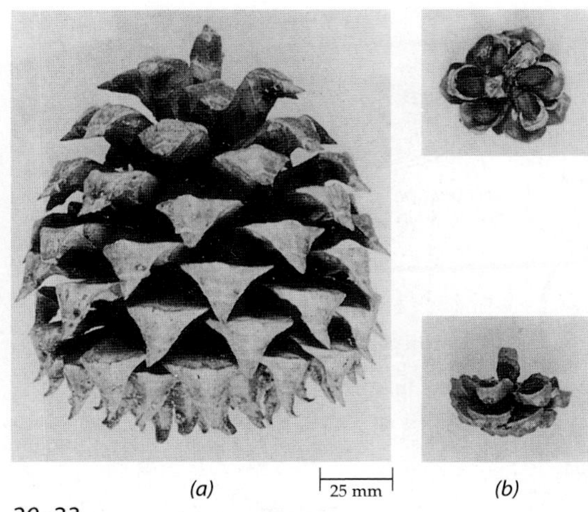

20–23
Relative sizes of some mature ovulate pine cones. ***(a)*** *Digger pine,* Pinus sabiniana. ***(b)*** *A pinyon pine,* Pinus edulis, *in top and side views. The wingless, edible seeds of this and certain other pines are called "pine nuts."* ***(c)*** *Sugar pine,* Pinus lambertiana. ***(d)*** *Yellow pine,* Pinus ponderosa. ***(e)*** *Eastern white pine,* Pinus strobus. ***(f)*** *Red pine,* Pinus resinosa.

The ovulate cones of pines are much larger and more complex in their structure than are the pollen-bearing cones (Figure 20–23). The **ovuliferous scales** (cone scales), which bear the ovules, are not simply megasporophylls. Instead, they are entire modified determinate branch systems properly known as **seed-scale complexes.** Each seed-scale complex consists of the ovuliferous scale—which bears two ovules on its upper surface—and a subtending sterile bract (Figure 20–24). The scales are arranged spirally around the axis of the cone. (The ovulate cone is, therefore, a compound structure, whereas the microsporangiate cone is a simple one, in which the microsporangia are directly attached to the microsporophylls.) Each ovule consists of a multicellular nucellus (the megasporangium) surrounded by a mas-

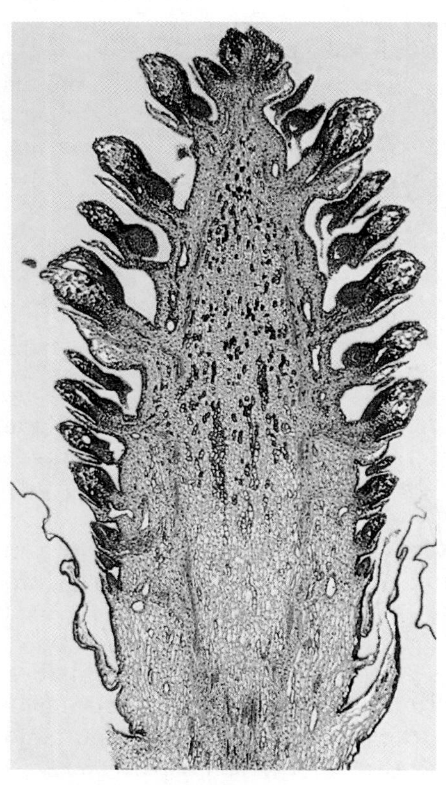

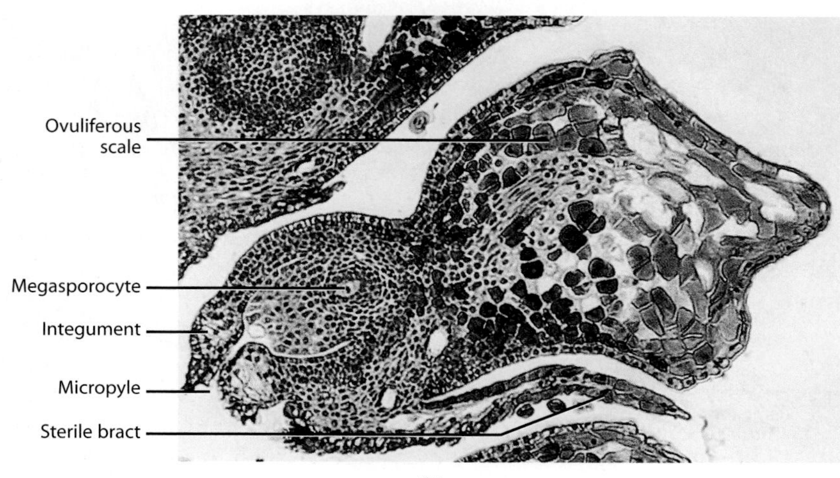

20–24
Pinus. ***(a)*** *Longitudinal view of young ovulate cone, showing its complex structure.* ***(b)*** *Detail of a portion of the cone. Note the megasporocyte (megaspore mother cell) surrounded by the nucellus. (See also the pine life cycle, Figure 20–22.)*

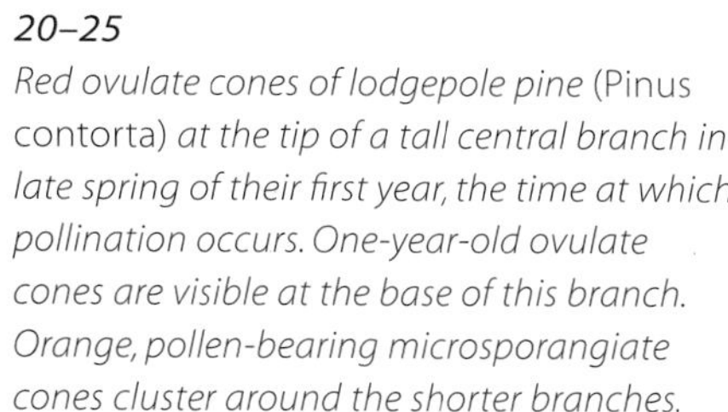

20–25
Red ovulate cones of lodgepole pine (Pinus contorta) *at the tip of a tall central branch in late spring of their first year, the time at which pollination occurs. One-year-old ovulate cones are visible at the base of this branch. Orange, pollen-bearing microsporangiate cones cluster around the shorter branches.*

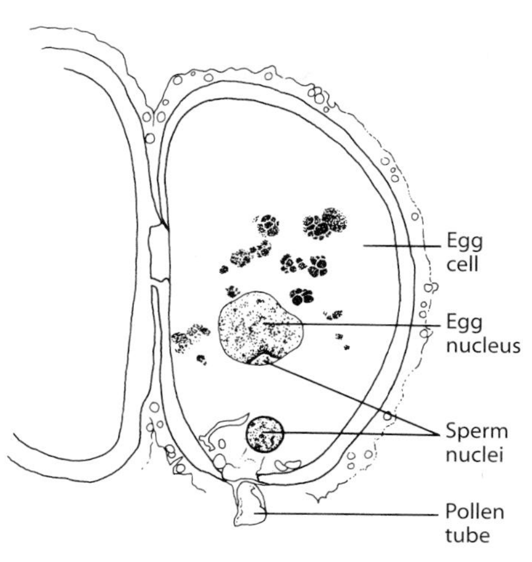

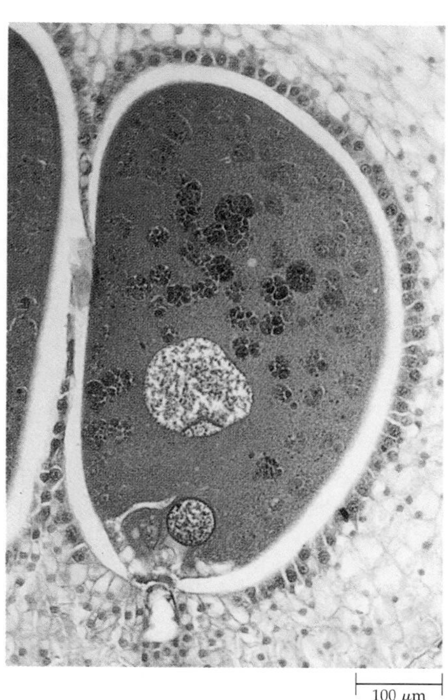

20–26
Pinus. *Fertilization: union of a sperm nucleus with the egg nucleus. The second sperm nucleus (below) is nonfunctional; it will eventually disintegrate.*

sive integument with an opening, the micropyle, facing the cone axis (Figure 20–24). Each megasporangium contains a single **megasporocyte,** or **megaspore mother cell,** which ultimately undergoes meiosis, giving rise to a linear series of four megaspores. However, only one of these megaspores is functional; the three nearest the micropyle soon degenerate.

Pollination in pines occurs in the spring (Figure 20–25). At this stage, the scales of the ovulate cone are widely separated. As the pollen grains settle on the scales, many adhere to pollination drops, which exude from the micropylar canals at the open ends of the ovules. As the pollination drops evaporate, they contract, carrying the pollen grains through the micropylar canal and into contact with the nucellus. At its micropylar end, the nucellus is slightly suppressed. The pollen grains come to lie in this shallow cavity. After pollination, the scales grow together and help protect the developing ovules. Shortly after the pollen grain comes in contact with the nucellus, it germinates to form a pollen tube. At this time, meiosis has not yet occurred in the megasporangium.

About a month after pollination, the four megaspores are produced, only one of which develops into a megagametophyte. The development of the megagametophyte is sluggish. It often does not begin until some six months after pollination, and even then may require another six months for completion. In the early stages of megagametophyte development, mitosis proceeds without immediate cell wall formation. About 13 months after pollination, when the female gametophyte contains some 2000 free nuclei, cell wall formation begins. Then, approximately 15 months after pollination, archegonia, usually two or three in number, differentiate at the micropylar end of the megagametophyte, and the stage is set for fertilization.

About 12 months earlier, the pollen grain germinated, producing a pollen tube that slowly digested its way through the tissues of the nucellus toward the developing megagametophyte. About a year after pollination, the generative cell of the four-celled male gametophyte undergoes division, giving rise to two kinds of cells—a **sterile cell** (stalk cell) and a **spermatogenous cell** (body cell). Subsequently, before the pollen tube reaches the female gametophyte, the spermatogenous cell divides, producing two sperm. The male gametophyte, or germinating pollen grain, is now mature. Recall that seed plants do not form antheridia.

Some 15 months after pollination, the pollen tube reaches the egg cell of an archegonium, where it discharges much of its cytoplasm and both of its sperm into the egg cytoplasm (Figure 20–26). One sperm nucleus unites with the egg nucleus, and the other degenerates.

Commonly, the eggs of all archegonia are fertilized and begin to develop into embryos (the phenomenon of polyembryony). Only one embryo usually develops fully, but about 3 to 4 percent of pine seeds contain more than one embryo and produce two or three seedlings upon germination.

During early embryogeny, four tiers of cells are produced near the lower end of the archegonium. Each of the four cells of the uppermost tier (that is, the tier farthest from the micropylar end of the ovule) begins to form an embryo. Simultaneously the four cells of the tier below the embryos, the suspensor cells, elongate greatly and force the four developing embryos through the wall of the archegonium and into the female gametophyte. Thus, a second type of polyembryony is found in the pine life cycle. Once again, however, usually only one of the embryos develops fully. During embryogeny, the integument develops into a seed coat.

The conifer seed is a remarkable structure, for it consists of a combination of two different diploid sporophytic generations—the seed coat (and remnants of the nucellus) and the embryo—and one haploid gametophytic generation (Figure 20–27). The gametophyte serves as a food reserve or nutritive tissue. The embryo consists of a hypocotyl-root axis, with a rootcap and apical meristem at one end and an apical meristem and several (generally eight) cotyledons, or seed leaves, at the other. The integument consists of three layers, of which the middle layer becomes hard and serves as the seed coat.

The seeds of pines are often shed from the cones during the autumn of the second year following the initial appearance of the cones and pollination. At maturity, the cone scales separate; the winged seeds of most species flutter through the air and are sometimes carried considerable distances by the wind. In some species of pines, such as lodgepole pines *(Pinus contorta),* the scales do not separate until the cones are subjected to extreme heat (see "The Great Yellowstone Fire" in Chapter 32). When a forest fire sweeps rapidly through the pine grove and burns the parent trees, most of the fire-resistant cones are only scorched. These cones open, releasing the seed crop accumulated over many years, and reestablish the species. In other species of pines, including the limber pine *(Pinus flexilis),* whitebark pine *(Pinus albicaulis),* and pinyon pines of western North America, as well as a few similar species of Eurasia, the wingless, large seeds are harvested, transported, and stored for later eating by large, crowlike birds called nutcrackers. The birds miss many of the seeds they store, aiding in the dispersal of the pines.

The pine life cycle is summarized in Figure 20–22.

Other Important Conifers Occur throughout the World

Although other conifers lack the needle clusters of pines, and may also differ in a number of relatively minor details of their reproductive systems, the living conifers form a fairly homogeneous group. In most conifers other than the pines, the reproductive cycle takes only a year; that is, the seeds are produced in the same season as the ovules are pollinated. In such conifers, the time between pollination and fertilization usually ranges from three days to three or four weeks, instead of 15 months or so.

Among the important genera of conifers other than

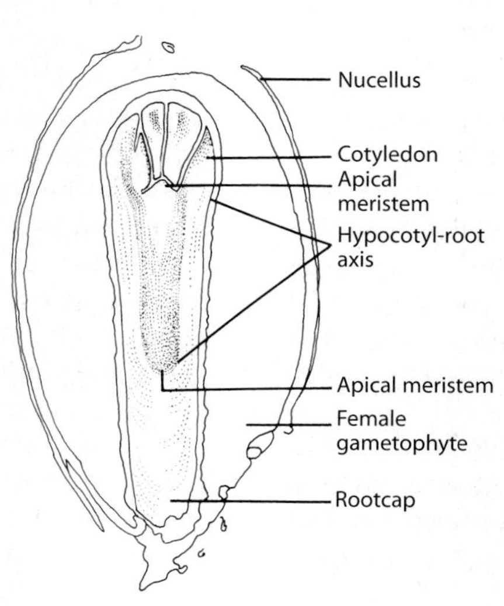

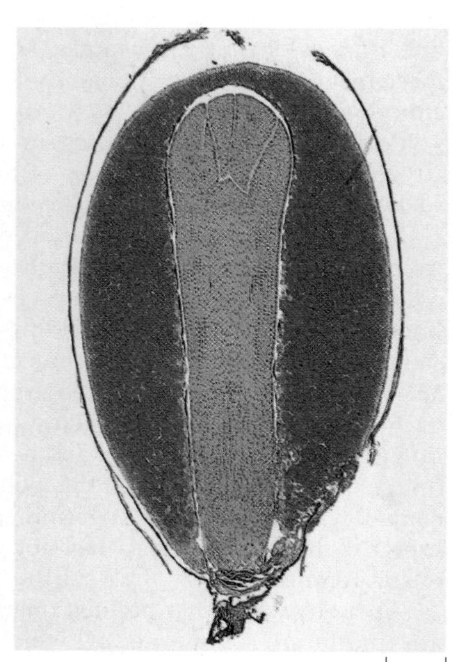

20–27
Pinus. *Longitudinal section of a seed. The hard protective seed coat (here removed) and embryo represent successive sporophyte* (2n) *generations, with a gametophyte generation intervening. A remnant of the nucellus (megasporangium) forms a papery shell around the gametophyte.*

(a)

(b)

20–28
Two genera of the pine family (Pinaceae). ***(a)*** *Ovulate cone of balsam fir* (Abies balsamea). *The upright cones, which are 5 to 10 centimeters long, do not fall to the ground whole, as they do in the pines. Instead, these cones shatter and fall apart while still attached, scattering the winged seeds.* ***(b)*** *European larch* (Larix decidua). *The needlelike leaves of larch, which are borne singly on both long shoots and short branch shoots, are spirally arranged. Unlike most conifers, the larches are deciduous; that is, they shed their leaves at the end of each growing season.*

pines are the firs (*Abies;* Figure 20–28a), larches (*Larix;* Figure 20–28b), spruces *(Picea),* hemlocks *(Tsuga),* Douglas firs *(Pseudotsuga),* cypresses (*Cupressus;* Figure 20–29), and junipers (*Juniperus;* Figure 20–30), often misleadingly called "cedars" in North America. In the yews (family *Taxaceae*), the ovules are not borne in cones but are solitary and surrounded by a fleshy, cuplike structure—the **aril** (Figure 20–31a).

One of the most interesting groups of conifers is the family *Araucariaceae,* the living members of which occur naturally only in the Southern Hemisphere. Until recently, only two surviving genera—*Agathis* and *Araucaria*—were known to exist (see the essay on page 488). A species of *Araucaria,* called Panama pine, is one of the most valuable timber trees in South America. Some species of *Araucaria,* such as the monkey-puzzle tree *(Araucaria araucana)* of Chile and the Norfolk Island pine *(Araucaria heterophylla),* are frequently cultivated in areas where the climate is mild (Figure 20–32). Seedlings of Norfolk Island pines are also grown as houseplants.

Another interesting group of conifers is the family *Taxodiaceae.* Taxodiaceous woods are found in the Triassic, and a number of fossils (leaves and cones) date from the Middle Jurassic (165–185 million years ago). The *Taxodiaceae* are represented today by widely scattered species that are the remnants of populations that were much more widespread during the Tertiary period (Figure 20–33). One of the most remarkable of these is the coast redwood, *Sequoia sempervirens,* the tallest living plant. The famous "big tree," *Sequoiadendron giganteum* (Figure 20–34), which forms spectacular, widely scattered groves along the western slope of the Sierra Nevada of California, and the bald cypresses *(Taxodium)* of the southeastern United States and Mexico (Figure 20–35) also belong to this family.

20–29
In the subglobose cones of the cypresses, the scales are crowded together, as in this Gowen cypress (Cupressus goveniana). *The small trees of this species—only about 6 meters tall at maturity—are extremely local, being found only near Monterey, California.*

20–30
The common juniper (Juniperus communis) *has spherical ovulate cones like those of the cypresses, but in juniper the scales are fleshy and fused together. Juniper "berries" give gin its distinctive taste and aroma.*

20–32
Norfolk Island pine (Araucaria heterophylla).

(a)

(b)

20–31
The conifers of the yew family (Taxaceae) *have seeds that are surrounded by a fleshy cup—the aril. The arils attract birds and other animals, which eat them and thus spread the seeds.* ***(a)*** *Members of the genus* Taxus, *the yews—which occur around the Northern Hemisphere—produce red, fleshy ovulate structures.* ***(b)*** *Sporophylls and microsporangia of the pollen-bearing cones of a yew. Ovulate cones and pollen-bearing cones are found on separate individuals. The seeds and leaves of yews contain a toxic substance and are an important cause of poisoning by plants in children in the United States, although fatalities are extremely rare.*

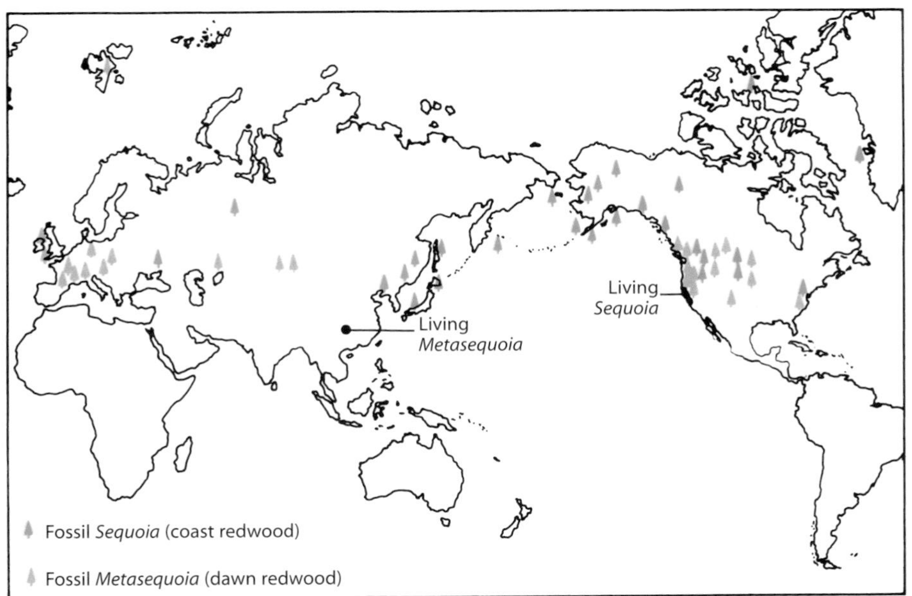

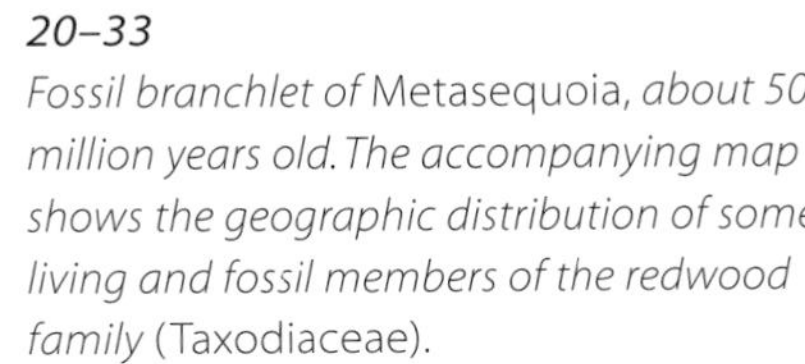

20–33
Fossil branchlet of Metasequoia, *about 50 million years old. The accompanying map shows the geographic distribution of some living and fossil members of the redwood family* (Taxodiaceae).

20–34
The "big trees" (Sequoiadendron giganteum) *of the western slope of California's Sierra Nevada are the largest of gymnosperms. The largest specimen, the General Sherman sequoia, is more than 80 meters high and estimated to weigh at least 2500 metric tons. The largest of living animals, the blue whale, pales by comparison. Blue whales rarely exceed 35 meters in length and 180 metric tons in weight.*

20–35
The bald cypress (Taxodium distichum) *is a deciduous member of the redwood family that grows in the swamps of the southeastern United States. Like the larch, it is one of the few conifers that sheds its leaves (actually leafy shoots) at the end of each growing season. In this autumn photograph, the leaves have begun to change color. Spanish "moss"* (Tillandsia usneoides), *actually a flowering plant related to the pineapple, is seen hanging in masses from the branches of these trees.*

20–36
The dawn redwood (Metasequoia glyptostroboides). *This tree, growing in Hubei Province in central China, is more than 400 years old.*

Like most of the living genera of *Taxodiaceae, Metasequoia,* the dawn redwood, was much more widespread in the Tertiary period than it is now (Figure 20–36). *Metasequoia* occurred widely across Eurasia and was the most abundant conifer in western and Arctic North America from the Late Cretaceous period to the Miocene epoch (from about 90 to about 15 million years ago). It survived both in Japan and in eastern Siberia until a few million years ago. The genus *Metasequoia* was first described from fossil material by the Japanese paleobotanist Shigeru Miki in 1941 (Figure 20–33a). Three years later the Chinese forester Tsang Wang, from the Central Bureau of Forest Research of China, visited the village of Mo-tao-chi in remote Sichuan Province in south-central China. There he discovered a huge tree of a kind he had never seen before. The natives of the area had built a temple around the base of the tree. Tsang collected specimens of the tree's needles and cones, and studies of these samples revealed that the fossil *Metasequoia* had "come to life." In 1948, paleobotanist Ralph Chaney of the University of California at Berkeley led an expedition down the Yangtze River and across three mountain ranges to valleys where dawn redwoods were growing, the last remnant of the once great *Metasequoia* forest. In 1980, about 8000 to 10,000 trees still existed in the *Metasequoia* valley, about 5000 of which had diameters over 20 centimeters. Unfortunately, the trees were not reproducing there, because the seeds were being harvested for cultivation and because of the lack of a suitable habitat in which seedlings could become established. Thousands of seeds have been distributed widely, however, and this "living fossil" can now be seen growing in parks and gardens all over the world.

The Other Living Gymnosperm Phyla: *Cycadophyta, Ginkgophyta,* and *Gnetophyta*

Cycads Belong to the Phylum *Cycadophyta*

The other groups of living gymnosperms are remarkably diverse and scarcely resemble one another at all. Among them are the cycads, phylum *Cycadophyta,* which are palmlike plants found mainly in tropical and subtropical regions. These unique plants, which appeared at least 250 million years ago during the Permian period, were so numerous in the Mesozoic era, along with the superficially similar *Bennettitales,* that this period is often called the "Age of Cycads and Dinosaurs." Living cycads comprise 11 genera, with about 140 species. *Zamia pumila,* found commonly in the sandy woods of Florida, is the only cycad native to the United States (Figure 20–37).

Most cycads are fairly large plants; some reach 18 meters or more in height. Many have a distinct trunk that is densely covered with the bases of shed leaves. The functional leaves characteristically occur in a cluster at the top of the stem; thus the cycads resemble palms. (Indeed, a common name for some cycads is "sago palms.") Unlike palms, however, cycads exhibit true, if sluggish, secondary growth from a vascular cambium; the central portion of their trunks consists of a great mass of pith. Cycads are often highly toxic, with both neurotoxins and carcinogenic compounds abundant. They harbor cyanobacteria and make important contributions of fixed nitrogen to the areas where they occur.

The reproductive units of cycads are more or less reduced leaves with attached sporangia that are loosely or

20–37
Male and female plants of Zamia pumila, *the only species of cycad native to the United States. The stems are mostly or entirely underground and, along with the storage roots, were used by Native Americans as food and as a source of starch. The two large gray cones in the foreground are ovulate cones; the smaller brown cones are microsporangiate cones.*

tightly clustered into conelike structures near the apex of the plant. The pollen and ovulate cones of cycads are borne on different plants (Figure 20–38). The pollen tubes formed by the male gametophytes of cycads are typically unbranched or only slightly branched. In most cycads, growth of the pollen tube results in significant destruction of the nucellar tissue. Before fertilization, the basal end of the male gametophyte swells and elongates, bringing the sperm close to the eggs. The basal end then ruptures, and the released multiflagellated sperm swim to the eggs (Figure 20–39). Each male gametophyte produces two sperm.

The role of insects in the pollination of cycads is of special interest. Beetles of several groups have frequently been found to be associated with the male cones, and less frequently with the female cones, of the members of several genera of cycads. For example, weevils (family *Curculionidae*) of the genus *Rhopalotria* carry out their entire life cycles upon and within male cones of *Zamia* and also visit the female cones. Pollen-eating beetles, although not weevils, have certainly been present throughout the history of the cycads. It seems reasonable to assume that there has been a long relationship between the members of the two groups. It is less certain that wind pollination plays an important role in the pollination of cycads, despite the general assumption that it does so.

(a)

(b)

20–38
(a) Encephalartos ferox, *a cycad native to Africa. Shown here is a female plant with ovulate cones.* **(b)** *A female plant of* Cycas siamensis. *Several megasporophylls have been removed to reveal seeds on the upper surfaces of other megasporophylls.*

20–39
Sexual reproduction in cycads and Ginkgo *is unusual in combining motile sperm with pollen tubes.* ***(a)*** *The sperm of the cycad* Zamia pumila, *shown here, swim by virtue of an estimated 40,000 flagella.* ***(b)*** *Sperm are transported to the vicinity of the egg cells in the ovule by means of a pollen tube (see Figure 20–12).*

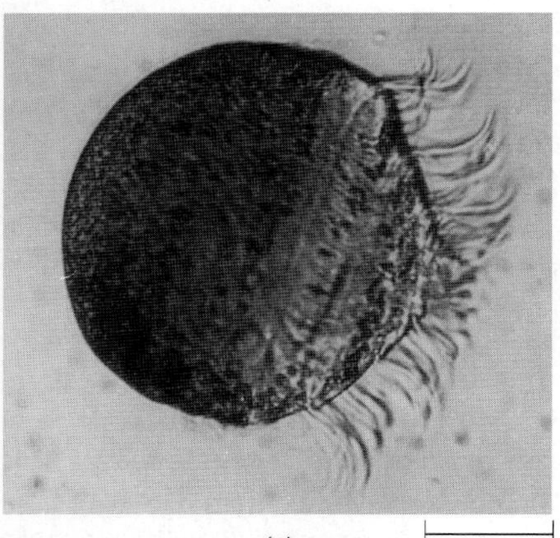

(a) 100 μm

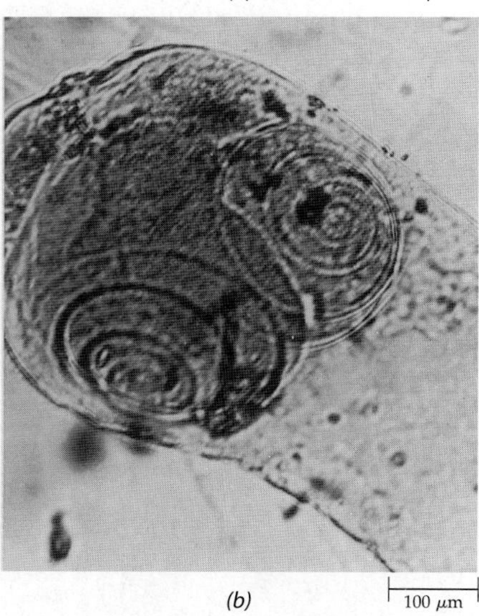

(b) 100 μm

Ginkgo biloba Is the Only Living Member of the Phylum *Ginkgophyta*

The maidenhair tree, *Ginkgo biloba,* is easily recognized by its fan-shaped leaves with their openly branched, dichotomous (forking) pattern of veins (Figure 20–40). It is an attractive and stately, but slow-growing, tree, which may reach a height of 30 meters or more. The leaves on the numerous slowly growing spur, or short, shoots of *Ginkgo* are more or less entire, whereas those on the long shoots and seedlings are often deeply lobed. Unlike most other gymnosperms, *Ginkgo* is deciduous; its leaves turn a beautiful golden color before falling in autumn.

Wollemia nobilis:

While loss of biological diversity is one of the world's most serious environmental problems, it is remarkable to discover surviving populations of an evolutionary line thought to be long extinct. This is more dramatic when the discovery is a tree up to 40 meters tall surviving in deep sandstone canyons within 150 kilometers of Sydney, Australia's largest city.

The "Wollemi Pine" was discovered in late 1994 and has been named **Wollemia nobilis,** ***commemorating the Wollemi National Park in which it occurs, as well as both David Noble, the Wildlife Officer who discovered it, and the stature of the trees. It is one of the world's rarest plant species, because fewer than 40 plants are known, growing in two small groves. It is a member of the*** **Araucariaceae,** ***an ancient family of conifers.***

Araucariaceae ***had a worldwide distribution and their greatest diversity in the Jurassic and Cretaceous periods, between 200 and 65 million years ago, but became extinct in the Northern Hemisphere in the Late Cretaceous at about the time of the extinction of the dinosaurs. The two other surviving genera of*** **Araucariaceae, Araucaria** ***and*** **Agathis,** ***have Southern Hemisphere distributions in lands that were formerly parts of the ancient supercontinent Gondwana.***

Wollemia ***is characterized by having the adult foliage leaves arranged in four rows, juvenile leaves that are very different from those of the adult, a distinctive bark type, and cones terminal on the branches of the same tree—the females on the upper branches and the males on the lower branches. The seeds are not winged. Undeveloped axillary buds buried under the bark may remain dormant for long periods before eventually developing as new lateral shoots or shoots from the base. These buds increase the chance of surviving environmental disasters such as storm, fire, or rock falls.***

Botanists at the Royal Botanic Gardens in Sydney are investigating features of **Wollemia** ***and are comparing gene sequences of the DNA of this and related groups, while horticultural researchers are studying its propagation from seeds, cuttings, and tissue cultures. Because of its rarity and since it is a handsome tree,***

A Newly Discovered Living Fossil

researchers aim to propagate Wollemia *and make it as widely available as possible in cultivation, first in other botanic gardens for conservation and eventually in general horticulture. It will, however, take years of work before it can be made widely available.*

Comparisons with fossils have involved detailed features of the leaf epidermis and the pollen, because these structures are often the best preserved parts of plant fossils. Such studies have confirmed that a pollen type found in southern Australia, New Zealand, and Antarctica, and originally thought to belong to an entirely different family, indeed represents the Wollemia *evolutionary line. That particular fossil pollen type, which was widespread from 94 to 30 million years ago, declined greatly at about the same time as rainforests became more restricted in Australia. Some plants apparently persisted, however, and the most recent pollen fossils have been recovered from marine oil drilling on the southern Australian continental shelf, in deposits only 2 million years old.*

As the third living genus of Araucariaceae, Wollemia *has provided DNA for studies of evolutionary relationships, clarifying aspects of the evolution of* Araucaria *and* Agathis *and their relationships to other conifers. The gene sequences confirm the conclusion that its closest relative is* Agathis, *but that it is appropriately recognized as a separate genus.*

Wollemia *reflects the dramatic worldwide history of floristic change on Earth. It is a member of an ancient plant family, an example of a lineage that survived the Late Cretaceous extinction of many plant and animal groups. Also, the distribution of* Araucariaceae *in Australia and South America reflects past contacts between the Southern Hemisphere land masses. Most recently, the habitats of* Wollemia*—together with its extreme rarity—appear to illustrate a late stage in the widespread replacement by flowering plants of large areas that were once dominated by conifers.*

As a rare species, Wollemia *requires conservation, and its survival shows the importance of national parks in conserving biodiversity. Its habitat is a rainforest of broad-leaved evergreen trees of* Ceratopetalum *and* Doryphora, *over which it rises as an emergent tall tree, an example of rainforest providing environmentally stable habitats that preserve species that retain many features of ancient plant groups.*

(a)

(b)

(c)

(d)

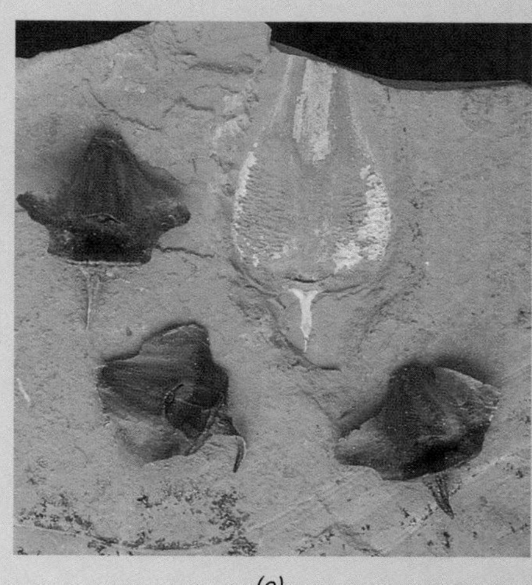

(e)

(a) *Tall trees of* Wollemia nobilis, *growing in a deep sandstone canyon.* **(b)** *Globular shaped ovulate cones and small, brown-colored male cones.* **(c)** *Ovulate cone and a branch with leaves arranged in four rows.* **(d)** *The mature leaves of* Wollemia *bear a close resemblance to the fossil leaves of the* Araucariaceae *from Jurassic deposits in eastern Australia.* **(e)** *Cone scales and seeds of* Wollemia *placed next to a fossil cone scale from the same Jurassic deposit.*

Ginkgo biloba is the sole living survivor of a genus that has changed little for more than 150 million years and is the only living member of the phylum *Ginkgophyta*. The living species shares features with other genera of gymnosperms that range back to the Early Permian period some 270 million years ago. There are probably no wild stands of *Ginkgo* anywhere in the world, but the tree was preserved in temple grounds in China and Japan. Introduced into other parts of the world, it has been an important feature of the parks and gardens of the temperate regions of the world for about 200 years. *Ginkgo* is especially resistant to air pollution and so is commonly cultivated in urban parks and along streets.

Like the cycads, *Ginkgo* bears its ovules and microsporangia on different individuals. The ovules of *Ginkgo* are borne in pairs on the end of short stalks and ripen to produce fleshy-coated seeds in autumn (Figure 20–40b). The rotting flesh of the *Ginkgo* seed coat is infamous for its vile odor, which is derived mainly from the presence of butanoic and hexanoic acids. These are the same fatty acids found in rancid butter and Romano cheese. The kernel of the seed (that is, the female gametophytic tissue and the embryo), however, has a fishy taste and is a prized delicacy in China and Japan.

In *Ginkgo*, fertilization within the ovules may not occur until after they have been shed from their parent tree. As in cycads, the male gametophyte forms an extensively branched, haustorium-like system that develops from an initially unbranched pollen tube (see Figure 20–12). Growth of the pollen tube within the nucellus is strictly intercellular, without any apparent damage to the adjacent nucellar cells. Eventually, the basal end of this system develops into a saclike structure that, at maturity, contains two large multiflagellated sperm (Figure 20–12b). Rupture of the saclike portion of the pollen tube releases these sperm to swim to the eggs within the female gametophyte of the ovule.

(a)

(b)

20–40
(a) Ginkgo biloba, *the maidenhair tree. This tree was given its English name because of the resemblance between its leaves and the leaflets of maidenhair fern.* **(b)** Ginkgo *leaves and fleshy seeds attached to spur shoots.*

The Phylum *Gnetophyta* Contains Members with Angiosperm-like Features

The gnetophytes comprise three living genera and about 70 species of very unusual gymnosperms: *Gnetum*, *Ephedra*, and *Welwitschia*. *Gnetum*, a genus of about 30 species, consists of trees and climbing vines with large, leathery leaves that closely resemble those of eudicotyledons (Figure 20–41). It is found throughout the moist tropics.

Most of the approximately 35 species of *Ephedra* are profusely branched shrubs with inconspicuous, small, scalelike leaves (Figure 20–42). With its small leaves and apparently jointed stems, *Ephedra* superficially resembles *Equisetum*. Most species of *Ephedra* inhabit arid or desert regions of the world.

(a)

(b)

(c)

20–41
The large, leathery leaves of the tropical gnetophyte Gnetum *resemble those of certain eudicots. The species of* Gnetum *grow as shrubs or woody vines in tropical or subtropical forests.* ***(a)*** *Megasporangiate inflorescences and leaves of* Gnetum gnemon. ***(b)*** *Microsporangiate inflorescence and leaves and* ***(c)*** *fleshy seeds with leaves of* Gnetum, *photographed in the Amazon Basin of southern Venezuela.*

(a)

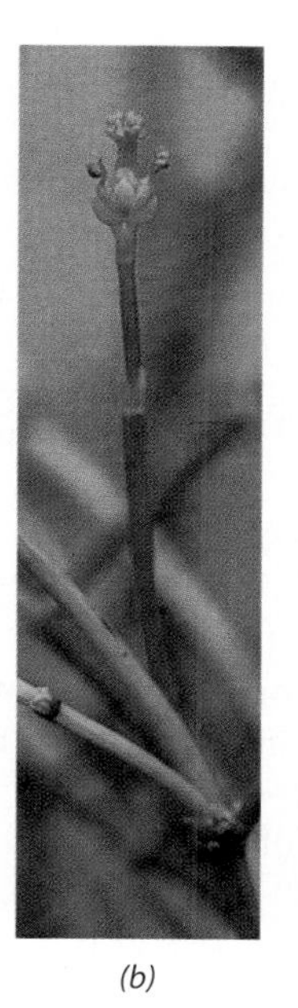

(b)

(d)

(c)

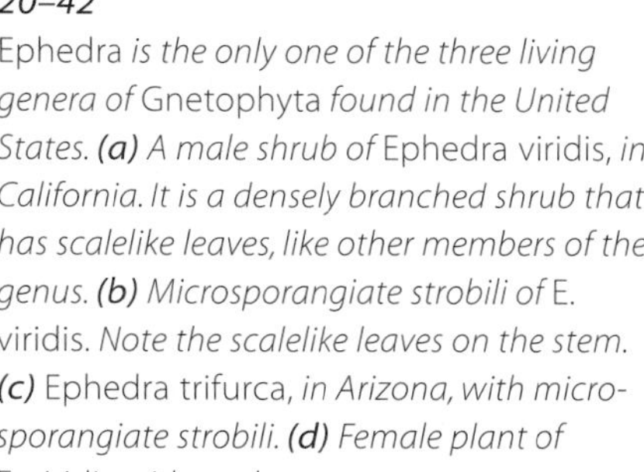

20–42
Ephedra *is the only one of the three living genera of* Gnetophyta *found in the United States.* ***(a)*** *A male shrub of* Ephedra viridis, *in California. It is a densely branched shrub that has scalelike leaves, like other members of the genus.* ***(b)*** *Microsporangiate strobili of* E. viridis. *Note the scalelike leaves on the stem.* ***(c)*** Ephedra trifurca, *in Arizona, with microsporangiate strobili.* ***(d)*** *Female plant of* E. viridis *with seeds.*

(a)

(b)

(c)

20–43
The gnetophyte Welwitschia mirabilis, *found only in the Namib Desert of Namibia and adjacent regions of southwestern Africa.* Welwitschia *produces only two leaves, which continue to grow for the life of the plant. As growth continues, the leaves break off at the tips and split lengthwise; thus older plants appear to have numerous leaves.* ***(a)*** *A large, seed-producing plant.* ***(b)*** *Microsporangiate strobili.* ***(c)*** *Ovulate strobili; the insect is a fire bug, sucking sap from the strobili.* Welwitschia *is dioecious (see page 501).*

Welwitschia is probably the most bizarre vascular plant (Figure 20–43). Most of the plant is buried in sandy soil. The exposed part consists of a massive, woody, concave disk that produces only two strap-shaped leaves that split lengthwise with age. The cone-bearing branches arise from meristematic tissue on the margin of the disk. *Welwitschia* grows in the coastal desert of southwestern Africa, in Angola, Namibia, and South Africa.

Although the genera of *Gnetophyta* are clearly related to one another and are appropriately placed together, they differ greatly in their characteristics. These genera do, however, have many angiosperm-like features, such as the similarity of their strobili to some angiosperm inflorescences (flower clusters), the presence of very similar vessels in their xylem, and the lack of archegonia in *Gnetum* and *Welwitschia*. Current analysis favors the idea that the last two features, while similar to those in angiosperms, were independently derived in *Gnetophyta* and angiosperms.

Recently, there has been increasing support, on the basis of careful morphologically based comparisons and macromolecular analyses, for the idea that gnetophytes are the group of gymnosperms most closely related to the angiosperms. In 1990, it was reported that double fertilization—a process involving the fusion of a second sperm nucleus with a nucleus of the female gametophyte—also occurs frequently in *Ephedra*. Thus double fertilization, hitherto considered unique to angiosperms, may actually have been present in the common ancestor of angiosperms and gnetophytes. Unlike flowering plants, however, where double fertilization produces a distinctive embryo-nourishing tissue called endosperm (in addition to an embryo), the second fertilization event in *Ephedra* (and also in *Gnetum*) produces extra embryos. As is the case with all gymnosperms, in *Ephedra* and *Gnetum* a large female gametophyte serves to nourish the developing embryo within the seed. None of the living gnetophytes, however, could possibly be an ancestor of any angiosperm—each of the three living genera of gnetophytes has its own unique specializations. Interestingly, the reproductive structures of at least some species of all three genera of gnetophytes produce nectar and are visited by insects. Wind pollination is clearly important—at least in *Ephedra*—but insects also play a role in the pollination of these plants.

Summary

Seed plants consist of five phyla with living representatives. One of these, the overwhelmingly successful angiosperms (phylum *Anthophyta*), is characterized by a unique set of features. The remaining four, which lack these specializations, are commonly grouped together as gymnosperms.

A Seed Develops from an Ovule

In addition to producing seeds, all seed plants bear megaphylls. The prerequisites of the seed habit include heterospory; retention of a single megaspore; development of the embryo, or young sporophyte, within the megagametophyte; and integuments. All seeds consist of a seed coat derived from the integument(s), an embryo, and stored food. In gymnosperms, the stored food is provided by the haploid female gametophyte.

Seed Plants Most Likely Evolved from the Progymnosperms

The oldest known seedlike structures occur in strata of the Late Devonian period, about 365 million years old. The likely progenitors of the gymnosperms and angiosperms are the progymnosperms, an extinct group of seedless Paleozoic vascular plants. Among the major extinct groups of gymnosperms are the seed ferns (phylum *Pteridospermophyta*), a diverse and unnatural group, and the cycadeoids, or *Bennettitales*, which had leaves resembling those of cycads but had very different reproductive structures.

All Gymnosperms Have the Same Basic Life Cycle

Living gymnosperms comprise four phyla: *Cycadophyta*, *Ginkgophyta*, *Coniferophyta*, and *Gnetophyta*. Their life cycles are essentially similar: an alternation of heteromorphic generations with large, independent sporophytes and greatly reduced gametophytes. The ovules (megasporangia plus integuments) are exposed on the surfaces of the megasporophylls or analogous structures. At maturity the female gametophyte of most gymnosperms is a multicellular structure with several archegonia. The male gametophytes develop inside pollen grains. Antheridia are lacking in all seed plants. In gymnosperms, the male gametes, or sperm, arise directly from the spermatogenous cell. Except for the cycads and *Ginkgo*, which have flagellated sperm, the sperm of seed plants are nonmotile.

Pollination and Pollen Tube Formation Eliminate the Need for Water for the Sperm to Reach the Egg

In seed plants, water is not necessary in order for the sperm to reach the eggs. Instead, the sperm are conveyed to the eggs by a combination of pollination and pollen tube formation. Pollination in gymnosperms is the transfer of pollen from microsporangium to mega-

Summary TABLE A Comparison of Some of the Main Features of the Gymnosperm Phyla with Living Representatives

PHYLUM	REPRESENTATIVE GENUS OR GENERA	TYPE OF TRACHEARY ELEMENT(S)	PRODUCE MOTILE SPERM?	POLLEN TUBE A TRUE SPERM CONVEYOR?	TYPE OF LEAVES PRODUCED	MISCELLANEOUS FEATURES
Cycadophyta (cycads)	*Cycas* and *Zamia*	Tracheids	Yes	No	Palmlike	Ovulate and microsporangiate cones simple and on separate plants
Ginkgophyta (maidenhair tree)	*Ginkgo*	Tracheids	Yes	No	Fan-shaped	Ovules and microsporangia on separate plants; fleshy-coated seeds
Coniferophyta (conifers)	*Abies*, *Picea*, *Pinus*, and *Tsuga*	Tracheids	No	Yes	Most needlelike or scalelike	Ovulate and microsporangiate cones on same plant; ovulate cones compound; pine needles in fascicles
Gnetophyta (gnetophytes)	*Ephedra*, *Gnetum*, and *Welwitschia*	Tracheids and vessel elements	No	Yes	*Ephedra:* small scalelike leaves; *Gnetum:* relatively broad, leathery leaves arranged in pairs; *Welwitschia:* two enormous, strap-shaped leaves	Ovulate and microsporangiate cones compound; borne on separate plants, except for some species of *Ephedra;* have several angiosperm-like features; leaves borne in opposite pairs

sporangium (nucellus). Fertilization occurs when one sperm of the male gametophyte (germinated pollen grain) unites with the egg, which in most gymnosperms is located in an archegonium. The second sperm has no apparent function (except perhaps in *Gnetum* and *Ephedra*), and it disintegrates. After fertilization in seed plants, each ovule develops into a seed.

There Are Four Phyla of Gymnosperms with Living Representatives

The conifers (phylum *Coniferophyta*) are the largest and most widespread phylum of living gymnosperms, with about 50 genera and some 550 species. They dominate many plant communities throughout the world, with pines, firs, spruces, and other familiar trees over wide stretches of the north. Living cycads (phylum *Cycadophyta*) consist of 11 genera and about 140 species, mainly tropical but extending away from the equator in warmer regions. Cycads are palmlike plants with trunks and sluggish secondary growth. There is only one living species of the phylum *Ginkgophyta,* the maidenhair tree *(Ginkgo biloba),* which is known only from cultivation. The three genera of the phylum *Gnetophyta* are the closest living relatives of the angiosperms (phylum *Anthophyta*).

Selected Key Terms

aril p. 483
determinate p. 474
generative cell p. 477
indeterminate p. 474
integument p. 468
megasporocytes, or **megaspore mother cells** p. 481
micropyle p. 468
microsporocytes, or **microspore mother cells** p. 477
nucellus p. 468
ovuliferous scales p. 480
pollen grain p. 472
pollen tube p. 472
pollination p. 472
polyembryony p. 472
prothallial cells p. 477
seed p. 468
seed coat p. 470
seed-scale complex p. 480
spermatogenous cell p. 481
sterile cell p. 481
tube cell p. 477

Questions

1. One of the most important evolutionary advances in the progymnosperms is the presence of a bifacial vascular cambium. What is a bifacial vascular cambium, and where is it found besides in the progymnosperms?
2. In what way do the *Bennettitales,* or cycadeoids, resemble cycads? How do they differ from the cycads?
3. The potential for polyembryony occurs twice in the pine life cycle. Explain.
4. Diagram and label the components of each of the following: a pine ovule with a mature megagametophyte; a mature pine microgametophyte (germinated pollen grain with sperm); and a mature pine seed.
5. Evidence exists in the cycads and *Ginkgo* that the first pollen tubes were haustorial structures and not true sperm conveyors. Explain.

Chapter 21

Introduction to the Angiosperms

21–1
A giant eucalyptus, or red tingle (Eucalyptus jacksonii), *growing in the Valley of the Giants in southwestern Australia. Note the man standing in the burnt-out base of this enormous angiosperm.*

OVERVIEW

Of all the plants, angiosperms—the flowering plants—are the ones that most directly affect our lives. The grains, fleshy fruits, and vegetables we eat are angiosperms, and the cotton and linen in the clothes we wear are from angiosperms as well.

The most obvious characteristic of a flowering plant is, of course, the flower. The flower contains the sexual reproductive parts of the plant and thus is of crucial importance not only for the formation of offspring but for the evolution of the species as a whole. This chapter focuses first on the basic structure of the flower. Emphasis then shifts to the formation of both the male and female gametophytes and an examination of pollination and fertilization, two processes that have features unique to angiosperms. Last to be discussed in the reproductive process are the conversions of the ovule into a seed and the surrounding ovary into a fruit. The chapter concludes with a comparison of outcrossing and self-pollination, including the different strategies angiosperms have evolved that ensure reproductive success.

CHECKPOINTS

By the time you finish reading this chapter, you should be able to answer the following questions:

1. What is a flower, and what are its principal parts?
2. What are some of the variations that exist in flower structure?
3. By what processes do angiosperms form microgametophytes (male gametophytes)? How are these processes both similar to and different from those that give rise to megagametophytes (female gametophytes)?
4. What is the structure, or composition, of the mature male gametophyte in angiosperms? Of the mature female gametophyte?
5. What is meant by "double fertilization" in angiosperms, and what are the products of this phenomenon?
6. What are some of the conditions that promote outcrossing in angiosperms, and under what circumstances might self-pollination be more advantageous than outcrossing?

Angiosperms—the flowering plants—make up much of the visible world of modern plants. Trees, shrubs, lawns, gardens, fields of wheat and corn, wildflowers, fruits and vegetables on the grocery shelves, the bright splashes of color in a florist's window, the geranium on a fire escape, duckweed and water lilies, eel grass and turtle grass, saguaro cacti and prickly pears—wherever you are, flowering plants are there also.

Diversity in the Phylum *Anthophyta*

Angiosperms make up the phylum *Anthophyta,* which includes about 235,000 species and is thus, by far, the largest phylum of photosynthetic organisms. In their vegetative features, angiosperms are enormously diverse. In size, they range from species of *Eucalyptus* trees well over 100 meters tall with trunks nearly 20 meters in girth (Figure 21–1), to some duckweeds, which are simple, floating plants often scarcely 1 millimeter long (Figure 21–2). Some angiosperms are vines that climb high into the canopy of the tropical rainforest, while others are epiphytes that grow in that canopy. Many angiosperms, such as cacti, are adapted for growth in extremely arid regions. For over 100 million years, the flowering plants have dominated the land.

In terms of their evolutionary history, the angiosperms are a group of seed plants with special characteristics: flowers, fruits, and distinctive life-cycle features that differ from those of all other plants. In this chapter, we outline these characteristics and place them in perspective, and in the following chapter, we discuss the evolution of the angiosperms. In Section 5, we consider in some detail the structure and development of the angiosperm plant body (sporophyte).

Angiosperms share so many features that do not exist in other plants that it is clear they were derived from a single common ancestor (a monophyletic group). Current investigations, based on an improved understanding of the fossil record, as well as better methods of analysis and comparisons of nucleic acid sequences, have led to a clearer recognition of the major evolutionary lines of angiosperms. Two of the major classes into which the group is divided are the *Monocotyledones* (monocots), with about 65,000 species (Figure 21–3), and the *Eudicotyledones* (eudicots), with about 165,000 species (Figure 21–4). Monocots are distinctive, including such familiar plants as the grasses, lilies, irises, orchids, cattails, and palms. The more diverse eudicots include almost all of the familiar trees and shrubs (other than the conifers) and many of the herbs (nonwoody plants). Other groups of archaic flowering plants that are neither monocots nor eudicots will be discussed in Chapter 22. Formerly these plants were grouped with the eudicots as "dicots," but we now know that this is an artificial system of classification that simply overemphasizes the dis-

(a)

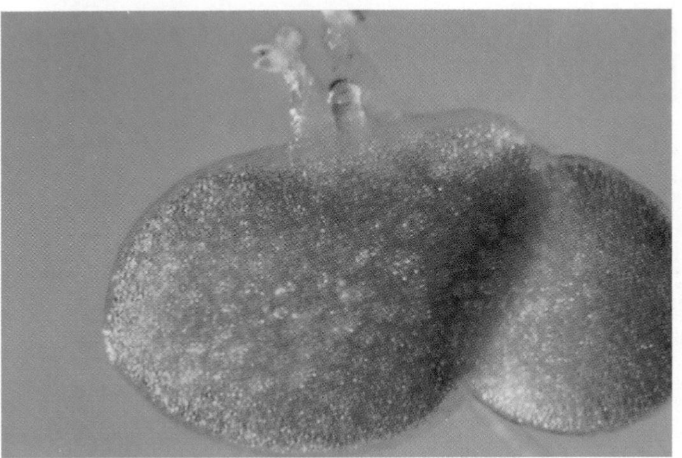

(c)

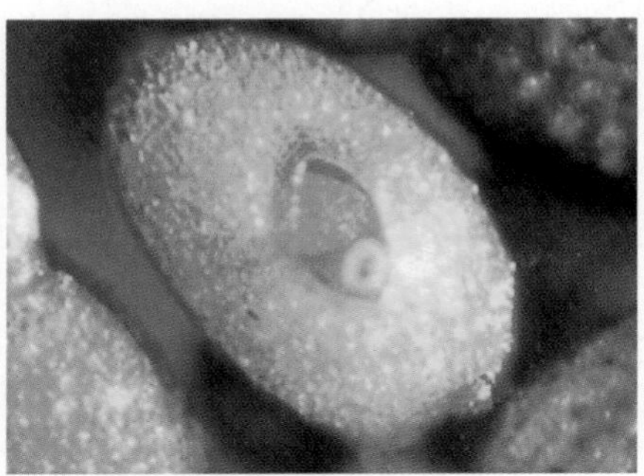

(b)

21–2
The duckweeds (family Lemnaceae*) are the smallest flowering plants.* ***(a)*** *A honeybee resting on a dense floating mat of three species of duckweed. The larger plants are* Lemna gibba, *about 2 to 3 millimeters long; the smaller ones are two species of* Wolffia, *up to 1 millimeter long.* ***(b)*** *A flowering plant of* Wolffia borealis *with a circular concave stigma (looking like a tiny doughnut) and a minute anther just above it, both protruding from a central cavity. The whole plant is less than 1 millimeter long.* ***(c)*** *Flowering plant of* Lemna gibba; *two stamens and a style protrude from a pocket on the upper surface of the plant.*

(a)

(b)

(c)

21–3

Monocots. ***(a)*** *A member of the palm family, the coconut palm* (Cocos nucifera), *growing in Tehuantepec, Oaxaca, Mexico. A coconut is a drupe, not a nut (see Chapter 22).* ***(b)*** *Flowers and fruits of the banana plant* (Musa × paradisiaca). *The banana flower has an inferior ovary, and the tip of the fruit bears a large scar left by the fallen flower parts.* ***(c)*** *Rice* (Oryza sativa) *is a member of the grass family.*

21–4

Eudicots. ***(a)*** *Saguaro cactus* (Carnegiea gigantea). *The cacti, of which there are about 2000 species, are almost exclusively a New World family. The thick, fleshy stems, which store water, contain chloroplasts and have taken over the photosynthetic function of the leaves.* ***(b)*** *Round-lobed hepatica* (Hepatica americana), *which flowers in deciduous woodlands in the early spring. The flowers*

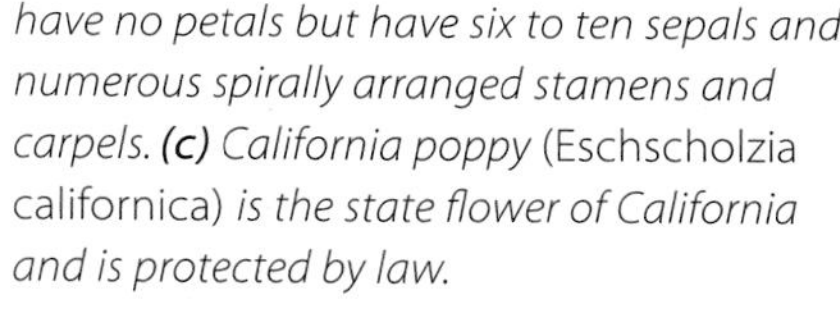
have no petals but have six to ten sepals and numerous spirally arranged stamens and carpels. ***(c)*** *California poppy* (Eschscholzia californica) *is the state flower of California and is protected by law.*

(a)

(b)

(c)

(a)

TABLE 21–1 Main Differences between Monocots and Eudicots

Characteristic	Eudicots	Monocots
Flower parts	In fours or fives (usually)	In threes (usually)
Pollen	Triaperturate (having three pores or furrows)	Monoaperturate (having one pore or furrow)
Cotyledons	Two	One
Leaf venation	Usually netlike	Usually parallel
Primary vascular bundles in stem	In a ring	Complex arrangement
True secondary growth, with vascular cambium	Commonly present	Rare

(b)

tinctiveness of monocots from other angiosperms. The major features of monocots and eudicots are indicated in Table 21–1.

In terms of their mode of nutrition, all but a few angiosperms are free-living, but both parasitic and myco-heterotrophic forms do exist (Figure 21–5). There are about 200 species of parasitic monocots and about 2800 species of parasitic eudicots, including the mistletoes (see Figure 22–44), *Cuscuta* (Figure 21–5a), and *Rafflesia* (Figure 21–5b). Parasitic flowering plants form specialized absorptive organs called haustoria that penetrate the tissues of their hosts. Myco-heterotrophs, in contrast, have obligate relationships with mycorrhizal fungi that are also associated with a green photosynthetic angiosperm. The fungus forms a bridge that transfers carbohydrates from the photosynthetic plant to the myco-heterotroph; Indian pipe (Figure 21–5c) is an example.

The Flower

The flower is a determinate (growth of limited duration) shoot that bears sporophylls—sporangium-bearing leaves (Figure 21–6). The name "angiosperm" is derived

(c)

21–5
Parasitic and myco-heterotrophic angiosperms. These plants have little or no chlorophyll; they obtain their food as a result of the photosynthesis of other plants. ***(a)*** *Dodder* (Cuscuta salina), *a parasite that is bright orange or yellow. Dodder is a member of the morning glory family* (Convolvulaceae). ***(b)*** *The world's largest flower,* Rafflesia arnoldii, *on Mount Sago, Sumatra. Plants of this genus are parasitic on the roots of a member of the grape family* (Vitaceae). ***(c)*** *Indian pipe* (Monotropa uniflora), *a myco-heterotroph that obtains its food from the roots of other plants via the fungal hyphae associated with its roots. There are more than 3000 species of parasitic angiosperms, representing 17 families.*

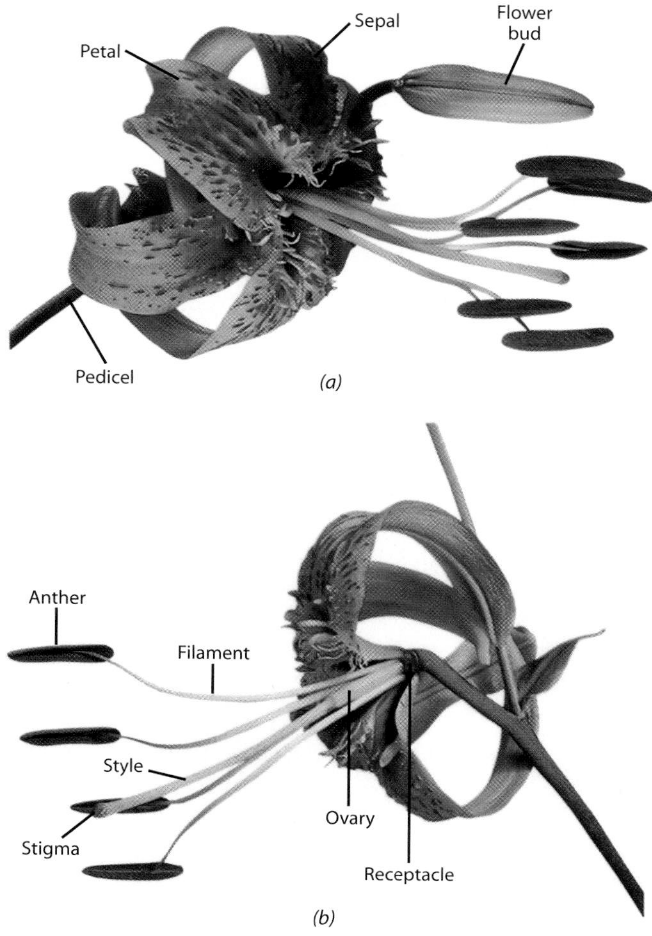

21–6
Parts of a lily (Lilium henryi) *flower.* ***(a)*** *An intact flower. In some flowers, such as lilies, the sepals and petals are similar to one another, and the perianth parts—the sepals and petals together—may then be referred to as tepals. Note that the sepals are attached to the receptacle below the petals.* ***(b)*** *Two tepals and two stamens have been removed to reveal the ovary. The gynoecium consists of the ovary, style, and stigma. The stamen consists of the filament and anther. Note that the sepals, petals, and stamens are attached to the receptacle below the ovary, which is composed, in the lily flower, of three fused carpels. Such a flower is said to be hypogynous.*

from the Greek words *angeion,* meaning "vessel," and *sperma,* meaning "seed." The definitive structure of the flower is the carpel—the "vessel." The **carpel** contains the ovules, which develop into seeds after fertilization.

Flowers may be clustered in various ways into aggregations called **inflorescences** (Figures 21–7 and 21–8). The stalk of an inflorescence or of a solitary flower is known as a **peduncle,** while the stalk of an individual flower in an inflorescence is called a **pedicel.** The part of the flower stalk to which the flower parts are attached is termed the **receptacle.**

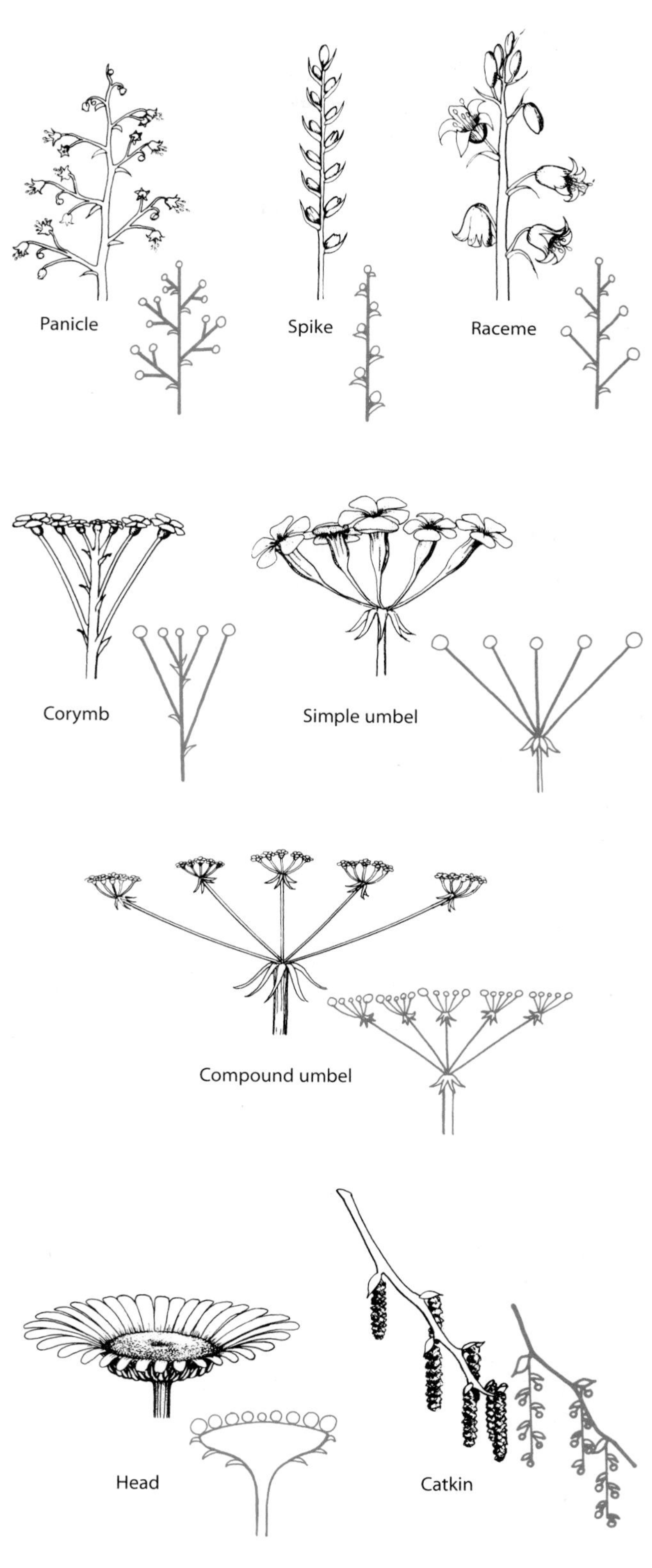

21–7
Illustrations of some of the common types of inflorescences found in the angiosperms, accompanied by simplified diagrams (in color).

(a) (b) (c) (d) (e)

21–8
Inflorescences of ***(a)*** *shooting star* (Dodecatheon meadia), ***(b)*** *butter-and-eggs* (Linaria vulgaris), ***(c)*** *lupine* (Lupinus diffusus), ***(d)*** *bluebells* (Mertensia virginica), *and* ***(e)*** *water hemlock* (Cicuta maculata). *Using Figure 21–7 as a guide, can you identify the types of inflorescences shown here?*

The Flower Consists of Sterile and Fertile, or Reproductive, Parts Borne on the Receptacle

Many flowers include two sets of sterile appendages, the **sepals** and **petals,** which are attached to the receptacle below the fertile parts of the flower, the **stamens** and **carpels.** The sepals arise below the petals, and the stamens arise below the carpels. Collectively, the sepals form the **calyx,** and the petals form the **corolla.** The sepals and petals are essentially leaflike in structure. Commonly the sepals are green and relatively thick, and the petals are brightly colored and thinner, although in many flowers the members of both whorls (a whorl is a circle of flower parts of one kind) are similar in color and texture. Together, the calyx (the sepals) and corolla (the petals) form the **perianth.**

The stamens—collectively the **androecium** ("house of man")—are microsporophylls. In all but a few living angiosperms, the stamen consists of a slender stalk, or **filament,** upon which is borne a two-lobed **anther** containing four microsporangia, or **pollen sacs,** in two pairs—a characteristic defining feature of angiosperms.

The carpels—collectively the **gynoecium** ("house of woman")—are megasporophylls that are folded lengthwise, enclosing one or more ovules. A given flower may contain one or more carpels, which may be separate or fused together, in part or entirely. Sometimes the individual carpel or the group of fused carpels is called a **pistil.** The word "pistil" comes from the same root as "pestle," the instrument with a similar shape that pharmacists use for grinding substances into a powder in a mortar.

In most flowers, the individual carpels or groups of fused carpels are differentiated into a lower part, the **ovary,** which encloses the ovules; a middle part, the **style,** through which the pollen tubes grow; and an upper part, the **stigma,** which receives the pollen. In some flowers, a distinct style is absent. If the carpels are fused, there may be a common style, or each carpel may retain a separate one. The common ovary of such fused carpels is generally (but not always) partitioned into two or more **locules**—chambers of the ovary that contain the ovules. The number of locules is usually related to the number of carpels in the gynoecium.

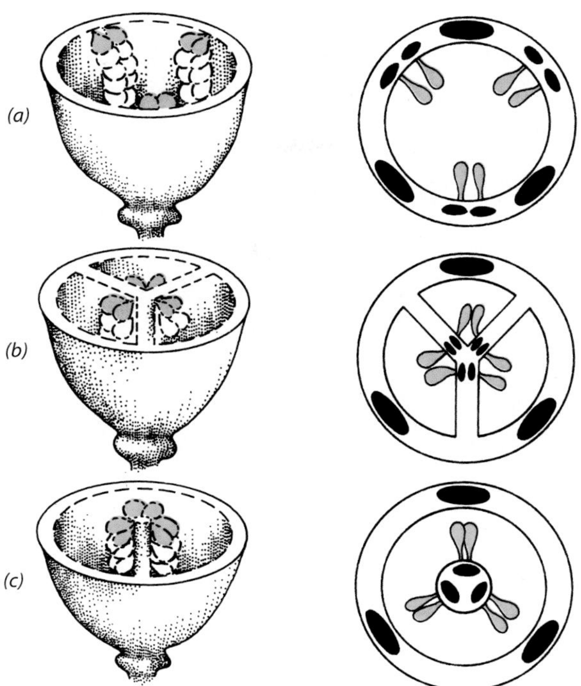

21–9
Types of placentation, with the ovules indicated in color. ***(a)*** *Parietal.* ***(b)*** *Axile.* ***(c)*** *Free central. Basal and apical placentation, a single ovule at the base or apex of a unilocular ovary, are not shown here. The vascular bundles are shown as solid structures in the ovary walls.*

The Ovules Are Attached to the Ovary Wall at the Placenta

The portion of the ovary where the ovules originate and to which they remain attached until maturity is called the **placenta.** The arrangement of the placentae—known as the **placentation**—and consequently of the ovules varies among different groups of flowering plants (Figure 21–9). In some, the placentation is *parietal;* that is, the ovules are borne on the ovary wall or on extensions of it. In other flowers, the ovules are borne on a central column of tissue in a partitioned ovary with as many locules as there are carpels. This is *axile* placentation. In still others, the placentation is *free central,* the ovules being borne on a central column of tissue not connected by partitions with the ovary wall. And finally, in some flowers a single ovule occurs at the base or apex of a unilocular ovary. This is known as *basal* or *apical* placentation. These differences are important in the classification of the flowering plants.

There Are Many Variations in Flower Structure

The majority of flowers include both stamens and carpels, and such flowers are said to be **perfect** (bisexual). If either stamens or carpels are missing, the flower is **imperfect** (unisexual) and, depending on the part that is present, the flower is said to be either **staminate** or **carpellate** (or pistillate) (Figure 21–10). If both staminate and carpellate flowers occur on the same plant, as in maize (see Figure 22–32a, b) and the oaks, the species is said to be **monoecious** (from the Greek words *monos,* "single," and *oikos,* "house"). If staminate and carpellate flowers are found on separate plants, the species is said to be **dioecious** ("two houses"), as in the willows and hemp *(Cannabis sativa).*

21–10
Staminate and carpellate flowers of a tan-bark oak (Lithocarpus densiflora), *of the family* Fagaceae. *Most members of this family, including the true oaks* (Quercus), *are monoecious, meaning that the staminate and carpellate flowers are separate but are borne on the same tree.*

Any one of the floral whorls—sepals, petals, stamens, or carpels—may be lacking in the flowers of certain groups. Flowers with all four floral whorls are called **complete** flowers. If any whorl is lacking, the flower is said to be **incomplete.** Thus an imperfect flower is also incomplete, but not all incomplete flowers are imperfect.

The particular arrangement of the floral parts may be spiral on a more or less elongated receptacle, or similar parts—such as petals—may be attached in a whorl. The parts may be united with other members of the same whorl *(connation)* or with members of other whorls *(adnation)*. An example of adnation is the union of stamens with the corolla (stamens adnate to the corolla), which is fairly common. When the floral parts of the same whorl are not joined, the prefixes *apo-* (meaning "separate") or *poly-* may be used to describe the condition. When the parts are connate, either *syn-* or *sym-* ("together") is used. For example, in an aposepalous or polysepalous calyx, the sepals are not joined; in a synsepalous one, they are.

In addition to this variation in arrangement of flower parts (spiral or whorled), the level of insertion of the sepals, petals, and stamens on the floral axis varies in relation to the ovary or ovaries (Figure 21–11). If the sepals, petals, and stamens are attached to the receptacle below the ovary, as they are in lily, the ovary is said to be **superior** (Figure 21–6). In other flowers the sepals, petals, and stamens apparently are attached near the top of the ovary, which is **inferior.** Intermediate conditions, in which part of the ovary is inferior, also occur in a number of kinds of plants.

In terms of their points of insertion, the perianth and stamens are said to be **hypogynous**—situated on the receptacle beneath the ovary and free from it and from the calyx (Figure 21–6); **epigynous**—arising from the top of the ovary (Figure 21–12); or **perigynous**—with the stamens and petals adnate to the calyx and thus forming a short tube *(hypanthium)* arising from the base of the ovary (Figure 21–13).

Finally, mention should be made of symmetry in flower structure. In some flowers, the members of the different whorls of the flower are made up of members of similar shape that radiate from the center of the flower and are equidistant from each other; they are radially symmetrical. Such flowers—tulips are an example—are said to be **regular,** or actinomorphic (from the Greek root *aktin-*, "ray"). In other flowers, one or more members of at least one whorl are different from other members of the same whorl, and such flowers are generally bilaterally symmetrical. Bilaterally symmetrical flowers—for example, snapdragons—are said to be **irregular,** or zygomorphic (Gk. *zygon*, "yoke" or "pair"). Some regular flowers have irregular color patterns, which give their visitors an image similar to that of a structurally irregular flower.

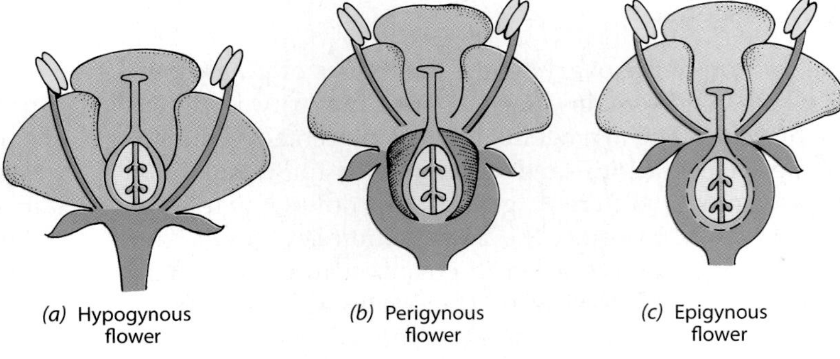

21–11
Types of flowers in common families of eudicots, showing differences in the position of the ovary. ***(a)*** *In* Ranunculaceae, *the buttercup family, for example, the sepals, petals, and stamens are attached below the ovary and there is no fusion; such flowers are said to be hypogynous.* ***(b)*** *In contrast, many* Rosaceae, *such as cherries, have superior ovaries, with the sepals, petals, and stamens fused together to form a cup-shaped extension of the receptacle called the hypanthium. Such flowers are said to be perigynous.* ***(c)*** *The flowers of other plants, for example* Apiaceae, *the parsley family, have inferior ovaries; that is, the sepals, petals, and stamens appear to be attached above the ovaries. Such flowers are said to be epigynous.*

(a)

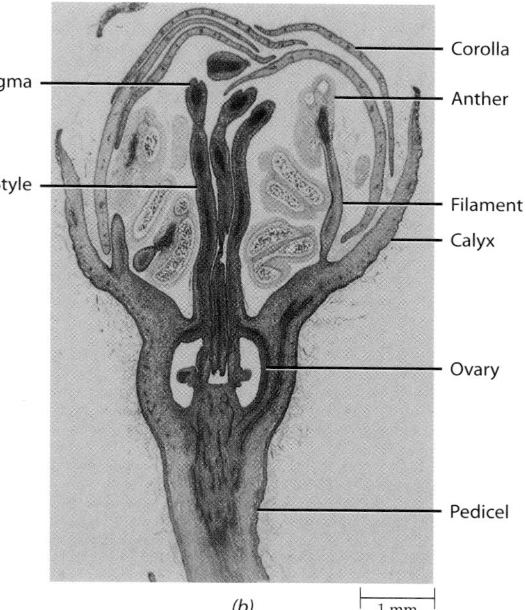

(b)

21–12
Apple (Malus sylvestris) *flowers,* ***(a)*** *and* ***(b)****, exhibit epigyny—their sepals, petals, and stamens apparently arise from the top of the ovary. In* ***(b)*** *the flower is nearly open, but the stamens are not yet erect.*

The Angiosperm Life Cycle

Angiosperm gametophytes are much reduced in size—more so than those of any other heterosporous plants, including the other seed plants (gymnosperms). The mature microgametophyte consists of only three cells. The mature megagametophyte (embryo sac), which is retained for its entire existence within the tissues of the sporophyte, or specifically of the ovule, consists of only seven cells in most kinds of angiosperms. Both antheridia and archegonia are lacking. Pollination is indirect; that is, pollen is deposited on the stigma, after which the pollen tube grows through or on the surface of tissues of the carpel to convey two nonmotile sperm to the female gametophyte. After fertilization, the ovule develops into a seed, which is enclosed in the ovary. At the same time, the ovary (and sometimes additional structures associated with it) develops into a fruit.

(a)

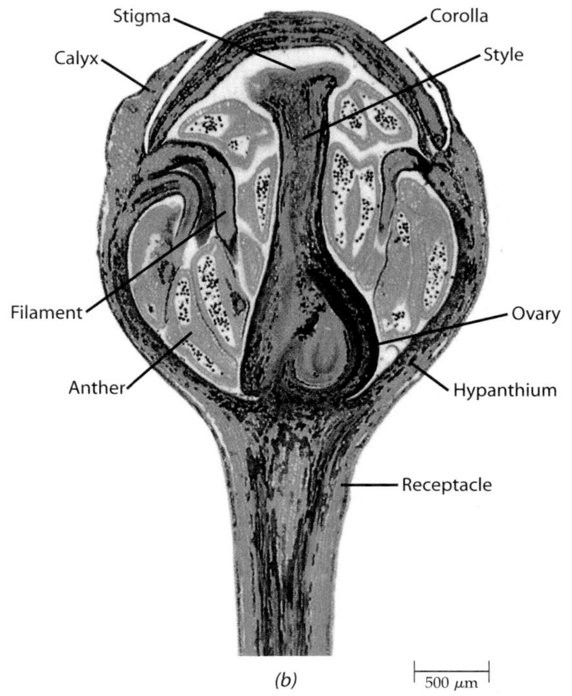

(b)

21–13
Cherry (Prunus) *flowers,* ***(a)*** *and* ***(b)****, exhibit perigyny—their sepals (calyx), petals (corolla), and stamens are attached to a hypanthium. In* ***(b)*** *the filaments of the stamens are bent and crowded in the hypanthium because the flower has not yet opened.*

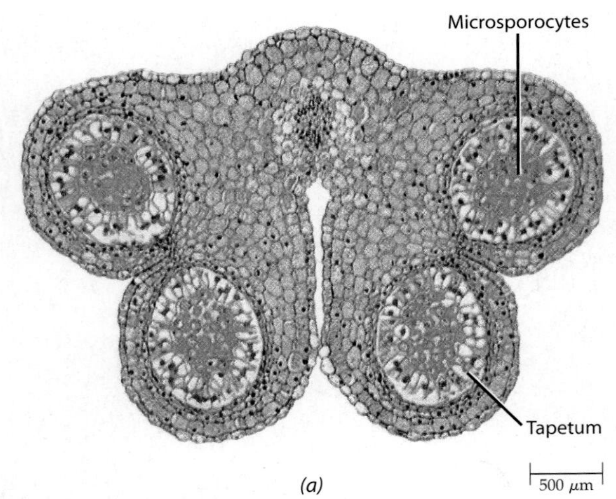

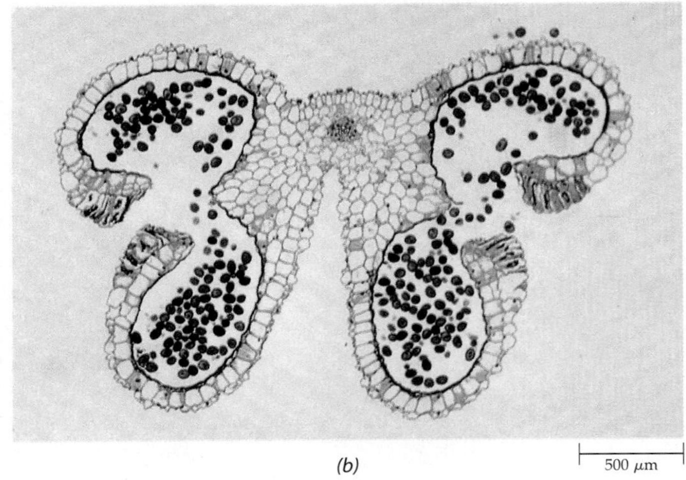

21–14
Two transverse sections of lily (Lilium) *anthers.* ***(a)*** *Immature anther, showing the four pollen sacs containing microsporocytes surrounded by the nutritive tapetum.* ***(b)*** *Mature anther containing pollen grains. The partitions between adjacent pollen sacs break down during dehiscence, as shown here.*

Microsporogenesis and Microgametogenesis Culminate in the Formation of Sperm

Two distinct processes—microsporogenesis and microgametogenesis—lead to the formation of the microgametophyte. Microsporogenesis is the formation of microspores (single-celled pollen grains) within the microsporangia, or pollen sacs, of the anther. Microgametogenesis is the development of the microgametophyte within the pollen grain up to the three-celled stage of development.

When it is first formed, the anther consists of a uniform mass of cells, except for the partly differentiated epidermis. Eventually, four groups of fertile, or **sporogenous,** cells become discernible within the anther. Each group is surrounded by several layers of sterile cells. The sterile cells develop into the wall of the pollen sac, which includes nutritive cells that supply food to the developing microspores. The nutritive cells constitute the **tapetum,** the innermost layer of the pollen sac wall (Figure 21–14a). The sporogenous cells become microsporocytes (pollen mother cells), which divide meiotically. Each diploid microsporocyte gives rise to a tetrad of haploid microspores. Microsporogenesis is completed with formation of the single-celled microspores.

During meiosis, each nuclear division may be followed immediately by cell wall formation, or the four microspore protoplasts may be walled off simultaneously after the second meiotic division. The first condition is common in monocots, the second in eudicots. Subsequently, the major features of the pollen grains are established (Figure 21–15). The pollen grains develop a resistant outer wall, the **exine,** and an inner wall, the **intine.** The exine is composed of the resistant substance **sporopollenin** (page 406), which apparently is derived primarily from the tapetum. This polymer is present in the spore walls of all plants. The intine, which is composed of cellulose and pectin, is laid down by the microspore protoplasts.

Microgametogenesis in angiosperms is uniform and begins when the microspore divides mitotically, forming two cells within the original microspore wall. The division forms a large **tube cell** and a small **generative cell,** which moves to the interior of the pollen grain. This two-celled pollen grain is an immature microgametophyte. In about two-thirds of flowering plant species, the microgametophyte is in this two-celled stage at the time the pollen grains are liberated from the anther (Figure 21–16). In the remainder, the generative nucleus divides prior to the release of the pollen grains, giving rise to two male gametes, or sperm (Figure 21–17). Mature pollen grains may be packed with starch or oils, depending on the group, and are a nutritious source of food for animals.

Pollen grains, like the spores of seedless plants, vary considerably in size and shape, ranging from less than 20 micrometers to more than 250 micrometers in diameter. They also differ in the number and arrangement of the apertures through which the pollen tubes ultimately grow. These apertures can be elongate (furrows), circular (pores), or a combination of the two. Nearly all families, many genera, and a fair number of species of flowering plants can be identified solely by their pollen grains, on the basis of such characteristics as size, number, and type of apertures, and exine sculpturing. In contrast to larger pieces of plants—such as leaves, flowers, and fruits—pollen grains, because of their tough, highly resistant exine, are very widely represented in the fossil

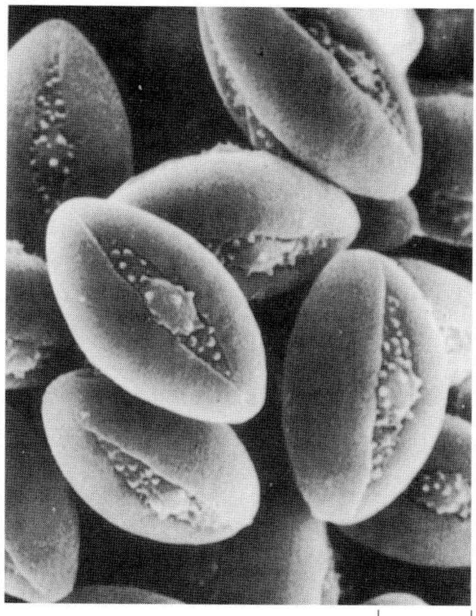

(a)

(b)

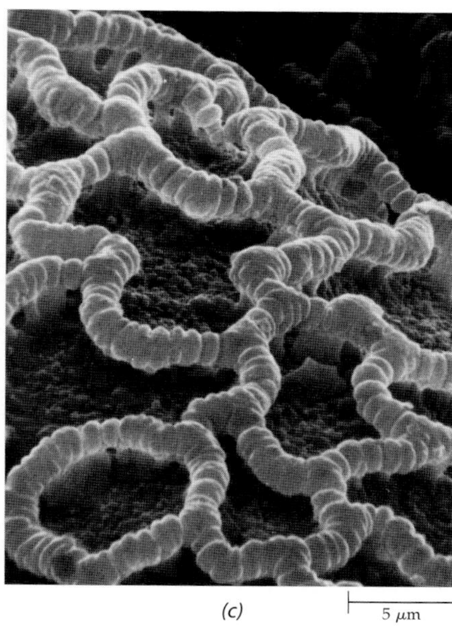

(c)

(d)

21–15

The wall of the pollen grain serves to protect the male gametophyte on its often hazardous journey from the anther to the stigma. The outer layer, or exine, is composed chiefly of a substance known as sporopollenin, which appears to be a polymer composed chiefly of carotenoids. The exine, which is remarkably tough and resistant, is often elaborately sculptured.

The sculpturing of pollen grain walls is distinctly different from one species to another, as revealed in these scanning electron micrographs of pollen grains. ***(a)*** *Pollen grains of the horse chestnut* (Aesculus hippocastanum). *The pore of each grain through which a pollen tube may emerge is visible in the furrow.* ***(b)*** *Pollen grains of a lily* (Lilium longiflorum). ***(c)*** *Detail of the surface of a lily* (L. longiflorum) *pollen grain.* ***(d)*** *Pollen grain of the western ragweed* (Ambrosia psilostachya). *The pollen of ragweed is an important cause of hay fever (see essay on page 508). Spiny pollen grains such as these are common among members of the sunflower family,* Asteraceae, *of which ragweed is a member.*

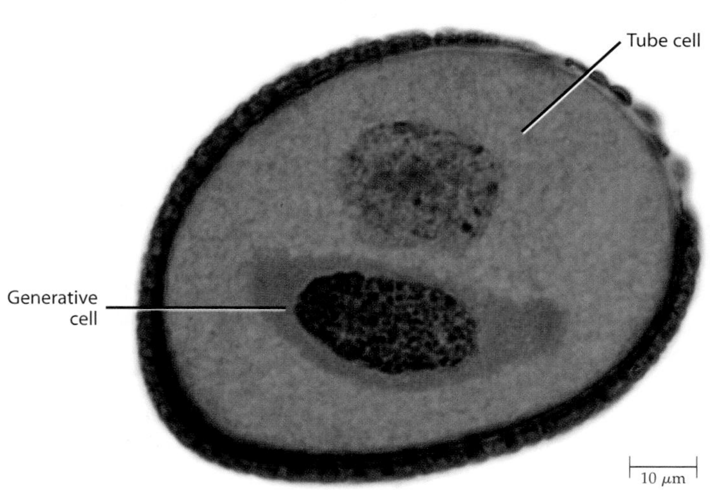

21–16

Mature pollen grain of Lilium, *containing a two-celled male gametophyte. The spindle-shaped generative cell will divide mitotically after the germination of the pollen grain. The larger tube cell, which contains the generative cell, will form the pollen tube. The round structure above the generative cell is the tube cell nucleus.*

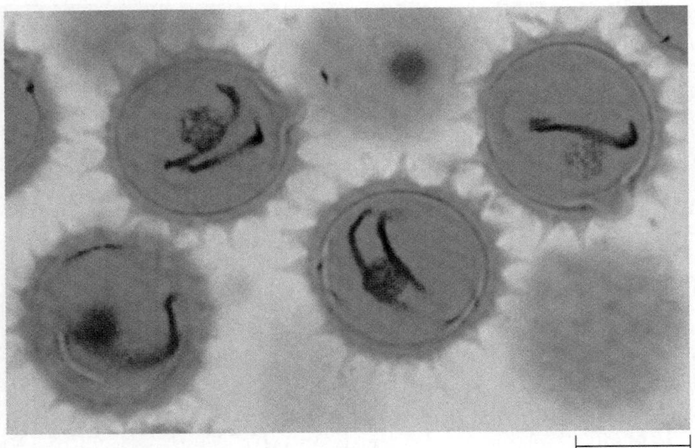

21–17
Mature pollen grains—three-celled male gametophytes—of the telegraph plant, Silphium *(family* Asteraceae*). Prior to pollination, each pollen grain contains two filamentous sperm cells, which are suspended in the cytoplasm of the larger tube cell. The pollen of* Silphium *is shed at the three-celled stage, whereas that of* Lilium, *shown in Figure 21–16, is shed at the two-celled stage.*

record. Studies of fossil pollen can provide valuable insights into the kinds of plants and plant communities, and thus to the nature of the climates, that existed in the past.

In contrast to the spores of most seedless plants, which are also products of meiosis, pollen grains undergo mitosis within the pollen sac in which they are produced. Pollen grains, therefore, have two or three nuclei when shed, whereas most spores have only one. In addition, spores germinate through a characteristic Y-shaped suture on their surface, whereas pollen grains germinate through their apertures. The distinctive structure and arrangement of these apertures allows most kinds of pollen grains to be distinguished from spores morphologically.

Megasporogenesis and Megagametogenesis Culminate in Formation of an Egg and Polar Nuclei

Two distinct processes—megasporogenesis and megagametogenesis—lead to the formation of the megagametophyte. Megasporogenesis is the process of megaspore formation within the nucellus (megasporangium); it takes place inside the ovule. Megagametogenesis is the development of the megaspore into the megagametophyte (female gametophyte, or embryo sac).

The ovule is a relatively complex structure, consisting of a stalk—the **funiculus**—bearing a nucellus enclosed by one or two integuments. Depending on the species, one to many ovules may arise from the placentae, or ovule-bearing regions of the ovary wall (Figure 21–9). Initially, the developing ovule is entirely nucellus, but soon it develops one or two enveloping layers, the integuments, which form a small opening, the **micropyle,** at one end of the ovule (Figure 21–18).

Early in the development of the ovule, a single megasporocyte arises in the nucellus (Figure 21–18a). The diploid megasporocyte divides meiotically (Figure 21–18b) to form four haploid megaspores, which are generally arranged in a linear tetrad. With this, megasporogenesis is completed. In most seed plants, three of the four megaspores disintegrate. The one farthest from the micropyle survives and develops into the megagametophyte.

The functional megaspore soon begins to enlarge at the expense of the nucellus, and the nucleus of the megaspore divides mitotically. Each of the resulting nuclei divides mitotically, followed by yet another mitotic division of the four resultant nuclei. At the end of the third mitotic division, the eight nuclei are arranged in two groups of four, one group near the micropylar end of the megagametophyte and the other at the opposite, or **chalazal,** end (Figure 21–18c). One nucleus from each group migrates into the center of the eight-nucleate cell; these two nuclei are then called **polar nuclei.** The three remaining nuclei at the micropylar end become organized as the **egg apparatus,** consisting of an **egg cell** and two short-lived cellular **synergids.** Cell wall formation also occurs around the three nuclei left at the chalazal end, forming the **antipodals.** The **central cell** contains the two polar nuclei. The eight-nucleate, seven-celled structure is the mature female gametophyte, or **embryo sac.**

The pattern of embryo sac development just described is the most common one. Other patterns occur in about a third of the flowering plants. In fact, one unusual pattern occurs in *Lilium,* which is the genus illustrated in Figure 21–18. In this case, there is no wall formation during megasporogenesis and all four megaspore nuclei participate in the formation of the embryo sac. Three of the nuclei move to the chalazal end of the embryo sac, while the remaining nucleus becomes situated at the micropylar end. This 3 + 1 arrangement of the nuclei represents the *first four-nucleate stage* in development of the embryo sac. What happens next is quite different at the two ends of the embryo sac. At the micropylar end of the sac, the single haploid nucleus undergoes mitosis, yielding two haploid nuclei. At the chalazal end, the mitotic spindles of the three sets of chromosomes unite and mitosis results in two nuclei that are $3n$ (triploid) in chromosome number. As a result of these events, a *second four-nucleate stage* is produced, with two haploid nuclei at the micropylar end of the embryo sac and two triploid nuclei at the chalazal end. Embryo sac development then proceeds in the manner described above for the more frequent kind of embryo sac formation, which has a single four-nucleate stage.

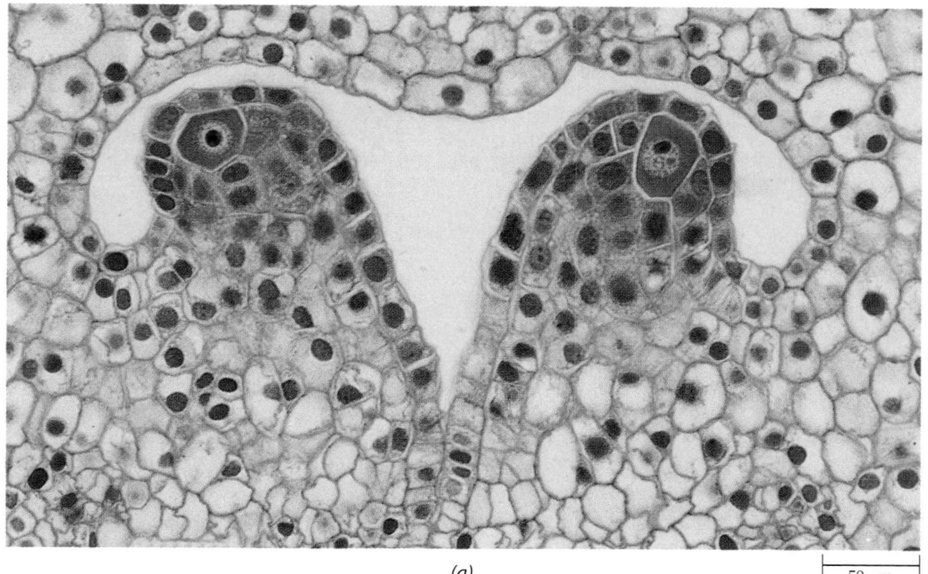

(a)

21–18

Lilium. *Some stages in development of an ovule and embryo sac.* ***(a)*** *Two young ovules, each with a single, large megasporocyte surrounded by the nucellus. Integuments have not begun to develop.* ***(b)*** *Ovule has now developed integuments with a micropyle. The megasporocyte is in the first prophase of meiosis.* ***(c)*** *Ovule with eight-nucleate embryo sac (only six of the nuclei can be seen here, four at the micropylar end and two at the opposite, chalazal end). The polar nuclei have not yet migrated to the center of the sac. The funiculus is the stalk of the ovule.*

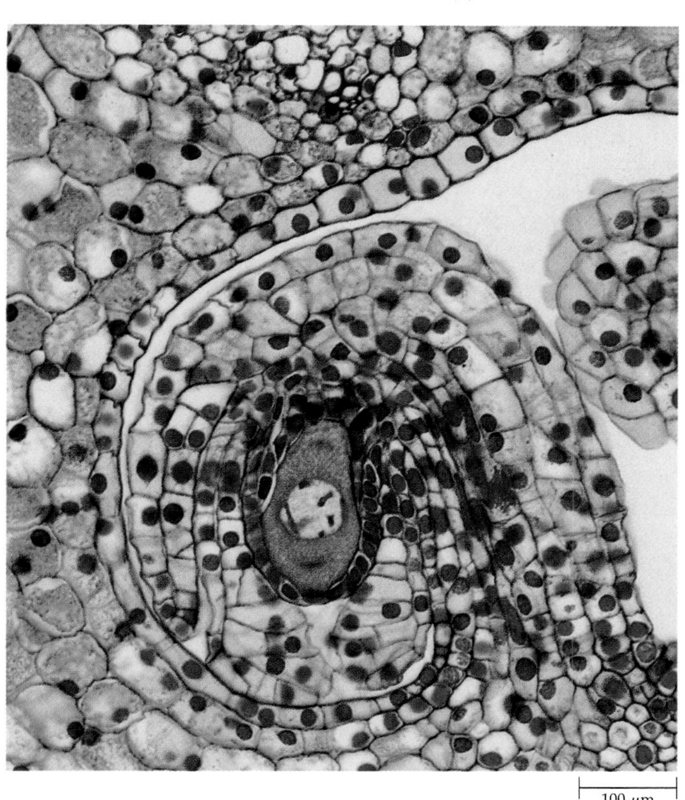

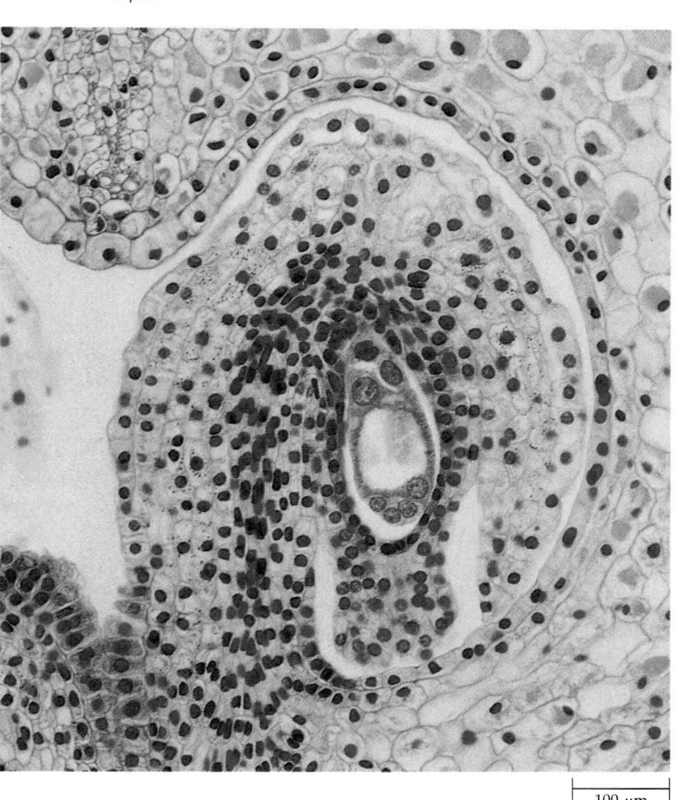

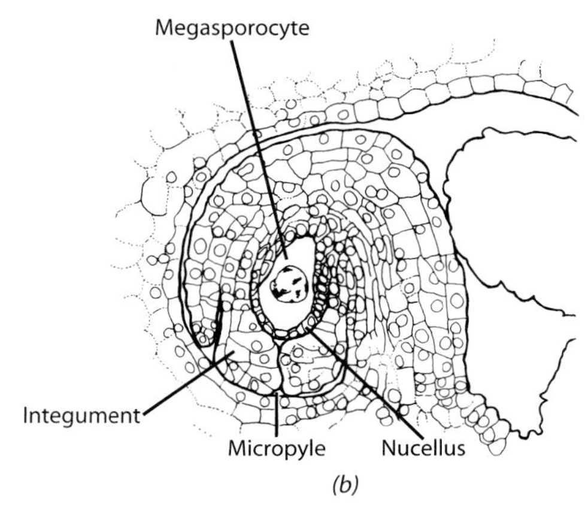

(b)

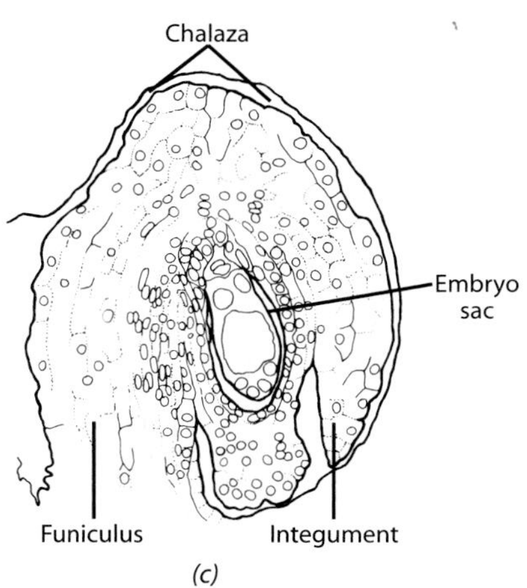

(c)

Hay Fever

In the temperate areas of the Northern Hemisphere, it is estimated that between 10 and 18 percent of all people will suffer from hay fever, which can be highly debilitating, at some point in their lives. Some of the proteins that occur in hollow spaces in pollen grain walls, and which can be released immediately following contact with a moist surface, are generally the culprits. Among these proteins are some that can act as very powerful allergens and antigens, provoking strong reactions by the human immune system. These are probably also involved in genetic self-incompatibility. Proteins may also be released in tiny particles of the tapetum, smaller than pollen grains, which may become airborne as the anther splits open.

Wind-borne pollen, such as that of grasses, birch trees, and ragweed, is particularly important as an agent of hay fever, because it is shed in large amounts directly into the air and is thus more likely to reach susceptible victims than are the larger pollen grains of insect-pollinated plants. The quantity of pollen inhaled seems to be the most important factor in determining whether there will be an allergic response, but surprisingly, some wind-borne pollen that is shed in huge amounts, such as that of maize and pines, rarely causes any difficulty. The scent of certain flowers can also cause reactions that resemble hay fever, perhaps in part by increasing the sensitivity of the nasal membranes.

In temperate North America, the hay fever season can be divided into three parts. In spring, most hay fever is associated with tree pollen from such sources as oaks, elms, poplars, pecans, and birches. In summer, grass pollen predominates, with Bermuda grass, timothy, and orchard grass being important in different regions. By fall, ragweed and grasses different from those that predominate in summer become major irritants. The susceptibility of individual people to different kinds of plants varies greatly.

When new kinds of plants, such as rapeseed, become widely cultivated, they may become important new causes of hay fever. For example, in the arid southwestern United States, the irrigation of large areas for lawns and golf courses, and the introduction of many kinds of weeds to the area, have made hay fever common where it was once virtually unknown.

The incidence of hay fever has been rising rapidly for more than 50 years, even though the pollen count is actually falling in many areas. Part of the increase is related to better detection of the problem, but there is clearly a genuine increase in hay fever. Understanding why this has occurred will require better knowledge of the human immune system.

Pollination and Double Fertilization in Angiosperms Are Unique

With **dehiscence** of the anther—that is, shedding of its contents—the pollen grains are transferred to the stigmas in a variety of ways (see Chapter 22). The process whereby this transfer occurs is called **pollination.** Once in contact with the stigma, the pollen grains take up water from the cells of the stigma surface. Following this hydration, the pollen grain germinates, forming a pollen tube. If the generative cell has not already divided, it soon does, forming the two sperm. The germinated pollen grain, with its tube nucleus and two sperm, is the mature microgametophyte (Figure 21–19).

The stigma and style are modified both structurally and physiologically to facilitate the germination of the pollen grain and growth of the pollen tube. The surface of many stigmas is essentially glandular tissue, known as **stigmatic tissue,** which secretes a sugary solution. The stigmatic tissue is connected with the ovule by **transmitting tissue,** which serves as a path through the style for the growing pollen tubes. Some styles contain open canals lined with transmitting tissue. In such styles, the pollen tubes grow either along or among the cells of the lining. In most angiosperms, however, the styles are solid, with one or more strands of transmitting tissue extending from the stigma to the ovules. The pollen tubes grow either between the cells of the transmitting tissue or within their thick walls, depending on the kind of plant.

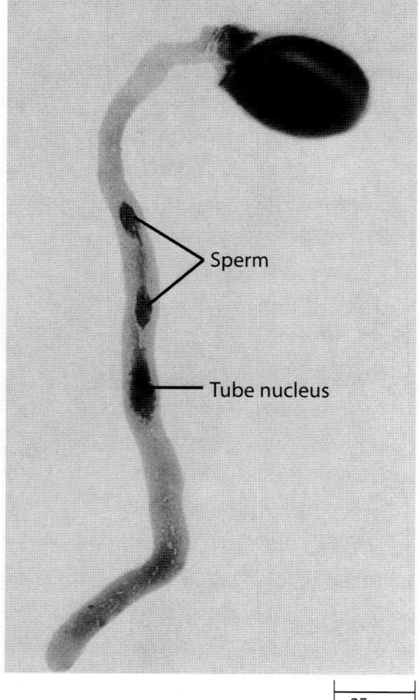

21–19
Microgametophyte, or mature male gametophyte, of Solomon's seal (Polygonatum). *The two sperm and the tube nucleus can be seen in the pollen tube.*

Commonly, the pollen tube enters the ovule through the micropyle and penetrates one of the two synergids next to the egg cell. The synergid begins to degenerate soon after pollination has taken place but before the pollen tube has reached the embryo sac. The two sperm and the tube nucleus are then released into the synergid through the pore that develops near the end of the pollen tube. Ultimately, one sperm nucleus enters the egg cell and the other enters the central cell, where it unites with the two polar nuclei (Figure 21–20). Recall that in most gymnosperms, only one of the two sperm is functional; one unites with the egg and the other degenerates. The involvement of both sperm in this process—the union of one with the egg and the other with the polar nuclei—is called **double fertilization.** It represents an unusual characteristic of the angiosperms, shared only with *Ephedra* and *Gnetum* (phylum *Gnetophyta*). In angiosperms with the most common type of embryo sac formation, the fusion of one of the sperm nuclei with the two polar nuclei, called **triple fusion,** results in a triploid ($3n$) **primary endosperm nucleus.** In *Lilium,* illustrated in Figures 21–18 and 21–20, in which one of the polar nuclei is triploid and the other haploid, triple fusion results in a pentaploid ($5n$) primary endosperm nucleus. Other situations occur in various groups of angiosperms. In any case, the tube nucleus degenerates during the process of double fertilization, and the remaining synergid and antipodals also degenerate near the time of fertilization or early in the course of differentiation of the embryo sac.

Significant progress has recently been made in our understanding of the origin of double fertilization and endosperm, features that have long been thought to be unique to flowering plants. These advances are based on the discovery of rudimentary double fertilization in *Ephedra* and *Gnetum,* two of the three genera of the phylum *Gnetophyta,* the closest living relatives of angiosperms (pages 490–492). The existence of a regular process of double fertilization in *Ephedra* and *Gnetum* as well as in the angiosperms suggests that a rudimentary type of double fertilization already existed in their common ancestor. Initially, however, the second fertilization product was diploid and yielded an extra embryo. Following the divergence of the angiosperm lineage from the common ancestor of this evolutionary line, the extra embryo apparently was modified to become a triploid embryo-nourishing endosperm. Thus endosperm may be regarded as a derivative of a second embryo, modified to perform a new function during the course of evolution.

The Ovule Develops into a Seed and the Ovary Develops into a Fruit

In double fertilization, several processes leading to development of the seed and fruit are initiated: the primary endosperm nucleus divides, forming the **endosperm;** the zygote develops into an embryo; the integuments develop into a seed coat; and the ovary wall and related structures develop into a fruit.

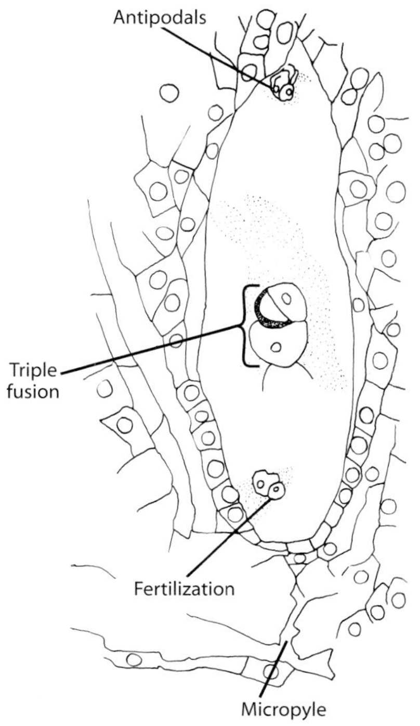

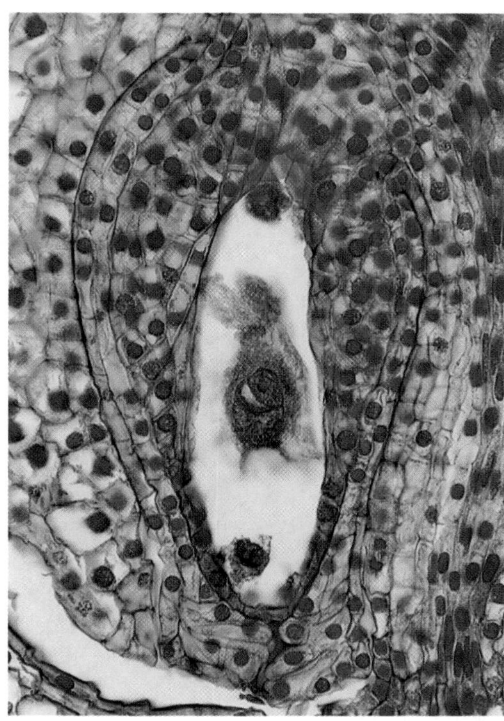

21–20
Double fertilization in Lilium. *Union of sperm and egg nuclei—"true" fertilization—can be seen in the lower half of the micrograph. Triple fusion of the other sperm nucleus and the two polar nuclei has taken place above. The three nuclei known as the antipodals can be seen at the chalazal end, opposite the micropyle, of the embryo sac.*

In contrast to the embryogeny (embryo development) of the majority of gymnosperms, which begins with a free-nuclear stage, embryogeny in angiosperms resembles that of the seedless vascular plants in that the first nuclear division of the zygote is accompanied by cell wall formation. In the early stages of development, the embryos of monocots undergo sequences of cell division similar to those of other angiosperms, and the embryo becomes a spherical ball of cells. It is with the formation of the cotyledons that monocot embryos become distinctive, forming only one cotyledon. Other angiosperm embryos form two cotyledons. The details of angiosperm embryogeny are presented in Chapter 23.

Endosperm formation begins with the mitotic division of the primary endosperm nucleus and usually is initiated prior to the first division of the zygote. In some angiosperms, a variable number of free nuclear divisions precedes cell wall formation; this process is known as nuclear-type endosperm formation. In other species the initial and subsequent mitoses are followed by cytokinesis, which is known as cellular-type endosperm formation. Although endosperm development may occur in a variety of ways, the function of the resulting tissue remains the same: to provide essential food materials for the developing embryo and, in many cases, the young seedling as well. In the seeds of some groups of angiosperms, the nucellus proliferates into a food-storage tissue known as **perisperm.** Some seeds may contain both endosperm and perisperm, as do those of the beet *(Beta)*. In many eudicots and some monocots, however, most or all of these storage tissues are absorbed by the developing embryo before the seed becomes dormant, as in peas or beans. The embryos of such seeds commonly develop fleshy, food-storing cotyledons. The principal food materials stored in seeds are carbohydrates, proteins, and lipids.

Angiosperm seeds differ from those of gymnosperms in the origin of their stored food. In four phyla of gymnosperms, the stored food is provided by the female gametophyte. In angiosperms, it is provided, at least initially, by endosperm, which is neither gametophytic nor sporophytic tissue. Another interesting difference is that the nutritious tissue is built up in angiosperms and *Gnetum* after fertilization occurs, whereas in other seed plants, the nutritious tissue is formed at least partly (in conifers) or entirely (other gymnosperms) *before* fertilization occurs.

Concomitantly with development of the ovule into a seed, the ovary (and sometimes other portions of the flower or inflorescence) develops into a fruit. As this occurs, the ovary wall, or **pericarp,** often thickens and becomes differentiated into distinct layers—the exocarp (outer layer), the mesocarp (middle layer), and the endocarp (inner layer), or exocarp and endocarp only. These layers are generally more conspicuous in fleshy fruits than in dry ones. Fruits are discussed in greater detail in Chapter 22.

An angiosperm life cycle is summarized in Figure 21–23 (pages 512 and 513).

A Variety of Conditions Promote Outcrossing in Angiosperms

Outcrossing is of critical importance for all eukaryotic organisms (pages 242–243). In plants, outcrossing is made possible by cross-pollination between individuals of the same species. In view of this relationship, it should therefore come as no surprise that angiosperms have evolved a variety of mechanisms that promote the transfer of pollen from one individual plant to another.

In dioecious plants, such as willows, the staminate and carpellate flowers occur on separate plants. In such plants, pollen must pass from one individual to another to achieve fertilization, and outcrossing is therefore inevitable. In monoecious plants, such as oaks, birches, and ragweed, there are separate staminate and carpellate flowers, but they occur together on the same individual. Their physical separation increases the chance that the wind, or an animal vector, will move pollen from one individual plant to another before depositing it on a receptive stigma. Maturation of the two kinds of flowers at different times further enhances the chance that outcrossing will occur.

Gymnosperms are also monoecious or dioecious, illustrating the importance of this condition in promoting outcrossing. Among them are *Ginkgo*, cycads, and junipers (conifers of the genus *Juniperus*). In all living gymnosperms, the ovule- and pollen-producing parts occur in different structures, as in different kinds of cones.

Another way in which angiosperms promote outcrossing is through **dichogamy,** a condition in which the stamens and carpels reach maturity at different times—even though they occur together in the same flowers. Those plants in which the stamens of an individual flower mature before the stigmas become receptive are said to be **protandrous** (see Figure 21–24a), and those in which the stigmas become receptive before the stamens mature are said to be **protogynous.** As a result of these conditions, at any given time a flower may be either effectively staminate or effectively carpellate. Dichogamy is widespread among flowering plants (Figure 21–21).

Another strategy by which plants achieve outcrossing is through physical separation of the stamens and stigma(s) within a flower (Figure 21–22). In this way, pollen derived from the stamens of a given flower will rarely reach the stigma(s) of that flower, even if the stamens and stigmas mature at the same time. Such a condition, then, greatly increases the likelihood of pollen being dispersed to another flower.

Genetic self-incompatibility also occurs widely in the angiosperms, with at least some, and often many, members of almost every plant family being genetically

(a)

(b)

21–21
These flowers of fireweed, Chamaenerion angustifolium, *illustrate dichogamy, in which the stamens and carpels on the same flower reach maturity at different times.* ***(a)*** *Flowers in the staminate phase, producing pollen.* ***(b)*** *Flowers in the carpellate phase. The flowers of fireweed open from the bottom of the long inflorescence upward. Soon after a flower opens, the anthers begin to shed pollen. About two days later, the style, which has been reflexed to one side, swings up into the center of the flower, the stigma opens, and the flower becomes carpellate. By this time, the anthers have completely shed their pollen. As a result, the lower flowers on a stem that has been blooming for more than a few days are always carpellate and the upper ones are staminate. The lower flowers have more abundant nectar, and bees fly to the bottom of the inflorescence first when they arrive at a new plant, then move upward. Thus they carry pollen first to the carpellate flowers and ultimately, as they move up, pick up additional pollen from the staminate flowers before moving on.*

self-incompatible. If a plant is genetically self-incompatible, it will be outcrossed even if its stamens and stigma come into contact regularly and mature at the same time. Two basic systems of genetic self-incompatibility occur in different groups of angiosperms.

In the more common system, *gametophytic self-incompatibility,* which is present in such economically important plant families as the grasses *(Poaceae)* and legumes *(Fabaceae)*, the behavior of the pollen is determined by its own (haploid) genotype. If the pollen grain's DNA carries an allele at its incompatibility locus that is identical to the allele that occurs at either of the two corresponding loci in the diploid stigma or style, then the entry of the pollen tube into the stigma, or its passage through the style, is barred. If the pollen grain carries at this locus an allele that is different from either of those present in the stigmatic tissue, then the pollen tube emerging from that grain is accepted.

In *sporophytic self-incompatibility,* present in the mustard family *(Brassicaceae)* and the sunflower family *(Asteraceae)*, the behavior of the pollen is determined by the genetics of the parent plant that produced the pollen, not by the alleles that occur in the pollen grain itself. In other words, the control system is based on the correspondence between the two kinds of diploid tissue. As in gametophytic self-incompatibility, the kind of "match" that occurs at the incompatibility locus determines acceptance or rejection.

21–22
This flower of Easter lily (Lilium longiflorum) *illustrates the wide separation of stigma and anthers that is characteristic of many plants.*

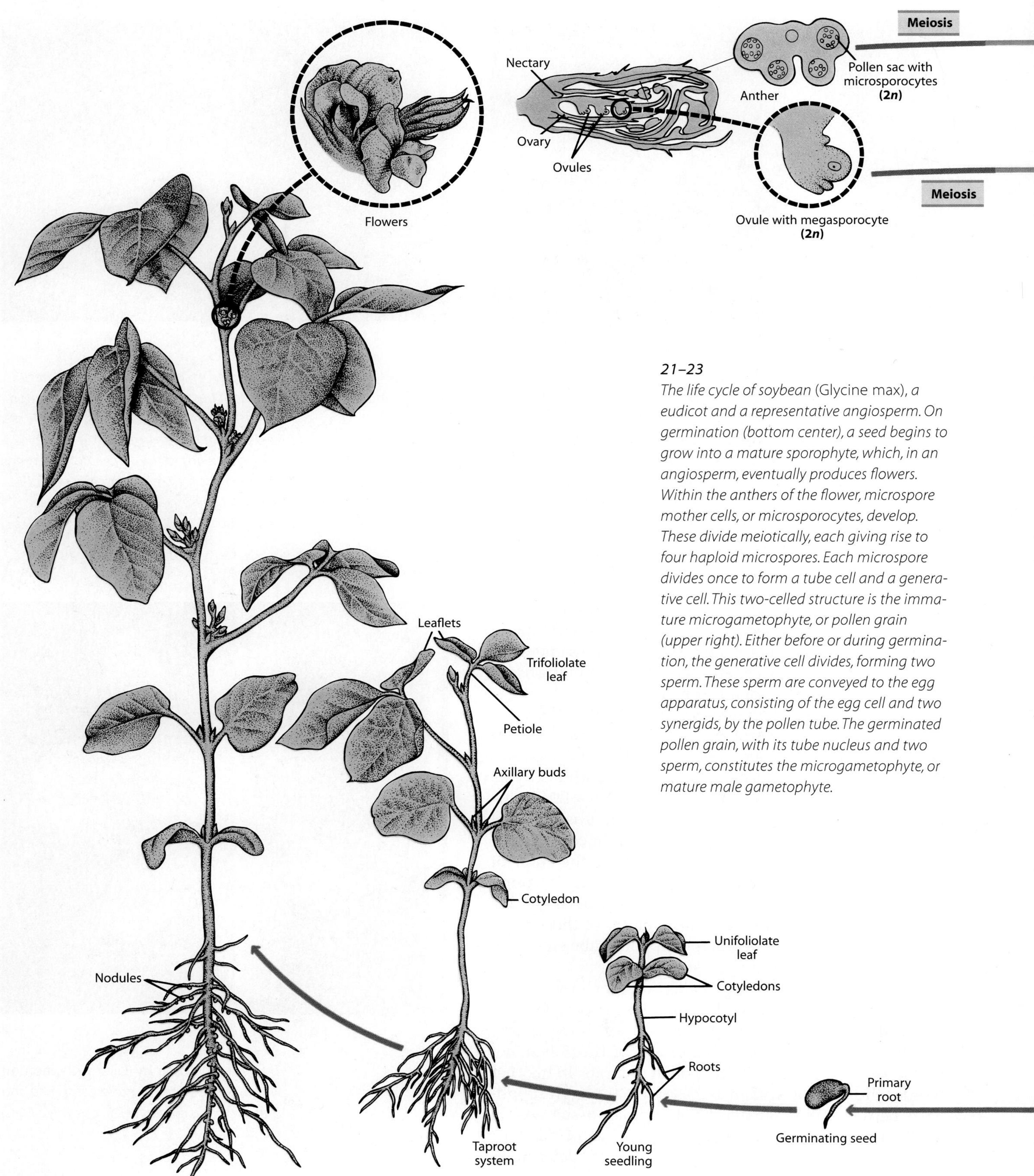

21–23

The life cycle of soybean (Glycine max), *a eudicot and a representative angiosperm. On germination (bottom center), a seed begins to grow into a mature sporophyte, which, in an angiosperm, eventually produces flowers. Within the anthers of the flower, microspore mother cells, or microsporocytes, develop. These divide meiotically, each giving rise to four haploid microspores. Each microspore divides once to form a tube cell and a generative cell. This two-celled structure is the immature microgametophyte, or pollen grain (upper right). Either before or during germination, the generative cell divides, forming two sperm. These sperm are conveyed to the egg apparatus, consisting of the egg cell and two synergids, by the pollen tube. The germinated pollen grain, with its tube nucleus and two sperm, constitutes the microgametophyte, or mature male gametophyte.*

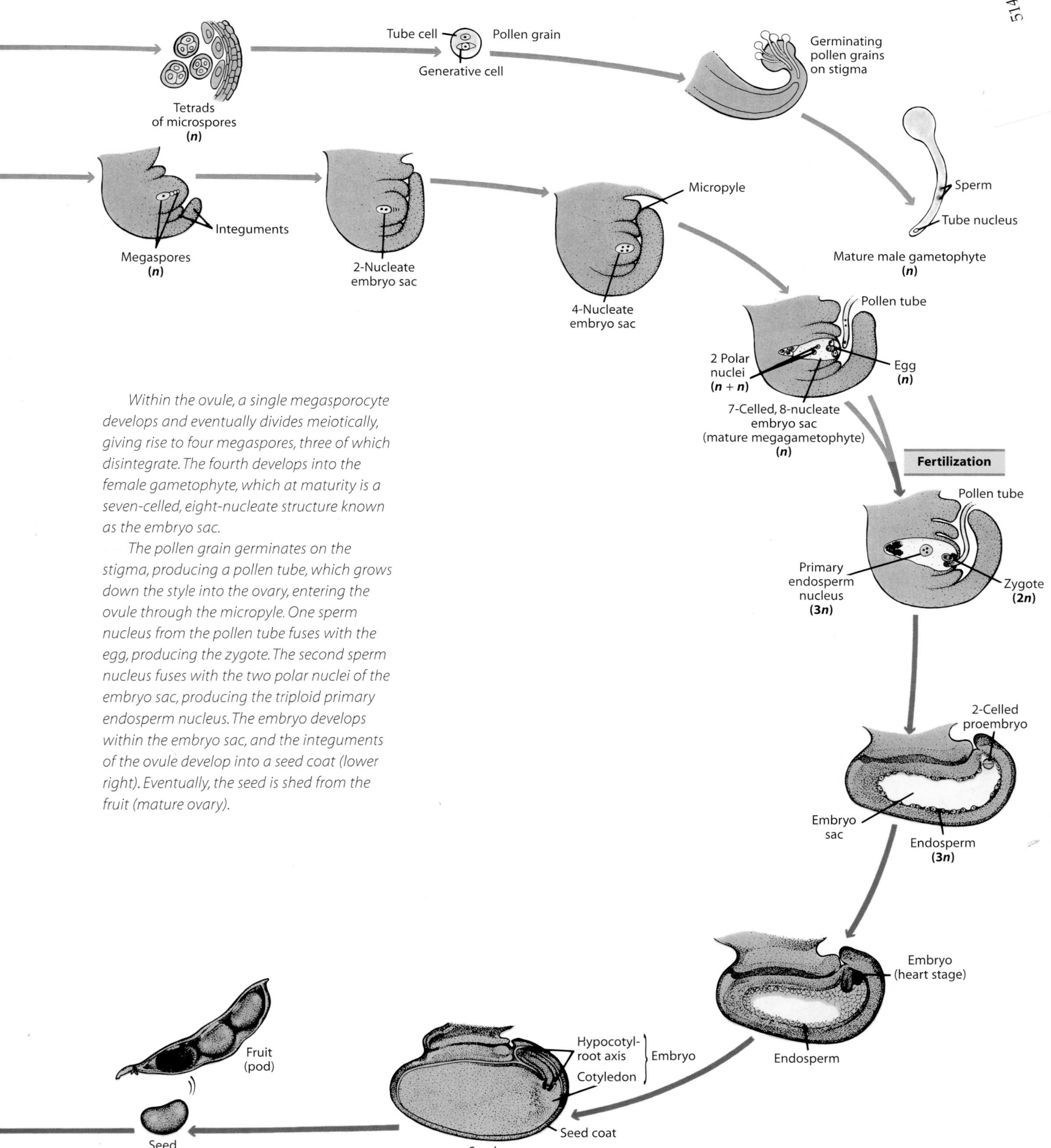

Within the ovule, a single megasporocyte develops and eventually divides meiotically, giving rise to four megaspores, three of which disintegrate. The fourth develops into the female gametophyte, which at maturity is a seven-celled, eight-nucleate structure known as the embryo sac.

The pollen grain germinates on the stigma, producing a pollen tube, which grows down the style into the ovary, entering the ovule through the micropyle. One sperm nucleus from the pollen tube fuses with the egg, producing the zygote. The second sperm nucleus fuses with the two polar nuclei of the embryo sac, producing the triploid primary endosperm nucleus. The embryo develops within the embryo sac, and the integuments of the ovule develop into a seed coat (lower right). Eventually, the seed is shed from the fruit (mature ovary).

Many Angiosperms Reproduce by Self-Pollination

In Chapter 22, we shall examine many of the specific plant–animal relationships that result in the accurate transfer of pollen between flowers. In most seed plants other than angiosperms, pollen is dispersed by the wind, as it is in a number of angiosperms. In submerged angiosperms, pollen is dispersed either through or on the surface of the water. All of these methods lead to outcrossing.

Many angiosperms, however, have adopted self-pollination as a regular mode of reproduction, despite the advantages of outcrossing. In temperate regions, for example, more than half of all species of flowering plants are self-pollinated. Self-pollinating plants often have smaller and less conspicuous flowers than outcrossing ones (Figure 21–24)—but then, the former have no need to attract animal visitors. In some self-pollinating plants, pollination occurs in the bud, after which the bud may or may not open. Often the bud simply falls off, leaving a ripening ovary behind. In others, pollination occurs only after the buds open—although pollen may also be dispersed by animals visiting such flowers.

The large numbers of angiosperms that are regularly self-pollinated clearly indicate that self-pollination is advantageous under certain circumstances. Populations of self-pollinating plants normally have higher proportions of genetically similar individuals than populations in which outcrossing predominates. Depending on their genotype, many or all of the self-pollinated individuals in a given population may be well suited to some specific habitat, such as the disturbed, open places where weeds, which are often self-pollinated, occur widely. A second, rather obvious advantage of self-pollination is the lack of dependency on animals or other vectors to achieve pollination. Whatever happens, short of physical damage or extreme drought, the self-pollinated plants will produce seed. This second advantage explains why self-pollinated plants are often relatively well represented in places where flower-visiting animals may be rare, for example on high mountains or in the Arctic.

Summary

Angiosperms, or Flowering Plants, Constitute the Phylum *Anthophyta*

The two largest classes of this phylum are the *Monocotyledones* (65,000 species) and the *Eudicotyledones* (165,000 species). Flowering plants differ from other seed plants in various distinctive characteristics, such as the presence of endosperm in their seeds; the fact that their ovules are enclosed within megasporophylls, the carpels; and their distinctive reproductive structure, the flower, which is characterized by the presence of carpels and distinctive microsporophylls, or stamens.

(a)

(b)

21–24
These two annual herbs are species of the genus Clarkia, *of the evening primrose family,* Onagraceae. *They grow in the foothills of California and are closely related, as shown by careful comparison of their nucleic acid components.* ***(a)*** Clarkia cylindrica *is an outcrossing species, although it is genetically self-compatible. It is protandrous, the eight anthers opening and shedding their pollen approximately two days before the stigma is receptive. In addition, as you can see, the stigma is widely separated from the anthers within the flower.* ***(b)*** *Sharply contrasting is* C. heterandra, *a self-compatible species in which the flowers are much smaller and paler. This species has only four stamens, and their anthers shed pollen directly onto the stigma.*

The Flower Is a Determinate Shoot That Bears Sporophylls

Individual flowers may have up to four whorls of appendages. From the outside in, the whorls are the sepals (collectively, the calyx); the petals (collectively, the corolla); the stamens (collectively, the androecium); and the carpels (collectively, the gynoecium). The sepals and petals are sterile, with the sepals often green and protective, covering the flower in bud, and the petals often colored and serving a function in attracting pollinators. Individual stamens are generally divided into a stalk, or filament, and an anther, containing four pollen sacs (two pairs). Carpels are usually differentiated into a swollen lower part, the ovary, and a slender upper part, the style, terminating in the receptive stigma. One or more of these whorls may be missing in the flowers of individual kinds of plants.

In Angiosperms, Pollination Is Followed by Double Fertilization

Pollination in angiosperms takes place by the transfer of pollen from anther to stigma. The male gametes, or sperm, of angiosperms are transmitted by means of a pollen grain, which is an immature male gametophyte. At the time of dispersal, such a gametophyte may contain either two or three cells. Initially, there is a tube cell and a generative cell, the latter dividing before or after dispersal to give rise to two sperm. The female gametophyte of an angiosperm is called an embryo sac. In many angiosperms, embryo sacs have eight nuclei at maturity, one of which is the egg (the number of cells varies in different groups). Both sperm function during angiosperm fertilization (double fertilization). One unites with the egg, producing a diploid zygote. The other unites with two polar nuclei, giving rise to the primary endosperm nucleus, which is usually triploid ($3n$). That nucleus divides, producing a unique kind of nutritive tissue, the endosperm, which may be absorbed by the embryo before the seed is mature or may persist in the mature seed. The angiosperms share double fertilization with the gnetophytes *Ephedra* and *Gnetum*, but in these gymnosperms, the process results in the formation of two embryos.

An Ovule Develops into a Seed, and an Ovary Develops into a Fruit

The ovaries (sometimes with some associated floral parts) develop into fruits, which enclose the seeds. Along with the flower from which it is derived, the fruit is a defining characteristic of the angiosperms.

A Variety of Conditions Promote Outcrossing in Angiosperms

Outcrossing in angiosperms is promoted by dioecism, in which staminate and carpellate flowers occur on different individual plants, and by monoecism, in which separate staminate and carpellate flowers occur on single individual plants. Outcrossing is also promoted by dichogamy, in which the stamens and carpels of a given flower mature at different times, or simply by the physical separation of these organs within an individual flower. Genetic self-incompatibility, which occurs widely in angiosperms, makes self-fertilization impossible even if the stamens and stigmas mature at the same time and come into contact with each other.

Many Angiosperms Reproduce by Self-Pollination

Self-pollination, resulting in the production of genetically uniform individuals, is characteristic of many angiosperms, including more than half of the species that occur in temperate regions. It clearly is favored over outcrossing under certain ecological conditions.

Selected Key Terms

anther p. 501
antipodals p. 506
carpel p. 499
central cell p. 506
chalazal p. 506
dichogamy p. 510
dioecious p. 501
egg apparatus p. 506
embryo sac p. 506
endosperm p. 509
epigynous p. 502
exine p. 504
filament p. 501
funiculus p. 506
generative cell p. 504
hypogynous p. 502
inflorescence p. 499
intine p. 504
irregular, or **bilaterally symmetrical** p. 502
locule p. 501
monoecious p. 501
pedicel p. 499
peduncle p. 499
perigynous p. 502
perisperm p. 510
petals p. 501
placenta p. 501
placentation p. 501
polar nuclei p. 506
pollen sac p. 501
primary endosperm nucleus p. 509
receptacle p. 499
regular, or **radially symmetrical** p. 502
sepals p. 501
stamens p. 501
stigmatic tissue p. 508
synergids p. 506
tapetum p. 504
transmitting tissue p. 508
tube cell p. 504

Questions

1. Distinguish among or between the following: calyx/corolla/perianth; stigma/style/ovary; complete/incomplete; perfect/imperfect; androecium/gynoecium.
2. Diagram and label as completely as possible a complete hypogynous flower, in which none of the floral parts are joined.
3. An imperfect flower is automatically incomplete, but not all incomplete flowers are imperfect. Explain.
4. Diagram and label completely a mature male gametophyte (germinated pollen grain) and a mature female gametophyte (embryo sac) of an angiosperm. Compare these gametophytes with their counterparts in pine.
5. Double fertilization followed by the formation of endosperm is unique to angiosperms. How does double fertilization in the gnetophytes *Ephedra* and *Gnetum* differ from that in angiosperms?
6. In some angiosperms, pollen production in the anthers occurs prior to or after the full development of the carpel of the same flower (dichogamy). What are the consequences of this shift in time frames?

Chapter 22

Evolution of the Angiosperms

22–1
The world's oldest flower. Imprint of Bevhalstia pebja *in Early Cretaceous rocks (about 130 million years old) in Surrey, England. This plant was about 25 centimeters high and probably lived in water. It combines a fernlike structure with small flowerlike reproductive structures. The actual width of the flower is about 7 mm.*

OVERVIEW

Reflect upon the evolutionary road we have traveled in the past few chapters of this book. One important trend concerns the mechanisms by which the sperm reach the eggs. In bryophytes and seedless vascular plants, sperm must swim through water to reach the eggs, whereas in gymnosperms the immature male gametophytes, or pollen grains, are largely carried by the wind to the nearby vicinity of the female gametophytes. There they germinate, producing pollen tubes and sperm. Hence, in the gymnosperms, water is no longer needed for the sperm to reach the eggs, and in two groups, the conifers and gnetophytes, the pollen tubes are true sperm conveyors, conveying nonmotile sperm more or less directly to the eggs. Nevertheless, gymnosperms are largely dependent upon wind for pollination. Angiosperms, by contrast, have evolved a set of features that attract a wide variety of pollinators—most notably insects—that assure a high degree of cross-pollination and evolutionary development.

In this chapter, we shall explore the angiosperms and learn why they have come to dominate the world's vegetation. It is a marvelous and unequaled success story. Let us begin with consideration of the origin of the angiosperms.

CHECKPOINTS

By the time you finish reading this chapter, you should be able to answer the following questions:

1. What are the current hypotheses on the origin of angiosperms, and what is the presumed relationship among the monocots, the eudicots, the woody magnoliids, and the paleoherbs?
2. What did the perianth (sepals and petals), stamens, and carpels of the earliest angiosperms look like? What are the four principal evolutionary trends among flowers?
3. What feature has evolved in angiosperms that has allowed them directed mobility in seeking a mate?
4. How do beetle-, bee-, moth-, bird-, and bat-pollinated flowers differ from one another?
5. What are some of the adaptations of fruits, in relation to their dispersal agents?
6. How, apparently, have secondary metabolites influenced angiosperm evolution?

In a letter to a friend, Charles Darwin once referred to the apparently sudden appearance of the angiosperms in the fossil record as "an abominable mystery." In the early fossil-bearing strata, about 400 million years old, one finds simple vascular plants, such as rhyniophytes and trimerophytes. Then there is a Devonian and Carboniferous proliferation of ferns, lycophytes, sphenophytes, and progymnosperms, which were dominant until about 300 million years ago. The early seed plants first appeared in the Late Devonian period and led to the gymnosperm-dominated Mesozoic floras. Finally, early in the Cretaceous period, and at least 130 million years ago (Figure 22–1), angiosperms appear in the fossil record, gradually achieving worldwide dominance in the vegetation by about 90 million years ago. By about 75 million years ago, many modern families and some modern genera of this phylum already existed (Figure 22–2).

Despite their relatively late appearance in the fossil record, why did the angiosperms rise to world dominance and then continue to diversify to such a spectacular extent? In this chapter, we shall attempt to answer this question, centering our discussion on the relationships of the angiosperms, their origin and diversification, the evolution of the flower, the evolution of fruits, and the role of certain chemical substances in angiosperm evolution. All five topics will illustrate some of the reasons for the evolutionary success of the group.

Relationships of the Angiosperms

Since the time of Darwin, scientists have attempted to understand the ancestry of the angiosperms. One approach has been to search for their possible ancestors in the fossil record. In this effort, particular emphasis has been placed on assessing the ease with which the ovule-bearing structures of various gymnosperms could be transformed into a carpel. Recently, phylogenetic analyses (cladistics) based on fossil, morphological, and molecular data have revitalized attempts to define the major natural groups of seed plants and to understand their interrelationships.

The most striking result from recent phylogenetic analyses is the support that they have provided for earlier ideas that the *Bennettitales* (page 472) and gnetophytes (page 490) are the seed plants most closely related to angiosperms. The term "anthophytes" (not to be confused with the use of the term *Anthophyta* here to refer to angiosperms) has been proposed to refer collectively to the *Bennettitales*, gnetophytes, and angiosperms. It emphasizes the shared possession of flowerlike reproductive structures by these three groups of seed plants. Two contrasting hypotheses have been proposed for the phylogenetic relationships among anthophytes. One hypothesis proposes that the gnetophytes are monophyletic, and the derived similarities that *Gnetum* (Figure 20–41) and *Welwitschia* (Figure 20–43) share with angiosperms are interpreted as examples of convergent evolution (Figure 22–3a). The second hypothesis considers the gnetophytes to be paraphyletic, with *Gnetum* and *Welwitschia* as sister groups to angiosperms (Figure 22–3b). The latter hypothesis interprets the derived similarities of *Gnetum*, *Welwitschia*, and angiosperms to be homologous.

It is significant that the *Bennettitales* and gnetophytes first appear in the fossil record in the Triassic period, about 225 million years ago. This seems to place some constraints on the possible earliest date of the appear-

(a)

(b)

22–2
Fossils of angiosperms from Late Cretaceous deposits (about 70 million years ago) in Wyoming. ***(a)*** *Leaf of a fan palm,* Sabalites montana, *an extinct species distantly related to the palmettos of the southeastern United States.* ***(b)*** *Leaf of an extinct member of the family* Platanaceae, *commonly known as the sycamores.*

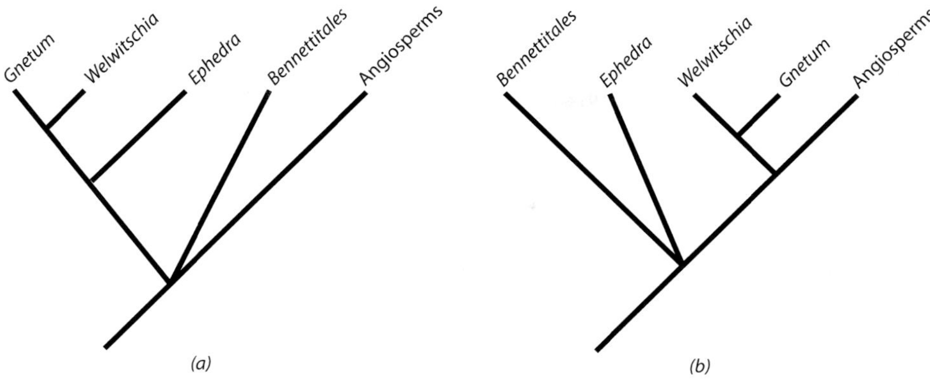

22–3
Two proposed hypotheses for the phylogenetic relationships among the anthophytes. ***(a)*** *One hypothesis states that the* Gnetales *are monophyletic, and the derived similarities of* Gnetum, Welwitschia, *and angiosperms are interpreted as convergent evolution. This hypothesis is consistent with current molecular data of both* rbcL *(the chloroplast-encoded gene sequence for the large subunit of Rubisco) and ribosomal RNA.* ***(b)*** *According to another hypothesis, the* Gnetales *are paraphyletic, with* Gnetum *and* Welwitschia *the sister group to the angiosperms. This hypothesis interprets the derived similarities of* Gnetum, Welwitschia, *and angiosperms as being homologous.*

ance of the angiosperms—that is, on the possibility that they could have arisen any earlier.

Origin and Diversification of the Angiosperms

The unique characteristics of the angiosperms include flowers, closed carpels, double fertilization leading to endosperm formation, a three-nucleate microgametophyte and an eight-nucleate megagametophyte, stamens with two pairs of pollen sacs, and the presence of sieve tubes and companion cells in the phloem (see Chapter 24). These similarities clearly indicate that the members of this phylum were derived from a single common ancestor. This common ancestor of the angiosperms ultimately would have been derived from a seed plant that lacked flowers, closed carpels, and fruits. The earliest known, clearly identifiable fossils of angiosperms are flowers and pollen grains up to 130 million years old, from the Early Cretaceous period (Figure 22–1). There are intriguing suggestions that much older fossils—up to 200 million years old—may have had some, but perhaps not all, of the characteristic features of angiosperms. Currently the interpretation of these fossils is enigmatic, and it appears most likely that the phylum did in fact originate in the Early Cretaceous (or perhaps uppermost Jurassic) period.

What were the earliest angiosperms like? Like gymnosperms, they clearly had pollen with a single aperture, as found in monocots and a few other groups of angiosperms, as well as cycads, *Ginkgo*, and other groups. This feature can therefore be considered an ancestral one that has been retained in the course of evolution. Angiosperms that produce single-aperture pollen, therefore, cannot be shown, on the basis of this character, to have had a common ancestor distinct from other angiosperms.

The Magnoliids Are Ancestral to Both Monocots and Eudicots

In Chapter 21, we discussed the two largest classes of angiosperms, the monocots and the eudicots, which between them comprise 97 percent of the members of the phylum. The monocots clearly had a common ancestor, as indicated by their single cotyledon and a number of other features. The same is true of the eudicots, which have a characteristic derived feature, their triaperturate pollen (pollen with three slits or pores, and also pollen types derived from the triaperturate group). The remaining 3 percent of living angiosperms, the **magnoliids,** include those with the most primitive features. These ancestors of both monocots and eudicots have traditionally been grouped with the eudicots as "dicots," but this is as illogical as grouping them with monocots. The evolutionary relationships of magnoliids are not well understood.

Although they have traditionally been regarded as dicots, all magnoliids, like the monocots, have pollen with a single aperture or some modification of this type. One of their characteristic features is the possession of oil cells with ethereal (ether-containing) oils, the basis of the characteristic scents of nutmeg, pepper, and laurel leaves. One group of magnoliids consists of plants known informally as the **woody magnoliids.** These plants have large, robust, bisexual flowers with many free parts, arranged spirally on an elongate axis, as in *Magnolia* (Figure 22–4). There are fewer than 20 families of this group with living members. Among them, the most familiar are the magnolia family *(Magnoliaceae)*, laurel family *(Lauraceae)*, and spicebush family *(Calycanthaceae)*.

One of the numerous fossil woody magnoliids is *Archaeanthus linnenbergeri*, which is probably very closely related to living *Magnolia* and the tulip tree *(Liriodendron)* (Figure 22–5). The stout flowers of *Archaeanthus* had an elongate axis with 100 to 130 spirally arranged carpels,

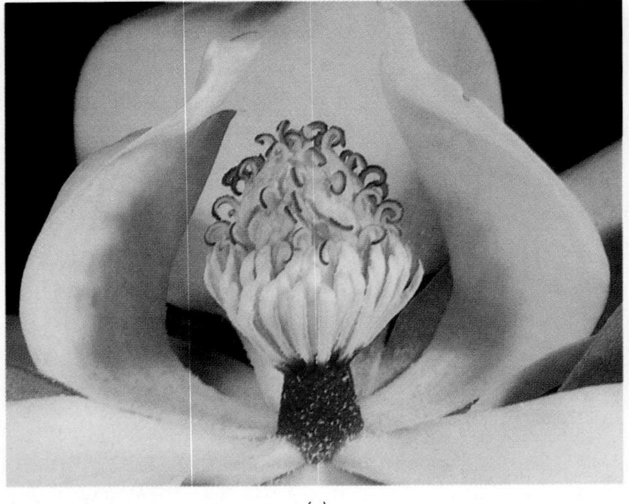

(a)

(b)

(c)

22–4
Flowers and fruits of the southern magnolia (Magnolia grandiflora), *a woody magnoliid. The cone-shaped receptacle bears numerous spirally arranged carpels from which curved styles emerge. Below the styles in* **(a)** *and* **(b)** *are the cream-white stamens.* **(a)** *The anthers have not yet shed their pollen, whereas the stigmas are receptive. In other words, the species is protogynous.* **(b)** *The floral axis of a second-day flower, showing stigmas that are no longer receptive and stamens that are shedding pollen.* **(c)** *Fruit, showing carpels and bright red seeds, each protruding on a slender stalk.*

(a)

(b)

(c)

(d)

22–5
Archaeanthus linnenbergeri, *an extinct angiosperm that generally resembles living magnolias.* **(a)** *Fossil reproductive axis.* **(b)** *Fossil leaf.* **(c)** *Reconstruction of a flowering branch.* **(d)** *Reconstruction of a fruiting branch. The reconstructions are based on the studies of David Dilcher of the University of Florida and Peter Crane of the Field Museum of Natural History in Chicago and were drawn by Megan Rohn. Careful studies by Dilcher, Crane, and others are revealing a great deal about the nature of early angiosperms and their flowers.*

22–6
Aristolochia grandiflora, *the Dutchman's pipe, belongs to the birthwort family* (Aristolochiaceae). Aristolochiaceae *are one of a few angiosperm families of living paleoherbs.*

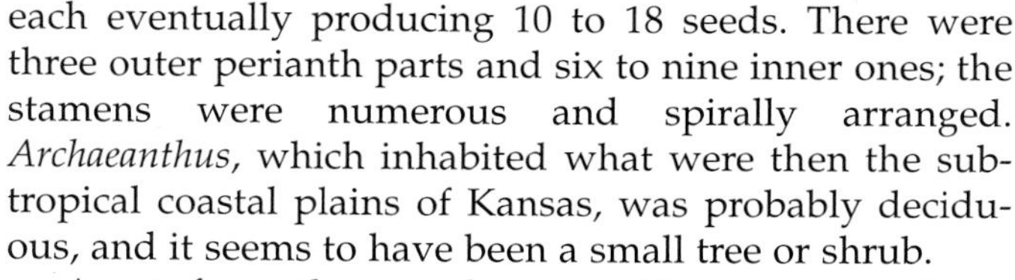

each eventually producing 10 to 18 seeds. There were three outer perianth parts and six to nine inner ones; the stamens were numerous and spirally arranged. *Archaeanthus*, which inhabited what were then the subtropical coastal plains of Kansas, was probably deciduous, and it seems to have been a small tree or shrub.

Apart from the woody magnoliids, the remaining magnoliids constitute a very diverse assemblage collectively called the **paleoherbs.** They are predominantly herbs with small flowers and often with relatively few flower parts; their flowers are often unisexual. Many paleoherbs, unlike the great majority of woody magnoliids, have carpels that are fused together. Among the living paleoherbs are families such as *Piperaceae*, the pepper family; *Aristolochiaceae*, the birthwort family (Figure 22–6); and *Nymphaeaceae*, the water lily family. A number of very early fossils, such as that shown in Figure 22–7, are also paleoherbs, although it is important to remember that this group basically includes all angiosperms that are not monocots, eudicots, or woody magnoliids. It is, therefore, a kind of evolutionary "grab bag" of angiosperms with primitive features.

The monocots arose from ancestors that may have been similar to the birthwort family, *Aristolochiaceae*, or to the pepper family, *Piperaceae*, judged from biochemical and morphological evidence. In any case, their ancestors were clearly not woody magnoliids. Although the fossil record of early monocots is rather poor, the first members of this class seem to have arisen as one of many diverse lines of paleoherbs, probably more than 120 million years ago. The major evolutionary lines of monocots were in existence at least 70 million years ago, before the close of the Cretaceous period.

22–7
A very early angiosperm, approximately 120 million years old, from fossil beds near Melbourne, Australia. This fossil, reported in 1990, probably represents an herbaceous plant that may have grown in marshy places. It has undivided leaves, and its flowers are subtended by bracts and are carpellate. This fossil paleoherb resembles members of the existing pepper family (Piperaceae).

The nature of the ancestors of the eudicots is less clear but is under active investigation. The triaperturate pollen that marks the members of this class appears in the fossil record in the Lower Cretaceous period, about 127 million years ago or perhaps a little earlier. The group became quite diverse over the subsequent 30 to 40 million years, by the mid-Cretaceous. The features of the flower of *Magnolia* have often been used to demonstrate what the features of a primitive angiosperm flower might have been. It seems clear now, however, that the *Magnolia* flower, which is large, with numerous, spirally arranged parts, is in fact an early specialization. The common ancestor of the angiosperms, and perhaps the eudicots as well, seems likely to have had relatively small, simple, perhaps green, and rather unattractive flowers, with their petals and sepals not clearly differentiated. These features are characteristic of some Early Cretaceous fossils, and some living plants as well.

Angiosperms Spread Rapidly throughout the World

The appearance and rapid diversification of the eudicots and the monocots led to the increasing domination of angiosperms throughout the world during the 35 million years of the upper Cretaceous period (100–65 million years ago). By approximately 90 million years ago, several existing orders and families of angiosperms had appeared (Figures 22–2 and 22–8), and the flowering plants had achieved dominance throughout the Northern Hemisphere. During the subsequent 10 million years, they achieved dominance in the Southern Hemisphere as well.

The early angiosperms possessed many adaptive traits that made them particularly resistant to drought and cold. Among these were tough leaves, commonly reduced in size; vessel elements (efficient water-conducting cells); and a tough, resistant seed coat that provided protection against the young embryo drying out. These features are not found in all of the flowering plants, nor are they restricted to the angiosperms, but they certainly have played major roles in the success of the phylum. The early evolution of the deciduous habit (the seasonal loss of leaves), which allows woody plants to lose their leaves and to become relatively inactive physiologically during periods of drought, extreme heat, and cold, may also have contributed to the evolutionary success of the group—especially during the past 50 million years when world climates have been undergoing active change.

A number of other factors seem to have been important in the early and continuing success of the angiosperms, and we shall discuss some of them in more detail in this chapter. The evolution of sieve-tube elements presumably made possible the more efficient conduction of sugars throughout the plant in the phloem, just as vessel elements are more efficient than tracheids in the xylem. Perhaps of even greater importance, the precise systems of pollination and specialized mechanisms of seed dispersal that became characteristic of the more advanced flowering plants allowed them to exist as widely scattered individuals in many different kinds of habitats. The enormous chemical diversity of the angiosperms, which includes the many kinds of defenses against diseases and herbivores, has likewise been of

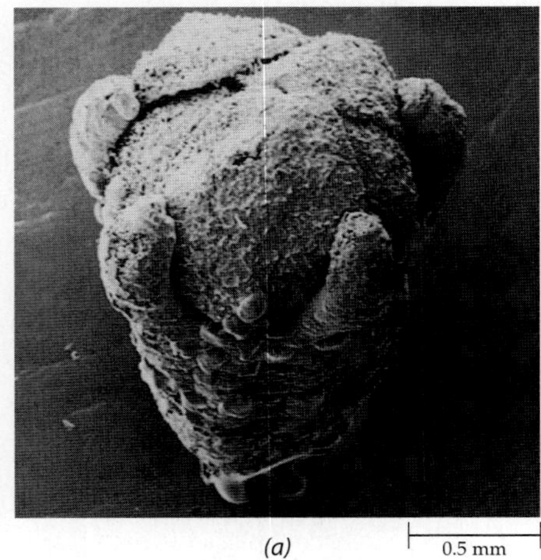

(a)

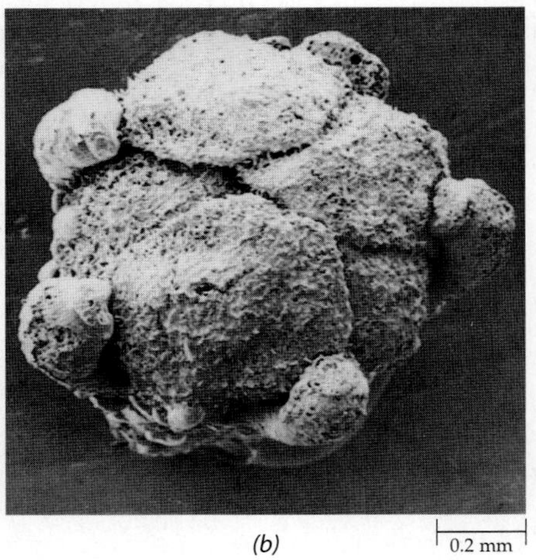

(b)

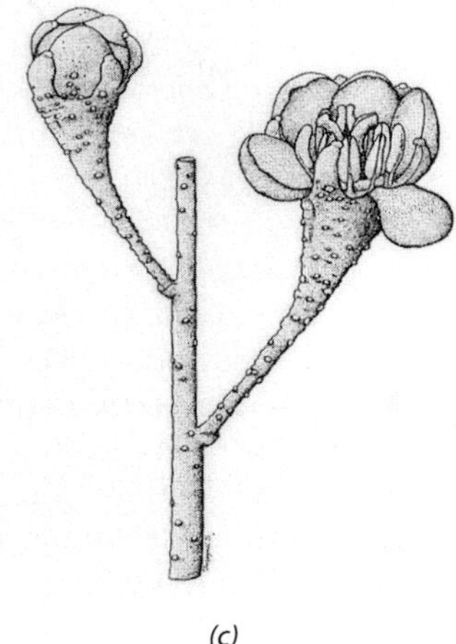
(c)

22–8
Silvianthemum suecicum, *a relatively specialized angiosperm that occurred in the Late Cretaceous period in southern Sweden, about 80 million years ago. Exquisitely preserved as fossil charcoal from ancient forest fires, these flowers are minute, perfect, and radially symmetrical, with five free sepals and five free petals, three fused carpels, disk-shaped nectaries, and minute, numerous seeds. Like two other genera that occurred with it,* Silvianthemum *is broadly related to the saxifrage family,* Saxifragaceae, *and especially to some of its woody relatives.* ***(a), (b)*** *Two views of the fossil flowers.* ***(c)*** *Reconstruction of a flower and bud. Else Marie Friis and her coworkers are making outstanding contributions to our knowledge of fossil angiosperms.*

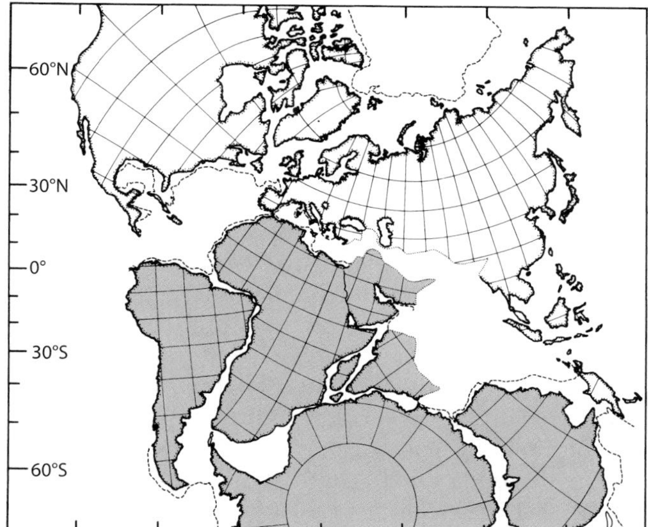

22–9
Relationship between the Earth's land areas at the time of the first appearance of the angiosperms in the fossil record (130 million years ago). In the middle of the Cretaceous period, about 100 million years ago, South America was still directly connected with Africa, Madagascar, and India and, via Antarctica, with Australia. These combined land masses, indicated here in color, formed the supercontinent of Gondwanaland.

great importance, as we shall see. These and other features bear directly on the fact that, compared with other plants, angiosperms are ecologically resilient and reproduce rapidly and efficiently.

About 130 million years ago, when the first definite angiosperm fossils appear in the fossil record, Africa and South America were directly connected with each other and also with Antarctica, India, and Australia in a great southern supercontinent called **Gondwanaland** (Figure 22–9). Africa and South America began to separate at about this time, forming the southern Atlantic Ocean, but they did not move completely apart in the tropical regions until about 90 million years ago. India began to move northward at about the same time, colliding with Asia starting about 65 million years ago and thrusting up the Himalayas in the process. Australia began to separate from Antarctica about 55 million years ago, but their separation did not become complete until more recently (Figure 22–10).

Within the central regions of West Gondwanaland, formed by what are now the continents of South America and Africa, habitats were arid to subhumid. Under these conditions, angiosperms and other kinds of organisms would have been challenged to produce new forms. With the final separation of these two continents, at about the time the angiosperms became abundant in the fossil record worldwide, the world climate changed greatly. This was especially true in these equatorial regions, which became milder, with fewer extremes of temperature and humidity. Magnoliid angiosperms—those with the most archaic features—have survived most abundantly and are best represented today in regions with relatively uniform, warm climates, such as those of southeast Asia and the South Pacific.

22–10
This forest of southern beech (Nothofagus menziesii), *growing in Fiordland, South Island, New Zealand, is a relict from the cool temperate forest that extended from southern South America across Antarctica into Australia and New Zealand from about 80 million years ago until perhaps 30 million years ago. Increasingly large gaps have developed between the land masses throughout that entire period of time up to the present.*

Evolution of the Flower

What were the flowers of the earliest angiosperms like? Of course, we do not know this from direct observation, but we can deduce their nature from what we know of certain living plants and from the fossil record. In general, the flowers of these plants were diverse both in the numbers of floral parts and in the arrangement of these parts. Most modern families of angiosperms tend to have more fixed floral patterns that do not vary much in their basic structural features within a particular family. We shall discuss the derivation of these patterns over the course of evolution in the following sections, which deal with the different whorls of the flower from the outside in, moving from the perianth inward to the androecium and the carpel.

The Parts of the Flower Provide Clues to Angiosperm Evolution

The Perianth of Early Angiosperms Did Not Have Distinct Sepals and Petals In the earliest angiosperms, the perianth, if present, was never sharply divided into calyx and corolla. Either the sepals and petals were identical, or there was a gradual transition in appearance between these whorls, as in modern magnolias and water lilies. In some angiosperms, including the water lilies, petals appear to have been derived from sepals. In other words, the petals can be viewed as modified leaves that have become specialized for attracting pollinators. In most angiosperms, however, petals were probably derived originally from stamens that lost their sporangia—becoming "sterilized"—and then were specially modified for their new role. Most petals, like stamens, are supplied by just one vascular strand. In contrast, sepals are normally supplied by the same number of vascular strands as are the leaves of the same plant (often three or more). Within sepals and petals alike, the vascular strands usually branch so that the number of strands that enter them cannot be determined from the number of veins in the main body of the structures.

Petal fusion has occurred a number of times during the evolution of the angiosperms, resulting in the familiar tubular corolla that is characteristic of many families (Figure 22–11c). When a tubular corolla is present, the stamens often fuse with it and appear to arise from it. In a number of evolutionarily advanced families, the sepals are also fused into a tube.

The Stamens of Early Angiosperms Were Diverse in Structure and Function The stamens of some families of woody magnoliids are broad, colored, and often scented, playing an obvious role in attracting floral visitors. In other archaic angiosperms, the stamens, although relatively small and often greenish, may also be fleshy. Many living angiosperms, in contrast, have stamens

(a)

(b)

22–11
Examples of specialized flowers. ***(a)*** *Wintergreen,* Chimaphila umbellata. *The sepals (not visible) and petals are reduced to five each, the stamens to ten, and the five carpels are fused into a compound gynoecium with a single stigma.* ***(b)*** *Lotus,* Nelumbo lutea. *The undifferentiated sepals and the numerous petals and stamens are spirally arranged; the carpels are embedded in a flat-topped receptacle.* ***(c)*** *Chaparral honeysuckle,* Lonicera hispidula. *The ovary is inferior and has two or three locules; the sepals are reduced to small teeth at its apex. The petals are fused into a corolla tube in the zygomorphic (bilaterally symmetrical) flower, and the five stamens, which protrude from the tube, are attached to its inner wall. The style is longer than the stamens, and the stigma is elevated above*

with generally thin filaments and thick, terminal anthers (for example, see Figures 21–6 and 21–22). In general, the stamens of monocots and eudicots seem to be much less diverse in structure and function than the stamens of magnoliid angiosperms.

In some specialized flowers the stamens are fused together. Their fused filaments may then form columnar structures, as in the members of the pea, melon, mallow (Figure 22–11d), and sunflower families, or they may be fused with the corolla, as in the phlox, snapdragon, and mint families.

In certain plant families, some of the stamens have become secondarily sterile: they have lost their sporangia and become transformed into specialized structures, such as nectaries. Nectaries are glands that secrete **nectar,** a sugary fluid that attracts pollinators and provides food for them. Most nectaries are not modified stamens but arose instead in other ways. During the course of evolution of the angiosperm flower, the sterilization of stamens, as noted above, also played an important role in the evolution of petals.

The Carpels of Many Early Angiosperms Were Unspecialized A number of woody magnoliids have generalized and sometimes leaflike carpels, with no specialized areas for the entrapment of pollen grains comparable to the specialized stigmas of most living angiosperms. The carpels of many woody magnoliids and other plants that retain archaic features are free from one another, instead of being fused together as in most contemporary angiosperms. In a few living woody magnoliids, the carpels are incompletely closed, although pollination is always indirect—the pollen does not contact the ovules directly. In the vast majority of living angiosperms, the carpels are closed (a condition that gave rise to the name of the phylum) and sharply differentiated into stigmas, styles, and ovaries. There is much variation in the arrangement of the ovules among contemporary groups of angiosperms, which often have fewer ovules than do some of the more generalized and archaic families of the phylum.

Four Evolutionary Trends among Flowers Are Evident

Insect pollination quite probably triggered the early evolution of angiosperms, both through the possibilities it provided for isolating small populations and with indirect pollination fostering competition between many pollen grains as they grew through the stigmatic tissue. The flowers of the earliest members of the phylum probably were bisexual, but unisexual flowers appeared very early in many different families. The undifferentiated perianth of early angiosperms soon gave rise to distinct petals and sepals. As the angiosperms continued to evolve, and relationships with specialized pollinators became more tightly linked, the number and arrangement

(c)

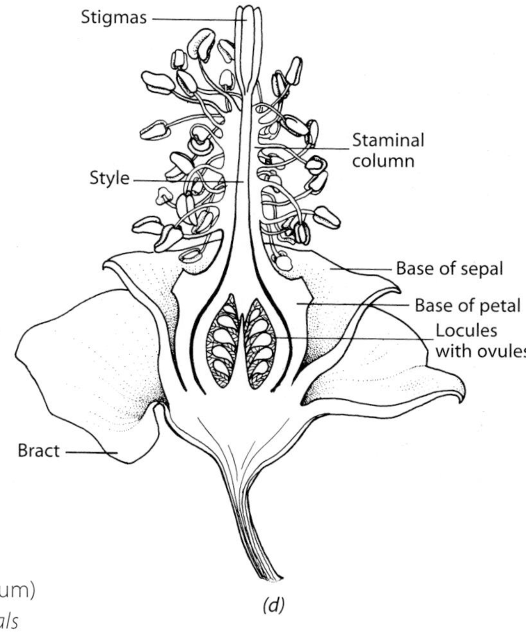

(d)

them. A pollinator visiting this flower would contact the stigma first, so that if it were carrying pollen from another flower, it would deposit that pollen on the stigma before reaching the anthers. Fruits of this species are shown in Figure 22–45c. ***(d)*** *A diagram of a longitudinal section of a cotton* (Gossypium) *flower, of the mallow family, with the sepals and petals removed and showing the column of stamens fused around the style.*

An Ambiguous Aquatic Plant

One of the great surprises revealed by contemporary studies of plant evolution concerns the relationship of the hornwort (Ceratophyllum), a unique aquatic angiosperm (a). Less than a decade ago, the perceptive studies of University of Connecticut botanist Donald Les, who contributed the material on which this essay is based, added a new dimension to the mystery of angiosperm evolution by offering a new hypothesis about the nature of Ceratophyllum. This small, odd genus, classified in a family of its own, had long thwarted the efforts of taxonomists to determine its relationships.

Ceratophyllum is a highly specialized plant that lacks roots entirely (even in the embryo), is drastically reduced in morphology and anatomy, and lives and reproduces entirely underwater (even pollination takes place below the water surface). Yet a number of its floral and vegetative features represent conditions that botanists regard as primitive for angiosperms. For many years, plant systematists have viewed Ceratophyllum as a highly modified descendant of water lilies, mainly because, like them, it is aquatic. However, a more critical evaluation of features indicated to Les that Ceratophyllum was not closely related either to the water lilies (Nymphaeaceae) or water lotus (Nelumbonaceae). In fact, Les's investigations of the morphological and anatomical features of Ceratophyllum indicated that it could be a living fossil descended from one of the earliest of angiosperm evolutionary lines, and that it might not be closely related to any known group of living angiosperms.

At first glance it is not easy to envision how a reduced, herbaceous, aquatic plant with simple flowers and water pollination could possibly represent an ancient angiosperm. But what is easy to overlook is the possibility that Ceratophyllum is so ancient that these unusual features have accumulated over a very long time and may represent poorly what its ultimate ancestor looked like. Many of its features, such as unisexuality, lack of vessels, lack of a perianth, lack of a pollen exine, a single integument (seed coat) layer, and branching pollen tubes, might have been expected in an ancient angiosperm, but they could also be interpreted as adaptations to an aquatic habit. Ceratophyllum might have become aquatic long ago—even though we usually think of aquatic plants as recently derived from terrestrial ancestors. Actually, fossilized Ceratophyllum-like fruits from the Lower Cretaceous period, some 115 million years ago, are among the oldest known angiosperm reproductive remains.

Recently, the base sequences in the DNA of Ceratophyllum have been studied. In a surprise to many critics of Les's original hypothesis, several analyses of chloroplast-encoded rbcL gene sequences (coding for the large subunit of ribulose bisphosphate carboxylase/oxygenase, or Rubisco) have indicated that Ceratophyllum is likely to be one of the most basal groups of all angiosperms (b). In the light of these additional findings, it is also interesting that Ceratophyllum possesses the simple flowers that botanists believe were characteristic of early paleoherbs. Although the data are not simple to analyze, largely because of the unique nature of Ceratophyllum, other molecular studies are tending to confirm the rbcL results. Ceratophyllum clearly illustrates the point that modern living descendants of ancient evolutionary lineages almost always possess a mosaic of both archaic and derived features and, taken as a whole, they may no longer resemble the ancestral plants from which they were derived.

(a)

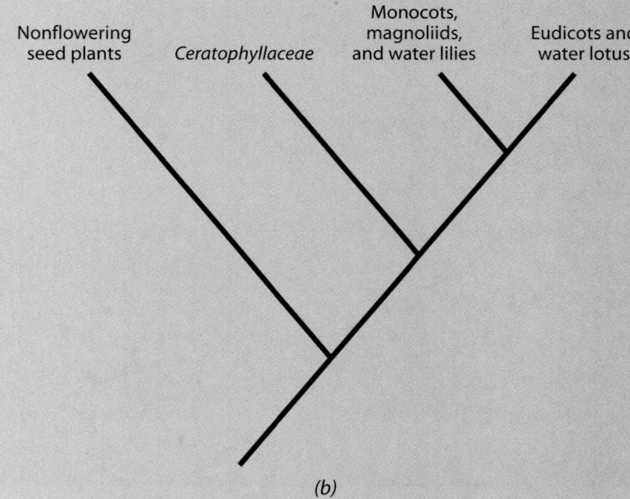

(b)

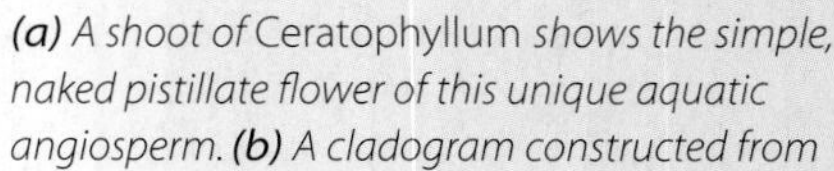

(a) A shoot of Ceratophyllum shows the simple, naked pistillate flower of this unique aquatic angiosperm. (b) A cladogram constructed from rbcL gene sequence data shows Ceratophyllum at the base of the flowering plants.

of floral patterns became more stereotyped. The following four trends are evident (Figure 22–11):

1. From flowers with few to many parts that are indefinite in number, flowers have evolved toward having few parts that are definite in number.
2. The number of floral whorls has been reduced from four in early flowers to three, two, or sometimes one in more advanced ones. The floral axis has become shortened so that the original spiral arrangement of parts is no longer evident. The floral parts often have become fused.
3. The ovary has become inferior rather than superior in position, and the perianth has become differentiated into a distinct calyx and corolla.
4. The radial symmetry (regularity), or actinomorphy, of early flowers has given way to bilateral symmetry (irregularity), or zygomorphy, in more advanced ones.

The *Asteraceae* and *Orchidaceae* Are Examples of Specialized Families

Among the most evolutionarily specialized of flowers are those of the family *Asteraceae* (*Compositae*), which are eudicots, and those of the family *Orchidaceae*, which are monocots. In number of species, these are the two largest families of angiosperms.

The Flowers of the *Asteraceae* Are Closely Bunched Together into a Head In the *Asteraceae* (the composites), the epigynous flowers are relatively small and closely bunched together into a head. Each of the tiny flowers has an inferior ovary composed of two fused carpels with a single ovule in one locule (Figure 22–12).

In composite flowers, the stamens are reduced to five in number and are usually fused to one another (coalescent) and to the corolla (adnate). The petals, also five in number, are fused to one another and to the ovary, and

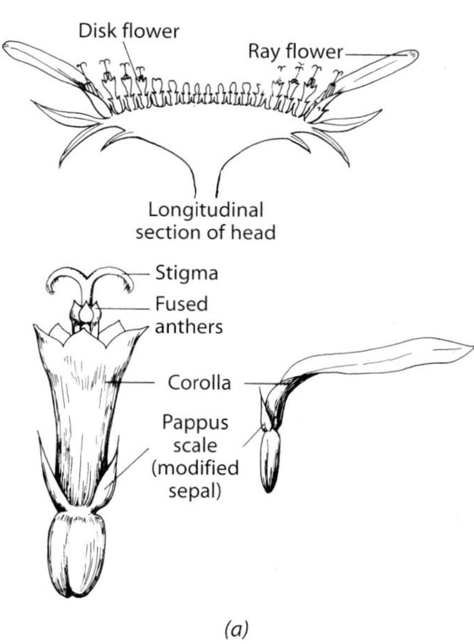

(a)

(b)

(c)

(d)

22–12
Composites (family Asteraceae*).* ***(a)*** *A diagram showing the organization of the head of a member of this family. The disk and ray flowers are subordinated to the overall display of the head, which functions as a single large flower in attracting pollinators.* ***(b)*** *Thistle,* Cirsium pastoris. *Members of the thistle tribe have only disk flowers. This particular species of thistle has bright red flowers and is regularly visited by hummingbirds, which are its primary agents of pollination.* ***(c)*** Agoseris, *a wild relative of the dandelion,* Taraxacum. *In the inflorescences of the chicory tribe (the group of composites to which dandelions and their relatives belong), there are no disk flowers. The marginal ray flowers, however, are often enlarged.* ***(d)*** *Sunflower,* Helianthus annuus.

the sepals are absent or reduced to a series of bristles or scales known as the **pappus.** The pappus often serves as an aid to dispersal by wind, as it does in the familiar dandelion, a member of the *Asteraceae* (Figure 22–12c; see also Figures 22–41 and 22–42). In other members of this family, such as beggar-ticks *(Bidens),* the pappus may be barbed, serving to attach the fruit to a passing animal and thus to enhance its chances of being dispersed from place to place. In many members of the family *Asteraceae,* each head includes two types of flower: (1) disk flowers, which make up the central portion of the aggregate, and (2) ray flowers, which are arranged on the outer periphery. The ray flowers are often carpellate, but sometimes they are completely sterile. In some members of the *Asteraceae,* such as sunflowers, daisies, and black-eyed Susans, the fused bilaterally symmetrical (zygomorphic) corolla of each ray flower forms a long strap-shaped "petal."

In general, the composite head has the appearance of a single large flower. Unlike many single flowers, however, the head matures over a period of days, with the individual flowers opening serially in an inward-moving spiral pattern. As a consequence, the ovules in a given head may be fertilized by several different pollen donors. The success of this plan as an evolutionary strategy is attested to by the great abundance of the members of the *Asteraceae* and also their great diversity, which, with about 22,000 species, makes them the second largest family of flowering plants.

Orchidaceae Is the Largest Angiosperm Family Another successful flower plan is that of the orchids *(Orchidaceae),* which, unlike the composites, are monocots. There are probably at least 24,000 species of orchids, making them the largest family of flowering plants. In contrast to the composites, however, individual species of orchids are rarely very abundant. Most species of orchids are tropical, and only about 140 are native to the United States and Canada, for example. In the orchids, the three carpels are fused and, as in the composites, the ovary is inferior (Figure 22–13). Unlike the composites, however, each orchid ovary contains many thousands of minute ovules. Consequently, each pollination event may result in the production of a huge number of seeds. Usually only one stamen is present (in one subfamily, the lady-slipper orchids, there are two), and this stamen is characteristically fused with the style and stigma into a single complex structure—the **column**. The entire contents of an anther are held together and dispersed as a unit—the **pollinium** (see Figure 22–25b). The three petals of orchids are modified so that the two lateral ones form wings and the third forms a cuplike lip that is often very

(a)

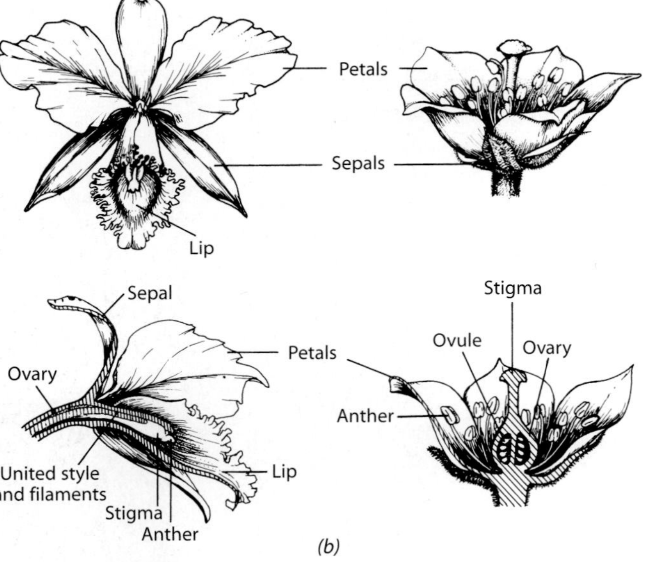

(b)

22–13
Orchids (family Orchidaceae*).* ***(a)*** *An orchid of the genus* Cattleya. *Orchids are one of the most specialized families of monocots.* ***(b)*** *A comparison of the parts of an orchid flower, shown on the left, with those of a radially symmetrical flower, shown on the right. The "lip" is a modified petal that serves as a landing platform for insects.*

(a)

(b)

22–14
Rhizanthella, *a Western Australian orchid, is a myco-heterotroph that grows entirely underground. Cracks in the soil form during the dry season, revealing the flowers, which never show above the soil surface.* ***(a)*** *A view from above, with leaves and debris cleared away, of the parted bracts of* Rhizanthella *through which the plant's pollinators (flies) enter.* ***(b)*** *A dozen flowers of* Rhizanthella *are seen here surrounded by the protective bracts.*

large and showy. The sepals, also three in number, are often colored and similar to petals in appearance. The flower is always bilaterally symmetrical and often bizarre in appearance.

Among the orchids are some species with flowers the size of a pinhead and others with flowers more than 20 centimeters in diameter. Several genera contain myco-heterotrophic species. Two Australian species grow entirely underground, their flowers appearing in cracks in the ground, where they are pollinated by flies (Figure 22–14). In the commercial production of orchids, the plants are cloned by making divisions of meristematic tissue, and thousands of identical plants can be produced rapidly and efficiently (see Chapter 28). There are more than 60,000 registered hybrids of orchids, many of them involving two or more genera. The seed pods of orchids of the genus *Vanilla* are the natural source of the popular flavoring of the same name (Figure 22–15).

22–15
Vanilla, *an orchid that is the source of the flavoring of the same name. Originally used by the Aztecs in what is now Mexico, vanilla is cultivated primarily on Madagascar and other islands in the western Indian Ocean and elsewhere in the Old World. Vanilla is extracted from the dried, fermented seed pods of this orchid. Chocolate is a blend of an extract from the pods of the cacao plant and vanilla. Synthetic vanilla flavoring (vanillin) is now used as the source of about 95 percent of all vanilla consumed.* ***(a)*** *Flowers of the vanilla orchid* (Vanilla planifolia). ***(b)*** *Hand pollination of vanilla plants in Mexico. This procedure is carried out, even in wild plants, to ensure a good crop of the seed pods from which vanilla is extracted.*

(a)

(b)

Animals Serve As Agents of Floral Evolution

Plants, unlike most animals, cannot move from place to place to find food or shelter or to seek a mate. In general, plants must satisfy those needs by growth responses and by the structures that they produce. Many angiosperms, however, have evolved a set of features that, in effect, allows them directed mobility in seeking a mate. This set of features is embodied in the flower. By attracting insects and other animals with their flowers, and by directing the behavior of these animals so that cross-pollination (and therefore cross-fertilization) will occur at a high frequency, the angiosperms have transcended their rooted condition. In this one respect, they have become just as mobile as animals. How was this achieved?

Flowers and Insects Have Coevolved The earliest seed-bearing plants were pollinated passively. Large amounts of pollen were blown about by the wind, reaching the vicinity of the ovules only by chance. The ovules, which were borne on the leaves or within cones, exuded sticky drops of sap from their micropyles. These drops served to catch the pollen grains and to draw them to the micropyle. As in most modern cycads (page 486) and gnetophytes, insects feeding on the pollen and other flower parts began returning to these new-found sources of food and thus transferred pollen from plant to plant. Such a system is more efficient than passive pollination by the wind. It allowed much more accurate pollination with many fewer pollen grains involved.

The more attractive the plants were to the insects (Figure 22–16), the more frequently they would be visited and the more seed they could produce. Any changes in the phenotype that made such visits more frequent or more efficient offered a selective advantage. Several important evolutionary developments followed. For example, plants that had flowers that provided special sources of food for their pollinators had a selective advantage. In addition to edible flower parts, pollen, and sticky fluid around the ovules, plants evolved floral nectaries. As previously mentioned, nectaries secrete the nutritious, sugary nectar that provides a source of energy for insects and other animals.

Attraction of insects to the naked ovules of these plants sometimes resulted in the loss of some of the ovules to the insects. The evolution of a closed carpel, therefore, gave certain seed plants—the ancestors of the angiosperms—a reproductive, and thus a selective, advantage. Further changes in the shape of the flower, such as the evolution of the inferior ovary, may have been additional means of protecting the ovules from being eaten by insects and other animals, thus providing a further reproductive advantage.

(a)

(b)

22–16

(a) A longhorn beetle (family Cerambycidae*), laden with pollen, visiting the flower of a member of the lily family* (Liliaceae) *in the mountains of northeastern Arizona.*

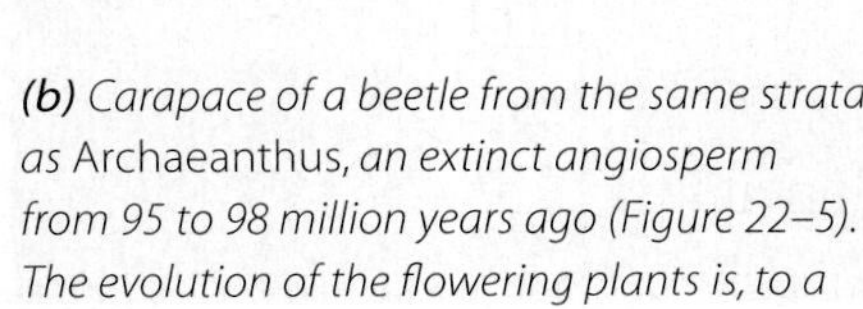

(b) Carapace of a beetle from the same strata as Archaeanthus*, an extinct angiosperm from 95 to 98 million years ago (Figure 22–5). The evolution of the flowering plants is, to a large extent, the story of increasingly specialized relationships between flowers and their insect pollinators, in which beetles played an important early role.*

(a)

(b)

22–17
Flower-visiting beetles. ***(a)*** *A pollen-eating beetle,* Asclera ruficornis, *at the open, bowl-shaped flowers of round-leaved hepatica* (Hepatica americana) *in spring in the woods of eastern North America. All species of this family* (Oedemeridae) *are obligate pollen-feeders as adults.* ***(b)*** *A cetoniine scarabid beetle,* Eupoecila australasiae, *visiting the flowers of* Angophora woodsiana, *near Brisbane, Australia. The cetoniine scarabs have membranous mouthparts used for drawing up nectar.*

Another important evolutionary development was the appearance of the bisexual flower. The presence of both carpels and stamens in a single flower (in contrast, for instance, to the separate microsporangiate and megasporangiate cones of living conifers) offers a selective advantage by making each visit by a pollinator more effective. The pollinator can both pick up and deliver pollen at each stop.

In the early part of the Tertiary period, 40 to 60 million years ago, such specialized groups of flower-visiting insects as bees and butterflies—which had been evolving with the angiosperms for about 50 million years at that point—became even more abundant and diverse. The rise and diversification of these groups of insects were directly related to the increasing diversity of angiosperms. In turn, the insects profoundly influenced the evolutionary course of angiosperms and contributed greatly to their diversification.

If a given plant species is pollinated by only one or a few kinds of visitors, selection favors specializations related to the characteristics of these visitors. Many of the modifications that have evolved in flowers promoted constancy of a specific type of visitor to that particular kind of flower. Some of the special modifications of flowers that came about during the course of their evolution in response to specific pollinators will be described in the following pages.

Beetle-Pollinated Flowers Typically Have a Dull Color but a Strong Odor Many modern species of angiosperms are pollinated solely or chiefly by beetles (Figure 22–17). The flowers of beetle-pollinated plants are either large and borne singly, such as those of magnolias, some lilies, California poppies, and wild roses, or small and aggregated in an inflorescence, such as those of dogwoods, elders, spiraeas, and many species of the parsley family *(Apiaceae)* (Figure 22–16a). Members of some 16 families of beetles are frequent visitors to flowers, although as a rule, these beetles derive most of their nourishment from other sources, such as sap, fruit, dung, and carrion. In beetles, the sense of smell is much more highly developed than the visual sense, and beetle-pollinated flowers are typically white or dull in color, with strong odors (Figure 22–18). These odors are usually fruity, spicy, or similar to the foul odors of fermentation. They are therefore distinct from the sweeter odors of flowers that are pollinated by bees, moths, and butterflies. Some beetle-pollinated flowers secrete nectar, which the beetles eat. In others, the beetles chew directly on the petals or on specialized food bodies (pads or clusters of cells on the surfaces of the various floral parts), and they also eat the pollen. Many beetle-pollinated flowers have inferior ovaries, with the ovules well buried in the floral tissues, out of reach of the chewing jaws of the beetles on which they depend for their pollination.

22–18
The foul-scented and often dark-colored flowers of many members of the milkweed family (Asclepiadaceae), *such as those of this African succulent plant,* Stapelia schinzii, *are pollinated by carrion flies.*

22–19
Bees have become as highly specialized as the flowers they have been associated with during the course of their evolution. Their mouthparts have become fused into a sucking tube containing a tongue. The first segment of each of the three pairs of legs has a patch of bristles on its inner surface. Those of the first and second pairs are pollen brushes that gather the pollen that sticks to the bee's hairy body. On the third pair of legs, the bristles form a pollen comb that collects pollen from these brushes and from the abdomen. From the comb, the pollen is forced up into pollen baskets, concave surfaces fringed with hairs on the upper segment of the third pair of legs. Shown here is a honeybee (Apis mellifera) *foraging in a flower of rosemary* (Rosemarinus officinalis). *In the rosemary flower, the stamens and stigma arch upward out of the flower, and both come into contact with the hairy back of any visiting bee of the proper size. Here the anthers can be seen depositing white pollen grains on the bee.*

Bee-Pollinated Flowers Are Usually Blue or Yellow with Distinctive Markings Bees, the most important group of flower-visiting animals, are responsible for the pollination of more species of plants than the members of any other animal group. Modern families of bees have existed for at least 80 million years, and they subsequently became diverse along with evolutionary radiation of the angiosperms. Both male and female bees live on nectar, and the females also collect pollen to feed the larvae. Bees have mouthparts, body hairs, and other appendages with special adaptations that make them suitable for collecting and carrying nectar and pollen (Figure 22–19). As Karl von Frisch and other investigators of insect behavior have shown, bees can learn quickly to recognize colors, odors, and outlines. The portion of the light spectrum that is visible to most insects, including bees, is somewhat different from the portion visible to humans. Unlike human beings, bees perceive ultraviolet as a distinct color; however, they don't perceive red, which therefore tends to merge with the background.

Many kinds of bees—especially solitary bees, which constitute a majority of species of the group (Figure 22–20)—are highly constant in their visits to flowers, confining their visits to one or a few plant species. Such constancy increases the efficiency of the particular species of bee—or of an individual bee—while it is visiting the flowers of one plant species. In relation to this specialization, bee species with narrowly restricted foraging habits often feature conspicuous morphological and physiological adaptations, such as coarse bristles in their pollen-collecting apparatus (if they visit plants with large pollen grains) or elongated mouthparts (if they take nectar from plants with long-tubed flowers). When they are constant to this degree, and have morphological and behavioral factors such that they actually bring about pollination, bees exert a powerful evolutionary force for specialization of the plants they visit. The process by which two or more species act as selective forces on one another and each undergoes evolutionary change is known as **coevolution.** There are some 20,000 species of bees, the great majority of which visit flowers for food.

Bee flowers—that is, flowers that coevolved with bees—have showy, brightly colored petals that are usually blue or yellow. They often have distinctive patterns by which bees can efficiently recognize them. Such patterns may include "honey guides," special markings that indicate the position of the nectar (Figure 22–21). Bee flowers are never pure red, and, as special photographic techniques have shown, they often have distinctive markings that are normally invisible to humans (Figure 22–22).

In bee flowers, the nectary is characteristically situated at the base of the corolla tube, where it is accessible only to specialized organs, such as the mouthpiece of bees, and not, for example, to the chewing mouthparts of beetles. Bee flowers also characteristically have a "landing platform" of some sort (Figures 22–13 and 22–21).

22–20
A sweat bee (family Halictidae*) gathering pollen from the stamens of a cactus* (Echinocereus) *in Baja California, Mexico. The tall structures in the center of the flower are the stigmas.*

22–21
"Honey guides" on the flowers of the foxglove (Digitalis purpurea) *serve as distinctive signals to insect visitors. The lower lip of the fused corolla serves as a landing platform of the kind that is commonly found in bee flowers.*

(a)

(b)

22–22
The color perception of most insects is somewhat different from that of human beings. To a bee, for example, ultraviolet light (which is invisible to humans) is seen as a distinct color. These photographs show a flower of marsh marigold (Caltha palustris) ***(a)*** *in natural light, showing the solid yellow color as the flower appears to humans, and* **(b)** *in ultraviolet light. The portions of the flower that appear light in* **(b)** *reflect both yellow and ultraviolet light, which combine to form a color known as "bee's purple," whereas the dark portions of the flower absorb ultraviolet and therefore appear pure yellow when viewed by a bee (see page 543).*

(a)

(b)

22–23
Bumblebees (Bombus). *These social bees are important pollinators of many genera of plants throughout the cooler parts of the Northern Hemisphere, and they have been introduced into regions where they are not native for the purpose of pollinating such plants as white clover* (Trifolium repens). ***(a)*** *A bumblebee gathering pollen from the flower of a California poppy* (Eschscholzia californica). ***(b)*** *Portion of the underground nest of a bumblebee colony, showing the cells in which the wormlike larvae complete their development. The bumblebees provision these cells with pollen and regurgitated nectar, which they obtain from flowers. Although a colony of bumblebees may visit a wide variety of flowers during the course of a season, an individual bee often visits only the flowers of one kind of plant on a single trip away from the nest.*

22–24
The beelike flowers of the orchid Ophrys speculum *attract male bees, which are so deceived by the resemblance to female bees of their species that they attempt to copulate with the flowers. In doing so, they often pick up a pollen sac (pollinium) from a flower and may then carry it to another flower of the same species. This orchid was photographed in Sardinia.*

Bumblebees are among the most familiar flower visitors in the North Temperate zone (Figure 22–23). They are social bees that live in colonies. The queens (sexual females) overwinter and, when they emerge in the spring, lay their eggs to establish a new colony. Bumblebees cannot fly until their wing muscles reach a temperature of about 32°C. To maintain this temperature, they must forage constantly on flowers with a copious supply of nectar. Many plants of the cool parts of North America and Eurasia, including lupines, larkspurs, and fireweed (see Figure 21–21), are regularly pollinated by bumblebees throughout their ranges.

Some of the evolutionarily more advanced flowers, in particular the orchids, have developed complex passageways and traps that force the bees that visit them to follow a particular route into and out of the flower. This ensures that both anther and stigma come into contact with the bee's body at a particular point and in the proper sequence (Figure 22–19).

An even more bizarre pollination strategy has been adopted by orchids of the genus *Ophrys*. The flower resembles a female bee, wasp, or fly (Figure 22–24). The males of these insect species emerge early in the spring, before the females. The orchids bloom early in the spring as well, and the male insects attempt to copulate with the orchid flower. During the course of its "sexual" visit, a pollinium may be deposited on the insect's body. When the insect visits another flower of the same species, the pollinium may be caught in the appropriate grooves on the stigma and thus bring about the pollination of that flower.

An additional range of flowers with different characteristics is pollinated by flies of various kinds, including mosquitoes. These insects feed on nectar but do not gather pollen or store food for their larvae. Examples of flowers pollinated by mosquitoes and flies are shown in Figures 22–18 and 22–25.

Flowers Pollinated by Moths and Butterflies Often Have a Long Corolla Tube Flowers that coevolved with butterflies and diurnal moths (those that are active during the day rather than at night) are similar in many respects to bee flowers. This is mainly because butterflies, moths, and bees are all guided to flowers by a combination of sight and smell (Figure 22–26). Some species of butterflies, however, are able to perceive red as a distinct color, and some butterfly-pollinated flowers are red and orange.

(a)

(c)

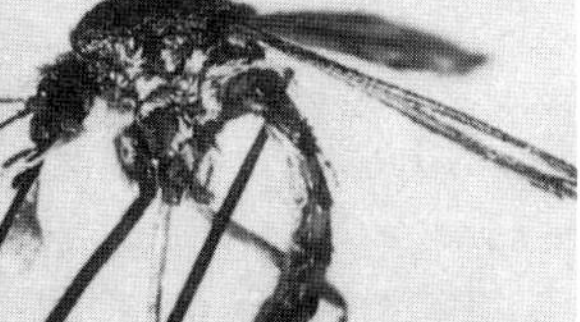

(b)

22–25

Pollination by mosquitoes and other flies. **(a)** *Some small-flowered orchids, such as* Habenaria elegans, *in which the flowers are white or green and relatively inconspicuous, are visited and pollinated by mosquitoes in North Temperate and Arctic regions. The mosquitoes obtain nectar from the flowers.* **(b)** *A female mosquito of the genus* Aedes *with an orchid pollinium attached to its head. Other small-flowered orchids, such as those of the genus* Spiranthes, *are pollinated by bees.* **(c)** *A fly on a flower of a lily* (Zigadenus fremontii). *Notice the conspicuous yellow nectaries.*

22–26
Copper butterfly (Lycaena gorgon) *sucking nectar from the flowers of a daisy. The long sucking mouthparts of moths and butterflies are coiled up at rest and extended when feeding. They vary in length from species to species. Only a few millimeters long in some of the smaller moths, they are 1 to 2 centimeters long in many butterflies, 2 to 8 centimeters long in some hawkmoths of the North Temperate zone, and as long as 25 centimeters in a few kinds of tropical hawkmoths.*

Most moths are nocturnal. The typical moth-pollinated flower—as seen, for example, in several species of tobacco *(Nicotiana)*—is white or pale in color and has a heavy fragrance, a sweet penetrating odor that often is emitted only after sunset. Well-known flowers that are pollinated by moths include the yellow-flowered species of evening primrose (*Oenothera*; see Figure 10–11b) and the pink-flowered amaryllis *(Amaryllis belladonna)*.

The nectary of a moth or butterfly flower is often located at the base of a long, slender corolla tube or a spur and is usually accessible only to the long sucking mouthparts of moths and butterflies. Hawkmoths, for instance, do not usually enter flowers, as bees do, but hover above them, inserting their long mouthparts into the floral tube. Consequently, hawkmoth flowers do not have the landing platforms, traps, and elaborate internal structural modifications seen in some of the bee flowers. Most moth–flower relationships typically involve smaller moths that do not use nearly as much energy as the hawkmoths. The flowers over which the moths scramble are often smaller than those pollinated by hawkmoths and have short corolla tubes. One of the most specialized moth–flower relationships is shown in Figure 22–27.

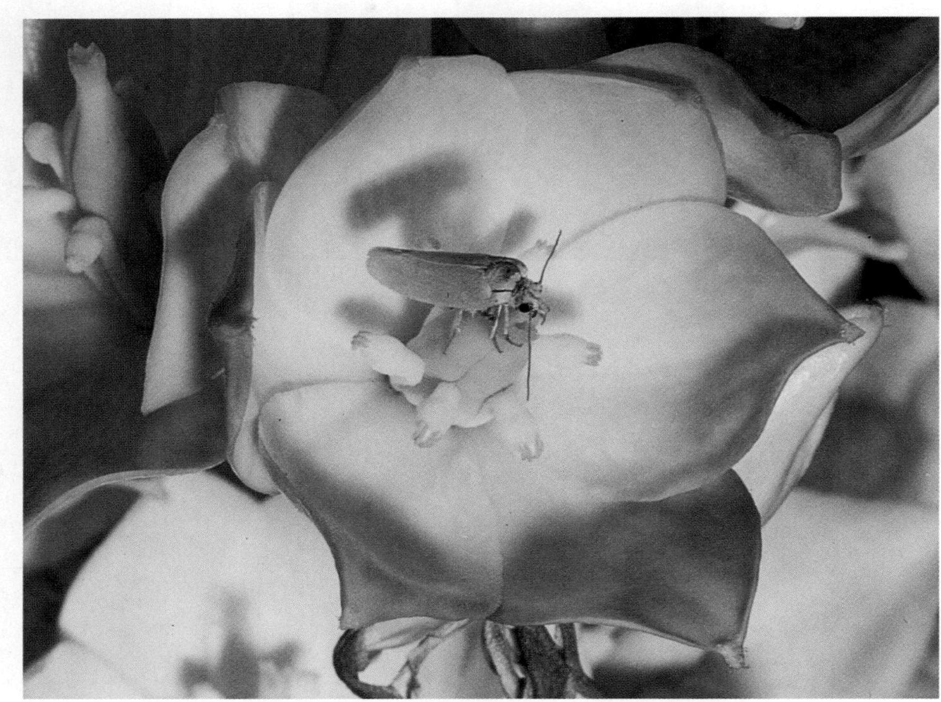

22–27
Yucca moth (Tegeticula yucasella) *scraping pollen from a yucca flower. The female moth visits the creamy white flowers by night and gathers pollen, which she rolls into a tight little ball and carries in her specialized mouthparts to another flower. In the second flower, the moth pierces the ovary wall with her long ovipositor and lays a batch of eggs among the ovules. She then packs the sticky mass of pollen through the openings of the stigma. Moth larvae and seeds develop simultaneously, with the larvae feeding on the developing yucca seeds. When the larvae are fully developed, they gnaw their way through the ovary wall and lower themselves to the ground, where they pupate until the yuccas bloom again. It is estimated that only about 20 percent of the seeds are usually eaten.*

22–28
A male Anna's hummingbird (Calypte anna) *at a flower of the scarlet monkey-flower* (Mimulus cardinalis) *in southern California. Note the pollen on the bird's forehead, which is in contact with the stigma of the flower.*

Bird-Pollinated Flowers Produce Large Amounts of Nectar and Are Often Red and Odorless Some birds regularly visit flowers to feed on nectar, floral parts, and flower-inhabiting insects. Many of these birds also serve as pollinators. In North and South America, the chief pollinators among the birds are hummingbirds (Figure 22–28). In other parts of the world, flowers are visited regularly by representatives of other specialized bird families (Figure 22–29).

Bird flowers have a copious, thin nectar (some actually drip with nectar when the pollen is mature) but usually have little odor because the sense of smell is poorly developed in birds. However, birds do have a keen color sense that is much like our own. It is not surprising, therefore, that most bird flowers are colorful, with red and yellow ones being the most common (Figure 22–12b). Bird-pollinated flowers include red columbine (Figure 22–30a), fuchsia, scarlet passion flower, eucalyptus, hibiscus, poinsettia (Figure 22–30b, c), and many members of the cactus, banana, and orchid families. Typically, these flowers are large or they form parts of large inflorescences, features that can be correlated with their importance as visual stimuli and their ability to hold large amounts of nectar.

22–29
A collared sunbird (Anthreptes collarii) *perching and feeding on a bird-of-paradise* (Strelitzia reginae) *flower in South Africa.*

(a)

(b)

(c)

22–30
Examples of bird-pollinated flowers. ***(a)*** *Columbine* (Aquilegia canadensis). *Alternating perianth segments are modified into nectar-filled tubes. Hummingbirds visit these hanging flowers, taking nectar from them on the wing. The nectar of columbine flowers is inaccessible to most other kinds of animals.* ***(b), (c)*** *Poinsettias* (Euphorbia pulcherrima). *In this familiar plant, a native of Mexico, the flowers are small, greenish, and clustered, but each cluster has a large, yellow nectary from which abundant nectar flows. Ants are seen feeding on the nectar in* ***(b).*** *Modified upper leaves, bright red in color, attract hummingbirds to the clusters of flowers.*

Bird and other animal pollinators usually restrict their visits to the flowers of a particular plant species, at least for a short period of time. This is not the only factor promoting outcrossing, however. The pollinator must not confine its visits to a single flower or to the flowers of a single plant. When flowers are visited regularly by large animals with a high rate of energy expenditure—such as birds, hawkmoths, or bats—the flowers must produce large amounts of nectar to support the metabolic requirements of the animals and keep them coming back. On the other hand, if an abundant supply of nectar is available to animals with a lower rate of energy expenditure, such as small bees or beetles, they will tend to remain at a single flower. Being satisfied there, these visitors will not move on to the flowers of other plants where they might bring about outcrossing. Consequently, species that are regularly pollinated by animals with a high rate of energy consumption, such as hummingbirds, have tended to evolve flowers with the nectar held in tubes or otherwise unavailable to smaller animals with lower rates of energy consumption. Similarly, the color red is a signal to birds but not to most insects. Birds, like ourselves, do not respond very strongly to odor clues. Thus, odorless, red flowers, being inconspicuous to insects, tend not to attract them. This adaptation is advantageous in view of the copious production of nectar by such flowers.

Bat-Pollinated Flowers Produce Copious Nectar and Have Dull Colors and Strong Odors Flower-visiting bats are found in tropical areas of both the Old World and the New World, and more than 250 species of bats—about a quarter of the total number of bat species—include at least some nectar, fruit, or pollen in their diet. These species of bats have slender, elongated muzzles and long, extensible tongues, sometimes with a brushlike tip, and their front teeth are often reduced in size or are missing altogether.

Bat flowers are similar in many respects to bird flowers, being large, robust flowers that produce copious nectar (Figure 22–31). Because bats feed at night, bat flowers are typically dull in color, and many of them open only at night. Many bat-pollinated flowers are tubular or structurally modified to protect their nectar in other ways. Some bat-pollinated flowers—and fruits that are regularly dispersed by bats—hang down on long stalks below the foliage, where the bats can fly more easily. Other bat-pollinated flowers are borne on the trunks of trees. Bats are attracted to the flowers largely through their sense of smell. Bat-pollinated flowers therefore characteristically have either very strong fermenting or fruitlike odors, or musty scents like those produced by bats to attract one another. Bats fly from tree to tree, eating pollen and other flower parts, and carrying pollen from flower to flower on their fur. At least 130 genera of angiosperms are pollinated by bats, or have their seeds dispersed by bats, or both. Included among bat-pollinated angiosperms are such economically important plants as bananas, mangoes, kapok, and sisal.

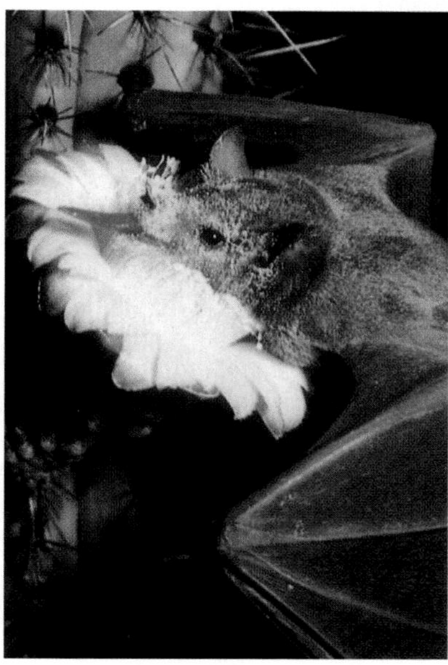

22–31
By thrusting its face into the tubular corolla of a flower of an organ-pipe cactus (Stenocereus thurberi), *this bat,* Leptonycteris curasoae, *is able to lap up nectar with its long, bristly tongue. Some of the pollen clinging to the bat's face and neck is transferred to the next flower it visits. This species of bat, which is one of the more specialized nectar-feeding bats, migrates from central and southern Mexico to the deserts of the south-western United States during late spring and early summer. Here it feeds on the nectar and pollen of organ-pipe and saguaro cacti and on the flowers of agaves.*

Some bats derive a significant portion of their dietary protein from the pollen they consume. As yet another example of coevolution, the pollen of the flowers they visit has been found to contain significantly higher levels of protein than does the pollen of insect-pollinated flowers.

Wind-Pollinated Flowers Produce No Nectar, Have Dull Colors, and Are Relatively Odorless About a century ago, botanists considered wind-pollinated angiosperms to be the most primitive members of the phylum. They thought that all other kinds of angiosperms had evolved from them. The conifers—thought by some scientists at the time to have been the direct ancestors of the angiosperms—have small, drab-colored, odorless, unisexual cones and are pollinated by the wind. Similarly, the many examples of wind-pollinated flowers have dull colors, are relatively odorless, and do not produce nectar. The petals of these flowers are either small or absent, and the sexes are often separated on the same plant. However, studies of other characteristics of these wind-pollinated angiosperms—in particular, their specialized wood and pollen—have convinced most botanists that all wind-pollinated angiosperms evolved not from the conifers but from insect-pollinated angiosperms.

According to current interpretations of the evidence, wind-pollinated angiosperms originated independently from several different ancestral stocks. They are best represented in temperate regions and are relatively rare in the tropics. In temperate climates, many individual trees of the same species are often found close together, and the dispersal of pollen by wind can occur readily in early spring, when the trees are leafless. In the tropics, on the other hand, many more kinds of trees are found in a given area, and the distance to the nearest individual of the same species may be quite great. Furthermore, in many tropical communities the trees are evergreen, and the dispersal of pollen by wind is more difficult than in temperate deciduous forests when the trees are leafless. Under these circumstances, pollination by insects and other animals that have the ability to seek out other individuals of the same plant species, sometimes over relatively great distances (20 kilometers or more, in some cases), is much more efficient than wind pollination.

Because wind-pollinated angiosperms do not depend on insects to transport their pollen from place to place, they devote no energy to the production of nutritious rewards for insect visitors. Wind pollination is very inefficient, however, and it is successful only where a large

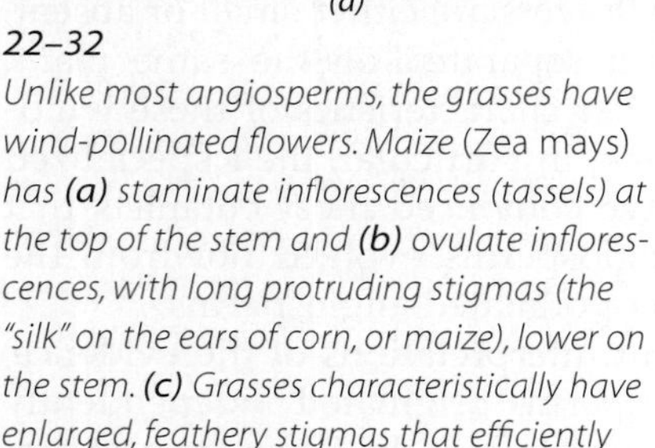

(a)

(b)

(c)

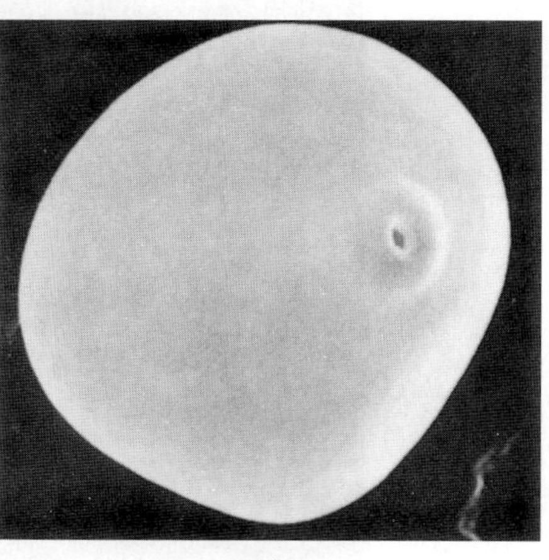

(d)

22–32

Unlike most angiosperms, the grasses have wind-pollinated flowers. Maize (Zea mays) *has* ***(a)*** *staminate inflorescences (tassels) at the top of the stem and* ***(b)*** *ovulate inflorescences, with long protruding stigmas (the "silk" on the ears of corn, or maize), lower on the stem.* ***(c)*** *Grasses characteristically have enlarged, feathery stigmas that efficiently catch the wind-blown pollen shed by the hanging anthers, as seen here in a grass of the genus* Agropyron. ***(d)*** *Scanning electron micrograph of a pollen grain of maize, showing the smooth pollen wall found in most wind-pollinated plants and the single aperture characteristic of monocots.*

number of individuals of the same species grow fairly close together. Nearly all wind-borne pollen falls to the ground within 100 meters of the parent plant. Thus, if the individual plants are widely scattered, the chance that a pollen grain will reach a receptive stigma is very slim. Many wind-pollinated plants are either dioecious (having male and female flowers on separate plants), such as the willows; monoecious (having separate male and female flowers on the same plant), such as the oaks (see Figure 21–10); or genetically self-incompatible, such as many grasses (page 511). Even though their pollen moves about somewhat randomly, these plants have still evolved devices that foster a high degree of outcrossing.

Wind-pollinated flowers usually have well-exposed stamens that can easily lose their pollen to the wind. In some, the anthers are suspended from long filaments hanging from the flower (Figures 22–32 and 22–33). The abundant pollen grains, which are generally smooth and small, do not adhere to one another as do the pollen grains of insect-pollinated species. The large stigmas are characteristically exposed, and they often have branches or feathery outgrowths adapted for intercepting wind-borne pollen grains. Most wind-pollinated plants have ovaries with single ovules (and hence single-seeded fruits) because each pollination event consists of the meeting of one pollen grain with one stigma and leads to the fertilization of one ovule for each flower. Thus, each oak flower produces only a single acorn and each grass flower produces only a single grain. To compensate, perhaps, plants with very small flowers tend to have them concentrated in large numbers into inflorescences (Figures 22–32 through 22–34).

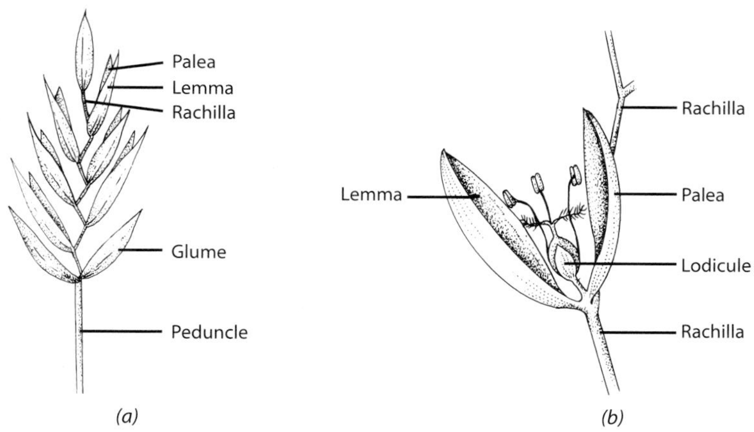

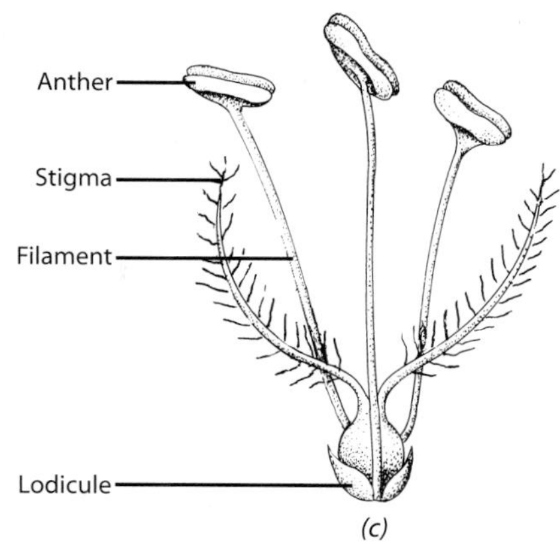

22–33
Grass flowers (florets) usually develop in clusters. ***(a)*** *As a cluster matures, a single pair of dry, chaffy bracts—the glumes —separate a little, exposing the elongating spikelet, with from one to many florets (depending on the species of grass) attached to a central axis, or rachilla.* ***(b)*** *Each floret is surrounded by two distinctive bracts of its own, the palea and the lemma. These are forced apart, exposing the inner parts of the flower* ***(c),*** *by the swelling of the lodicules—small, rounded bodies at the base of the carpel—and are spread wide when the grass is in flower. The stamens, usually three in number, have slender filaments and long anthers, and the stigmas are typically long and feathery and so are efficient at intercepting the wind-borne pollen.*

22–34
Most common species of trees in temperate regions are wind-pollinated. The staminate flowers of the paper birch (Betula papyrifera) *hang down in catkins—flexible, thin tassels several centimeters long. These catkins are whipped by passing breezes, and the pollen, when mature, is scattered about by the wind.*

Some Submerged Aquatic Angiosperms Are Pollinated via Water A few angiosperms—about 79 families and 380 genera—are submerged aquatic plants, living in marine and freshwater habitats. In 18 of these genera, the pollen is either transported underwater or floats from one plant to another across the surface of the water. In some of these plants, pollen grains are either threadlike (filiform), thus increasing their chances of coming into contact with receptive stigmas, or linked into chains, with similar effect (Figure 22–35a). The filiform pollen of some aquatic genera is unusual in lacking an exine. In other genera of submerged aquatic plants, the pollen is dispersed across the surface of the water. For example, in the freshwater eel grass *Vallisneria*, the entire staminate flower is released under water and floats to the surface where its three stamens become erect and function like a sail. Floating just below the tension zone at the

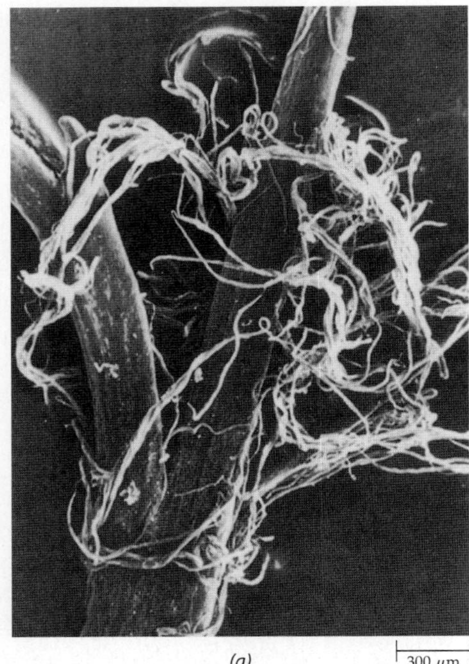

(a)

(b)

22–35

Unique pollination systems in submerged aquatic angiosperms. ***(a)*** *The slender, branching stigmas of the sea-nymph,* Amphibolis, *seen here in a scanning electron micrograph, have captured many grains of the threadlike pollen released by staminate plants of this species.* ***(b)*** *Tiny staminate flowers of eel grass,* Vallisneria, *float into the depression in the surface-tension layer of the water that surrounds the much larger carpellate flower. The individual pollen grains are visible on the staminate flowers.*

water surface, the carpellate flower creates a small depression into which the staminate flowers fall (Figure 22–35b).

In contrast to these plants with very specialized pollination systems, which are clearly advanced in an evolutionary sense, most aquatic angiosperms are either wind-pollinated or insect-pollinated, like their terrestrial ancestors. Their flowers are held above the surface of the water.

The Most Important Pigments in Floral Coloration Are the Flavonoids

Color is one of the most conspicuous features of angiosperm flowers—a characteristic by which members of the phylum are easily recognized. The varied colors of different kinds of flowers evolved in relation to their pollination systems and, in general, are advertisements for particular kinds of animals, as we have just seen.

The pigments that are responsible for the colors of angiosperm flowers are generally common in vascular plants other than angiosperms. It is the way in which they are concentrated in angiosperm flowers, and particularly in their corollas, that is a special characteristic of the flowering plants. Surprisingly, all flower colors are produced by a small number of pigments. Many red, orange, or yellow flowers owe their color to the presence of carotenoid pigments similar to those that occur in leaves (and in all plants, in green algae, and in some other organisms, as well). The most important pigments in floral coloration, however, are **flavonoids,** which are compounds with two six-carbon rings linked by a three-carbon unit. Flavonoids probably occur in all angiosperms, and they are sporadically distributed among the members of other groups of plants. In leaves, flavonoids block far-ultraviolet radiation, which is destructive to nucleic acids and proteins. They usually selectively admit light of blue-green and red wavelengths, which are important for photosynthesis.

Pigments belonging to one major class of flavonoids, the **anthocyanins,** are major determinants of flower color (Figure 22–36). Most red and blue plant pigments are anthocyanins, which are water-soluble and are found in vacuoles. By contrast, the carotenoids are oil-soluble and are found in plastids. The color of an anthocyanin pigment depends on the acidity of the cell sap of the vacuole. Cyanidin, for example, is red in acid solution, violet in neutral solution, and blue in alkaline solution. In some plants, the flowers change color after pollination, usually because of the production of large amounts of anthocyanins, and then become less conspicuous to insects.

The **flavonols,** another group of flavonoids, are very commonly found in leaves and also in many flowers. A number of these compounds are colorless or nearly so,

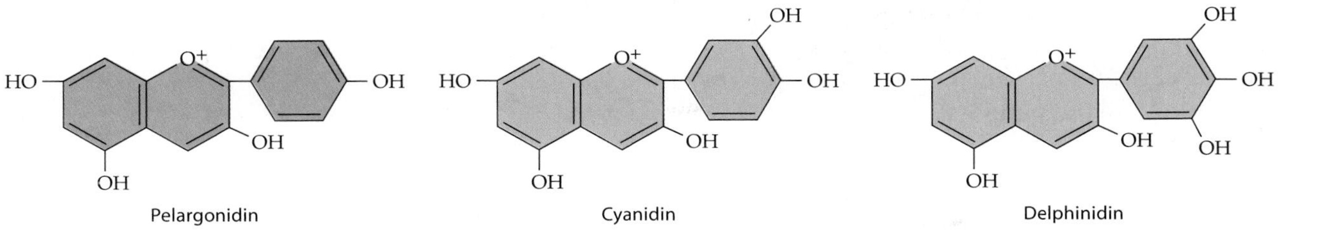

Pelargonidin Cyanidin Delphinidin

22–36
Three anthocyanin pigments, the basic pigments on which flower colors in many angiosperms depend: pelargonidin (red), cyanidin (violet), and delphinidin (blue). Related compounds known as flavonols are yellow or ivory, and the carotenoids are red, orange, or yellow. Betacyanins (betalains) are red pigments that occur in one group of eudicots. Mixtures of these different pigments, together with changes in cellular pH, produce the entire range of flower color in the angiosperms. Changes in flower color provide "signals" to pollinators, telling them which flowers have opened recently and are more likely to provide food.

but they may contribute to the ivory or white hues of certain flowers.

For all flowering plants, different mixtures of flavonoids and carotenoids (as well as changes in cellular pH) and differences in the structural, and thus the reflective, properties of the flower parts produce the characteristic colors. The bright fall colors of leaves come about when large quantities of colorless flavonols are converted into anthocyanins as the chlorophyll breaks down. In the all-yellow flowers of the marsh marigold *(Caltha palustris),* the ultraviolet-reflective outer portion is colored by carotenoids, whereas the ultraviolet-absorbing inner portion is yellow to our eyes because of the presence of a yellow chalcone, one of the flavonoids. To a bee or other insect, the outer portion of the flower appears to be a mixture of yellow and ultraviolet, a color called "bee's purple," whereas the inner portion appears pure yellow (Figure 22–22). Most, but not all, ultraviolet reflectivity in flowers is related to the presence of carotenoids, and thus ultraviolet patterns are more common in yellow flowers than in others.

In the goosefoot, cactus, and portulaca families and in other members of the order *Chenopodiales (Centrospermae),* the reddish pigments are not anthocyanins or even flavonoids but a group of more complex aromatic compounds known as **betacyanins** (or betalains). The red flowers of *Bougainvillea* and the red color of beets are due to the presence of betacyanins. No anthocyanins occur in these plants, and the families characterized by betacyanins are closely related to one another.

Evolution of Fruits

Just as flowers have evolved in relation to their pollination by many different kinds of animals and other agents, so have fruits evolved for dispersal in many different ways. Fruit dispersal, like pollination, is a fundamental aspect of the evolutionary radiation of the angiosperms. Before we consider this subject in more detail, however, we must present some basic information about fruit structure.

A fruit is a mature ovary, which may or may not include some additional flower parts. A fruit in which such additional parts are retained is known as an **accessory fruit.** Although fruits usually have seeds within them, some—**parthenocarpic fruits**—may develop without seed formation. The cultivated strains of bananas are familiar examples of this exceptional condition.

Fruits are generally classified as simple, multiple, or aggregate, depending on the arrangement of the carpels from which the fruit develops. **Simple fruits** develop from one carpel or from several united carpels. **Aggregate fruits,** such as those of magnolias, raspberries, and strawberries, consist of a number of separate carpels of one gynoecium. The individual parts of aggregate fruits are known as **fruitlets;** they can be seen, for example, in the magnolia fruit shown in Figure 22–4c. **Multiple fruits** consist of the gynoecia of more than one flower. The pineapple, for example, is a multiple fruit consisting of an inflorescence with many previously separate ovaries fused on the axis on which the flowers were borne (the other flower parts being squeezed between the expanding ovaries).

Simple fruits are by far the most diverse of the three groups. When ripe, they may be soft and fleshy, dry and woody, or papery. There are three main types of fleshy fruits—berries, drupes, and pomes. In **berries**—examples of which are tomatoes, dates, and grapes—there may be one to several carpels, each of which is typically many-seeded. The inner layer of the fruit wall is fleshy. In **drupes,** there may also be one to several carpels, but each carpel usually contains only a single seed. The inner layer of the fruit is stony and usually tightly adherent to the seed. Peaches, cherries, olives, and plums are familiar drupes. Coconuts are drupes whose outer layer is fibrous rather than fleshy, but in temperate regions we usually see only the coconut seed with the adherent stony inner layer of the fruit (Figure 22–37). **Pomes** are highly specialized fleshy fruits that are characteristic of one subfamily of the rose family. The pome is derived from a compound inferior ovary in which the fleshy portion comes largely from the enlarged base of the perianth. The inner portion, or endocarp, of a pome resembles a tough membrane, as you know from eating apples and pears, the two most familiar examples of this kind of fruit.

Dry simple fruits are classified as either dehiscent (Figures 22–38 and 22–39) or indehiscent (Figure 22–40). In **dehiscent fruits,** the tissues of the mature ovary wall (the pericarp) break open, freeing the seeds. In **indehiscent fruits,** on the other hand, the seeds remain in the fruit after the fruit has been shed from the parent plant.

There are several kinds of dehiscent simple dry fruits. The **follicle** is derived from a single carpel that splits down one side at maturity, as in the columbines and milkweeds (Figure 22–38a). Follicles were also characteristic of the extinct, Middle Cretaceous plant *Archaeanthus* (Figure 22–5), and they are also found in magnolias (Figure 22–4c). In the pea family *(Fabaceae)*, the characteristic fruit is a **legume.** Legumes resemble follicles, but they split along both sides (Figure 22–39). In the mustard family *(Brassicaceae)*, the fruit is called a **silique** and is formed of two fused carpels. At maturity, the two sides of the fruit split off, leaving the seeds attached to a persistent central portion (Figure 22–38c). The most common sort of dehiscent simple dry fruit is the **capsule**, which is formed from a compound ovary in plants with either a superior or an inferior ovary. Capsules shed their seeds in a variety of ways. In the poppy family *(Papaveraceae)*, the seeds are often shed when the capsule splits longitudinally, but in some members of this family they are shed through holes near the top of the capsule (Figure 22–38b).

Indehiscent simple dry fruits are found in many different plant families. Most common is the **achene,** a small, single-seeded fruit in which the seed lies free in

(a)

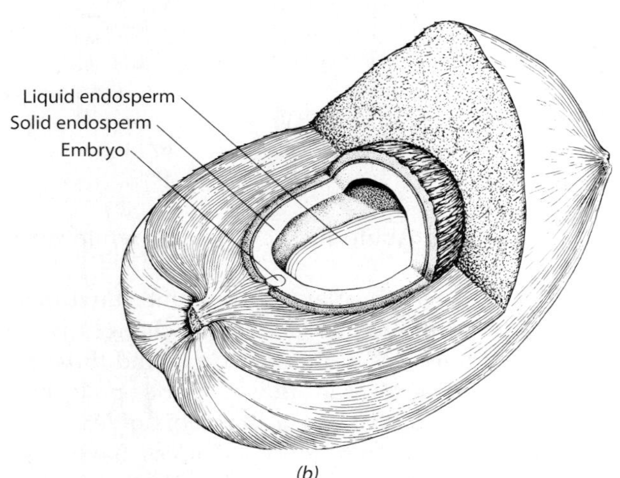

(b)

22–37
The coconut palm (Cocos nucifera) *has a very wide range on ocean shores throughout the world because its fruits are able to float for long periods and then germinate when they reach land.* ***(a)*** *A coconut germinating on the beach in Florida.* ***(b)*** *Diagram of a coconut fruit. The coconut milk is liquid endosperm. Cell walls form around the nuclei in the liquid endosperm, which becomes solid by the time of germination. When coconuts are transported commercially, their husks are usually removed first, so that in temperate countries people usually see only the stony inner shell of the fruit surrounding the seed.*

(a)

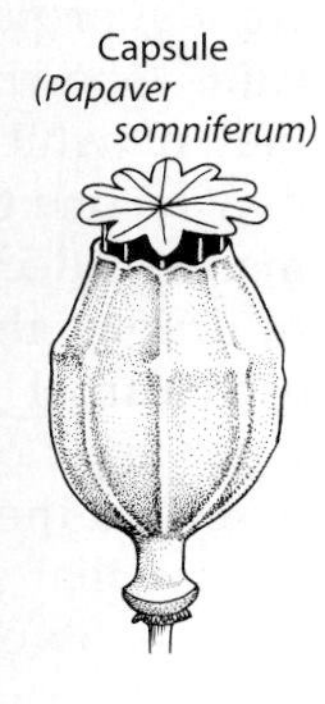

(b)

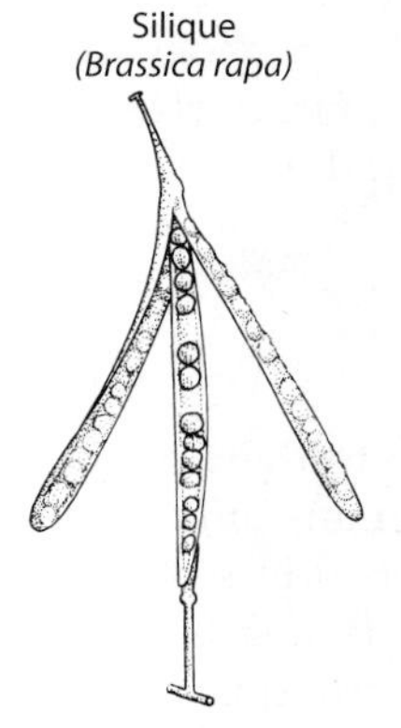

(c)

22–38
Dehiscent fruits. ***(a)*** *Bursting follicles of a milkweed* (Asclepias). ***(b)*** *In some members of the poppy family* (Papaveraceae), *such as poppies (genus* Papaver*), the capsule sheds its seeds through pores near the top of the fruit.* ***(c)*** *Plants of the mustard family* (Brassicaceae) *have a characteristic fruit known as a silique, in which the seeds arise from a central partition, and the two enclosing valves fall away at maturity.*

(a)

(b)

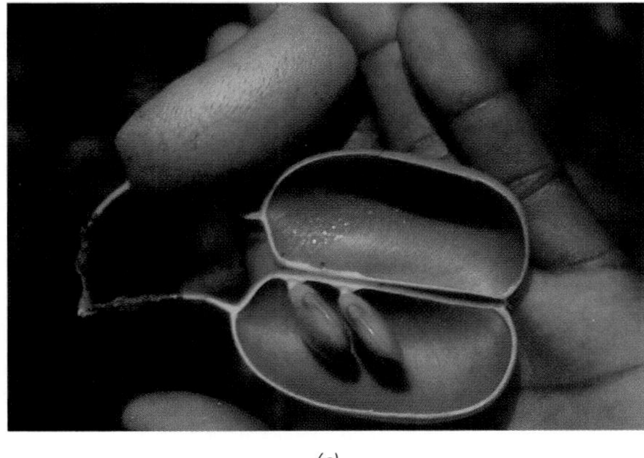

(c)

22–39
The legume, a kind of fruit that is usually dehiscent, is the characteristic fruit of the pea family, Fabaceae *(also called* Leguminosae*). With about 18,000 species,* Fabaceae *is one of the largest families of flowering plants. Many members of the family are capable of nitrogen fixation because of the presence of nodule-forming bacteria of the genus* Rhizobium *on their roots (see Chapter 30). For this reason, these plants are often the first colonists on relatively infertile soils, as in the tropics, and they may grow rapidly there. The seeds of a number of plants of this family, such as peas, beans, and lentils, are important foods.* ***(a)*** *Legumes of the garden pea,* Pisum sativum. ***(b)*** *Legumes of* Albizzia polyphylla, *growing in Madagascar. Each seed is in a separate compartment of the fruit.* ***(c)*** *Legume of* Griffonia simplicifolia, *a West African tree. The two valves of the legume are split apart, revealing the two seeds inside.*

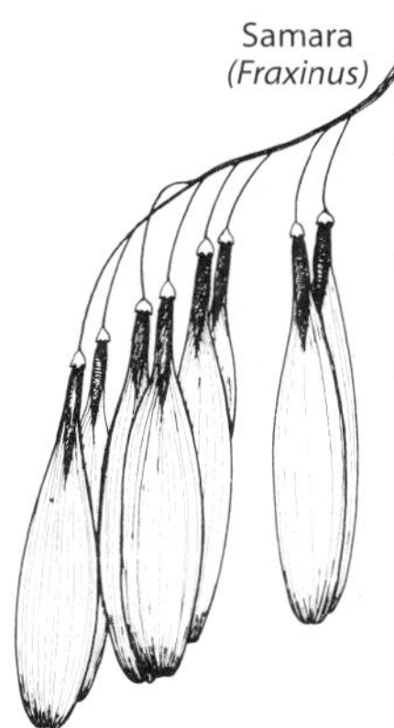

22–40
The samara, a winged indehiscent fruit characteristic of ashes (Fraxinus) *and elms* (Ulmus), *retains its single seed at maturity. Samaras are dispersed by wind.*

the cavity except for its attachment by the funiculus. Achenes are characteristic of the buttercup family *(Ranunculaceae)* and the buckwheat family *(Polygonaceae)*. Winged achenes, such as those found in elms and ashes, are commonly known as **samaras** (Figure 22–40). The achenelike fruit that occurs in grasses *(Poaceae)* is known as a **caryopsis,** or grain; in it, the seed coat is firmly united to the fruit wall. In the *Asteraceae,* the complex, achenelike fruit is derived from an inferior ovary; technically, it is called a **cypsela** (see Figure 22–42). Acorns and hazelnuts are examples of **nuts,** which resemble achenes but have a stony fruit wall and are derived from a compound ovary. Finally, in the parsley family *(Apiaceae)* and the maples *(Aceraceae),* as well as a number of other, unrelated groups, the fruit is a **schizocarp,** which splits at maturity into two or more one-seeded portions (Figure 22–41a).

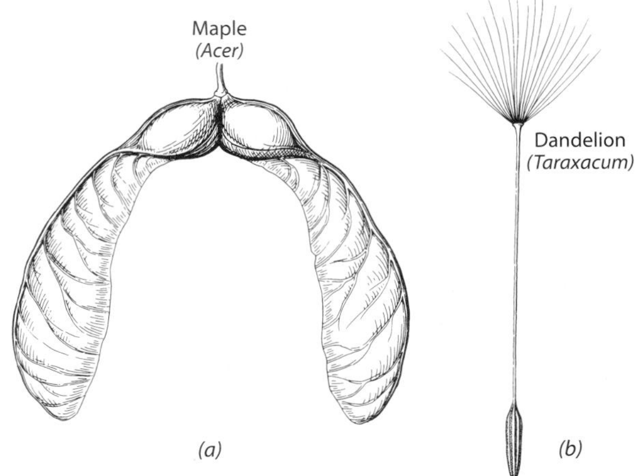

22–41
Wind-dispersed fruits. ***(a)*** *In maples* (Acer), *each half of the schizocarp has a long wing.* ***(b)*** *The fruits of the dandelion* (Taraxacum) *and many other composites have a modified calyx, called the pappus, which is adherent to the mature cypsela and may form a plume-like structure that aids in wind dispersal.*

Fruits and Seeds Have Evolved in Relation to Their Dispersal Agents

Just as flowers have evolved according to the characteristics of the pollinators that visit them regularly, so have fruits evolved in relation to their dispersal agents. In both coevolutionary systems, there have, in general, been many changes in relation to different dispersal agents within individual families and a great deal of convergent evolution toward similar-appearing structures with similar functions. We shall review some of the adaptations of fruits here, in relation to their dispersal agents.

Many Plants Have Wind-Borne Fruits and Seeds Some plants have light fruits or seeds that are dispersed by the wind (Figures 22–38a, 22–40, 22–41). The dustlike seeds of all members of the orchid family, for example, are wind-borne. Other fruits have wings, which are sometimes formed from perianth parts, that allow them to be blown from place to place. In the schizocarps of maples, for example, each carpel develops a long wing (Figure 22–41a). The two carpels separate and fall when mature. Many members of the *Asteraceae*—dandelions, for example—develop a plumelike pappus, which aids in keeping the light fruits aloft (Figures 22–41b and 22–42). In some plants, the seed itself, rather than the fruit, bears the wing or plume. The familiar butter-and-eggs *(Linaria vulgaris)* has a winged seed, and both fireweed *(Chamaenerion)* and milkweed (*Asclepias;* Figure 22–38a) have plumed seeds. In willows and poplars (family *Salicaceae*), the seed coat is covered with woolly hairs. In tumbleweeds *(Salsola),* the whole plant (or a portion of it) is blown along by the wind, scattering seeds as it moves (Figure 22–43).

Other plants shoot their seeds aloft. In touch-me-not *(Impatiens),* the valves of the capsules separate suddenly, throwing seeds for some distance. In the witch hazel *(Hamamelis),* the endocarp contracts as the fruit dries, discharging the seeds so forcefully that they sometimes travel as far as 15 meters from the plant. Another example of self-dispersal is shown in Figure 22–44. In contrast to these active methods of dispersal, the seeds or fruits of many plants simply drop to the ground and are dispersed more or less passively (or sporadically, such as by rainwater or floods).

Fruits and Seeds Adapted for Floating Are Dispersed by Water The fruits and seeds of many plants, especially those growing in or near water sources, are adapted for floating, either because air is trapped in some part of the fruit or because the fruit contains tissue that includes large air spaces. Some fruits are especially adapted for dispersal by ocean currents. Notable among these is the coconut (Figure 22–37), which is why almost every newly formed Pacific atoll quickly acquires its own co-

22–42
The familiar small, indehiscent fruits of dandelions, which are technically known as cypselas (but often loosely called achenes), have a plumelike, modified calyx (the pappus) and are spread by the wind. This photograph shows the fruiting heads of a plant of the genus Agoseris, *which is closely related to the dandelions.*

22–43
In tumbleweeds (Salsola), *the whole plant breaks off and is blown across open country, scattering its seeds as it tumbles along. So many tumbleweeds blew into Mobridge, South Dakota, on November 8, 1989, that the town was besieged. The tumbleweeds are natives of Eurasia, but they are widely naturalized as weeds in North America and elsewhere.*

(a)

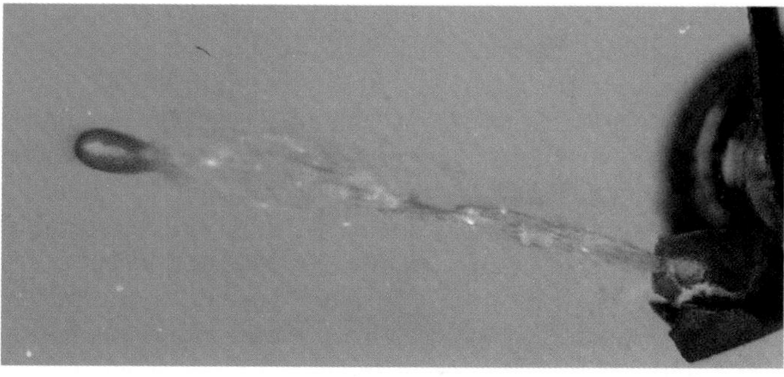

(b)

22–44
Dwarf mistletoe (Arceuthobium), *a parasitic eudicot that is the most serious cause of loss of forest productivity in the western United States.* ***(a)*** *A plant growing on a pine branch in California.* ***(b)*** *Seed discharge. Very high hydrostatic pressure builds up in the fruit and shoots the seeds as much as 15 meters laterally. The seeds have an initial velocity of about 100 kilometers per hour. This is one of the ways in which the seeds are spread from tree to tree, although they are also sticky and can be carried from one tree to another over much longer distances by adhering to the feet or feathers of birds.*

conut tree. Rain, also a common means of fruit and seed dispersal, is particularly important for plants that live on hillsides or mountain slopes.

Fruits and Seeds That Are Fleshy or Have Adaptations for Attachment Are Dispersed by Animals The evolution of sweet and often highly colored, fleshy fruits was clearly involved in the coevolution of animals and flowering plants. The majority of fruits in which much of the pericarp is fleshy—bananas, cherries, raspberries, dogwoods, grapes—are eaten by vertebrates. When such fruits are eaten by birds or mammals, the seeds the fruits contain are spread by being passed unharmed through the digestive tract or, in birds, by being regurgitated at a distance from the place where they were ingested (Figure 22–45). Sometimes, partial digestion aids the germination of seeds by weakening their seed coats.

(a)

(b)

(c)

22–45
The seeds of fleshy fruits are usually dispersed by vertebrates that eat the fruits and either regurgitate the seeds or pass them as part of their feces. Examples of vertebrate-dispersed fruits are shown here. ***(a)*** *Strawberries* (Fragaria), *an example of an aggregate fruit. The achenes are borne on the surface of a fleshy receptacle. Immature strawberries, like the immature stages of many bird- or mammal-dispersed fruits, are green, but they become red when the seeds are mature and thus ready for dispersal.* ***(b)*** *The berries of many cacti, such as this prickly pear* (Opuntia) *growing in southern Mexico, are conspicuous at maturity.* ***(c)*** *Berries of chaparral honeysuckle,* Lonicera hispidula. *These berries develop from inferior ovaries, and they therefore have fused portions of the outer floral whorls incorporated in them. A flower of this species is shown in Figure 22–11c.*

When fleshy fruits ripen, they undergo a series of characteristic changes, mediated by the hormone ethylene, which will be discussed in Chapter 28. Among these changes are a rise in sugar content, a softening of the fruit caused by the breakdown of pectic substances, and often a change in color from inconspicuous, leaflike green to bright red (Figure 22–45a), yellow, blue, or black. The seeds of some plants, especially tropical ones, often have fleshy appendages, or arils, with the bright colors characteristic of fleshy fruits and, like these fruits, are aided in their dispersal by vertebrates. The arils of yews (*Taxus*; see Figure 20-31) are not homologous to the structures called arils in angiosperms, because yew arils are outgrowths of the seed, not of the fruit. Nevertheless, they play a similar role in seed dispersal.

Unripe fruits are often green or colored in such a way that they are inconspicuous among the plant's green leaves, and thus they are somewhat concealed from birds, mammals, and insects. They may also be disagreeable to the taste—such as unripe cherries *(Prunus)*, which have a strong, acidic taste—thereby discouraging animals from eating them before the seeds are ripe. The changes in color that accompany ripening are the plant's "signal" that the fruit is ready to be eaten—the seeds are ripe and ready for dispersal (Figure 22–45). It is no coincidence that red is such a predominant color among ripe fruits. Red fruit is inconspicuous to insects, blending with the background of green leaves. Insects are too small to disperse the large seeds of fleshy fruits effectively, and such concealment is therefore advantageous to these plants. At the same time, the red fruits are very conspicuous to vertebrates, which then eat the fruits and help disperse their ripe seeds.

A number of angiosperms have fruits or seeds that are dispersed by adhering to fur or feathers (Figure 22–46). These fruits and seeds have hooks, barbs, spines, hairs, or sticky coverings that allow them to be transported, often for great distances, attached to the bodies of animals.

Other important agents of seed dispersal in some plants are ants (Figure 22–47). Such plants have evolved a special adaptation on the exterior of their seeds, called an elaiosome, a fleshy pigmented appendage that contains lipids, protein, starch, sugars, and vitamins. The ants usually carry such seeds back to their nests, where the elaiosomes are consumed by other workers or larvae and the seeds are left intact. The seeds readily germinate in this location, and seedlings often become established, protected from their predators and perhaps benefiting from nutrient enrichment as well. Up to a third of the species in some plant communities, such as the herb understory communities in the deciduous forests of the central and eastern United States, are dispersed by ants in this way. These plants include such familiar species as spring beauty *(Claytonia virginiana)*, Dutchman's-

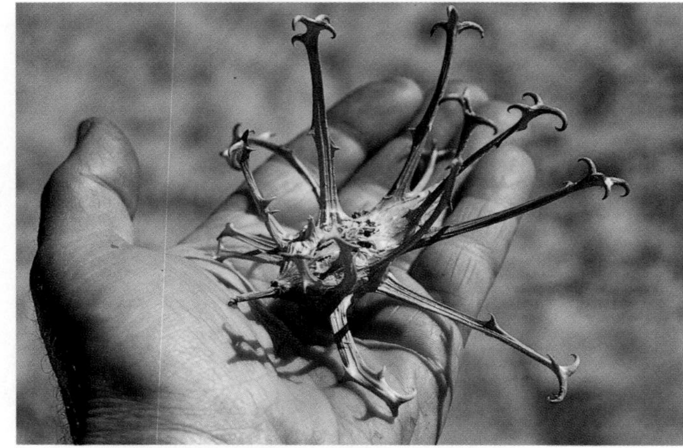

(a)

(b)

22–46

(a) *The fruits of the African plant* Harpagophytum, *a member of the sesame family* (Pedaliaceae), *are equipped with "grappling hooks," which catch in the fur on the legs of large mammals. In this way, the fruits are spread from place to place.* ***(b)*** *Mature inflorescences of cocklebur* (Xanthium), *which attach themselves to passing animals and are dispersed. In this case, the entire inflorescence is the dispersal unit, rather than the fruits alone, as in* Harpagophytum. *Cocklebur is a member of the family* Asteraceae.

22–47
An ant (Rhytidoponera metallica) *taking the seed of an acacia back to its nest in the Australian bushland. Note the large, yellow elaiosome.*

breeches *(Dicentra cucullaria),* bloodroot *(Sanguinaria canadensis),* many species of violets *(Viola),* and *Trillium* (see Figure 1–6b).

Biochemical Coevolution

Also important in the evolution of angiosperms are the so-called *secondary metabolites,* or *products* (page 32). Once thought of as waste products, these include an array of chemically unrelated compounds, such as alkaloids, quinones, essential oils (including terpenoids), glycosides (including cyanogenic substances and saponins), flavonoids, and even raphides (needlelike crystals of calcium oxalate). The presence of certain of these compounds can characterize whole families, or groups of families, of flowering plants (Figure 22–48).

In nature, these chemicals appear to play a major role either in restricting the palatability of the plants in which they occur or in causing animals to avoid the plants altogether (Figure 22–49). When a given family of plants is characterized by a distinctive group of secondary plant products, those plants are apt to be eaten only by insects belonging to certain families. The mustard family *(Brassicaceae),* for example, is characterized by the presence of mustard-oil glycosides and associated

Calactin

Sinigrin

Caffeine

Nicotine

Theobromine

22–48
Secondary metabolites: sinigrin, from black mustard, Brassica nigra; *calactin, a cardiac glycoside, from the milkweed* Asclepias curassavica; *nicotine, from tobacco,* Nicotiana tabacum, *a member of the nightshade family; caffeine, from* Coffea arabica *of the madder family; and theobromine, a prominent alkaloid in coffee, tea* (Thea sinensis), *and cocoa* (Theobroma cacao). *Nicotine, caffeine, and theobromine are alkaloids, members of a diverse class of nitrogen-containing ring compounds that are physiologically active in vertebrates.*

22–49
Poison ivy (Toxicodendron radicans) *produces a secondary plant product, 3-pentadecanedienyl catechol, which causes an irritating rash on the skin of many people. The ability to produce this alcohol presumably evolved under the selective pressure exerted by herbivores. Fortunately, the plant is easily identifiable by its characteristic compound leaves with their three leaflets.*

enzymes that break down these glycosides to release the pungent odors associated with cabbage, horseradish, and mustard. Plant-eating insects of most groups ignore plants of the mustard family and will not feed on them even if they are starving. However, certain groups of true bugs and beetles, and the larvae of some groups of moths, feed only on the leaves of plants of the mustard family. The larvae of most of the members of the butterfly subfamily *Pierinae* (which includes the cabbage butterflies and orange-tips) also feed only on these plants. The same chemicals that act as deterrents to most groups of insect herbivores often act as feeding stimuli for these narrowly restricted feeders. For example, certain moth larvae that feed on cabbage will extrude their mouthparts and go through their characteristic feeding behavior when presented with agar or filter paper containing juices pressed from these plants.

Clearly the ability to manufacture these mustard-oil glycosides and to retain them in their tissues is an important evolutionary step that protects the plants of the mustard family from most herbivores. From the standpoint of herbivores in general, such protected plants represent an unexploited food source for any group of insects that can tolerate or break down the poisons manufactured by the plant. The main evolutionary development of the butterfly group *Pierinae* probably occurred after its ancestors had acquired the ability to feed on plants of the mustard family by breaking down these toxic molecules.

Herbivorous insects that are narrowly restricted in their feeding habits to groups of plants with certain secondary plant products are often brightly colored. This coloration serves as a signal to their predators that they carry the noxious chemicals in their bodies and hence are unpalatable. For example, the assemblage of insects found feeding on a milkweed plant on a summer day might include bright green chrysomelid beetles, bright red cerambycid beetles and true bugs, and orange and black monarch butterflies, among others (see Figure 2–28). Milkweeds *(Asclepiadaceae)* are richly endowed with alkaloids and cardiac glycosides, heart poisons that have potent effects in vertebrates, the main potential predators of these insects. If a bird ingests a monarch butterfly, severe gastric distress and vomiting will follow, and orange-and-black patterns like that of the monarch will be avoided by the predator in the future. Other insects, such as the viceroy butterfly (which also has an orange-and-black pattern), have evolved similar coloration and markings. Thus, they escape predation by capitalizing on their resemblance to the poisonous monarch. This phenomenon, known as **mimicry,** is ultimately dependent upon the plant's chemical defenses. Various drugs and psychedelic chemicals, such as the active ingredients in marijuana *(Cannabis sativa)* and the opium poppy *(Papaver somniferum),* among others, are also secondary plant products that in nature presumably play a role in discouraging the attacks of herbivores (Figure 22–50).

Still more complex systems are known. When the leaves of potato or tomato plants are wounded, as by the Colorado potato beetle, the concentration of proteinase inhibitors, which interfere with the digestive enzymes in the guts of the beetle, rapidly increases in the wounded tissues. Other plants manufacture molecules that resemble the hormones of insects or other predators and thus interfere with the predators' normal growth and development. One of these natural products that resembles a human hormone is a complex molecule called diosgenin, which is obtained from wild yams. Diosgenin is only two simple chemical steps away from 16-dehydropregnenolone, or 16D, the main active ingredient in many oral contraceptives, and wild yams were once a major source for the manufacture of 16D. Unfortunately, these plants grow slowly, and the supply of wild yams was soon largely exhausted. Species of *Solanum* are being cultivated as alternative sources of molecules suitable for simple conversion to 16D.

(a) (b) (c) (d)

22–50
Some plants that produce hallucinogenic and medicinal compounds. ***(a)*** *Mescaline, from the peyote cactus* (Lophophora williamsii), *is used ceremonially by many Native American groups of northern Mexico and the southwestern United States.* ***(b)*** *Tetrahydrocannabinol (THC) is the most important active molecule in marijuana* (Cannabis sativa). ***(c)*** *Quinine, a valuable drug used in the treatment and prevention of malaria, is derived from tropical trees and shrubs of the genus* Cinchona. ***(d)*** *Cocaine, a drug that has recently been abused to an unparalleled extent, is derived from coca* (Erythroxylum coca), *a cultivated plant of northwestern South America. A Peruvian woman is shown here harvesting the leaves of cultivated coca. The secondary metabolites identified in these plants presumably protect them from the depredations of insects, but they are also physiologically active in vertebrates, including humans.*

As mentioned previously, pollination and fruit-dispersal systems have developed particular coevolutionary patterns in which many of the possible variants have evolved not once but several times within a particular plant family or even genus. The resulting array of forms gives the angiosperms an extremely wide variety of pollination and fruit-dispersal mechanisms. In the case of biochemical relationships, however, the evolutionary steps appear to have been large and definitive, and whole families of plants can be characterized biochemically and associated with major groups of plant-eating insects. These biochemical relationships appear to have played a key role in the success of the angiosperms, which have a vastly more diverse array of secondary plant products than any other group of organisms.

Summary

The Angiosperms' Closest Relatives Are Thought to Be the *Bennettitales* and Gnetophytes

The ancestry of the angiosperms, long a subject of debate, has been investigated by phylogenetic analyses. Such methods have defined seed plants as one evolutionary line, or clade, and the *Bennettitales* and gnetophytes as the seed plants most closely related to angiosperms. All three groups possess flowerlike reproductive structures and collectively are referred to as "anthophytes." Only the angiosperms, however, belong to the phylum *Anthophyta.*

Several Factors Help Explain the Worldwide Success of Angiosperms

The earliest definite angiosperm remains are from the Early Cretaceous period, about 130 million years ago; they include both flowers and pollen. The flowering plants became dominant worldwide between 80 and 90 million years ago. Possible reasons for their success include various adaptations for drought resistance, including the evolution of the deciduous habit, as well as the evolution of efficient and often specialized mechanisms for pollination and seed dispersal. The primary radiation of angiosperms took place when the southern continents were united in the great land masses of Gondwanaland, which have separated progressively during the history of the group.

The Magnoliids, the Living Relatives of Early Angiosperms, Fall into Two Groups

The woody magnoliids have large, robust, bisexual flowers, often with many, spirally arranged floral parts. The remaining magnoliids, the paleoherbs, have relatively few flower parts and often unisexual flowers. Both of these groups, which together comprise about 3 percent of living angiosperm species, are characterized by pollen with a single pore (or furrow), as are the monocots, which include about 22 percent of living angiosperms. The eudicots, with pollen that has three apertures (pores or furrows), comprise about three-quarters of the species of angiosperms and are an enormously successful group.

The Four Whorls of Flower Parts Have Evolved in Different Ways

Typical angiosperm flowers consist of four whorls. The outermost whorl consists of sepals, which are specialized leaves that protect the flower in bud. In contrast, the petals of most angiosperms have evolved from stamens that have lost their sporangia during the course of evolution. Stamens with anthers that comprise two pairs of pollen sacs are one of the diagnostic features of angiosperms. In the course of evolution, differentiation between the anther and the slender filament seems to have increased. Carpels are somewhat leaflike structures that have been transformed during the course of evolution to enclose the ovules. In most plants, the carpels have become specialized and differentiated into a swollen, basal ovary, a slender style, and a receptive terminal stigma. The loss of individual floral whorls and fusion within and between adjacent whorls have led to the evolution of many specialized floral types, which are often characteristic of particular families.

Angiosperms Are Pollinated by a Variety of Agents

Pollination by insects is basic in the angiosperms, and the first pollinating agents were probably beetles. The closing of the carpel, in an evolutionary sense, may have contributed to protection of the ovules from visiting insects. Pollination interactions with more specialized groups of insects seem to have evolved later in the history of the angiosperms, and wasps, flies, butterflies, and moths have each left their mark on the morphology of certain angiosperm flowers. The bees, however, are the most specialized and constant of flower-visiting insects and have probably had the greatest effect on the evolution of angiosperm flowers. Each group of flower-visiting animals is associated with a particular group of floral characteristics related to the animals' visual and olfactory senses. Some angiosperms have become wind-pollinated, shedding copious quantities of small, non-sticky pollen and having well-developed, often feathery stigmas that are efficient in collecting pollen from the air. Water-pollinated plants have either filamentous pollen grains that float to submerged flowers or various ways of transmitting pollen through or across the surface of the water.

Various Factors Affect the Relationship between Plant and Pollinator

Flowers that are regularly visited and pollinated by animals with high energy requirements, such as hummingbirds, hawkmoths, and bats, must produce large amounts of nectar. These sources of nectar must then be protected and concealed from other potential visitors with lower energy requirements. Such visitors might satiate themselves with nectar from a single flower (or from the flowers of a single plant) and therefore fail to move on to another plant of the same species to effect cross-pollination. Wind pollination is most effective when individual plants grow together in large groups, whereas insects, birds, or bats can carry pollen great distances from plant to plant.

Flower Colors Are Determined Mainly by Carotenoids and Flavonoids

Carotenoids are yellow, oil-soluble pigments that occur in the chloroplasts and act as accessory pigments in photosynthesis. Flavonoids are water-soluble ring compounds present in the vacuole. Anthocyanins, which are blue or red pigments that constitute one major class of flavonoids, are especially important in determining the colors of flowers and other plant parts.

Fruits Are Basically Mature Ovaries

Fruits are just as diverse as the flowers from which they are derived, and they can be classified either morphologically, in terms of their structure and development, or functionally, in terms of their methods of dispersal. Fruits are basically mature ovaries, but if additional flower parts are retained in their mature structure, they are said to be accessory fruits. Simple fruits are derived from one carpel or from a group of united carpels, aggregate fruits from the free carpels of one flower, and multiple fruits from the fused carpels of several or many flowers. Dehiscent fruits split open to release the seeds, and indehiscent fruits do not.

Fruits and Seeds Are Dispersed by Wind, Water, or Animals

Wind-borne fruits or seeds are light and often have wings or tufts of hairs that aid in their dispersal. The fruits of some plants expel their seeds explosively. Some seeds or fruits are borne away by water, in which case they must be buoyant and have water-resistant coats. Others are disseminated by birds or mammals and frequently have fleshy coverings that are tasty or hooks, spines, or other devices that adhere to the coats of mammals or to feathers. Ants disperse the seeds and fruits of many plants; such dispersal units typically have an oily appendage, an elaiosome, which the ants consume.

Secondary Metabolites Are Important in the Evolution of Angiosperms

Biochemical coevolution has been an important aspect of the evolutionary success and diversification of the angiosperms. Certain groups of angiosperms have evolved various secondary products, or secondary metabolites, such as alkaloids, which protect them from most foraging herbivores. However, certain herbivores (normally those with narrow feeding habits) are able to feed on those plants and are regularly found associated with them. Potential competitors are excluded from the same plants because of their inability to handle the toxins. This pattern indicates that a stepwise pattern of coevolutionary interaction has occurred, and it appears likely that the early angiosperms also may have been protected by their ability to produce some chemicals that functioned as poisons for herbivores.

Selected Key Terms

accessory fruit p. 543
achene p. 544
aggregate fruits p. 543
anthocyanins p. 542
berries p. 543
betacyanins p. 543
capsule p. 544
caryopsis p. 545
coevolution p. 532
cypsela p. 545
dehiscent fruits p. 544
drupes p. 543
flavonoids p. 542
flavonols p. 542
follicle p. 544
fruitlets p. 543
Gondwanaland p. 523
indehiscent fruits p. 544
legume p. 544
magnoliids p. 519
mimicry p. 550
multiple fruits p. 543
nectar p. 525
nut p. 545
paleoherbs p. 521
pappus p. 528
parthenocarpic fruits p. 543
pollinium p. 528
pomes p. 544
samara p. 545
schizocarp p. 545
silique p. 544
simple fruits p. 543
woody magnoliids p. 519

Questions

1. What concept is embodied in the term "anthophyte" (not to be confused with *Anthophyta*)?
2. What unique characteristics of *Anthophyta* (angiosperms) indicate that the members of this phylum were derived from a single common ancestor?
3. Evolutionarily, petals apparently have been derived from two different sources. What are they?
4. Explain what is meant by coevolution, and provide two examples involving different insects and flowers.
5. Why are wind-pollinated angiosperms best represented in temperate regions and relatively rare in the tropics?
6. Distinguish among simple, aggregate, and multiple fruits, and give an example of each.

SECTION 5

The Angiosperm Plant Body: Structure and Development

*Expanding shoot of the shagbark hickory (*Carya ovata*). Over winter, this shoot was greatly telescoped and existed as a terminal bud. As the bud expanded, the protective bud scales (below) separated and folded back. After the shoot is fully expanded, a new terminal bud will form and pass through a dormant period before it will be capable of expanding and repeating the cycle.*

Chapter 23

Early Development of the Plant Body

OVERVIEW

Depending upon your point of view, when you watch a seedling emerge from the soil you are witnessing the beginning, middle, or end of a process. To a gardener, the emergence of the seedling is the first visible sign of life—a harbinger of the flower, fruit, or leaf that is the gardener's goal. But we know from Chapter 21 that the life of the new plant begins much earlier, with the fusion of sperm and egg to form a zygote within the embryo sac. Thus, to the botanist, the emergence of the seedling is merely an intermediate step in plant development. In the present chapter we take yet a third perspective: the emergence of the seedling is the culmination of embryonic development.

Embryonic development begins once the zygote divides; the two daughter cells constitute the first cells of the young embryo. In this chapter we follow the embryo as it progresses through a typical series of stages, eventually forming a structurally mature embryo consisting of a specific arrangement of embryonic tissues and organs within the seed.

But as gardeners know, the embryonic root and shoot will not emerge from the seed in the process of seed germination until the seed is exposed to the proper external conditions, including adequate moisture, temperature, and oxygen. Even if these conditions are met, some seeds still will not germinate because conditions within the seed are inadequate. Such seeds are said to be dormant. We discuss the factors affecting seed germination and dormancy and then describe the emerging young seedling.

CHECKPOINTS

By the time you finish reading this chapter, you should be able to answer the following questions:

1. How is polarity important in embryonic development of plants?
2. What are the three primary meristems of plants, and to which tissues do they give rise?
3. Through what sequence of stages of development do the embryos of eudicots progress? How does embryo development in monocots differ from that in eudicots?
4. How have mutations helped us understand embryo development?
5. What are the main parts of a mature eudicot or monocot embryo? What additional structures are present in grass embryos?

In the previous section, we traced the long evolutionary development of the angiosperms beginning with their presumed ancestor, a relatively complex, multicellular green alga. As we pointed out, the forked axes of early vascular plants were the forerunners of the shoots and roots of most modern vascular plants.

In this section, we shall be concerned with the structure and development of the angiosperm plant body, or sporophyte, which is a result of that long period of evolutionary specialization. This chapter begins with the formation of the embryo, a process known as **embryogenesis.** Embryogenesis establishes the body plan of the plant, consisting of two superimposed patterns: an **apical-basal pattern** along the main axis and a **radial pattern** of concentrically arranged tissue systems (Figure 23–1). Embryogenesis is accompanied by seed development. The seed, with its mature embryo, stored food, and protective seed coat, confers significant selective advantages over plants lacking seeds: the seed enhances the plant's ability to survive adverse environmental conditions and facilitates dispersal of the species.

As we follow the development of the angiosperm plant body, keep in mind the evolution of the vascular plants that we explored in the previous section. Developmental and evolutionary biologists are particularly excited over what can be learned about the evolution of developmental patterns. Great strides have been made by studying highly conserved genes—genes with similar DNA sequences in distantly related organisms—that regulate key developmental pathways. Much of what we know about this regulation comes from studying mutations that disrupt normal embryo development.

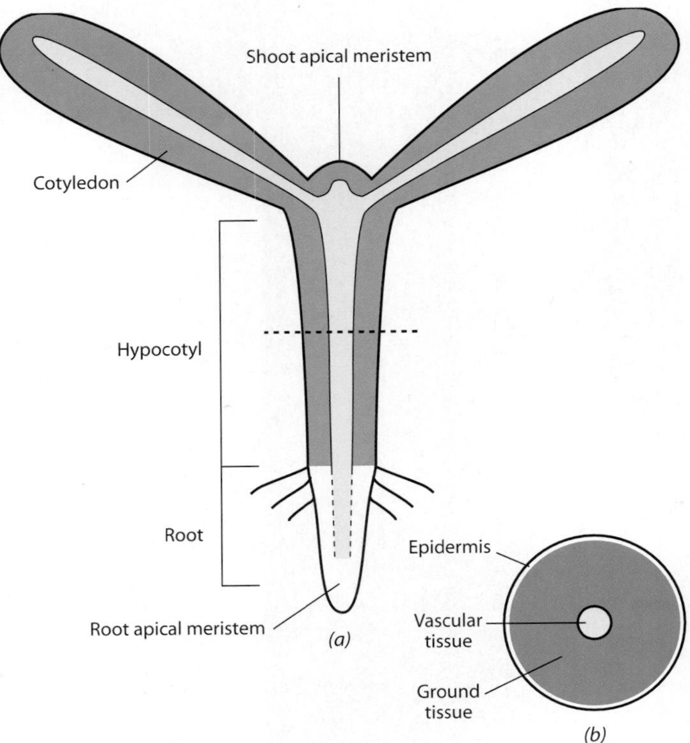

23–1
Diagrammatic representation of the body plan of an Arabidopsis *seedling.* ***(a)*** *The apical-basal pattern consists of an axis with a shoot tip at one end and a root tip at the other.* ***(b)*** *A transverse section through the hypocotyl reveals the radial pattern, consisting of the three tissue systems, as represented by the epidermis, ground tissue, and vascular tissue.*

Formation of the Embryo

The early stages of embryogenesis are essentially the same in magnoliids, eudicotyledons, and monocotyledons (Figures 23–2 and 23–3). Formation of the embryo begins with division of the zygote within the embryo sac of the ovule. In most flowering plants, the first division of the zygote is asymmetrical and transverse with regard to the long axis of the zygote (Figures 23–2a and 23–3a). With this division, the **polarity** of the embryo is established. The upper (chalazal) pole, consisting of a small *apical cell,* gives rise to most of the mature embryo. The lower (micropylar) pole, consisting of a large *basal cell,* produces a stalklike **suspensor** that anchors the embryo at the micropyle, the opening in the ovule through which the pollen tube enters.

Polarity is a key component of biological pattern formation. The term arises by analogy with a magnet, which has plus and minus poles. "Polarity" means simply that whatever is being discussed—a plant, an animal, an organ, a cell, or a molecule—has one end that is different from the other. Polarity in plant stems is a familiar phenomenon. In plants that are propagated by stem cuttings, for example, roots will form at the lower end of the stem, with leaves and buds at the upper end.

The establishment of polarity is an essential first step in the development of all higher organisms, because it fixes the structural **axis** of the body, the "backbone" on which the lateral appendages will be arranged. In some angiosperms, polarity is already established in the egg cell and zygote, where the nucleus and most of the cytoplasmic organelles are located in the upper portion of the cell, and the lower portion is dominated by a large vacuole.

Through an orderly progression of divisions, the embryo eventually differentiates into a nearly spherical structure—the **embryo proper**—and the suspensor (Figures 23–2d and 23–3b, c). Before this stage is reached, the developing embryo is frequently referred to as the **proembryo.**

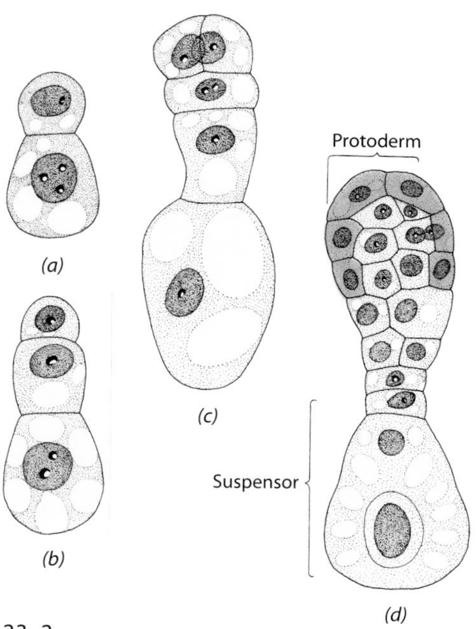

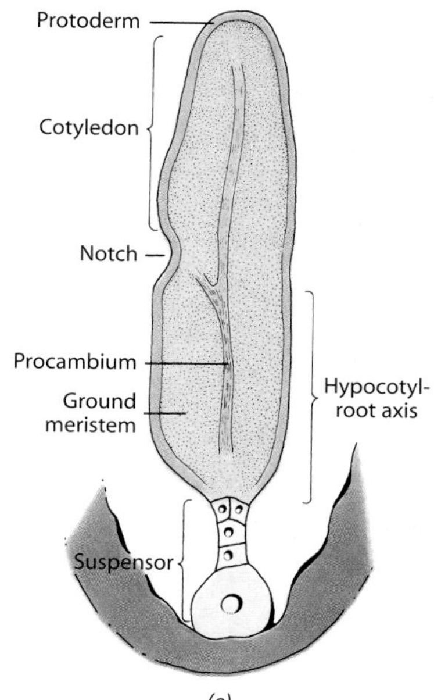

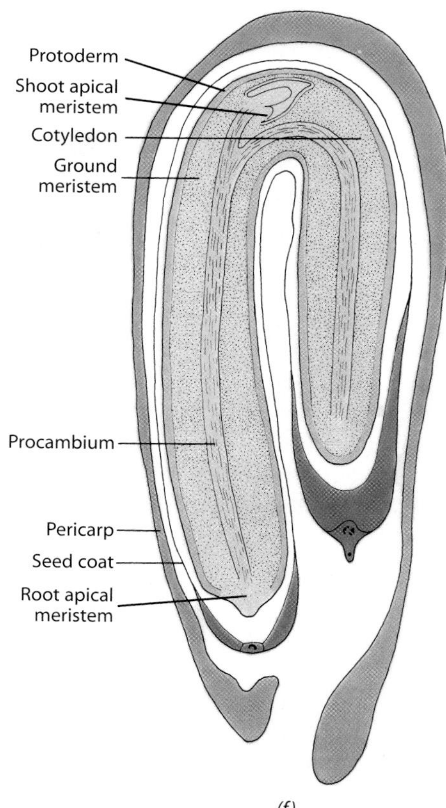

23–2
Stages in the development of the embryo of arrowhead (Sagittaria), *a monocot. Early stages:* **(a)** *The two-celled stage, resulting from transverse division of the zygote.* **(b)** *The three-celled proembryo.* **(c)** *Disregarding the large basal cell, the proembryo is now at the four-celled stage. All four of these cells, through a series of divisions, contribute to formation of the embryo proper.* **(d)** *The protoderm has been initiated at the terminal end of the embryo proper. At this stage, the suspensor consists of only two cells, one of which is the large basal cell. Late stages:* **(e)** *A depression, or notch (the site of the future apical meristem of the shoot), has formed at the base of the emerging cotyledon.* **(f)** *The cotyledon curves, and the embryo is approaching maturity. The suspensor has disappeared.*

The Protoderm, Procambium, and Ground Meristem Are the Primary Meristems

When first formed, the embryo proper consists of a mass of relatively undifferentiated cells. Soon, however, changes in the internal structure of the embryo proper result in the initial development of the tissue systems of the plant. The future epidermis, the **protoderm,** is formed by periclinal divisions—divisions parallel to the surface—in the outermost cells of the embryo proper (Figures 23–2d and 23–3c). Subsequently, vertical divisions within the embryo proper result in the initial distinction between **procambium** and **ground meristem** (Figure 23–3d, e). The ground meristem, the precursor of the **ground tissue,** surrounds the procambium, the precursor of the vascular tissues, xylem and phloem. The protoderm, ground meristem, and procambium—the so-called **primary meristems,** or primary meristematic tissues—extend into the other regions of the embryo as embryogenesis continues (Figures 23–2e, f and 23–3e, f).

Embryos Typically Progress through a Sequence of Developmental Stages

The stage of embryo development preceding cotyledon development—that is, when the embryo proper is spherical—is often referred to as the *globular stage.* Development of the cotyledons, the first leaves of the plant, may begin either during or after the time at which the procambium becomes discernible. As it develops, the globular embryo in eudicots gradually assumes a two-lobed form, or heart shape, and this stage is often called the *heart stage* (Figure 23–3d). The globular embryos of monocots, which form only one cotyledon, become cylindrical in shape (Figure 23–2e). It is during the transition between the globular stage of development and the emergence of the cotyledon(s) that the apical-basal pattern of the embryo becomes discernible, with the partitioning of the axis into shoot meristem, cotyledon(s), hypocotyl (the stemlike axis below the cotyledon(s)), embryonic root, and root meristem.

23–3
*Stages in the development of the embryo of shepherd's purse (*Capsella bursa-pastoris*), a eudicot.* ***(a)*** *Two-celled stage, resulting from transverse division of the zygote into an upper apical cell and a lower basal cell.* ***(b)*** *Six-celled proembryo. The suspensor is now distinct from the two terminal cells, which develop into the embryo proper. Endosperm provides food for the developing embryo.* ***(c)*** *The embryo proper is globular and has a protoderm, which will develop into the epidermis. The large cell near the bottom is the basal cell of the suspensor.* ***(d)*** *Embryo at the heart stage, when the cotyledons, the first leaves of the plant, begin to emerge.* ***(e)*** *The embryo seen here is at the torpedo stage. In* Capsella *the embryos curve. Ground meristem, the precursor of the ground tissue, surrounds the procambium, which will develop into the vascular tissues, xylem and phloem.* ***(f)*** *Mature embryo. The part of the embryo below the cotyledons is the hypocotyl. At the lower end of the hypocotyl is an embryonic root, or radicle.*

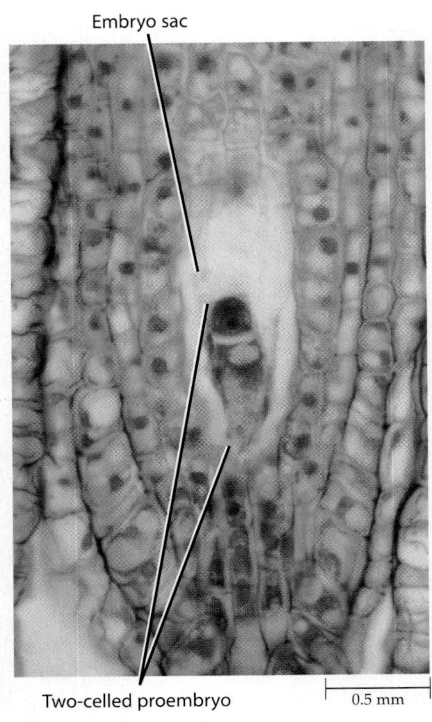

(a)

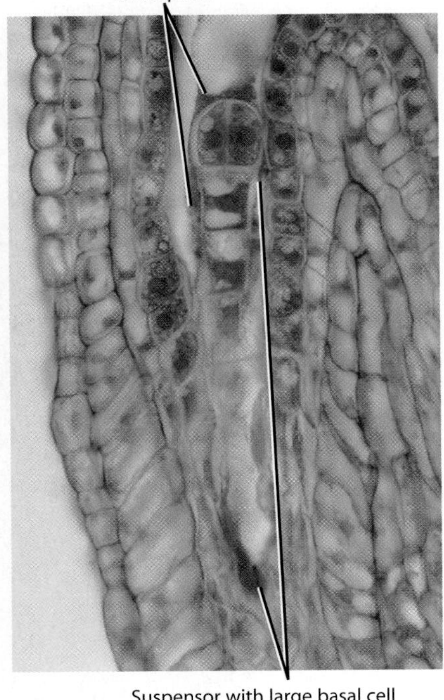

(b)

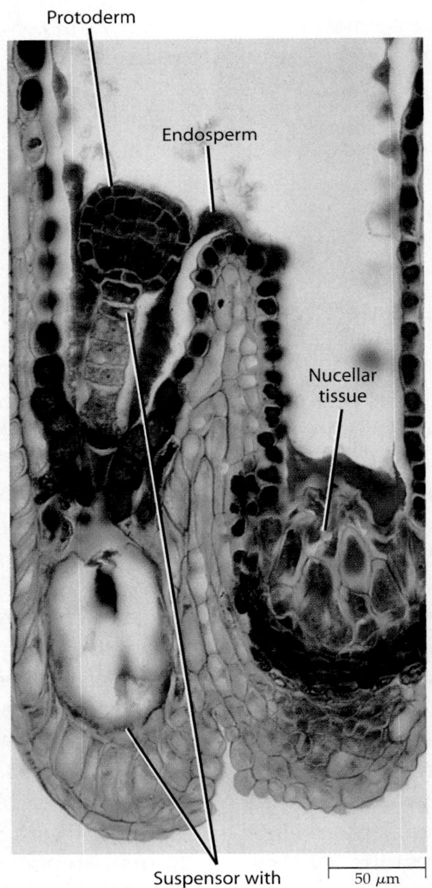

(c)

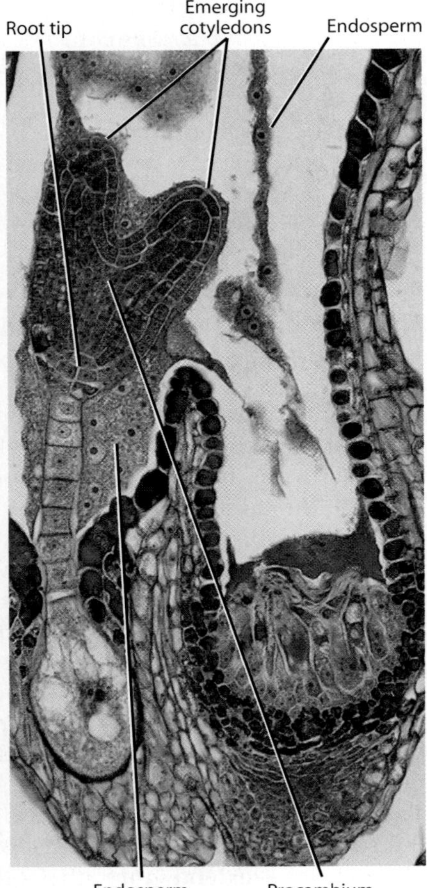

(d)

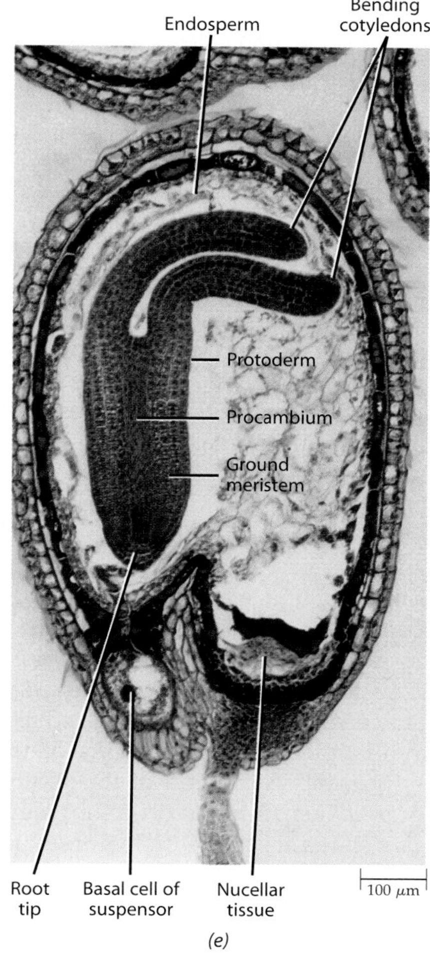

(e)

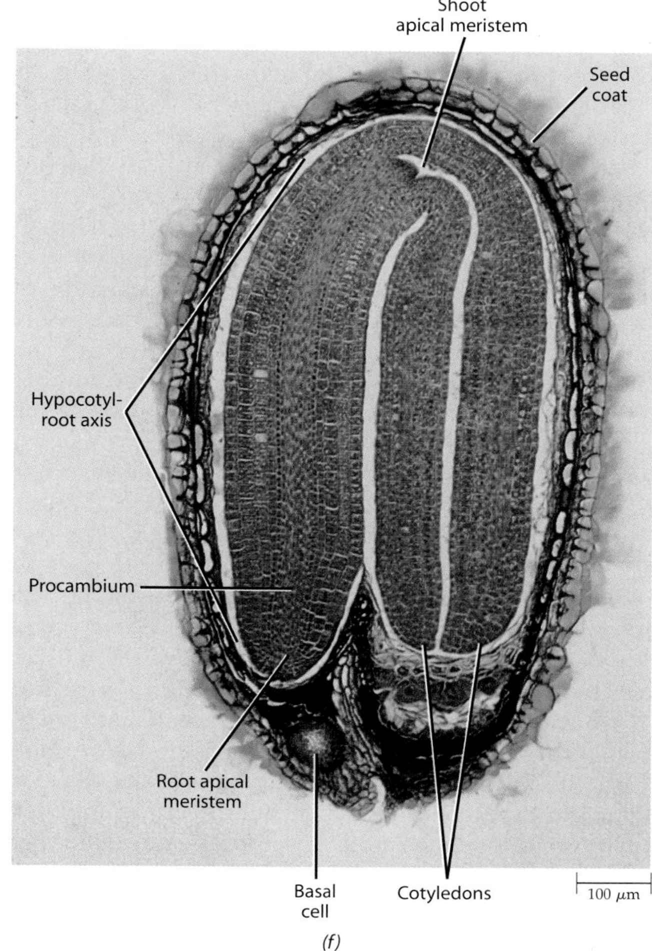

(f)

As embryo development continues, the cotyledon(s) and axis elongate (the so-called *torpedo stage* of embryo development), and the primary meristems extend along with them (Figures 23–2f and 23–3e). During elongation, the embryo either remains straight or becomes curved. The single cotyledon of the monocot often becomes so large in comparison with the rest of the embryo that it is the dominating structure (see Figure 23–7c).

During the early stages of embryogenesis, cell division takes place throughout the young sporophyte. However, as the embryo develops, the addition of new cells gradually becomes restricted to the apical meristems of the shoot and root. As discussed previously (page 8), apical meristems are found at the tips of all shoots and roots and are composed of cells that are capable of repeated division. In magnoliid and eudicot embryos, the apical meristem of the shoot arises between the two cotyledons (Figure 23–3f). In monocot embryos, on the other hand, the shoot apical meristem arises on one side of the cotyledon and is completely surrounded by a sheathlike extension from the base of the cotyledon (Figure 23–2f). The apical meristems of both shoot and root are of great importance, because these tissues are the source of virtually all of the new cells responsible for the development of the seedling and the adult plant.

The Suspensor Plays a Supporting Role in Development of the Embryo Proper

The suspensors associated with the embryos of *Selaginella*, a seedless vascular plant (page 442), and pine (page 482) function merely to push the developing embryos into nutritive tissues. In the past, it was believed that the suspensors of angiosperms played a similar role. We now know, however, that angiosperm suspensors are metabolically active. They play a role in supporting early development of the embryo proper by providing it with nutrients and growth regulators, particularly gibberellins. In some embryos, the suspensor produces proteinaceous materials, which are then absorbed by the rapidly growing embryo. The suspensor is short-lived, undergoing programmed cell death (page 576) at the torpedo stage of development. It is therefore not present in the mature seed (Figure 23–2f).

Several lines of evidence indicate that the normally developing embryo proper limits the growth and differentiation of the suspensor, the cells of which have the potential to generate an embryo. Such evidence is provided by several *Arabidopsis* embryo-defective mutants, such as *raspberry*1, *sus,* and *twn,* in which development of the embryo proper is disrupted. In all of these mutants, disrupted embryo development leads directly to proliferation of suspensor cells. Some of these suspensor cells acquire characteristics normally restricted to cells of the embryo proper, including the production of food reserves. The *twn* mutants are the most striking of the *Arabidopsis* embryo-defective mutants. In these mutants, the cells of the suspensor undergo embryogenic transformation, resulting in the formation of seeds with viable twin and occasionally triplet embryos (Figure 23–4). The results of these studies reveal that interactions occur between the suspensor and embryo proper. It has been suggested that during normal development, the embryo proper transmits specific inhibitory signals to the suspensor that suppress the embryonic pathway.

Genes Have Been Identified That Determine Major Events in Embryogenesis

The description of embryo formation tells us how the primary body plan of the plant develops, but it reveals little about the underlying mechanisms. As mentioned in the *Arabidopsis thaliana* essay on page 228, large populations of mutagen-treated *Arabidopsis* plants are being systematically screened for mutations that have an effect on plant development. Using this procedure, it is possible to identify the genes that govern plant development, a first step in determining how the genes function.

Using this method, very promising results have been obtained in identifying genes responsible for major events in *Arabidopsis* embryogenesis. (One approach to generating mutations that affect embryogenesis in *Arabidopsis* is outlined in Figure 23–5.) Over 50 distinct genes governing embryonic pattern formation have thus far been identified. Some of these regulatory genes affect the apical-basal pattern of the embryo and seedling. Mutations in these genes delete different regions of the apical-basal pattern (Figure 23–6). Another group of genes is involved in determining the radial pattern of tissue differentiation. Mutations in one of these genes, for example, prevents formation of the protoderm. Still another group of genes is involved in regulating the changes in cell shape that give the embryo and seedling their characteristic elongated shape.

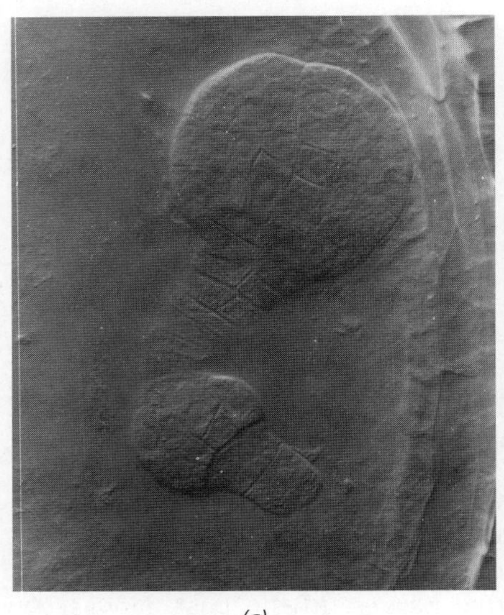

(a)

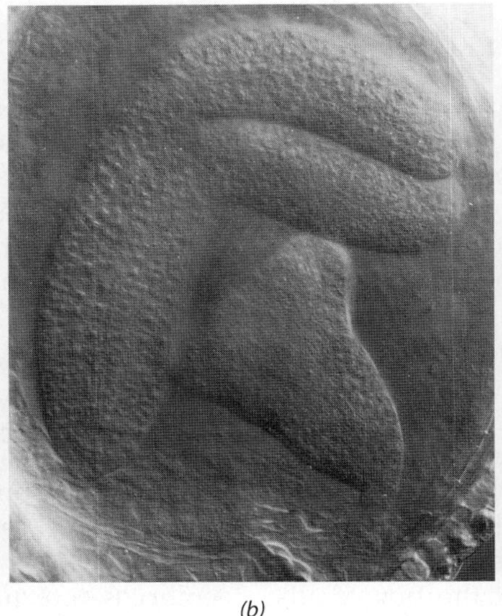

(b)

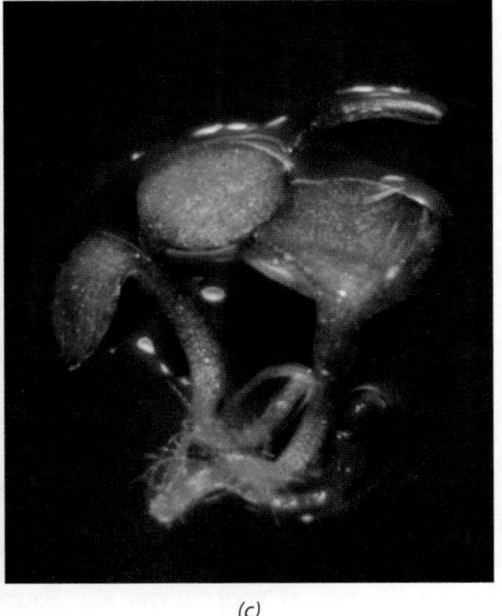

(c)

23–4
Twin embryo development in the twn *mutant of* Arabidopsis thaliana, *a eudicot.* ***(a)*** *A secondary embryo can be seen developing from the suspensor of the larger, primary embryo. Both embryos are at the globular stage.* ***(b)*** *The cotyledons of the primary embryo are partly developed. Cotyledon development in the secondary embryo is at an early stage.* ***(c)*** *Twin seedlings from a germinated seed. The seedling on the left resembles the wild-type (nonmutant) seedling. Its sibling has one large cotyledon.*

Appendix D

Classification of Organisms

There are several different ways to classify organisms. The one presented here follows the overall scheme described in Chapter 13, in which organisms are divided into three domains: *Archaea*, *Bacteria*, and *Eukarya*. *Archaea* and *Bacteria* are distinct lineages of prokaryotic organisms. *Eukarya*, which consists entirely of eukaryotic organisms, includes four kingdoms: *Protista*, *Animalia*, *Fungi*, and *Plantae*. The chief taxonomic categories are domain, kingdom, phylum, class, order, family, genus, species.

The classification that follows includes the phyla of *Protista*, except those considered *Protozoa*, as well as the *Fungi* and *Plantae*. Certain classes given prominence in this book are also included, but the listings are far from complete. The number of species given for each group is the estimated number of living species that have been described and named. Only groups that include living species are described. Viruses are not included in this appendix, but are discussed in Chapter 14.

Domain *Archaea*

Archaea are prokaryotic cells. They lack a nuclear envelope, plastids, mitochondria, and other membrane-bounded organelles, and 9-plus-2 flagella. They are unicellular but sometimes aggregate into filaments or other superficially multicellular bodies. Their predominant mode of nutrition is absorption, but one group of genera obtain their energy by metabolizing sulfur, and another genus, *Halobacterium*, does so through the operation of a proton pump. Many archaea are methanogens, generators of methane. Others are among the most "salt-loving" (extreme halophiles) and "heat-loving" (extreme thermophiles) of all known prokaryotes. Reproduction is asexual, by fission; genetic recombination has not been observed. They are diverse morphologically, being motile flagellated or nonmotile rods, cocci, and spirilla. *Archaea* differ fundamentally from *Bacteria* in the base sequences of their ribosomal RNAs and lipid composition of their plasma membranes. They also differ from *Bacteria* in lacking peptidoglycans in their cell walls. There are fewer than 100 named species.

Domain *Bacteria*

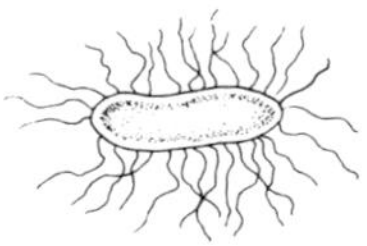

Like *Archaea*, *Bacteria* lack a nuclear envelope, plastids, mitochondria, and other membrane-bounded organelles, and 9-plus-2 flagella. They are unicellular, but many form aggregates. Their predominant mode of nutrition is absorption, but some groups are photosynthetic or chemosynthetic. Reproduction is predominantly asexual, by fission or budding, but portions of DNA molecules may also be exchanged between cells under certain circumstances. They are motile by simple flagella or by gliding, or they may be nonmotile.

About 2600 species of bacteria are recognized at present, but this is probably only a small fraction of the actual number. The recognition of species is not comparable with that in eukaryotes and is based largely upon metabolic features. One group, the class *Rickettsiae*—very small bacteria—occurs widely as parasites in arthropods and may consist of tens of thousands of species, depending upon the classification criteria used; they are not included in the estimate given here.

The *Bacteria* can be divided into twelve major lineages, or kingdoms. Among these, cyanobacteria are an ancient group that is abundant and important ecologically. Formerly and misleadingly called "blue-green algae," cyanobacteria have a type of photosynthesis that is based on chlorophyll *a*. Cyanobacteria, like the red algae, also have accessory pigments called phycobilins. Many cyanobacteria can fix atmospheric nitrogen, often in specialized cells called heterocysts. Some cyanobacteria form complex filaments or other colonies. Although some 7500 species of cyanobacteria have been described, a more reasonable estimate puts the number of these specialized bacteria at about 200 distinct nonsymbiotic species.

Domain *Eukarya*

Kingdom *Fungi*

Eukaryotic multicellular or rarely unicellular organisms in which the nuclei occur in a basically continuous mycelium; this mycelium becomes septate in certain groups and at certain stages of the life cycle. Fungi are heterotrophic; they obtain their nutrition by absorption. Members of all but one phylum *(Chytridiomycota)* form important symbiotic relationships with the roots of plants, called mycorrhizae. Reproductive cycles typically include both sexual and asexual phases. There are over 70,000 valid species of fungi to which names have been given, and many more will eventually be found. Some have been named two or more times; this is particularly so for fungi that may be classified both as ascomycetes and as members of the deuteromycetes. The major characteristics of the phyla of *Fungi* are provided in Table 15–1.

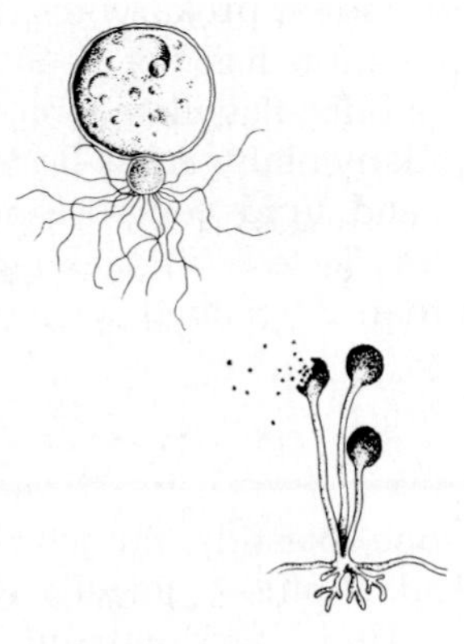

Phylum *Chytridiomycota:* Chytrids. Predominantly aquatic heterotrophic organisms with motile cells characteristic of certain stages in their life cycle. The motile cells of most have a single, posterior, whiplash flagellum. Their cell walls are composed of chitin, but other polymers may also be present, and they store their food as glycogen. There are about 790 species.

Phylum *Zygomycota:* Terrestrial fungi with the hyphae septate only during the formation of reproductive bodies; chitin is predominant in the cell walls. Zygomycetes can usually be recognized by their profuse, rapidly growing hyphae. The class includes about 1060 described species, some of which occur as components of the endomycorrhizae that are found in about 80 percent of all vascular plants.

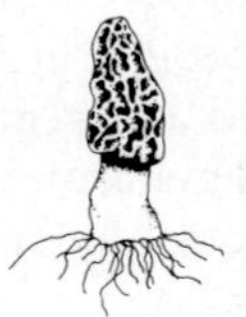

Phylum *Ascomycota:* Terrestrial and aquatic fungi with the hyphae septate but the septa perforated; complete septa cut off the reproductive bodies, such as spores or gametangia. Chitin is predominant in the cell walls. Sexual reproduction involves the formation of a characteristic cell—the ascus—in which meiosis takes place and within which ascospores are formed. The hyphae in many ascomycetes are packed together into complex "bodies" known as ascomata. There are about 32,300 species of ascomycetes.

Phylum *Basidiomycota:* Terrestrial fungi with the hyphae septate but the septa perforated; complete septa cut off reproductive bodies, such as spores. Chitin is predominant in the cell walls. Sexual reproduction involves formation of basidia, in which meiosis takes place and on which the basidiospores are borne. *Basidiomycota* are dikaryotic during most of their life cycle, and there is often complex differentiation of "tissues" within their basidiomata. They are the fungal components of most ectomycorrhizae. There are some 22,300 described species.

Class *Basidiomycetes:* Includes the hymenomycetes and gasteromycetes. The hymenomycetes produce basidiospores in a hymenium exposed on a basidioma; the mushrooms, coral fungi, and shelf, or bracket, fungi. The gasteromycetes produce basidiospores inside basidiomata, where they are completely enclosed for at least part of their development; puffballs, earthstars, stinkhorns, and their relatives. The basidia of most *Basidiomycetes* are aseptate (internally undivided).

Class *Teliomycetes:* Consists of fungi commonly referred to as rusts. Unlike the *Basidiomycetes,* the rusts do not form basidiomata, and they have septate basidia.

Class *Ustomycetes:* Commonly referred to as smuts. Like the *Teliomycetes,* they do not form basidiomata, and they form septate basidia.

Yeasts: A yeast, by definition, is simply a unicellular fungus that reproduces primarily by budding. The yeasts are not a formal taxonomic group. The yeast growth form is exhibited by a broad range of unrelated fungi encompassing the *Zygomycota, Ascomycota,* and *Basidiomycota.* Most yeasts are ascomycetes, but at least a quarter of the genera are *Basidiomycota.*

Deuteromycetes: The deuteromycetes are an artificial assemblage of about 15,000 distinct species of fungi for which only the asexually reproductive state is known or in which the sexual reproductive features are not used as the basis of classification. They are often referred to as the "Fungi Imperfecti."

Lichens: A lichen is a mutualistic symbiotic association between a fungal partner and a certain genus of green algae or of cyanobacteria. The fungal component of the lichen is called the mycobiont, and the photosynthetic component is called the photobiont. About 98 percent of the mycobionts belong to the *Ascomycota,* the remainder to the *Basidiomycota.* About 13,250 species of lichen-forming fungi have been described.

Kingdom *Protista*

Eukaryotic unicellular or multicellular organisms. Their modes of nutrition include ingestion, photosynthesis, and absorption. True sexuality is present in most phyla. They move by means of 9-plus-2 flagella or are nonmotile. Fungi, plants, and animals are specialized multicellular groups derived from *Protista.* The phyla treated in this book are categorized as heterotrophic protists (slime molds and water molds; the first three phyla listed below) and photosynthetic protists (the algae). The characteristics of the phyla of *Protista* are outlined in Tables 16–1 and 17–1.

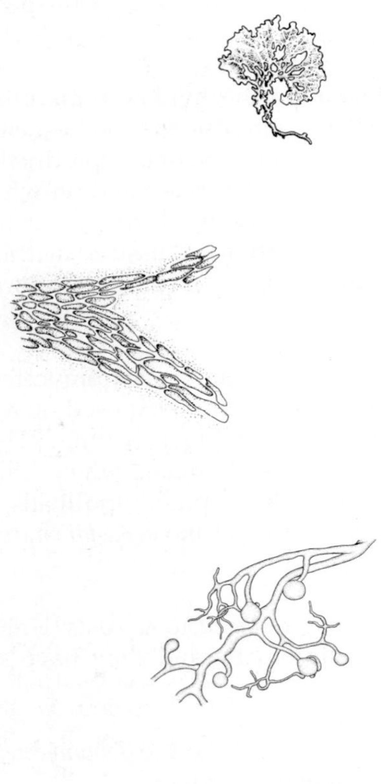

Phylum *Myxomycota*: Plasmodial slime molds. Heterotrophic amoeboid organisms that form a multinucleate plasmodium that creeps along as a mass and eventually differentiates into sporangia, each of which is multinucleate and eventually gives rise to many spores. Sexual reproduction is occasionally observed. The predominant mode of nutrition is by ingestion. There are about 700 species.

Phylum *Dictyosteliomycota*: Cellular slime molds, or dictyostelids. Heterotrophic organisms that exist as separate amoebas (called myxamoebas). Eventually, the myxamoebas swarm together to form a pseudoplasmodium, within which they retain their individual identities. Ultimately, the pseudoplasmodium differentiates into a fruiting body. Sexual reproduction involves structures known as macrocysts. Pairs of amoebas first fuse, forming zygotes. Subsequently, these zygotes attract and then engulf nearby amoebas. The principal mode of nutrition is by ingestion. There are about 50 known species in four genera.

Phylum *Oomycota*: Water molds and related organisms. Aquatic or terrestrial organisms with motile cells characteristic of certain stages of their life cycle. The flagella are two in number—one tinsel and one whiplash as is characteristic of heterokonts. Their cell walls are composed of cellulose or celluloselike polymers, and they store their food as glycogen. There are about 694 species.

Phylum *Euglenophyta*: Euglenoids. About a third of the approximately 40 genera of euglenoids have chloroplasts, with chlorophylls *a* and *b* and carotenoids; the others are heterotrophic and essentially resemble members of the phylum *Zoomastigina,* within which they would probably best be included. They store food as paramylon, an unusual carbohydrate. Euglenoids usually have two apical flagella and a contractile vacuole. The flexible pellicle is rich in proteins. Sexual reproduction is unknown. There are some 900 species, most of which occur in fresh water.

Phylum *Cryptophyta*: Cryptomonads. Photosynthetic organisms that possess chlorophylls *a* and *c* and carotenoids; in addition some cryptomonads contain a phycobilin, either phycocyanin or phycoerythrin. They are rich in polyunsaturated fatty acids. In addition to a regular nucleus the cryptomonads contain a reduced nucleus called a nucleomorph. There are 200 known species.

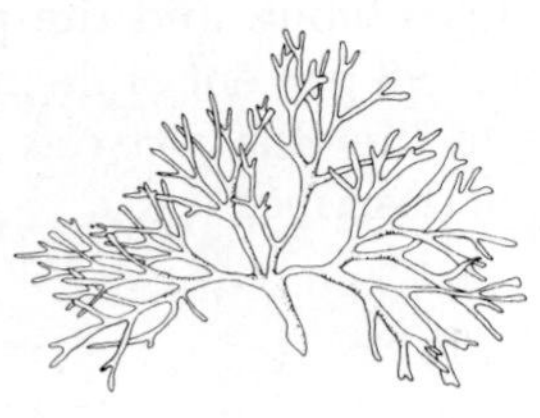

Phylum *Rhodophyta*: Red algae. Primarily marine algae characterized by the presence of chlorophyll *a* and phycobilins. Particularly abundant in tropical and warm waters. Their carbohydrate food reserve is floridean starch, and the cell walls are composed of cellulose or pectins, with calcium carbonate in many. No motile cells are present at any stage in the complex life cycle. The vegetative body is built up of closely packed filaments in a gelatinous matrix and is not differentiated into roots, leaves, and stems. It lacks specialized conducting cells. There are some 4000 to 6000 known species.

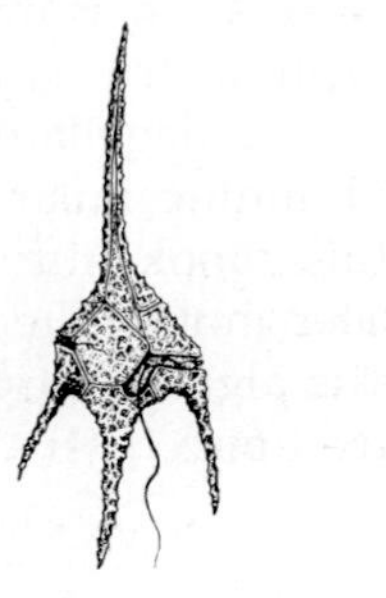

Phylum *Dinophyta*: Dinoflagellates. Autotrophic organisms, about half of which possess chlorophylls *a* and *c* and carotenoids; the other half lack a photosynthetic apparatus and hence obtain their nutrition either by ingesting solid food particles or by absorbing dissolved organic compounds. Food is stored as starch. A layer of vesicles—often containing cellulose—lies beneath the plasma membrane. This phylum contains some 2000 to

4000 known species, mostly biflagellated organisms. These all have lateral flagella, one of which beats in a groove that encircles the organism. Sexual reproduction is generally isogamous, but anisogamy is also present. The mitosis of dinoflagellates is unique. Many—in a form called zooxanthellae—are symbiotic in marine animals, and they make important contributions to the productivity of coral reefs.

Phylum *Haptophyta:* Haptophytes. Mostly photosynthetic organisms that contain chlorophyll *a* and some variation of chlorophyll *c*. Some have the accessory pigment fucoxanthin. The most distinctive feature of the haptophytes is the haptonema, a threadlike structure that extends from the cell along with two flagella. There are about 300 known species of haptophytes.

Phylum *Chrysophyta:* Chrysophytes. Primarily unicellular or colonial organisms that possess chlorophylls *a* and *c* and carotenoids, mainly fucoxanthin. Food is stored as the water-soluble carbohydrate chrysolaminarin. The cell walls are absent or consist of cellulose that may be impregnated with minerals; some are covered with silica scales. There are about 1000 known living species.

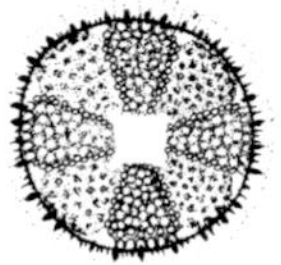

Phylum *Bacillariophyta:* Diatoms. Unicellular or colonial organisms with two-part siliceous cell walls, the two halves of which fit together like a Petri dish. Diatoms have chlorophylls *a* and *c*, as well as fucoxanthin. Reserve storage materials include lipids and chrysolaminarin. Diatoms lack flagella, except on some male gametes. It is estimated that there are at least 100,000 living species, plus thousands of extinct ones.

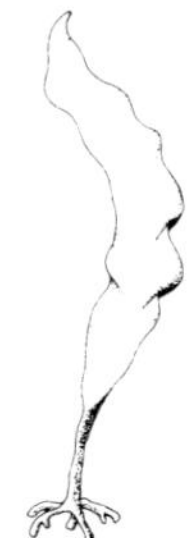

Phylum *Phaeophyta:* Brown algae. Multicellular, nearly entirely marine algae characterized by the presence of chlorophylls *a* and *c* and fucoxanthin. The carbohydrate food reserve is laminarin, and the cell walls have a cellulose matrix containing algin. Motile cells are biflagellated, with one forward flagellum of the tinsel type and one trailing flagellum of the whiplash type. A considerable amount of differentiation is found in some of the kelps (some of the large brown algae of the order *Laminariales*), with specialized conducting cells for transporting the products of photosynthesis to the regions of the body that receive little light. There is, however, no differentiation into roots, leaves, and stems, as in the vascular plants. Although there are only about 1500 species, the brown algae dominate rocky shores throughout the cooler regions of the world.

Phylum *Chlorophyta:* Green algae. Unicellular or multicellular photosynthetic organisms characterized by the presence of chlorophylls *a* and *b* and various carotenoids. The carbohydrate food reserve is starch; only green algae and plants, which are clearly descended from the green algae, store their reserve food inside their plastids. The cell walls of green algae are formed of polysaccharides, sometimes cellulose. Motile cells generally have two apical whiplash flagella. True multicellular genera do not exhibit complex patterns of differentiation. Multicellularity has arisen at least twice. There are about 17,000 known species.

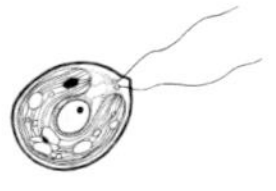

Class *Chlorophyceae:* Green algae in which the unique mode of cell division involves a phycoplast—a system of microtubules parallel to the plane of cell division. The nuclear envelope persists throughout mitosis, and chromosome division occurs within it. Motile cells, if present, are symmetrical and possess two, four, or many flagella that are apical and directed forward.

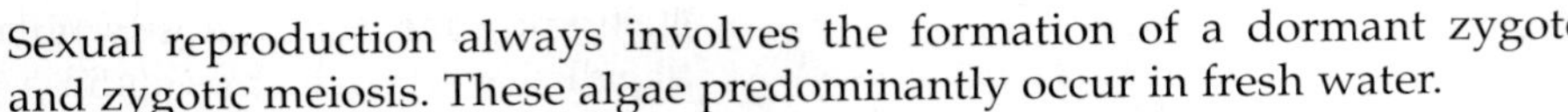

Sexual reproduction always involves the formation of a dormant zygote and zygotic meiosis. These algae predominantly occur in fresh water.

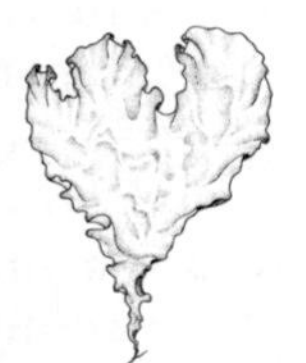

Class *Ulvophyceae:* Green algae with a closed mitosis in which the nuclear envelope persists; the spindle is persistent through cytokinesis. Motile cells, if present, are symmetrical and possess two, four, or many flagella that are apical and directed forward. Sexual reproduction often involves alternation of generations and sporic meiosis, and dormant zygotes are rare. These are predominantly marine algae.

Class *Charophyceae:* Unicellular, few-celled, filamentous, or parenchymatous green algae in which cell division involves a phragmoplast—a system of microtubules perpendicular to the plane of cell division. The nuclear envelope breaks down during the course of mitosis. Motile cells, if present, are asymmetrical and possess two flagella that are subapical and extend laterally at right angles from the cell. Sexual reproduction always involves the formation of a dormant zygote and zygotic meiosis. Certain members of this class resemble plants more closely than do any other organisms. These algae predominantly occur in fresh water.

Kingdom *Plantae*

The plants are autotrophic (some are derived heterotrophs), multicellular organisms possessing advanced tissue differentiation. All plants have an alternation of generations, in which the diploid phase (sporophyte) includes an embryo and the haploid phase (gametophyte) produces gametes by mitosis. Their photosynthetic pigments and food reserves are similar to those of the green algae. Plants are primarily terrestrial. Summaries of some of the characteristics of the various phyla are found in the Summary Tables of Chapters 18 to 20 and in Table 21–1.

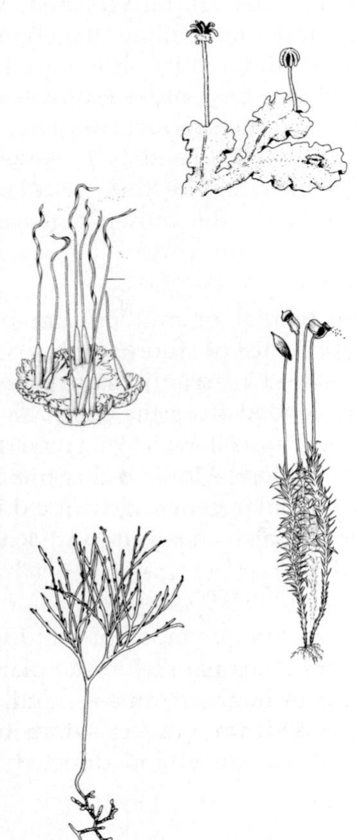

Phylum *Hepatophyta:* Liverworts. *Hepatophyta* and the two following phyla, all of which constitute the bryophytes, have multicellular gametangia with a sterile jacket layer; their sperm are biflagellated. In all three phyla, most photosynthesis is carried out in the gametophyte, upon which the sporophyte is dependent. Liverworts lack specialized conducting tissue (with possibly a few exceptions) and stomata; they are the simplest of all living plants. The gametophytes are thallose or leafy, and the rhizoids are single-celled. There are about 6000 species.

Phylum *Anthocerophyta:* Hornworts. Bryophytes with thallose gametophytes; the sporophyte grows from a basal intercalary meristem for as long as conditions are favorable. Stomata are present on the sporophyte; there is no specialized conducting tissue. There are about 100 species.

Phylum *Bryophyta:* Mosses. Bryophytes with leafy gametophytes; the sporophytes have complex patterns of dehiscence. Specialized conducting tissue is present in both gametophytes and sporophytes of some species. Rhizoids are multicellular. Stomata are present on the sporophytes. There are about 9500 species.

Phylum *Psilotophyta:* Psilotophytes. The *Psilotophyta* and the three following phyla constitute the living phyla of seedless vascular plants. Psilotophytes are homosporous. There are two genera, one of which has leaflike appendages on the stem; both genera have extremely simple sporophytes, with no differentiation between root and shoot. The sperm are motile. There are several species.

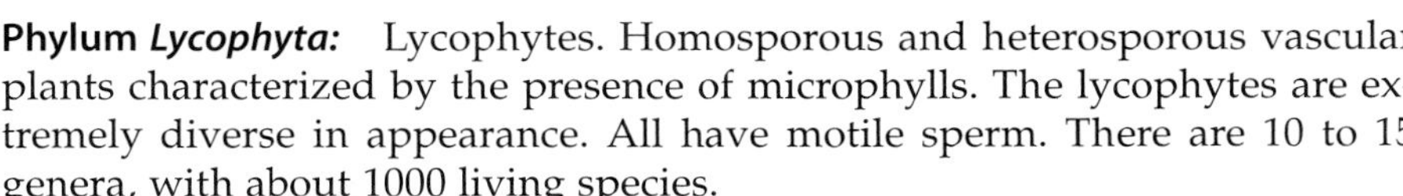

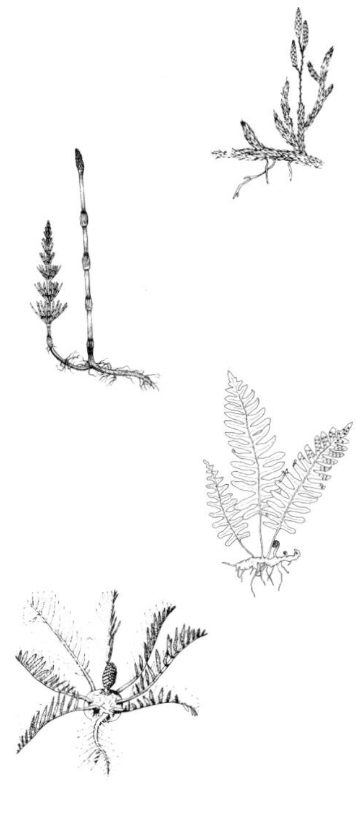

Phylum *Lycophyta:* Lycophytes. Homosporous and heterosporous vascular plants characterized by the presence of microphylls. The lycophytes are extremely diverse in appearance. All have motile sperm. There are 10 to 15 genera, with about 1000 living species.

Phylum *Sphenophyta:* Horsetails. A single genus of homosporous vascular plants, *Equisetum,* with jointed stems marked by conspicuous nodes and elevated siliceous ribs. Sporangia are borne in a strobilus at the apex of the stem. Leaves are scalelike. Sperm are motile. There are 15 living species of horsetails.

Phylum *Pterophyta:* Ferns. Mostly homosporous, although some are heterosporous. All possess a megaphyll. The gametophyte is more or less free-living and usually photosynthetic. Multicellular gametangia and free-swimming sperm are present. There are about 11,000 species.

Phylum *Cycadophyta:* Cycads. This and the following three phyla make up the gymnosperms. Cycads have sluggish cambial growth and pinnately compound, palmlike or fernlike leaves; ovules and seeds are exposed. The sperm are flagellated and motile but are carried to the vicinity of the ovule in a pollen tube. There are 11 genera with about 140 species.

Phylum *Ginkgophyta:* *Ginkgo.* Gymnosperm with considerable cambial growth and fan-shaped leaves with open dichotomous venation; ovules and seeds exposed; seed coats fleshy. Sperm are carried to the vicinity of the ovule in a pollen tube but are flagellated and motile. There is only one species.

Phylum *Coniferophyta:* Conifers. Gymnosperms with active cambial growth and simple leaves; ovules and seeds exposed; sperm nonflagellated. The most familiar group of gymnosperms. There are some 50 genera with about 550 species.

Phylum *Gnetophyta:* Gnetophytes. Gymnosperms with many angiospermlike features, such as vessels; the gnetophytes are the only gymnosperms in which vessels occur. They are the group of gymnosperms most closely related to angiosperms. Motile sperm are absent. There are three very distinctive genera with about 70 species.

Phylum *Anthophyta:* Flowering plants; angiosperms. Seed plants in which ovules are enclosed in a carpel and seeds are borne within fruits. The angiosperms are extremely diverse vegetatively but are characterized by the flower, which is basically insect-pollinated. Other modes of pollination, such as wind pollination, have been derived in a number of different lines. The gametophytes are much reduced, with the female gametophyte often consisting of only seven cells at maturity. Double fertilization involving the two sperm of the mature microgametophyte gives rise to the zygote (sperm and egg) and to the primary endosperm nucleus (sperm and polar nuclei); the former becomes the embryo and the latter becomes a special nutritive tissue called the endosperm. There are about 235,000 species.

Class *Monocotyledones:* Monocots. Flower parts usually in threes; leaf venation usually parallel; primary vascular bundles in the stem are scattered; true secondary growth is not present; one cotyledon. There are about 65,000 species.

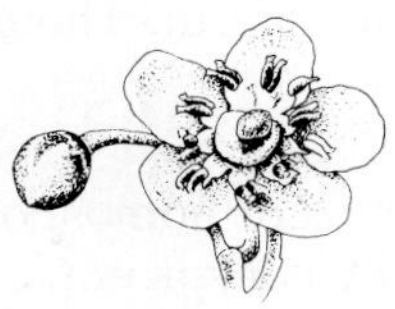

Class *Eudicotyledones:* Eudicots. Flower parts usually in fours or fives; leaf venation usually netlike; primary vascular bundles in the stem are in a ring; many with a vascular cambium and true secondary growth; two cotyledons. There are about 165,000 species.

Together, the monocots and eudicots represent about 97 percent of angiosperms. The remaining 3 percent of living angiosperms are the magnoliids, those angiosperms with the most primitive features and the ancestors of both monocots and eudicots.

Some other common terms used to describe major groups of plants deserve mention here. In systems in which the algae and fungi are regarded as plants, they are often grouped as a subkingdom, *Thallophyta,* the thallophytes: organisms with no highly differentiated tissues, such as root, stem, or leaf, and no vascular tissues (xylem and phloem). The bryophytes and vascular plants are then grouped into a second subkingdom, *Embryophyta,* in which the zygote develops into a multicellular embryo still encased in an archegonium or an embryo sac. All embryophytes are marked by an alternation of heteromorphic generations. The term "embryophyte" commonly is used as a synonym for plants.

Although they are no longer used in formal schemes of classification, terms such as "algae," "thallophytes," "vascular plants," and "gymnosperms" are still sometimes useful in an informal sense. An even earlier scheme divided all plants into "phanerogams," those with flowers, and "cryptogams," those lacking flowers; these terms are still occasionally seen.

Glossary

A

Å: *See* ångstrom.

a- [Gk. *a-*, not, without]: Prefix that negates the succeeding part of the word; "an-" before vowels and "h".

abscisic acid [L. *abscissus,* to cut off]: A plant hormone that brings about dormancy in buds, maintains dormancy in seeds, and brings about stomatal closing, among other effects.

abscission (ăb•sizh´ŭn): The dropping off of leaves, flowers, fruits, or other plant parts, usually following the formation of an abscission zone.

abscission zone: The area at the base of a leaf, flower, or fruit, or other plant part containing tissues that play a role in the separation of a plant part from the plant body.

absorption spectrum: The spectrum of light waves absorbed by a particular pigment.

accessory bud: A bud generally located above or on either side of the main axillary bud.

accessory cell: *See* subsidiary cell.

accessory fruit: A fruit, or assemblage of fruits, whose fleshy parts are derived largely or entirely from tissues other than the ovary. An example is the strawberry, whose receptacle is fleshy and whose fruits (achenes) are embedded in its surface.

accessory pigment: A pigment that captures light energy and transfers it to chlorophyll *a.*

acclimation: The process by which numerous physical and physiological processes prepare a plant for winter.

achene: A simple, dry, one-seeded indehiscent fruit in which the seed coat is not adherent to the pericarp.

acid: A substance that dissociates in water, releasing hydrogen ions (H^+) and thus causing a relative increase in the concentration of these ions; having a pH in solution of less than 7; a proton donor; the opposite of "base."

acid growth hypothesis: The hypothesis that acidification of the cell wall leads to hydrolysis of restraining bonds within the wall and, consequently, to cell elongation driven by the turgor pressure of the wall.

actin filament: A helical protein filament, 5 to 7 nanometers thick, composed of globular actin molecules; a major constituent of the cytoskeleton of all eukaryotic cells; also called microfilament.

actinomorphic [Gk. *aktis,* ray of light, + *morphē,* form]: Pertaining to a type of flower that can be divided into two equal halves in more than one longitudinal plane; also called radially symmetrical or regular; *see also* zygomorphic.

action spectrum: The spectrum of light waves that elicits a particular reaction.

active site: The region of an enzyme surface that binds the substrate during the reaction catalyzed by the enzyme.

active transport: Energy-requiring transport of a solute across a membrane in the direction of increasing concentration (against the concentration gradient).

ad- [L. *ad-*, toward, to]: Prefix meaning "toward" or "to."

adaptation [L. *adaptare,* to fit]: A peculiarity of structure, physiology, or behavior that aids in fitting an organism to its environment.

adaptive radiation: The evolution from one kind of organism to several divergent forms, each specialized to fit a distinct and diverse way of life.

adenine (ăd´e•nēn): A purine base present in DNA, RNA, and nucleotide derivatives, such as ADP and ATP.

adenosine triphosphate (ATP): A nucleotide consisting of adenine, ribose sugar, and three phosphate groups; the major source of usable chemical energy in metabolism. On hydrolysis, ATP loses one phosphate to become adenosine diphosphate (ADP), releasing usable energy.

adhesion [L. *adhaerere,* to stick to]: The sticking together of unlike objects or materials.

adnate [L. *adnatus,* grown together]: Said of fused unlike parts, as stamens and petals; *see also* connate.

ADP: *See* adenosine triphosphate.

adsorption [L. *ad-,* to, + *sorbere,* to suck in]: The adhesion of a liquid, gaseous, or dissolved substance to a solid, resulting in a higher concentration of the substance.

adventitious [L. *adventicius,* not properly belonging to]: Referring to a structure arising from an unusual place, such as buds at other places than leaf axils, or roots growing from stems or leaves.

aeciospore (ē´sĭ•o•spor) [Gk. *aikia,* injury, + *spora,* seed]: A binucleate spore of rust fungi; produced in an aecium.

aecium, *pl.* **aecia:** In rust fungi, a cuplike structure in which aeciospores are produced.

aerobic [Gk. *aer,* air, + *bios,* life]: Requiring free oxygen.

aerobic respiration: *See* respiration.

after-ripening: Term applied to the metabolic changes that must occur in some dormant seeds before germination can occur.

agar: A gelatinous substance derived from certain red algae; used as a solidifying agent in the preparation of nutrient media for growing microorganisms.

aggregate fruit: A fruit developing from the several separate carpels of a single flower.

akinete: A vegetative cell that is transformed into a thick-walled resistant spore in cyanobacteria.

albuminous cell: Certain ray and axial parenchyma cells in gymnosperm phloem that are spatially and functionally associated with the sieve cells; also called Strasburger cells.

aleurone [Gk. *aleuron,* flour]: A proteinaceous material, usually in the form of small granules, occurring in the outermost cell layer of the endosperm of wheat and other grains.

alga, *pl.* **algae** (ăl'ga, ăl'je): Traditional term for a series of unrelated groups of photosynthetic eukaryotic organisms lacking multicellular sex organs (except for the charophytes); the "blue-green algae," or cyanobacteria, are one of the groups of photosynthetic bacteria.

algin: An important polysaccharide component of brown algal cell walls; used as a stabilizer and emulsifier for some foods and for paint.

alkali [Arabic *algili,* the ashes of the plant saltwort]: A substance with marked basic properties.

alkaline: Pertaining to substances that release hydroxyl ions (OH^-) in water; having a pH greater than 7.

alkaloids: Bitter-tasting nitrogenous compounds that are basic (alkaline) in their chemical properties; includes morphine, cocaine, caffeine, nicotine, and atropine.

allele (ă•lēl′) [Gk. *allēlōn,* of one another, + *morphē,* form]: One of the two or more alternative forms of a gene.

allelopathy [Gk. *allelon,* of one another, + *pathos,* suffering]: The inhibition of one species of plant by chemicals produced by another plant.

allopatric speciation [Gk. *allos,* other, + *patra,* fatherland, country]: Speciation that occurs as the result of the geographic separation of a population of organisms.

allopolyploid: A polyploid formed from the union of two separate chromosome sets and their subsequent doubling.

allosteric interaction [Gk. *allos,* other, + *steros,* shape]: A change in the shape of a protein resulting from the binding to the protein of a nonsubstrate molecule; in its new shape, the protein typically has different properties.

alternate phyllotaxy: Leaf arrangement in which there is one bud or one leaf at a node.

alternation of generations: A reproductive cycle in which a haploid (n) phase, the gametophyte, produces gametes, which, after fusion in pairs to form a zygote, germinate, producing a diploid ($2n$) phase, the sporophyte. Spores produced by meiotic division from the sporophyte give rise to new gametophytes, completing the cycle.

amino acids [Gk. *Ammon,* referring to the Egyptian sun god, near whose temple ammonium salts were first prepared from camel dung]: Nitrogen-containing organic acids, the units, or "building blocks," from which protein molecules are built.

ammonification: Decomposition of amino acids and other nitrogen-containing organic compounds, resulting in the production of ammonia (NH_3) and ammonium ions (NH_4^+).

amoeboid [Gk. *amoibe,* change]: Moving or eating by means of pseudopodia (temporary cytoplasmic protrusions from the cell body).

amphi- [Gk. *amphi-,* on both sides]: Prefix meaning "on both sides," "both," or "of both kinds."

amylase (ăm′ĭ•lās): An enzyme that breaks down starch into smaller units.

amyloplast: A leucoplast (colorless plastid) that forms starch grains.

an- [Gk. *an-,* not, without]: Prefix equivalent to "a-," meaning "not" or "without"; used before vowels and "h."

anabolism [Gk. *ana-,* up, + *-bolism* (as in metabolism)]: The constructive part of metabolism; the total chemical reactions involved in biosynthesis.

anaerobic [Gk. *an-,* without, + *aer,* air, + *bios,* life]: Referring to any process that can occur without oxygen, or to the metabolism of an organism that can live without oxygen; strict anaerobes cannot survive in the presence of oxygen.

analogous [Gk. *analogos,* proportionate]: Applied to structures similar in function but different in evolutionary origin, such as the phyllodes of an Australian *Acacia* and the leaves of an oak.

anaphase [Gk. *ana,* away, + *phasis,* form]: A stage in mitosis in which the chromatids of each chromosome separate and move to opposite poles; similar stages in meiosis in which chromatids or paired chromosomes move apart.

anatomy: The study of the internal structure of organisms; morphology is the study of their external structure.

andro- [Gk. *andros,* man]: Prefix meaning "male."

androecium [Gk. *andros,* man, + *oikos,* house]: (1) The floral whorl that comprises the stamens; (2) in leafy liverworts, a packetlike swelling containing the antheridia.

aneuploid: A chromosomal aberration in which the chromosome number differs from the normal chromosome number for the species by a small number.

angiosperm [Gk. *angion,* a vessel, + *sperma,* a seed]: Literally, a seed borne in a vessel (carpel); thus one of a group of plants whose seeds are borne within a mature ovary (fruit).

ångstrom [after A. J. Ångstrom, a Swedish physicist, 1814–74]: A unit of length equal to 10^{-10} meter; abbreviated Å.

anion [Gk. *anienae,* to go up]: A negatively charged ion.

anisogamy [Gk. *aniso,* unequal, + *gamos,* marriage]: The condition of having dissimilar motile gametes.

annual [L. *annulus,* year]: A plant whose life cycle is completed in a single growing season.

annual ring: In wood, the growth layer formed during a single year; *see also* growth layer.

annulus [L. *anus,* ring]: In ferns, a row of specialized cells in a sporangium; in gill fungi, the remnant of the inner veil forming a ring on the stalk.

antenna complex: The portion of a photosystem that consists of pigment molecules (antenna pigments) that gather light and "funnel" it to the reaction center.

anterior: Situated before or toward the front.

anther [Gk. *anthos,* flower]: The pollen-bearing portion of a stamen.

antheridiophore [Gk. *anthos,* flower, + *phoros,* bearing]: In some liverworts, a stalk that bears antheridia.

antheridium: A sperm-producing structure that may be multicellular or unicellular.

anthocyanin [Gk. *anthos,* flower, + *kyanos,* dark blue]: A water-soluble blue or red pigment found in the cell sap.

Anthophyta: The phylum of angiosperms, or flowering plants.

anthophytes: Collectively the *Bennettitales,* gnetophytes, and angiosperms, all with a shared possession of flowerlike reproductive structures; not to be confused with the *Anthophyta,* the angiosperm phylum.

antibiotic [Gk. *anti,* against or opposite, + *biotikos,* pertaining to life]: Natural organic substances that retard or prevent the growth of organisms; generally used to designate substances formed by microorganisms that prevent growth of other microorganisms.

anticlinal: Perpendicular to the surface.

anticodon: In a tRNA molecule, the three-nucleotide sequence that base-pairs with the mRNA codon for the amino acid carried by that particular tRNA; the anticodon is complementary to the mRNA codon.

antipodals: Three (sometimes more) cells of the mature embryo sac, located at the end opposite the micropyle.

apical dominance: The influence exerted by a terminal bud in suppressing the growth of lateral, or axillary, buds.

apical meristem: The meristem at the tip of the root or shoot in a vascular plant.

apomixis [Gk. *apo,* separate, away from, + *mixis,* mingling]: Reproduction without meiosis or fertilization; vegetative reproduction.

apoplast [Gk. *apo,* away from, + *plastos,* molded]: The cell wall continuum of a plant or organ; the movement of substances via the cell walls is called apoplastic movement or transport.

apoptosis (ā-po-to-sis)**:** Programmed cell death.

apothecium [Gk. *apotheke,* storehouse]: A cup-shaped or saucer-shaped open ascoma.

arch-, archeo- [Gk. *arche,* archos, beginning]: Prefix meaning "first," "main," or "earliest."

Archaea: A phylogenetic domain of prokaryotes consisting of the methanogens, most extreme halophiles and hyperthermophiles, and *Thermoplasma.*

archegoniophore [Gk. *archegonos*, the first of a race, + *phoros*, bearing]: In some liverworts, a stalk that bears archegonia.

archegonium, *pl.* **archegonia:** A multicellular structure in which a single egg is produced; found in the bryophytes and some vascular plants.

aril (ăr′ĭl) [L. *arillus*, grape, seed]: An accessory seed covering, often formed by an outgrowth at the base of the ovule; often brightly colored, which may aid in dispersal by attracting animals that eat it and, in the process, carry the seed away from the parent plant.

artifact [L. *ars*, art, + *facere*, to make]: A product that exists because of an extraneous, especially human, agency and does not occur in nature.

artificial selection: The breeding of selected organisms to produce strains with desired characteristics.

ascogenous hyphae [Gk. *askos*, bladder, + *genous*, producing]: Hyphae containing paired haploid male and female nuclei; they develop from an ascogonium and eventually give rise to asci.

ascogonium: The oogonium or female gametangium of the ascomycetes.

ascoma, *pl.* **ascomata:** A multicellular structure in ascomycetes lined with specialized cells called asci, in which nuclear fusion and meiosis occur. Ascomata may be open or closed. Also called an ascocarp.

ascospore: A spore produced within an ascus; found in ascomycetes.

ascus, *pl.* **asci:** A specialized cell, characteristic of the ascomycetes, in which two haploid nuclei fuse to produce a zygote that immediately divides by meiosis; at maturity, an ascus contains ascospores.

aseptate [Gk. *a-*, not, + L. *septum*, fence]: Nonseptate; lacking cross walls.

asexual reproduction: Any reproductive process, such as fission or budding, that does not involve the union of gametes.

assimilate stream: The flow of assimilates, or food materials, in the phloem; moves from source to sink.

atom [Gk. *atomos*, indivisible]: The smallest unit into which a chemical element can be divided and still retain its characteristic properties.

atomic nucleus: The central core of an atom, containing protons and neutrons, around which electrons orbit.

atomic number: The number of protons in the nucleus of an atom.

atomic weight: The weight of a representative atom of an element relative to the weight of an atom of carbon ^{12}C, which has been assigned the value 12.

ATP: *See* adenosine triphosphate.

ATP synthase: An enzyme complex that forms ATP from ADP and phosphate during oxidative phosphorylation in the inner mitochondrial membrane.

auto- [Gk. *autos*, self, same]: Prefix meaning "same" or "self-same."

autoecious [Gk. *autos*, self, + *oikia*, dwelling]: In some rust fungi, completing the life cycle on a single species of host plant.

autopolyploid: A polyploid formed from the doubling of a single genome.

autoradiograph: A photographic print made by a radioactive substance acting upon a sensitive photographic film.

autotroph [Gk. *autos*, self, + *trophos*, feeder]: An organism that is able to synthesize the nutritive substances it requires from inorganic substances in its environment; *see also* heterotroph.

auxin [Gk. *auxein*, to increase]: A class of plant hormones that control cell elongation, among other effects.

axial system: In secondary xylem and secondary phloem; the term applied collectively to cells derived from fusiform cambial initials. The long axes of these cells are oriented parallel with the main axis of the root or stem. Also called longitudinal system and vertical system.

axil [Gk. *axilla*, armpit]: The upper angle between a twig or leaf and the stem from which it grows.

axillary: Term applied to buds or branches occurring in the axil of a leaf.

B

bacillus, *pl.* **bacilli** (ba•sĭl′ŭs) [L. *baculum*, rod]: A rod-shaped bacterium.

backcross: The crossing of a hybrid with one of its parents or with a genetically equivalent organism; a cross between an individual whose genes are to be tested and one that is homozygous for all of the recessive genes involved in the experiment.

Bacteria: The phylogenetic domain consisting of all prokaryotes that are not members of the domain *Archaea*.

bacteriophage [Gk. *bakterion*, little rod, + *phagein*, to eat]: A virus that parasitizes bacterial cells.

bacterium, *pl.* **bacteria:** A prokaryotic organism. *See also Bacteria.*

bacteroid: An enlarged, deformed *Rhizobium* or *Bradyrhizobium* cell found in root nodules; capable of nitrogen fixation.

bark: A nontechnical term applied to all tissues outside the vascular cambium in a woody stem; *see also* inner bark *and* outer bark.

basal body: A self-reproducing, cylinder-shaped cytoplasmic organelle from which cilia or flagella arise; identical in structure to the centriole, which is involved in mitosis and meiosis in most animals and protists.

base: A substance that dissociates in water, causing a decrease in the concentration of hydrogen ions (H^+), often by releasing hydroxyl ions (OH^-); bases have a pH in solution of more than 7; the opposite of "acid."

basidioma, *pl.* **basidiomata:** A multicellular structure, characteristic of the basidiomycetes, within which basidia are formed.

basidiospore: A spore of the basidiomycota, produced within and borne on a basidium following nuclear fusion and meiosis. Also called a basidiocarp.

basidium, *pl.* **basidia:** A specialized reproductive cell of the *Basidiomycota*, often club-shaped, in which nuclear fusion and meiosis occur.

berry: A simple fleshy fruit that includes a fleshy ovary wall and one or more carpels and seeds; examples are the fruits of grapes, tomatoes, and bananas.

bi- [L. *bis*, double, two]: Prefix meaning "two," "twice," or "having two points."

biennial: A plant that normally requires two growing seasons to complete its life cycle, flowering and fruiting in its second year.

bilaterally symmetrical: *See* zygomorphic.

biological clock [Gk. *bios*, life, + *logos*, discourse]: The internal timing mechanism that governs the innate biological rhythms of organisms.

biomass: Total dry weight of all organisms in a particular population, sample, or area.

biome: A complex of terrestrial communities of very wide extent, characterized by its climate and soil; the largest ecological unit.

biosphere: The zone of air, land, and water at the surface of the Earth that is occupied by organisms.

biotechnology: The practical application of advances in hormone research and DNA biochemistry to manipulate the genetics of plants.

biotic: Relating to life.

bisexual flower: A flower that has at least one functional stamen and one functional carpel.

bivalent [L. *bis*, double, + *valere*, to be strong]: A pair of synapsed homologous chromosomes. Also called a tetrad.

blade: The broad, expanded part of a leaf; the lamina.

body cell: Vegetative or somatic cell; *see also* spermatogenous cell.

bordered pit: A pit in which the secondary wall arches over the pit membrane.

bract: A modified, usually reduced leaflike structure.

branch root: *See* lateral root.

bryophytes (brī′o•fīts): The members of the phyla of nonvascular plants; the mosses, hornworts, and liverworts.

bud: (1) An embryonic shoot, often protected by young leaves; (2) a vegetative outgrowth of yeasts and some bacteria as a means of asexual reproduction.

bulb: A short underground stem covered by enlarged and fleshy leaf bases containing stored food.

bulk flow: The overall movement of water or some other liquid induced by gravity, pressure, or an interplay of both.

bulliform cell: A large epidermal cell present, with other such cells, in longitudinal rows in grass leaves; also called motor cell. Believed to be involved with the mechanism of rolling and unrolling of the leaves.

bundle scar: Scar or mark left on leaf scar by vascular bundles broken at the time of leaf fall, or abscission.

bundle sheath: Layer or layers of cells surrounding a vascular bundle; may consist of parenchyma or sclerenchyma cells, or both.

bundle-sheath extension: A group of cells extending from a bundle sheath of a vein in the leaf mesophyll to either upper or lower epidermis or both; may consist of parenchyma, collenchyma, or sclerenchyma.

C

C_3 pathway: *See* Calvin cycle.

C_3 plants: Plants that employ only the Calvin cycle, or C_3 pathway, in the fixation of CO_2; the first stable product is the three-carbon compound 3-phosphoglycerate.

C_4 pathway: The set of reactions through which carbon dioxide is fixed to a compound known as phosphoenolpyruvate (PEP) to yield oxaloacetate, a four-carbon compound.

C_4 plants: Plants in which the first product of CO_2 fixation is a four-carbon compound (oxaloacetate); both the Calvin cycle (C_3 pathway) and C_4 pathway are employed by C_4 plants.

callose: A complex branched carbohydrate that is a common wall constituent associated with the sieve areas of sieve elements; may develop in reaction to injury in sieve elements and parenchyma cells.

callus [L. *callos,* hard skin]: Undifferentiated tissue; a term used in tissue culture, grafting, and wound healing.

calorie [L. *calor,* heat]: The amount of energy in the form of heat required to raise the temperature of one gram of water 1°C. In making metabolic measurements, the kilocalorie (kcal), or Calorie—the amount of heat required to raise the temperature of one kilogram of water 1°C—is generally used.

Calvin cycle: The series of enzymatically mediated photosynthetic reactions during which carbon dioxide is reduced to 3-phosphoglyceraldehyde and the carbon dioxide acceptor, ribulose 1,5-bisphosphate, is regenerated. For every three molecules of carbon dioxide entering the cycle, a net gain of one molecule of glyceraldehyde 3-phosphate results.

calyptra [Gk. *kalyptra,* covering for the head]: The hood or cap that partially or entirely covers the capsule of some species of mosses; it is formed from the expanded archegonial wall.

calyx (kā′lĭks) [Gk. *kalyx,* a husk, cup]: The sepals collectively; the outermost flower whorl.

CAM: *See* crassulacean acid metabolism.

cambial zone: A region of thin-walled, undifferentiated meristematic cells between the secondary xylem and secondary phloem; consists of cambial initials and their recent derivatives.

cambium [L. *cambiare,* to exchange]: A meristem that gives rise to parallel rows of cells; commonly applied to the vascular cambium and the cork cambium, or phellogen.

capsid: The protein coat of a virus particle.

capsule: (1) In angiosperms, a dehiscent, dry fruit that develops from two or more carpels; (2) a slimy layer around the cells of certain bacteria; (3) the sporangium of bryophytes.

carbohydrate [L. *carbo,* ember, + *hydro,* water]: An organic compound consisting of a chain of carbon atoms to which hydrogen and oxygen are attached in a 2:1 ratio; examples are sugars, starch, glycogen, and cellulose.

carbon cycle: Worldwide circulation and utilization of carbon atoms.

carbon fixation: The conversion of CO_2 into organic compounds during photosynthesis.

carbon-fixation reactions: In photosynthetic cells, the light-independent enzymatic reactions concerned with the synthesis of glucose from CO_2, ATP, and NADPH; also called light-independent reactions and dark reactions.

carnivorous: Feeding upon animals, as opposed to feeding upon plants (herbivorous); also refers to plants that are able to utilize proteins obtained from trapped animals, chiefly insects.

carotene (kăr′o•tēn) [L. *carota,* carrot]: A yellow or orange pigment belonging to the carotenoid group.

carotenoids (kă•rŏt′e•noids): A class of fat-soluble pigments that includes the carotenes (yellow and orange pigments) and the xanthophylls (yellow pigments); found in chloroplasts and chromoplasts of plants. Carotenoids act as accessory pigments in photosynthesis.

carpel [Gk. *karpos,* fruit]: One of the members of the gynoecium, or inner floral whorl; each carpel encloses one or more ovules. One or more carpels form a gynoecium.

carpellate: Pertaining to a flower with one or more carpels but no functional stamens; also called pistillate.

carpogonium [Gk. *karpos,* fruit, + *gonos,* offspring]: In red algae, the female gametangium.

carposporangium [Gk. *karpos,* fruit, + *spora,* seed, + *angeion,* vessel]: In red algae, a carpospore-containing cell.

carpospore: In red algae, the single diploid protoplast found within a carposporangium.

carriers: Transport proteins that bind specific solutes and undergo conformational change in order to transport the solute across the membrane.

caryopsis [Gk. *karyon,* a nut, + *opsis,* appearance]: Simple, dry, one-seeded indehiscent fruit with the pericarp firmly united all around the seed coat; a grain characteristic of the grasses (family *Poaceae*).

Casparian strip [after Robert Caspary, German botanist]: A bandlike region of primary wall containing suberin and lignin; found in anticlinal—radial and transverse—walls of endodermal and exodermal cells.

catabolism [Gk. *katabolē,* throwing down]: Collectively, the chemical reactions resulting in the breakdown of complex materials and involving the release of energy.

catalyst [Gk. *katalysis,* dissolution]: A substance that accelerates the rate of a chemical reaction but is not used up in the reaction; enzymes are catalysts.

category [Gk. *katēgoria,* category]: In a hierarchical classification system, the level at which a particular group is ranked.

cation [Gk. *katienai,* to go down]: A positively charged ion.

catkin: A spikelike inflorescence of unisexual flowers; found only in woody plants.

cDNA: *See* complementary DNA.

cell [L. *cella,* small room]: The structural unit of organisms; in plants, cells consist of the cell wall and the protoplast.

cell division: The division of a cell and its contents, usually into two roughly equal parts.

cell plate: The structure that forms at the equator of the spindle in the dividing cells of plants and a few green algae during early telophase.

cell sap: The fluid contents of the vacuole.

cellular respiration: *See* respiration.

cellulase: An enzyme that hydrolyzes cellulose.

cellulose: A carbohydrate; the chief component of the cell wall in plants and some protists; an insoluble complex carbohydrate formed of microfibrils of glucose molecules attached end to end.

cell wall: The rigid outermost layer of the cells found in plants, some protists, and most prokaryotes.

central mother cells: Relatively large vacuolate cells in a subsurface position in apical meristems of shoots.

centriole [Gk. *kentron,* center, + L. *-olus,* little one]: A cytoplasmic organelle found outside the nuclear envelope, and identical in structure to a basal body; centrioles are found in the cells of most eukaryotes other than fungi, red algae, and the nonflagellated cells of plants. Centrioles divide and organize spindle fibers during mitosis and meiosis.

centromere [Gk. *kentron,* center, + *meros,* a part]: Region of constriction of chromosome that holds sister chromatids together.

chalaza [Gk. *chalaza,* small tubercle]: The region of an ovule or seed where the funiculus unites with the integuments and the nucellus.

channel proteins: Transport proteins that form water-filled pores that extend across cellular membranes; when open, the channel proteins allow specific solutes to pass through them.

chemical potential: The activity or free energy of a substance; it is dependent upon the rate of motion of the average molecule and the concentration of the molecules.

chemical reaction: The making or breaking of chemical bonds between atoms or molecules.

chemiosmotic coupling: Coupling of ATP synthesis to electron transport via an electrochemical H^+ gradient across a membrane.

chemoautotrophic: Refers to prokaryotes that are able to manufacture their own basic foods by using the energy released by specific inorganic reactions; *see also* autotroph.

chiasma (kī•ă z'ma) [Gk. *chiasma,* a cross]: The X-shaped figure formed by the meeting of two nonsister chromatids of homologous chromosomes; the site of crossing-over.

chitin (kī'tĭn) [Gk. *chiton,* tunic]: A tough, resistant, nitrogen-containing polysaccharide forming the cell walls of certain fungi, the exoskeleton of arthropods, and the epidermal cuticle of other surface structures of certain protists and animals.

chlor- [Gk. *chloros,* green]: Prefix meaning "green."

chlorenchyma: Parenchyma cells that contain chloroplasts.

chlorophyll [Gk. *chloros,* green, + *phyllon,* leaf]: The green pigment of plant cells, which is the receptor of light energy in photosynthesis; also found in algae and photosynthetic bacteria.

chloroplast: A plastid in which chlorophylls are contained; the site of photosynthesis. Chloroplasts occur in plants and algae.

chlorosis: Loss or reduced development of chlorophyll.

chroma- [Gk. *chroma,* color]: Prefix meaning "color."

chromatid [Gk. *chroma,* color, + L. *-id,* daughters of]: One of the two daughter strands of a duplicated chromosome, which are joined at the centromere.

chromatin: The deeply staining complex of DNA and proteins that forms eukaryotic chromosomes.

chromatophore [Gk. *chroma,* color, + *phorus,* a bearer]: In some bacteria, a discrete vesicle delimited by a single membrane and containing photosynthetic pigments.

chromoplast: A plastid containing pigments other than chlorophyll, usually yellow and orange carotenoid pigments.

chromosome [Gk. *chroma,* color, + *soma,* body]: The structure that carries the genes. Eukaryotic chromosomes are visualized as threads or rods of chromatin, which appear in contracted form during mitosis and meiosis and otherwise are enclosed in a nucleus; each eukaryotic chromosome contains a linear DNA molecule; prokaryotes typically have a single chromosome consisting of a circular DNA molecule.

chrysolaminarin: The storage product of the chrysophytes and diatoms.

cilium, *pl.* **cilia** (sĭl'ē•ŭm) [L. *cilium,* eyelash]: A short, hairlike flagellum, usually numerous and arranged in rows.

circadian rhythms [L. *circa,* about, + *dies,* a day]: Regular rhythms of growth and activity that occur on an approximately 24-hour basis.

circinate vernation [L. *circinare,* to make round, + *vernare,* to flourish]: As in ferns, the coiled arrangement of leaves and leaflets in the bud; such an arrangement uncoils gradually as the leaf develops further.

cisterna, *pl.* **cisternae** [L. *cistern,* a reservoir]: A flattened or saclike portion of the endoplasmic reticulum or a Golgi body (dictyosome).

clade: One evolutionary line of organisms.

cladistics: System of arranging organisms following an analysis of their primitive and advanced features so that their phylogenetic relationships will be reflected accurately.

cladogram: A line diagram that branches repeatedly, suggesting phylogenetic relationships among organisms.

cladophyll [Gk. *klados,* shoot, + *phyllon,* leaf]: A branch resembling a foliage leaf.

clamp connection: In the *Basidiomycota,* a lateral connection between adjacent cells of a dikaryotic hypha; ensures that each cell of the hypha will contain two dissimilar nuclei.

class: A taxonomic category between phylum and order in rank. A class contains one or more orders, and belongs to a particular phylum.

cleistothecium [Gk. *kleistos,* closed, + *thekion,* small receptacle]: A closed, spherical ascoma.

climax community: The final stage in a successional series; its nature is determined largely by the climate and soil of the region.

cline: A graded series of changes in some characteristics within a species, often correlated with a gradual change in climate or another geographical factor.

clone [Gk. *klon,* twig]: A population of cells or individuals derived by asexual division from a single cell or individual; one of the members of such a population.

cloning: Producing a cell line or culture all of whose members are characterized by a specific DNA sequence; a key element in genetic engineering.

closed vascular bundle: A vascular bundle in which a cambium does not develop.

coalescence [L. *coalescere,* to grow together]: The union of floral parts of the same whorl, as petals to petals.

coccus, *pl.* **cocci** (kŏk'ŭs) [Gk. *kokkos,* a berry]: A spherical bacterium.

codon (kō'dŏn): Sequence of three adjacent nucleotides in a molecule of DNA or mRNA that form the code for a single amino acid, or for the termination of a polypeptide chain.

coenocytic (se•nō•sĭ'tic) [Gk. *koinos,* shared in common, + *kytos,* a hollow vessel]: A term used to describe an organism or part of an organism that is multinucleate, the nuclei not separated by walls or membranes; also called siphonaceous, siphonous, or syncytial.

coenzyme: An organic molecule, or nonprotein organic cofactor, that plays an accessory role in enzyme-catalyzed processes, often by acting as a donor or acceptor of electrons; NAD^+ and FAD are common coenzymes.

coevolution [L. *co,* together, + *e-,* out, + *volvere,* to roll]: The simultaneous evolution of adaptations in two or more populations that interact so closely that each is a strong selective force on the other.

cofactor: One or more nonprotein components required by enzymes in order to function; many cofactors are metal ions, while others are called coenzymes.

cohesion [L. *cohaerrere,* to stick together]: The mutual attraction of molecules of the same substance.

cold hardiness: The ability of a plant to survive the extreme cold and drying effects of winter weather.

coleoptile (kō′lē•op′till) [Gk. *koleos,* sheath, + *ptilon,* feather]: The sheath enclosing the apical meristem and leaf primordia of the grass embryo; often interpreted as the first leaf.

coleorhiza (kō′lē•o•rī′za) [Gk. *koleos,* sheath, + *rhiza,* root]: The sheath enclosing the radicle in the grass embryo.

collenchyma [Gk. *kolla,* glue]: A supporting tissue composed of collenchyma cells; common in regions of primary growth in stems and in some leaves.

collenchyma cell: Elongated living cell with unevenly thickened nonlignified primary cell wall.

colloid (kŏl′oid): A permanent suspension of fine particles.

community: All the organisms inhabiting a common environment and interacting with one another.

companion cell: A specialized parenchyma cell associated with a sieve-tube element in angiosperm phloem and arising from the same mother cell as the sieve-tube element.

competition: Interaction between members of the same population or of two or more populations to obtain a resource that both require and that is available in limited supply.

complementary DNA (cDNA): A single-stranded molecule of DNA that has been synthesized from an mRNA template by reverse transcription.

complete flower: A flower having four whorls of floral parts—sepals, petals, stamens, and carpels.

complex tissue: A tissue consisting of two or more cell types; epidermis, periderm, xylem, and phloem are complex tissues.

compound: A combination of atoms in a definite ratio, held together by chemical bonds.

compound leaf: A leaf whose blade is divided into several distinct leaflets.

compression wood: The reaction wood of conifers; develops on the lower sides of leaning trunks or limbs.

concentration gradient: The concentration difference of a substance per unit distance.

cone: *See* strobilus.

conidiophore: Hypha on which one or more conidia are produced.

conidium, *pl.* **conidia** [Gk. *konis,* dust]: An asexual fungal spore not contained within a sporangium; it may be produced singly or in chains; most conidia are multinucleate.

conifer: A cone-bearing tree.

conjugation: The temporary fusion of pairs of bacteria, protozoa, and certain algae and fungi during which genetic material is transferred between the two individuals.

conjugation tube: A tube formed during the process of conjugation to facilitate the transfer of genetic material.

connate (kōn′āt): Said of similar parts that are united or fused, as petals fused in a corolla tube; *see also* adnate.

consumer: In ecology, an organism that derives its food from another organism.

continuous variation: Variation in traits to which a number of different genes contribute; the variation often exhibits a "normal" or bell-shaped distribution.

contractile vacuole: A clear, fluid-filled vacuole in some groups of protists that takes up water within the cell and then contracts, expelling its contents from the cell.

convergent evolution [L. *convergere,* to turn together]: The independent development of similar structures in organisms that are not directly related; often found in organisms living in similar environments.

cork: A secondary tissue produced by a cork cambium; made up of polygonal cells, nonliving at maturity, with suberized cell walls, which are resistant to the passage of gases and water vapor; the outer part of the periderm. Also called phellem.

cork cambium: The lateral meristem that forms the periderm, producing cork (phellem) toward the surface (outside) of the plant and phelloderm toward the inside; common in stems and roots of gymnosperms and woody angiosperms. Also called phellogen.

corm: A thickened underground stem, upright in position, in which food is accumulated, usually in the form of starch.

corolla [L. *corona,* crown]: The petals collectively; usually the conspicuously colored flower whorl.

corolla tube: A tubelike structure resulting from the fusion of the petals along their edges.

cortex: Ground-tissue region of a stem or root bounded externally by the epidermis and internally by the vascular system; a primary-tissue region; also used to refer to the peripheral region of a cell protoplast.

cotransport: Membrane transport in which the transfer of one solute depends on the simultaneous or sequential transfer of a second solute.

cotyledon (kŏt′ĭ•lē′dŭn) [Gk. *kotyledon,* cup-shaped hollow]: Seed leaf; generally absorbs food in monocotyledons and stores food in other angiosperms.

coupled reactions: Reactions in which energy-requiring chemical reactions are linked to energy-releasing reactions.

covalent bond: A chemical bond formed between atoms as a result of the sharing of two electrons.

crassulacean acid metabolism, or **CAM:** A variant of the C_4 pathway; phosphoenolpyruvate fixes CO_2 in C_4 compounds at night and then, during the daytime, the fixed CO_2 is transferred to the ribulose bisphosphate of the Calvin cycle within the same cell. Characteristic of most succulent plants, such as cacti.

cristae, *sing.* **crista:** The enfoldings of the inner mitochondrial membrane, which form a series of crests or ridges containing the electron-transport chains involved in ATP formation.

crop rotation: The practice of growing different crops in regular succession to aid in the control of insects and diseases, to increase soil fertility, and to decrease erosion.

cross-fertilization: The fusion of gametes formed by different individuals; the opposite of self-fertilization.

crossing-over: The exchange of corresponding segments of genetic material between the chromatids of homologous chromosomes at meiosis.

cross-pollination: The transfer of pollen from the anther of one plant to the stigma of a flower of another plant.

cross section: *See* transverse section.

cryptogam: An archaic term for all organisms except the flowering plants (phanerogams), animals, and heterotrophic protists.

cultivar: A variety of plant found only under cultivation.

cuticle: Waxy or fatty layer on outer wall of epidermal cells, formed of cutin and wax.

cutin [L. *cutis,* skin]: Fatty substance deposited in many plant cell walls and on outer surface of epidermal cell walls, where it forms a layer known as the cuticle.

cyclic electron flow: In chloroplasts, the light-induced flow of electrons originating from and returning to photosystem I.

cyclosis (sī•klō′sis) [Gk. *kyklosis,* circulation]: The streaming of cytoplasm within a cell.

-cyte, cyto- [Gk. *kytos,* hollow vessel, container]: Suffix or prefix meaning "pertaining to the cell."

cytochrome [Gk. *kytos,* container, + *chroma,* color]: Heme proteins serving as electron carriers in respiration and photosynthesis.

cytokinesis [Gk. *kytos,* hollow vessel, + *kinesis,* motion]: Division of the cytoplasm of a cell following nuclear division.

cytokinin [Gk. *kytos,* hollow vessel, + *kinesis,* motion]: A class of plant hormones that promotes cell division, among other effects.

cytology: The study of cell structure and function.

cytoplasm: The living matter of a cell, exclusive of the nucleus; the protoplasm.

cytoplasmic ground substance: *See* cytosol.

cytosine: One of the four pyrimidine bases found in the nucleic acids DNA and RNA.

cytoskeleton: The flexible network within cells, composed of microtubules and actin filaments, or microfilaments.

cytosol: The cytoplasmic matrix of the cytoplasm in which the nucleus, various organelles, and membrane systems are suspended.

D

day-neutral plants: Plants that flower without regard to daylength.

de- [L. *de-*, away from, down, off]: Prefix meaning "away from," "down," or "off"; for example, dehydration means removal of water.

deciduous [L. *decidere*, to fall off]: Shedding leaves at a certain season.

decomposers: Organisms (bacteria, fungi, heterotrophic protists) in an ecosystem that break down organic material into smaller molecules that are then recirculated.

dehiscence [L. *de*, down, + *hiscere*, split open]: The opening of an anther, fruit, or other structure, which permits the escape of reproductive bodies contained within.

dehydration synthesis: The synthesis of a compound or molecule involving the removal of water; also called a condensation reaction.

denitrification: The conversion of nitrate to gaseous nitrogen; carried out by a few genera of free-living soil bacteria.

deoxyribonucleic acid (DNA): Carrier of genetic information in cells; composed of chains of phosphate, sugar molecules (deoxyribose), and purines and pyrimidines; capable of self-replication as well as determining RNA synthesis.

deoxyribose [L. *deoxy*, loss of oxygen, + *ribose*, a kind of sugar]: A five-carbon sugar with one less atom of oxygen than ribose; a component of deoxyribonucleic acid.

dermal tissue system: The outer covering tissue of the plant; the epidermis or the periderm.

desmotubule [Gk. *desmos*, to bind, + L. *tubulus*, small tube]: The tubule traversing a plasmodesmatal canal and uniting the endoplasmic reticulum of the two adjacent cells.

determinate growth: Growth of limited duration, characteristic of floral meristems and of leaves.

deuterium: Heavy hydrogen; a hydrogen atom, the nucleus of which contains one proton and one neutron. (The nucleus of most hydrogen atoms consists of only a proton.)

dichotomy: The division or forking of an axis into two branches.

dicotyledon: Obsolete term used to refer to all angiosperms other than monocotyledons; characterized by having two cotyledons; *see also* eudicotyledons *and* magnoliids.

dictyosome: *See* Golgi body.

differentiation: A developmental process by which a relatively unspecialized cell undergoes a progressive change to a more specialized cell; the specialization of cells and tissues for particular functions during development.

diffuse-porous wood: A wood in which the pores, or vessels, are fairly uniformly distributed throughout the growth layers or in which the size of pores changes only slightly from early wood to late wood.

diffusion [L. *diffundere*, to pour out]: The net movement of suspended or dissolved particles from a more concentrated region to a less concentrated region as a result of the random movement of individual molecules; the process tends to distribute such particles uniformly throughout a medium.

digestion: The conversion of complex, usually insoluble foods into simple, usually soluble forms by means of enzymatic action.

dikaryon (Gk. *di*, two, + *karyon*, a nut]: In fungi, mycelium with paired nuclei, each usually derived from a different parent.

dikaryotic: In fungi, having pairs of nuclei within cells or compartments.

dimorphism [Gk. *di*, two, + *morphē*, form]: The condition of having two distinct forms, such as sterile and fertile leaves in ferns, or sterile and fertile shoots in horsetails.

dioecious [Gk. *di*, two, + *oikos*, house]: Unisexual; having the male and female (or staminate and ovulate) elements on different individuals of the same species.

diploid: Having two sets of chromosomes; the $2n$ (diploid) chromosome number is characteristic of the sporophyte generation.

disaccharide [Gk. *di*, two, + *sakcaron*, sugar]: A carbohydrate formed of two simple sugar molecules linked by a covalent bond; sucrose is an example.

disk flowers: The actinomorphic, tubular flowers in *Asteraceae;* contrasted with flattened, zygomorphic ray flowers. In many *Asteraceae*, the disk flowers occur in the center of the inflorescence, the ray flowers around the margins.

distal: Situated away from or far from the point of reference (usually the main part of body); opposite of proximal.

DNA: *See* deoxyribonucleic acid.

DNA sequencing: Determination of the order of nucleotides in a DNA molecule.

domain: The taxonomic category above the kingdom level; the three domains are *Archaea*, *Bacteria*, and *Eukarya*.

dominant allele: One allele is said to be dominant with respect to an alternative allele if the homozygote for the dominant allele is indistinguishable phenotypically from the heterozygote; the other allele is said to be recessive.

dormancy [L. *dormire*, to sleep]: A special condition of arrested growth in which the plant and such plant parts as buds and seeds do not begin to grow without special environmental cues. The requirement for such cues, which include cold exposure and a suitable photoperiod, prevents the breaking of dormancy during superficially favorable growing conditions.

double fertilization: The fusion of the egg and sperm (resulting in a $2n$ fertilized egg, the zygote) and the simultaneous fusion of the second male gamete with the polar nuclei (typically resulting in a $3n$ primary endosperm nucleus); a unique characteristic of all angiosperms.

doubling rate: The length of time required for a population of a given size to double in number.

drupe [Gk. *dryppa*, overripe olive]: A simple, fleshy fruit, derived from a single carpel, usually one-seeded, in which the inner fruit coat is hard and may adhere to the seed.

druse: A compound, more or less spherical crystal with many component crystals projecting from its surface; composed of calcium oxalate.

E

early wood: The first-formed wood of a growth increment; it contains larger cells and is less dense than the subsequently formed late wood; replaces the term "spring wood."

eco- [Gk. *oikos*, house]: Prefix meaning "house" or "home."

ecology: The study of the interactions of organisms with their physical environment and with one other.

ecosystem: A major interacting system that involves both living organisms and their physical environment.

ecotype [Gk. *oikos*, house, + L. *typus*, image]: A locally adapted variant of an organism, differing genetically from other ecotypes.

edaphic [Gk. *edaphos,* ground, soil]: Pertaining to the soil.

egg: A nonmotile female gamete, usually larger than a male gamete of the same species.

egg apparatus: The egg cell and synergids located at the micropylar end of the female gametophyte, or embryo sac, of angiosperms.

elater [Gk. *elater,* driver]: (1) An elongated, spindle-shaped, sterile cell in the sporangium of a liverwort sporophyte (aids in spore dispersal); (2) clubbed, hygroscopic bands attached to the spores of the horsetails.

electrochemical gradient: The driving force that causes an ion to move across a membrane due to the difference in the electric charge across the membrane in combination with the difference in the ion's concentration on the two sides of the membrane.

electrolyte: A substance that dissociates into ions in aqueous solution and so makes possible the conduction of an electric current through the solution.

electromagnetic spectrum: The entire spectrum of radiation, which ranges in wavelength from less than a nanometer to more than a kilometer.

electron: A subatomic particle with a negative electric charge equal in magnitude to the positive charge of the proton, but with a mass of 1/1837 of that of the proton. Electrons orbit the atom's positively charged nucleus and determine the atom's chemical properties.

electron-dense: In electron microscopy, not permitting the passage of electrons and so appearing dark.

electron transport: The movement of electrons down a series of electron-carrier molecules that hold electrons at slightly different energy levels; as electrons move down the chain, the energy released is used to form ATP from ADP and phosphate. Electron transport plays an essential role in the final stage of cellular respiration and in the light-dependent reactions of photosynthesis.

element: A substance composed of only one kind of atom; one of more than 100 distinct natural or synthetic types of matter that, singly or in combination, compose virtually all materials of the universe.

embryo [Gk. *en,* in, + *bryein,* to swell]: A young sporophytic plant, before the start of a period of rapid growth (germination in seed plants).

embryogenesis: Development of an embryo from a fertilized egg, or zygote; also called embryogeny.

embryophytes: The bryophytes and vascular plants, both of which produce embryos; a synonym for plants.

embryo sac: The female gametophyte of angiosperms, generally an eight-nucleate, seven-celled structure; the seven cells are the egg cell, two synergids and three antipodals (each with a single nucleus), and the central cell (with two nuclei).

endergonic: Describing a chemical reaction that requires energy to proceed; opposite of exergonic.

endo- [Gk. *endo,* within]: Prefix meaning "within."

endocarp [Gk. *endo,* within, + *karpos,* fruit]: The innermost layer of the mature ovary wall, or pericarp.

endocytosis [Gk. *endon,* within, + *kytos,* hollow vessel]: The uptake of material into cells by means of invagination of the plasma membrane; if solid material is involved, the process is called phagocytosis; if dissolved material is involved, it is called pinocytosis.

endodermis [Gk. *endon,* within, + *derma,* skin]: A single layer of cells forming a sheath around the vascular region in roots and some stems; the endodermal cells are characterized by a Casparian strip within radial and transverse walls. In roots and stems of seed plants, the endodermis is the innermost layer of the cortex.

endogenous [Gk. *endon,* within, + *genos,* race, kind]: Arising from deep-seated tissues, as in the case of lateral roots.

endomembrane system: Collectively, the cellular membranes that form a continuum (plasma membrane, tonoplast, endoplasmic reticulum, Golgi bodies, and nuclear envelope).

endoplasmic reticulum: A complex, three-dimensional membrane system of indefinite extent present in eukaryotic cells, dividing the cytoplasm into compartments and channels. Those portions that are densely coated with ribosomes are called rough endoplasmic reticulum, and other portions with fewer or no ribosomes are called smooth endoplasmic reticulum.

endosperm [Gk. *endon,* within, + *sperma,* seed]: A tissue, containing stored food, that develops from the union of a male nucleus and the polar nuclei of the central cell; it is digested by the growing sporophyte either before or after the maturation of the seed; found only in angiosperms.

energy: The capacity to do work.

energy of activation: The energy that must be possessed by atoms or molecules in order to react.

energy-transduction reactions: *See* light reactions.

entrainment: The process by which a periodic repetition of light and dark, or some other external cycle, causes a circadian rhythm to remain synchronized with the same cycle as the modifying, or entraining, factor.

entropy: A measure of the randomness or disorder of a system.

enzyme: A protein that is capable of speeding up specific chemical reactions by lowering the required activation energy, but is unaltered itself in the process; a biological catalyst.

epi- [Gk. *epi,* upon]: Prefix meaning "upon" or "above."

epicotyl: The upper portion of the axis of an embryo or seedling, above the cotyledons (seed leaves) and below the next leaf or leaves.

epidermis: The outermost layer of cells of the leaf and of young stems and roots; primary in origin.

epigeous [Gk. *epi,* upon, + *ge,* the Earth]: Type of seed germination in which the cotyledons are carried above ground level.

epigyny [Gk. *epi,* upon, + *gyne,* woman]: A pattern of floral organization in which the sepals, petals, and stamens apparently grow from the top of the ovary; *see also* of hypogyny.

epiphyte (ĕp'ī•fīt): An organism that grows upon another organism but is not parasitic on it.

epistatic [Gk. *epistasis,* a stopping]: Term used to describe a gene the action of which modifies the phenotypic expression of a gene at another locus.

essential elements: Chemical elements essential for normal plant growth and development; also referred to as essential minerals and essential inorganic nutrients.

ethylene: A simple hydrocarbon that is a plant hormone involved in the ripening of fruit; $H_2C = CH_2$.

etiolation (e'tĭ•o•lā'shŭn) [Fr. *etioler,* to blanch]: A condition involving increased stem elongation, poor leaf development, and lack of chlorophyll; found in plants growing in the dark or with a greatly reduced amount of light.

etioplast: Plastid of a plant grown in the dark and containing a prolamellar body.

eudicotyledons: One of two major classes of angiosperms, *Eudicotyledones;* formerly grouped with the magnoliids, a diverse group of archaic flowering plants, as "dicots"; abbreviated as eudicot.

Eukarya: The phylogenetic domain containing all eukaryotic organisms.

eukaryote [Gk. *eu,* good, + *karyon,* kernel]: A cell that has a membrane-bounded nucleus, membrane-bounded organelles, and chromosomes in which the DNA is associated with proteins; an organism composed of such cells. Plants, animals, fungi, and protists are the four kingdoms of eukaryotes.

eusporangium: A sporangium that arises from several initial cells and, before maturation, forms a wall of more than one layer of cells.

eustele [Gk. *eu-*, good, + *stele*, pillar]: A stele in which the primary vascular tissues are arranged in discrete strands around a pith: typical of gymnosperms and angiosperms.

evolution: The derivation of progressively more complex forms of life from simple ancestors; Darwin proposed that natural selection is the principal mechanism by which evolution takes place.

exergonic [L. *ex*, out, + Gk. *ergon*, work]: Energy-yielding, as in a chemical reaction; applied to a "downhill" process.

exine: The outer wall layer of a spore or pollen grain.

exocarp [Gk. *exo*, without, + *karpos*, fruit]: The outermost layer of the mature ovary wall, or pericarp.

exocytosis [Gk. *ex*, out of, + *kytos*, vessel]: A cellular process in which particulate matter or dissolved substances are enclosed in a vesicle and transported to the cell surface; there, the membrane of the vesicle fuses with the plasma membrane, expelling the vesicle's contents to the outside.

exodermis: The outer layer, one or more cells in depth, of the cortex in some roots; these cells are characterized by Casparian strips within the radial and transverse walls. Following development of Casparian strips, a suberin lamella is deposited on all walls of the exodermis.

exon [Gk. *exo*, outside]: A segment of DNA that is both transcribed into RNA and translated into protein; exons are characteristic of eukaryotes. *See also* intron.

eyespot: A small, pigmented structure in flagellate unicellular organisms that is sensitive to light; also called a stigma.

F

F_1: First filial generation. The offspring resulting from a cross. F_2 and F_3 are the second and third generations resulting from such a cross.

facilitated diffusion: Passive transport with the assistance of carrier proteins.

family: A taxonomic group between order and genus in rank; the ending of family names in animals and heterotrophic protists is *-idae;* in all other organisms it is *-aceae.* A family contains one or more genera, and each family belongs to an order.

fascicle (făs′ĭ•kûl) [L. *fasciculus*, a small bundle]: A bundle of pine leaves or other needlelike leaves of gymnosperms; an obsolete term for a vascular bundle.

fascicular cambium: The vascular cambium originating within a vascular bundle, or fascicle.

fat: A molecule composed of glycerol and three fatty acid molecules; the proportion of oxygen to carbon is much less in fats than it is in carbohydrates. Fats in the liquid state are called oils.

feedback inhibition: Control mechanism whereby an increase in the concentration of some molecule inhibits the further synthesis of that molecule.

fermentation: The extraction of energy from organic compounds without the involvement of oxygen.

ferredoxin: Electron-transferring proteins of high iron content; some are involved in photosynthesis.

fertilization: The fusion of two gamete nuclei to form a diploid zygote.

fiber: An elongated, tapering, generally thick-walled sclerenchyma cell of vascular plants; its walls may or may not be lignified; it may or may not have a living protoplast at maturity.

fibril: Submicroscopic threads composed of cellulose molecules, which constitute the form in which cellulose occurs in the cell wall.

field capacity: The percentage of water a particular soil will hold against the action of gravity; also called field moisture capacity.

filament: (1) The stalk of a stamen; (2) a term used to describe the threadlike bodies or segments of certain algae or fungi.

fission: Asexual reproduction involving the division of a single-celled individual into two new single-celled individuals of equal size. Also applied to the division of plastids.

fitness: The genetic contribution of an organism to future generations, relative to the contributions of organisms living in the same environment that have different genotypes.

flagellum, *pl.* **flagella** [L. *flagellum*, whip]: A long threadlike organelle that protrudes from the surface of a cell. The flagella of bacteria are capable of rotary motion and consist of a single protein fiber each; eukaryotic flagella, which are used in locomotion and feeding, consist of an array of microtubules with a characteristic internal 9 + 2 microtubule structure and are capable of a vibratory, but not rotary, motion. A cilium is a small eukaryotic flagellum.

flavonoids: Phenolic compounds; water-soluble pigments present in the vacuoles of plant cells; those found in red wines and grape juice have been reported to lower cholesterol levels in the blood.

flavoprotein: A dehydrogenase that contains a flavin and often a metal and plays a major role in oxidation; abbreviated FP.

floral tube: A cup or tube formed by the fusion of the basal parts of the sepals, petals, and stamens; floral tubes are often found in plants that have a superior ovary.

floret: One of the small flowers that make up the composite inflorescence or the spike of the grasses.

florigen [L. *flor-*, flower, + Gk. *-genes*, producer]: A hypothetical plant hormone that promotes flowering.

flower: The reproductive structure of angiosperms; a complete flower includes calyx, corolla, androecium (stamens), and gynoecium (carpels), but all flowers contain at least one stamen or one carpel.

fluid-mosaic model: Model of membrane structure, with the membrane composed of a lipid bilayer in which globular proteins are embedded.

follicle [L. *folliculus*, small ball]: A dry, dehiscent simple fruit derived from a single carpel and opening along one side.

food chain, food web: A chain of organisms existing in any natural community such that each link in the chain feeds on the one below and is eaten by the one above; there are seldom more than six links in a chain, with autotrophs on the bottom and the largest carnivores at the top.

fossil [L. *fossils*, dug up]: The remains, impressions, or traces of an organism that has been preserved in rocks found in the Earth's crust.

fossil fuels: The altered remains of once-living organisms that are burned to release energy; oil, gas, and coal.

founder cells: The group of cells in the peripheral zone of the apical meristem involved with the initiation of a leaf primordium.

founder effect: Type of genetic drift that occurs as the result of the founding of a population by a small number of individuals.

FP: *See* flavoprotein.

free energy: Energy available to do work.

frond: The leaf of a fern; any large, divided leaf.

fruit: In angiosperms, a mature, ripened ovary (or group of ovaries), containing the seeds, together with any adjacent parts that may be fused with it at maturity; sometimes applied informally, and misleadingly, as in "fruiting body," to the reproductive structures of other kinds of organisms.

fucoxanthin (fū′kō•zăn′thĭn) [Gk. *phykos*, seaweed, + *xanthos*, yellowish-brown]: A brownish carotenoid found in brown algae and chrysophytes.

fundamental tissue system: *See* ground tissue system.

funiculus [L. *funiculus*, small rope or cord]: The stalk of the ovule.

fusiform initials [L. *fusus*, spindle]: The vertically elongated cells in the vascular cambium that give rise to the cells of the axial system in the secondary xylem and secondary phloem.

G

gametangium, *pl.* **gametangia** [Gk. *gamein,* to marry, + L. *tangere,* to touch]: A cell or multicellular structure in which gametes are formed.

gamete [Gk. *gamete,* wife]: A haploid reproductive cell; gametes fuse in pairs, forming zygotes, which are diploid.

gametic meiosis: Meiosis resulting in the formation of haploid gametes from a diploid individual; the gametes fuse to form a diploid zygote that divides to form another diploid individual.

gametophore [Gk. *gamein,* to marry, + *phoros,* bearing]: In the bryophytes, a fertile stalk that bears gametangia.

gametophyte: In plants, which have an alternation of generations, the haploid (n), gamete-producing generation, or phase.

gel: A mixture of substances having a semisolid or solid constitution.

gemma, *pl.* **gemmae** (jĕm′ă) [L. *gemma,* bud]: A small mass of vegetative tissue; an outgrowth of the thallus, for example, in liverworts or certain fungi; it can develop into an entire new plant.

gene: A unit of heredity; a sequence of DNA nucleotides that codes for a protein, tRNA, or rRNA molecule, or regulates the transcription of such a sequence.

gene flow: The movement of alleles into and out of a population.

gene frequency: The relative occurrence of a particular allele in a population.

gene pool: All the alleles of all the genes of all the individuals in a population.

generative cell: (1) In many gymnosperms, the cell of the male gametophyte that divides to form the sterile and spermatogenous cells; (2) in angiosperms, the cell of the male gametophyte that divides to form two sperm.

genetic code: The system of nucleotide triplets (codons) in DNA and RNA that dictates the amino acid sequence in proteins; except for three "stop" signals, each codon specifies one of 20 amino acids.

genetic drift: Evolution (change in allele frequencies) owing to chance processes.

genetic engineering: The manipulation of genetic material for practical purposes; also referred to as recombinant DNA technology.

genetic recombination: The occurrence of gene combinations in the progeny that are different from the combinations present in the parents.

genome: The totality of genetic information contained in the nucleus, plastid, or mitochondrion.

genomic library: A library encompassing an entire genome either in the nucleus or in the nucleoid of an organelle (mitochondrion, plastid) in eukaryotes, or in the nucleoid of a prokaryote.

genotype: The genetic constitution, latent or expressed, of an organism, as contrasted with the phenotype; the sum total of all the genes present in an individual.

genus, *pl.* **genera:** The taxonomic group between family and species in rank; genera include one or more species.

geotropism: *See* gravitropism.

germination [L. *germinare,* to sprout]: The beginning or resumption of growth by a spore, seed, bud, or other structure.

gibberellins (jĭb•ĕ•rĕ′lĭns) [*Gibberella,* a genus of fungi]: A class of plant hormones, the best known effect of which is to increase the elongation of plant stems.

gill: The strips of tissue on the underside of the cap in *Basidiomycetes.*

girdling: The removal from a woody stem of a ring of bark extending inward to the cambium; also called ringing.

glucose: A common six-carbon sugar ($C_6H_{12}O_6$); the most common monosaccharide in most organisms.

glycerol: A three-carbon molecule with three hydroxyl groups attached; glycerol molecules combine with fatty acids to form fats or oils.

glycogen [Gk. *glykys,* sweet, + *gen,* of a kind]: A carbohydrate similar to starch that serves as the reserve food in bacteria, fungi, and most organisms other than plants.

glycolysis: The anaerobic breakdown of glucose to form two molecules of pyruvate, resulting in the net liberation of two molecules of ATP; catalyzed by enzymes in the cytosol.

glyoxylate cycle: A variant of the Krebs cycle; present in bacteria and some plant cells, for net conversion of acetate into succinate and, eventually, new carbohydrate.

glyoxysome: A peroxisome containing enzymes necessary for the conversion of fats into carbohydrates; glyoxysomes play an important role during the germination of seeds.

Golgi body (gôl′jē): In eukaryotes, a group of flat, disk-shaped sacs that are often branched into tubules at their margins; serve as collecting and packaging centers for the cell and concerned with secretory activities; also called dictyosomes. The term "Golgi complex" is used to refer collectively to all the Golgi bodies, or dictyosomes, of a given cell.

grafting: A union of different individuals in which a portion, called the scion, of one individual is inserted into a root or stem, called the stock, of the other individual.

grain: *See* caryopsis; also a term used to refer to the alignment of wood elements.

grana, *sing.* **granum:** Structures within chloroplasts, seen as green granules with a light microscope and as a series of stacked thylakoids with an electron microscope; the grana contain the chlorophylls and carotenoids and are the sites of the light reactions of photosynthesis.

gravitropism [L. *gravis,* heavy, + Gk. *tropes,* turning]: The response of a shoot or root to the pull of the Earth's gravity; also called geotropism.

ground meristem [Gk. *meristos,* divisible]: The primary meristem, or meristematic tissue, that gives rise to the ground tissues.

ground tissue: Tissues other than the vascular tissues, the epidermis, and the periderm; also called fundamental tissue.

ground tissue system: All tissues other than the epidermis (or periderm) and the vascular tissues; also called fundamental tissue system.

growth layer: A layer of growth in the secondary xylem or secondary phloem; *see also* annual ring.

growth ring: A growth layer in the secondary xylem or secondary phloem, as seen in transverse section; may be called a growth increment, especially where seen in other than transverse section.

guanine [Sp. from Quechua, *huanu,* dung]: A purine base found in DNA and RNA. Its name is based on the fact that guanine is abundant in the form of a white crystalline base in guano and other kinds of animal excrement.

guard cells: Pairs of specialized epidermal cells surrounding a pore, or stoma; changes in the turgor of a pair of guard cells cause opening and closing of the pore.

guttation [L. *gutta,* a drop]: The exudation of liquid water from leaves; caused by root pressure.

gymnosperm [Gk. *gymnos,* naked, + *sperma,* seed]: A seed plant with seeds not enclosed in an ovary; the conifers are the most familiar group; not a monophyletic group.

gynoecium [Gk. *gyne,* woman, + *oikos,* house]: The aggregate of carpels in the flower of a seed plant.

H

habit [L. *habitus,* condition, character]: Characteristic form or appearance of an organism.

habitat [L. *habitare,* to inhabit]: The environment of an organism; the place where it is usually found.

hadrom: The central strand of water-conducting cells found in the axes of some moss gametophytes and sporophytes.

haploid [Gk. *haploos*, single]: Having only one set of chromosomes (n), in contrast to diploid ($2n$).

hardwood: A name commonly applied to the wood of a magnoliid or eudicot tree.

Hardy–Weinberg law: The mathematical expression of the relationship between the relative frequencies of two or more alleles in a population. It demonstrates that the frequencies of alleles and genotypes will remain constant in a random-mating population in the absence of inbreeding, selection, or other evolutionary forces.

haustorium, *pl.* **haustoria** [L. *haustus*, from *haurire*, to drink, draw]: A projection of a fungal hypha that functions as a penetrating and absorbing organ; in parasitic angiosperms, a modified root capable of penetrating and absorbing materials from host tissues.

heartwood: Nonliving and commonly dark-colored wood in which no water transport occurs; it is surrounded by sapwood.

helical phyllotaxy: Leaf arrangement in which there is one leaf at a node, forming a helical pattern around the stem; also called alternate phyllotaxy.

heliotropism [Gk. *helios*, sun]: *See* solar tracking.

hemicellulose (hĕm′ē•sĕl′ū•lōs): A polysaccharide resembling cellulose but more soluble and less ordered; found particularly in cell walls.

herb [L. *herba*, grass]: A nonwoody seed plant with a relatively short-lived aerial portion.

herbaceous: An adjective referring to nonwoody plants.

herbarium: A collection of dried and pressed plant specimens.

herbivorous: Feeding upon plants.

heredity [L. *heredis*, heir]: The transmission of characteristics from parent to offspring through the gametes.

hermaphrodite [Gk. for Hermes and Aphrodite]: An organism possessing both male and female reproductive organs.

hetero- [Gk. *heteros*, different]: Prefix meaning "other" or "different."

heterocyst [Gk. *heteros*, different, + *cystis*, a bag]: A transparent, thick-walled, nitrogen-fixing cell that forms in the filaments of certain cyanobacteria.

heteroecious (hĕt′er•ē′shŭs) [Gk. *heteros*, different, + *oikos*, house]: As in some rust fungi, requiring two different host species to complete the life cycle.

heterogamy [Gk. *heteros*, other, + *gamos*, union or reproduction]: Reproduction involving two types of gametes.

heterokaryotic [Gk. *heteros*, other, + *karyon*, kernel]: In fungi, having two or more genetically distinct types of nuclei within the same mycelium.

heterokonts: Organisms with one long, ornamented (tinsel) flagellum and one shorter and smooth (whiplash) flagellum; includes oomycetes, chrysophytes, diatoms, brown algae, and certain other groups.

heteromorphic [Gk. *heteros*, different, + *morphē*, form]: A term used to describe a life history in which the haploid and diploid generations are dissimilar in form.

heterosis [Gk. *heterosis*, alteration]: Hybrid vigor, the superiority of the hybrid over either parent in any measurable character.

heterosporous: Having two kinds of spores, designated as microspores and megaspores.

heterothallic [Gk. *heteros*, different, + *thallus*, sprout]: A term used to describe a species, the haploid individuals of which are self-sterile or self-incompatible; two compatible strains or individuals are required for sexual reproduction to take place.

heterotroph [Gk. *heteros*, other, + *trophos*, feeder]: An organism that cannot manufacture organic compounds and so must feed on organic materials that have originated in other plants and animals; *see also* autotroph.

heterozygous: Having two different alleles at the same locus on homologous chromosomes.

Hill reaction: The oxygen evolution and photoreduction of an artificial electron acceptor by a chloroplast preparation in the absence of carbon dioxide.

hilum [L. *hilum*, a trifle]: (1) Scar left on seed after separation of seed from funiculus; (2) the part of a starch grain around which the starch is laid down in more or less concentric layers.

histone: The group of five basic proteins associated with the chromosomes of all eukaryotic cells.

holdfast: (1) Basal part of a multicellular alga that attaches it to a solid object; may be unicellular or composed of a mass of tissue; (2) cuplike structure at the tips of some tendrils, by means of which they become attached.

homeo-, homo- [Gk. *homos*, same, similar]: Prefix meaning "similar" or "same."

homeostasis (hō′mē•ō•stā′sĭs) [Gk. *homos*, similar, + *stasis*, standing]: The maintaining of a relatively stable internal physiological environment within an organism, or a steady-state equilibrium in a population or ecosystem. Homeostasis usually involves feedback mechanisms.

homeotic mutation: A mutation that changes organ identity so that the wrong structures appear in the wrong place or at the wrong time.

homokaryotic [Gk. *homos*, same, + *karyon*, kernel]: In fungi, having nuclei with the same genetic makeup within a mycelium.

homologous chromosomes: Chromosomes that associate in pairs in the first stage of meiosis; each member of the pair is derived from a different parent. Homologous chromosomes are also called homologs.

homology [Gk. *homologia*, agreement]: A condition indicative of the same phylogenetic, or evolutionary, origin, but not necessarily the same in present structure and/or function.

homosporous: Having only one kind of spore.

homothallic [Gk. *homos*, same, + *thallus*, sprout]: A term used to describe a species in which the individuals are self-fertile.

homozygous: Having identical alleles at the same locus on homologous chromosomes.

hormogonium, *pl.* **hormogonia:** A portion of a filament of a cyanobacterium that becomes detached and grows into a new filament.

hormone [Gk. *hormaein*, to excite]: An organic substance produced usually in minute amounts in one part of an organism, from which it is transported to another part of that organism on which it has a specific effect; hormones function as highly specific chemical signals between cells.

host: An organism on or in which a parasite lives.

humus: Decomposing organic matter in the soil.

hybrid: Offspring of two parents that differ in one or more heritable characteristics; offspring of two different varieties or of two different species.

hybridization: The formation of offspring between unlike parents.

hybrid vigor: *See* heterosis.

hydrocarbon [Gk. *hydro*, water, + L. *carbo*, charcoal]: An organic compound that consists only of hydrogen and carbon atoms.

hydrogen bond: A weak bond between a hydrogen atom attached to one oxygen or nitrogen atom and another oxygen or nitrogen atom.

hydroids: The water-conducting cells of the moss hadrom; they resemble the tracheary elements of vascular plants except that they lack specialized wall thickenings.

hydrolysis [Gk. *hydro*, water, + *lysis*, loosening]: Splitting of one molecule into two by addition of the H^+ and OH^- ions of water.

hydrophyte [Gk. *hydro*, water, + *phyton*, a plant]: A plant that depends on an abundant supply of moisture or that grows wholly or partly submerged in water.

hydrotropism: The growth of roots toward regions of greater water potential, or in response to a moisture gradient.

hydroxyl group: An OH^- group; a negatively charged ion formed by the dissociation of a water molecule.

hymenium [Gk. *hymen,* a membrane]: The layer of asci on an ascoma, or of basidia on a basidioma, together with any associated sterile hyphae.

hyper- [Gk. *hyper,* above, over]: Prefix meaning "above" or "over."

hypertonic: Refers to a solution that has a concentration of solute particles high enough to gain water across a selectively permeable membrane from another solution.

hypha, *pl.* **hyphae** [Gk. *hyphe,* web]: A single tubular filament of a fungus, oomycete, or chytrid; the hyphae together comprise the mycelium.

hypo- [Gk. *hypo,* less than]: Prefix meaning "under" or "less."

hypocotyl: The portion of an embryo or seedling situated between the cotyledons and the radicle.

hypocotyl-root axis: The embryo axis below the cotyledon or cotyledons, consisting of the hypocotyl and the apical meristem of the root or the radicle.

hypodermis [Gk. *hypo,* under, + *derma,* skin]: One or more layers of cells beneath the epidermis, which are distinct from the underlying cortical or mesophyll cells.

hypogeous [Gk. *hypo,* under, + *ge,* the Earth]: Type of seed germination in which the cotyledons remain underground.

hypogyny [Gk. *hypo,* under, + *gyne,* woman]: Floral organization in which the sepals, petals, and stamens are attached to the receptacle below the ovary; *see also* epigyny.

hypothesis [Gk. *hypo,* under, + *tithenai,* to put]: A temporary working explanation or supposition based on accumulated facts and suggesting some general principle or relation of cause and effect; a postulated solution to a scientific problem that must be tested by experimentation and, if disproved or shown to be unlikely, is discarded.

hypotonic: Refers to a solution that has a concentration of solutes low enough to lose water across a selectively permeable membrane to another solution.

I

IAA: *See* indoleacetic acid.

imbibition (ĭm•bĭ•bĭsh′ŭn): Adsorption of water and swelling of colloidal materials because of the adsorption of water molecules onto the internal surfaces of the materials.

imperfect flower: A flower lacking either stamens or carpels.

imperfect fungi: The deuteromycetes, or conidial fungi, which reproduce only asexually, or in which the sexual cycle has not been observed; most deuteromycetes are ascomycetes.

inbreeding: The breeding of closely related plants or animals; in plants, it is usually brought about by repeated self-pollination.

incomplete flower: A flower lacking one or more of the four kinds of floral parts, that is, lacking sepals, petals, stamens, or carpels.

indehiscent (ĭn′de•hĭs′ĕnt): Remaining closed at maturity, as are many fruits (samaras, for example).

independent assortment: *See* Mendel's second law.

indeterminate growth: Unrestricted or unlimited growth, as with a vegetative apical meristem that produces an unrestricted number of lateral organs indefinitely.

indoleacetic acid (IAA): Naturally occurring auxin, a kind of plant hormone.

indusium, *pl.* **indusia** (ĭn•dŭ′zi•ŭm) [L. *indusium,* a woman's undergarment]: Membranous growth of the epidermis of a fern leaf that covers a sorus.

inferior ovary: An ovary that is completely or partially attached to the calyx; the other floral whorls appear to arise from the ovary's top.

inflorescence: A flower cluster, with a definite arrangement of flowers.

initial: (1) In the meristem, a cell that remains within the meristem indefinitely and at the same time, by division, adds cells to the plant body; (2) a meristematic cell that eventually differentiates into a mature, more specialized cell or element.

inner bark: In older trees, the living part of the bark; the bark inside the innermost periderm.

integral proteins: Transmembrane proteins and other proteins that are tightly bound to the membrane.

integument: The outermost layer or layers of tissue enveloping the nucellus of an ovule; develops into the seed coat.

inter- [L. *inter,* between]: Prefix meaning "between," or "in the midst of."

intercalary [L. *intercalare,* to insert]: Descriptive of meristematic tissue or growth not restricted to the apex of an organ; that is, growth in the regions of the nodes.

interfascicular cambium: The vascular cambium arising between the fascicles, or vascular bundles, from interfascicular parenchyma.

interfascicular region: Tissue region between vascular bundles in a stem; also called pith ray.

internode: The region of a stem between two successive nodes.

interphase: The period between two mitotic or meiotic cycles; the cell grows and its DNA replicates during interphase.

intine: The inner wall layer of a spore or pollen grain.

intra- [L. *intra,* within]: Prefix meaning "within."

intron [L. *intra,* within]: A portion of mRNA as transcribed from eukaryotic DNA that is removed by enzymes before the mRNA is translated into protein. *See also* exon.

ion: An atom or molecule that has lost or gained one or more electrons, and thus is positively or negatively charged.

irregular flower: A flower in which one or more members of at least one whorl differ in form from other members of the same whorl.

iso- [Gk. *isos,* equal]: Prefix meaning "similar," "alike." Is- before a vowel.

isogamy: A type of sexual reproduction in which the gametes (or gametangia) are alike in size; found in some algae and fungi.

isomer [Gk. *isos,* equal, + *meros,* part]: One of a group of compounds identical in atomic composition but differing in structural arrangement; for example, glucose and fructose.

isomorphic [Gk. *isos,* equal, + *morphē,* form]: Identical in form.

isotonic: Having the same osmotic concentration.

isotope: One of several possible forms of a chemical element that differ from other forms in the number of neutrons in the atomic nucleus, but not in chemical properties.

K

karyogamy [Gk. *karyon,* kernel, + *gamos,* marriage]: The union of two nuclei following fertilization, or plasmogamy.

kelp: A common name for any of the larger members of the order *Laminariales* of the brown algae.

kinetin [Gk. *kinetikos,* causing motion]: A purine that probably does not occur in nature but that acts as a cytokinin in plants.

kinetochore [Gk. *kinetikos,* putting in motion, + *chorus,* chorus]: Specialized protein complexes, which develop on each centromere, and to which spindle fibers are attached during mitosis or meiosis.

kingdom: One of the seven chief taxonomic categories; for example, *Fungi* or *Plantae.*

Kranz anatomy [Ger. *Kranz,* wreath]: The wreathlike arrangement of mesophyll cells around a layer of large bundle-sheath cells, forming two concentric layers around the vascular bundle; typically found in the leaves of C_4 plants.

Krebs cycle: The series of reactions that results in the oxidation of pyruvate to hydrogen atoms, electrons, and carbon dioxide. The electrons, passed along electron-carrier molecules, then go through the oxidative phosphorylation and terminal oxidation processes. Also called the tricarboxylic acid cycle or TCA cycle.

L

L_1, L_2, L_3 layers: The outer cell layers of angiosperm apical meristems with a tunica-corpus organization.

lamella (la•mĕl'a) [L. *lamella,* thin metal plate]: Layer of cellular membranes, particularly photosynthetic, chlorophyll-containing membranes; *see also* middle lamella.

lamina: The blade of a leaf.

laminarin: One of the principal storage products of the brown algae; a polymer of glucose.

lateral meristems: Meristems that give rise to secondary tissue; the vascular cambium and cork cambium.

lateral root: A root that arises from another, older root; also called a branch root, or secondary root, if the older root is the primary root.

late wood: The last part of the growth increment formed in the growing season; it contains smaller cells and is denser than the early wood; replaces the term "summer wood."

leaching: The downward movement and drainage of minerals, or inorganic ions, from the soil by percolating water.

leaf: The principal lateral appendage of the stem; highly variable in both structure and function; the foliage leaf is specialized as a photosynthetic organ.

leaf buttress: A lateral protrusion below the apical meristem; represents the initial stage in the development of a leaf primordium.

leaf gap: Region of parenchyma tissue in the primary vascular cylinder above the point of departure of the leaf trace or traces in ferns.

leaflet: One of the parts of a compound leaf.

leaf primordium [L. *primordium,* beginning]: A lateral outgrowth from the apical meristem that will eventually become a leaf.

leaf scar: A scar left on a twig when a leaf falls.

leaf trace: That part of a vascular bundle extending from the base of the leaf to its connection with a vascular bundle in the stem.

leaf trace gap: Region of parenchyma tissue in the primary vascular cylinder of a stem above the point of departure of the leaf trace or traces in seed plants.

legume [L. *legumem,* leguminous plant]: (1) A member of the *Fabaceae,* the pea or bean family; (2) a type of dry simple fruit that is derived from one carpel and opens along both sides.

lenticels (lĕn'tĭ•sĕls) [L. *lenticella,* a small window]: Spongy areas in the cork surfaces of stem, roots, and other plant parts that allow interchange of gases between internal tissues and the atmosphere through the periderm; occur in vascular plants.

leptoids: The food-conducting cells associated with the hydroids of some moss gametophytes and sporophytes; they resemble the sieve elements of some seedless vascular plants.

leptom: The food-conducting tissue consisting of leptoids; surrounds the hadrom in the axes of some moss gametophytes and sporophytes.

leptosporangium: A sporangium that arises from a single initial cell and whose wall is composed of a single layer of cells.

leucoplast (lū'kō•plăst) [Gk. *leuko,* white, + *plasein,* to form]: A colorless plastid; leucoplasts are commonly centers of starch formation.

liana [F. *liane,* from *lier,* to bind]: A large, woody vine that climbs on other plants.

life cycle: The entire sequence of phases in the growth and development of any organism from time of zygote formation to gamete formation.

ligase: An enzyme that joins together (ligates) two molecules in an energy-dependent process; DNA ligase, for example, is essential for DNA replication, catalyzing the covalent bonding of the 3′ end of a new DNA fragment to the 5′ end of a growing chain.

light reactions: The reactions of photosynthesis that require light and cannot occur in the dark; also called light-dependent reactions and energy-transduction reactions.

lignin: One of the most important constituents of the secondary wall of vascular plants, although not all secondary walls contain lignin; after cellulose, lignin is the most abundant plant polymer.

ligule [L. *ligula,* small tongue]: A minute outgrowth or appendage at the base of the leaves of grasses and those of certain lycophytes.

linkage: The tendency for certain genes to be inherited together owing to the fact that they are located on the same chromosome.

lipid [Gk. *lipos,* fat]: One of a large variety of nonpolar organic molecules that are insoluble in water (which is polar) but dissolve readily in nonpolar organic solvents; lipids include fats, oils, steroids, phospholipids, and carotenoids.

loam soils: Soils containing sand, silt, and clay in proportions that result in ideal agricultural soils.

locule (lŏk'ūl) [L. *loculus,* small chamber]: A cavity within a sporangium or a cavity of the ovary in which ovules occur.

locus, *pl.* **loci:** The position on a chromosome occupied by a particular gene.

long-day plants: Plants that must be exposed to light periods longer than some critical length for flowering to occur; they flower in spring or summer.

lumen [L. *lumen,* light, an opening for light]: The space bounded by the plant cell wall; the thylakoid space in chloroplasts; the narrow, transparent space of endoplasmic reticulum.

lysis [Gk. *lysis,* a loosening]: A process of disintegration or cell destruction.

lysogenic bacteria: Bacteria-carrying viruses (phages) that eventually break loose from the bacterial chromosome and set up an active cycle of infection, producing lysis in their bacterial hosts.

lysosome [Gk. *lysis,* loosening, + *soma,* body]: An organelle, bounded by a single membrane, and containing hydrolytic enzymes that are released when the organelle ruptures and are capable of breaking down proteins and other complex macromolecules.

M

macrocyst: A flattened, irregular structure, encircled by a thin membrane, in which zygotes are formed during the life cycle of cellular slime molds.

macroevolution: Evolutionary change on a grand scale, involving major evolutionary trends.

macrofibril: An aggregation of microfibrils, visible with the light microscope.

macromolecule [Gk. *makros,* large]: A molecule of very high molecular weight; refers specifically to proteins, nucleic acids, polysaccharides, and complexes of these.

macronutrients [Gk. *makros,* large, + L. *nutrire,* to nourish]: Inorganic chemical elements required in large amounts for plant growth, such as nitrogen, potassium, calcium, phosphorus, magnesium, and sulfur.

magnoliids: The 3 percent of living angiosperms with the most primitive features; the ancestors of both monocots and eudicots.

major veins: The larger leaf vascular bundles, which are associated with ribs; they are largely involved with the transport of substances into and out of the leaf.

maltase: An enzyme that hydrolyzes maltose to glucose.

mannitol: One of the storage molecules of the brown algae; an alcohol.

mating type: A particular genetically defined strain of an organism that is incapable of sexual reproduction with another member of the same strain but capable of such reproduction with members of other strains of the same organism.

matrotrophy: Pertaining to a form of nutrition provided by the maternal gametophyte as, for example, in the case of a moss gametophyte providing nutrients to the zygote and developing sporophyte.

mega- [Gk. *megas,* large]: Prefix meaning "large."

megagametophyte [Gk. *megas,* large, + *gamos,* marriage, + *phyton,* plant]: In heterosporous plants, the female gametophyte; located within the ovule of seed plants.

megaphyll [Gk. *megas,* large, + *phyllon,* leaf]: A generally large leaf with several to many veins; its leaf trace or traces is/are associated with a leaf gap (in ferns) and a leaf trace gap (in seed plants); in contrast to microphyll.

megasporangium, *pl.* **megasporangia:** A sporangium in which megaspores are produced; *see also* nucellus.

megaspore: In heterosporous plants, a haploid (*n*) spore that develops into a female gametophyte; in most groups, megaspores are larger than microspores.

megaspore mother cell: A diploid cell in which meiosis will occur, resulting in the production of four megaspores; also called a megasporocyte.

megasporocyte: *See* megaspore mother cell.

megasporophyll: A leaf or leaflike structure bearing a megasporangium.

meiosis (mī•ō′sĭs) [Gk. *meioun,* to make smaller]: The two successive nuclear divisions in which the chromosome number is reduced from diploid (2*n*) to haploid (*n*) and segregation of the genes occurs; gametes or spores (in organisms with an alternation of generations) may be produced as a result of meiosis.

meiospores: Spores that arise through meiosis and are therefore haploid.

membrane potential: The voltage difference across a membrane due to the differential distribution of ions.

Mendel's first law: The factors for a pair of alternate characteristics are separate, and only one may be carried in a particular gamete (genetic segregation).

Mendel's second law: The inheritance of one pair of characteristics is independent of the simultaneous inheritance of other traits, such characteristics assorting independently as though there were no others present (later modified by the discovery of linkage).

meristem [Gk. *merizein,* to divide]: The undifferentiated, perpetually young plant tissue from which new cells arise.

meso- [Gk. *mesos,* middle]: Prefix meaning "middle."

mesocarp [Gk. *mesos,* middle, + *karpos,* fruit]: The middle layer of the mature ovary wall, or pericarp, between the exocarp and endocarp.

mesocotyl: The internode between the scutellar node and the coleoptile in the embryo and seedling of grasses (*Poaceae*).

mesophyll: The ground tissue (parenchyma) of a leaf, located between the layers of epidermis; mesophyll cells generally contain chloroplasts.

mesophyte [Gk. *mesos,* middle, + *phyton,* a plant]: A plant that requires an environment that is neither too wet nor too dry.

messenger RNA (mRNA): The class of RNA that carries genetic information from the gene to the ribosomes, where it is translated into protein.

metabolism [Gk. *metabolē,* change]: The sum of all chemical processes occurring within a living cell or organism.

metaphase: The stage of mitosis or meiosis during which the chromosomes lie in the equatorial plane of the spindle.

metaxylem [Gk. *meta,* after, + *xylon,* wood]: The part of the primary xylem that differentiates after the protoxylem; the metaxylem reaches maturity after the portion of the plant part in which it is located has finished elongating.

micro- [Gk. *mikros,* small]: Prefix meaning "small."

microbody: *See* peroxisome.

microevolution: Evolutionary change within a population over a succession of generations.

microfibril: A threadlike component of the cell wall, composed of cellulose molecules, visible only with the electron microscope.

microfilament: *See* actin filament.

microgametophyte [Gk. *mikros,* small, + *gamos,* marriage, + *phyton,* plant]: In heterosporous plants, the male gametophyte.

micrometer: A unit of microscopic measurement convenient for describing cellular dimensions; 1/1000 of a millimeter; its symbol is µm.

micronutrients [Gk. *mikros,* small, + L. *nutrire,* to nourish]: Inorganic chemical elements required only in very small, or trace, amounts for plant growth, such as iron, chlorine, copper, manganese, zinc, molybdenum, nickel, and boron.

microphyll [Gk. *mikros,* small, + *phyllon,* leaf]: A small leaf with one vein and one leaf trace not associated with either a leaf gap or leaf trace gap; in contrast to megaphyll. Microphylls are characteristic of the lycophytes.

micropyle: In the ovules of seed plants, the opening in the integuments through which the pollen tube usually enters.

microsporangium: A sporangium within which microspores are formed.

microspore: In heterosporous plants, a spore that develops into a male gametophyte.

microspore mother cell: A cell in which meiosis will occur, resulting in four microspores; in seed plants, often called a pollen mother cell; also called a microsporocyte.

microsporocyte: *See* microspore mother cell.

microsporophyll: A leaflike organ bearing one or more microsporangia.

microtubule [Gk. *mikros,* small, + L. *tubulus,* little pipe]: Narrow (about 25 nanometers in diameter), elongate, nonmembranous tubule of indefinite length; microtubules occur in the cells of eukaryotes. Microtubules move the chromosomes in cell division and provide the internal structure of cilia and flagella.

middle lamella: The layer of intercellular material, rich in pectic compounds, cementing together the primary walls of adjacent cells.

mimicry [Gk. *mimos,* mime]: The superficial resemblance in form, color, or behavior of certain organisms (mimics) to other more powerful or more protected ones (models), resulting in protection, concealment, or some other advantage for the mimic.

mineral: A naturally occurring chemical element or inorganic compound.

minor veins: The small leaf vascular bundles, which are located in the mesophyll and enclosed by a bundle sheath; they are involved with distribution of the transpiration stream and the uptake of the products of photosynthesis.

mitochondrion, *pl.* **mitochondria** [Gk. *mitos,* thread, + *chondrion,* small grain]: A double-membrane-bounded organelle found in eukaryotic cells; contains the enzymes of the Krebs cycle and the electron-transport chain; the major source of ATP in nonphotosynthetic cells.

mitosis (mī•tō′sĭs) [Gk. *mitos,* thread]: A process during which the duplicated chromosomes divide longitudinally and the daughter chromosomes then separate to form two genetically identical daughter nuclei; usually accompanied by cytokinesis.

mitotic spindle: The array of microtubules that forms between the opposite poles of a eukaryotic cell during mitosis.

mole: The name for a gram molecule. The number of particles in 1 mole of any substance; always equal to Avogadro's number: 6.022×10^{23}.

molecular weight: The relative weight of a molecule when the weight of the most frequent kind of carbon atom is taken as 12; the sum of the relative weights of the atoms in a molecule.

molecule: Smallest possible unit of a compound, consisting of two or more atoms.

mono- [Gk. *monos,* single]: Prefix meaning "one" or "single."

monocotyledon: A plant whose embryo has one cotyledon; one of the two great classes of angiosperms, *Monocotyledones;* often abbreviated as monocot.

monoecious (mō•nē'shŭs) [Gk. *monos,* single, + *oikos,* house]: Having the anthers and carpels produced in separate flowers on the same individual.

monokaryotic [Gk. *monos,* one, + *karyon,* kernel]: In fungi, having a single haploid nucleus within one cell or compartment.

monomers [Gk. *monos,* single, + *meros,* part]: Small, repeating units that can be linked together to form polymers.

monophyletic: Pertaining to a taxon descended from a single ancestor.

monosaccharide [Gk. *monos,* single, + *sakcaron,* sugar]: A simple sugar, such as five-carbon and six-carbon sugars, that cannot be dissociated into smaller sugar particles.

-morph, morph- [Gk. *morphē,* form]: Suffix or prefix meaning "form."

morphogenesis: The development of form.

morphology [Gk. *morphē,* form, + *logos,* discourse]: The study of form and its development.

mRNA: *See* messenger RNA.

mucigel: A slime sheath covering the surface of many roots.

multigene family: A collection of related genes on a chromosome; most eukaryotic genes appear to be members of multigene families.

multiple epidermis: A tissue composed of several cell layers derived from the protoderm; only the outer layer assumes characteristics of a typical epidermis.

multiple fruit: A cluster of mature ovaries produced by a cluster of flowers, as in the pineapple.

mutagen [L. *mutare,* to change, + Gk. *genaio,* to produce]: An agent that increases the mutation rate.

mutant: A mutated gene or an organism carrying a gene that has undergone a mutation.

mutation: Any change in the hereditary state of an organism; such changes may occur at the level of the gene (gene mutations, or point mutations) or at the level of the chromosome (chromosome mutations).

mutualism: The living together of two or more organisms in an association that is mutually advantageous.

myc-, myco- [Gk. *mykes,* fungus]: Prefix meaning "pertaining to fungi."

mycelium [Gk. *mykes,* fungus]: The mass of hyphae forming the body of a fungus, oomycete, or chytrid.

mycology: The study of fungi.

mycorrhiza, *pl.* **mycorrhizae:** A symbiotic association between certain fungi and plant roots; characteristic of most vascular plants.

N

NAD⁺: *See* nicotinamide adenine dinucleotide.

NADP⁺: *See* nicotinamide adenine dinucleotide phosphate.

nanoplankton (năn'ō•plăngk'tŏn) [Gk. *nanos,* dwarf, + *planktos,* wandering]: Plankton with dimensions of less than 70 to 75 micrometers.

nastic movement: A plant movement that occurs in response to a stimulus, but whose direction is independent of the direction of the stimulus.

natural selection: The differential reproduction of genotypes based on their genetic constitution.

nectary [Gk. *nektar,* the drink of the gods]: In angiosperms, a gland that secretes nectar, a sugary fluid that attracts animals to plants.

netted venation: The arrangement of veins in the leaf blade that resembles a net; characteristic of the leaves of magnoliids and eudicots; also called reticulate venation.

neutron [L. *neuter,* neither]: An uncharged particle with a mass slightly greater than that of a proton, found in the atomic nucleus of all elements except hydrogen, in which the nucleus consists of a single proton.

niche: The role played by a particular species in its ecosystem.

nicotinamide adenine dinucleotide (NAD⁺): A coenzyme that functions as an electron acceptor in many of the oxidation reactions of respiration.

nicotinamide adenine dinucleotide phosphate (NADP⁺): A coenzyme that functions as an electron acceptor in many of the reduction reactions of biosynthesis; similar in structure to NAD⁺ except that it contains an extra phosphate group.

nitrification: The oxidation of ammonium ions or ammonia to nitrate; a process carried out by a specific free-living soil bacterium.

nitrogen fixation: The incorporation of atmospheric nitrogen into nitrogen compounds; carried out by certain free-living and symbiotic bacteria.

nitrogen-fixing bacteria: Soil bacteria that convert atmospheric nitrogen into nitrogen compounds.

nitrogenous base: A nitrogen-containing molecule having basic properties (tendency to acquire an H atom); a purine or pyrimidine; one of the building blocks of nucleic acids.

node [L. *nodus,* knot]: The part of a stem where one or more leaves are attached; *see also* internode.

nodules: Enlargements or swellings on the roots of legumes and certain other plants inhabited by symbiotic nitrogen-fixing bacteria.

noncyclic electron flow: The light-induced flow of electrons from water to NADP⁺ in oxygen-evolving photosynthesis; it involves both photosystems I and II.

nonseptate: *See* aseptate.

nucellus (noo•sĕl'ŭs) [L. *nucella,* a small nut]: Tissue composing the chief part of the young ovule, in which the embryo sac develops; equivalent to a megasporangium.

nuclear envelope: The double membrane surrounding the nucleus of a cell.

nucleic acid: An organic acid consisting of joined nucleotide complexes; the two types are deoxyribonucleic acid (DNA) and ribonucleic acid (RNA).

nucleoid: A region of DNA in prokaryotic cells, mitochondria, and chloroplasts.

nucleolar organizer region: A special area on a certain chromosome associated with the formation of the nucleolus.

nucleolus, *pl.* **nucleoli** (noo•klē'ō•lŭs) [L. *nucleolus,* a small nucleus]: A small, spherical body found in the nucleus of eukaryotic cells, which is composed chiefly of rRNA in the process of being transcribed from copies of rRNA genes; the site of production of ribosomal subunits.

nucleoplasm: The ground substance of a nucleus.

nucleosome: A complex of DNA and histone proteins that forms the fundamental packaging unit of eukaryotic DNA; its structure resembles a bead on a string.

nucleotide: A single unit of nucleic acid, composed of a phosphate, a five-carbon sugar (either ribose or deoxyribose), and a purine or a pyrimidine.

nucleus, *pl.* **nuclei:** (1) A specialized body within the eukaryotic cell bounded by a double membrane and containing the chromosomes; (2) the central part of an atom of a chemical element.

nut: A dry, indehiscent, hard, one-seeded simple fruit, generally produced from a gynoecium of more than one fused carpel.

O

obligate anaerobe: *See* strict anaerobe.

-oid [Gk. *oid,* like, resembling]: Suffix meaning "like" or "similar to."

Okazaki fragments [after R. Okazaki, Japanese geneticist]: In DNA replication, the discontinuous segments in which the 3′ to 5′ strand (the lagging strand) of the DNA double helix is synthesized; typically 1000 to 2000 nucleotides long in prokaryotes, and 100 to 200 nucleotides long in eukaryotes.

ontogeny [Gk. *on,* being, + *genesis,* origin]: The development, or life history, of all or part of an individual organism.

oo- [Gk. *oion,* egg]: Prefix meaning "egg."

oogamy: Sexual reproduction in which one of the gametes (the egg) is large and nonmotile, and the other gamete (the sperm) is smaller and motile.

oogonium (ŏ′o•gŏ′nē•ŭm): A unicellular female sex organ that contains one or several eggs.

oospore: The thick-walled zygote characteristic of the oomycetes.

open vascular bundle: A vascular bundle in which a vascular cambium develops.

operator: A segment of DNA that interacts with a repressor protein to regulate the transcription of the structural genes of an operon.

operculum (o•pûr′ku•lŭm) [L. *operculum,* lid]: In mosses, the lid of the sporangium.

operon [L. *opus, operis,* work]: In the bacterial chromosome, a segment of DNA consisting of a promoter, an operator, and a group of adjacent structural genes; the structural genes, which code for products related to a particular biochemical pathway, are transcribed onto a single mRNA molecule, and their transcription is regulated by a single repressor protein.

opposite phyllotaxy: Term applied to leaves occurring in pairs at a node.

order: A category of classification between the rank of class and family; classes contain one or more orders, and orders in turn are composed of one or more families.

organ: A structure composed of different tissues, such as root, stem, leaf, or flower parts.

organelle (or′găn•el) [Gk. *organella,* a small tool]: A specialized, membrane-bounded part of a cell.

organic: Pertaining to living organisms in general, to compounds formed by living organisms, and to the chemistry of compounds containing carbon.

organism: Any individual living creature, either unicellular or multicellular.

osmosis (ŏs•mō′sĭs) [Gk. *osmos,* impulse or thrust]: The diffusion of water, or any solvent, across a selectively permeable membrane; in the absence of other forces, the movement of water during osmosis will always be from a region of greater water potential to one of lesser water potential.

osmotic potential: The change in free energy or chemical potential of water produced by solutes; carries a negative (minus) sign; also called solute potential.

osmotic pressure: The potential pressure that can be developed by a solution separated from pure water by a selectively permeable membrane; in the absence of other forces, the movement of water during osmosis will always be from a region of greater water potential to one of lesser water potential.

osmotic potential: The change in free energy or chemical potential of water produced by solu permeable or semipermeable membrane; it is an index of the solute concentration of the solution.

outcrossing: Cross-pollination between individuals of the same species.

outer bark: In older trees, the dead part of the bark; the innermost periderm and all tissues outside it; also called rhytidome.

ovary [L. *ovum,* an egg]: The enlarged basal portion of a carpel or of a gynoecium composed of fused carpels; a mature ovary, sometimes with other adherent parts, is a fruit.

ovule (ō•vūl) [L. *ovulum,* a little egg]: A structure in seed plants containing the female gametophyte with egg cell, all being surrounded by the nucellus and one or two integuments; when mature, an ovule becomes a seed.

ovuliferous scale: In certain conifers, the appendage or scalelike shoot to which the ovule is attached.

oxidation: The loss of an electron by an atom or molecule. Oxidation and reduction (gain of an electron) take place simultaneously, because an electron that is lost by one atom is accepted by another. Oxidation-reduction reactions are an important means of energy transfer in living systems.

oxidative phosphorylation: The formation of ATP from ADP and inorganic phosphate; oxidative phosphorylation takes place in the electron-transport chain of the mitochondrion.

P

pairing of chromosomes: Side-by-side association of homologous chromosomes.

paleobotany [Gk. *palaios,* old]: The study of fossil plants.

paleoherbs: Apart from the woody magnoliids, collectively the remaining magnoliids; constitute a very diverse assemblage; include all angiosperms that are not woody magnoliids, eudicots, or monocots.

palisade parenchyma: A leaf tissue composed of columnar chloroplast-bearing parenchyma cells with their long axes at right angles to the leaf surface.

panicle (păn′ĭ•kûl) [L. *panicula,* tuft]: An inflorescence, the main axis of which is branched, and whose branches bear loose flower clusters.

para- [Gk. *para,* beside]: Prefix meaning "beside."

paradermal section [Gk. *para,* beside, + *derma*, skin]: Section cut parallel to the surface of a flat structure, such as a leaf.

parallel evolution: The development of similar structures having similar functions in two or more evolutionary lines as a result of the same kinds of selective pressures.

parallel venation: The pattern of venation in which the principal veins of the leaf are parallel or nearly so; characteristic of monocots.

paramylon: The storage molecule of euglenoids.

paraphyletic: Pertaining to a taxon that excludes species that share a common ancestor with species included in the taxon.

paraphysis, *pl.* **paraphyses** [Gk. *para,* beside, + *physis,* growth]: In certain fungi, a sterile filament growing among the reproductive cells in the fruiting body; also applied to the sterile filaments growing among the gametangia and sporangia of certain brown algae and among the antheridia and archegonia of mosses.

parasexual cycle: The fusion and segregation of heterokaryotic haploid nuclei in certain fungi to produce recombinant nuclei.

parasite: An organism that lives on or in an organism of a different species and derives nutrients from it; the association is beneficial to the parasite and harmful to the host.

parenchyma (pa•rĕng′kĭ•ma) [Gk. *para,* beside, + *en,* in, + *chein,* to pour]: A tissue composed of parenchyma cells.

parenchyma cell: Living, generally thin-walled cell of variable size and form; the most abundant kind of cell in plants.

parthenocarpy [Gk. *parthenos,* virgin, + *karpos*, fruit]: The development of fruit without fertilization; parthenocarpic fruits are usually seedless.

passage cell: Endodermal cell of root that retains a thin wall and Casparian strip when other associated endodermal cells develop thick secondary walls.

passive transport: Non-energy-requiring transport of a solute across a membrane down the concentration or electrochemical gradient by either simple diffusion or facilitated diffusion.

pathogen [Gk. *pathos,* suffering, + *genesis,* beginning]: An organism that causes a disease.

pathogenic: Disease-causing.

pathology: The study of plant or animal diseases, their effects on the organism, and their treatment.

PCR: See polymerase chain reaction.

pectin: A highly hydrophilic polysaccharide present in the intercellular layer and primary wall of plant cell walls; the basis of fruit jellies.

pedicel (ped′ĭ•sĕl): The stalk of an individual flower in an inflorescence.

peduncle (pe•dŭng′kûl): The stalk of an inflorescence or of a solitary flower.

pentose phosphate cycle: The pathway of oxidation of glucose 6-phosphate to yield pentose phosphates.

peptide: Two or more amino acids linked by peptide bonds.

peptide bond: The type of bond formed when two amino acid units are joined end to end by the removal of a molecule of water; the bonds always form between the carboxyl ($-COOH$) group of one amino acid and the amino ($-NH_2$) group of the next amino acid.

perennial [L. *per,* through, + *annuus,* a year]: A plant in which the vegetative structures live year after year.

perfect flower: A flower having both stamens and carpels; hermaphroditic flower.

perfect stage: That phase of the life history of a fungus that includes sexual fusion and the spores associated with such fusions.

perforation plate: Part of the wall of a vessel element that is perforated.

peri- [Gk. *peri,* around]: Prefix meaning "around" or "about."

perianth (pĕr′ĭ•anth) [Gk. *peri,* around, + *anthos,* flower]: (1) The petals and sepals taken together; (2) in leafy liverworts, a tubular sheath surrounding an archegonium, and later, the developing sporophyte.

pericarp [Gk. *peri,* around, + *karpos,* fruit]: The fruit wall, which develops from the mature ovary wall.

periclinal: Parallel to the surface.

pericycle [Gk. *peri,* around, + *kykos,* circle]: A tissue characteristic of roots that is bounded externally by the endodermis and internally by the phloem.

periderm [Gk. *peri,* around, + *derma,* skin]: Outer protective tissue that replaces epidermis when it is destroyed during secondary growth; includes cork, cork cambium, and phelloderm.

perigyny [Gk. *peri,* around, + *gyne,* female]: A form of floral organization in which the sepals, petals, and stamens are attached to the margin of a cup-shaped extension of the receptacle; superficially, the sepals, petals, and stamens appear to be attached to the ovary.

perisperm [Gk. *peri,* around, + *sperma,* seed]: Food-storing tissue derived from the nucellus that occurs in the seeds of some flowering plants.

peristome (pĕr′ĭ•stōm) [Gk. *peri,* around, + *stoma,* a mouth]: In mosses, a fringe of teeth around the opening of the sporangium.

perithecium: A spherical or flask-shaped ascoma.

permanent wilting percentage: The percentage of water remaining in a soil when a plant fails to recover from wilting even if placed in a humid chamber.

permeable [L. *permeare,* to pass through]: Usually applied to membranes through which liquid substances may diffuse.

peroxisome: A spherical, single membrane-bounded organelle, ranging in diameter from 0.5 to 1.5 micrometers; some peroxisomes are involved in photorespiration, and others (called glyoxysomes) with the conversion of fats to sugars during seed germination; also called microbodies.

petal: A flower part, usually conspicuously colored; one of the units of the corolla.

petiole: The stalk of a leaf.

pH: A symbol denoting the relative concentration of hydrogen ions in a solution; pH values run from 0 to 14, and the lower the value the more acidic a solution, that is, the more hydrogen ions it contains; pH 7 is neutral, less than 7 is acidic, and more than 7 is alkaline.

phage: *See* bacteriophage.

phagocytosis: *See* endocytosis.

phellem: *See* cork.

phelloderm (fĕl′o•durm) [Gk. *phellos,* cork, + *derma,* skin]: A tissue formed inwardly by the cork cambium, opposite the cork; inner part of the periderm.

phellogen (fĕl′o•jĕn): *See* cork cambium.

phenolics: A broad range of compounds, all of which have a hydroxyl group ($-OH$) attached to an aromatic ring (a ring of six carbons containing three double bonds); includes flavonoids, tannins, lignins, and salicylic acid.

phenotype: The physical appearance of an organism; the phenotype results from the interaction between the genetic constitution (genotype) of the organism and its environment.

phloem (flō′ĕm) [Gk. *phloos,* bark]: The food-conducting tissue of vascular plants, which is composed of sieve elements, various kinds of parenchyma cells, fibers, and sclereids.

phloem loading: The process by which substances (primarily sugars) are actively secreted into the sieve tubes.

phosphate: A compound formed from phosphoric acid by replacement of one or more hydrogen atoms.

phospholipids: A phosphorylated lipid; similar in structure to a fat, but with only two fatty acids attached to the glycerol backbone, with the third space occupied by a phosphorus-containing molecule; important components of cellular membranes.

phosphorylation (fŏs′fo•rĭl•ā′shŭn) [Gk. *phosphoros,* bringing light]: A reaction in which phosphate is added to a compound; for example, the formation of ATP from ADP and inorganic phosphate.

photo-, -photic [Gk. *photos,* light]: Prefix or suffix meaning "light."

photobiont [Gk. *photos,* light, + *bios,* life]: The photosynthetic component of a lichen.

photolysis: The light-dependent oxidative splitting of water molecules that takes place in photosystem II of the light reactions of photosynthesis.

photon [Gk. *photos,* light]: The elementary particle of light.

photoperiodism: Response to duration and timing of day and night; a mechanism evolved by organisms for measuring seasonal time.

photophosphorylation [Gk. *photos,* light, + *phosphoros,* bringing light]: The formation of ATP in the chloroplast during photosynthesis.

photorespiration: The oxygenase activity of Rubisco combined with the salvage pathway, consuming O_2 and releasing CO_2; occurs when Rubisco binds O_2 instead of CO_2.

photosynthesis [Gk. *photos,* light, + *syn,* together + *tithenai,* to place]: The conversion of light energy to chemical energy; the production of carbohydrates from carbon dioxide and water in the presence of chlorophyll by using light energy.

photosystem: A discrete unit of organization of chlorophyll and other pigment molecules embedded in the thylakoids of chloroplasts and involved with the light-dependent reactions of photosynthesis.

phototropism [Gk. *photos,* light, + *trope,* turning]: Growth in which the direction of the light is the determining factor, as the growth of a plant toward a light source; turning or bending in response to light.

phragmoplast: A spindle-shaped system of fibrils, which arises between two daughter nuclei at telophase and within which the cell plate is formed during cell division, or cytokinesis. The fibrils of the phragmoplast are composed of microtubules. Phragmoplasts are found in all green algae except the members of the class *Chlorophyceae* and in plants.

phragmosome: The layer of cytoplasm that forms across the cell where the nucleus becomes located and divides.

phycobilins: A group of water-soluble accessory pigments, including phycocyanins and phycoerythrins, which occur in the red algae and cyanobacteria.

phycology [Gk. *phykos,* seaweed]: The study of algae.

phycoplast: A system of microtubules that develops between the two daughter nuclei parallel to the plane of cell division. Phycoplasts occur only in green algae of the class *Chlorophyceae.*

phyllo-, phyll- [Gk. *phyllon,* leaf]: Prefix meaning "leaf."

phyllode (fĭl′ōd): A flat, expanded, photosynthetic petiole or stem; phyllodes occur in certain genera of vascular plants, where they replace leaf blades in photosynthetic function.

phyllotaxy: The arrangement of the leaves on a stem; also called phyllotaxis.

phylogeny [Gk. *phylon,* race, tribe]: Evolutionary relationships among organisms; the developmental history of a group of organisms.

phylum, *pl.* **phyla** [Gk. *phylon,* race, tribe]: A taxonomic category of related, similar classes; a higher-level category below kingdom and above class; until recently called division by botanists.

physiology: The study of the activities and processes of living organisms.

phyto-, -phyte [Gk. *phyton,* plant]: Prefix or suffix meaning "plant."

phytoalexin: [Gk. *phyton,* plant, + *alexein,* to ward off, protect]: A chemical substance produced by a plant to combat infection by a pathogen (fungi or bacteria).

phytochrome: A phycobilinlike pigment found in the cytoplasm of plants and a few green algae that is associated with the absorption of light; photoreceptor for red and far-red light; involved in a number of timing processes, such as flowering, dormancy, leaf formation, and seed germination.

phytoplankton [Gk. *phyton,* + *planktos,* wandering]: Aquatic, free-floating, microscopic, photosynthetic organisms.

pigment: A substance that absorbs light, often selectively.

pileus [L. *pileus,* a cap]: The caplike part of mushroom basidiomata and of certain ascomata.

pinna, *pl.* **pinnae** (pĭn′a) [L. *pinna,* feather]: A primary division, or leaflet, of a compound leaf or frond; may be divided into pinnules.

pinocytosis: *See* endocytosis.

pistil [L. *pistillum,* pestle]: A term sometimes used to refer to an individual carpel or a group of fused carpels.

pistillate: *See* carpellate.

pit: A recessed cavity in a cell wall where the secondary wall does not form.

pit membrane: The middle lamella and two primary cell walls between two pits.

pit-pair: Two opposite pits plus the pit membrane.

pith: The ground tissue occupying the center of the stem or root within the vascular cylinder; usually consists of parenchyma.

pith ray: *See* interfascicular region.

placenta, *pl.* **placentae** [L. *placenta,* cake]: The part of the ovary wall to which the ovules or seeds are attached.

placentation: The manner of ovule attachment within the ovary.

plankton [Gk. *planktos,* wandering]: Free-floating, mostly microscopic, aquatic organisms.

plaque: Clear area in a sheet of cells resulting from the killing or lysis of contiguous cells by viruses.

-plasma, plasmo-, -plast [Gk. *plasma,* form, mold]: Prefix or suffix meaning "formed" or "molded"; examples are protoplasm, "first-molded" (living matter), and chloroplast, "green-formed."

plasma membrane or **plasmalemma:** Outer boundary of the cytoplasm, next to the cell wall; consists of a single membrane; also called cell membrane, or ectoplast.

plasmid: A relatively small fragment of DNA that can exist free in the cytoplasm of a bacterium and can be integrated into and then replicated with a chromosome. Plasmids make up about 5 percent of the DNA of many bacteria, but are rare in eukaryotes.

plasmodesma, *pl.* **plasmodesmata** [Gk. *plasma,* form, + *desma,* bond]: The minute cytoplasmic threads that extend through openings in cell walls and connect the protoplasts of adjacent living cells.

plasmodium (plăz•mō′dĭ•ŭm): Stage in life cycle of myxomycetes (plasmodial slime molds); a multinucleate mass of protoplasm surrounded by a membrane.

plasmogamy [Gk. *plasma,* form, + *gamos,* marriage]: Union of the protoplasts of gametes that is not accompanied by union of their nuclei.

plasmolysis (plăz•mŏl′ĭ•sĭs) [Gk. *plasma,* form, + *lysis,* a loosening]: The separation of the protoplast from the cell wall because of the removal of water from the protoplast by osmosis.

plastid (plăs′tĭd): Organelle in the cells of certain groups of eukaryotes that is the site of such activities as food manufacture and storage; plastids are bounded by two membranes.

pleiotropy [Gk. *pleros,* more, + *trope,* a turning]: The capacity of a gene to affect more than one phenotypic characteristic.

plumule [L. *plumula,* a small feather]: The first bud of an embryo; the portion of the young shoot above the cotyledons.

pneumatophores [Gk. *pneuma,* breath, + *-phoros,* carrying]: Negatively gravitropic extensions of the root systems of some trees growing in swampy habitats; they grow upward and out of the water and probably function to ensure adequate aeration.

point mutation: An alteration in one of the nucleotides in a chromosomal DNA molecule; an allele of a gene changes, becoming a different allele; also called gene mutation.

polar molecule: A molecule with positively and negatively charged ends.

polar nuclei: Two nuclei (usually), one derived from each end (pole) of the embryo sac, which become centrally located; they fuse with a male nucleus to form the primary (typically $3n$) endosperm nucleus.

pollen [L. *pollen,* fine dust]: A collective term for pollen grains.

pollen grain: A microspore containing a mature or immature microgametophyte (male gametophyte); pollen grains occur in seed plants.

pollen mother cell: *See* microspore mother cell.

pollen sac: A cavity in the anther that contains the pollen grains.

pollen tube: A tube formed after germination of the pollen grain; carries the male gametes into the ovule.

pollination: In angiosperms, the transfer of pollen from an anther to a stigma. In gymnosperms, the transfer of pollen from a pollen-producing cone directly to an ovule.

poly- [Gk. *polys,* many]: Prefix meaning "many."

polyembryony: Having more than one embryo within the developing seed.

polygenic inheritance: The inheritance of quantitative characteristics determined by the combined effects of multiple genes.

polymer (pŏl′ĭ•mer): A large molecule composed of many similar molecular subunits.

polymerase chain reaction (PCR): A technique for amplifying specific regions of DNA by multiple cycles of DNA polymerization, utilizing special primers, DNA polymerase molecules, and nucleotides; each cycle is followed by a brief heat treatment to separate complementary strands.

polymerization: The chemical union of monomers such as glucose or nucleotides to form polymers such as starch or nucleic acid.

polynucleotide: A single-stranded DNA or RNA molecule.

polypeptide: A molecule composed of amino acids linked together by peptide bonds, not as complex as a protein.

polyphyletic: Pertaining to a taxon whose members are derived from two or more ancestors not common to all members of the taxon.

polyploid (pŏl′ĭ•ploid): Referring to an organism, tissue, or cell with more than two complete sets of chromosomes.

polysaccharide: A polymer composed of many monosaccharide units joined in a long chain, such as glycogen, starch, and cellulose.

polysome or **polyribosome:** An aggregation of ribosomes actively involved in the translation of the same RNA molecule, one after another.

pome (pōm) [Fr. *pomme,* apple]: A simple fleshy fruit, the outer portion of which is formed by the floral parts that surround the ovary and expand with the growing fruit; found only in one subfamily of the *Rosaceae* (apples, pears, quince, pyracantha, and so on).

population: Any group of individuals, usually of a single species, occupying a given area at the same time.

P-protein: Phloem-protein; a proteinaceous substance found in cells of angiosperm phloem, especially in sieve-tube elements; also called slime.

preprophase band: A ringlike band of microtubules, found just beneath the plasma membrane, that delimits the equatorial plane of the future mitotic spindle of a cell preparing to divide.

primary endosperm nucleus: The result of the fusion of a sperm nucleus and the two polar nuclei.

primary growth: In plants, growth originating in the apical meristems of shoots and roots, as contrasted with secondary growth.

primary meristem or **primary meristematic tissue** (mĕr′ĭ•ste•măt′ĭk): A tissue derived from the apical meristem; of three kinds: protoderm, procambium, and ground meristem.

primary metabolites: Molecules that are found in all plant cells and are necessary for the life of the plant; examples are simple sugars, amino acids, proteins, and nucleic acids.

primary pit-field: Thin area in a primary cell wall through which plasmodesmata pass, although plasmodesmata may occur elsewhere in the wall as well.

primary plant body: The part of the plant body arising from the apical meristems and their derivative meristematic tissues; composed entirely of primary tissues.

primary root: The first root of the plant, developing in continuation of the root tip or radicle of the embryo; the taproot.

primary tissues: Cells derived from the apical meristems and primary meristematic tissues of root and shoot; as opposed to secondary tissues derived from a cambium; primary growth results in an increase in length.

primary cell wall: The wall layer deposited during the period of cell expansion.

primordium, *pl.* **primordia** [L. *primus,* first, + *ordiri,* to begin to weave]: A cell or organ in its earliest stage of differentiation.

pro- [Gk. *pro,* before]: Prefix meaning "before" or "prior to."

procambium (prō•kăm′bē•ŭm) [L. *pro,* before, + *cambiare,* to exchange]: A primary meristematic tissue that gives rise to primary vascular tissues.

proembryo: Embryo in early stages of development, before the embryo proper and suspensor become distinct.

prokaryote [Gk. *pro,* before, + *karyon,* kernel]: A cell lacking a membrane-bounded nucleus and membrane-bounded organelles; *Bacteria* and *Archaea.*

prolamellar body: Semicrystalline body found in plastids arrested in development by the absence of light.

promoter: A specific segment of DNA to which RNA polymerase attaches to initiate transcription of mRNA from an operon.

prophase [Gk. *pro,* before, + *phasis,* form]: An early stage in nuclear division, characterized by the shortening and thickening of the chromosomes and their movement to the metaphase plate.

proplastid: A minute self-reproducing body in the cytoplasm from which a plastid develops.

prop roots: Adventitious roots arising from the stem above soil level and helping to support the plant: common in many monocots, for example, maize *(Zea mays).*

prosthetic group: A heat-stable metal ion or an inorganic group (other than an amino acid) that is bound to a protein and serves as its active group.

protease (prō′tē•ās): An enzyme that digests protein by the hydrolysis of peptide bonds; proteases are also called peptidases.

protein [Gk. *proteios,* primary]: A complex organic compound composed of many (100 or more) amino acids joined by peptide bonds.

prothallial cell [Gk. *pro,* before, + *thallos,* sprout]: The sterile cell or cells found in the male gametophytes, or microgametophytes, of vascular plants other than angiosperms; believed to be remnants of the vegetative tissue of the male gametophyte.

prothallus (prō•thăl′ŭs): In homosporous vascular plants, such as ferns, the more or less independent, photosynthetic gametophyte; also called the prothallium.

protist: A member of the kingdom *Protista.*

proto- [Gk. *protos,* first]: Prefix meaning "first"; for example, *Protozoa,* "first animals."

protoderm [Gk. *protos,* first, + *derma,* skin]: Primary meristematic tissue that gives rise to epidermis.

proton: A subatomic, or elementary, particle, with a single positive charge equal in magnitude to the charge of an electron and a mass of 1; the basic component of every atomic nucleus.

protonema, *pl.* **protonemata** [Gk. *protos,* first, + *nema,* a thread]: The first stage in development of the gametophyte of mosses and certain liverworts; protonemata may be filamentous or platelike.

protoplasm: A general term for the living substance of all cells.

protoplast: The protoplasm of an individual cell; in plants, the unit of protoplasm inside the cell wall.

protostele [Gk. *protos,* first, + *stele,* pillar]: The simplest type of stele, consisting of a solid column of vascular tissue.

protoxylem: The first part of the primary xylem, which matures during elongation of the plant part in which it is found.

protracheophyte: An organism with branched axes and multiple sporangia but with water-conducting cells similar to the hydroids of modern mosses rather than to the tracheary elements of vascular plants; an intermediate stage in the evolution of vascular plants, or tracheophytes.

proximal (prŏk′sĭ•măl) *[L. proximus,* near]: Situated near the point of reference, usually the main part of a body or the point of attachment; opposite of distal.

pseudo- [Gk. *pseudes,* false]: Prefix meaning "false."

pseudoplasmodium: A multicellular mass of individual amoeboid cells, representing the aggregate phase in the cellular slime molds.

pulvinus, *pl.* **pulvini:** Jointlike thickening at the base of the petiole of a leaf or petiolule of a leaflet, and having a role in the movements of the leaf or leaflet.

pumps: Transport proteins driven by either chemical energy (ATP) or light energy; in plant and fungal cells, they typically are proton pumps.

punctuated equilibrium: A model of evolutionary change that proposes that there are long periods of little or no change punctuated with brief intervals of rapid change.

purine (pū′rēn): The larger of the two kinds of nucleotide bases found in DNA and RNA; a nitrogenous base with a double-ring structure, such as adenine or guanine.

pyramid of energy: Energy relationships among various feeding levels involved in a particular food chain; autotrophs (at the base of the pyramid) represent the greatest amount of available energy; herbivores are next; then primary carnivores, secondary carnivores, and so forth. Similar pyramids of mass, size, and number also occur in natural communities.

pyrenoid [Gk. *pyren,* the stone of a fruit, + *oides,* like]: Differentiated regions of the chloroplast that are centers of starch formation in green algae and hornworts.

pyrimidine: The smaller of the two kinds of nucleotide bases found in DNA and RNA: a nitrogenous base with a single-ring structure, such as cytosine, thymine, or uracil.

Q

quantasome [L. *quantus,* how much, + Gk. *soma,* body]: Granules located on the inner surfaces of chloroplast lamellae; believed to be involved in the light-dependent reactions of photosynthesis.

quantum: The ultimate unit of light energy.

quiescent center: The relatively inactive initial region in the apical meristem of a root.

R

raceme [L. *racemus,* bunch of grapes]: An indeterminate inflorescence in which the main axis is elongated but the flowers are borne on pedicels that are about equal in length.

rachis (rā′kĭs) [Gk. *rachis,* a backbone]: Main axis of a spike; the axis of a fern leaf (frond), from which the pinnae arise; in compound leaves, the extension of the petiole corresponding to the midrib of an entire leaf.

radially symmetrical: *See* actinomorphic.

radial micellation: The radial orientation of cellulose microfibrils in guard cell walls; plays a role in the movement of guard cells.

radial section: A longitudinal section cut parallel to the radius of a cylindrical body, such as a root or stem; in the case of secondary xylem, or wood, and secondary phloem, parallel to the rays.

radial system: In secondary xylem and secondary phloem, the term applied to all the rays, the cells of which are derived from ray initials; also called the horizontal system, or ray system.

radicle [L. *radix,* root]: The embryonic root.

radioisotope: An unstable isotope of an element that decays or disintegrates spontaneously, emitting radiation; also called a radioactive isotope.

raphe [Gk. *raphe,* seam]: (1) Ridge on seeds, formed by the stalk of the ovule, in those seeds in which the stalk is sharply bent at the base of the ovule; (2) groove on the frustule of a diatom.

raphides (răf′ĭ•dēz) [Gk. *rhaphis,* a needle]: Fine, sharp, needlelike crystals of calcium oxalate found in the vacuoles of many plant cells.

ray flowers: *See* disk flowers.

ray initial: An initial in the vascular cambium that gives rise to the ray cells of secondary xylem and secondary phloem.

reaction center: The complex of proteins and chlorophyll molecules of a photosystem capable of converting light energy to chemical energy in the photochemical reaction.

reaction wood: Abnormal wood that develops in leaning trunks and limbs; *see also* compression wood *and* tension wood.

receptacle: That part of the axis of a flower stalk that bears the floral organs.

recessive: Describing a gene whose phenotypic expression is masked in the heterozygote by a dominant allele; heterozygotes are phenotypically indistinguishable from dominant homozygotes.

recombinant DNA: DNA formed either naturally or in the laboratory by the joining of segments of DNA from different sources.

reduction [L. *reductio,* a bringing back; originally "bringing back" a metal from its oxide]: Gain of an electron by an atom; reduction takes place simultaneously with oxidation (the loss of an electron by an atom), because an electron that is lost by one atom is accepted by another.

regular: *See* actinomorphic.

regulator gene: A gene that prevents or represses the activity of the structural genes in an operon.

replicate: Produce a facsimile or a very close copy; used to indicate the production of a second molecule of DNA exactly like the first molecule or the production of a sister chromatid.

replication fork: In DNA synthesis, the Y-shaped structure formed at the point where the two strands of the original molecule are being separated and the complementary strands are being synthesized.

repressor: A protein that regulates DNA transcription; this occurs because RNA polymerase is prevented from attaching to the promoter and transcribing the gene. *See also* operator.

resin duct: A tubelike intercellular space lined with resin-secreting cells (epithelial cells) and containing resin.

resonance energy transfer: The transfer of light energy from an excited chlorophyll molecule to a neighboring chlorophyll molecule, exciting the second molecule and allowing the first one to return to its ground state.

respiration: An intracellular process in which molecules, particularly pyruvate in the Krebs cycle, are oxidized with the release of energy. The complete breakdown of sugar or other organic compounds to carbon dioxide and water is termed aerobic respiration, although the first steps of this process are anaerobic.

restriction enzymes: Enzymes that cleave the DNA double helix at specific nucleotide sequences.

reticulate venation: *See* netted venation.

reverse transcription: The process by which an RNA molecule is used as a template to make a single-stranded copy of DNA.

rhizobia [Gk. *rhiza,* root, + *bios,* life]: Bacteria of the genera *Rhizobium* or *Bradyrhizobium,* which may be involved with leguminous plants in a symbiotic relationship that results in nitrogen fixation.

rhizoids [Gk. *rhiza,* root]: (1) Branched rootlike extensions of fungi and algae that absorb water, food, and nutrients; (2) root-hairlike structures in liverworts, mosses, and some vascular plants, which occur on free-living gametophytes.

rhizome: A more or less horizontal underground stem.

ribonucleic acid (RNA): Type of nucleic acid formed on chromosomal DNA and involved in protein synthesis; composed of chains of phosphate, sugar molecules (ribose), and purines and pyrimidines; RNA is the genetic material of many kinds of viruses.

ribose: A five-carbon sugar; a component of RNA.

ribosomal RNA: Any of a number of specific molecules that form part of the structure of a ribosome and participate in the synthesis of proteins.

ribosome: A small particle composed of protein and RNA; the site of protein synthesis.

ring-porous wood: A wood in which the pores, or vessels, of the early wood are distinctly larger than those of the late wood, forming a well-defined ring in cross sections of the wood.

RNA: *See* ribonucleic acid.

root: The usually descending axis of a plant, normally below ground, which serves to anchor the plant and to absorb and conduct water and minerals into it.

rootcap: A thimblelike mass of cells that covers and protects the growing tip of a root.

root hairs: Tubular outgrowths of epidermal cells of the root; greatly increase the absorbing surface of the root.

root pressure: The pressure developed in roots as the result of osmosis, which causes guttation of water from leaves and exudation from cut stumps.

rRNA: *See* ribosomal RNA.

Rubisco: RuBP carboxylase/oxygenase, the enzyme that catalyzes initial reaction of the Calvin cycle, involving the fixation of carbon dioxide to ribulose 1,5-bisphosphate (RuBP).

runner: *See* stolon.

S

samara: Simple, dry, one-seeded or two-seeded indehiscent fruit with pericarp-bearing, winglike outgrowths.

sap: (1) A name applied to the fluid contents of the xylem or the sieve elements of the phloem; (2) the fluid contents of the vacuole are called cell sap.

saprophyte [Gk. *sapros,* rotten, + *phyton,* plant]: An organism that secures its food directly from nonliving organic matter; also called a saprobe.

sapwood: Outer part of the wood of stem or trunk, usually distinguished from the heartwood by its lighter color, in which conduction of water takes place.

satellite DNA: A short nucleotide sequence repeated in tandem fashion many thousands of times; this region of the chromosome has a distinctive base composition and is not transcribed.

savanna: Grassland containing scattered trees.

scarification: The process of cutting or softening a seed coat to hasten germination.

schizo- [Gk. *schizein,* to split]: Prefix meaning "split."

schizocarp (skĭz'ō•karp): Dry simple fruit with two or more united carpels that split apart at maturity.

sclereid [Gk. *skleros,* hard]: A sclerenchyma cell with a thick, lignified secondary wall having many pits. Sclereids are variable in form but typically not very long; they may or may not be living at maturity.

sclerenchyma (skle•rĕng'kĭ•ma) [Gk. *skleros,* hard, + L. *enchyma,* infusion]: A supporting tissue composed of sclerenchyma cells, including fibers and sclereids.

sclerenchyma cell: Cell of variable form and size with more or less thick, often lignified, secondary walls; may or may not be living at maturity; includes fibers and sclereids.

scutellum (sku•tĕl'ŭm) [L. *scutella,* a small shield]: The single cotyledon of a grass embryo, specialized for absorption of the endosperm.

secondary growth: In plants, growth derived from secondary or lateral meristems, the vascular and cork cambiums; secondary growth results in an increase in girth, and is contrasted with primary growth, which results in an increase in length.

secondary plant body: The part of the plant body produced by the vascular cambium and the cork cambium; consists of secondary xylem, secondary phloem, and periderm.

secondary root: *See* lateral root.

secondary tissues: Tissues produced by the vascular cambium and cork cambium.

secondary cell wall: Innermost layer of the cell wall, formed in certain cells after cell elongation has ceased; secondary walls have a highly organized microfibrillar structure.

secondary metabolites: Molecules that are restricted in their distribution, both within the plant and among different plants; important for the survival and propagation of the plants that produce them; there are three major classes—alkaloids, terpenoids, and phenolics. Also called secondary products.

second messenger: A small molecule that is formed in or released into the cytosol in response to an external signal; it relays the signal to the interior of the cell; examples are calcium ions and cyclic AMP.

seed: A structure formed by the maturation of the ovule of seed plants following fertilization.

seed coat: The outer layer of the seed, developed from the integuments of the ovule.

seedling: A young sporophyte, which develops from a germinating seed.

segregation: The separation of the chromosomes (and genes) from different parents at meiosis; *see* Mendel's first law.

selectively permeable [L. *seligere,* to gather apart, + *permeare,* to go through]: Applied to membranes that permit passage of water and some solutes but block the passage of others; semipermeable membranes permit the passage of water but not solutes.

sepal (sē'păl) [L. *sepalum,* a covering]: One of the outermost flower structures, a unit of the calyx; sepals usually enclose the other flower parts in the bud.

septate [L. *septum,* fence]: Divided by cross walls into cells or compartments.

septum, *pl.* **septa:** A partition or cross wall.

sessile (sĕs'ĭl) [L. *sessilis,* of or fit for sitting, low, dwarfed]: Attached directly by the base; referring to a leaf lacking a petiole or to a flower or fruit lacking a pedicel.

seta, *pl.* **setae** (sē'ta) [L. *seta,* bristle]: In bryophytes, the stalk that supports the capsule, if present; part of the sporophyte.

sexual reproduction: The fusion of gametes followed by meiosis and recombination at some point in the life cycle.

sheath: (1) The base of a leaf that wraps around the stem, as in grasses; (2) a tissue layer surrounding another tissue, such as a bundle sheath.

shoot: The above-ground portions, such as the stem and leaves, of a vascular plant.

short-day plants: Plants that must be exposed to light periods shorter than some critical length for flowering to occur; they usually flower in autumn.

shrub: A perennial woody plant of relatively low stature, typically with several stems arising from or near the ground.

sieve area: A portion of the sieve-element wall containing clusters of pores through which the protoplasts of adjacent sieve elements are interconnected.

sieve cell: A long, slender sieve element with relatively unspecialized sieve areas and with tapering end walls that lack sieve plates; found in the phloem of gymnosperms.

sieve element: The cell of the phloem that is involved in the long-distance transport of food substances; sieve elements are further classified into sieve cells and sieve-tube elements.

sieve plate: The part of the wall of sieve-tube elements bearing one or more highly differentiated sieve areas.

sieve tube: A series of sieve-tube elements arranged end-to-end and interconnected by sieve plates.

sieve-tube element: One of the component cells of a sieve tube; found primarily in flowering plants and typically associated with a companion cell; also called sieve-tube member.

signal transduction: The process by which a cell converts an extracellular signal into a response.

silique [L. *siliqua,* pod]: The fruit characteristic of the mustard family; two-celled, the valves splitting from the bottom and leaving the placentae with the false partition stretched between. A small, compressed silique is called a silicle.

simple fruit: A fruit derived from one carpel or several united carpels.

simple leaf: An undivided leaf; as opposed to a compound leaf.

simple pit: A pit not surrounded by an overarching border of secondary wall; as opposed to a bordered pit.

simple tissue: A tissue composed of a single cell type; parenchyma, collenchyma, and sclerenchyma are simple tissues.

siphonaceous [Gk. *siphon,* a tube, pipe]: In algae, multinucleate cells without cross walls; coenocytic; also called siphonous.

siphonostele [Gk. *siphon,* pipe, + *stele,* pillar]: A type of stele containing a hollow cylinder of vascular tissue surrounding a pith.

slime: *See* P-protein.

softwood: A name commonly applied to the wood of a conifer.

solar tracking: The ability of the leaves and flowers of many plants to move diurnally, orienting themselves either perpendicular or parallel to the sun's direct rays; also called heliotropism.

solute: A molecule dissolved in a solution.

solute potential: *See* osmotic potential.

solution: Usually a liquid, in which the molecules of the dissolved substance, the solute (for example, sugar), are dispersed between the molecules of the solvent (for example, water).

somatic cells [Gk. *soma,* body]: All cells except the gametes and the cells from which the gametes develop.

soredium, *pl.* **soredia** [Gk. *soros,* heap]: A specialized reproductive unit of lichens, consisting of a few cyanobacterial or green algal cells surrounded by fungal hyphae.

sorus, *pl.* **sori** (sō′rŭs) [Gk. *soros,* heap]: A group or cluster of sporangia or spores.

specialized: (1) Of organisms, having special adaptations to a particular habitat or mode of life; (2) of cells, having particular functions.

speciation: The origin of new species in evolution.

species, *pl.* **species** [L. *species,* kind, sort]: A kind of organism; species are designated by binomial names written in italics.

specific epithet: The second part of a species name; for example, *mays* of *Zea mays,* which is maize.

specificity: Uniqueness, as in proteins in given organisms or enzymes in given reactions.

sperm: A mature male gamete, usually motile and smaller than the female gamete.

spermatangium, *pl.* **spermatangia** (spûr′mă•tăn′jĭ•ŭm) [Gk. *sperma,* sperm, + L. *tangere,* to touch]: In the red algae, the structure that produces spermatia.

spermatium, *pl.* **spermatia** [Gk. *sperma,* sperm]: In the red algae and some fungi, a minute, nonmotile male gamete.

spermatogenous cell: The cell of the male gametophyte, or pollen grain, of gymnosperms, which divides mitotically to form two sperm.

spermatophyte [Gk. *sperma,* seed, + *phyton,* plant]: A seed plant.

spermogonium, *pl.* **spermogonia** (spûr′mă•gō′nĭ•ŭm) [Gk. *sperma,* sperm, + *gonos,* offspring]: In the rust fungi, the structure that produces spermatia.

spike [L. *spica,* head of grain]: An indeterminate inflorescence in which the main axis is elongated and the flowers are sessile.

spikelet: The unit of inflorescence in grasses; a small group of grass flowers.

spindle fibers: Bundles of microtubules, some of which extend from the kinetochores of the chromosomes to the poles of the spindle.

spine: A hard, sharp-pointed structure; usually a modified leaf, or part of a leaf.

spirillum, *pl.* **spirilli** [L. *spira,* coil]: A long coiled or spiral bacterium.

spongy parenchyma: A leaf tissue composed of loosely arranged, chloroplast-bearing cells.

sporangiophore (spo•răn′jĭ•o•fōr′) [Gk. *spora,* seed, + *pherein,* to carry]: A branch bearing one or more sporangia.

sporangium, *pl.* **sporangia** (spo•răn′jĭ•ŭm) [Gk. *spora,* seed, + *angeion,* a vessel]: A hollow unicellular or multicellular structure in which spores are produced.

spore: A reproductive cell, usually unicellular, capable of developing into an adult without fusion with another cell.

spore mother cell: A diploid ($2n$) cell that undergoes meiosis and produces (usually) four haploid cells (spores) or four haploid nuclei.

sporic meiosis: Meiosis resulting in the formation of haploid spores by a diploid individual, or sporophyte; the spores give rise to haploid individuals, or gametophytes, which eventually produce gametes that fuse to form diploid zygotes; the zygotes, in turn, develop into sporophytes; this kind of life cycle is known as an alternation of generations.

sporophyll (spō′rō•fĭl): A modified leaf or leaflike organ that bears sporangia; applied to the stamens and carpels of angiosperms, fertile fronds of ferns, and other similar structures.

sporophyte (spō′rō•fīt): The spore-producing, diploid ($2n$) phase in a life cycle characterized by alternation of generations.

sporopollenin: The tough substance of which the exine, or outer wall, of spores and pollen grains is composed; a cyclic alcohol highly resistant to decay.

stalk cell: *See* sterile cell.

stamen (stā′mĕn) [L. *stamen,* thread]: The part of the flower producing the pollen, composed (usually) of anther and filament; collectively, the stamens make up the androecium.

staminate (stăm′ĭ•nat): Pertaining to a flower having stamens but no functional carpels.

starch [M.E. *sterchen,* to stiffen]: A complex insoluble carbohydrate; the chief food storage substance of plants; composed of a thousand or more glucose units.

statoliths [Gk. *statos,* stationary, + *lithos,* stone]: Gravity sensors; starch-containing plastids (amyloplasts) or other bodies in the cytoplasm.

stele (stē′le) [Gk. *stele,* a pillar]: The central cylinder, inside the cortex, of roots and stems of vascular plants.

stem: The part of the axis of vascular plants that is above ground, as well as anatomically similar portions below ground, such as rhizomes or corms.

stem bundle: Vascular bundle belonging to the stem.

sterigma, *pl.* **sterigmata** [Gk. *sterigma,* a prop]: A small, slender protuberance of a basidium, bearing a basidiospore.

sterile cell: One of two cells produced by division of the generative cell in developing pollen grains of gymnosperms; it is not a gamete, and eventually degenerates.

stigma: (1) The region of a carpel that serves as a receptive surface for pollen grains and on which they germinate; (2) a light-sensitive, pigmented structure; see eyespot.

stigmatic tissue: The pollen-receptive tissue of the stigma.

stipe: A supporting stalk, such as the stalk of a gill fungus or the leaf stalk of a fern.

stipule (stĭp′yūl): An appendage, often leaflike, that occurs on either side of the basal part of a leaf, or encircles the stem, in many kinds of flowering plants.

stolon (stō′lŏn) [L. *stolo,* shoot]: A stem that grows horizontally along the ground surface and may form adventitious roots, such as the runners of a strawberry plant.

stoma, *pl.* **stomata** (stō′ma) [Gk. *stoma,* mouth]: A minute opening bordered by guard cells in the epidermis of leaves and stems through which gases pass; also used to refer to the entire stomatal apparatus—the guard cells plus their included pore.

stratification: The process of exposing seeds to low temperatures for an extended period before attempting to germinate them at warm temperatures.

strict anaerobe: an anaerobe that is killed by oxygen; can live only in the absence of oxygen.

strobilus, *pl.* **strobili** (strōb′ĭ•lŭs) [Gk. *strobilos,* a cone]: A reproductive structure consisting of a number of modified leaves (sporophylls) or ovule-bearing scales grouped terminally on a stem; a cone. Strobili occur in many kinds of gymnosperms, lycophytes, and sphenophytes.

stroma [Gk. *stroma,* anything spread out]: The ground substance of plastids.

structural gene: Any gene that codes for a protein; in distinction to regulatory genes.

style [Gk. *stylos,* column]: A slender column of tissue that arises from the top of the ovary and through which the pollen tube grows.

sub- [L. *sub,* under, below]: Prefix meaning "under" or "below"; for example, subepidermal, "underneath the epidermis."

suberin (sū'ber•ĭn) [L. *suber,* the cork oak]: Fatty material found in the cell walls of cork tissue and in the Casparian strip of the endodermis.

subsidiary cell: An epidermal cell morphologically distinct from other epidermal cells and associated with a pair of guard cells; also called an accessory cell.

subspecies: The primary taxonomic subdivision of a species. Varieties are used as equivalent to subspecies by some botanists, or subspecies may be divided into varieties.

substrate [L. *substratus,* strewn under]: The foundation to which an organism is attached; the substance acted on by an enzyme.

substrate phosphorylation: Phosphorylation—the formation of ATP from ADP and inorganic phosphate—that takes place during glycolysis.

succession: In ecology, the orderly progression of changes in community composition that occurs during the development of vegetation in any area, from initial colonization to the attainment of the climax typical of a particular geographic area.

succulent: A plant with fleshy, water-storing stems or leaves.

sucker: A sprout produced by the roots of some plants that gives rise to a new plant; an erect sprout that arises from the base of stems.

sucrase (sū'krās): An enzyme that hydrolyzes sucrose into glucose and fructose; also called invertase.

sucrose (sū'krōs): A disaccharide (glucose plus fructose) found in many plants; the primary form in which sugar produced by photosynthesis is translocated.

superior ovary: An ovary that is free and separate from the calyx.

suspension: A heterogeneous dispersion in which the dispersed phase consists of solid particles sufficiently large that they will settle out of the fluid dispersion medium under the influence of gravity.

suspensor: A structure at the base of the embryo in many vascular plants. In some plants, it pushes the embryo into nutrient-rich tissue of the female gametophyte.

symbiosis (sĭm'bi•ō'sĭs) [Gk. *syn,* together with, + *bios,* life]: The living together in close association of two or more dissimilar organisms; includes parasitism (in which the association is harmful to one of the organisms) and mutualism (in which the association is advantageous to both).

sympatric speciation [Gk. *syn,* together with, + *patra,* fatherland, country]: Speciation that occurs without geographic isolation of a population of organisms; usually occurs as the result of hybridization accompanied by polyploidy; may occur in some cases as a result of disruptive selection.

symplast [Gk. *syn,* together with, + *plastos,* molded]: The interconnected protoplasts and their plasmodesmata; the movement of substances in the symplast is called symplastic movement, or symplastic transport.

sympodium, *pl.* **sympodia:** A stem bundle and its associated leaf traces.

syn-, sym- [Gk. *syn,* together with]: Prefix meaning "together."

synapsis [Gk. *synapsis,* a contract or union]: The pairing of homologous chromosomes that occurs prior to the first meiotic division; crossing-over occurs during synapsis.

synergids (sĭ•nûr'jĭds): Two short-lived cells lying close to the egg in the mature embryo sac of the ovule of flowering plants.

syngamy [Gk. *syn,* together with, + *gamos,* marriage]: *See* fertilization.

synthesis: The formation of a more complex substance from simpler ones.

systematics: Scientific study of the kinds and diversity of organisms and of the relationships between them.

T

taiga: The northern coniferous forest.

tangential section: A longitudinal section cut at right angles to the radius of a cylindrical structure, such as a root or stem; in the case of secondary xylem, or wood, and secondary phloem, at right angles to the rays.

tapetum (ta•pē'tŭm) [Gk. *tapes,* a carpet]: Nutritive tissue in the sporangium, particularly an anther.

taproot: The primary root of a plant formed in direct continuation with the root tip or radicle of the embryo; forms a stout, tapering main root from which arise smaller, lateral roots.

taxon: General term for any one of the taxonomic categories, such as species, class, order, or phylum.

taxonomy [Gk. *taxis,* arrangement, + *nomos,* law]: The science of the classification of organisms.

teliospore (tē'lĭ•ō•spōr): In the rust fungi, a thick-walled spore in which karyogamy and meiosis occur and from which basidia develop.

telium, *pl.* **telia:** In the rust fungi, the structure that produces teliospores.

telomere: The end of a chromosome; has repetitive DNA sequences that help counteract the tendency of the chromosome otherwise to shorten with each round of replication.

telophase: The last stage in mitosis and meiosis, during which the chromosomes become reorganized into two new nuclei.

template: A pattern or mold guiding the formation of a negative or complement; a term applied especially to DNA duplication, which is explained in terms of a template hypothesis.

tendril [L. *tendere,* to extend]: A modified leaf or part of a leaf or stem modified into a slender coiling structure that aids in support of the stems; tendrils occur only in some angiosperms.

tension wood: The reaction wood of magnoliids and eudicots; develops on the upper side of leaning trunks and limbs.

tepal: One of the units of a perianth that is not differentiated into sepals and petals.

test cross: A cross of a dominant with a homozygous recessive; used to determine whether the dominant is homozygous or heterozygous.

tetrad (tĕ t'răd): A group of four spores formed from a spore mother cell by meiosis; *see also* bivalent.

tetraploid (tĕt'ra•ploid) [Gk. *tetra,* four, + *ploos,* fold]: Twice the usual, or diploid ($2n$), number of chromosomes (that is, $4n$).

tetrasporangium, *pl.* **tetrasporangia** [Gk. *tetra,* four, + *spora,* seed, + *angeion,* vessel]: In certain red algae, a sporangium in which meiosis occurs, resulting in the production of tetraspores.

tetraspore [Gk. *tetra,* four, + *spora,* seed]: In certain red algae, the four spores formed by meiotic division in the tetrasporangium of a spore mother cell.

tetrasporophyte [Gk. *tetra,* four, + *spora,* seed, + *phyton,* plant]: In certain red algae, a diploid individual that produces tetrasporangia.

texture: Of wood, refers to the relative size and amount of variation in size of elements within the growth rings.

thallophyte: A term previously used to designate fungi and algae collectively, now largely abandoned.

thallus (thăl'ŭs) [Gk. *thallos,* a sprout]: A type of body that is undifferentiated into root, stem, or leaf; the word *thallus* was used commonly when fungi and algae were considered to be plants, to distinguish their simple construction, and that of certain gametophytes, from the differentiated bodies of plant sporophytes and the elaborate gametophytes of the bryophytes.

theory [Gk. *theorein,* to look at]: A well-tested hypothesis; one unlikely to be rejected by further evidence.

thermodynamics [Gk. *therme,* heat, + *dynamis,* power]: The study of energy exchanges, using heat as the most convenient form of measurement of energy. The first law of thermodynamics states that in all processes, the total energy of the universe remains constant. The second law of thermodynamics states that the entropy, or degree of randomness, tends to decrease.

thermophile: An organism with an optimal growth temperature between 45 and 80°C.

thigmomorphogenesis: The alteration of plant growth patterns in response to mechanical stimuli.

thigmotropism [Gk. *thigma,* touch]: A response to contact with a solid object.

thorn: A hard, woody, pointed branch.

thylakoid [Gk. *thylakos,* sac, + *oides,* like]: A saclike membranous structure in cyanobacteria and the chloroplasts of eukaryotic organisms; in chloroplasts, stacks of thylakoids form the grana; chlorophylls are found within the thylakoids.

thymine: A pyrimidine occurring in DNA but not in RNA; *see also* uracil.

Ti plasmid: A circular plasmid of *Agrobacterium tumifaciens* that enables the bacterium to infect plant cells and produce a tumor (crown gall tumor); a powerful tool in the transfer of foreign genes into plant genomes.

tissue: A group of similar cells organized into a structural and functional unit.

tissue culture: A technique for maintaining fragments of plant or animal tissue alive in a medium after removal from the organism.

tissue system: A tissue or group of tissues organized into a structural and functional unit in a plant or plant organ. There are three tissue systems: dermal, vascular, and ground, or fundamental.

tonoplast [Gk. *tonos,* stretching, tension, + *plastos,* formed, molded]: The cytoplasmic membrane surrounding the vacuole in plant cells; also called vacuolar membrane.

torus, *pl.* **tori:** The central thickened part of the pit-membrane in the bordered pits of conifers and some other gymnosperms.

totipotent: The potential of a plant cell to develop into an entire plant.

tracheary element: The general term for a water-conducting cell in vascular plants; tracheids and vessel elements.

tracheid (trā'kē•ĭd): An elongated, thick-walled conducting and supporting cell of xylem. It has tapering ends and pitted walls without perforations, as contrasted with a vessel element. Found in nearly all vascular plants.

tracheophyte: A vascular plant.

transcription: The enzyme-catalyzed assembly of an RNA molecule complementary to a strand of DNA.

transduction: The transfer of genes from one organism to another by a virus.

transfer cell: Specialized parenchyma cell with wall ingrowths that increase the surface area of the plasma membrane; apparently functions in the short-distance transfer of solutes.

transfer RNA (tRNA): Low-molecular-weight RNA that becomes attached to an amino acid and guides it to the correct position on the ribosome for protein synthesis; there is at least one tRNA molecule for each amino acid.

transformation: The transfer of naked DNA from one organism to another; also called "gene transfer." Transposons are often used as vectors in transformation when it is carried out in the laboratory.

transgenic organism: An organism whose genome contains DNA—from the same or a different species—that has been modified by the methods of genetic engineering.

transition region: The region in the primary plant body showing transitional characteristics between structures of root and shoot.

translation: The assembly of a protein on the ribosomes; mRNA is used to direct the order of the amino acids.

translocation: (1) In plants, the long-distance transport of water, minerals, or food; most often used to refer to food transport; (2) in genetics, the interchange of chromosome segments between nonhomologous chromosomes.

transmembrane proteins: Globular proteins that traverse the lipid bilayer of cellular membranes. Some extend across the lipid bilayer as a single alpha helix and others as multiple alpha helices.

transmitting tissue: A tissue similar to stigmatic tissue and serving as a path for the pollen tube in the style.

transpiration [Fr. *transpirer,* to perspire]: The loss of water vapor by plant parts; most transpiration occurs through stomata.

transpiration stream: The flow of water through the xylem, from the roots to the leaves.

transport protein: A specific membrane protein responsible for transferring solutes across membranes; grouped into three broad classes: pumps, carriers, and channels.

transposon [L. *transponere,* to change the position of something]: A DNA sequence that carries one or more genes and is flanked by sequences of bases that confer the ability to move from one DNA molecule to another; an element capable of transposition, which is the changing of a chromosomal location.

transverse section: A section cut perpendicular, or at right angles, to the longitudinal axis of a plant part.

tree: A perennial woody plant generally with a single stem (trunk).

tricarboxylic acid cycle or **TCA cycle:** *See* Krebs cycle.

trichogyne [Gk. *trichos,* a hair, + *gyne,* female]: In the red algae and certain ascomycetes and *Basidiomycota,* a receptive protuberance of the female gametangium for the conveyance of spermatia.

trichome [Gk. *trichos,* hair]: An outgrowth of the epidermis, such as a hair, scale, or water vesicle.

triglyceride: Glycerol ester of fatty acids; the main constituent of fats and oils.

triose [Gk. *tries,* three, + *ose,* suffix indicating a carbohydrate]: Any three-carbon sugar.

triple fusion: In angiosperms, the fusion of the second male gamete, or sperm, with the polar nuclei, resulting in formation of a primary endosperm nucleus, which is triploid ($3n$) in most groups.

triploid [Gk. *triploos,* triple]: Having three complete chromosome sets per cell ($3n$).

tritium: A radioactive isotope of hydrogen, ^{3}H. The nucleus of a tritium atom contains one proton and two neutrons, whereas the more common hydrogen nucleus consists only of a proton.

tRNA: *See* transfer RNA.

-troph, tropho- [Gk. *trophos,* feeder]: Suffix or prefix meaning "feeder," "feeding," or "nourishing"; for example, autotrophic, "self-nourishing."

trophic level: A step in the movement of energy through an ecosystem, represented by a particular set of organisms.

tropism [Gk. *trope,* a turning]: A response to an external stimulus in which the direction of movement is usually determined by the direction from which the most intense stimulus comes.

tube cell: In male gametophytes, or pollen grains, of seed plants, the cell that develops into the pollen tube.

tuber [L. *tuber,* swelling]: An enlarged, short, fleshy underground stem, such as that of the potato.

tundra: A treeless circumpolar region, best developed in the Northern Hemisphere and most found north of the Arctic Circle.

tunica-corpus: The organization of the shoot apex of most angiosperms and a few gymnosperms, consisting of one or more peripheral layers of cells (the tunica layers) and an interior (the corpus). The tunica layers undergo surface growth (by anticlinal divisions), and the corpus undergoes volume growth (by divisions in all planes).

turgid (tûr′jĭd) [L. *turgidus,* to be swollen]: Swollen, distended; referring to a cell that is firm due to water uptake.

turgor pressure [L. *turgor,* a swelling]: The pressure within the cell resulting from the movement of water into the cell.

tylose [Gk. *tylos,* a lump]: A balloonlike outgrowth from a ray or axial parenchyma cell through the pit in a vessel wall and into the lumen of the vessel.

type specimen: Usually a dried plant specimen housed in an herbarium; selected by a taxonomist to serve as the basis for comparison with other specimens in determining whether they are members of the same species or not.

U

umbel (ŭm′bĕl) [L. *umbella,* sunshade]: An inflorescence, the individual pedicels of which all arise from the apex of the peduncle.

unicellular: Composed of a single cell.

unisexual: Usually applied to a flower lacking either stamens or carpels; a perianth may be present or absent.

unit membrane: A visually definable, three-layered membrane, consisting of two dark layers separated by a lighter layer, as seen with an electron microscope.

uracil (ū′ra•sĭl): A pyrimidine found in RNA but not in DNA; *see also* thymine.

uredinium, *pl.* **uredinia** [L. *uredo,* a blight]: In rust fungi, the structure that produces urediniospores.

urediniospore [L. *uredo,* a blight, + *spora,* spore]: In rust fungi, a reddish, binucleate spore produced in summer.

V

vacuolar membrane: *See* tonoplast.

vacuole [L. *vacuus,* empty]: A space or cavity within the cytoplasm filled with a watery fluid, the cell sap; a lysosomal compartment.

variation: The differences that occur within the offspring of a particular species.

variety: A group of plants or animals of less than species rank. Some botanists view varieties as equivalent to subspecies, and others consider them divisions of subspecies.

vascular [L. *vasculum,* a small vessel]: Pertains to any plant tissue or region consisting of or giving rise to conducting tissue; for example, xylem, phloem, vascular cambium.

vascular bundle: A strand of tissue containing primary xylem and primary phloem (and procambium if still present) and frequently enclosed by a bundle sheath of parenchyma or fibers.

vascular cambium: A cylindrical sheath of meristematic cells, the division of which produces secondary phloem and secondary xylem.

vascular plant: A plant that has xylem and phloem; also called tracheophyte.

vascular rays: Ribbonlike sheets of parenchyma that extend radially through the wood, across the cambium, and into the secondary phloem; they are always produced by the vascular cambium.

vascular system: All the vascular tissues in their specific arrangement in a plant or plant organ.

vector [L., a bearer, carrier, from *vehere,* to carry]: (1) A pathogen that carries a disease from one organism to another; (2) in genetics, any virus or plasmid DNA into which a gene is integrated and subsequently transferred into a cell.

vegetative: Of, relating to, or involving propagation by asexual processes; also referring to nonreproductive plant parts.

vegetative reproduction: (1) In seed plants, reproduction by means other than by seeds; apomixis; (2) in other organisms, reproduction by vegetative spores, fragmentation, or division of the somatic body. Unless a mutation occurs, each daughter cell or individual is genetically identical to its parent.

vein: A vascular bundle forming a part of the framework of the conducting and supporting tissue of a leaf or other expanded organ.

velamen [L. *velumen,* fleece]: A multiple epidermis covering the aerial roots of some orchids and aroids; also occurs on some terrestrial roots.

venation: Arrangement of veins in the leaf blade.

venter [L. *venter,* belly]: The enlarged basal portion of an archegonium containing the egg.

vernalization [L. *vernalis,* spring]: The induction of flowering by cold treatment.

vessel [L. *vasculum,* a small vessel]: A tubelike structure of the xylem composed of elongate cells (vessel elements) placed end to end and connected by perforations. Its function is to conduct water and minerals through the plant body. Found in nearly all angiosperms and a few other vascular plants (for example, gnetophytes).

vessel element: One of the cells composing a vessel; also called vessel member.

viable [L., *vita,* life]: Able to live.

volva [L. *volva,* a wrapper]: A cuplike structure at the base of the stalk of certain mushrooms.

W

wall pressure: The pressure of the cell wall exerted against the turgid protoplast; opposite and equal to the turgor pressure.

water potential: The algebraic sum of the solute potential and the pressure potential, or wall pressure; the potential energy of water.

water vesicle: An enlarged epidermal cell in which water is stored; a type of trichome.

weed [O.E. *weod,* used at least since the year 888 in its present meaning]: Generally an herbaceous plant not valued for use or beauty, growing wild, and regarded as using ground or hindering the growth of useful vegetation.

whorled phyllotaxy: Arrangement of three or more leaves or floral parts in a circle at a node.

wild type: In genetics, the phenotype or genotype that is characteristic of the majority of individuals of a species in a natural environment.

wood: Secondary xylem.

woody magnoliids: Magnoliids with large, robust, bisexual flowers with many free parts, arranged spirally on an elongated axis, or receptacle.

X

xanthophyll: (zăn′thō•fĭl) [Gk. *xanthos,* yellowish-brown, + *phyllon,* leaf]: A yellow chloroplast pigment; a member of the carotenoid class.

xerophyte [Gk. *xeros,* dry, + *phyton,* a plant]: A plant that has adapted to arid habitats.

xylem [Gk. *xylon,* wood]: A complex vascular tissue through which most of the water and minerals of a plant are conducted; characterized by the presence of tracheary elements.

Z

zeatin (zē′ă•tĭn): Plant hormone; a natural cytokinin isolated from maize.

zooplankton [Gk. *zoe*, life, + *plankton*, wanderer]: A collective term for the nonphotosynthetic organisms present in plankton.

zoosporangium: A sporangium bearing zoospores.

zoospore (zō′o•spōr): A motile spore, found among algae, oomycetes, and chytrids.

zygomorphic [Gk. *zygo*, pair, + *morphē*, form]: A type of flower capable of being divided into two symmetrical halves only by a single longitudinal plane passing through the axis; also called bilaterally symmetrical.

zygosporangium, *pl.* **zygosporangia:** A sporangium containing one or more zygospores.

zygospore: A thick-walled, resistant spore that develops from a zygote, resulting from the fusion of isogametes.

zygote (zī′gōt) [Gk. *zygotos*, paired together]: The diploid ($2n$) cell resulting from the fusion of male and female gametes.

zygotic meiosis: Meiosis by a zygote to form four haploid cells, which divide by mitosis to produce either more haploid cells or a multicellular individual that eventually gives rise to gametes.

Illustration Credits

All photographs not credited herein are by Ray F. Evert. All section openers are by Rhonda Nass.

Chapter 1
Opener, p. xvi, Rhonda Nass; **1.1** Tokai University Research Center; **1.2** © Stanley M. Awramik, University of California/Biological Photo Service; **1.3** Richard E. Dickerson, "Chemical Evolution and the Origin of Life," *Scientific American*, vol. 239(3), pages 70–86, 1978; **1.4** Michael Durham/Ellis Nature Photography; **1.5** Sidney W. Fox; **1.6(a)** Gary Breckon; **1.6(b)** Larry West/Bruce Coleman, Inc.; **1.7** Harlan P. Banks; **1.8** Anne Wertheim/Animals Animals; **1.9** Dr. Jeremy Burgess/Science Photo Library/Photo Researchers Inc.; **1.10** After W. Troll, *Vergleichende Morphologie der Hoheren Pflanzen*, vol. 1, pt. 1, Verlage von Gebruder Borntraeger, Berlin, 1937; **1.11(a)** Dr. Anne La Bastille/Photo Researchers, Inc.; **1.11(b)** B. C. Alexander/Photo Researchers, Inc.; **1.11(c)** Martin Harvey/The Wildlife Collection; **1.11(d)** Jack Swenson/The Wildlife Collection; **1.11(e)** Fred Hirschmann; **1.11(f)** Stephen P. Parker/Photo Researchers, Inc.; **1.12** From H. Curtis and N. Sue Barnes, *Invitation to Biology*, 5th ed., Worth Publishers, New York, 1994; **1.13** Line art adapted from: P. Ehrlich, A. Ehrlich, and Holdren, *Ecoscience: Population, Resources, Environment*, W. H. Freeman and Company, New York, 1977; photos (left to right): Jean Pragen/Tony Stone Images, Ken Biggs/Tony Stone Images, John S. Lough

Chapter 2
2.4(c) L. M. Beidler; **2.8** M. Kruatrachue and R. Evert, *American Journal of Botany*, vol. 64, pages 310–325, 1977; **2.9, 2.10** From H. Curtis and N. Sue Barnes, *Invitation to Biology*, 5th ed., Worth Publishers, New York, 1994; **2.12** B. E. Juniper; **2.17–2.19** From H. Curtis and N. Sue Barnes, *Invitation to Biology*, 5th ed., Worth Publishers, New York, 1994; **2.25(a)** Dr. Jeremy Burgess/Science Photo Library/Photo Researchers, Inc.; **2.25(b)** Dr. Morley Read/Science Photo Library/Photo Researchers, Inc.; **2.25(c)** Gerry Ellis/Ellis Nature Photography; **2.25(d)** ©Bill Strode/Woodfin Camp & Associates; **2.26(b)** Marcel Isy-Schwart/Image Bank; **2.27** Gerry Ellis/Ellis Nature Photography; **2.28(a)** Frans Lanting/Minden Pictures; **2.28(b)** Dwight Kuhn/Bruce Coleman, Inc.; **2.29** Albert F. W. Vick, Jr./National Wildlife Research Center; **2.30** Katherine Esau; **2.31(b)** Steve Solum/Bruce Coleman, Inc.

Chapter 3
3.1(a) Dr. Jeremy Burgess/Science Photo Library/Photo Researchers, Inc.; **3.1(b)** Courtesy of the National Library of Medicine; **3.2** A. Ryter; **3.3(photo)** N. J. Lang, *Journal of Phycology*, vol. 1, pages 127–134, 1965; **3.4(photo)** George Palade; **3.5(photo)** Michael A. Walsh; **3.11** Myron C. Ledbetter; **3.12** D.S. Neuberger; **3.13** R. R. Dute; **3.15** David Stetler; **3.16(photo)** K. Esau; **3.17** K. Esau; **3.18(a)** Mary Alice Webb; **3.19** M. Kruatrachue and R. Evert, *American Journal of Botany*, vol. 64, pages 310–325, 1977; **3.21** Peter K. Hepler, from M. M. McCauley and P. K. Hepler, *Development*, vol. 109 (4), pages 753–764, 1990; **3.22(a, b)** R. R. Dute; **3.27(b)** M. V. Parthasarathy, from M. V. Parthasarathy et al., *American Journal of Botany*, vol. 72, pages 1318–1323, 1985; **3.28** R. R. Powers; **3.29(b)** Lewis Tilney; **3.30(a)** Brian Wells and Keith Roberts in Alberts et al., *Molecular Biology of the Cell*, 2nd ed., Garland Publishing, Inc., New York, 1989; **3.30(b)** After Alberts et al., *op. cit.*, 1989; **3.32** After K. Esau, *Anatomy of Seed Plants*, 2nd ed., John Wiley and Sons, Inc., New York, 1977; **3.34(a)** H. A. Core, W. A. Côté, and A. C. Day, *Wood Structure and Identification*, 2nd ed., Syracuse University Press, Syracuse, NY, 1979; **3.34b** After R. D. Preston, in A. W. Robards (ed.), *Dynamic Aspects of Plant Ultrastructure*, McGraw-Hill Book Company, New York, 1974; **Page 47** After Richard E. Williamson, *Plant Physiology*, vol. 82, pp. 631-634, 1986, and B. Alberts, D. Bray, J. Lewis, M. Raff, K. Roberts, and J. D. Watson, *Molecular Biology of the Cell*, 2nd ed., Garland Publishing, Inc., New York, 1989

Chapter 4
4.4–4.7 From H. Curtis and N. Sue Barnes, *Invitation to Biology*, 5th ed., Worth Publishers, New York, 1994; **4.8** After A. L. Lehninger, *Biochemistry*, 2nd ed., Worth Publishers, New York, 1975; **Page 80,** Doug Wechsler/Earth Scenes; **4.14** Birgit Satir; **4.16(a–c)** David G. Robinson; **4.19(a–c)** E. B. Tucker, *Protoplasma*, vol. 113, pages 193–201, 1982

Chapter 5
5.2–5.7 From H. Curtis and N. Sue Barnes, *Invitation to Biology*, 5th ed., Worth Publishers, New York, 1994; **5.8(a, b)** Thomas A. Steitz; **5.9, 5.11, 5.14, 5.15** From H. Curtis and N. Sue Barnes, *op. cit.*

Chapter 6
6.2, 6.10, 6.14, 6.15(a) From H. Curtis and N. Sue Barnes, *Invitation to Biology*, 5th ed., Worth Publishers, New York, 1994; **6.15(b)** John N. Telfold; **6.16** From H. Curtis and N. Sue Barnes, *op. cit.*; **6.18(b)** ©1989 Egyptian Expedition of the Metropolitan Museum of Art, Rogers Fund, 1915 (15.5.19e); **Page 122,** M. P. Price/Bruce Coleman, Inc.

Chapter 7
7.2 Paul W. Johnson/Biological Photo Service; **7.3** Colin Milkins/Oxford Scientific Films; **7.4** From Peter Gray, *Psychology*, Worth Publishers, New York, 1991; **7.5** From H. Curtis and N. Sue Barnes, *Invitation to Biology*, 5th ed., Worth Publishers, New York, 1994; **7.6** Prepared by Govindjee; **7.7** L. E. Graham; **7.12** D. Branton; **Page 137,** After Alberts et al., *Molecular Biology of the Cell*, 2nd ed., Figure 9–52, Garland Publishing, Inc., New York, 1989; **7.15** Adapted from B. Alberts et al., *Molecular Biology of the Cell*, 2nd ed., Garland Publishing, Inc., New York, 1989, page 378; **7.17** Dr. Jeremy Burgess/Science Photo Library/Photo Researchers, Inc.; **7.27** After J. A. Teeri and L. G. Stowe, *Oecologia*, vol. 23, pages 1–12, 1976; **7.28** Art adapted from N. A. Campbell, *Biology*, 4th ed., Benjamin/Cummings Publishing Company, Inc., Menlo Park, CA, 1996; **7.28(a)** Leonard L. Rue/Bruce Coleman, Inc.; **7.28(b)** Phil Degginger/Bruce Coleman, Inc.

Chapter 8
8.1 From H. Curtis and N. Sue Barnes, *Invitation to Biology*, 5th ed., Worth Publishers, New York, 1994; **8.3** James S. Busse; **Page 159 (a–d)** Susan Wick, from Russell H. Goddard, Susan M. Wick, Carolyn D. Silflow, and D. Peter Snustad, *Plant Physiology*, vol. 104, pages 1–6, 1994; **8.4** After Catharina J. Venverloo and K. R. Libbenga, 1987; **8.6(a–f)** W. T. Jackson, *Physiologia Plantarum*, vol. 20, pages 20–29, 1967; **8.10** J. Cronshaw; **8.11(a–c)** P. K. Hepler, *Protoplasma*, vol. 111, pages 121–133, 1982

Chapter 9
9.5 B. John; **9.7** W. Tai; **9.8(a, b)** P. B. Moens; **9.9** G. Ostergren; **Page 180, (a)** G. I. Bennard/Oxford Scientific Films; **(b, c)** Heather Angel/Biofotos; **Page 182,** Table adapted from A. Griffiths et al., *An Introduction to Genetic Analysis*, 6th ed., W. H. Freeman and Company, New York, 1996

Chapter 10
10.1 Moravian Museum, Brno, Czechoslovakia; **10.2** From H. Curtis and N. Sue Barnes, *Invitation to Biology*, 5th ed., Worth Publishers, New York, 1994; **10.3** Adapted from K. von Frisch, *Biology*, translated by Jane Oppenheimer, Harper and Row Publishers, Inc., New York, 1964; **10.4–10.9** From H. Curtis and N. Sue Barnes, *op. cit.;* **10.10** After Figure 5-15 in: Anthony J. F. Griffiths, Jeffrey H. Miller, David T. Suzuki, Richard C. Lewontin, William M. Gelbart, *An Introduction to Genetic Analysis*, 6th ed., W. H. Freeman and Company, New York, 1996; **10.11(a)** C. G. G. J. van Steenis; **10.11(b)** Warren L. Wagner, Smithsonian, Washington, D.C.; **10.12** Nik Kleinberg; **10.13** From H. Curtis and N. Sue Barnes, *op. cit.;* **10.15** Dr. Max–B. Schroder and Hannelore Oldenburg, *Flora*, vol. 184, pages 131–136, 1990; **10.16** Heather Angel; **10.17** From H. Curtis and N. Sue Barnes, *op. cit.;* **10.18(a)** A. Barington Brown/Science Source/Photo Researchers, Inc., from J. D. Watson, *The Double Helix*, Atheneum Publishers, New York, 1968; **10.18(b)** Will and Deni McIntyre/Photo Researchers, Inc.; **10.19–10.21** From H. Curtis and N. Sue Barnes, *op. cit.*; **10.22** A. B. Blumenthal, H. J. Kreigstein, and D. S. Hognes, *Cold Spring Harbor Symposium on Quantitative Biology*, vol. 38, page 205, 1973; **10.23** From H. Curtis and N. Sue Barnes, *Biology*, 5th ed., Worth Publishers, New York, 1989; **Page 188,** from H. Curtis and N. Sue Barnes, *op. cit.*, 1994

Chapter 11
11.1 Computer Graphics Laboratory, University of California at San Francisco; **11.2** From H. Curtis and N. Sue Barnes, *Invitation to Biology*, 5th ed., Worth Publishers, New York, 1994; **11.9** Hans Ris; **11.10** After W. M. Becker, J. B. Reece, M. F. Poenie, *The World of the Cell*, 3rd ed., Benjamin/Cummings Publishing Company, Inc., Menlo Park, CA, 1996, page 586; **11.11–11.14** From H. Curtis and N. Sue Barnes, op. cit.; **11.15(a)** John Bova/Photo Researchers, Inc.; **11.15(b)** Arnold Sparrow/ Brookhaven National Laboratory; **11.16(a)** Victoria Foe; **11.17** Adapted from B. Alberts et al., *Molecular Biology of the Cell*, 2nd ed., Garland Publishing, Inc., New York, 1989; **11.18(a)** James German; **11.18** Line art adapted from P. Chambon, "Split Genes," *Scientific American*, May 1981, pages 60–71; **11.19** Adapted from R. Lewin, *Science*, vol. 212, pages 28–32, 1981; **11.20** Adapted from B. Alberts et al., *op. cit.*; **11.21** From H. Curtis and N. Sue Barnes, *op. cit.*; **11.22** Adapted from Neil A. Campbell, *Biology*, 4th ed., Benjamin/ Cummings Publishing Company, Inc., Menlo Park, CA, 1996; **Page 228 (a–d)** Anthony Bleecker, from Linda L. Hensel, Vojislava Grbic, David A. Baumgarten, and Anthony Bleecker, *The Plant Cell*, vol. 5, pages 553–564, 1993; drawing from S. Ross-Craig, *Drawings of British Plants*, Part III, *Cruciferae*, Bell, London, 1949; **11.23** Adapted from a photo by John T. Fiddles and Howard M. Goodman; **11.24** From H. Curtis and N. Sue Barnes, *op. cit.*; **11.26** Vysis; **11.27** Redrawn from H. Curtis and N. Sue Barnes, *Biology*, 5th ed., Worth Publishers, New York, 1989

Chapter 12
12.1 By permission of Mr. G. P. Darwin, courtesy of The Royal College of Surgeons of England; **12.2** From H. Curtis and N. Sue Barnes, *Invitation to Biology*, 5th ed., Worth Publishers, New York, 1994; **12.3** After D. Lack, *Darwin's Finches*, Harper and Row, Publishers, Inc., New York, 1961; **12.4(a)** Mervin W. Larson/Bruce Coleman; **12.4(b)** Frans Lanting/Minden Pictures; **12.6(a)** E. R. Degginger/Bruce Coleman, Inc.; **12.6(b)** Sidney Ash; **12.6(c)** Patrick Herendeen, specimen from the Smithsonian Institution; **12.7** A. J. Griffiths; **12.8, 12.9** From H. Curtis and N. Sue Barnes, *op. cit.*; **12.10(a)** M&C Photography/Peter Arnold, Inc.; **12.10(b)** Kent and Donna Dannen/Photo Researchers, Inc.; **12.11(a)** Heather Angel/Biofotos; **12.11(b)** Robert Ornduff, University of California, Berkeley; **12.12** J. Antonovics/Visuals Unlimited; **12.13** After Jens Clausen and William H. Hiesey, *Experimental Studies on the Nature of Species IV, Genetic Structure of Ecological Races*, Carnegie Institution of Washington Publication 615, Washington, D.C., 1958; **12.15** Adapted from E. O. Wilson and W. H. Burrert, *A Primer of Population Biology*, Sinauer Associates, Sunderland, MA, 1971; **Pages 250, 251,** Gerald Carr; leaf drawings after R. H. Robichaux; **12.16, 12.17** From H. Curtis and N. Sue Barnes, *op. cit.*; **12.18** In Marion Ownbey, "Natural hybridization and amphiploidy in the genus *Tragopogon*," *American Journal of Botany*, 37 (7), 487–499, 1950; **12.19(a, b)** Heather Angel/Biofotos; **12.19(c–e)** C. J. Marchant; **12.21(a)** Thase Daniel/Bruce Coleman, Inc.

Chapter 13
13.1 Corbis/Bettmann; **13.2(a)** Larry West; **13.2(b)** L. Campbell/ NHPA; **13.2(c)** Imagery; **Table 13.1(top)** G. R. Roberts; **Table 13.1 (bottom)** David Thomas/Oxford Scientific Films; **Page 266(a–c)** E. S. Ross; **13.3** University of Wisconsin Herbarium; **13.4** Reproduced from Michael Neushul, *Botany*, Hamilton Publishing Company (John Wiley), Santa Barbara, CA, 1974; **13.5** From Curtis and Barnes, *Biology*, 5th ed., Worth Publishers, New York, 1989; **13.7** From Curtis and Barnes, *Invitation to Biology*, 5th ed., Worth Publishers, New York, 1994; **13.8** Modified from Carl R. Woese; **13.9** W. Jack Jones, U.S. Environmental Protection Agency, Athens, GA; **13.10(a)** L. V. Leak, *Journal of Ultrastructural Research*, vol. 21, pages 61–74, 1967; **13.10(b)** Helmut Konig and Karl Stetter; **13.10(c)** K. Esau; **13.11** Adapted from Christian de Duve, *Scientific American*, vol. 274(4), pages 50–57, 1996; **13.12(a)** L. Wingren; **13.12(b)** L. E. Graham; **13.13(a)** Matt Meadows/ Peter Arnold, Inc.; **13.13(b)** L. E. Graham; **13.13(c)** Kim Taylor/Bruce Coleman, Inc.; **13.13(d)** G. I. Bernard/Oxford Scientific Films/Animals Animals; **13.13(e)** E. V. Gravé; **13.14(a)** Photo Researchers, Inc.; **13.14(b)** L. West; **13.14(c)** K. B. Sandved; **13.14(d)** John A. Lynch/Photo NATS; **13.15(a)** L. West/Bruce Coleman, Inc.; **13.15(b)** J. W. Perry; **13.15(c)** E. S. Ross; **13.15(d)** Robert Carr/Bruce Coleman, Inc.; **13.15(e)** John Shaw/Bruce Coleman, Inc.; **13.15(f–h)** J. Dermid; **13.15(i)** Steve Solum 1985/Bruce Coleman, Inc.; **13.15(j)** E. Beals

Chapter 14

14.1 MSU Instructional Media Center/R. Hammerschmidt; **14.2** Paul Chesley/Photographers Aspen; **14.3(a)** Jack L. Pate, from Michael T. Madigan, John M. Martino, and Jack Parker, *Brock Biology of Microorganisms,* Prentice-Hall, Upper Saddle River, N.J., 1997; **14.3(b)** T. D. Brock and S. F. Conti, *ibid.*; **14.4** D. A. Cuppels and A. Kelman, *Phytopathology,* vol. 70, pages 1110–1115, 1980; **14.5** C. C. Brinton, Jr., and John Carnahan; **14.6(a)** USDA; **14.6(b)** David Phillips/Visuals Unlimited; **14.6(c)** Richard Blakemore; **14.7** Hans Reichenbach; **14.8** P. Gerhardt; **14.9** E. J. Ordal; **14.10(photo)** M. Jost; **14.11(a)** E. V. Gravé; **14.11(c)** Winston Patnode/Photo Researchers, Inc.; **14.12**(top) Fred Bavendam/Peter Arnold, Inc.; **14.12(bottom)** After M. R. Walter, *American Scientist,* vol. 65, pages 563–571, 1977; **14.13(a)** Robert D. Warmbrodt; **14.13(b)** Paul W. Johnson/Biological Photo Service; **14.14** Robert and Linda Mitchell; **14.15** Germaine Cohen-Bazire; **14.16** T. D. Pugh and E. H. Newcomb; **14.17** J. F. Worley/USDA; **14.18(a)** M. V. Parthasarathy; **14.18(b)** Henry Donselman; **14.19** After G. N. Agrios, *Plant Pathology,* 2nd ed., Academic Press, Inc., New York, 1978; **14.20** USGS Eros Data Center; **14.21** R. Robinson/Visuals Unlimited; **14.22** Leonard Lessin/Peter Arnold, Inc.; **14.23** Redrawn from Michael T. Madigan, John M. Martino, and Jack Parker, *Brock Biology of Microorganisms,* Prentice Hall, Upper Saddle River, N.J., 1997; **14.24(a)** K. Esau; **14.24(b)** Jean-Yves Sgro; **14.25(a)–(d)** Jean-Yves Sgro; **14.26(a)** G. Gaard and R. W. Fulton; **14.26(b), (c)** G. Gaard and G. A. deZoeten; **14.27** M. Dollet and R. G. Milne, from M. Dollet et al., *Journal of General Virology,* vol. 67, pages 933–937, 1986; **14.29** D. Maxwell; **14.30(a)** K. Maramorosch; **14.30(b)** E. Shikata; **14.30(c)** E. Shikata and K. Maramorosch; **14.31** T. White; **14.32** Th. Koller and J. M. Sogo, Swiss Federal Institute of Technology, Zurich.

Chapter 15

15.1 R. M. Meadows/Peter Arnold Inc.; **15.2** Charles M. Fitch/Taurus Photos; **15.3(a)** M. Powell; **15.3(b)** E. S. Ross; **15.3(c)** Thomas Volk; **15.3(d)** E. S. Ross; **15.4** R. J. Howard; **15.5** M. D. Coffey, B. A. Palevitz, and P. J. Allen, *The Canadian Journal of Botany,* vol. 50, pages 231–240, 1972; **15.6** E. C. Swann and C. W. Mims, from E. C. Swann and C. W. Mims, *Canadian Journal of Botany,* vol. 69, pages 1655–1665, 1991; **15.7** After N. A. Campbell, *Biology,* 4th ed., Benjamin/Cummings Publishing Company, Menlo Park, CA, 1996; **15.8** M. Powell; **15.9** John W. Taylor; **Page 315, (b)** John Hogdin, from A. H. R. Buller, *Researches on Fungi,* vol. 6, Longman, Inc., New York; **15.13(a)** Thomas Volk; **15.13(b)** G. J. Breckon; **15.15(a, b)** J. C. Pendland and D. G. Boucias; **15.16(a)** C. Bracker; **15.16(b)** D. S. Neuberger; **15.16(c)** Bryce Kendrick; **15.19(a)** David J. McLaughlin, Alan Beckett, and Kwon S. Yoon, *Botanical Journal of the Linnaean Society,* vol. 91, page 253, 1985; **15.19(b)** D. J. McLaughlin and A. Beckett; **15.20, 15.21(b)** Haisheng Lu and D. J. McLaughlin, *Mycologia,* vol. 83, page 320, 1991; **15.22(a)** C. W. Perkins/Earth Scenes; **15.22(b)** Thomas Volk; **15.22(c)** Peter Katsaros/Photo Researchers, Inc.; **15.22(d)** J. W. Perry; **15.24** E. S. Ross; **15.25** Walter Hodge/Peter Arnold, Inc.; **15.26(a, b)** R. Gordon Wasson, Botanical Museum of Harvard University; **15.27(a)** Jane Burton/Bruce Coleman, Inc.; **15.27(b)** Thomas Volk; **15.27(c), (d)** Jeff Lepore/Photo Researchers Inc.; **15.28** E. S. Ross; **15.31(a, b)** Stephen J. Kron; **15.32** Grant Heilman Photography; **15.33(a)** Andrew McClenaghan/Photo Researchers Inc.; **15.33(b)** John Durham/Science Photo Library/Photo Researchers Inc.; **15.34(a, b)** G. L. Barron; **Page 333, (a, b)** G. L. Barron, University of Guelph; **(c)** N. Allin and G. L. Barron, University of Guelph; **Page 335,** John Webster, University of Exeter; **15.35(a, b)** E. Imre Friedmann; **15.36(a)** E. S. Ross; **15.36(b)** Meredith Blackwell/Louisiana State University; **15.36(c)** Robert A. Ross; **15.37(a, b)** E. S. Ross; **15.37(c)** Stephen Sharnoff/National Geographic Society; **15.37(d)** Larry West; **15.39(a–c)** V. Ahmadjian and J. B. Jacobs; **15.40** S. A. Wilde; **15.41(a, b)** Bryce Kendrick; **15.42** William J. Yawney and Richard C. Schultz, Iowa State University; **15.43** D. J. Read; **15.44** B. Zak, U. S. Forest Service; **15.45(a)** R. D. Warmbrodt; **15.45(b)** R. L. Peterson and M. L. Farquhar, *Mycologia,* vol. 86, pages 311–326, 1994; **15.46** Thomas N. Taylor, Ohio State University

Chapter 16

16.1 G. Vidal and T. D. Ford, *Precambrian Research,* 1985; **16.2** D. P. Wilson/Science Source/Photo Researchers, Inc.; **16.3(a)** Biophoto Associates/Photo Researchers, Inc.; **16.4** Linda Graham; **16.5** D. S. Neuberger; **16.7(a)** Victor Duran; **16.7(b)** Ed Reschke/Peter Arnold, Inc.; **16.8(a, b)** K. B. Raper **16.8(c), (d)** London Scientific Films/Oxford Scientific Films; **16.8(e–g)** Robert Kay; **16.9** K. B. Raper; **16.10** Linda Graham; **16.11** Geoff McFadden and Paul Gilson; **16.12** Kim Taylor/Bruce Coleman, Inc.; **16.13(a)** D. P. Wilson/Eric and David Hosking Photography; **16.13(b)** E. S. Ross; **16.13(c)** R. C. Carpenter; **16.13(d)** Jean Baxter/Photo NATS; **16.14(a, b)** M. Littler and D. Littler, Smithsonian Institution; **16.15(a)** Linda Graham; **16.15(b)** J. Waaland, University of Washington; **16.16** Ronald Hoham, Colgate University; **16.17** (photo) C. Pueschel and K. M. Cole, *American Journal of Botany,* vol. 69, pages 703–720, 1982; **16.20(a, b)** D. P. Wilson/Science Source/Photo Researchers, Inc.; **16.20(c)** Florida Department of Natural Resources; **Page 364(a)** Erik Freeland/Matrix; **(b)** Florida Department of Natural Resources; **16.21** Robert F. Sisson/National Geographic Society; **16.22(a–c)** J. Burkholder, North Carolina State University; **16.23** J. Burkholder, North Carolina State University; **16.24(a)** Elizabeth Venrick, Scripps Institution of Oceanography, University of California, San Diego; **16.24(b)** Holly Kunz/Institute of Marine Sciences, courtesy of Peggy Hughes and David Garrison

Chapter 17

17.1 Doug Wechsler/Animals Animals; **17.3(a)** A. W. Barksdale; **17.3(b)** A.W. Barksdale, *Mycologia,* vol. 55, pages 493–501, 1963; **17.5** After J. H. Niederhauser and W. C. Cobb, *Scientific American,* vol. 200, pages 100–112, 1959; **17.6(a)** M. I. Walker/Science Source/Photo Researchers, Inc.; **17.6(b)** F. Rossi; **17.6(c)** Biophoto Associates/Science Source/Photo Researchers, Inc.; **17.6(d)** Dr. Ann Smith/SPL/Photo Researchers, Inc.; **17.8(a, b)** C. Sandgren; **17.9(a)** G. R. Roberts; **17.9(b), (c)** D. P. Wilson/Eric and David Hosking Photography; **Page 380 (a)** Bob Evans/Peter Arnold, Inc.; **(b)** W. H. Hodge/Peter Arnold, Inc.; **(c)** Kelco Communications; **17.10(b)** Oxford Scientific Films; **17.11** C. J. O'Kelly; **17.12(a, b)** R. Evert and John West; **17.15(a–f)** Ronald Hoham; **17.16** After G. L. Floyd; **17.17** After K. R. Mattox and K. D Stewart, in *Systematics of the Green Algae,* D. E. G. Irvine and D. M. John (eds.), 1984; **17.18** L. E. Graham; **17.19** W. L. Dentler/Biological Photo Service/U. of Kansas; **17.21(a–c)** L. Graham; **17.22** David L. Kirk, *Science,* vol. 231, page 51, 1986; **17.23** J. Robert Waaland/Biological Photo Service; **17.24(a)** L. E. Graham; **17.24(b)** M. I. Walker/Science Source/Photo Researchers, Inc.; **17.25** L. E. Graham; **17.26(a, b)** Larry Hoffman; **17.27** Gary Floyd; **17.28** After R. T. Skagel, R. J. Bandoni, G. E. Rouse, W. B. Schofield, J. R. Stein, and T. M. C. Taylor, *An Evolutionary Survey of the Plant Kingdom,* Wadsworth Publishing Company, Inc., Belmont, CA, 1966; **17.29(a)** James Graham; **17.29(b, c)** K. Esser, *Cryptograms,* Cambridge University Press, Cambridge, 1982; **17.29(d)** E. S. Ross; **17.30** D. P. Wilson/Eric and David Hosking Photography; **17.32(a)** Robert A. Ross; **17.32(b)** Grant Heilman Photography; **17.32(c)** L. R. Hoffman; **17.32(d)** V. Paul; **17.33(a, b)** Richard W. Greene; **17.34(a-d)** M. I. Walker/Science Source/Photo Researchers, Inc.; **17.35(a–c)** Lee W. Wilcox; **17.35(d)** Spike Walker; **17.36(a, b)** L. E. Graham; **17.37(a, b)** William H. Amos/Bruce Coleman, Inc.

Chapter 18

18.1 T. S. Elias; **18.2** Brent Mishler; **18.3** D. R. Given; **18.4** After Brent D. Mishler, *Annals of the Missouri Botanical Garden,* vol. 81, pages 451–483, 1994; 18.5(b) R. E. Magill, Botanical Research Institute, Pretoria; **18.6** L. Graham; **18.7(a, b)** D. S. Neuberger; **18.10** Karen Renzaglia; **18.11** Alan S. Heilman, D. K. Smith, Paul G. Davison, Ken D. McFarland, Department of Botany, University of Tennessee, Knoxville; **18.12(a)** John Wheeler; **18.13(a)** Field Museum of Natural History; **18.13(b)** Dr. G. J. Chafaris/Dr. E. R. Degginger; **18.15(a)** Harold Taylor AB IPP/Oxford Scientific Films; **18.16(a,b)** J. J. Engel, *Fieldiana: Botany* (New Series), vol. 3, pages 1–229, 1980; **18.16(c)** J. J. Engel; **18.18** K. B. Sandved; **18.19(a)** Andrew Drinnan; **18.19(d)** D. S. Neuberger; **18.21(a)** Larry West; **18.22** Martha Cook; **18.23(a)** M. C. F. Proctor; **18.24** D. S. Neuberger; **18.25(a–c)** C. Hebant, *Journal of the Hattori Botanical Laboratory,* vol. 39, pages 235–254, 1975; **18.26(a, b)** D. S. Neuberger; **18.28** Rod Planck/Photo Researchers, Inc.; **18.29** Fred D. Sack and D. J. Paolillo, Jr., *American Journal of Botany,* vol. 70, pages 1019–1030, 1983; **18.30(b)** R. E. Magill, Missouri Botanical Garden, St. Louis; **18.31(a)** A. E. Staffen; **18.31(b)** E. S. Ross

Chapter 19

19.1 Dianne Edwards; **19.2** M. K. Rasmussen and Stuart A. Naquin; **19.3** After A. S. Foster and E. M. Gifford, Jr, *Comparative Morphology of Vascular Plants,* 2nd ed., W. H. Freeman & Company, New York, 1974; **19.4** After Katherine Esau, *Plant Anatomy,* 2nd ed., John Wiley & Sons,

New York, 1965; **19.5(a, d)** After K. K. Namboodiri and C. Beck, *American Journal of Botany*, vol. 55, pages 464–472, 1968; **19.5(b, c)** After K. Esau, *Plant Anatomy*, 2nd ed., John Wiley & Sons, Inc., New York, 1965; **19.7** After G. M. Smith, *Cryptogamic Botany*, vol. 2, in *Bryophytes and Pteridophytes*, 2nd ed., McGraw-Hill Book Company, New York, 1955; **19.10(a)** After David S. Edwards, *Review of Palaeobotany and Palynology*, vol. 29, pages 177–188, 1980; **19.10(b)** After J. Walton, *Phytomorphology*, vol. 14, pages 155–160, 1964; **19.10(c)** After F. M. Heuber, *International Symposium on the Devonian System*, vol. 2, D. H. Oswald (ed.), Alberta Society of Petroleum Geologists, Calgary, Alberta, Canada, 1968; **19.11** Field Museum of Natural History; **19.12** After David S. Edwards, *Botanical Journal of the Linnean Society*, vol. 93, pages 173–204, 1986; **19.13** Specimen provided by Ripon Microslides, Ripon, WI; **19.15(a)** D. S. Neuberger; **19.16(a)** David Johnson, Big Bend National Park; **19.16(b, c)** D. S. Neuberger; **19.16(d)** Fletcher and Baylis/Photo Researchers, Inc.; **19.19** W. H. Wagner; **19.20** After Kristine Rasmussen and Stuart Naquin; **Page 456, (a)** Interpreted by M. K. Rasmussen and Stuart Naquin from Tom L. Phillips and William A. DiMichelle, *Annals of the Missouri Botanical Garden*, vol. 9, pages 560–588, 1992; **Page 457, (b)** After W. N. Stewart and T. Delevoryas, *Botanical Review*, vol. 22, pages 45–80, 1956; **Page 457, (c)** After H. P. Banks, *Evolution and Plants of the Past*, Wadsworth Publishing Company, Inc., Belmont, CA, 1970; **19.22(a)** R. L. Peterson, M. J. Howarth, and D. P. Whittier, *Canadian Journal of Botany*, vol. 59, pages 711–720, 1981; **19.22(b)** Heather Angel; **19.23(a)** D. Cameron; **19.23(b)** R. Schmid; **19.25(a)** R. Carr; **19.25(b)** Gerry Ellis/Ellis Nature Photography; **19.27** Dr. Jeremy Burgess/Science Photo Library/Photo Researchers, Inc.; **19.30(a, c, e)** W. H. Wagner; **19.30(b)** James L. Castner; **19.30(d)** Nancy A. Murray; **19.30(f)** David Johnson; **19.30(g)** Geoff Bryant/Photo Researchers, Inc.; **19.32(a)** W. H. Wagner; **19.32(b)** John D. Cunningham/Visuals Unlimited; **19.34** Bill Ivy/Tony Stone Images, Inc.; **19.35(a–d)** C. Neidorf; **19.37(a–c)** D. Farrar; **19.39(a)** D. S. Neuberger; **19.39(b, c)** W. H. Wagner

Chapter 20
20.1 M. K. Rasmussen and Stuart A. Naquin; **20.4(a, b)** After J. M. Pettitt and C. B. Beck, *Contributions from the Museum of Paleontology*, University of Michigan Press, vol. 22, pages 139–54, 1968; **20.4(c)** J. M. Pettitt and C. B. Beck, *Science*, vol. 156, pages 1727–1729, 1967; **20.6** Charles B. Beck; **20.7** After S. E. Scheckler, *American Journal of Botany*, vol. 62, pages 923–934, 1975; **20.8, 20.9** M. K. Rasmussen and Stuart A. Naquin; **20.10** After Peter R. Crane, *Annals of the Missouri Botanical Garden*, vol. 72, pages 716–793, 1985; **20.11(a)** Field Museum of Natural History, Transparency # B83046c; **20.11(b)** After P. Crane, *op. cit.*; **20.12** After W. E. Friedman, *Trends in Ecology and Evolution*, vol. 8, pages 15–20, 1993; **20.13, 20.14(a)** J. Dermid; **20.15** E. S. Ross; **20.16** J. Kummerow; **20.19** B. Haley; **20.21(c)** Gary J. Breckon; **20.25** N. Fox-Davies/Bruce Coleman Ltd.; **20.28(a)** W. H. Hodge/Peter Arnold, Inc.; **20.28(b)** E. S. Ross; **20.29** H. H. Iltis; **20.30** Grant Heilman Photography; **20.31(a)** Larry West; **20.31(b)** J. Burton/Bruce Coleman, Inc.; **20.32** Geoff Bryant/Photo Researchers, Inc.; **20.33 (photo)** Carolina Biological Supply Company; **20.34** Mark Wetter; **20.35** Gene Ahrens/Bruce Coleman, Inc.; **20.36** Sichuan Institute of Biology; **20.37** D. A. Steingraeber; **20.38(a)** Knut Norstog; **20.38(b)** D. T. Hendricks and E. S. Ross; **20.39** Knut Norstog; **Page 489 (a, b)** W. Jones; **(c–e)** Jaime Plaza; **20.40(a)** J. W. Perry; **20.40(b)** Runk and Schoenberger/Grant Heilman Photography; **20.41(a)** Gerald D. Carr; **20.41(b, c)** G. Davidse; **20.42(a)** E. S. Ross; **20.42(b, d)** J. W. Perry; **20.42(c)** K. J. Niklas; **20.43(a)** C. H. Bornman; **20.43(b, c)** E. S. Ross

Chapter 21
21.1 E. S. Ross; **21.2(a–c)** W. P. Armstrong; **21.3(a, b)** G. J. Breckon; **21.3(c)** E. S. Ross; **21.4(a)** E. S. Ross; **21.4(b)** E. R. Degginger/Earth Scenes; **21.4(c)** T. Davis/Photo Researchers, Inc.; **21.5(a, b)** E. S. Ross; **21.5(c)** G. Carr; **21.8(a, c)** Larry West; **21.8(b, e)** J. H. Gerard; **21.8(d)** Grant Heilman Photography; **21.10** E. S. Ross; **21.12(a)** Larry West; **21.12(b)** Specimen provided by Rudolf Schmid; **21.13(a)** Runk and Schoenberger/Grant Heilman Photography; **21.15(a)** P. Echlin; **21.15(b, c)** J. Heslop-Harrison and Y. Heslop-Harrison; **21.15(d)** J. Mais; **21.21(a, b)** P. Hoch; **21.22** James L. Castner; **21.24(a, b)** C. S. Webber

Chapter 22
22-1 Chris R. Hill, courtesy of E. A. Jarzembowski, Maidstone Museum and Art Gallery; **22.2** E. Dorf; **22.3** After Peter R. Crane, Else Marie Friis, and Kaj Raunsgaard Pedersen, *Nature*, vol. 374, pages 27–33, 1995; **22.4(a–c)** James L. Castner; **22.5(a–d)** David L Dilcher, (Reconstructions by Megan Rohn in consultation with D. Dilcher); **22.6** Geoff Bryant/Photo Researchers, Inc.; **22.7 (photo)** D. W. Taylor and L. J. Hickey, *Science*, vol. 247, page 702, 1990; **22.8(a, b)** E. M. Friis; **22.10** Mike Andrews/Animals Animals/Earth Scenes; **22.11(a, c)** E. S. Ross; **22.11(b)** J. W. Perry; **Page 526(a)** Donald H. Les; **22.12(b, c)** E. S. Ross; **22.12(d)** A. Sabarese; **22.13(a)** E. S. Ross; **22.14(a, b)** J. A. L. Cooke; **22.15(a, b)** W. H. Hodge; **22.16(a)** E. S. Ross; **22.16(b)** D. L. Dilcher; **22.17(a)** Larry West; **22.17(b)** T. J. Hawkeswood; **22.18, 22.19, 22.20** E. S. Ross; **22.21** Larry West; **22.22(a, b)** T. Eisner; **22.23, 22.24, 22.25(a, c)** E. S. Ross; **22.25(b)** L. B. Thein; **22.26** E. S. Ross; **22.27** M. P. L. Fogden/Bruce Coleman, Inc.; **22.28** R. A. Tyrell; **22.29** Oxford Scientific Films/G. I. Bernard; **22.30(a–c)** E. S. Ross; **22.31** D. J. Howell; **22.32(a, b)** T. Hovland/Grant Heilman Photography; **22.32(c)** E. S. Ross; **22.32(d)** J. F. Skvarla, University of Oklahoma; **22.33(b,c)** After D. B. Swingle, *A Textbook of Systematic Botany*, McGraw-Hill Book Company, New York, 1946; **22.34** D. S. Neuberger; **22.35(a)** John M. Pettitt and R. Bruce Knox, *Scientific American*, vol. 244, pages 134–144, 1981; **22.35(b)** Sean Morris/Oxford Scientific Films; **22.37(a)** J. L. Castner; **22.38(a)** E. S. Ross; **22.37(b), 22.38(b)** After R. T. Skagel, R. J. Bandoni, G. E. Rouse, W. B. Schofield, J. R. Stein, and T. M. C. Taylor, *An Evolutionary Survey of the Plant Kingdom*, Wadsworth Publishing Company, Inc., Belmont, CA, 1966; **22.38(c)** After L. Benson, *Plant Classification*, D. C. Heath and Company, Boston, 1957; **22.39(a)** E. S. Ross; **22.39(b,c)** K. B. Sandved; **22.40** After Skagel et al., op. cit.; **22.42** E. S. Ross; **22.43** Larry Atkinson/Mobridge Tribune; **22.44(a)** E. S. Ross; **22.44(b)** U.S. Forest Service; **22.45(a–c)** E. S. Ross; **22.46(a, b)** E. S. Ross; **22.47** J. Kendrick; **22.49** Robert and Linda Mitchell; **22.50(a, c, d)** T. Plowman; **22.50(b)** E. S. Ross

Chapter 23
23.3 After A. S. Foster and E. M. Gifford, Jr., *Comparative Morphology of Vascular Plants*, 2nd ed., W. H. Freeman & Company, New York, 1974; **23.4(a–c)** Daniel M. Vernon; **23.5** After Alberts et al., *op. cit.*, 1989; **23.6(a–e)** Gerd Juergens, Lehrstuhl fuer Entwicklungsgenetik, Universitaet Tuebingen; **Page 563** Werner H. Muller/Peter Arnold, Inc.; **23.9** Tom McHugh/Photo Researchers, Inc.

Chapter 24
24.7 M. C. Ledbetter and K. B. Porter, *Introduction to the Fine Structure of Plant Cells*, Springer-Verlag, Inc., New York, 1970; **24.14(a, b)** H. A. Core, W. A. Côté, and A. C. Day, *Wood: Structure and Identification*, 2nd ed., Syracuse University Press, Syracuse, NY, 1979; **24.15** I. B. Sachs, Forest Products Laboratory, U.S.D.A.; **24.17, 24.22, 24.27** After K. Esau, *Anatomy of Seed Plants*, 2nd ed., John Wiley & Sons, New York, 1977; **24.20(a)** M. A. Walsh; **24.23** R. F. Evert, "Dicotyledons," in H.-D. Behnke and R. D. Sjolund (eds.), *Sieve Elements: Comparative Structure, Induction, and Development*, Springer-Verlag, Berlin, 1990; **24.24** J. S. Pereira; **24.26** Randall Brand; **24.28** T. Vogelman and G. Martin; **24.29** M. David Marks and Kenneth A. Feldman

Chapter 25
25.1, 25.2 After J. E. Weaver, *Root Development of Field Crops*, McGraw-Hill Book Company, New York, 1926; **25.4(a, b)** Margaret E. McCully; **25.7** F. A. L. Clowes; **25.8** After K. Esau, *Plant Anatomy*, 2nd ed., John Wiley & Sons, Inc., New York, 1965; **25.9(a)** Robert Mitchell/Earth Scenes; **25.13** H. T. Bonnett, Jr., *Journal of Cell Biology*, vol. 37, pages 199-205, 1968; **25.14** After W. Braune, A. Leman, and H. Taubert, *Pflanzenanatomisches Prakticum*, VEB Gustav Fischer Verlag, Jena, 1967; **25.15** Robert D. Warmbrodt, *New Phytologist*, vol. 102, pages 175–192, 1986; **25.19** E. R. Degginger/Bruce Coleman, Inc.; **Page 604 (c, d)** M. E. Galway, J. D. Masucci, A. M. Lloyd, V. Walbot, R. W. Davis, and J. W. Schiefelbein, *Developmental Biology*, vol. 166, pages 740–754, 1994; **25.20** Robert and Linda Mitchell; **25.22(a, b)** D. A. Steingraeber

Chapter 26
26.5(a–d) Mary Ellen Gerloff; **Page 617(a)** H. C. Jones, Tennessee Valley Authority; **(b)** J. S. Jacobson and A. C. Hill (eds.), *Recognition of Air Pollution Injury to Vegetation: A Pictorial Atlas*, Air Pollution Control Association, Pittsburgh, PA, 1970; **(c)** After T. H. Maugh, II, *Science*, vol. 226, pages 1408–1410, 1984; **26.12(c)** W. Eschrich; **26.14, 26.15** After K. Esau, *Anatomy of Seed Plants*, 2nd ed., John Wiley & Sons, New York, 1977; **Page 626**, After P. A. Deschamp and T. J. Cooke, *Science*, vol. 219, pages 505–507, 1983; **26.17, 26.19** Rhonda Nass/Ampersand; **26.18(a)** James W. Perry; **26.24(a)** Michele McCauley; **26.25** William A. Russin; **26.31** Daniel J. Barta; **26.32** Joanne M. Dannenhoffer; **26.33(a, b)**

Raymon Donahue and Greg Martin; **26.36(a–i)** Shirley C. Tucker; **26.37(a)** Leslie Sieburth; **26.39(a, b)** Leslie Sieburth; **6.39(c)** Mark Running; **26.40** James W. Perry; **26.41** E. R. Degginger/Earth Scenes; **26.42** David A. Steingraeber; **26.43(a, b)** James W. Perry; **26.44** G. R. Roberts

Chapter 27
27.19 S. Gutierrez/Photo Researchers, Inc.; **27.23** I. B. Sachs, Forest Products Laboratory, U.S.D.A.; **27.26** H.A. Core, W. A. Côté, and A. C. Day, *Wood Structure and Identification,* 2nd ed., Syracuse University Press, Syracuse, NY, 1979; **27.28(a)** Galen Rowell 1985/Peter Arnold, Inc.; **27.28(b)** C. W. Ferguson, Laboratory of Tree-Ring Research, University of Arizona; **27.30** Regis Miller; **27.32** After R. B. Hoadley, *Understanding Wood,* The Taunton Press, Newton, CT, 1980; **Page 668 (a,b)** After Peter M. Ray, Taylor A. Steeves, and Sara A. Fultz, *Botany,* Saunders College Publishing, Philadelphia, 1983

Chapter 28
28.4 Roni Aloni, Tel Aviv University; **28.6** Runk and Schoenberger/Grant Heilman Photography; **28.8(a–c)** Bruce Iverson; **28.9(a)** E. Webber/Visuals Unlimited; **28.9(b)** F. Skoog and C. O. Miller, *Symposia of the Society for Experimental Biology,* vol. 11, pages 118–131, 1957; **28.11** J. D. Goeschl; **28.13** D. R. McCarty; **28.15** S. W. Wittwer; **28.16(a)** After M. B. Wilkins (ed.), *Advanced Plant Physiology,* Pitman Publishing Ltd., London, 1984; **28.16(b)** J. E. Varner; **28.17** J. van Overbeck, *Science,* vol. 152, pages 721–731, 1966; **28.18** Carolina Biological Supply Co.; **28.19** Abbott Laboratories; **28.25** After illustration by Hilleshög, Laboratory for Cell and Tissue Culture, Research Division, Landskrona, Sweden; **28.26** Phillip A. Harrington/Fran Heyl Associates; **28.27** Eugene W. Nester; **28.30(a)** Sara Patterson; **28.30(b–d)** Harry J. Klee; **28.30(e)** Richard M. Amasino; **28.31** Keith Wood, University of California, San Diego

Chapter 29
29.3(a, b) Department of Botany, University of Wisconsin; **29.4(a, b)** Randy Moore/Biophot; **29.5** After B. E. Juniper, *Annual Review of Plant Physiology,* vol. 27, pages 385–406, 1976; **29.6** Stephen A. Parker/Photo Researchers, Inc.; **29.7(a, b)** Jack Dermid; **29.8** After A. W. Galston, *The Green Plant,* Prentice-Hall, Inc., Upper Saddle River, NJ, 1968; **29.9(a)** Biophoto Associates/Science Source/Photo Researchers, Inc.; **29.9(b, c)** After B. Sweeney, *Rhythmic Phenomena in Plants,* Academic Press, Inc., New York, 1969; **29.10** Steve A. Kay and A. Millar; **29.11** After A. W. Naylor, *Scientific American,* vol. 186, pages 49–56, 1952; **29.12** After P. M. Ray, *The Living Plant,* Holt, Rinehart, & Winston, Inc., New York, 1963; **29.13(a, b)** Department of Botany, University of Wisconsin; **29.14(a–d)** U.S. Department of Agriculture; **29.18** Breck P. Kent/Earth Scenes; **29.19** U.S. Department of Agriculture; **29.20** A. Lang, M. Kh. Chailakhyan, and I. A. Frolova, *Proceedings of the National Academy of Sciences,* vol. 74, pages 2412–2416, 1977; **29.21** R. Amasino; **29.23** After A. W. Naylor, *Scientific American,* vol. 286, pages 49–56, 1952; **29.25(a, b)** Robert L. Dunne/Bruce Coleman, Inc.; **29.26(a, b)** Runk and Schoenberger/Grant Heilman Photography; **29.27** Janet Braam; **29.28(a)** J. Ehleringer and I. Forseth, University of Utah; **29.28(b)** Gene Ahrens/Bruce Coleman, Inc.; **29.29** After J. Ehleringer and I. Forseth, *Science,* vol. 210, pages 1094–1098, 1980

Chapter 30
30.1 Liphatech, Inc., Milwaukee, WI; **30.2(a)** Donald Specker/Earth Scenes; **30.2(b)** Biological Photo Service; **30.3(a, b)** University of Wisconsin, Madison, Department of Soil Science; **30.4** E. Crichton/Bruce Coleman, Ltd.; **Page 737 (a)** Dwight Kuhn/Bruce Coleman, Inc.; **(b)** L. West, Bruce Coleman, Inc.; **30.6** After B. Gibbons, *National Geographic,* vol. 166, pages 350–388, 1984 (Ned M. Seidler, artist); **30.7** After F. B. Salisbury and C. W. Ross, *Plant Physiology,* 4th ed., Figure 5–12b, Wadsworth Publishing Company, Belmont, CA, 1992; **30.8** From H. Curtis and N. Sue Barnes, *Invitation to Biology,* 5th ed., Worth Publishers, Inc., New York, 1994; **30.9(a, b)** B. F. Turgeon and W. D. Bauer, *Canadian Journal of Botany,* vol. 60, pages 152–161, 1982; **30.9(c)** Ann Hirsch, University of California, Los Angeles; **30.9(d)** E. H. Newcomb and S. R. Tandon; **30.10** J. M. L. Selker and E. H. Newcomb, *Planta,* volume 165, pages 446–454, 1985; **30.11(b)** H. E. Calvert; **30.13** From H. Curtis and N. Sue Barnes, *Invitation to Biology,* 5th ed., Worth Publishers, Inc., New York, 1994; **Page 745(a)** Grant Heilman Photography; **(b)** J. H. Troughton and L. Donaldson, *Probing Plant Structures,* McGraw-Hill Book Company, New York, 1972; **Page 746** Foster/Bruce Coleman, Inc.

Chapter 31
31.1 Stephen Hale; **31.2(a, b)** J. H. Troughton; **31.3** After D. E. Aylor, J. Y. Parlange, and A. D. Krikorian, *American Journal of Botany,* vol. 60, pages 163–171, 1973; **31.4** After M. Richardson, *Translocation in Plants,* Edward Arnold Publishers, Ltd., London, 1968; **31.5** After A. C. Leopold, *Growth and Development,* McGraw-Hill Book Company, New York, 1964; **31.6** After M. Richardson, *Translocation in Plants,* Edward Arnold Publishers, Ltd., London, 1968; **31.7** After L. Taiz and E. Zeiger, *Plant Physiology,* The Benjamin/Cummings Publishing Company, Inc., Redwood City, CA, 1991; **31.9** After P. F. Scholander, H. T. Hammel, E. D. Bradstreet, and E. A. Hemmingsen, *Science,* vol. 148, pages 239–246, 1965; **31.10, 31.11** After M. H. Zimmermann, *Scientific American,* vol. 208, pages 132–142, 1963; **31.12(a)** G. R. Roberts; **31.14** After M. Richardson, *Translocation in Plants,* Edward Arnold Publishers, Ltd., London, 1968; **31.15** H. Reinhard/Bruce Coleman, Inc.; **31.16** After E. Hausermann and A. Frey-Wyssling, *Protoplasma,* vol. 57, pages 37–80, 1963; **31.17** After Todd E. Dawson, *Oecologia,* vol. 95, pages 565–574, 1993; **31.19** After J. S. Pate, *Transport in Plants,* I., *Phloem Transport,* M. H. Zimmermann and J. A. Milburn (eds.), Springer-Verlag, Berlin, 1975; **31.22** After Malpighii, *Opera Posthuma,* London, 1675; **31.23(a, b)** E. Fritz; **31.24(a, b)** M. H. Zimmermann; **31.26** After Neil A. Campbell, *Biology,* 4th ed., The Benjamin/Cummings Publishing Company, Inc., Menlo Park, CA, 1996

Chapter 32
32.1(a) N. H. Cheatham/DRK Photo; **32.1(b)** Michael Fogden/DRK Photo; **32.3** Harry Taylor/Oxford Scientific Films; **Page 778(a)** Kevin Byron/Bruce Coleman, Inc.; **(b)** John Shaw/Bruce Coleman, Inc.; **32.4** C. H. Muller; **32.5(a, b)** Australian Department of Lands; **Page 781(b)** R. T. Smith/Ardea; **32.6** G. E. Likens; **32.11(a)** Fred Bavendam/Peter Arnold, Inc. **32.11(b)** Wendell Metzen/Peter Arnold, Inc.; **32.11(c)** James H. Carmichael/Bruce Coleman, Inc.; **32.11(d)** L. West/Bruce Coleman, Inc.; **32.12** Jane Burton/Bruce Coleman, Inc.; **32.13** Bruce Coleman, Inc.; **32.14(a)** J. Dermid; **32.14(b)** E. S. Ross; **Page 790, (a)** U.S. Geological Survey; **Page 791, (b)** Jeff Henry/Peter Arnold, Inc.; **32.15** John Marshall; **32.16(a)** Keith Wendt/UW Arboretum; **32.16(b)** W. R. Jordan/UW Arboretum

Chapter 33
33.1(a) Jack Wilburn/Earth Scenes; **33.1(b)** Ardea Photographics; **33.3** After A. W. Kiichler; **33.4** Dr. Gene Feldman/NASA/Goddard Space Flight Center; **33.5** C. D. MacNeill, Jepson Herbarium, University of California, Berkeley; **Page 802,** Institut für Auslandbeziehungen, Stuttgart, Germany, painting by F. G. Weitsch (1806); **33.8(a)** Michael Fogden/DRK photo; **33.8(b)** Martin Wendler/NHPA; **33.8(c)** E. S. Ross; **33.9** Tom Bean/Allstock; **33.10** Frans Lanting/Minden Pictures; **33.11** Peter Ward/Bruce Coleman, Inc.; **33.12** J. Dermid; **33.13(a)** Martin Wendler/NHPA; **33.13(b)** Max Thompson/NAS/Photo Researchers, Inc.; **33.13(c)** Ric Ergenbright; **33.13(d)** Greg Vaughn/Tom Stack and Associates; **33.14** F. C. Vasck; **Page 809, (a)** Robert Hitchman/Bruce Coleman, Inc., **(b)** Ronald F. Thomas/Bruce Coleman, Inc.; **33.15(a, b)** J. Reveal; **33.16(a)** Jeff Foott; **33.16(b)** Pat Caulfield; **33.18** Soil Conservation Service; **33.19(a)** Rod Planck/Tom Stack and Associates; **33.19(b)** P. White; **33.20** J. H. Gerard; **Page 815,** Tom and Pat Leeson; **33.21** R. Burda/Taurus Photos, Inc.; **33.22** E. S. Ross; **33.23** E. Beals; **33.24** D. Brokaw; **33.25** E. S. Ross; **33.26** J. Bartlett and D. Bartlett/Bruce Coleman, Inc.; **33.27(a, b)** W. D. Billings

Chapter 34
34.1 Irven de Vore/Anthrophoto; **34.2(a, b)** E. S. Ross; **34.3(a, b)** E. S. Ross; **34.4** E. S. Ross; **34.5** W. H. Hodge/Peter Arnold, Inc.; **34.6(a)** James P. Blair, © 1983, National Geographic Society; **34.6(b)** Susan Pierres/Peter Arnold, Inc.; **34.7** E. S. Ross; **34.8** G. R. Roberts; **Page 829,** John Doebley; **34.9(a)** E. Zardini; **34.9(b)** C. B. Heiser, Jr.; **34.9(c)** A. Gentry; **34.9(d)** M. K. Arroyo; **34.10** E. S. Ross; **34.11(a)** C. F. Jordan; **34.11(b)** M. J. Plotkin; **34.12** M. J. Plotkin; **34.13** K. B. Sandved; **34.14** W. H. Hodge/Peter Arnold, Inc.; **34.15** E. S. Ross; **34.16** W. H. Hodge/Peter Arnold, Inc.; **34.17** Harvey Lloyd/Peter Arnold, Inc.; **34.18** New York Public Library Picture Collection; **34.20** R. Abernathy; **34.21** C. A. Black; **34.22** AP/Wide World Photos; **34.23** W. H. Hodge/Peter Arnold, Inc.; **34.24** Agricultural Research Service, U.S.D.A.; **34.25(a)** Agricultural Research Service, U.S.D.A; **34.25(b)** University of Wisconsin; **34.26** Robert and Linda Mitchell; **34.27** ©1982 Angelina Lax/Photo Researchers, Inc.; **34.28(a, b)** J. Aronson; **Page 845,** Gary Braasch/Tony Stone Images; **34.30** M. J. Plotkin; **34.31** Michael J. Balick

Index

Numbers in **boldface** indicate figures and tables.

A

ABA (*see* abscisic acid)
Abies, 483 (*see also* fir)
  balsamea, **483, 788**
  concolor, 788, **789**
  lasiocarpa, 777, 791
abscisic acid (ABA), **674t**, 683–84, **684**
  cell division and, **687t**
  cell wall extensibility and, 689
  dormancy and, 683–84
  molecular basis of action of, 688
  seed development and, 684
  stomatal movement and, **692**, 692–93, 753
  water relations and, 684
abscisin, 683
abscission, 636
  auxin and, 683
  ethylene and, 683
abscission zone, 636, **637, 657**
absolute temperature, 864
absorption
  of food by cotyledons, 563, 569
  by haustoria, 310, **311**
  by roots, 590, 595–600, 603, 605–6, 608
absorption spectrum, 131, **131**
Acacia (acacia), **10, 239, 549**, 590
  bull's-horn, 642
  cornigera, 775, **775**
  mutualism between ants and, 775, **775**
ACC (1-aminocyclopropane-1-carboxylic acid), 682, **682**
accessory buds, **657**
accessory fruit, 543
accessory pigments, 133 (*see also* carotenoids, chlorophylls, phycobilins)
acclimation, dormant condition in buds and, 718–19, **719**
ACC synthase, **682, 698**
Acer, **545**, 624, **637, 718**
  negundo, **657**
  saccharinum, **625**
  saccharum, **341, 625, 669t, 762**
Aceraceae, 545
Acetabularia, 395, **395**
Acetobacter diazotrophicus, 747
acetyl CoA, 114, **114**, 122
  in Krebs cycle, 115, **115**
acetyl group (CH_3CO), 114, **114**, 115
acetylsalicylic acid (aspirin), **36**
Achemilla vulgaris, **761**
achenes, 544–45
Achlya, 373, **373**
  ambisexualis, **373**
  bisexualis, 26
Achnanthes, 376
acid growth hypothesis, 689
acid rain, 616, **617**, 744
acids, 862–63
  fatty, 22–24, **23, 24**, 122, **124**
acorns, 545
acritarchs, **347**, 348
acropetal transport of auxin, 676
actin, 60
  myosin and, 47
actin filaments, **59**, 60, **60, 71t**
  cell division and, 158, 163, 166
actinomorphy, 502, 527
actinomycetes, **281**, 284–85
  nitrogen-fixing, 741
action spectrum, 131, **131**
active site, 101, **101**, 104
active transport, **82**, 83–84, **84**
adaptation, as result of natural selection, 245–47
adaptive radiation, 249, 250
adder's tongue, 454
Addicott, Frederick T., 683
adenine, 30, **30, 31**, 198, **198**, 199, **200**
  in ATP, **31**, 204
  in DNA, **30**, 198, 199, **200**, 201, 204
  in NAD, 102
  in RNA, **209t**, 223, **223**
adenosine diphosphate (*see* ADP)
adenosine triphosphate (*see* ATP)
Adiantum, **455**
adnation, 502
ADP (adenosine diphosphate), 31, 106
  in glycolysis, 111, **111**
  structure of, **105**
adventitious roots, 566, **568**
  auxin and, 678
aecia, aeciospores, 327, **329**
Aedes, **535**
aerial roots, 603
aerobes, 287
aerobic, 6
aerobic pathway in respiration, 113–22
Aeschynomene hispida, 670
Aesculus
  hippocastanum, **505, 613**, 648, **657**
  pavia, **627**
aethalia, 352, **353**
aflatoxins, 334
Africa, 523, 827
after-ripening, 565
agar, 358, **358**, 380
Agaricaceae, **265t**
Agaricales, **265t**
Agaricus
  bisporus, **265t**, 325
  campestris, 325
Agathis, 483, 488, 489
Agave (agave; century plant), 644, 808
  shawii, **808**
Agent Orange, 679
ageotropum (mutant of *Pisum sativum*), 706
aggregate fruits, 543
Aglaophyton major, 312, 434, **434**
Agoseris, **527, 546**
agricultural revolution, 824–35
agriculture, 11, 823, 824–35
  beginnings of, 824–25, 840
  domestication of plants, **825**, 826, **827–31**, 832, **833–35**
  in Fertile Crescent, 825
  future of, 837–47
    genetic engineering, 847
    genetic variability of crop plants, 840–41
    improvement of existing crops, 838–43, **843**
    medicinal drugs, **833**, 843, 846, **846**
    new crops, 841–43, 846–47
    research efforts, 839
    sustainability, 844–45
  as global phenomenon, 833–35
  in the New World, 828–32
  population growth and, 835–37
  soils and, 744–45
  sustainable, 699
Agrobacterium, 294, **294**
  tumefaciens, **696**, 696–98, **698**
Agropogon cristatum, **175**
Agropyron, **540**
Agrostis tenuis, 146, 244–45, **245, 595**
A horizon (topsoil), 732, **732**
AIDS, 296, 308
air bubbles
  in tracheids and vessel elements, 577
  water transport in the xylem and, **756**, 756–58, **757**
air currents, **797**
  transpiration and, 754
air plants, 585
air pollution, 616
  lichens and, 340
air roots, 605, **605**
air seeding, 757
air spaces (*see* intercellular spaces)
akinetes, **289, 290**, 290–91
alanine, **28**
Albizzia polyphylla, **545**
albuminous cells, **579**, 582–83, **587t**
alcohol fermentation, 123, **123**
alcohol(s), 883
  ethyl (ethanol), 123, **123**, 308
  mannitol, 382
alder, 741
  red, **577, 669t**
Alectoria sarmentosa, **337**
aleurone layer, 563, **564**
  gibberellins and, 685, **686**, 688
alfalfa, 614, 619, **620**, 729, 738, 739
algae, 11, **275t**, 348, **349t, 372t** (*see also by name of group*)
  accessory pigments of, **349t**, 365, **372t**, 381, 383
  bioluminescence of, **363**, 366
  blooms of, 350, **364**, 365, **367**, 378–79, 391
  blue-green (*see* cyanobacteria)
  brown (*see* brown algae)
  carbon cycle and, 350
  cell walls of, **349t, 372t**
  chlorophylls of, **349t**, 365, 367, **372t**, 378, 383
  chrysophytes (*Chrysophyta*) (*see* chrysophytes)
  classification of, **275t**
  diatoms (*see* diatoms)
  ecology of, 348–50
  economic importance of, 350, 361, **364**, 366, **367**, 373–75, 379, 381–82
  evolution of, 348, 352, 360, 383
  green (*Chlorophyta*), (*see* green algae)
  edible, 350, 361, **380**
  food reserves of, **349t, 372t**
  habitats of, **349t, 372t**
  haptophyte (*Haptophyta*) (*see* haptophyte algae)
  red (*Rhodophyta*) (*see* red algae)
  life cycle of, **171, 362, 374, 377, 384–86, 394**
  snow, 383, **386**
  symbiotic, 351, 356–57, **357**, 358, **365**, 378, 383
algin, 382
alginates, 380
alkaloids, **32**, 32–33, 549, **549**, 550
  fungal endophytes as producers of, 335
Allard, H. A., 709, 716
alleles
  crossing-over and, 191, **191**
  defined, 186
  dominant, 185, 186
  evolution and, 239
    gene flow, 241
    genetic drift, 241
  frequency of, 239, 240
  heterozygous/homozygous, 186
  interactions among
    of different genes, 195
    that affect the phenotype, 194
  multiple, 194
  principle of independent assortment and, 189, **189**, 190
  principle of segregation and, 186, **186**
  recessive
    diploidy and, 243, **243t**
    genetic variability and, 243
allelopathy, 33, 779
Allium, **162**
  cepa, **562**, 567, **567, 593, 644**
  porrum, **341**
allolactose, 217, 218
Allomyces, **308t**
  arbusculus, **314**
  life cycle of, **314**
  reproduction in, 313
allopatric speciation, 249
  punctuated equilibrium model and, 257
allopolyploidy (allopolyploids), **252**, 252–53
allosteric enzymes, 104
allspice, 833
Alnus, 741
  rubra, **577, 669t**
alpha-galacturonic acid, **22**
alpha-glucose, **19, 20**, 22
alpha helix, 28, **28, 29**
alpine penny cress, 747
alpine tundra, **798–99, 817**
alternation of generations, **171**, 172
  in algae, **171**, 360–61, 383, **384**, 393, **394**
  in *Allomyces*, 313, **314**
  in bryophytes, 402, **410–11**
  in chytrids, 313
  in vascular plants, 430
  isomorphic/heteromorphic, 172
altitude, distribution of organisms and, 801, **801**, 802
aluminum (Al)
  acid rain and, 616
  agriculture and, 747
Amanita, 325
  muscaria, **323**
  virosa, 325
amaranths, grain, 842–43
Amaranthus, 842
  retroflexus, **126**
Amaryllis (amaryllis), 536
  belladonna, **536**
Ambrosia psilostachya, **505**
American elm, **657, 663**
amino acids
  active site of enzymes and, 101, **101**
  coding sequences, 688, **688**

essential, 26
in food plants, 26
genetic code and, **209t**
molecular structure of, 26–29, **28, 29**
molecular systematics and, **269,** 269–70
nitrogen and, 26
proteinoid microspheres and, **4**
protein synthesis and, 208, 209, **209,** 210
sequence of, 28
in transfer RNA (tRNA), 210
aminoacyl (A) site, 211–13
aminoacyl-tRNA, 210
aminoacyl-tRNA synthetases, 210, 211
1-aminocyclopropane-1-carboxylic acid (ACC), 682
ammonia (NH_3), 3, **3,** 26
ammonification, 736, **736,** 737
ammonium (NH_4^+), nitrogen cycle and, **736,** 737–38
ammonium ions, 737, 738
amoebas, 365, 872 (*see also* myxamoebas)
of cellular slime molds, **354**
of plasmodial slime molds, **353**
Amorphophallus titanum, 192
AMP (adenosine monophosphate), **31,** 106
structure of, **105**
Amphibolis, **542**
α-amylase, 685, 686, **688**
amylopectin, 20, **21**
amyloplasts, **23,** 51, **51,** 591 (*see also* plastids)
gravitropism and, 704–5, **705**
amylose, 20, **21**
Anabaena, **272,** 290, **290,** 741, **741**
azollae, **44,** 462
cylindrica, **288**
anabolism, 123–24, **124**
anaerobes, 287
facultative, 287
strict, 287
anaerobic pathways, in respiration, 122–23
anaerobic processes, 6
analogy (analogous features), systematics and, 267
anaphase
meiotic, **174, 175,** 176, 177
mitotic, **160, 161,** 163
anatomy of plants, 11
Andreaea, 416
rothii, **415**
Andreaeidae, 412, 416
Andreaeobryum, 416
androecium, 501
of liverworts, **409,** 412
Anethum graveolens, 833
aneuploidy, 193
Aneurophyton-type progymnosperms, 471, **471**
angiosperms (flowering plants), **275t,** 470, 495–516, 875 (*see also Anthophyta;* flowers)
ancestors of, **519–23**
aquatic, 541–42, **542**
cytoplasmic inheritance in, 196
diversity of, 496–98
embryogeny, **556–61**
evolution of, 428, 431, 432, 496, 517–53 (*see also* flowers, evolution of; fruits, evolution of)
biochemical coevolution, 549–51
hypotheses of relationships among anthophytes, 518–19, **519**
magnoliids, 519, **520,** 521
origin and diversification, 519–23
resistance to drought and cold, 522
single common ancestor, 519
fossil, **517–18, 520, 522**
gametophytes of, 431
life cycle of, 503–14, **512–13**
development of the seed and fruit, 509–10
double fertilization, 509, **509**
megasporogenesis and megagametogenesis, 506
microsporogenesis and microgametogenesis, 504–6
outcrossing, 510–11
pollination, 508–9, **530–42**
self-pollination, 514
parasitic, **498, 547**
phylogenetic relationships with other embryophytes, **472**
saprophytic, **498**
specialized families of, **527–29**
vessel elements of, 428
Angophora woodsiana, **531**
angstrom (Å), 42
Animalia (animal kingdom), 276
animals (*see also* birds; insects; *and specific animals*)
dispersal of fruits and seeds by, **547,** 547–49
distribution of (*see specific biomes*)
domestication of, 826, **826,** 831
as fruit and seed distributors, **547–49**
germination and, 565
secondary plant metabolites and, 549
anions, 735, 856
anise, 833
anisogamy, **372**
Anna's hummingbird, **537**
annual rings, 664
annuals, 648
annular scars, in *Oedogonium*, 392, **392**
annulus, 454
Antarctica, 801, 821
lichens in, 336
antenna complex, 135, **136**
antenna pigments, 135
antheridia, 372
of algae, **385,** 392, **392**
of ancestor of plants, 398, **402**
of ascomycetes, 320
of bryophytes, **402,** 404, 405
of club mosses (*Lycopodiaceae*), **437,** 438
of *Equisetum*, **449, 451**
evolution of vascular plants and, **430,** 431
of ferns, 455, 458–59, **460–61**
of hornworts, 412
of *Marchantia*, **405, 408, 411**
of liverworts, 408, **408, 409**
of *Lycopodium*, **437**
of mosses, 414, **417, 419,** 420, **420**
of oomycetes, **374**
of *Psilotum*, **444, 447**
of seedless vascular plants, 431
of *Selaginella*, **441,** 442
of *Takakia*, **407**
antheridiol, 26
antheridiophores, 408, **408**
anthers, **184, 499,** 501, 504
dehiscence of, **504,** 508
of lily, **504**
of orchids, 528
outcrossing and, **511**
of wind-pollinated flowers, 540, **541**
anthesin, 715
Anthocerophyta, 874 (*see also* bryophytes; hornworts)
Anthoceros, 412, **413**
anthocyanins, 35, **35,** 54–55, 542, 543, **543**
Anthophyta, **265t,** 470, 875 (*see also* angiosperms)
diversity in, 496–98
anthophytes, 518, **519**
anthracnose diseases, 334
anthrax, 287
Anthreptes collarii, **537**
antibiotics, 287
deuteromycetes and, 334
anticlinal divisions, 612, **612**
of fusiform initials, **649**
anticodons, 210, **211–13**
antiparallel strands, 199, **200**
antipodals, 506, 509, **509**
antiporters, 83, **84**
Antirrhinum majus, 639
ants, **538,** 606, 642
mutualism between acacia trees and, 775, **775**
seed dispersal by, 548
añu, **830**
A_o, A_1, 138
apex (apices)
apical-basal pattern of, 556, **556,** 557, 560
floral, **638,** 639
reproductive, 639
root, **479,** 591, 611
shoot, **436, 440, 446, 450, 460, 479, 566,** 610–12, **612,** 614, 635, **635**
transition to flowering, 639
aphids, 298, 766, **767**
Ap horizon, 732
Apiaceae, 264, **502,** 531, 545, 833
apical cell, 556, **558**
apical dominance, 678, **678**
apical meristems, 8, 157, **157,** 425, 426, 559, **562,** 571, **571**
of embryo, **559, 561, 564**
functions of, 570
of pines, 474, **475,** 482
of roots, **8,** 559, 562, **570,** 591–92, 604, 605
quiescent center, 594
types of organization of, 592, **593**
of shoots, **8, 556, 557,** 557, **558,** 559, **561,** 562, **562,** 568, **570,** 611–14
apical placentation, 501, **501**
Apis mellifera, **532**
Apium
graveolens, **643**
petroselinum, **752**
Apocynales, 34
apomixis, **180,** 255–56
apoplast, 87
apoplastic pathway, 759, **760**
apoplastic transport, 87
apoptosis, 355, 578
apple, **241, 587, 649,** 655
flowers, **503**
appressoria, 339
aquaporins, 83
aquatic plants, **397,** 401, 407, 526 (*see also* algae; water molds; *and specific plants*)
angiosperms, 541–42, **542**
leaf dimorphism in, 626, **626**
Aquilegia canadensis, **538**
Arabidopsis, **157, 556,** 682, 698, 704
thaliana, 47, 228, **228,** 560, **560, 561, 586,** 692, 708, 714
embryo-defective mutants of, 560, **560, 561**
floral organ identity in, 639–40, **640,** 641
flowering, 716, **717**
roots of, 604, **604**
thigmomorphogenesis, 721–22, **722**
trichomes on the epidermis of, 585
Araceae, 37
Araucaria, 483, 488, 489
araucana, 483
heterophylla, 483, 484
Araucariaceae, 483, 488
arbuscules, 341, **341, 344**
Arceuthobium, **547**
archaea, 6, 270, **270,** 282, 294–96
extreme halophilic, 295
extreme thermophilic, 296, **296**
methane-producing (methanogens), 295, **295,** 296
Archaea (domain), **272,** 294–96, 869
Archaeanthus, 521, **530,** 544
linnenbergeri, 519, **520**
Archaeopteris, **471**
macilenta, **471**
Archaeopteris-type progymnosperms, 471
Archaeosperma arnoldii, 468, **469**
archegonia
of ancestors of plants, 398, **402**
of bryophytes, **402,** 404–6
of club mosses (*Lycopodiaceae*), **437,** 438
of *Equisetum*, **449, 451**
evolution of vascular plants and, 431
of ferns, 455, 458–59, **460**
of hornworts, 412
of liverworts, 408, **408, 409**
of *Lycopodium*, **437**
of *Marchantia*, 449
of mosses, 414, **417, 418,** 420
of pines, 481, 482
of *Psilotum*, **444, 447**
of seedless vascular plants, 431
of *Selaginella*, **441,** 442
archegoniophores, 408, **408**
Arctic, 801
Arctic tundra, **798–99, 819,** 819–21
Arctophila fulva, **819**
Arctostaphylos, 565
viscida, **565**
Arcyria nutans, **353**
areoles, **631**
arginine, **27**
Argyroxiphium sandwicense (silversword), 250, **250,** 251
arils, 483, **484**
seed dispersal and, 548
Aristida purpurea, **591**
Aristolochia, **428, 654**
grandiflora, **521**
Aristolochiaceae, 521, **521**
tridentata, **808**
Aristotle, **733**
Armillaria
gallica, 307
ostoyae, 307
arrowhead, **558–59**
Artemisia dracunculus, 833
tridentata, 808
Arthrobotrys anchonia, **333**
Arthuriomyces peckianus, **311**
artificial selection, 238, **238**
Asarum canadense, **762**
asci, **318,** 318–19, **319**
in yeasts, 330, **331**
Asclepiadaceae, **266, 531,** 550 (*see also* milkweed)
Asclepias, 34, **544,** 546
curassavica, **549**
Asclera ruficornis, **531**
ascocarp (*see* ascomata)
Ascodesmis nigricans, **319**
ascogenous hyphae, 320
ascogonia, 320
ascomata, 319, 338
ascomycetes (*Ascomycota*), 275, **308t,** 309, **309,** 310, **310,** 312, **317,** 317–20, 870
life cycle of, 318, **318,** 320
ascospores, 318, **318, 319,** 319, 320, 330, 331, **331**
aseptate hyphae, 310
asexual reproduction (vegetative reproduction), 166, 170, 179, **179, 180**
in algae, 352, 360, 371, 373, **374, 377,** 379, **389–90,** 391–92, 396
in ascomycetes, 318, **318**
in bryophytes, 404
in *Chlamydomonas*, 388
in diatoms, 376, **376, 377,** 378
in euglenoids, 352
in fungi, 311–12
genetic variability and, 242
in mosses, 414, 421
in slime molds, 355–56
in vascular plants, 255, **256**
in yeasts, 330
ash, 545, **755**
green, **627**
white, **669t**
Asia, 826–27
A (aminoacyl) site, 211, **212, 213**
asparagine, **27,** 740
asparagus, 642
Asparagus officinalis, 642
aspartate (aspartic acid), **27,** 144, **144**
aspen, quaking, **788**
Aspergillus, **332,** 333
flavus, 334
fumigatus, **332**
oryzae, 334
parasiticus, 334
soyae, 334
aspirin, **36**
Asplenium
rhizophyllum, **180**
septentrionale, **453**
viride, **800**

assimilate stream of the phloem, 764, **764**
nutrients exchanged between transpiration stream and, **763,** 763–64
assimilate transport, 764–69
and aphids, 766, **767**
phloem loading, 767–68, **769**
phloem unloading, 768–69
pressure-flow hypothesis, 766–69, **768**
radioactive tracer experiments, 765–66
in sieve tubes, 765–67
Asteraceae (Compositae), 256, 264, **505, 506,** 511, **527,** 527–28, 545
aster yellows, 293
athlete's foot, 334
atmosphere, 844
atmospheric gases, in early Earth, 3–4
atom, 863 *(see also specific elements)*
atomic mass, 863
atomic mass units (daltons), 863
atomic number, 863
atomic structure
of familiar elements, **864t**
models of, **853,** 865–67, **866, 867**
atomic weight, 863
chemical reactivity of, 867
ATP (adenosine triphosphate), 31, **31,** 93
alcohol fermentation and, 123
chemiosmotic coupling and, 120, **121**
in electron transport chain, 118
energy supplied (released) by, **31,** 31–32, 105, **105,** 106, 120–22
in glycolysis, 111, **111,** 112, **112,** 113, **113**
in oxidative phosphorylation, 119, 120
in photosynthesis, **134,** 135, **136,** 137
cyclic photophosphorylation, 139, **139**
in respiration, 108, **108,** 109, **110**
role of, 105–6
structure of, **105**
ATPases, 47, 105, **105**
ATP synthase, 119, 120, **120, 121,** 137, **138**
Atriplex, 585, 744, **744**
Atriplex nummularia, **843**
atropine, 33
Aureococcus, 379
aureomycin, 287
Auricularia auricula, **322**
autoecious parasites, 327
autopolyploidy (autopolyploids), 252, **252**
autoradiography, 864–65
autotrophic cells, 124
autotrophs, 286, 287, 783
chemosynthetic, 287
photosynthetic, 5, 287
autumn crocus plant *(Colchicum autumnale)*, 193
auxiliary cell, 361
auxins, 66, **674t,** 675–79, **676, 679, 687t,** 696 (*see also* indoleacetic acid (IAA); naphthaleneacetic acid (NAA))
abscission and, 683
Agent Orange and, 679
apical dominance and, **674t,** 678, **678**
cambial reactivation and, 650
cell differentiation and, 676, **677**
cell division and, 678, 681, **687t**
cell elongation and, 66, **675, 687t**
cell wall extensibility and, 689
circinate vernation and, 458
differentiation of vascular tissue and, 676
ethylene and, **674t,** 682–83
fruit growth and, **674t,** 678, **679**
gibberellins and, 689
gravitropism and, 704–5, **705**
leaf development and, **674t**
phototropism and, **702,** 703
ratio of cytokinin to, 681
root growth and, **674t, 678**
synthetic, 679
in tissue culture, 680, **681**
transport of, 676, 677
vascular cambium and, 678
in Went's experiment, 676, **702**
auxospores, **376, 377**
Avena sativa, 146, 563, **589,** 675
Avicennia germinans, 605
avocado, 683
Avogadro's number, 871
awns, wheat with, 196
axial core, 175, **176**
axial system, 648
axil, leaf, **8,** 611, 626
axile placentation, 501, **501**
axis
of plant body, 556
stemlike (*see* epicotyl; hypocotyl-root axis)
Azolla, 290, 462, 741, **741**
filiculoides, **741**
Azotobacter, 741
Azotococcus, 741

B

Bacillariophyta (diatoms), **275t,** 349, 371, **372t, 375,** 375–78, 873 (*see also* algae)
heterotrophic, 378
pennate, **276**
resting stages of, 376, 378
bacilli (rods), 284, **285**
Bacillus
cereus, **286,** 779
thuringiensis, 697
bacteria, 287–94 (*see also* prokaryotes *and by name*)
cell walls of, 283, **283**
commercial uses of, 287
diseases caused by, 287, 293–94
endospores of, 286, **286**
evolution of mitochondria and chloroplasts from, 53
gram-negative, 283, **283**
gram-positive, 283, **283**
green, 291
nitrifying, 738
photosynthetic, 127–28, 133, 291 (*see also* cyanobacteria; green bacteria; prochlorophytes; purple bacteria)
as prokaryotes, 6
purple, 282, 283
photosynthesis in, 291
purple nonsulfur, 291, **291**
purple sulfur, **127,** 128, 291
thermophilic, **282**
Bacteria (domain), 270, **270, 272,** 282, 287–94, 869–70
major lineages (kingdoms) of, 287
bacterial chromosome, 156, **156**
bacteriochlorophyll, 133, 291
bacteriology, 12
bacteriophages, 286, 297, **297**
bacteroids, 739, **739**
badnaviruses, 300
Baker, Kathleen Drew, 361
bald cypresses, 483
balsa, 659
balsam fir, **483, 788**
banana, **497,** 827
banyan tree, 603
bar (unit of pressure), 77
Barbarea vulgaris, **728**
barberry *(Berberis)*, 327, **328**
bark, 647, 650, 654-59, **658**
definition of, 654
inner, **655,** 658, 659
outer, **655,** 658
ring, 658
scale, 658, **658**
barley, **635,** 685, **686,** 688, 825, **825**
bran, 563
foxtail, **278**
barley malt, 685
barrel cactus, 809, **809**
basal bodies, 61, **387**
basal cell, 556, **557, 558**
basal placentation, 501, **501**
bases, 862–63
basidia, **321,** 324, **324**
basidiocarp (*see* basidiomata)
basidiomata, 320, **320,** 322, 325
Basidiomycetes, **265t, 275t,** 309, 322, **323,** 323–26, 871
life cycle of, **320–21**
Basidiomycota, **265t,** 275, **308t,** 309, 311, 312, 320–30, 871 (*see also* fungi)
classes of, 322
life cycle of, **328–29**
mycelium of, 322
basidiospores, 320, **320–321,** 322–24, **324,** 325, 326, **326,** 327, **328**
in *Ustomycetes* (smuts), 330
basil, 833
basipetal transport of auxin, 676
Bassham, James A., 140
basswood (linden; *Tilia americana*), **575, 577, 580, 657, 669t**
bark of, 655
stems of, 614, 615, 618, 651, 652, **652**
bast fibers, 575
Bateson, William, *190*
Batrachospermum
moniliforme, 357
sirodotia, **360**
bats, pollination by, 538–39, **539**
Bauhin, Caspar, 262
bay leaves, 833
Beagle, HMS, 236, **236**
bean golden mosaic, 300
beans, 738
broad, **8,** 738, **766**
castor, **562,** 563, **566,** 567, **578**
contender, **685**
garden, **562,** 563, **566,** 567
beech, **762, 813**
American, **669t**
family, 342
bees, pollination by, **532,** 532–35, **533, 534**
bee's purple, **533,** 543
beetles, 550
Colorado potato, 550
longhorn, **530**
pollination by, **530,** 531, **531**
pollination of cycads and, 487
beets, **835**
sugar-, **272,** 606, **607,** 834, 835, **835**
beet yellows virus, 300, **300**
beggar-ticks, 528
Beijerinckia, 741
Belt, Thomas, 775
Beltian bodies, 775, **775**
Bennettitales (cycadeoids), 472, **473,** 518
Benson, Andrew A., 140
bent grass *(Agrostis tenuis)*, 244–45, **245, 595**
benzylamino purine (BAP), **680**
Berberis, 327
berries, 543
beta-carotene, 133
betacyanins (betalains), 543, **543**
beta-fructose, **20**
beta-glucose, **19,** 22
beta oxidation, 122
beta pleated sheet, 28, **29**
Beta vulgaris, 606, **607, 835**
Betula
alleghaniensis, **343, 669t**
papyrifera, **541, 658**
Betulaceae, 342
Bevhalstia pebja, **517**
B horizon (subsoil), 732, **732**
bicarbonate ion (HCO_3^-), 144, **144, 149**
Bidens, 528
biennials, 648
bifacial vascular cambium, 470
binary fission, 156, 285
bindweed, field, **599**
binomial names, 262–64
binomial system of nomenclature, 262
biochemical coevolution, 549–51
biodiversity, 844, 845 (*see also* genetic variability)
biogeochemical cycles, 736
biological clocks, **12, 709**
circadian rhythms and, 706–9
biological species concept, 248
bioluminescence, 122, 699, 708
of dinoflagellates, 366
biomass, 784
pyramids of, 785, **785**
biomes, 9, 796, **820t** (*see also* forests; *specific types of biomes)*
definition of, 797
distribution of, 797, **798–99,** 802
map, **798–99**
patterns of wind and rainfall and, **801,** 802
relationship between altitude and latitude and, 800, **801**
biosphere, 5
biotechnology, 231, 693–99 (*see also* genetic engineering; recombinant DNA)
birches, 342
ectomycorrhiza of, **343**
paper, **10, 541**
yellow, **669t**
bird-of-paradise, **537**
birds
DDT and, **781**
in fruit and seed dispersal, **547,** 548
germination and, 565
pollination by, 537–38, **538**
bird's-nest fungi, 323, 326, **326**
birth-control pills, 846, **846**
birthwort family, 521, **521**
bisexual gametophytes, 430, 431, **436,** 438, 444, **444,** 449, 449, **450**
bivalents (tetrads), 172, **173**
black ironwood, 670
black locust, **627, 656, 657,** 659
Blackman, F. F., 128
black oak, 256, **257, 658**
black pepper, 832
black stem rust of wheat, 327
life cycle of, **328–29**
black walnut, **669t**
black wart disease, 313
bladderwort, **737**
blazing star, **591**
blights, 293, **294**
blister rust, **327**
bloodroot, 549
blooms
of algae, 350, **363, 364,** 365, **367,** 378–79, 391
of chrysophytes, 379
of cyanobacteria, 289, 290
of dinoflagellates, 364
of extreme halophilic archaea, **295**
of haptophytes, 367
bluebells, **500**
blueberry, lowbush, **762**
bluegrass
annual, **633**
Kentucky *(Poa pratensis)*, 146, 244
apomixis in, 255–56
blue-green algae (*see* cyanobacteria)
body cell (*see* spermatogenous cell)
body plan, 556, **556**
Bohr, Niels, 865
bolting, gibberellin and, 686, **687**
bond energy, 875
bonds, chemical, 105, 106, 868–70
Bonnemaisonia hamifera, **358**
Bonner, James, 711, 715
Bordeaux mixture, 373
bordered pits (pit-pairs), **64,** 659, 660, **661, 662**
boreal forests (taiga), 150, **818,** 818–19
Borlaug, Norman, **839**
boron (B), as essential element, 727, **730t**
Borthwick, Harry A., 712
Boston ivy, 641
botany (plant biology), 2, 11
areas of study in, 11-12
importance of knowledge of, 12–14
Botrychium, 453, 454–55
parallelum, **454**
virginianum, **59**
bottleneck effect, 242
botulism, 287
Bougainvillea, 543
box elder, **657**
Brachythecium, 421
bracken fern, **458**
bracket fungi (*see* shelf fungi)
Bradyrhizobium, 738–40
japonicum, 739, **739**

bran, 563
branches of lycophyte trees, 456
branching in progymnosperms, 471, **471**
branch traces, 622, **624**
Brassica
  juncea, 747
  napus, 698
    var. *napobrassica,* 763
  nigra, **549,** 833
  oleracea, 47, **238**
    var. *capitata,* 643, 686
    var. *caulorapa,* 643, **644**
  rapa, **544**
Brassicaceae, **228,** 340, 511, 544, **544,** 549
brassinolides, 675, **675t**
brassins, 26
Brazil, **803**
Briggs, Winslow, 703, **703**
bristlecone pine, **664,** 664–65
British soldier lichens, **337, 339**
broad beans *(Vicia faba),* **8,** 738, **766**
broccoli, **238**
Bromeliaceae, 149, **337**
bromeliads, 804, **804**
brown algae *(Phaeophyta),* **7, 171, 275t,** 370, 371, **372t,** 379–83, 873 (*see also* algae)
  conducting tissues in, **382**
  edible, 380
  fucoxanthin in, 381
  life cycles of, 383
  thalli, 381
Browne, John, 845
brown rot of stone fruits, 317
brown spot of corn, 313
brown tides, 379
brussels sprouts, **238**
Bryidae ("true mosses"), 412, 416–22 (*see also* bryophytes; mosses)
  "cushiony" or "feathery" growth patterns in, 421, **422**
  sexual reproduction in, 416, **417–18,** 418–21
  specialized tissues for water and food conduction, 416, **416, 417**
Bryophyta (*see* bryophytes; mosses)
bryophytes, 172, **275t,** 277, 400–423 (*see also* hornworts; liverworts; mosses)
  comparative structure of, **403,** 403–4
  comparative summary of characteristics of, **422t**
  life cycles in, **410–11, 418–19**
  matrotrophy in, 404–6
  phyla of, 407
  phylogenetic relationships with other embryophytes, **472**
  relationships to other groups, 401–4, **402**
  reproduction in, 404, **405,** 406
  sporopollenin-walled spores of, 402, 406–7
  symbiosis and, 403
  vascular plants compared to, 425
BT gene, 697, **698**
buckeye, red, **627**
buckwheat family, 545
budding, of yeasts, 330, 331, **331**
bud primordia, 611, **611, 634**
buds, **8, 611, 613,** 622
  accessory, **657**
  auxin-kinetin ratio and, 681, **681**
  dormancy of, 717, **718,** 718–19, **719**
  hormones and, **678**
  lateral (axillary), **657,** 678, **678**
  in mosses, **416**
  terminal, **657**
  on woody stems, **657**
bud scale(s), **613**
  dormancy and, 718, **718**
  scars, **657**
bugs (*see also* insects)
  true, 550
bulbs, 643, **644**
  forcing, 719
bulk flow, 77
bulliform cells, 632, **633**
bull kelp, **379**
bull's-horn acacia, 642, 775, **775**
bumblebees, 534, **534**
bundle scars, **657**
bundle-sheath cells, 144, **145,** 146
  of grasses, 632, **633**
bundle-sheath extensions, 632
bundle sheaths, **628, 629,** 632
  in grasses, 632, **632, 633**
butter-and-eggs, **500,** 546
buttercup, **596, 598,** 614, 619, **620**
  family, **502,** 545
  water, **197**
butterflies, 550
  monarch, 34, **35,** 550
  pollination by, 535–36, **536**
  viceroy, 550
butternut, **657**
butterwort, 737
buttonwood, **658**

C

C_3–C_4 intermediates, 147
C_3 pathway (*see* Calvin cycle)
C_4 pathway (C_4 photosynthesis; Hatch-Slack pathway), 144, **144, 145,** 146, 147, **148,** 149
  in grasses, 632, **633**
C_3 plants, 144
  global warming and, 151
C_4 plants, 144, **144, 145,** 146, 147, **147, 148,** 149
  global warming and, 151
cabbages, **238,** 550, 643
  bolting and flowering, 686, **687**
  family, **73**
cacao, 834, **834**
Cactaceae, **266**
cacti, 266, **266, 497,** 543
  barrel, 809, **809**
  cladophylls of, 642, **642**
  dispersal of, **547**
  family, 266
  fishhook, **278**
  organ-pipe, **539**
  peyote, **551**
  prickly-pear, 779, **780**
  saguaro, **10, 539, 808**
  water storage in, 644
Cactoblastis cactorum, 779, **780**
caffeine, 32–33, **33, 549,** 833
calactin, **549**
calamites (giant horsetails), 445, 456–57, **466**
calcification
  algae and, 350
  of red algae, 358–59
calcite, 731
calcium (Ca), **76**
  cycling, 782–83
  deposits (stromatolites), **289**
  as essential element, 727, **730t**
  gravitropism and, 704, 705
calcium carbonate (calcite), 731
calcium ions (Ca^{2+})
  halophytes and, 744
  hormone action and, 691, **692**
  as second messengers, 87
calcium oxalate, **575**
  in vacuoles, 54, **54**
California, chaparral of, 817, **817**
California poppy, **497, 534**
Callitriche heterophylla, 626
Callixylon, 471
  newberryi, **470**
callose, **579,** 580, **580, 583**
  definitive, 580
  wound, 580
callus, 681, **681**
calmodulin, 87, 704
Calocedrus decurrens, **278,** 788
calorie, 864
Calostoma cinnabarina, **326**
Calothrix, **289**
Caltha palustris, **533,** 543
Calvin, Melvin, 139
Calvin cycle (C_3 pathway), **134,** 139–43, **140, 141,** 147, **148,** 149
  C_4 pathway and, 144
  photorespiration and, 143
Calycanthaceae, 519
Calypte anna, **537**
calyptra, 406, **410, 411,** 419, 421 (*see also* venter)
calyx, 501, 502, 527
calyx tube, **638**
cambial initials (*see* fusiform initials; ray initials)
cambial zone, 649
cambium, 443
  cork (phellogen), 8, 427, 587, **587,** 600, 602, 647, 648, 653–55, **655,** 658, **659**
  fascicular, **619,** 650–51
  interfascicular, **619,** 650–51
  of *Isoetes,* 443
  reactivation of, 649–50, 678
  storied/nonstoried, **649**
  supernumerary, 606, **607**
  vascular (*see also* vascular cambium), 8, 427, 576, **576,** 647–50, **649, 650**
Cambrian period, 877
  Lower, 312
Camellia sinensis, 833
cAMP (cyclic adenosine monophosphate; cyclic AMP), 354
  cellular slime molds and, 355
CAM photosynthesis, 147, **148,** 149
CAM plants, 147, **148**
  stomatal movements and, 754
Campylopus, **401**
canary grass, 675
Candida albicans, 331
Candolle, Augustin-Pyramus de, 264
Cannabis sativa, 550, **551**
canola, 698
cap (pileus), 324
Capparidaceae, **263**
Capsella bursa-pastoris, **557**
Capsicum, 833
  frutescens, **299**
capsids, 298, **299**
capsules (*see also* sporangia)
  of angiosperms, 544
  of bacteria, 284
  of bryophytes, **405,** 406, 408, **408, 409,** 412, 414, **414,** 416, **418, 419,** 420, 421, **421, 422**
  of mosses, 414, 416, **418, 419,** 420, 421, **421, 422**
caraway, 833
carbohydrates, 18, **19–22, 38t** (*see also by name*)
  in carbon cycle, 150
  formation of, **20–22**
  molecular structure of, 18–22
  oxidation of, 109–24
  in plasma membrane, 74–75
carbon (C), 18 (*see also* carbon cycle)
  covalent bonds of, 869
  as essential element, 727
  fixation of (*see* carbon-fixation reactions)
carbon balance, global (*see also* greenhouse effect)
  forests and, 150
  prokaryotes and, 287
carbon cycle, 150–51, **151**
  algae and, 150, 350
  peatlands and, 415
carbon dioxide (CO_2) (*see also* carbon cycle)
  in atmosphere, 3, 150–51, 845
  diffusion of, 78
  from fermentation, **110,** 122, **123**
  from glucose oxidation, 109
  Krebs cycle and, 109, **115,** 116, **121, 124**
  in oceans, 150
  in photosynthesis, 751 (*see also* carbon-fixation reactions)
  from pyruvate oxidation and decarboxylation, 114
  stomatal movement and concentration of, 753
carbon-fixation reactions, 133, 134, **134,** 135, 139–51
  advantages and disadvantages of, 149
  Calvin cycle (C_3 pathway), 139–43
  crassulacean acid metabolism (CAM), 147, **148, 149,** 149
  four-carbon pathway (C4 pathway; Hatch-Slack pathway), 144, **144, 145,** 146, 147, **148,** 149
  in grasses, 632, **633**
Carboniferous period, 431, 435, 445, **456,** 877
  plants of, 456–57, **456–57**
  swamps of, **433**
carbon to nitrogen (C/N) ratio, 746
carbonyl groups (—C=O), 18
carboxyl groups (—COOH)
  in amino acids, 26, **27**
  in fatty acids, **23**
cardamom, 832
cardiac glycosides, 34, 550
Carex aquatilis, **819**
carinal canals, 445, **448**
carnauba wax, 25
Carnegiea gigantea, **497, 808**
carnivorous plants, 642, 737, **737**
carotene, 133, **134**
carotenoids (carotenoid pigments), 49–51, 133, **134,** 135, 542, 543, **543**
carpels, 499, **499,** 501, 521, 528, **638**
  development of, **638,** 639
  of early angiosperms, **524,** 525, 527, 530
  fruits and, 543
  insect pollination and, 530, 531
  outcrossing and, 510, **511**
carpogonia, **359,** 360
carposporangia, **362**
carpospores, **359,** 360, **362**
carposporophytes, 361
carrageenan, 358, 380
carrier proteins, **82,** 83
carrion flies, **531**
carrot, **62,** 606, **681,** 686, 695
Carum carvi, 833
Carya
  cordiformis, **669t**
  ovata, **554, 627, 647, 658**
caryopsis, 545
Casparian strips, **594,** 598, **598, 599,** 600
cassava, 831
castings, **733**
castor bean, **562,** 563, **566,** 567, **578**
catabolism, 123–24, **124**
catalysts, 863
  enzymes as, 30, 100
categories, taxonomic, 264
caterpillars
  of the monarch butterfly, 34, **35**
  resistance to damage from, 697, **698**
Catharanthus roseus, **846**
cation exchange, 735
cations, 856
catkins, **499, 541**
Cattleya, **528**
cauliflower, **238**
caulimoviruses, 300
cavitation, 756
cDNA (complementary DNA) libraries, 226
Ceanothus, 741
cedar
  incense, **278,** 788
  western red, **669t**
celery, 643
cell cycle, 155, 157, **157**
  controls that regulate, **157,** 158
  phases of, 157, **157**
cell death, programmed, 355, 578
cell differentiation, 214, 572
  hormones and, 687, **687t**
cell division, 155, 156, **157, 159** (*see also* cell cycle; cytokinesis; meiosis; mitosis)
  in algae, **376,** 376–78, **377,** 386–88
  embryogenesis and, 556, 557, **557, 558–59,** 559
  in eukaryotes, 156
  growth and, 571
  hormones and, **674t,** 687, **687t**
    auxins, 676
    cytokinins, 680, 681
    gibberellins, 684

in plants, 158, 160–66
in prokaryotes, 156, **156**
region of, 594, **594**
reproduction of the organism and, 166
cell elongation, hormones and, **674t,** 675, **675t,** 685, 686
cell enlargement, 571
cell expansion, 59, **66,** 166, 634
direction of, 690
gravitropism and, 704
hormones and, 681, 682, 687, **687, 689,** 689–90
cell membrane (*see* plasma membrane)
cell plate, 163, **164,** 165, **165,** 166
cells (*see also specific types and parts of cells*)
components of, 41, 43, 45–46, **46t, 70–71t**
diffusion and, 78–79
eukaryotic (*see* eukaryotic cells)
evolution of, 4–6
expansion of (*see* cell expansion)
files (lineages) of, **570,** 572, 591, 604
first use of the term, **40**
forerunners of the first, 4–5
fossils, 2, **2,** 43
hormonal influences on basic processes of, **687t**
prokaryotic (*see* prokaryotes)
structure of, 40–71
totipotency and, 688, 695
cell sap, 54
cell theory, 41, 65
cell-to-cell communication, 86–89
plasmodesmata and, 87
signal transduction and, 86–87, **87**
cellular differentiation (*see* cell differentiation)
cellular membranes (*see* membranes, cellular)
cellular respiration (*see* respiration)
cellular slime molds (*Dictyosteliomycota;* dictyostelids), 275, 348, **349t,** 354–56 (*see also* plasmodial slime molds)
cellular-type endosperm formation, 510
cellulose, 21, **21,** 22, **38t**
in cell walls, **62,** 62–63, **63,** 689, **689,** 690, **690** (*see also* cells walls)
cellulose microfibrils (*see* microfibrils)
cellulose plates of dinoflagellates, 363, **363**
cellulose synthase, 66, **67**
cell walls, 21, 22, 43, **44,** 45, **45, 63, 70t**
of angiosperms, 504, 506, 510
cell division and, 156, 158, 163, **165,** 166
cellulose in, **62,** 62–63, **63**
of chytrids, 312
of diatoms (frustules), 376, **376, 377,** 378
direction of cell expansion and, 690, **690**
extensibility of, 63, 689
of fungi, 310
growth of, 66
of guard cells, 752, **753**
lignin in, 36, 63
noncellulosic molecules in, 63
oomycetes and, 371, **372**
overview of, 61–62, **62**
of pines, 481
plasmolysis and, 81
primary, **62,** 64
of prokaryotes, 283–84
of red algae, 358
secondary, 64–65, **65**
pits in, 66
of slime molds, 355, **356**
suberized, 25
turgor pressure and, 81
Cenozoic era, 877
central cell, 506
central dogma of molecular biology, 209
central mother cell zone, 612, 613
centric diatoms, **375,** 376, 378
centrioles, 158
in fungi, 311
centripetal differentiation, 435
centromeres, 160, **160, 162**
in meiosis, 176
centrosome, 158
century plant (*Agave*), 644, 808
shawii, **808**
Cerambycidae, **530**
Ceratium, **363**
tripos, **363**
Ceratophyllum, 526, **526**
Cercis canadensis, **54**
cereals (grains), 825, 843
cetoniine scarabid beetle, **531**
CFCs, 844
Chaetoceros, **375**
Chailakhyan, M. Kh., 715
chalazal, 506, **507,** 556
Chamaenerion, 546
angustifolium, **511, 792**
Chaney, Ralph, 486
channel proteins, **82,** 83
chaparral, 565, **565, 796,** 817, **817**
chaparral honeysuckle, **524**
dispersal of, **547**
Chara, **47, 387,** 398
Charales, 397, 398, 402
Chargaff, Erwin, 198
Charophyceae (charophytes), 387, **387,** 396–98, 874 (*see also* algae; green algae)
bryophytes compared to, 401–2, 407
checkpoints, **157,** 158
cheese, deuteromycetes and, 334
chemical compounds, 868
chemical equations, 871
chemical equilibrium, 871–72
chemical formulas, 870
chemical reactions, 871
energy and, 875–76
chemical reactivity, basis of, 867
chemiosmotic coupling, 120
in chloroplasts and mitochondria, 137, 138
in photophosphorylation, **138**
chemoorganotrophs, 295
chemosynthetic autotrophs, 287, 738
chemotaxis, 355
Chenopodiales (*Centrospermae*), 543
Chenopodium quinoa, 830, **830**
cherries, **502**
cherry
black, **669t**
flowers, **503**
chestnut, horse, **505, 613**
chestnut blight, 317
chiasmata, 172–73, **173,** 176
chicken pox, 296
chicory tribe, **527**
Chihuahuan Desert, **808**
Chimaphila umbellata, **524**
chimeral meristems, 634
China, 826
chitin, 22, **38t,** 310
Chlamydia pneumoniae, 287
Chlamydomonas, 44, **44,** 45, **61, 171,** 383, **387, 388,** 388–89, **389**
life cycle of, **389**
nivalis, **386**
Chlorella, **274**
chloride (Cl^-), **76**
chloride ions, stomatal movement and, **692,** 693
chlorine (Cl)
as essential element, 727, **730t**
chlorobium chlorophyll, 133
Chlorococcum, 390–91, **391**
echinozygotum, **391**
chlorofluorocarbons (CFCs), 844
Chloromonas, 383
brevispina, **386**
granulosa, **386**
Chlorophyceae (chlorophytes), 386, **386, 387,** 388–92, 873–74 (*see also* algae, green algae)
filamentous and parenchymatous, 392
life cycle of, 388, **389, 390**
motile colonial, 389–90, **390**
motile unicellular, 388–89
nonmotile colonial, **391**
nonmotile unicellular, 390–**91**
chlorophyll *a,* 133 (*see also* P_{680},P_{700} chlorophyll molecules)
in light reactions, 135, 136
in prochlorophytes, 292
chlorophyll *a/b*-binding protein (CAB), 708, **709**
chlorophyll *b,* 133
in prochlorophytes, 292
chlorophyll *c,* 133
chlorophylls, 2, 49, 131, 133
absorption spectrum of, 131
magnesium and, 729
molecular structure of, **133**
in photosystems, 135
synthesis of, 634
Chlorophyta, 873 (*see also* algae; green algae)
chloroplast endoplasmic reticulum, 357, 367
chloroplasts, **44–45,** 48, **49, 50, 70t,** 132, **132**
of bryophytes, 401–2, 421
chemiosmotic coupling in, 137, 138
of cryptomonads, 357
cytoplasmic inheritance involving, 196
of euglenoids, 351
evolutionary origin of, 53, 272, **273**
genome of, **273,** 274
of green algae, 351
in photosynthesis, 49–50, **126, 143,** 144, 146, **146**
of red algae, 358
structure of, 49
chlorosis, 729, **729, 730t**
Choanephora, 317
choanoflagellates, 277, 312, **312**
cholera, 287
cholesterol, 25, **25**
bran and, 563
Chondromyces crocatus, **285**
Chondrus crispus, **358**
C horizon (soil base), 732, **732**
chromatids
in meiosis, 172, **173**
in mitosis, 160, **160–162,** 163
chromatin, 47, **48,** 219, **220**
chromophore, 713
chromoplasts, 48, 50–51, **70t** (*see also* plastids)
chromosome mutations, 192–94
chromosomes, 41 (*see also* DNA)
bacterial, 156, **156**
cell division and, 156, **156,** 157, **157,** 158
prophase, 160, **160**
changes in number of (aneuploidy), 193
chemical structure of, 197
daughter, **161**
duplication of, 193
eukaryotic, 218–22 (*see also* DNA (deoxyribonucleic acid), of eukaryotic chromosomes)
homologous (homologs), **170,** 170–77, **173**
crossing-over and, 172, 191, **191**
linkage of genes and, 190, 191, **191**
in meiosis I, 174–77
in meiosis II, 177
inversions of, 193
linkage of genes and, 190, 191, **191**
in meiosis, 169
Mendel's laws and, 190, **190**
in mitosis, 160, **160,** 161, **161, 162,** 163
number of, 170 (*see also* diploid; haploid)
prokaryotic, 216–18
translocations of, 193
chrysanthemum, **710**
Chrysanthemum, 780
indicum, 715
chrysanthemum stunt viroid (CSVd), 303
Chrysochromulina, 367
chrysolaminarin, 378
Chrysophyta (chrysophytes), **275t,** 349, 371, **372t, 378,** 378–79, 873 (*see also* algae)
Chytridiomycota (chytrids), 275, **308t,** 870
Chytridium confervae, **313**
chytrids (*Chytridiomycota*), 309, **309,** 312–13, **313, 314**
evolutionary origin of, 312
Cicuta maculata, **500**
ciguatera, 364
cilia, 60 (*see also* flagella)
Cinchona, **551, 833**
Cinnamomum zeylanicum, 832
cinnamon, 832
circadian rhythms, 706–9, **707**
circinate vernation, 458
Cirsium pastoris, **527**
cisternae, 56
citrate (citric acid), 115
citrate synthase, 747
citric acid cycle (*see* Krebs cycle)
citrus stubborn disease, 292, **292**
clades, 267
cladistics (phylogenetic analysis), 267
cladograms, 267–68, **268, 432, 472, 519, 526**
Cladonia
cristatella, 337, **339**
subtenuis, **337**
Cladophora, 393, **393**
cladophylls, 642, **642**
Cladosporium herbarum, 307
clamp connections, 322, **322,** 333
Clarkia, **514**
cylindrica, **514**
heterandra, **514**
Clasmatocolea humilis, **409**
classes, taxonomic, 264
classification (*see* systematics; taxonomy)
clathrin, 58, **58,** 86
Clausen, Jens, 246
Clavariaceae, **277**
Clavibacter, 294, **294**
Claviceps purpurea, 335, **335**
clay soils, 732, 734, 735, **735**
Claytonia virginiana, 548
cleavage furrow, 386
cleistothecium, 319, **319**
climacteric, 683
climatic changes (*see* global warming)
climax community, 792
clines, 246
clonal analysis, 632, 634
clonal propagation (micropropagation), 693–95, **694**
clonal reproduction, competition and, 777
clones, 695
cloning, DNA (gene cloning), 225, 226, **226** (*see also* recombinant DNA)
Clostridium, 741
botulinum, **285,** 286
clover, 719, 738, 740
white, 244, **534**
cloves, 832
club mosses (*Lycopodiaceae*), 429, **435,** 435–38, **436** (*see also* seedless vascular plants)
gametophytes of, 431, **436, 437,** 438
life cycle of, 435, **436–37,** 438
suspensor in, 442
coal age plants, **456–57, 466–67**
coated pits, 86, **86**
coated vesicles, 58, **58,** 86, **86**
cobalt (Co), 729
coca, **551**
cocaine, 32, **32, 551**
cocci, 284, **285**
coccolithophorids, 367, **367**
coccoliths, 367
Cochliobolus heterostrophus, 840, **840**
Cochrane, Theodore S., **263**
cocklebur, **548,** 681, 709, **709,** 710, **710,** 711
cocoa, **549**
coconut, 543–44
dispersal by ocean currents, 546
fruit, **544**
lethal yellowing of, 293, **293**
palm, **497, 544,** 834
coconut cadang-cadang viroid (CCCVd), 303
coconut milk, **544,** 695
growth factor in, 679–80
Cocos nucifera, **293, 497, 544,** 679, 834
coding sequences, 688, **688**
Codium, 394, **395**
fragile, 394
magnum, 383
codons, 209, **209**
initiation, 212
stop, 214
Coelomomyces, **308t**

coenocytic hyphae, 310
coenzyme A (CoA), 114, **114**
coenzyme Q (CoQ; ubiquinone), **117,** 118, **118**
coenzymes, 102
coevolution, 247, 532
of animals and flowering plants, 547
biochemical, 549–51
cofactors, 102–3
Coffea arabica, **549,** 827–28, **833**
coffee, **549,** 827, 833, **833**
cohesion-tension theory of water transport, 754–58, **756, 757**
air bubbles and, **756,** 756–57, **757**
tensile strength of water and, 757
test of, **757,** 757–58
Colacium, 351
colchicine, 193, 840
cold (*see also* temperature)
dormancy and, 717, 719
flowering response and, 719
cold hardiness, 719
colds (viral), 296
Cole, Lamont, 785
Coleochaetales, 397, 402
Coleochaete, **387,** 397, **397**
coleoptile, **562,** 563, **564,** 568, 675, 676, **686, 702**
phototropism and, 703, **703**
coleorhiza, **562,** 563, **564,** 568
Coleus, 624, **678**
blumei, **611, 612, 634**
collared sunbird, **537**
collenchyma (cells or tissue), **571,** 572, **573,** 574, **574–575**
of leaves, **628,** 632
of stems, 618
Colocasia esculenta, 827
Colorado potato beetle, 550
color perception, in bees, 532, **533**
colors (*see also* pigments)
of flowers
bat-pollinated flowers, 539
bee-pollinated flowers, 532
beetle-pollinated flowers, 531
pigments responsible for, 542–43, **543**
wind-pollinated flowers, 539
of wood, 667
columbine, **538**
columella, 591, 604
gravitropism and, 704, **705**
column, 528
comets, 4, **4**
Commelina communis, 35
Commelinales, **265t**
commelinin, **35**
communication, cell-to-cell, 86–89
plasmodesmata and, 87
signal transduction and, 86–87, **87**
communities, 9, **10,** 11, 786 (*see also* biomes)
climax, 792
definition of, 774
companion cells, **580–582,** 582–83, **583, 587t,** 651
competition, 776–79
clonal reproduction and, 777
defined, 776
growth rate and, 776
by inhibiting the growth of others, 779
competitive exclusion, principle of, 776–77
complementary DNA (cDNA) libraries, 226
complementary strands, 199
complex tissues, 573
Compositae, 264 (*see* Asteraceae)
composites, 256, **527,** 527–28
compost, 746
compression wood, 666, **666**
Comptonia, 741
concentration gradient, diffusion and, 78
conceptacles, **385**
Conchocelis, 361
conchospores, **359**
conducting tissue (*see also* phloem; xylem)
in algae, **382**
of Anthoceros, **413**
bryophytes and vascular plants compared, 402
of mosses, 416, **417**
cones (*see also* strobili)
cone scales (*see* ovuliferous scales)
conidia, 311, 318, **318, 319,** 320, **332**
conidial fungi (deuteromycetes), 332, **332,** 871
predaceous, **333**
conidiogenous cells, 311
conidiophores, 318
Coniferophyta (*see* conifers)
coniferous forests, 802, 816, 819 (*see also* taiga)
northern (taiga), **818,** 818–19
conifers *(Coniferophyta),* **275t, 278,** 470, 472–86, 875 (*see also* pines)
of Carboniferous period, 457
commercial uses of, **669t**
compression wood of, 666
wood of, 659–61, **660, 661**
Coniochaeta, 319
conjugation, **284,** 285–86
conjugation tubes, 396, **396**
connation, 502
contender beans, **685**
continuous variation, 195, **195**
contraceptive, oral, 78
contractile vacuoles, 80, 351
convergent evolution, 266, **266,** 267, 428, 518, **519**
Convolvulaceae, 498
Convolvulus arvensis, **599**
Cooksonia, **6, 424, 426,** 434
pertoni, 434
copepods, 366
co-pigmentation, 35, **35**
copper (Cu), as essential element, 727, **730t**
copper butterfly, 536
Coprinus, **324**
cinereus, **321**
CoQ (coenzyme Q), **117–19, 121**
coral fungi, 323
coralline algae *(Corallinaceae),* **358,** 359, **359,** 361 (*see also* algae; red algae)
coral reefs, **358,** 359
dinoflagellates and, 365, **365**
Cordaites, **456,** 457, **457**
Coreopsis, 811
corepressors, 217, **217,** 218
coriander, 833
Coriandrum sativum, 833
cork (phellem), 7, **40,** 587, 602, 647, 653
commercial, 658–59, **659**
cork cambium (phellogen), 8, 427, 587, **587,** 600, 602, 647, 648, 653–55
cork oak, 658, **659**
cork tissue, 600
corms, 442, **443,** 643
corn (*see* maize)
corn smut, 330, **330**
corn streak, 300
corn stunt disease, **292**
Cornus florida, 334
corolla, 501, 502, 527, 528
corolla tube, 524, **524,** 532, 535–36, **539**
corpus, 612, **612,** 613
Correns, Carl, 189
cortex, **8,** 573, **573**
of lichens, **338**
of roots, 592, **593, 597,** 598–600, 602
of stems, 613, 614, **615,** 618
cortisone, 846
corymb, **499**
Costa Rica, **803**
cotranslational import, 214, **214, 215**
cotransport systems, carrier proteins as, 83
cotton, 827, 828, **837**
flowers, **525**
cottonwood, 68, 342, **583, 631, 669t**
cotyledons (seed leaves), **8,** 482, **562,** 563 (*see also* eudicots; monocots; scutellum)
of angiosperms, 510
development of, 557, **557,** 559, **560**
germination and, **566,** 567–68
of gymnosperms, **478–79, 482**
notch at the base of, **558**
coupled reactions, 105
covalent bonds, 868–70, **869**
polar, 869–70, **870**
crabgrass *(Digitaria sanguinalis),* 146–47
Crane, Peter, **520**
Crassulaceae, 147
crassulacean acid metabolism (CAM), 147, **148, 149,** 149
Crataegus, **643**
creosote bush, **664, 808**
Cretaceous period, 470, 488, inside front cover
Early, **517,** 519
Lower, 522
upper, 522
Crick, Francis, 197–99, **199,** 209
cristae, 52, **52,** 114
Crocus sativus, 833
cross-grained wood, 667
crossing-over, 172, **173,** 176, 191, **191,** 193
cross-pollination, **184** (*see also* outcrossing)
Croton (crotons), **610, 624,** 634
crown gall, 294, **294**
crown-gall tumors, 696, **696**
crown wart of alfalfa, 313
crozier, 320
Crucibulum laeve, **326**
crustose coralline red algae, **358,** 359, **359**
crustose lichens, **336,** 338
Cryphonectria parasitica, 317
cryptomonads *(Cryptophyta),* **275t,** 348, **349t, 356,** 356–57, **357,** 872 (*see also* algae)
crystal violet, 283
cucumber, 676, **677, 752**
ethylene and sex expression in, 683
Cucumis sativus, 676, **677,** 683, **752**
Cucurbita
maxima, 567, **576, 581, 582**
pepo, **600**
cucurbits *(Cucurbitaceae),* ethylene and sex expression in, 683
Culver's-root, 624, **625**
Cupressus, 483
goveniana, 483
cupules, **469**
Curculionidae, 487
Curtis Prairie, 792–93, **793**
Cuscuta, 498
salina, **498**
cuticles, 7, 25, **25,** 597
as barrier to water loss, 752
of bryophytes, 403
of epidermal cells, 584, 597
cuticular wax, 25
cutin, 24, 25, 63
cyanidin, 542, **543**
Cyanidium, 357
cyanobacteria (blue-green algae), 43, **44,** 282, **288,** 288–91, **289,** 403, 870 (*see also* bacteria)
akinetes of, **289, 290,** 290–91
chloroplast evolution, and 290
environmental conditions and, 288
gas vesicles of, 289
glycogen and, 288
heterocysts of, 290, **290**
importance of, 288
in lichens, 340
marine, 289
membranes of, **288**
motile, 288
mucilaginous sheaths of, 288
nitrogen-fixing, 290
photosynthesis and, 288, 290
pigments, 288
planktonic, 289–90
in stromatolites, 289, **289**
structure of, **288–90**
symbiotic, 290 (*see also* lichen)
Cyathea, 453, **453**
australis, **445**
cycadeoids *(Bennettitales),* 472, **473,** 518
cycads *(Cycadophyta),* **239, 275t,** 470, 472, 473, 486–87, **487,** 875
Cycas siamensis, **487**
cycles
nutrient (*see* nutrient cycles)
seasonal growth, 648
water, **734**
cyclic adenosine monophosphate (*see* cAMP)
cyclic AMP (*see* cAMP)
cyclic electron flow, 139, **139**
cyclic photophosphorylation, 139, **139**
cyclosis (cytoplasmic streaming), 46
in giant algal cells, 47, **47**
cyclosporin, 308
Cyclotella meneghiniana, **375**
cylinders (*see* steles)
Cymbidium orchids, **278**
Cyperaceae, 340
cypresses, 483, **483**
bald, 483, **485**
Gowen, **483**
Cypripedium calceolus, **260**
cypsela, 545, **545, 546**
Cyrtomium falcatum, **459**
cysteine, **27,** 29
Cystopteris bulbifera, **278**
cysts, resting, 365
cytochrome c, 269, **269**
cytochromes, 116, **117**
cytokinesis, 156, 157, **157,** 158, 163, **164, 165**
in *Chlorophyceae,* 386, **387**
cytokinins, 590, **674t,** 676, 679–82, **680,** 696
genetic engineering and, 698, **698**
leaf senescence delayed by, 681–82
cytology, 11, 190
cytoplasm, 41, **44,** 45, **70–71t**
division of (*see* cytokinesis)
of prokaryotes, 282
cytoplasmic inheritance, 196, **196**
cytoplasmic male sterility, 196
cytoplasmic sleeve, 87, **88**
cytoplasmic streaming (cyclosis), 46
in giant algal cells, 47, **47**
cytosine, 198, **198,** 199, **200**
cytoskeleton, 43, 58, **71t, 273**
cytosol, 46

D

Dactylis glomerata, 244
daisy, **536**
daltons, 863
damping-off diseases, 374
dandelion *(Taraxacum officinale),* **132,** 180, **278,** 528, **545, 777**
fruits of (cypselas), **546**
darkness, flowering and, 711
dark reactions (*see* light-independent reactions)
dark reversion, **713**
Darwin, Charles, 3, 35, **235,** 236–38, 248, 257, 518, 675, **707**
Darwin, Francis, 675
Dasycladus, **395**
dATP (deoxyadenosine triphosphate), 204 (see also ATP)
Daucus carota, **62,** 606, **681,** 686, 695
daughter cells, 156
daughter chromosomes, **161,** 163
Dawsonia superba, **417**
daylength
dormancy in buds and, 718–19, **719**
flowering and, **709,** 709–14 (*see also* photoperiodism)
day-neutral plants, 710
DDT, 781, **781**
Death Valley, **808,** 810
deciduous forests
temperate, 812–14, **813, 814**
tropical, 806
deciduous trees, 522
decomposers, 287, 783
fungi as, 307
decussate phyllotaxy, **611,** 624
deficiencies, nutrient, 727, 729, 730t, 731
definitive callose, 580
dehiscence
of anthers, **504,** 508
of spores, 412, **414–15,** 416
dehiscent fruits, 544, **544**
dehydration synthesis, 20, **20,** 22
amino acids and, 28

16-dehydropregnenolone (16D), 550
deletions, 192–93
delphinidin, **543**
denaturation, 104
denaturation of proteins, 30
dendrochronologists, 664–65
dendrometer, **758**
denitrification, **736,** 738
Dennstaedtia punctilobula, **458**
density of wood, 668, 670
deoxyribose, 30, **30**
deoxyribonucleic acid (*see* DNA)
derivatives, 571
dermal tissue system, 426, **427,** **571,** 572, 583–86, **586t**
dermatophytes, 334
desert plants, **10**
  convergent evolution of, 266
  germination of seeds of, 565
deserts, **10, 798–99,** 807–10, **808**
  cold (ice), 807, 821
desmids, 397, **397**
desmotubules, 68, **68,** 87, **88**
deuteromycetes (conidial fungi), 332, **332,** 871
  predaceous, **333**
development, 571 (*see also* cell differentiation; growth; morphogenesis)
  of flowers, **638,** 639, **640,** 641
  of fruits, 509–10
    gibberellins and, 686
  hormones and (*see* hormones)
developmental plasticity, 246
Devonian period, 425, **426,** 427, 428, 431, 435, 443, 471, 877
  Early (Lower), 312, 339, **426,** 434
  Late (Upper), 312, 431, 434, 445, **456,** 468, **469,** 470
  Middle, **426,** 427
de Vries, Hugo, 189, 192, **192,** 252
diabetes, **846**
diatomaceous earth, 378
diatoms *(Bacillariophyta),* **275t,** 349, 371, **372t, 375,** 375–78, 873 (*see also* algae)
  cell wall of, 376
  centric, 376
  heterotrophic, 378
  life cycle of, **376–377**
  pennate, **276,** 376
  resting stages of, 376, **378**
Dicentra cucullaria, 549
2,4-dichlorophenoxyacetic acid (2,4-D), **676,** 679
dichogamy, 510, **511**
Dicksonia, **455**
dicots, 496 (*see also* eudicots; magnoliids)
Dictyophora duplicata, **326**
dictyosomes (*see* Golgi complex)
Dictyosteliomycota (dictyostelids), 884 (*see also* cellular slime molds; *Dictyostelium*)
Dictyostelium
  discoideum, 354, **354,** 355
  life cycle of, **354**
  mucoroides, **356**
Diener, Theodor O., 302
differentiation, cell (*see* cell differentiation)
diffuse-porous woods, **662,** 665
diffusion, **77,** 77–79, **79**
  concentration gradient and, 78
  facilitated, **82,** 83
  osmosis as special case of, 78–79
  passive transport and, 82–83
  simple, **82,** 82–83
digger pine, **480**
Digitalis, 34
  purpurea, **533**
Digitaria, **300**
digitoxin, 34
digoxin, 34
dihybrid crosses, 185
dihydroxyacetone phosphate, 111, **111,** 142
dikaryon, 312
dikaryotic cells of ascomycetes, 320
dikaryotic mycelium (or hyphae), 322, **322**
dilated rays, 652, **652,** 655
Dilcher, David, **520**
dill, 833
Dinobryon, 378, **378**
dinoflagellates *(Dinophyta),* **275t,** 348, 349, **349t, 350,** 361–66, **363,** 872–73 (*see also* algae)
  feeding by, 365
  red tides and toxic blooms caused by, 364
  resting cysts formed by, 365
  as symbionts, 365
  toxic or bioluminescent compounds produced by, 366
Dinophyta, 872–73 (*see also* dinoflagellates)
dinucleotides, 102
dioecious plants, 501
Dionaea muscipula, 721, **721,** 737
Dioscorea, 834, 846, **846**
diosgenin, 550
Diospyros, **67**
  virginiana, **669t**
dioxin, 679
Diphasiastrum, 438
  complanatum, **435**
  digitatum, **278**
diptheria, 287
diploid number, 47, 170
diploidy, 171, 172
  preservation of variability and, 243
disaccharides, 18, **38t**
Dischidia rafflesiana, 606, **606**
Discula destructiva, 334
diseases (*see also specific diseases and types of diseases)*
  bacterial, 293–94
  fungal, 307–8
    ascomycetes, 317
    chytrids, 313
  mycoplasmalike organisms (phytoplasmas) as cause of, 292–93
  oomycetes as cause of, 373–74
  viral, 296–97, 301–2
  viroid, 302–3
disk flowers, **527,** 528
dispersal agents, evolution of fruits and, 546–49
  biochemical coevolution and, 549–51
  vertebrates, **547,** 547–49
  water, 546–47
  wind, 546
Distichlis palmeri, 744
distichous phyllotaxy, 624
disulfide bridges, 29
diterpenoids, 33, 34
diurnal movements, 706, **707**
divisions, taxonomic, 264
DNA (deoxyribonucleic acid), **30,** 31, **31,** 41
  amount of, in selected bacterial and eukaryotic genomes, **221t**
  antiparallel strand of, 199
  base pairing of, 199, **200**
  cell division and, 156, **157,** 158
  of *Ceratophyllum,* 526
  chloroplasts and, 50
  in chromatin, 47
  composition of, in various species, **198**
  double helix, **198**
  of eukaryotic chromosomes, 218–22
    condensation of DNA, 220–21
    histones and nucleosomes, 219–20
    introns and exons, 222, **222**
    repetition of nucleotide sequences, 221–22
    single-copy DNA, 222
    specific binding proteins, 221
  in genetic engineering, 224–231
  information-carrying capacity of, 204
  interspersed repeated, 222
  lagging strands of, 202, 203
  leading strands of, 202, 203
  of mitochondria, 53
  mutations and, 192
  in nucleolus, 48
  nucleotides in (*see* nucleotides)
  Okazaki fragments, 202, **203**
  polymerases, 201, **203**
  of prokaryotes, 216–18, 282
  recombinant (*see* recombinant DNA)
  replication of, **201,** 201–4, 208–9, **209**
    as bidirectional, 201–2
    energetics of, 203–4
    errors in, 203
    initiation of (origin of replication), 201, 202
    overview of, **202**
    summary of, **203**
    telomeres, 203
    templates, 201
  RNA compared to, 208, **208**
  RNA synthesis and, 208
  simple-sequence repeated, 222
  single-copy, 222
  spacer, **219,** 222
  structure of, 197–200, **198–200, 207**
    complementary nature of strands, 199
  tandemly repeated, 221–22
  of viruses, 297, 300
  Watson-Crick model, 197, **198–199**
DNA cloning (gene cloning), 225, 226, **226–27** (*see also* recombinant DNA)
DNA ligase, 202, **203,** 224
DNA polymerases, 201, 202, **203**
  polymerase chain reaction (PCR) and, 226
  replication errors and, 203
DNA sequencing, **229,** 229–30
  molecular systematics and, 270
dodder, **498**
Dodecatheon meadia, **500**
Doebley, John, 829
dogbane, 34
dogs, 826, 828
dolipore septa, 322, **322**
domains, looped, 220
domains, symplastic, 88
domains, taxonomic, 881–82
  major distinguishing features of, 271, **271t**
  nucleotide sequences as evidence for three, 270, **270**
  representatives of, **272**
dominance (dominant characteristics), 185
  incomplete, 194
donor organism, 224
dormancy, 717–19
  acclimation and, 718–19
  breaking of, 684
  of buds, 717, **718,** 718–19, **719**
  gibberellin and, 685
  of seeds, 565, 685, 717–18
dormin, 683
double bonds, 857
double fertilization, 492, 509, **509**
Douglas fir, 483, **669t**
downy mildew in grapes, 373
Drepanophycus, **426**
Drew Baker, Kathleen 361
Drosera rotundifolia, **737**
drought, in Mediterranean climates, 817
drugs, plants as sources of, 843, 846
drupes, 543
druses, **54**
Dryopteris
  crassirhizoma, 680
  marginalis, **458**
Dubautia, 250
  reticulata, **250**
  scabra, 251
duckweeds, **496**
duplication of chromosomes (polyploidy), 193 (*see also* allopolyploidy; autopolyploidy)
  speciation and, 249, 252–54
duplications, 193
DuPraw, E. J., 219
Durvillea antarctica, **379**
"dust bowl" conditions, 812, **812**
Dutch elm disease, 317
Dutchman's breeches, 548–49
Dutchman's pipe, **428, 521, 654**
dwarf mutants, gibberellin and, 684–85, **685**
dyes, lichen, 338
dynamic instability, 59

E

early wood, 665
Earth, **2**
  biological productivity of, **800**
earthstars, 323, **326**
earthworms, **733**
Easter lily, **511**
Ebola virus, 296
Echinocereus, **266, 533**
ecology, 11, 773 (*see also* ecosystems; interactions between organisms)
  definition of, 774
  global, 796–822 (*see also* biomes)
  restoration, 792–93, **793**
economic botany, 11
*Eco*RI, **225**
ecosystems (ecological systems), 9, 773, 774
  development of communities and, 786–89, 792–93
  flow of energy through, 784–86
  nutrient recycling and, 782–83
  pesticides and, 781
  trophic levels and, 783–86
ecotypes, 246–47, **247**
Ectocarpus, 382, **382,** 383
  siliculosus, **382**
ectomycorrhizae, 341, 342, **342,** 343, 774–75
Ecuador, **803**
eel grass, 541, **542**
effector site, 104
egg apparatus, 506
egg cell, 506 (*see also* gametes)
  of algae, **362,** 371, **372, 374, 377,** 383, **384–85,** 392, **398**
  of angiosperms, 506, 509, **513**
  of bryophytes, **405, 410, 419**
  of gymnosperms, 472, **473,** 474, **479, 481, 482, 487, 488**
  of oomycetes, 371, **374**
  of seedless vascular plants, 442, 445, 449
eggplants, **843**
Eichhornia crassipes, **787**
Einstein, Albert, 130
Elaeis guineensis, 834
elaiosome, 548, **549**
Elaphoglossum, **453**
elaters
  of *Equisetum,* 448–49, **449, 450**
  of *Marchantia,* 408, **409,** 449
elderberry, **573,** 614, **618–19,** 650, **651–653**
Eldredge, Niles, 257
electrical energy, 2
electrical gradient, 82
electric potential (voltage), membranes and, 74
electrochemical gradient, 82, 119, **121**
electromagnetic spectrum, 129
electronegativity, 867
electron carriers, 116, **117–19, 121,** 122, 137, **138–39**
electron(s), 863
  energy level of, 865, **865**
  in oxydation-reduction, 98–99
  in photosynthesis, 131, 132, 134, 135–38, **136,** 138 (*see also* cyclic electron flow; noncyclic electron flow)
electron transport chain, 108, 109, **110,** 116, **116,** 117, 118, **119,** 122
  in photosynthesis, 135, 137
electrophoresis, 229, **229**
electroporation, 699
elements, 18, 851 *(see also specific types)*
  essential, 727, **728t,** 729, **730t,** 731
Elettaria cardamomum, 832
elevation (*see* altitude)
elicitors, 780, 782
Elkinsia polymorpha, 468, **469**
elm, **545, 623, 669t**
  American, **657, 663**
Elodea, 81, **81, 128**
elongation
  cell
    gibberellins and, 685, 686
    hormones and, **674t,** 675, **675t**
  internodal, **611,** 613, **613,** 614, 618, 682
  region of, **594,** 595

embolism, water transport and, **756,** 756–57, **757**
embryogenesis, 556–61, 571
in angiosperms, 509, 510
genes and, 560, **561**
primary meristems and, 557
stages in, 557, **557, 558–59,** 559
suspensors and, 559–60
Embryophyta (embryophytes), 876
phylogenetic relationships between major groups of, 472
as synonym for plants, **402,** 406
embryos (young sporophytes), 9, 468, 470 (*see also* embryogenesis)
of angiosperms, 509, 510, **513,** 555–68
asexually produced, **180,** 255–56
in bryophytes, **402,** 404, **405,** 406, **410, 418**
of Capsella, **558–59**
dormancy of, 565
eudicot, 557–59, **558–64,** 565, **566–68**
evolution of seed and, 510, 562
formation of, 556–561, **557, 588–61**
of gnetophytes, 492
of gymnosperms, 472
maize, 562, **567, 568**
mature, 562
monocot, **557**–59, **562**–63, 566, **567, 568**
mutant of *Arabidopsis,* 560, **560, 561**
of pines, 482, **482,** 559
proper, 556, 557, **557**
of *Sagittaria,* **557**
of seedless vascular plants, 438, 442, 449, 559
vascular transition in, 636, **637**
wheat, **564**
embryo sac, 506, **507, 513**
Emiliania huxleyi, 367, **367**
enations, **430**
Encephalartos ferox, **487**
endangered species, 13, 805, **815,** 818, 844
endergonic processes (reactions), 96, **96,** 876
ATP and, 105
endocarp, 510
endocytosis, 85, **85, 86**
receptor-mediated, 86
endodermis, 448, 475, 598, **599,** 600, 603
endomembrane system, **57, 71t** (*see also* plasma membrane)
evolutionary origin of, 272, **273**
mobility of cellular membranes exemplified by, 58
endomycorrhizae, 312, 341, **341, 344,** 774 (*see also* mycorrhizae)
endophytes, 308, 335
endoplasmic reticulum (ER), 47, **48,** 53, **55, 56, 57, 71t**
cisternal, **55, 56**
cortical, 56, **56**
plasmodesmata and, **67, 68**
polypeptide targeting and sorting and, 214, **214, 215**
ribosomes and **55,** 56, **57**
rough, **55,** 56, **57,** 58
smooth, 56, **57**
tubular, 56, **56**
endosperm, 509, 510, **557, 558,** 562, **562,** 563
gibberellins and, 685–86, **686**
liquid, **544**
endospores, **285,** 286, **286**
endosymbionts, 272, **273,** 274, 290, 351, 356–57, **357,** 358, **365,** 378, 383
in *Vorticella,* 274, **274**
energy, 93–107 (*see also* ATP; respiration; photosynthesis)
of activation, 99–100, **100,** 105, 863
bond, 863
concept of, 95
ecosystems and, 9
electrical, 2
first law of thermodynamics, 95
flow of, **94**
through ecosystems, 784–86
free, 864
living organisms' need for a steady input of, 97–98
in nucleotides, 31–32
from organic molecules, 4–5
photosynthesis and, 2
potential
of electrons, 95, 853
osmotic (solute), 79, **79**
of water, 76–77
pyramid of, 785, **785**
second law of thermodynamics, 95–97
solar (radiant), 2, 5, 94
storage as fats and oils, 22, 24
energy level of electrons, 865, **865**
energy-transduction reactions, 133, **134**
Engelmann, T. W., 131
Engelmann spruce, 777, 791
Entogonia, **375**
Entomophthorales, 317
Entomophthora muscae, **277**
entrainment, 707, 708, **708**
entropy, 97
environment (environmental factors) (*see also specific factors*)
clines and ecotypes as adaptations to, 246–47, **247**
lichens used to monitor the, 340
stomatal movement and, 753
transpiration and, 754
enzymes (enzymatic reactions), 30, 93, 99–106, **100,** 863
active site of, 101, **101**
allosteric, 104
as catalysts, 100
cofactors and, 102–3
metabolic pathways and, 103, **103**
pH of the surrounding solution and, 104
regulation of activity of, 103–5
regulatory, 104
restriction (restriction endonucleases), 224–25, **225,** 696
DNA sequencing techniques and, 229
substrates and, 100–101
temperature and, 103–4, **104**
Eocene epoch, inside front cover
Ephedra, 490, **491,** 492
double fertilization in, 492, 509
trifurca, **491**
viridis, **491**
epicotyl, 562, **562, 566,** 567
epicuticular wax, 25, 584, **584**
epidermis, 7, **8, 571,** 572, **573,** 583–86, **585**
of grasses, 632, **633**
of leaves, 583–86, **584–86,** 627, **628, 629,** 630, 632, **633, 634**
of roots, 595–97, **596, 597,** 602
of stems, 618
epigeous germination, **566,** 567, **567**
epigynous flowers (epigyny), 502, **502, 503,** 527
Epiphyllum, **642**
epiphytes, 421, 435, 445, 449, **453**
roots of, 605, 606, **606**
in tropical rainforests, 804, **804**
epistasis, 195
epithem, **761**
equatorial plane, **160–61,** 162
equilibrium, chemical, 871–72
Equisetites, 445
Equisetum, 445, **448,** 456, **463t** (*see also* horsetails)
arvense, **448–9,** 680
ferrissii, 255, **255, 256**
hyemale, 57, **256**
laevigatum, **256**
life cycle of, 449, **450–51**
sylvaticum, **278**
telemateia, **728**
ergosterol, **25**
ergot, 335, **335**
ergotism, 335
Ericaceae, 343
Eriophrum
angustifolium, **819**
scheuchzeri, **819**
erosion, 736, **838**
Erwinia, 293, **294**
amylovora, **285,** 293
tracheiphila, 294
Erysiphe aggregata, **319**
Erythroxylon coca, **551**
Escherichia coli (E. coli), 26, **43, 225, 226,** 229
conjugating, **284**
DNA of, 209, 216
operons in, 218
in protein synthesis studies, 210
ribosomes in, **211**
Eschscholzia californica, **497, 534**
E (exit) site, 212, **212, 213**
essential elements (essential minerals; essential inorganic nutrients), 727–31, **728t**
criteria for, 727
functions of, 729–31, **730t**
essential oils, 34, 549
estrogen, 26
ethylene, 548, **674t, 682,** 682–83
abscission and, 683
cell expansion and, 682
cell wall extensibility and, 689
fruit ripening and, 683
genetic engineering and, 697–98, **698**
orientation of cellulose microfibrils and, 690, **690**
sex expression in cucurbits and, 683
ethyl methanosulfonate, **561**
etiolation, 713–14
etioplasts, 51–52, **52**
Euastrum affine, **397**
Eucalyptus (eucalyptus), 342
cloeziana, 25
giant (red tingle), **495**
globulus, **584**
jacksonii, **495**
Eucheuma, 380
euchromatin, 220, 221
eudicots (eudicotyledons; *Eudicotyledones*), **497,** 504 (*see also* magnoliids)
embryos of, 557, **557,** 559, **562,** 563
evolution of, 519, 522
major features of, **498t**
parasitic, 498
seeds of, **562**
types of flowers in, **502**
Eudicotyledones, 496, 876 (*see also* eudicots)
Eudorina, **390**
Eugenia aromatica, 832
Euglena, 80, 351, **351**
Euglenophyta (euglenoids), **275t,** 348–52, **349t,** 872 (*see also* algae)
Eukarya (domain), 270, **270, 272,** 274
kingdoms in, 275–79
eukaryotes, **44,** 53 (*see also Eukarya*)
chromosomes of, 218–20 (*see also* DNA (deoxyribonucleic acid), of eukaryotic chromosomes)
DNA replication in, 202, **202,** 203
evolutionary origin of, **272–73,** 272–74, 348
mRNA in, 222–23, **223**
regulation of gene expression in, 220–21
transcription in, 221, 223, **223**
eukaryotic cells, 41–68 (*see also* eukaryotes)
first appearance of, 6
prokaryotic cells compared to, 41, 43, **44, 45t,** 272
Euphorbia (euphorbias), **266,** 644
corollata, **793**
pulcherrima, **538**
Euphorbiaceae, 34, **266,** 834
Eupoecila australasiae, **531**
Europa, 3
European larch, **483**
Eurystoma angulare, **469**
eusporangia, **453,** 455
eusporangiate ferns, **453,** 453–55
eusteles, 428, **428,** 429, **463t,** 471
evapotranspiration, 734
evening primrose, 192, **192,** 252, 536
family, **514**
Oenothera glazioviana, 192, **192**
Oenothera lamarckiana, 252
evergreen species, 648, 812
evergreen wood fern, **458**
evolution, 235–58 (*see also* individual organisms, natural selection; phylogenetic trees; speciation; systematics)
agents of change in, 240–42
gene flow, 241
genetic drift, 241–42
mutations, 193–94, 240–41
nonrandom mating (inbreeding), 242
clockface of, **12**
coevolution, 247, 532
of animals and flowering plants, 547
biochemical, 549–51
convergent, 266, **266,** 267
of C_4 pathway, 147
Darwin's theory of, 236–38
of eukaryotes, **272–73,** 272–74
gradualism model of, 257, 258
Hardy-Weinberg equilibrium and, 240
of human beings, 11–12, 824
of major groups of organisms, 256–58
photorespiration and, 143
of plants, 2–11 (*see also* angiosperms, evolution of; bryophytes; vascular plants, evolution of)
aggregations of molecules, 4–5
chemical precursors, 3–4
communities of plants, 9, **10,** 11
of Devonian period, 425, **426,** 427
on Earth, 2–3
photosynthesis, 5–7
transition to land, 7–11
punctuated equilibrium model of, 257–58
excited state, 131, 853
exergonic processes (reactions), **96,** 96–97, 864
ATP and, 105
exine, 504, **505**
exocarp, 510
exocytosis, 58, 85, **85**
exodermis, **597,** 600
of roots, **596**
exons, 222, **222,** 223, **223**
expansins, 689
explants, 695
extensins, 63, 689
extinction of plant species, present and future, 844, 847
extinct plants
gymnosperms, 472
seedless vascular plants, 425
extreme halophilic archaea, 295
extreme thermophilic archaea, 296, **296**
Exuviaella, **363**
eyelash cup, **317**
eyespot (*see* stigmas (eyespot))

F

F_1, F_2 (filial) generations, **185, 187, 189, 190**
Fabaceae, 264, 511, 544, **545** (*see also* legumes)
facilitated diffusion, **82,** 83
FAD (flavin adenine dinucleotide), 116, **116, 119**
$FADH_2$, 116, **117, 119,** 122
Fagaceae, 342, **502**
Fagus
americana, **669t**
grandifolia, **762**
fairy rings, 325, **325**
false annual rings, 664
false Solomon's seal, **762**
families, taxonomic, 264, 266
fan palm, **518**
fasciation, **294**
fascicles, 474, **475**
fascicular cambium, 650
fats, 22, **23, 38t**
respiration and, 122
saturated, 24
unsaturated, 24
fatty acids, 22, **23, 38t**
respiration and, 122
feedback inhibition, 104, **104**
Fd (*see* ferredoxin),
feeder roots, 590, 596
fennel, 833
fermentation, 109, **110**

  alcohol, 123, **123**
  lactate, 122, 123
ferns, **55, 59, 108, 275t,** 424, 431, 449, **452, 453,** 453–55, 457–63, 804 (see also *Pterophyta*)
  bracken, **458**
  Carboniferous period, 457
  cladogram of, 268, **268**
  in cladograms showing phylogenetic relationships, **268**
  eusporangiate, **453,** 453–55
  evergreen wood, **458**
  Filicales, **453,** 454, 455, 458–62
  filmy, **453**
  grape, 454
  heterosporous, 454, 462 (*see also* water ferns)
  homosporous, 430, **459** (see also *Filicales*)
  leptosporangiate, 453, **453,** 454, 455, 460, 462–63
  life cycle of, **460–61**
  maidenhair, 455, **458**
  ostrich, **458**
  seed, **456,** 457, **457**
  species of, 449
  sweet, 741
  tree, 453, **455,** 457
  walking, **180**
  water, **462,** 462–63
  whisk, 443
Ferocactus
  acanthodes, 809, **809**
  melocactiformis, **643**
ferredoxin (Fd), **136,** 138, **139**
Fertile Crescent, 825
fertilization, **169,** 170, **170,** 171, **171,** 180, **184,** 312 (*see also* pollination; sexual reproduction; *and specific plants*)
  in algae, 360, 372, 396, **396**
  in angiosperms, 509
  in bryophytes, 400, 402, 406, **410,** 414, **419**
  in chytrids, **313, 314**
  in cycads and *Ginkgo,* 473, 490
  double, 492, 509, **509**
    in gnetophytes, 492
  in fruit development, 679
  in fungi, 311–12
  in *Lycopodiaceae,* **437,** 438
  in oomycetes, 373, **374**
  in pines, **481,** 482
  in seedless vascular plants, 438, 442, 449
fertilization tubes, **373, 374**
fertilizers, 745, 838
  nitrogen, 738, 741–42, 745
fescue, tall, 335
Festuca arundinacea, 335
F_1, F_2 generations, 185
fibers, 572, 575, **575, 577**
  bast, 575
  gelatinous, 666
  phloem, 618, 650, 651
  xylem, 579
fiber tracheid, 576–77
fibrous proteins, 29
fibrous root system, 590, **591**
Ficus benghalensis, 603
fiddleheads, 455, **458** (*see also* leaves, of ferns)
  uncoiling of, 458
field capacity of soils, 734
field hypothesis of phyllotaxy, 624
figure, of wood, 667
filaments, (angiosperm), **499,** 501, **503**
  fused, 525
  of wind-pollinated flowers, 540, **541**
filaments (fungal), 310
Filicales, **453,** 454, 455, 458–62
  life cycle of, 459, **460–61**
filmy ferns, **453**
fimbriae, 284, **284**
fir, 483
  balsam, **483, 788**
  Douglas, 483, **669t**
  subalpine, 777, 791
  white, 788, **789**
fire blight, 293
fireflies, 698
fires
  dormancy of seeds and, 565, **565**
  forest, 788–91, **789–791**
  grasslands and, 811–12
  Mediterranean-type vegetation and, 817
  in taigas, 819
fireweed, **511,** 546, **792**
fir mosses, 438
first available space hypothesis, 624
fish, acid rain and, 616
Fissidens, **400**
fitness of an individual, 239
flagella, **44,** 52, **60,** 60–61, **61**
  of algae, **60, 349t, 372t,** 383, 387–93, 396, 398
  of chytrids, 312, 313
  of cryptomonads, 356
  of cycad sperm, **488**
  of diatoms, 376
  of dinoflagellates, 361
  emergent, **351**
  of euglenoids, 351, **351**
  of eukaryotes, 60–61
  of haptophytes, 366, **367**
  of heterokonts, 371
  nonemergent, **351**
  of oomycetes, 371
  of prokaryotes, 284
  of slime molds, **353,** 354
  structure of, 60, **61**
  tinsel, **60, 349t,** 371, **372t,** 376
  whiplash, **60,** 313, **349t,** 354, **371, 372t,**
flagellar roots, 386, 387, **387**
flavin adenine dinucleotide (FAD), 116, **116,** 119
flavin mononucleotide (FMN), **116, 117,** 118, **119**
flavins, 704
flavones, 35, **35**
flavonoids, 34–35, 542, 543, 549
flavonols, 35, 542–43, **543**
flax, 575, 825
Fleming, Sir Alexander, 334
flies, 529, **529, 531,** 535, **535**
floral apex, **638,** 639
floral stimulus (flowering hormone), 715
florets, **541**
Florey, Howard, 334
floridean starch, 358
florigen, 715, 716
flour, 563
flour corns, **828**
flowering
  cold and, 719
  daylength and (photoperiodism), **709,** 709–14
  genetic control of, 716
  hormonal control of, 715–16
  inhibitors and promoters of, 715, **716**
  phytochrome and, 712
flowering hormone (floral stimulus), 715
flowering plants (*see* angiosperms)
flowers, 495, 498–502 (*see also* inflorescences; pollination)
  bee, 532
  bisexual, 531
  bolting and, 686, **687**
  carpellate (or pistillate), 501, **511**
  colors of, 50, 531, 532, 539, 542–43, **543**
  complete, 502
  definition of, 498
  development of, **638,** 639–41
    floral organ identity, 639–40, **640,** 641
  dioecious, 501
  disk, **527,** 528
  double, 639
  epigynous, 502, **502,** 527
  of eudicots, **497, 498t, 502**
  evolution of, 524–43 (*see also* angiosperms, evolution of)
    animals as agents of, 530–39
    carpels, 525, 527
    evolutionary trends, 527
    most specialized flowers, 527–29
    perianth, 524
    pollination mechanisms, 525, 530–40
    sepals and petals, 524
    stamens, 524–25
  hormones and, 715–16
  hypogynous, **499,** 502, **502**
  imperfect (unisexual), 501
  incomplete, 502
  irregular (bilateral; zygomorphic), 502
  of magnoliids, 519, **520,** 521, 522
  of monocots, **497, 498t**
  monoecious, 501, **502**
  odors of, 531, 532, 535, 537–39
  parts of, 499, **499,** 501–2
  perfect (bisexual), 501
  perigynous, 502, **502**
  placentation of, 501, **501**
  regular (actinomorphic; radial), 502, 527
  senescence or wilting of, genetic engineering and, 698, **698**
  staminate, 501, **511**
  symmetry in, 502
  terminology, 499–502
  underground, 529, **529**
  unisexual, 501
  vasculature of, 524
flow of energy through ecosystems, 784–86
fluid-mosaic model of membranes, 74, **75**
fluorescence, 131
fluorides, as air pollutant, 616
fly agaric, **323**
fMet (*see* methionine)
FMN (flavin mononucleotide), **116, 117,** 118, **119**
Foeniculum vulgare, 833
foliose lichens, **336,** 338
follicles, 544
food (*see* nutrients, inorganic; nutrition, plant)
food bodies, 531
food chains, 783, 785
food-conducting cells, tissues, 8, 382, 402, 403, 425, 579–583 (*see also* leptoids; phloem)
food reserves, 20, 22, **349t, 372t** (*see also by type*)
food storage
  in fleshy roots, 606, **607**
  in leaves, 643
  in seeds, 470, 482, 510, 563, **564**
  in stems, 642–43
food transport (*see also* assimilate transport; leptoids; phloem)
  in bryophytes, 402, 403
  in vascular plants, 425, 764–69
food web, 783, **784**
foolish seedling disease, 684
foot
  of bryophytes, **405,** 406, **410, 413**
  of seedless vascular plants, 445
Foraminifera, 378
foreign genes, 696–99
forests
  carbon cycle and, 150
  coniferous, 816, 819
  fires in, 482, 788–91, **789–791**
  monsoon, 806, 807
  old-growth, 815
  rainforests, 489, **798–800, 803–5**
    temperate, 804
    tropical, **10, 803,** 803–5
  subtropical mixed, 806, **807**
  taiga, **818,** 818–19
  temperate
    deciduous, **10,** 812–14, **813, 814**
    mixed, 815–16, **816**
    pyramid of numbers for, **786**
  tropical mixed, 806
forestry, sustainable, 815
formulas, chemical, 870
Forsythia, **50**
fossil fuels, 150, 845
fossils, 238, **239**
  of algae, 348, 359
  of angiosperms, 470, **517,** 518, **518,** 519, 521, **521,** 522
  of bryophytes, 403
  earliest known, 2, **3**
  eukaryotic, 348
  of flowers, **517,** 519, **520–22**
  of fungi, 312
  "gaps" in fossil record, 257
  of gymnosperms, 457, 472, 488–89, **489**
  of lichens, 339
  of megaspore, **469**
  of *Metasequoia,* **485**
  of multicellular organisms, 6, **6**
  of mycorrhizae, 343, **344**
  of photosynthetic organisms, 5
  of pollen, 504, 506, 519
  of prokaryotes, 2, **3,** 282
  radiocarbon dating of, 852
  of seeds, 468
  of vascular plants, **424,** 427, 429, 432, 434
  of wood, 470
founder cells, 634
founder effect, 241–42, **242**
Fox, Sidney W., 4
foxglove, 34, **533**
Fragaria, 547
  ananassa, 679, **679**
  virginia, **762**
Franklin, Rosalind, 198, **198**
Fraxinus, **545, 755**
  americana, **669t**
  pennsylvanica var. *subintegerrima,* **627, 657**
free central placentation, 501, **501**
free energy, 864
free-energy change, 97
Friis, Else Marie, 522
Frisch, Karl von, 532
Fritschiella, **387,** 392, **392**
Frolova, I. A., 715
fronds, 455, 457, 458 (*see also* leaves, of ferns)
frost, reducing susceptibility to, 699
fructans, 20
fructose, 19, **19**
fructose 1,6-bisphosphate, **111, 113**
fructose 6-phosphate, 111, **111, 113**
fruit drop, 683
fruitlets, 543
fruits, 503, 543
  accessory, 543
  aggregate, 543
  auxin and formation of, 679
  climacteria, 683
  colors of, 50–51, 683
  definition of, 543
  dehiscent, 544, **544**
  development and growth of, 509–10, 686
  dispersal of 546–51 (*see also* seeds, dispersal of)
  dry, 544–45
  evolution of, 543–49
  fleshy, 543, 547–48, 683
  indehiscent, 544–45
  lenticels on, 654
  multiple, 543
  parthenocarpic, 543
  ripening of, 683, 697, **698**
    genetic engineering and, 697, **698**
  simple, 543, 544
Frullania, **409,** 412
frustules, 376, **376, 377,** 378
fruticose lichens, 338
Fucales, 379, 381, **381**
fuchsia, 537
fucoxanthin, 365, 367, 378, 381–82 (*see also* carotenoids; xanthophylls)
Fucus, **171,** 383, **385** (*see also* brown algae)
  life cycle of, **171, 385**
  vesiculosus, **379**
Funaria hygrometrica, **56, 420**
functional groups (chemical), 871, **871t**
functional phloem, 659
fundamental tissue system (*see* ground tissue system)
fungi, 5, **5,** 11, 306–46 (*see also* mushrooms; *specific types of fungi*)
  cell walls of, 310
  classification of, 277, 309
  coenocytic, 310, 313
  conidial, 332–34

as decomposers, 307
destructiveness of, 307–8
diseases caused by, 307–8, 313
endophytic, 335
evolutionary origin of, 277, 312
fossil, 312
as heterotrophic absorbers, 310
hyphae of, 310
importance of, 307–9
life cycles of, **314, 316, 318, 321, 328–329**
mitosis and meiosis in, 310–12
mycorrhizal (*see* mycorrhizae)
parasitic, 310, 317, 334, 339
pathogenic, 307–8, 313, 317
phototropism in, 315, **315**
predaceous, 333, **333**
reproduction in, 311–12
saprophytic, 310, 312, 313
spores of, 308, 311–13, 315, **315, 316,** 326
symbiotic relationships of, 308, 334–44 (*see also* lichens; mycorrhizae)
Fungi (kingdom), **265t**, 275, 277, **277,** 882
Fungi Imperfecti (*see* deuteromycetes)
funiculus, 506, **507,** 562
furrowing, of cell, 386–87, **387**
Fusarium, 780
fusiform initials, 648, 649, **649, 650,** 667
fynbos, 817

G

G_0, G_1, G_2 phases, 157, **157,** 158
GA_3 (gibberellic acid) 626, **674t**, 684, 689
galactans, **349t**
β-galactosidase, **218**
α-galacturonic acid, **22**
Galápagos Islands, 236, 237, **237**
Galápagos tortoises, 236, **237**
gametangia, 312, **316,** 317 (*see also* antheridia; archegonia)
of algae, 383, **384, 385**
of bryophytes, 402, **405**
of chytrids, **314**
of mosses, 416, **417,** 420
plurilocular, 383
of vascular plants, 402
gametes, 169, 170, **170, 171 372** (*see also* egg cell; sperm)
of algae, 360, 383, **384, 385,** 388–89, **389,** 394
allopolyploidy and, **252**
of angiosperms, 504, 506, **506**
autopolyploidy and, **252**
of bryophytes, 400
of chytrids, **314**
defined, 170
evolution of, 8
of fungi, 312, **313, 314**
Mendel's experiments with peas and, 184, 186, **186, 187,** 188, **189,** 190, **190**
nonmotile, **372**
number of chromosomes in, 47
of oomycetes, 371, 372
plasmodium formation and, 354
segregation of alleles in formation of, 186, **186**
of slime molds, 352, **353**
gametic meiosis, 171, **171**
gametophores, 408
gametophytes, **171,** 172 (*see also* embryo sac; megagametophytes; microgametophytes; pollen)
of algae, **359,** 360, 361, 383, **384**
of angiosperms, 431, 503 (*see also* pollen)
bisexual, 430, 431, **436,** 438, 444, **444,** 449, 449, **450**
of bryophytes, 403, 404, 425
of chytrids, **314**
of club mosses, 431, **436, 437,** 438
of cycads and *Ginkgo,* 473, **473,** 474
of *Equisetum,* 449, **449, 450**
evolution of, 430, 431
of ferns, 455, 458, 459, **459–461,** 462
of gymnosperms, 472–73
of heterosporous and homosporous plants, 430, 431
of hornworts, 412, **413**
of liverworts, 407, 408, **408–411**
of *Lycopodium,* **436–37**
of *Marchantia,* **410–11**
of mosses, 414, **414,** 416, **416, 418, 419,** 420, **420,** 421, **422**
and mycorrhizae, 438, 445
of pines, 472, 477, **477, 478**
of *Polypodium,* 460–61
of *Psilotum,* 444, **444,** 445, **446, 447**
of rhyniophytes, 434
of *Selaginella,* **440, 441,** 442
of vascular plants, 430–31
gametophytic self-incompatibility, 511
gamma rays, **129**
Ganoderma applanatum, 323
gap junctions, 87
garbanzos, 825
garden bean, **562,** 563, **566,** 567
Garner, 716
Garner, W. W., 709
gas exchange, lenticels and, 654
Gassner, Gustav, 719
gasteromycetes, 323, 325, 326, **326**
gas vesicles, 289
gating, 83
Geastrum saccatum, **326**
gel electrophoresis, 688
geminiviruses, 300, **300**
gemma cups, 408, **409**
gemmae, 404, 408, **409**
gene cloning (*see* DNA cloning)
gene expression
biological clock and, 708, **709**
chromosome condensation and, 220–21
in eukaryotes, 220–21
hormones and, 687, **687t,** 688, **688**
regulation of (*see* gene regulation)
RNA and, 208
specific binding proteins and, 221
temperature and, 197
gene flow, 241
gene monitoring, 698
gene mutations (point mutations), 192
gene pool, 239–40
genera, origin of, 256
General Sherman sequoia, **485**
generation, filial (F_1, F_2), 185–89
generative cells, 196
of angiosperms, 504, **505,** 508, **512**
of pines, 477, **477, 479,** 481
gene regulation, 214
in eukaryotes, 220–21
in prokaryotes, 216
genes, 31, 184, 197 (*see also* alleles; chromosomes; DNA)
cell differentiation and, 214, 572
coding sequences of, 222, 688, **688**
dominant/recessive, 185–87
embryogenesis and, 560, **561**
epistasis of, 195
eukaryotic, 218, 221
expression (*see* gene expression)
foreign, 696–99
frequency, 239–40
homeotic, 639
interactions among alleles of different, 195
linkage of, 190–91, **191**
movable (transposons), 193
noncoding sequences, 221, 222
pleiotropic, 195
prokaryotic, 210, 216–18
regulator, 216, **217**
regulatory sequences of, 688, **688**
reporter, 698–99
sequencing, 229–30
structural, 216, **217**
traits controlled by several, 195
transfer of, 696–99
genetic code, **209,** 209–10
as nearly identical in all organisms, 210
genetic drift, 241–42
genetic engineering, 13, 696–99 (*see also* recombinant DNA)
Agrobacterium tumefaciens and, **696,** 696–97
benefits and risks of, 699
food quality improved by, 698
hormone production and, 697–98
nitrogen-fixing bacteria and, 747
genetic isolation (reproductive isolation), 248, 256
genetic linkage maps (*see* linkage maps)
genetic recombination, 172
genetics (heredity), 11, 183–204 (*see also* alleles; DNA; genes; genotypes; phenotypes)
chemical basis of, 197
continuous variation, 195, **195**
Darwin's theory of evolution and, 238
epistasis, 195
incomplete dominance, 194, **194**
linkage of genes and, 190–91, **191**
Mendel's experiments in (*see* Mendelian genetics)
mutations, 192–94
polygenic inheritance, 195
genetic self-incompatibility, 510–11
genetic variability (genetic diversity), 239
diploidy and, 243
meiosis and, 170, 177–79
natural selection and, 243–45
preservation and promotion of, 242–43
as protection against pathogens, 840–41
sexual reproduction and, 242
genome, 53, 196, 297
genomic libraries, 226
Genomosperma
kidstonii, **469**
latens, **469**
genotypes, 186, 197
genus (genera), 262
geologic eras, inside front cover
geotropism (*see* gravitropism)
germination, 54
of coconuts, **544**
cold treatment and, 717
dormancy and, 565, 717–18
epigeous, **566,** 567, **567**
in eudicots, 566
of fungal spores, 317, 330
hormones and, 684, 685
hypogeous, **566,** 567, **567**
imbibition and, 80, **80**
light and, 564, 711–12, **712,** 717
photoperiodism (phytochrome) and, 711–12, **712, 713**
requirements for, 564–65, 717–18
of soybeans, **512**
of sporocarps, **462**
temperature and, 564–65, 717
water and, 564–65
giant cell, 356, **356**
Gibberella
acuminata, **310**
fujikuroi, 684
gibberellic acid (GA_3), 626, **674t,** 684, 689
gibberellins, 560, **674t,** 684–87, **685, 686**
bolting and, 686, **687**
commercial uses of, 686–87
dormancy and, 685–86, **686,** 719
dwarf mutants and, 684–85, **685**
flowering and, 715, 719
fruit development and, 686, **687**
molecular basis of action of, 688, 690
orientation of cellulose microfibrils and, 690, **690**
in roots, 590
in seeds, 685
in sex expression, 683
structure of, 685
Gibbs, Josiah Willard, 97
gill fungi, 325
gills of mushrooms, 324, **324**
ginger, **762,** 832
Ginkgo biloba, 60, **473,** 488, **488, 490,** 490
Ginkgophyta, 60, **275t,** 470, 472, 473, **473,** 488, **488, 490,** 490, 887
Givnish, Thomas, 778
glades, 811, **811**
gladiolus, 643, **644**
Gladiolus grandiflorus, **644**
gleba, 326
Gleditsia triacanthos, **627**
GL1 (*GLABROUS1*) gene, 585–86
global ecology, 796–822 (*see also* biomes)
global warming, 150–51, 415, 845
peatland carbon and, 415
globular proteins, 29
globular stage, 557, **560**
Glomus, **308t,** 317
etunicatum, **341**
versiforme, **341**
glucose, 19, **19,** 20, 22
fermentation of, 123
oxidation of, 99, 109, **110, 121** (*see also* glycolysis)
as product of photosynthesis, 128, 142
glucose 6-phosphate, 111, **111, 113**
glumes, **541**
glutamate (glutamic acid), **27, 28**
glutamine, **27**
glutamine synthetase–glutamate synthase pathway, 742, **742**
gluten, **835**
glyceraldehyde, **19**
glyceraldehyde 3-phosphate (3-phosphoglyceraldehyde; PGAL), 111–13, **111–13,** 140–42, **140, 141**
glycerol, **38t**
glycine, **28**
Glycine max, **51,** 698, 715, 738, 739, **740,** 826, **826**
life cycle, **512–13**
glycocalyx, 284
glycogen, 20, 22, **38t,** 284
glycolipids, **38t,** 75
glycolysis, 108, 109–13, **110,** 122
anaerobic, 122, 123, **123**
steps of, 111–12, **111–12**
summary of, 112–13, **113**
glycoproteins, 58, 63, 74
glycosides, **549,** 549–50
glycosidic bond, **20**
glyoxysomes, 54
glyphosate, resistance to, 697
Gnetales, **519**
Gnetophyta (gnetophytes), **275t,** 470, 472–74, 490–92, **491,** 492, **492,** 518, 887
relationship to angiosperms, 492, 509
Gnetum, 490, **491,** 492, 518, **519**
double fertilization in, 492, 509
goat's beard *(Tragopogon),* 253, **253**
goldenrods, **762,** 778, **778**
goldeye lichen, **337**
Golgi complex (Golgi bodies), **57,** 57–58, **71t**
Gondwanaland, 488, 523, **523**
Gonium, **390**
gonorrhea, 287
Gonyaulax
polyedra, **363, 708**
tamarensis, 364
goosefoot, 543
Gossypium, **525,** 828, **828**
Gould, Stephen Jay, 257
Gowen cypress, **483**
gradualism model of evolution, 257, 258
grain of wood, 667
grains (cereal), 195, **564, 568**
Gram, Hans Christian, 283
gram-negative and gram-positive bacteria, 283, **283**
grana (grana thylakoids), 49, **50, 126, 132**
granite mosses *(Andreaeidae),* 412, 416
grape ferns, 454
grapes, 373, 825
gibberellic acid and, 686–87, **687**
tendrils of, 641, **642**
grapevine, **672**
grasses, 253, 256, 511 (*see also names of specific grasses*)
Agrostis tenuis, 244–45, **245**
C_3 and C_4, 146, **147**
caryopsis of, 545
eel, 541, **542**
embryos of, 563
epidermis of, 632, **633**
fungal endophytes in, 335

Kentucky bluegrass *(Poa pratensis)*, 146, 244, 255–56
leaves of, 632, **633**
orchard, 244
wind-pollinated flowers of, **540, 541**
grasslands, **798–99**, 810–12, **811, 812**
biome map of, 798–99
pyramid of numbers for, **786**
gravitropism, 591, 704–5, **705, 723t**
grazing mammals, 812, 826
Great Basin Desert, 807, **808**
green algae *(Chlorophyta)*, **171, 275t**, 348, 349, 371, **372t**, 383–98, **386, 402**, 873–74 (*see also* algae; *Charophyceae* (charophytes); *Chlorophyceae* (chlorophytes); *Ulvophyceae* (ulvophytes))
cell division in, 386–87, 387
classification of, 383, 386
cytoplasmic streaming in, 47, **47**
economic uses of, 380
as endosymbionts, 274, **274**
food reserves in, **372t**, 383
life cycles of, **389, 394**
relationship to plants, 383, 396–98, 401–2
sitosterol in, 25, **25**
green ash, **627, 657**
green bacteria, 291
greenhouse effect, 13, 150 (*see also* global warming)
greenhouse gases, 844–45
Green Revolution, 839, 847
green snow, **386**
Griffonia simplicifolia, **545**
ground meristem, 557, **557, 570, 571** (*see also* primary meristems)
of shoots, **611**, 613, 614
ground tissue system, 426, **427**, 557, **557, 571**, 572–73, **573**, 573–76, **586t**
growth, 571 (*see also* meristems)
of bud, **611**, 612, **613**, 681, 718–19
deficiency symptoms of, **730t**
of embryo, 556–60
hormones and (*see* hormones)
increments (rings), 664–66
indeterminate, 571
intercalary, 634
nutritional requirements for (*see* nutrition, plant)
phytochrome and, 712–14
primary, 8, 426–27, 571
in stems, 611–14
regulators of, **675t** *(see also by name)*
in roots, 591–95
secondary (*see* secondary growth)
in stems, 611–14
thigmomorphogenesis and, 721–22
growth rings, 664–66, **666**
Grypania, 348
guanine, 198, **198**, 199, **200**
guanosine triphosphate (GTP), 212
guard cells, 583, **584**, 584–85, **585**
radial orientation of cellulose microfibrils in, 752
stomatal movement and, 692, **692**, 693
turgor pressure in, 752
guayule, 842, **842**
GUS gene, 698
guttation, 761, **761**
Guzmán, Rafael, 829
Gymnodinium
breve, 364, **364**
catenella, 364
costatum, **363**
neglectum, **363**
gymnosperms, 467–94 (*see also* conifers; cycads; *Ephedra; Ginkgophyta; Gnetophyta; Welwitschia;* pines)
evolution of, 431
extinct, 472
living, 472–74
main features of, **493t**
phylogenetic relationships with other embryophytes, **472**
gynoecium, **499**, 501
gypsy moths, 782

H

Habenaria elegans, **535**
Haberlandt, Gottlieb, 695
hadrom, 416
Haeckel, Ernst, **264**
Haemanthus katherinae, **161**
Hagemann, Wolfgang, 65
hairs (*see also* trichomes)
glandular (secretory), 585
leaf, 34, **140**, 250, **585**, 585–86, **586**, 627, **628, 634**
root (*see* root hairs)
Hale-Bopp comet, 4, **4**
Hales, Stephen, 751
half-life of radioisotopes, 864
Halimeda, 395, **395**
hallucinogenic compounds, **551**
Halobacterium halobium, 295
halophilic archaea, extreme, 295
halophytes, 744
Hamamelis, 546
Hamner, Karl C., 711, 715
Hantaan virus, 296
haploid cells, **171**, 172, 173, 175, *175*, 177, 179, 331, 332 (*see also* meiosis)
haploid generation, 172, 279 (*see also* gametophytes; sporophytes)
haploidization, in deuteromycetes, 333
haploid number, 47, **169**, 170 (*see also* meiosis)
haploid organisms, 171, 243
haploid phase of life cycle, **171**
haploid spores, **313**, 314, 316, 318, **324**, 329, 330, 354
haplophase, **353**
Haplopappus gracilis, 47
haptonema, 366, **367**
haptophyte algae *(Haptophyta)*, **275t**, 348, 349, **349t**, 366–67, **367**, 885 (*see also* algae)
hardwoods, 659 *(see also by name)*
Hardy, G. H., 239, 240
Hardy-Weinberg law (equilibrium), 240–41, 877–78
hares, snowshoe, 782
Harpagophytum, **548**
Hartig net, 342, **343**
Hatch, M. D., 144
Hatch-Slack pathway (C4 pathway), 144, **144, 145**, 146, 147, **148**, 149
in grasses, 632, **633**
haustoria, 310, **311**, 339, **339**, 498
Hawaiian Islands, 250–51, **250–51**
hawkmoths, 536, **536**
Hawksworth, David L., 339
hawthorn, **643**
hay fever, 508
hazelnuts, 545
heads (flower), **499**
heart disease, 287
heart stage, 557, **557**
heartwood, **655**, 665, 667
heat content, 864
heather family, mycorrhizae in, 343
Hedera helix, 603
helanalin, 780
Helenium, 780
Helianthus, 255, **255**
annuus, 255, **255, 527, 722**, 831, **831**
anomalus, 255, **255**
petiolaris, 255, **255**
helical phyllotaxy, 624
helicases, 201
Helicobacter pylori, 287
Heliconia irrasa, **803**
heliotropism, **722**, 722–23, **723t**
heliotropsim, **723t**
helix
alpha, 28, **28**, 29
double, **198–200**, 201–2
heme, **117**
hemicellulose cross bridges, **689**
hemicelluloses, 63, **62**, 66
Hemitrichia serpula, **353**
hemlock, 483
eastern, **669t**
water, **500**
western, **342**
hemorrhagic fevers, 296
hemp, 575
henbane, 710, **710**, 711, 719
hepatica, **497**, 531
Hepatica americana, **497, 531**
hepatics (*see* liverworts)
Hepatophyta (*see* liverworts)
hepatitus, 296
herbaceous plants, 619–21, 648, 812–13, 814
Herbaspirillum
rubrisubalbicans, 747
seropedicae, 747
herbicides, 679, 697, 781
herbivores, 783 (*see also* animals; mammals)
interactions between plants and, 779–82, **780**
herbs, 832, 833
heredity (*see* genetics; inheritance)
Hericium coralloides, **323**
herpes, 296
Herpothallon sanguineum, **277**
heterochromatin, 220–21
heterocysts, 290, **290**
heteroecious, 327
heterokaryosis, 333
heterokonts, 371
heteromorphic generations, alternation of, 172
Heterosigma, 379
heterospory, 431, 442, 443, 468
in angiosperms, 503
in ferns, 454, 462
in gymnosperms, 503
in *Isoetes*, 443
in progymnosperms, 471
in *Selaginella*, 442
heterothallic organisms, 317, 373
heterotrophs, 5, 124, 286, 783
heterozygotes, 242–43
heterozygous, 186, 187
Hevea, 834, 842
brasiliensis, 34, **34**
hexokinase, **101**, 111, **111**
hexoses, 18
hickory
bitternut, **669t**
shagbark, **554, 627, 647**
hierarchical taxonomic categories, 264
Hiesey, William, 246
Hildebrandt, A. C., 695
Hill, Robin, 128
Hill reaction, 128
hilum, 563
*Hind*III, **225**
histidine, 26, **27**
histones, 41, 158, 219–21
Hodgkin's disease, **846**
Holcus lanatus, **762**
holdfast, 382
homeotic genes, 639
homeotic mutations, 639, 640, **641**
homologous chromosomes (homologs), **170**, 170–77, **173**
crossing-over and, 172, 191, **191**
linkage of genes and, 190, 191, **191**
in meiosis I, 174–77
in meiosis II, 177
homology (homologous features), 266
Homo sapiens, 824
homospory, 430
in *Equisetum*, 448
in ferns, 454, 458
in *Lycopodium*, 438
in *Psilotum*, 444
homothallic organisms, 317, 373
homozygotes, 242–43
homozygous, 186, 187, **189**, 190, **190**
honey guides, 532, **533**
honey locust, **627**
honeysuckle, 624
chaparral, **524, 547**
Hoodia, **266**
hook
fungal, 320
seedling, 567
Hooke, Robert, **40**, 41
Hordeum
jubatum, **278**
vulgare, 563, **635**, 685, 825, **825**
horizons, soil, 731–32, **732**
hormogonia, 288
hormones, 86, 673–701, **674t** (*see also specific hormones*)
cellular processes and, **687t**
defined, 674
flowering and, 715–16
genetic engineering and production of, 697–98
groups of, 675
molecular basis of action of, 687–93
cell expansion, **689**, 689–90
gene expression, 687, 688, **688**
receptors, 690–91, **691**
second messengers, 691, **691, 692**
stomatal movement, **692**, 692–93
sensitivity to, 675
in tissue culture, 680–81, **681**, 693
hornworts *(Anthocerophyta)*, **275t**, 400, 407, 412, 526, **526**, 886 (*see also* bryophytes)
comparative structure of, 403, 404
horse chestnut, **505, 613**, 648, **657**
horsetails, **57**, 255, **256, 275t, 278**, 680, **728**, 875 (see also *Equisetum;* seedless vascular plants; *Sphenophyta*)
giant (calamites), 456–57, **466**
host cell, ancestral, 272
Hubbard Brook Experimental Forest, 782–83, **783**
human(s)
and agriculture, 824–25
culture and society, 824, 826, 836–37
evolution of, 11–12, 824
impact on environment, 824, 826, 832, 837, 844–45
impact on nutrient cycles, 743–45
Industrial Revolution and, 836
populations, 835–37
Humboldt, Alexander von, 802, **802**
humidity, transpiration and, 754
hummingbirds, 537, **537, 538**
humus, 732, **732**
hunger and poverty, 847
hunter-gatherers, **823**, 824–25
Huperzia, 438
lucidula, **438**
Hutchinson, G. E., 774
hyacinths, water, **787**
hybridization, 243, 252, 253, 838–39, 847
recombination speciation and, 255
somatic, 693, 695
hybridization, nucleic acid, 230
hybrids, 243
fertile, 253
of maize, 243, 829, 839
sterile, 252, 255–56
of wheat, 254
hybrid vigor, 243
hydathodes, 761, **761**
hydraulic lift, 761, **762**
Hydrodictyon, 391, **391**
reticulatum, **391**
hydrogen (H), 3, **3**, 18
as essential element, 727
isotopes of, 864, **865t**
hydrogen bonds, 873, **873t**, 874
hydrogen sulfide (H_2S), **127**, 128
hydroids, 416, **417**, 420
hydrolysis, 20, **20**, 871
hydrolytic enzymes, gibberellins and, 685, **686**
hydrophilic groups, molecules, 18, 874
hydrophobic interactions, 874
hydrophobic molecules, 22, 874
hydrophytes, leaves of, 626, 627, **629**
hydroponics, 693
hydrostatic pressure, 77, 80 (*see also* turgor pressure)
hydrostatic pressure hypothesis, 705
hydrotropism, 706, **723t**
hydroxyl groups (—OH), 18, 859

Hygrocybe aurantiosplendens, **309**
hymenium (hymenial layer), 319, **321,** 323, **323,** 324
hymenomycetes, **320,** 323, **323,** 324 (see also *Basidiomycota*)
Hymenophyllum tunbrigense, **453**
Hyoscyamus niger, 710, **710,** 719
hypanthium, 502, **502, 503**
hyperaccumulators, 746–47
hypertonic solutions, 79
hyphae, 310 (*see also* conidiophores; haustoria; rhizoids)
  of ascomycetes, 318, 320
  coenocytic, 310, 313, 315
  dikaryotic, monokaryotic, **322**
  of predaceous fungi, 333, **333**
  receptive, 327
  septate, 310, 318, 322
hypocotyl, **157, 556, 557, 566**
hypocotyl-root axis, 482, **482, 513,** 562, 563, **637**
hypodermis of pine leaves (needles), 474
hypogeous germination, **566,** 567, **567**
hypogynous flowers, 502, **502**
hypotonic solutions, 79

I

IAA (indoleacetic acid), 676, **676,** 681, **681,** 695, 703
i^6Ade (isopentenyladenine), **680**
ice desert, **798–99,** 821
  biome map of, **798–99**
ice plant, 644
igneous rocks, 731
Iltis, Hugh, 829
imbibition (hydration), 80, 81, 565
immunofluorescence microscopy, 159
Impatiens, 546
inbreeding, 242
incense cedar, 788
incomplete dominance, 194, **194**
indehiscent fruits, 544–45
independent assortment, principle of, 188–89, **189,** 190
indeterminate growth, 571
Indian pipe, 498, **498**
indoleacetic acid (*see* IAA)
induced-fit hypothesis, 101, **101**
inducers, 217
induction in signal transduction, 87, **87**
indusia, 458, **458, 459**
Industrial Revolution, 836
infanticide, 835
infection (*see* diseases)
infection threads, 739
inflorescences (flower clusters), **192, 253, 293,** 491, 499, **499, 501** (*see also* flowers)
  of wind-pollinated flowers, 540
influenza, 296
infared radiation, **129**
Ingenhousz, Jan, 127
inheritance, 195–96 (*see also* genetics)
initials, 157, 453, 571 (*see also* apical meristems; cambium)
  defined, 157
  fusiform, 648, 649, **649, 650,** 667
  ray, 648, 649, **650**
initiation codons, 212
initiation stage of translation, 212, **212, 213**
inky cap, **321**
inorganic nutrients (*see* essential elements)
insects (*see also* pesticides; *names of specific insects*)
  biochemical coevolution and, 549–51
  carnivorous plants and, 642, **737**
  juvenile hormones and, 780
  pollination by, **530,** 530–35
    bees, **532,** 532–35, **533, 534**
    beetles, 531, **531**
    in bisexual flowers, 531
    in cycads, 487
    evolution of angiosperms and, 525
    in gnetophytes, 492, **492**
    mosquitoes and flies, 535, **535**
    moths and butterflies, 535–36, **536**
    nectar, 530
    wind pollination and, 539
  resistance to insecticides, 781
  secondary plant products and, 549, 550
  symbiotic relationships between fungi and, 308
  trichomes as defense against, 585
  viruses transmitted by, 298
integral proteins, 74
integuments, 470, 507
  of angiosperms, 506, **507,** 509, **513**
  evolution of, 468, **469**
  of pines, 481, 482
interactions between organisms, 774–82
  competition, 776–79
  mutualism, 774–75
  phytoalexins and, 780, 782
  plant-herbivore and plant-pathogen, 779–82, **780**
  tannins and, 782
intercalary growth, 634
intercalary meristems, 614
intercellular spaces, **573**
  ethylene and, 682
  in leaves, **629,** 630, 631, 632, 634
  in root cortex, 598
  transpiration and, 752
interfascicular cambium, 650
interfascicular parenchyma, 614, 619
interlocked grain, 667
intermediate filament, 58
International Code of Botanical Nomenclature, 262, 264
International Panel on Climate Change, 845
International Rice Research Institute, **827**
internodal elongation, 613, **613,** 614, 618, 682
internodes, **8,** 445, **450,** 568, **610,** 611, **611,** 613
interphase
  meiotic, 172
  mitotic, 157, **157,** 158, **165**
interspersed repeated DNA, 222
intine, 504
introns, 222–23, **222, 223**
inversions, chromosomal, 193
ionic interactions (ionic bonds), 868, **868**
ion(s), 868 (*see also specific elements*)
  active transport of, 82–84
  channels, 83
  as cofactors, 102–3
  functions of, 729, **730t**
  phloem-immobile, phloem mobile, 729
  transport in plants, 729, 762–63, **763,** 764
ipecac, 846
Ipomoea batatas, 606, **607,** 831
Iridaceae (irises), 833
Irish moss, **358,** 412
Irish potato famine (1846–1847), 373, 841
iron (Fe)
  as essential element, 727, **730t**
  in electron carriers, 116
  oxide, 732
iron-sulfur proteins, 116, **117**
ironwood, black, 670
irrigation, 837, **837, 843**
isidia, 338
islands, speciation and, 249, **249**
isocitrate (isocitric acid), 115, **115**
Isoetes (quillworts), **23, 55,** 149, 442–43, **443**
  muricata, 23, **55**
  storkii, **442**
Isoetaceae, 442–43, **463t**
isogametes (isogamy), 312, **372, 394**
isolation, reproductive (genetic), 248, 256
isoleucine, 26, **27**
isomers, 858
isomorphic generations, alternation of, 172
isopentenyl adenine (i^6Ade), **680**
isoprene, 33, **33**
isotonic solutions, 79
isotopes, 851–53
isozymes, 103
ivy, 603
  poison, **550**
  Swedish, **585**

J

jacket layer (sterile), 402, **405, 461**
Jacob, François, 216
Jaffe, M. J., 706, 721
Janzen, Daniel, 775
jasmonates, 675, **675t**
Jefferson, Thomas, 802
Jeffrey pines, **244**
jelly fungi, 323, **324**
jojoba, **841,** 841–42
Joshua tree, 808
Juglans
  cinerea, **657**
  nigra, **669t,** 779
junipers (*Juniperus*), 483
  communis (common), 484, **484**
  osteosperma, **810**
juniper savannas, **798–99,** 810, **810**
Jurassic period, 483, 488, 877
jute, 575

K

Kalanchoë daigremontiana, **180**
kale, **238**
kaolinite, 731
kapok, 539
Kaplan, Donald, 65
karyogamy, 312, **318, 321, 353, 389**
  in *Chlamydomonas*, 389, **389**
  in mushrooms, **321**
  in plasmodial slime molds, **353**
  in wheat rust, **328–29**
Keck, David, 246
kelps, **7, 370,** 371, 379, **380,** 382–83 (*see also* brown algae)
  commercial uses of, 380, 382
  internal structure of, 382
kelvins (K), 876
Kentucky bluegrass (*Poa pratensis*), 146, 244
  apomixis in, 255–56
kernel (grain), 195, **564, 568**
α-ketoglutarate (α-ketoglutaric acid), **115**
kinases, 106
  protein, 691
kinetin, 680, **680,** 681, **681,** 695
kinetochore microtubules, 161, 162, **162,** 163
  meiotic, 176
kinetochores, **160,** 161, 162
King Clone, **664, 808**
kingdoms, taxonomic, 264, 271
  in the domain *Eukarya*, **275t,** 275–79
Klebsormidium, **387**
Knop, W., 693
knots in wood, 668, **668**
kohlrabi, **238,** 643, **644**
kombu, 380
Krakatau, 789
Kranz anatomy, 144, 632
Krebs, Sir Hans, 115
Krebs cycle, 108, 109, **110,** 114, 115, **115,** 122, 124
Küchler, A. W., **798**
Kuehneola uredinis, **327**
Kurosawa, E., 684

L

Labiatae, 624
Labrador tea, 819
lactate (lactic acid), 122–123
lactate fermentation, 122, 123
lactose, 217, 218
lactose (*lac*) operon, 218, **218**
Lactuca sativa, **164, 635,** 711
lacZ gene, 226
lady's mantle, **761**
lady's slipper, yellow, **260**
lagging strands, 202, 203
Laguncularia racemosa, 605, **605**
Lamarck, Jean Baptiste de, 236
Lamiaceae, 833
lamina, 429, 624
Laminaria, 379, **379,** 381, 382 (*see also* kelps)
  life cycle of, **384**
laminarin, **372t,** 381
land, transition of plants to, 7–11
land-form leaves, 626, **626**
landing platform, on bee flowers, 532, **533,** 536
Lang, Anton, 715
larches, 483, **818**
  European, **483**
Larix, 483, **818**
  decidua, **483**
Larrea divaricata, **664, 808**
late blight of potatoes, 373, **375**
lateral (axillary) buds, **657,** 678, **678**
lateral meristems, 8, 648 (*see also* cork cambium; vascular cambium)
lateral roots (branch roots), 590, **592,** 603, **603**
late wood, 665
latex, 34
Lathyrus odoratus, 190–91
Latin names, 262–63
latitude, distribution of organisms and, 800–801, **801,** 802
Lauraceae (laurel family), 519, 833
Laurus nobilis, 833
LDA (lysergic acid amide), 335
leaching, 735, 738
lead (P^6), 244–45, 616
leading strands, 202, 203
leaf axil, **8,** 611, 626, 642
leaf blade, 624, **625,** 626, **628,** 635, **635**
leaf buttress, 634, **634**
leaf gaps, **428,** 429, **429**
leafhoppers, 298
leaflets, **8,** 626
leaf primordia, 611, **611, 614,** 622, 634, **634,** 635, **635**
  phyllotaxy and, 624
leaf scars, 636, **657**
leaf sheath, 624, **625**
leaf spots, **294**
leaf stalk (*see* petiole)
leaf-stem relation, 622–24
leaf trace gaps, 429, 622
leaf traces, 429, 622, **623, 624**
leaves, 7, 426 (*see also* cotyledons; shoots; *specific plants*)
  abscission of, 636, 683
  air pollution and, 616
  of angiosperms, 624–36
  of aquatic plants, 626, **626**
  of bryophytes, **414,** 415, 416
  of CAM plants, 149
  coloration, 543
  composting, 746
  compound, 626, **627**
  of cordaites, 457
  of C_3, C_4 plants, 144, **145,** 146, 632
  of cycads, 486
  of desert plants, **266,** 644
  development of, 632–35, **634, 635**
  dimorphic, **626**
  of *Ephedra*, **491**
  epidermis of, 583–86, **584–86,** 627, **628, 629,** 630, 632, **633, 634**
  of *Equisetum*, 448, **448**
  of eudicots, 624, **625,** 626, 627, 631
  evolution of, **428,** 429–30
  of ferns, 454, **454,** 455, 458, 462, **462,** 463 (*see also* fronds)
  food storage in, 643
  function of, 611, 630
  gas exchange by, 144, 627, 692, 752 (*see also* stomata)
  of *Gnetum*, **491**
  of *Ginkgo*, 488, 490, **490**
  of grasses, 632, **633**
  hairs (trichomes) of, 34, **140,** 250, **585,** 585–86, **586,** 627, **628, 634**
  heliotropism of, 722–23
  of hydrophytes, 626, 627
  of *Isoetes*, 442, **442–43**
  of larch, **483**
  of *Lycophyta*, 438, **438**
  of Magnoliids, 624, 626, 627, **629,** 631
  of mesophytes, 626, 630
  modifications of, 641–44, **644**

of monocots, 624, **625,** 631, 635
morphology of, 624–26
movements of, 707, 719–23
phyllotaxy of (arrangement on stem), 624
of pines (needles), 474, **475,** 476
scalelike, 643
seed (*see* cotyledons)
of *Selaginella,* 439, **439**
senescence of, 681–82, 698
sessile, 624, **625**
simple, **625,** 626
stomata on, **584,** 584–85, **585,** 626, 627, **628, 629,** 630, **630**
structure of, 626–32
succulent, 644, 807
sun and shade, 636, **636**
vascular tissues of stems and, 622–24
veins of (*see* veins)
water storage in, 644
of *Welwitschia,* **492**
of *Wollemia nobilis,* 488, **489**
of xerophytes, 626, 627, **629,** 631
of yews, **484**
lectins, 75, 740
leghemoglobin, 740
Legionnaires' disease, 287
legumes, 511, 544, **545,** 641, 825
cultivation of, 825–26, **826,** 827, 828
nitrogen-fixing bacteria and, 738–41
Leguminosae, 264
lemma, **541**
Lemna
gibba, **496,** 776, **776**
polyrhiza, 776, **776**
Lemnaceae, **496**
lenticels, 586, 602, **653,** 654, **654, 657, 658**
Lentinula edodes, 325
Lepidodendron, **456, 466**
leptoids, 416, **417,** 420
leptom, 416
Leptonycteris curasoae, **539**
leptosporangia (*see* sporangia, of ferns)
leptosporangiate ferns, 453, **453,** 454, 455, 460, 462–63
Les, Donald, 526
lethal yellowing of coconut, 293, **293**
lettuce, **164, 635,** 711–12, **712**
leucine, **27**
leucoplasts, 48, 51, **51, 70t** (*see also* plastids)
leukemia, **846**
lianas, **803,** 804, **804**
Liatris
punctata, **591**
pycnostachya, **793**
lichen acids, 339–40
lichens, 334–40, 883
algal or cyanobacterial component of, 340, 334–35, **338,** 339
commercial uses of, 340
crustose, **336,** 338
ecological importance of, 339–40
evolution of, 334
foliose, **336,** 338
fruticose, **337,** 338
fungal component of, 334, 338, 339
marine, 336
mycobiont-photobiont relationship in, 339, **339**
pollution and, 340
reproduction in, 338
structure of, 338, **338**
survival of, 338–39
water content of, 338–39
weathering of rocks and, 786
Licmophora flagellata, **375**
lid (*see* operculum)
life cycles, 172 (*see also specific groups or organisms*)
types of, 171, **171**
light (*see also* energy; phototropism)
absorption of, 130–33 (*see also* pigments)
competition for, 776–78
flowering and, **709,** 709–14
leaf development and, 636
nature of, 128–30, **129, 130**
in photosynthesis, 2, 130–39
plant responses to, 709–14
plastid formation and, 51–52, **52**
red, far-red, **129,** 711–14
in seed germination, 564, 711–12, **712,** 717
stomatal movement and, 753–54
ultraviolet (ultraviolet radiation), 6, 12, 129, 192, 203
bees' perception of, 532, **533,** 543
carotenoids and, 543
flavonoids and, 35, 542
visible, 129, **129,** 130
wavelengths of, 129, **129**
light-independent reactions, 134 (*see also* carbon-fixation reactions)
light reactions (light-dependent reactions), 133–39
lignification, 36
lignins, 36–37, 63, 428, 572
evolution of plants and, 425
ligule, 439, **440, 443, 625**
lilac, **570, 628**
Liliaceae, **196, 530**
lilies
African blood, **161**
Easter, **511**
parts of, **499**
pollen grains of, **505**
voodoo, 37
water, 521, **575, 629, 787**
Lilium, **176, 504, 505,** 506, **506**
double fertilization in, 509, **509**
henryi, **499**
longiflorum, **505, 511**
limber pine, 482
limestone (calcium carbonate), 150, 731
Linaria vulgaris, **500,** 546
linden, **575, 577, 762** (*see also* basswood)
Lindera benzoin, **762**
linkage maps, 191, **191**
linked genes, 190–91, **191**
Linnaeus, Carl, 262, **262**
linolenic acid, **23**
Linum usitatissimum, 825
lip cells, 454
lip of flower, 528, **528,** 529
lipid bilayer, 24, **24,** 74, **75,** 76
lipids, 18, 22–26, **38t** (*see also* fats; glycolipids; oils; phospholipids; steroids; *and specific types of lipids*)
lipopolysaccharide, 283
Liriodendron, 519
tulipifera, **662, 669t**
Lithocarpus densiflora, **502**
liverworts (*Hepatophyta*), **275t,** 400, 407–12, 886 (*see also* bryophytes)
comparative structure of, 403
leafy, **409,** 409–12, **412**
life cycle of, **410–11**
sporophytes of, 406
thalloid, 407–8, **408**
loam soils, 732, 735, **735**
Lobaria verrucosa, **338**
locules, 501
locus (on chromosome), 186, 190, **191**
locust
black, **627, 669t**
honey, **627**
lodgepole pine, **481,** 482, 791
lodicules, **541**
London plane tree, 248
long-day plants, 710, **710,** 711
Longistigma caryae, **767**
long shoot, **483,** 488
Lonicera, 624
hispidula, **524, 547**
looped domains, 220
loose smut of oats, 330
Lophophora williamsii, **551**
lorica, **351**
Los Angeles, smog in, 616
lotus, **524**
sacred, 718
luciferase, 698, **699,** 708, **709**
luciferin, 698–99
lumen of endoplasmic reticulum, 56
lupine, **500, 722,** 830
arctic, 718
white, 764
Lupinus
albus, 764
arcticus, 718
arizonicus, **722**
diffusus, **500**
Lycaena gorgon, **536**
Lycophyta (lycophytes), **275t, 426,** 429–31, **432,** 435–43, **436t,** 887 (*see also* club mosses)
club mosses (*Lycopodiaceae*), 435–38
Isoetaceae, 442–43
Isoetes, 442, **442,** 443
life cycles of, **436–37, 440–41**
Lycopodium, **436–37,** 438, **438**
Selaginellaceae, 439–42
lycophyte trees, 435, 456, **456,** 457, **466**
Lycopodiaceae (*see* club mosses)
Lycopodiella, 438
Lycopodium, 435, 438 (*see also* club mosses; *Lycophyta*)
lagopus, **427, 436,** 438, **438**
Lyell, Charles, 238
Lygodium, 453, **453**
Lymantria dispar, 782
lymphoma, **846**
lysergic acid amide (LDA), 335
lysine, 26, **27**
lysosomes, 55

M

McClintock, Barbara, 193, 203
mace, **832**
MacMillan, J., 684
Macrocystis, 380, 382
integrifolia, 382
pyrifera, **380**
macrocysts, 356, **356**
macroevolution, 257
macrofibrils, **63**
macromolecules, 18
macronutrients, 727, **728t, 730t**
Madagascar, **805**
madder family, **833**
Madia, 251
magnesium (Mg/Mg^{2+}), **76**
in chlorophyll molecule, **133**
as cofactor, 102
as essential element, 727, 729, **730t**
Magnolia, 519, 522
grandiflora, **520**
Magnoliaceae, 519
magnolias, 544
southern, **520**
magnoliids, 519, **520,** 521, 523, 525
embryos of, 559
maidenhair ferns, **455, 458**
maidenhair tree (*Ginkgo biloba*), 60, **275t,** 470, 472, 473, **473,** 488, **488, 490,** 490, 888
Mairan, Jean-Jacques de, 706
maize (*Zea mays*), 45, **45, 46, 50, 145,** 829
actin filaments in, **60**
aflatoxins in, 334
Black Mexican, **195**
classification of, 265
continuous variation in, **195**
cultivation of, 828, **828,** 829, **829,** 838, 839
domestication of, 829, **829**
embryo of, **562**
endocytosis in, **86**
flowers, **540**
germination in, **567,** 568, **568**
hybrid, 243, 829, 839
leaf stomata, **584**
leaves of, **625**
photosynthesis in, **134, 145,** 146, **146**
pollen grain, **540**
roots, 590, **592, 593,** 594, **594, 597,** 600, 603, **603**
sieve-tube elements, **581**
southern leaf blight of, 840, **840**
stems, 614, 621, **621**
viviparous, 684, **684**
maize streak, 300
malaria, **551**
malate (malic acid), 144, **144, 145,** 147, **149,** 693
male sterility, cytoplasmic, 196
Malloch, D. W., 343
Malpighi, Marcello, **765**
Malthus, Thomas, 237
Malus sylvestris, **503, 587, 649,** 655
mammals (*see also* animals *and by name*)
soil dwelling, **733**
Mammillaria microcarpa, **278**
manganese (Mn/Mn^{2+}), 137, **730t**
Mangifera indica (mango), 827
mangroves, 603, **605**
Manihot esculenta, 831, **831**
Manila hemp, 575
manioc, 831, **831,** 834
mannitol, 382
mantle of ectomycorrhizae, 342, **342**
manzanita, 565, **565**
maples, 545, **545,** 624, **637, 718, 813**
red, **10**
schizocarps of, 546
sugar, **625, 669t**
map units, 191
maquis, **796,** 817
Marasmius oreades, 325
Marattiales, 454–55, **463t**
marble, 731
Marchantia, **278,** 403, **403, 405,** 408, 409 (*see also* bryophytes)
life cycle of, **410–11**
margo, **662**
mariculture, 350
marijuana, 550, **551**
Mars, 3
Marsilea, 462
polycarpa, **462**
Marsileales, 454, 462, **463t**
maternal inheritance, 196
maternity plant (*Kalanchoë daigremontiana*), **180**
mating strains, types, **316, 321, 328,** 388, **389, 394, 396**
matorral, 817
matrix of mitochondrion, 114, **114,** 118, **119,** 120, **120, 121,** 122
matrotrophy, in bryophytes, 404–6
Matteuccia struthiopteris, **458**
Matthaei, Heinrich, 209
maturation, region of, **594,** 595, 596
Maxwell, James Clerk, 128–29
May apple, **762**
meadow rue, **762**
measles, 296
measurements, in microscopy, **42t**
mechanical stimulation, plant responses to, 721–22
Medicago sativa, 614, 619, **620,** 729, 738
medicinal compounds, plants, 550, **551,** 843, 846, **846**
medicine and botany, 11, 843, 846
Mediterranean climates, **10,** 816, 817
Mediterranean scrub, **796, 798–99, 817**
Medullosa noei, **456, 457**
megagametogenesis, 506
megagametophytes (female gametophytes), 431
of angiosperms, 506
evolution of seeds and, 468
of gymnosperms, 470
of pines, **478, 479,** 481, 482
of *Selaginella,* **440, 441,** 442
megapascal (MPa), 77
megaphylls, **429,** 470
evolution of, 429, 430, **430**
of ferns, 455
megasporangia, 431
evolution of seeds and, 468
of *Isoetes,* 443, **443**
in ovules, **468**
of pines and other conifers, 476
of *Selaginella,* **440, 441,** 442
megaspore mother cells (megasporocytes), **480,** 481, 506, **513**

megaspores, 431, **513**
of angiosperms, 506
evolution of seeds and, 468
in ovules, **468, 469**
of pines, 481
of *Selaginella*, **440, 441,** 442
megasporocytes (megaspore mother cells), **480,** 481, 506, **513**
megasporogenesis, 506
megasporophylls, 442, 443, **443, 487,** 501
meiocytes, 173
meiosis, **169,** 169–78 (*see also* sexual reproduction)
in *Agropogon*, **175**
in algae, **362,** 383, **384, 385, 389, 394**
in angiosperms, 504
in *Chlamydomonas*, 389, **389**
in chytrids, 312
in diatoms, **376, 377**
in *Equisetum*, **450**
in *Fucus*, **385**
in fungi, 310–12
gametic, 171, **171**
genetic variability and, 170, 177–79
in gymnosperms, **478,** 481
haploid, diploid numbers and, 170–71
in hymenomycetes, 324
life cycles and, 171, **171**
in liverworts, **410**
in *Lycopodium*, **436**
mitosis compared and, **178,** 179
in mosses, **418, 419**
in oomycetes, **374**
phases of, 173–78
in plasmodial slime molds, **353**
in *Psilotum*, **446**
sporic, **171,** 172
in *Ulvophyceae*, 393, **394**
in vascular plants, **430**
in wheat rust, **328**
in yeasts, 330
zygotic, 171, **171, 379,** 389, 391, 392, 397
meiospores, 172
Melampsora lini, **311**
Melilotus alba, **301**
melon, 525, 679
membrane potential, 82
membrane proteins (*see* integral proteins; transmembrane proteins)
membranes, cellular, **73** (*see also* endomembrane system; plasma membrane; *and specific membranes*)
fluid-mosaic model of, 74, **75**
solute transport across, 82–84
structure of, 74–76
unit membrane model of, 74
water transport across (*see* water transport (movement))
Mendel, Gregor, **183,** 184–90, 195
Mendel's first law, 185–87
Mendel's second law, 188–89
Mendelian genetics, 184–90, **185, 185t,** 242
chromosomal basis of Mendel's laws, 189–90, **190**
laws of probability, 188
method of, 184–85
pleiotropy, 195
principle of independent assortment, 188–89, **189,** 190
principle of segregation, 185–87
Mentha, 833
meristematic cap, 614, **614**
meristematic cells, 572, **572**
meristem culture, pathogen-free plants produced by, 695
meristems, 8, 571 (*see also* cambium)
apical (*see* apical meristems)
cell cycle and, 157, 166
chimeral, 634
ground, 557, **557, 570, 571, 611,** 613, 614 (*see also* primary meristems)
of hornworts, 412
intercalary, 614
lateral, 8, 427
peripheral, 613
pith, 613
primary, 557, 559, 570, **570, 571, 611,** 614 (*see also* ground meristem; procambium; protoderm)
Mertensia virginica, **500**
mescaline, **551**
Mesembryanthemum crystallinum, 644
mesocarp, 510
mesocotyl, **564,** 568
mesophyll, **8, 132, 573, 585**
of grass leaves, 632, **632, 633**
photosynthesis and, 630–31
of pine leaves (needles), 474, 475
mesophyll cells, 49, **126,** 144, **144, 145,** 146, **146**
phloem loading and, 767
somatic hybrids and, 693
transpiration and, 752, **752, 764**
of Venus flytrap, 721
mesophytes, 626
Mesozoic era, 431, 472, 486, 877
mesquite, 590
messenger RNA (*see* mRNA)
mestome sheath, 632, **633**
metabolic pathways, 103, **103**
feedback inhibition of, **104**
metabolism, 123–24
metabolites (secondary plant products), **549,** 549–51, **551**
metal ions, as cofactors, 102
metamorphic rocks, 731
metaphase
meiotic, **174, 175,** 176, 177
mitotic, 159, **160,** 161, **161,** 162, **162,** 163, **165**
plate, **161,** 162
metaphloem, 621 (*see also* primary phloem)
Metasequoia (dawn redwood), **485,** 486
glyptostroboides, **486**
metaxylem, 600, **619,** 621, **622, 623** (*see also* primary xylem)
methane, 3, **3**
Methanococcus jannaschii, 230, 270, **271**
methanogens, 295, **295,** 296
Methanothermus fervidus, **272**
methionine (fMet), 26, **27,** 212, 682, **682**
metric table, 879
Mexico, agriculture in, 828–30
micelles, 62, **63**
Micrasterias
radiosa, **397**
thomasiana, **397**
microbodies (*see* peroxisomes)
Micrococcus luteus, **285**
microcysts, 354
microenvironment, 777
microevolution, 240
microfibrils, 46, **59,** 62, **62, 63, 67, 165** (*see also* cell walls)
growth of cell wall and, 66, **66**
hormones and, 689, **689,** 690, **690**
orientation of, **66,** 159, **165,** 690, **690**
microfilaments (*see* actin filaments)
microgametogenesis, 504
microgametophytes, 431, 508, **512** (*see also* pollen)
of angiosperms, 504
of *Selaginella*, **440, 441,** 442
microhabitat isolation, **257**
micronutrients (trace elements), 727, **728t, 730t**
microphylls, **429, 442**
of club mosses, 435, **438**
of *Equisetum*, 445, **448**
evolution of, 429, **430**
fertile (*see* sporophylls)
of lycophytes, 435, 436, **438,** 439, **442–43**
micropropagation (clonal propagation), 693–95, **694**
micropyle, 468, **468, 469,** 470, **473, 479,** 481, 506, **507,** 509, **509,** 556, **562,** 563
microscopes, 40, **42–43,** 159
microspheres, proteinoid, 4
microsporangia, 431, 442, **443** (*see also* pollen sacs)
of pines and other conifers, 476, 477
of *Selaginella*, **440**
of yews, **484**
microsporangiate cones
of cycads, 487
of *Ephedra*, **491**
of pines, 476, 477, **477, 478,** 481, **481**
microspore mother cells (*see* microsporocytes)
microspores, 431, 468, **473**
of angiosperms, 504, **512**
of pines, 477, **479**
of *Selaginella*, **440, 441,** 442
microsporocytes (microspore or pollen mother cells)
of angiosperms, 504
of pines, 477, **478**
microsporogenesis, 504
microsporophylls, 442, 443, **443,** 501
of pines, 477, **477**
microtubule organizing centers, 59, 163, 311 (*see also* polar rings)
microtubules, 29, **59, 71t**
cell cycle and cell division and, 158–62, **161, 162,** 165, **165**
cell plate formation and, 165, **165,** 166
cell wall formation and, 66, **67, 165**
in flagella and cilia, 61, **61,** 105
functions of, 59
hormones and orientation of, 690
kinetochore, 161, 162, **162,** 163
phragmoplast, 165
polar, 163
spindle, 161
structure of, 58–59
middle lamella, **22,** 64, 166
midrib, **628,** 631, 635
midvein, 628, 631, **633,** 635
Miki, Shigeru, 486
mildews, 317, 373 (*see also* molds)
milkweed, 34, **35, 544,** 546, **549,** 550
family, 266, **266, 531**
Miller, Carlos O., 680
Miller, Stanley L., 3, **3**
millets, 826
mimicry, 550
Mimosa pudica, **36,** 720, **720**
Mimulus cardinalis, **537**
minerals, 6
absorption by roots, 762–63
cycles (*see* nutrient cycles)
as nutrients, 731 (*see also* essential elements)
transport in plants, 729, 762–64
mint family, 34, **245,** 624, 833
miso, 334
Mississippian period, 471
mistletoe, 498, **547**
Mitchell, Peter, 118, 120
mitochondria, **44, 45, 52,** 52–53, **71t**
cell cycle and, 158
chemiosmotic coupling in, 137, 138
cytoplasmic inheritance and, 196
DNA of, 53
electron transport chain in, 116, **117,** 118
evolution of, 53, **272–73,** 272–74
functions of, 52, **108,** 113–14, **143**
matrix of, 114, **114,** 118, **119,** 120, **120, 121,** 122
structure of, 114, **114**
mitosis, 156, 157, **157,** 158, **160, 161,** 166
in algae, 360, 386–87, 388–98
in chytrids, 312
in dinoflagellates, 363
duration of, 163
in euglenoids, 352
in fungi, 310–11
meiosis compared and, **178,** 179
in oomycetes, 371, **374**
phases of, 160–63, **162**
in prokaryotes, 285–86
in slime molds, 352, 355
mitotic spindle, 158–60, **160,** 161, **162,** 163, **165**
mixotrophs, 348
Mnium, **417**
moa plant, **444**
Mojave Desert, **808**
molds, 315, **316,** 317
slime (*see* cellular slime molds; plasmodial slime molds)
water, **171,** 275, 373, 872
molecular biology, 11
molecular chaperones, 29
molecular clock, 270
molecular formulas, 19, 870
molecular genetics, 197
molecular mass, 870
molecular systematics, 268–71
molecular weights, 870–71
molecules, 18, 868–71
biologically important, **38t**
diffusion of, 76–78
energy and, 97–99, 105–6, 875–76
excited, 131–32, 135, 865
organic, 3–4, 18, 869–70 (*see also* specific types)
representations of, 19, **19**
moles (units), 97, 871
molybdenum (Mo/MoO_4^{2-}), as essential element, 727, **730t**
monarch butterfly, 34, **35,** 550
Monilinia fructicola, 317
monkeyflower, scarlet, **537**
monkey-puzzle tree, 483
monocots (monocotyledons), **497,** 498, 504, 887
cotyledon of, 559, 563
embryos of, 557, **558–59,** 559, 563
eudicots compared and, **498t**
evolution of, 519, 521
genetic engineering and, 699
leaf development in, 635
major features of, **498t**
parasitic, 498
seeds of, **562**
Monocotyledones, **265t,** 496, 887
Monod, Jacques, 216
monoecious plants, 501, **502**
monohybrid crosses, 185
monokaryotic (uninucleate) cells, 322
monomers, 18
monooxygenases, 780
monophyletic taxa, 265
monosaccharides, 18, **19**
monospores, 360
monoterpenoids, 33, 34
Monotropa uniflora, **498**
monsoon forests, **798–99,** 806, 807
Monterey pines, **475, 477**
Morchella, **308t, 320,** 342
morels, **309,** 317, **320,** 342
Moricandia, **625**
arvensis, **73**
morphine, 32, **32**
morphogenesis, 166, 572
hormones and, 687
morphological species concept, 248
Morus alba, 624, **625**
mosaics (mosaic diseases), 301
mosquitoes, pollination by, 535, **535**
moss, Irish, **358**
mosses (*Bryophyta*), **56, 275t,** 400, 407, 412–22, **432,** 886 (*see also* bryophytes; *and specific names of mosses*)
asexual reproduction in, 414, 421
cladogram of, 267–68, **268, 402**
classification of, 412
club (*see* club mosses)
comparative structure of, 403
comparative summary of characteristics of, **422t**
economic uses of, 415
fir, 438
granite (*Andreaeidae*), 412, 416
habitats of, 401, **401**
Irish (algae), 412
life cycles of, 414, 416, **418–19**
liverworts compared and, 412
peat (*Sphagnidae*), 412, **414,** 414–15
protonemata of, 407
reindeer (lichens), 412
scale (liverworts), 412
sea (algae), 412
Spanish, 412

sporophytes of, 406
"true," *(Bryidae)*, 412, 416–22
conducting tissues of, 416, **416, 417**
"cushiony," "feathery," 421, **422**
sexual reproduction in, 416, **417–18,** 418–21
weathering of rocks and, 786
moths, 550, 782
pollination by, 535–36, **536**
motility
of algae, 388–90, 393
of chytrids, 313
of diatoms, 376
of dinoflagellates, 361, 363–65
of euglenoids, 351
of oomycetes, 371, 373
of prokaryotes, 284, 288–89, 292–93
of slime molds, 352, **354–55**
motor proteins, 163
mountain forests, **798–99**
mountain lilacs, 741
mountains, precipitation and, 801, **801,** 802
Mount St. Helens, 789, 792, **792**
movement proteins, 300, 302
M phase, 157, **157**
mRNA (messenger RNA), 208, 209, **688**
in eukaryotes, 222–23, **223**
gene regulation in prokaryotes and, 216, **216**
protein synthesis and, 210–14
mRNA-binding sites, 211
mucigel, 591, **592,** 597
mucilage, 618
mucilaginous sheaths, 288, 397
mulberry, 624, **625**
mullein, **585**
Müller, Hermann J., 181
multicellular organisms, evolution of, **6, 7,** 11, **12,** 14, 45
multilayered structure, 387, **387**
Multinational Coordinated *Arabidopsis thaliana* Genome Research Project, 228
multipass transmembrane proteins, 75, **75,** 82
multiple fruits, 543
mumps, 296
Münch, Ernst, 767
Musa × *paradisiaca*, **497,** 827
mushrooms, **277, 308t, 309,** 323 *(see also Basidiomycetes)*
commercially cultivated, 325
edible, classification of, 265
fairy rings, 325, **325**
hallucinogenic, 325, **325**
inky cap, **321**
life cycle of, 320, **320–21**
oyster, 333, **333**
parts of, 324–25
poisonous, 325
reproduction in, 324
mustard, 833
black, **549**
family, **228, 238,** 340, 511, 544, **544,** 549, 550, **728**
Indian, 747
mustard-oil glycosides, 549–50
mutagens, 192
mutants
dwarf, gibberellin and, 684–85, **685**
viviparous, 684, **684**
mutations, 172, 192–94, 240, **241**
chromosome, 192–94
embryogenesis and, 560, **560, 561**
evolutionary change and, 193–94, 240–41
gene (point), 192
homeotic, 639, 640, **641**
of prokaryotes, 285–86
mutualism (mutualistic relationships), 334, 774–75 *(see also* lichens; mycorrhizae; symbiosis)
of mycobiont and photobiont, 339, **339**
mycelia, **306,** 310, **320,** 325
of ascomycetes, 318, 320
of *Basidiomycota*, 322
dikaryotic, monokaryotic, 322
heterokaryotic, 333
of mushrooms, 324–25
septate, 310, 318, 322
Mycena
haematopus, **309**
lux-coeli, **122**
mycobionts, 334
photobionts' relationship with, 339, **339**
myco-heterotroph, **498, 529**
mycologists, 348
mycology, 12, 310
mycoplasmalike organisms (MLOs; phytoplasmas), 292–93, **293**
mycoplasmas, 292
mycorrhizae, 308, 312, 340–44, 431, 597, 762
of club mosses *(Lycopodiaceae)*, 431, **436, 437,** 438
evolution of, 343
fossil, 343, **344**
mutualism and, 774–75
tree nutrition and, 340, **340,** 341
vesicular-arbuscular (V/A), 341
mycorrhizal fungus, 340–44, 762, 774
of angiosperms, 342–43
of ferns, 455
of *Lycopodiaceae*, 431, 436
of *Psilotum*, 431, 444, 445
mycotoxins, 307, 334
myosin, actin and, 47
Myrica gale, 741
Myristica fragrans, 832
myxamoebas (amoeba-like cells), **354, 356**
myxobacteria, 285, **285**
myxomatosis, 777
Myxomycota (myxomycetes), 884 *(see also* plasmodial slime molds)
life cycle of, **353–55**
myxospores, 285, **285**

N

NAA (naphthalenacetic acid), **676, 678**
NAD^+/NADH, **102,** 102–3, **117**
in alcohol fermentation, 123
in electron transport chain, 118
in glycolysis, 109, 112, **112,** 113, **113,** 122
in Krebs cycle, 114, **114,** 116
$NADP^+$/NADPH, 134, **134,** 135, **136,** 138, 141
naphthaleneacetic acid (NAA), **676, 678**
nastic movements, 719–21, **723t**
natural disasters, 789
natural gas, 295
Naturalist in Nicaragua, The, 775
natural selection, 236, 237–38, 240
adaptation as result of, 245–47
genetic variability and, 243–45
rapid changes in natural populations and, 244–45
neck canal cells, 404, **405**
necrosis, 729
nectar, 525
of bat-pollinated flowers, 538, 539, **539**
of bird flowers, 537, **538**
insect pollination and, 530, 531, **531**
nectaries, **522,** 525, 530
insect pollination and, 532, **535,** 536
bee flowers, 532
needles, pine *(see* leaves, of pines)
Neljubov, Dimitry, 682
Nelumbo
lutea, **524**
nucifera, 718
nematodes
mycorrhizae and, 340
predaceous fungi and, **333**
neomycin, 287
Nepeta cataria, 262
Neptunia pubescens, **638**
Nereocystis, 382
Nerium oleander, **629**
netted venation, 631, **631**
nettle, stinging, 778
neurological illness, 364
Neurospora, **308t,** 708
neutrons, 851
Newton, Sir Isaac, 128
N-formyl methionine (fMet), 212, **213**
nickel (Ni/Ni^{2+}), as essential element, 727, **730t**
Nicolson, Garth, 74
Nicotiana (see also tobacco)
glauca, **696**
silvestris, 715, **716**
tabacum, 715, **716**
nicotinamide adenine dinucleotide *(see* NAD^+/NADH)
nicotine, 33, **33, 549**
nif genes, 740
nightshade family, 695, 831
Nilson-Ehle, H., 195
Nirenberg, Marshall, 209
Nitella, **47, 76**
nitrate ions (NO_3^-), 738
nitrates, 6, 26, 738
assimilation of, 742
nitrification, 736, **736,** 738
nitrifying bacteria, 738
nitrite ions (NO_2^-), 738
nitrites, 26, 738
Nitrobacter, 738
nitrogen (N), 18
amino acids and, 26
assimilation of, 742, **742**
in atmosphere, 3, 736, **736**
as essential element, 727, 729, **730t**
in prokaryote metabolism, 287, 290, 295, 737–41
nitrogenase, 290, 738
nitrogen cycle, 736–42
human impact on, 743
stages of, 736, **736**
nitrogen dioxide, in photochemical smog, 616
nitrogen fertilizers, 738, 741–42, 745
nitrogen fixation
biological, 290, **736,** 738–42
genetic manipulation of, 747
industrial, 741
manipulation of, 747
nitrogen-fixing bacteria, 287, 290, 738–41, 747
nonsymbiotic, 741
nitrogenous bases, 30, **31,** 198–199, 200
nitrogen oxides, as air pollutants, 616, **617**
Nitrosococcus nitrosus, **285**
Nitrosomonas, 738
Nobel, Park, 809
Noble, David, 488
Noctiluca scintillans, **363**
nodD genes, 740
nodes, **8,** 445, 611, 613
nod genes, 740
nodules, root, 739, 740, **740**
nodulins, 740
Nomuraea rileyi, **319**
noncyclic electron flow, photophosphorylation, **136,** 138
nondisjunction, 173
nonfunctional phloem, **656,** 659
non-heme iron proteins, 116, **117**
nonpolar covalent bonds, 870
nonrandom mating, 242
Norfolk Island pine, 483, **484**
nori, 380
northern spotted owl, 815, **815**
Northwestern coniferous forest, **798–99**
Nostoc, 335, 412
commune, **289**
Nothofagus, 342
menziesii, **523**
no-tillage cropping system, **838**
nucellus, 468, **468,** 470, **480,** 510
of angiosperms, 506
of pines, **482**
nuclear division *(see* meiosis, mitosis)
nuclear envelope, 41, **44, 45,** 47, **48**
endoplasmic reticulum and, 56
of euglenoids, 352
of fungi, 311
nuclear genome, 53
nuclear pores, 47, 48, **48**
nuclear-type endosperm formation, 510
nucleic acid probe, 230, **230**
nucleic acids, 18, 30–32, **31** *(see also* DNA; *specific compounds)*
sequencing of, **229,** 229–30
molecular systematics and, 270–71
nucleoids, 41, **43, 44,** 52, **52,** 53
nucleolar organizer regions, 48
nucleoli, **45,** 48, 160, **160**
nucleomorphs, 357
nucleoplasm, 47
nucleosides, 102
nucleosomes, **219,** 219–20
nucleotides, 30, **30, 38t,** 102 *(see also specific compounds)*
in ATP, 31
in DNA, 209
mutations and, 192–93
replication of DNA and, 203–4
structure of, 198, **198,** 199
genetic code and, 209
protein synthesis and, 209, 210, **211**
in RNA, 208, 209
sequencing of, 229–30, 270–71
in viral genomes, 297–98
nucleus (atomic), 863–67
nucleus (of cell), 41, 45, **45,** 47–48, **70t**
diploid, haploid, 47, 170–73, 175, 177, 179, 331, 332
division of, 156, **157,** 158, **159,** 160, **160,** 161, 163 *(see also* meiosis; mitosis)
evolutionary origin of, **273,** 274
functions of, 47
generative, 504
polar, 506, **507,** 509
tube cell, **505**
Nudaria, **380**
nudibranch, 395
numbers, pyramid of, 786, **786**
nutcrackers, 482
nutmeg, 832, **832**
nutrient cycles, 735–36
carbon, 150–51, **151,** 350, 415
human impact on, 743–44
nitrogen, **736,** 736–43
phosphorus, 742, **743**
nutrient deficiencies, 727, 729, **730t,** 731
nutrients, inorganic *(see also* essential elements; macronutrients; micronutrients; nutrition, plant; soils; *specific nutrients)*
concentration of, in plants, 727, **728t**
exchanged between transpiration and assimilate streams, **763,** 763–64
functions of, 729, **730t,** 731
uptake by roots, 762–64, **763**
nutrition, plant, 726–49 *(see also* nutrient cycles)
agriculture and, 744–45
of angiosperms, 498
essential elements, 727–31
general requirements, 727–31
mycorrhizae and, 340–41, 762, 774
research on, 745–47
soils and, 731–35
nuts, 545
nyctinastic movements, 719–21 *(see also* tropisms)
nyctinasty, **723t**
Nymphaeaceae, 521
Nymphaea odorata, **575, 629, 787**

O

oaks, **577** *(see also Quercus)*
black, 256, **257, 658**
cladogram of, 268, **268**
cork, 658, **659**
red, 65, 577
scarlet, 256, **257**
white, **657, 666, 667, 669t**
oat, 146, 675
bran, 563
loose smut of, 330
root system, **589**
oca, **830**
ocean currents
dispersal of fruits by, 546
global warming and, 845
oceans

  carbon dioxide in, 150
  evolution of photosynthetic organisms in, 6–7
  primitive, 4
Ochroma lagopus, 659
Ocimum vulgare, 833
odors, of flowers, 531, 539
Oedemeridae, **531**
Oedogonium, 392
  cardiacum, **392**
  foveolatum, **392**
Oenothera, **536**
  gigas, 252
  glazioviana, 192, **192**
  lamarckiana, 252
O horizon of soil, **732**
oil bodies, **23, 51,** 55, **55, 71t** (*see also* plastoglobuli)
oil cells, 519
oil palms, 834
oils, 22, **23, 38t**
  essential, 34, 549
Okazaki fragments, 202, **203**
okra, 827
Olea capensis, 670
oleander, **629**
oleic acid, **23**
Oligocene epoch, 877
oligosaccharides, 74
oligosaccharins, 22, 62
Oliver, F. W., 457
olives, 825
O'Mara, J. G., 840
Onagraceae, **514**
Oncidium sphacelatum, **606**
onion, 162, **562, 593**
  germination in, **567,** 567–68
Onoclea sensibilis, **458**
oogamy, 371–72, **372,** 430
oogonia, 372, **374, 385**
oomycetes *(Oomycota)*, **171,** 348, 371–74, **372t,** 884
  life cycle of, **374**
  water molds, **171,** 275, 373, 884
oospores, 372, **374**
opaline silica, 376
Oparin, A. I., 3
open system, the Earth as, 98, **98**
operator sequence, 216
operculum, 414, **414**
operons, 216, **217,** 218, **218**
Ophioglossales, 454, 455, **463t**
Ophioglossum, 47, **454,** 454–55
  reticulatum, 455
Ophiostoma ulmi, 317
Ophrys, **535**
  speculum, **534**
opines, 696
opine-synthesizing enzyme, 696
opium poppy, **32,** 550
opposite phyllotaxy, 624
Opuntia, **547**
  inermis, 779, **780**
orange snow, **386**
orbitals, 855
orchard grass, 244
Orchidaceae, **260,** 343, 527, **528,** 528–29
orchids, **528,** 528–29
  commercial production of, 529
  Cymbidium, **278–79**
  epiphytic, 804
  lady slipper, **260**
  mycorrhizae in, 343
  pollination of, **534,** 534–35, **535**
  roots of, 605, **606**
  saprophytic, 529
  underground, 529, **529**
  vanilla, 529, **529,** 833
orders, taxonomic, 264
oregano, 833
organelles, 41, **44** *(see also by specific type)*
organic molecules, 3–4, 18, 869–70 *(see also by name)*
organismal theory, 65
organisms *(see also by name)*
  classification of, 262–71, 881–88
  interactions between, 774–82
organ-pipe cactus, **539**
organs, 86 (*see also* leaves; roots; stems)
organ transplants, 308
Origanum vulgare, 833
origin of replication, 201, 202
Origin of Species, On the, 248, 257
Orthotrichum, 421
Oryza sativa, 146, **497,** 563, 684, 826, **827**
Oscillatoria, **289**
osmoregulatory mechanisms, 744
osmosis, 78–81, **79**
  in living organisms, 80–81
osmotic potential (solute potential), 79, **79**
  of guard cells, 693
osmotic pressure, 79
ostrich fern, **458**
outcrossing (outbreeding), 256, 477
  in angiosperms, 510–11
  in gymnosperms, 510
  mechanisms that promote, 242–43
outgroups, 267
ovaries, **499,** 501, 503, **513**
  beetle pollination and, 531
  development into fruit, 509–10, 686
  of early angiosperms, **524,** 527
  inferior, superior, 502
  of orchids, 528
ovary walls, 501, **501,** 506, 510
ovulate cones
  of balsam fir, **483**
  of pines, 476–77, **477, 478,** 480, **480, 481,** 481, **483**
ovules, **184, 468, 469**
  of angiosperms, 499, 501, 506, **507,** 525
  evolution of, 468, 525
  of *Ginkgo*, 490
  of gymnosperms, 472
  insect pollination and, 530
  of pines, **477–479,** 480–82
  of wind-pollinated flowers, 530, 540
ovuliferous scales (cone scales), 480, 481
owl, northern spotted, 815, **815**
Oxalis, **707**
  tuberosa, **830**
oxaloacetate (oxaloacetic acid), 115, 144, **145, 149**
oxidation, 98, **99**
  beta, 122
  of glucose, 99, 109, **110, 121** (*see also* glycolysis)
  of pyruvate, **114,** 122
oxidation-reduction (redox) reactions, 98–99, **99**
oxidative phosphorylation, 108, 109, **110,** 118–20, **121** (*see also* phosphorylation)
Oxydendrum, **813**
oxygen (O), 18 (*see also* respiration)
  in atmosphere, 5–6, 9, 11, 844
  diffusion of, 78
  as essential element, 727, **730t**
  germination and, 565
  nitrogen fixation and, 740
  photosynthesis and, 5–6, 9, 11
  prokaryotes and, 287
Oxyria digyna, 247, 249
oyster mushroom, 333, **333**
ozone (O_3), 616, **617,** 844
  in atmosphere, 6
ozone layer, 844

P

P_{680}, P_{700} chlorophyll molecules, 135, **136,** 138, 139, **139,** 152
palea, **541**
paleobotany, 11
paleoherbs, 521, **521,** 526
Paleozoic era, **469,** 470
Palhinhaea, 438
palisade parenchyma, **628, 629,** 630–31, 636
palmately compound leaves, 626, **627**
Palmer's grass, 744
palms, 648
  coconut, **497, 544**
  fan, 518
  oil, 834
  sago, 486
PAN (peroxyacetyl nitrate), 616
Panama pine, 483
Pandorina, **390**
panacle, **499**
Panicum, 827
Papaveraceae, 544
Papaver somniferum, **32, 544,** 550
paper birch, **10, 541, 658**
pappus, 528, **545,** 546
paradermal section, **628**
paraheliotropism, 723
parallel venation, **625,** 631
paramylon, **351,** 352
paraphyletic taxa, 266
paraphyses, **417**
parapodia, **396**
Para rubber trees, 842
parasexuality, in deuteromycetes, 333–34
parasitic organisms
  angiosperms, 498, **498**
  bacteria, 293
  fungi, 310, 317, 334, 339
  oomycetes, 373
parasitic relationships, 334
parenchyma (parenchyma cells), **48, 571,** 572–73, **573,** 574, 582, 583, 602
  in algae, 381, 392
  auxin and, 676
  of bryophytes, **417**
  interfascicular, 614, 619
  of *Isoetes*, 443
  of leaves, 628
   palisade and spongy, **628, 629,** 630–31, 636
   pine leaves (needles), 474, 475
  of phloem, **579–80,** 582–83
  of pulvinus, 720, **720**
  ray, 573, 648
  of stems, 618, 619, **619, 620**
  storage, 606, **607**
  of xylem, 579, **579**
parenthesomes, 322, **322**
parietal placentation, 501, **501**
Parmelia perforata, **336**
parsimony, principle of, 268
parsley, **752,** 833
  family, 264, **502,** 531, 833
   schizocarp, 545
Parthenium argentatum, 842, **842**
parthenocarpic fruits, 543
Parthenocissus
  quinquefolia, 641
  tricuspidata, 641
particle bombardment, 699
particle model of light, 130
passage cells, 600
passion flower, 537
passive transport, 82
Pasternak, D., **843**
patch-clamp technique, for studying ion channels, 83
pathogen(s) (*see* diseases; plant diseases)
Pauling, Linus, 198, 199
PCR (polymerase chain reaction), 226
pea(s), **545,** 563, **566,** 567, 738, 780, 825, **825**
  ethylene and, 682, **683**
  family *(Fabaceae)*, 264, 544, **545,** 636
  garden, **545**
  leaves of, **624**
  Mendel's experiments with, 184–90, **185, 185t,** 242 (*see also* Medelian genetics)
  seed germination in, 566
  tendrils of, 641, 706
peach, 543, 719
peanut(s), 828
pear, **62, 576,** 655
pear decline, 293
peat mosses *(Sphagnidae)*, 412, **414,** 414–15
pectic acid, **22**
pectins, 22, **22,** 63, 572
  in cell walls, **62**
  growth of the cell wall and, 66
pedicel, 499
peduncle
  of dinoflagellates, 365, **366**
  of flowers, 499
pelargonidin, **543**
pellicle, 351, **351**
penicillin, 334, 779
penicillinases, 779
Penicillium, **332,** 333, 334
  camemberti, 334
  chrysogenum, 779
  notatum, **332**
  roqueforti, 334
pennate diatoms, **375,** 376, 378
Pennisetum, 827
Pennsylvanian period, 877
3-pentadecanedienyl catechol, **550**
pentoses, 18
PEP (phosphoenolpyruvate), 112, **112,** 144, **144,** 146
PEP carboxylase, 144, **144,** 146
Peperomia, 644
peppers, 521, **521,** 832, 833
peptide bond, 28
peptidoglycans, 283, **283**
peptidyl transferase, 212
perennials, 648
perforation plate, 576, **577**
perforations of vessel elements, 576, 577, **577,** 578
perianth, **409,** 501, 502
  of early angiosperms, 524
  of liverworts, 412
pericarp, 510, 547, **564, 557**
  in algae, **362**
periclinal divisions, 557, 612, **612, 649**
pericycle, 600, 601, **601,** 602, 603
periderm, 427, 476, 572, 586–87, **587,** 600, **601,** 602, **653,** 653–54
  gas exchange through, 654
peridinin, 365
peridium, 325, **326**
perigynous flowers, 502, **502**
perigyny, **503**
peripheral proteins, 74, **75**
peripheral zones, 613
periphyses, 327
perisperm, 510, 563
peristome, 421, **421**
perithecium, 319, **319**
periwinkle, **846**
permafrost, 819, 820
permanent wilting percentage (point) of soils, 735, **735**
permeability, 78, 79
Permian period, 429, 431, 457, 474
Peronosporales, 373
peroxisomes, **53,** 53–54, **71t**
  in photorespiration, **143**
peroxyacetyl nitrate (PAN), 616
Persea americana, 683
persimmon, **67, 669t**
pesticides, 781, **837**
pest management systems, integrated, 781
petal primordia, **638**
petals, **499,** 501
  development of, **638,** 639
  of early angiosperms, 524, **524,** 527
  fusion of, 524
  of orchids, **528,** 528–29
petiole, **8,** 624, **625,** 626
petiolule, 626
Petroselinum crispum, 833
petunia, **698**
peyote cactus, **551**
Pfiesteria piscicida, 364, 366, **366**
PGA (3-phosphoglyceric acid), 140, **140,** 142, **143**
PGAL (glyceraldehyde 3-phosphate; 3-phosphoglyceraldehyde), **111,** 111–13, 140–42
pH, 875
  of acid rain, 616–17
  scale, 875, **875t**
  of soils, 735
Phaeocystis, 367

  pouchetii, **367**
Phaeophyta (*see* brown algae)
phages (*see* bacteriophages)
phagocytes, evolution of, 272, **273,** 274
phagocytosis, 85, **85**
Phalaris canariensis, 675
Phanerochaete chrysosporium, 308
Phaseolus, **58,** 738
  vulgaris, **562, 566,** 567, 678, 684, **685, 705**
phellem, 587 (*see also* cork)
phelloderm, 587, **587,** 602, 653, 654, **654**
phellogen (*see* cork cambium)
phenolics, 34–37
phenotypes, 186, 194–96
  environment and, 197
phenotypic ratio(s), 185–89
phenylalanine, 26, **27, 28**
pheophytin, 137
Phlegmariurus, 435, 438
phloem, 8, **8,** 426, **427,** 572, **576,** 579–83, **587t**
  assimilate transport in, 764–69
    aphids and, 766, **767**
    phloem loading, unloading, 767–69, **769**
    pressure-flow hypothesis of, 766–69, **768**
    radioactive tracer experiments, 765–66
  in brown algae, 382
  cell types in, **579t,** 580–83
  development of, 580–582, **583, 594,** 600, 614–21
  florigen transport in, 715–16
  functional, nonfunctional, **656,** 659
  hormones and, 676
  primary, **571, 573,** 600
    of roots, 601, **601,** 602
    of stems, 618, **619, 620**
  secondary, 427, 470
    in pines and other conifers, 476, **476**
    in seedless vascular plants, **435, 442, 444–45, 448, 455**
  transfer cells and, 574, **574**
  of veins, 631
  viral movement in, 300, **300**
  xylem interchange and, 767, **768**
phloem-immobile ions, 729, 764
phloem loading and unloading, 767–68, **769**
phloem-mobile ions, 729, 764
phloem unloading, 768–69
phlox, 525
Phoenix dactylifera, 825
phosphate ($H_2PO_4^-$; HPO_4^{2-})
  absorption of, 742
  in ATP, **105,** 106
  in nucleic acids, 30, **30,** 31, **31**
  inorganic (P_i), 106, 112, 120, 138
  in phospholipids, 24, **24**
  mycorrhizae and, 340
phosphoenolpyruvate (PEP), 112, **112,** 144, **144,** 146
phosphofructokinase, **111**
phosphoglucoisomerase, 111, **111**
3-phosphoglyceraldehyde (PGAL), **111,** 111–13, **140,** 140–42
1,3-bisphosphoglycerate, 112, **112**
3-phosphoglycerate (3-phosphoglyceric acid; PGA), 140, **140,** 142, **143**
3-phosphoglycerate kinase, 140
phosphoglycolate, 142, **142,** 143, **143**
phospholipid bilayer, 24, **24,** 74, **75,** 76
phospholipids, **24,** 24–25, **38t,** 74
  in cellular membranes, 74
phosphorus (P), 18
  as essential element, 727, **730t**
  in fertilizers, 745
  mycorrhizae and, 340
phosphorus cycle, 742, **743**
phosphorylation, 106 (*see also* photophosphorylation)
  oxidative, 108, 109, **110,** 118–20, **121**
  substrate-level, 112, **112**
Photinus pyralis, 698
photobionts, 334–35, **338,** 339
photochemical smog, 616
photoconversion reactions, 712
photoelectric effect, 129
photolysis, 137
photomorphogenic responses (photomorphogenesis), 714, **714**
photons, 130, 135
photoperiodism, 709–14
  in animals, 709
  chemical basis of, 711–14, **712, 713**
  dormancy and, 718–19
  flowering and, 709–11
photophosphorylation, 137–38, **138**
  cyclic, 139, **139**
  noncyclic, **136,** 138
photoreceptors (photoreceptor proteins), 703, 704 (*see also* phytochrome)
photorespiration, 54, **142,** 142–44
photosynthates, 631, 632
photosynthesis, 2, 126–51
  action spectrum for, 131, **131**
  in algae, 350, 395
  CAM, 147, **148,** 149, 443
  carbon cycle and, 150–51, **151**
  carbon-fixation reactions of (*see* carbon-fixation reactions)
  chloroplasts and, 49–50
  competition and, 776
  cotyledons and, 567
  in C_3 plants, 140–42, 144, 146–47
  in C_4 plants, 144–49
  efficiency of, 138, 142–44, 146, 149
  evolution of plants and, **5,** 5–6
  historical perspective on, 127–28
  in lichens, 334, 338–39
  light reactions of, 133–39
  mesophyll and, 630–31
  overall reaction (equation) for, 127
  overview of, **134**
  photochemical smog and, 616
  pigments in (*see* accessory pigments; carotenoids; chlorophylls; phycobilins)
  products of, 142
  in prokaryotes, 127–28, 133, 283, 291
  redox reactions during, 99
  stomatal movements and, 751–54
  transpiration and, 751
photosynthetic autotrophs, 287
photosynthetic bacteria, 127–28, 133, 291 (*see also* cyanobacteria; green bacteria; prochlorophytes; purple bacteria)
photosynthetic organisms (*see also* autotrophs)
  earliest, 5
  ecosystems and, 9, 11
  seashore environment and, 6–7
Photosystems I and II, 135, **136,** 138, **138,** 139
photosystems, 135, **135, 136**
  in purple and green bacteria, 291
phototropism, 675, **702,** 703–4, **723t**
  in fungi, 315, **315**
phragmoplasts, 159, **159,** 165, **165,** 387, **387**
phragmosome, 158, **159**
phycobilins, 133, 288, 291, 357, 358
phycocyanin, 288, 357
phycoerythrin, 288, 357
phycology, 12
phycoplasts, 386
phyllotaxy, 624
phylogenetic analysis (cladistics), 267
phylogenetic trees, **264,** 265
  traditional, 267, **267**
phylogeny (phylogenetic relationships), 265
  selected characters used in analyzing, **268t**
phylum (phyla), taxonomic, 264
Physarum, **276, 352**
Physoderma
  alfalfae, 313
  maydis, 313
phytoalexins, 32, 62, 780, 782
phytochrome, **712, 713**
  in *Chlorophyta*, 383
  dark reversion of, **713**
  photoconversion of, **713**
  photoperiodism and, **712,** 712–14, **713**
phytohormones (*see* hormones)
phytomeres, 611, **611**
Phytophthora, 373
  cinnamomi, 373
  infestans, 373, **375,** 841
phytoplankton, **785,** 786
  concentration of, **800**
  marine, 348, 349–50, **350**
phytoplasmas, 292–293
Picea, 483
  engelmannii, 777, 791
  glauca, **788, 818**
  rubens, **669t**
pickles, 287
pickleweed, 744
picoplankton, 294
Pierinae, 550
pigments, 130–33 (*see also by name*)
  absorption spectrum of, 131, **131**
  accessory, 133, 288, 291, **349t, 372t**
  action spectrum, 131, **131**
  antenna, 135
  anthocyanins, 35, **35,** 54–55, 542, 543, **543**
  in bacteria, 282, 288, 291, **291,** 292
  carotenoids, 49–51, 133, **134,** 135, 542, 543, **543**
  colors of angiosperm flowers and, 542–43
  P_{680}, P_{700}, 135, **136,** 138, **139**
  photoperiodism and, 712–714, **713**
  photosynthetic (*see* accessory pigments; carotenoids; chlorophylls; phycobilins)
  P_r, P_{fr}, 712–714
  stomatal movements and, 753
  vacuoles and, 54
pigweed, **126**
pileus (cap), 324
pili, 284, **284,** 285
Pilobolus, 315, **315**
Pimenta officinalis, 833
Pinaceae, 342, **483** (*see also* pines)
pineapple, **148,** 149, 674
  family, 337
  fruit, 543
pine nuts, **480**
pine-oak chapparal, **10**
pines, 474–82, 559 (*see also* conifers; *Pinus*)
  bristlecone, 475, **475, 664,** 664–65
  cladogram of, 268, **268**
  digger, **480**
  eastern white, **480**
  family, 342, **483**
  leaves (needles) of, 474, **475, 476**
  life cycle of, 476–77, **478–79,** 482
  limber, 482
  loblolly, **669t**
  lodgepole, **342, 481,** 482, 791
  longleaf, **474, 475**
  Monterey, **475, 477**
  Norfolk Island, 483, **484**
  Panama, 483
  pinyon, **475, 480,** 482
  ponderosa, **669t**
  red, **480**
  slash, **669t**
  sugar, **278, 480, 669t,** 789
  western white, **669t**
  white, 340, **480,** 660, **660–662, 668, 669t**
  whitebark, 482
  yellow, **480**
Pinguicula grandiflora, 737
pinnae, 455
pinnately compound leaves, 626
pinocytosis, **85,** 85–86
Pinus, 659
  albicaulis, 482
  contorta, **342, 481,** 482, 791
  ectomycorrhiza of, **342, 343**
  edulis, **475, 480**
  elliottii, 659, **669t, 807**
  flexilis, 482
  lambertiana, **278, 480, 669t,** 788, **789**
  life cycle of, **478–479**
  longaeva, 475, **475, 664,** 664–65
  monticola, **669t**
  palustris, **474, 475**
  ponderosa, **480, 669t**
  radiata, **475, 477**
  resinosa, **480**
  sabiniana, **480**
  strobus, **340, 480, 660–662, 668**
  taeda, **669t**
pinyon-juniper savannas and woodlands, **810**
pinyon pine, **475, 480,** 482
Piperaceae, 521, **521**
Piper nigrum, 832
Pirozynski, K. A., 343
pisatin, 780
pistil, 501
Pisum sativum (*see* pea(s))
pitcher plant, 642
pit connections, in red algae, 360, **360**
pit-fields, primary, 64, **64, 573**
pith, 428, 573, **573**
  hollow, 445
  of roots, 595, **597**
  of stems, 613, 614, **615,** 618
pith meristems, 613
pith rays, 618, **651**
pit-pairs, **64,** 66, 577, 659–60, **661, 662**
pit(s), **64,** 66, 576, **576, 577**
  apertures, 660
  bordered, **64,** 66, **661, 662**
  cavity, 66
  coated, 86, **86**
  membranes, **64,** 66, 577, **662**
  ramiform, **576**
  simple, 66, **576**
placenta, 501
  of bryophytes, 404–6
placentation, 501, **501**
Placobranchus ocellatus, 395
Plagiogyria, **453**
plainsawn boards, 667, **667**
plankton, 289, 348, **785** (*see also* phytoplankton; zooplankton)
Plantae (plant kingdom), **265t,** 277, **278,** 279, 886 (*see also* bryophytes; vascular plants)
plantains, 827
plant body, 8, 425–30
  internal organization of, 572–73
  primary, 427, 571
  secondary, 427
plant breeding, 837–43, 846
plant morphology, 11
plant physiology, 11
plant(s)
  and adaptations to life on land, 797, 800–2
  ancestors of, 401–3
  biotechnology (*see* biotechnology)
  C_3, C_4, 144–47, **148,** 149
  CAM, 147–49
  classification of, **275**
  domestication of, 824–32
  nutrition and soils, 726–49
  people and, 823–49
  tissues, 573–587
plant diseases, 287, 292–94, 296–303
plant-herbivore interactions, 779–80, 782
plant-pathogen interactions, 779–80, 782
plant pathogens, 292–94, **292–94**
plasma membrane (plasmalemma), 41, **44, 45, 70t, 75, 599** (*see also* endomembrane system; membranes, cellular)
  in algae, 386
  of bacteria, 282–83
  cell-to-cell communication and, 87
  central control hypothesis, 705
  diffusion across, 78–79, **79**
  endomembrane system's evolution from, 272, **273,** 274
  functions of, 46, 74
  growth of cell walls and, 66
  plasmolysis and, 81, **81**
  of prokaryotes, 282–83
  solute transport across, 82–84, 574
  three-layered appearance of, 46, **46**
  transfer cells and, 574
  turgor pressure and, 80–81
  vesicle-mediated transport across, 85
plasmids
  mutations and, 193
  recombinant DNA (DNA cloning) and, 224–26, **225, 226**
  Ti (tumor-inducing), 696, **696,** 697

plasmodesmata, 56, 64, **64,** 65, 66, **67,** 68, **68, 88, 164,** 166, 398, 404, 582, **583**
cell-to-cell communication and, 87–89
movement of viruses via, 300, **300**
size exclusion limits of, 88
plasmodia, 352, **352, 353,** 354
plasmodial slime molds (*Myxomycota;* myxomyocetes), 275, **275t, 276,** 348, **349t,** 352–54, 884
life cycle of, 352, **353**
plasmodiocarp, 352, **353**
plasmogamy, 312, **318**
in *Chlamydomonas*, 389, **389**
in mushrooms, **321**
in plasmodial slime molds, **353**
in wheat rust, **329**
plasmolysis, 81, **81**
Plasmopara viticola, 373
plasticity, developmental, 246
plastids, 48–52, **70t,** 581, **581** (*see also* amyloplasts; chloroplasts; chromoplasts; leucoplasts)
of bryophytes, 404
cell cycle and, 158
cytoplasmic inheritance involving genes in, 196
developmental cycle of, **51**
of euglenoids, 352
plastocyanin (PC), 137, **138, 139**
plastoglobuli, 49
plastoquinones (PQ, Qa, Q_b), **136, 138**
Platanaceae, **518**
Platanus
occidentalis, 248, **248, 658, 669t**
orientalis, 248, **248**
x *hybrida*, 248, **248**
Plectranthus, **585**
pleiotropy, 195–96
Pleistocene epoch, inside front cover
Pleopeltis polypodioides, **453**
Pleurotus ostreatus, 333, **333**
Pliny the Elder, 327, 772
Pliocene epoch, inside front cover
plum, 543
plumule, 562, **562,** 563, **564,** 567, 568
plurilocular gametangia, 383
plurilocular sporangia, **382,** 383
pneumatophores (air roots), 605, **605**
Pneumocystis carinii, 308
pneumonia, bacterial, 287
Poa
annua, **633**
pratensis, 146, 244
Poaceae, 256, **265t,** 511 (*see also* grasses)
pod (fruit), 513 (*see also* legume)
Podandrogyne formosa, **263**
Podophyllum peltatum, **762**
poi, 827
poinsettias, **538**
point mutations (gene mutations), 192
poison ivy, **550**
polar covalent bonds, 869–70, **870**
polar desert, 820
polarity, 874
of embryos, 556
of water molecules, 872–73, **872, 873**
polar microtubules, 163
polar molecules, **24,** 869, **870**
polar nuclei, 506, **507,** 509
polar rings, 358
polio, 296
pollen (pollen grains), **184,** 431, 504–6 (*see also* pollination)
of angiosperms, 504, 506, **506,** 508
single-aperture, 519
walls of, **505**
bats and, 539
of cycads, 487
of eudicots, **506**
germination of, 473, 481, **508**
of gymnosperms, 472
hay fever and, 508
of monocots, **505, 540**
of pines, 477, **477, 478–79,** 481
spores compared to, 506
of wind-pollinated flowers, 540
of *Wollemia nobilis*, 489
pollen basket, 532
pollen sacs, 501, 504, **504**
pollen tube, 467
pollen tubes, **184,** 431
of angiosperms, 501, 503, 504, **505,** 508, **508,** 509, **512, 513**
of cycads, 487, 488
of *Ginkgo*, **473,** 490
of gymnosperms, 472, 473, **473,** 474
of pines, **477, 479,** 481
pollination, **184,** 431, 508–9
in angiosperms, 503
by bats, 538–39, **539**
by birds, 537–38, **538**
cross-, **184,** 530 (*see also* outcrossing)
of cycads, 487
evolution of angiosperms and, 525
in gnetophytes, **492**
in gymnosperms, 472
by insects, **530,** 530–36
in pines, 481
self-, **184,** 514
homozygotes and, 242
inhibition of, 243, **243**
wind, 530, 539–40, **540, 541**
pollinium, 528
pollution, **616–17**
bryophytes and, 401
Ginkgo and, 490
lichens and, 340, 401
poly-A tail, 223
poly-β-hydroxybutyric acid, 284
polyembryony, 472, 482
polygenic inheritance, 195
Polygonaceae, 545
Polygonatum, **508**
polymerase chain reaction (PCR), 226
polymerization, 18
polymers, 18
polynomial names, 262
polypeptides (polypeptide chains)
of proteins, 28, **28, 29,** 30
protein synthesis and, 210, 212, **212, 213,** 214
targeting and sorting of, 214, **214, 215**
Polyphagus euglenae, 309, **313**
polyphyletic taxa, 266
polyploid cells, 170
polyploidy (polyploids), 193 (*see also* allopolyploidy; autopolyploidy)
speciation and, 249, 252–54
Polypodium, life cycle of, **460–61**
polypores, 323, 324
Polyporus arcularis, **323**
polyribosomes (polysomes), 212, **212**
polysaccharides, 18, **38t** (*see also by name*)
in bacteria, 283, **283,** 284, 288
in cell walls, 62–3, **349t, 372t**
in fungi, 310
as storage forms of sugar, 20
structural role of, 20, 21
Polysiphonia, life cycle of, **362**
polysomes (polyribosomes), **48, 55,** 55–56
polysporangiates, **432**
Polytrichum
juniperinum, **422**
life cycle of, **418–19**
piliferum, **420**
pomegranates, 825
pomes, 544
ponderosa pine, **669t**
pondweed, **128**
popcorn, **828**
poplar, 342, 546
poppy
California, **497, 534**
capsules, 544
family, **544**
opium, 32, 550
population genetics, 239
populations, 774
allopatric speciation and geographic separation of, 249
defined, 239
Hardy-Weinberg law and, 239–40
human, 835–37
natural selection within, 245
rapid evolutionary changes in, 244–45
Populus
deltoides, **68, 583, 631, 669t**
tremuloides, **788**
pores
of bryophytes, 403, **403**
epidermal, **761**
fungal, **310, 319, 322, 323,** 324
in pollen grains (*see* pollen)
in sieve elements, 579–80, **581–83**
stomatal (*see* stomata)
of wood, 665
pore space of soils, 732, 734–35
Porolithon craspedium, **358**
Porphyra, 359, 361, 380
nereocystis, **359**
porphyrin ring, **117**
portulaca, 543
position effects, 193
Postelsia palmiformis, **276**
post-sieve-tube transport, 769
posttranslational import, 214, **214, 215**
postzygotic isolating mechanisms, 256
potassium (K), **76**
cycling, 783
as essential element, 727, **730t**
in fertilizers, 745
leaf movements and, 753–54
potassium ions (K^+), stomatal movement and, **692,** 693
potato(es), **21,** 550, 698, **830,** 830–31
cultivation of, **830,** 834, **840**
dormancy of, 719
eye (bud), 642, 719
genetic diversity of, 841
late blight of, 373, **375**
ring rot, **294**
scab, **294**
seed bank for, **840**
soft rot, 293
spindle tuber disease of, 302
sweet, 831, 834
white (Irish), 642–43, 831
wilt, **285,** 294
potato famine in Ireland (1846–1847), 373, 841
potato spindle tuber viroid (PSTVd), 302
potential (potential energy)
of electrons, 95, 865
osmotic, **79**
osmotic (solute), 79
of water, 76–77
Potentilla glandulosa, 246, **247,** 249
Poterioochromonas, 378
P-protein (slime), **580, 581,** 581–82, **583**
P-protein bodies (slime bodies), 582, **582, 583**
PQ (plastoquinones), **138**
prairies (*see* grasslands)
Precambrian era, inside front cover
precipitation (*see also* rainfall)
global warming and, 845
mountains and, 801, **801,** 802
predaceous fungi, 333, **333**
preovules, 468
preprophase bands, 158, 159, 165, 166
of bryophytes, 404
preprophase spindle, 160
pressure, osmotic, 79
pressure chamber, cohesion-tension theory of water transport and, 757
pressure-flow hypothesis, 767–69, **768**
prezygotic isolating mechanisms, 256
prickles, 642
prickly pear, dispersal of, 547
Priestley, Joseph, 127
primary active transport, 84, **84**
primary consumers, 783 (*see also* herbivores)
primary endosperm nucleus, 509, **513**
primary growth, 8, 426–27, 570–88
in roots, 591–95
in stems, 611–14
primary meristems, 557, 559, 570, **570, 571** (*see also* ground meristem; procambium; protoderm)
of roots, 591–92, **593, 594**
of shoots, **611,** 614
primary metabolites, definition of, 32
primary phloem, **571, 573,** 579–83
of roots, 601, **601,** 602
of stems, 618, **619, 620**
primary phloem fibers, 618, 650, 651
primary pit-fields, 64, **64, 573**
primary plant body, 427, 570–87
primary producers, 783
primary roots, 566, 590 (*see also* radicle; taproot)
primary structure
comparison in root and stem, 595
of roots, 595–600
of stems, 610–46
primary structure of proteins, 28, **28, 29**
primary tissues, 427, 573–87
primary vascular tissues, 576–83
primary xylem, **571, 573,** 576–79
of roots, 600, 601, **601,** 602
of stems, 618, **619, 620**
vascular cylinder and, 600
primordia
bud, 611, **611, 634**
flower, **638, 639**
leaf, 611, **611, 614,** 622, 634, **634,** 635, **635**
phyllotaxy and, 624
root, **603**
primrose, **243**
evening, 192, **192,** 252, 536
family, **514**
giant, 252
principle of independent assortment, 188–89
principle of segregation, 185–87
probability, laws of, 188
procambial strands, 614, 618, **618,** 619, **619,** 621, **621,** 622, **634**
procambium, 557, **557, 570, 571,** 576 (*see also* primary meristems)
of leaves, 634
of shoots, **611,** 613, 614, **614**
of stems, 618
Prochlorococcus, 292
Prochloron, 292, **292**
prochlorophytes, 292, **292**
Prochlorothrix, 292
product rule of probability, 188
proembryo, 556, **557, 558**
progesterone, 846
progymnosperms (*Progymnospermophyta*), 429, 431, 443, **470,** 470–72
phylogenetic relationships with other embryophytes, **472**
prokaryotes, 6, 11, **43, 44,** 281–305 (see also archaea; bacteria)
autotrophic, 286, 287, 783
chemosynthetic, 287
photosynthetic, 5, 287
cell division in, **156**
cell wall of, 283–84
characteristics of, 282–84
chromosomes of, 216–18
commercial uses of, 287
disease-causing, 287
DNA replication in, 202, 203
eukaryotes compared and, 41, 43, 44, **45t,** 272
fimbriae and pili of, 284, **284**
flagella of, 284
forms (shapes) of, 284–85, **285**
heterotrophic, 5, 286, 783
inclusion bodies (storage granules) of, 284
metabolic diversity of, 286–87
overview of, 282
oxygen need or tolerance of, 287
plasma membrane of, 282–83
regulation of gene expression in, 216
reproduction of, 285
world ecosystem and, 287
prolamellar bodies, 51
proline, **27**
promoters, 210

propagation (*see* asexual reproduction; sexual reproduction)
prophase, 158, 160, **160,** 163
  meiotic, **174,** 174–76, **175,** 177
  mitotic, 160
proplastids, **51,** 51–52, **70t**
prop roots, 603, **603**
Prosopis juliflora, 590
prosthetic groups, 103
protandrous plants, 510, **514**
protective layer, 636, **637**
proteinase inhibitors, 550
protein kinases, 691
proteinoid microspheres, 4
proteins, 18, **38t** (*see also* amino acids; *specific types of proteins*)
  carrier, **82,** 83
  channel, **82,** 83
  denaturation of, 30
  fibrous, 29
  in food, 825, 838
  globular, 29
  hay fever and, 508
  integral, 74
  iron-sulfur, 116, **117**
  molecular chaperones, 29
  molecular structure of, 26–30
    primary structure, 28, **28, 29**
    quaternary structure, **29,** 30
    secondary structure, 28, **29**
    tertiary structure, 29, **29,** 30
  movement, 300, 302
  peripheral, 74, **75**
  respiration and, 122
  synthesis of, 208, 210–14
    DNA strands used as templates, 210
    RNA synthesis from DNA templates, 210
    rRNA (ribosomal RNA), 211–12
    stages in, 212, **212**
    targeting and sorting of polypeptides, 214, **214, 215**
    transfer RNA (tRNA), 210–11
    translation of mRNA into proteins, 212–14
  transmembrane, 74, 75, **75**
    multipass, 75, **75,** 82
  transport, 82
  viral, 297–98
prothallial cells, 477, **477, 479**
prothallus, 458–59
Protista (protists; kingdom), 270, 271, 275, **275t,** 275–76, **276,** 347–99, 883
  comparative summary of characteristics of, **349t, 372t**
  endosymbionts in, 274
  in evolution, 348
  heterotrophic, 351, 356, 365, 367, 378
  photosynthetic, 351, 356, 357, 365, 367, 378, 381, 383
protoderm, 557, **557, 558, 570, 571, 593** (*see also* primary meristems)
  of shoots, **611,** 613, 614, **618**
protofilaments, 58, **59**
Protogonyaulax tamarensis, 364
protogynous plants, 510
Protolepidodendron, **426**
proton acceptors, 862
proton donors, 862
protonemata, 407, **416, 418**
  of *Sphagnidae*, 414
proton gradient, **84, 110,** 118, **119,** 120, **136,** 137, 138, 152
proton pump, 83–84
protons, 863, 874
  in photosynthesis, 135, **136,** 137
protophloem, 579, 595, 614, 621, **622, 623** (*see also* primary phloem)
protoplasm, 45
protoplast fusion, 693, 695
protoplasts, 45
  of mature sieve elements, 580, 581
Protosphagnales, 414
protosteles, **427, 428, 429, 435,** 439, **442, 444, 463t**
protoxylem, 577, 595, **596,** 614, 618, **619,** 621, **622, 623** (*see also* primary xylem)
protoxylem lacuna, **622**
protoxylem poles, 600, 601, **601**
protozoa, classification of, 275
protracheophytes, 416, 434, **434**
Prunella vulgaris, **245**
Prunus, **503**
  persica var. *nectarina*, 264
  persica var. *persica*, 264
  serotina, **669t**
Prymnesium, 367
  farvum, **367**
Psaronius, 457
Pseudolycopodiella, 438
Pseudomonas, 293, **294**
  marginalis, **284**
  solanacearum, **285,** 294
  syringa, 699
Pseudomyrmex ferruginea, 775, **775**
pseudoplasmodia (slugs), **354,** 355
pseudopodium, 414
Pseudotrebouxia, 335
Pseudotsuga, 483
  menziesii, **669t**
Psilocybe mexicana, **325**
psilocybin, 325
Psilophyton, **426, 433**
  princeps, **433**
Psilotophyta (psilotophytes), **275t,** 430, 431, 443–45, **463t,** 886 (*see also Psilotum; Tmesipteris*)
Psilotum, 443–45, **463t**
  life cycle of, **446–47**
  nudum, **444**
P (peptidyl) site, 212, **212, 213**
psychedelic chemicals, 550, **551**
psychrophiles, 287
Pteridium aquilinum, **458**
pteridophytes, phylogenetic relationships with other embryophytes, **472**
Pteridospermales, 456
Pteridospermophyta, 472
pteridosperms (seed ferns), **456,** 457, **457,** 472
Pterophyta, 431, 449, 452–63, 457–63, **463t,** 887 (*see also* ferns)
Ptichodiscus brevis, 364
Puccinia graminis, life cycle of, 327, **328–29**
puffballs, 322, 323, 326, **326**
pulvini, 720, **720**
pumpkins, 828
pumps, **82,** 83 (*see also by name*)
punctuated equilibrium model of evolution, 257–58
Punnett square, 187, 188, **189**
purines, 198, **198,** 199, **199**
purple bacteria, 282, 283
  nonsulfur, **291**
  photosynthesis in, 291
  sulfur, **127,** 128, 291
purple blazing star, **793**
purple sage, **779**
Puya raimondii, 648
pyramids of biomass, energy, numbers, **785, 786**
pyrenoids, **44,** 352, 358, 388, **388**
pyrethrum, 780
pyrimidines, 198, **198,** 199
pyrophosphate bridge, 102
Pyrus communis, **62, 576,** 655
pyruvate (pyruvic acid), 113–15
  in alcohol fermentation, 123, **123**
  in glycolysis, 109, **110,** 113
  in Krebs cycle, 115
  oxidation and decarboxylation of, 114, **114**
  in photosynthesis, 144, **145, 149**
Pythium, 374

Q

Q_a, Q_b (plastoquinones), **136, 138**
quaking aspen, **788**
quantum, of light, 130
quartersawed boards, 667, **667**
quartz (SiO_2), 731
quartzite, 731
Quaternary period, inside front cover
quaternary structure of proteins, **29,** 30
Quercus, **502, 577,** 624
  alba, **657, 666, 667, 669t**
  coccinea, 256, **257**
  rubra, **65, 577, 625, 655,** 662, **662, 663, 669t**
  suber, 658, **659**
  velutina, 256, **257, 658**
quiescent center, 594, **594,** 604
quillworts (*Isoetes*), **23, 55,** 149, **442–43, 443**
  muricata, 23, **55**
  storkii, **442**
quinine, **551, 833**
quinoa, 830, **830**
quinones, 118, 137 (*see also* coenzyme Q)

R

rabies, 296
raceme, **499**
rachilla, **541**
rachis, 453, 455, 626, 825
radial files, 648, 662
radial micellation, 753, **753**
radial pattern, 556, **556**
radial section (surface), **655, 660**
radial system, 648
radiant energy (*see* solar energy)
radicle, **557,** 562, **562,** 563, **564,** 566, 568
radioactive decay, 864
radioactive isotopes (radioisotopes), 864
radioactive tracers, **755,** 765, **766**
radiocarbon dating, 864
radio waves, 129
radish, **570, 595, 759t**
Rafflesia, 498
  arnoldii, **498**
ragweed, western, **505**
Raillardiopsis, 251
rainfall, **797** (*see also* precipitation)
  in deserts, 807
  dispersal of fruits and seeds by, 547
rainforests, 489, **798–800, 803–5**
  biome map of, **798–99**
  temperate, 804
  tropical, **10,** 803–5
ramiform pits, **576**
Ranunculaceae, **502,** 545
Ranunculus, **596, 598,** 614, 619, **620**
  peltatus, **197**
rapeseed, **508**
Raphanus sativus, **570, 595, 759t**
raphides, **54,** 549
Ratibida pinnata, **793**
rattlesnake plantain, 778, **778**
ray flowers, **527,** 528
ray initials, 648, 649, **650**
ray(s), 573, 648, **650, 655, 661, 663**
  of angiosperm woods, 662, **662**
  of conifers, 476
  dilated, **652**
  parenchyma, **661**
  phloem, 476, **652, 655, 656**
  pith, **651**
  vascular, 648, 649, **649,** 650, **650**
  xylem, 476, **652, 655**
ray tracheid, **661**
*rbc*L gene, 271, 526
reaction center, 135
reaction wood, 666, **666**
Reboulia hemisphaerica, **406**
Recent epoch, inside front cover
receptacle (algal), **385**
receptacle (flower), 499, **499,** 501
reception of signals, 87, **87**
receptor-mediated endocytosis, **85,** 86
receptors, 87
  hormones and, 690–91, **691**
recessive characteristics, 185, 186
recognition sequences, 224
recombinant DNA, 224–31 (*see also* genetic engineering)
  Arabidopsis thaliana as model experimental organism, 228, **228**
  DNA libraries and, 226
  DNA sequencing techniques and, **229,** 229–30
  nucleic acid hybridization and, 230
  plasmids and, 224–26, **225, 226**
  polymerase chain reaction (PCR) and, 226
  restriction enzymes and, 224–25, **225**
  screening processes in, 225–26
  techniques for locating genes of interest, 231
recombination, genetic, 172
recombination nodules, 176, **176**
recombination speciation, 255
red alder, **577, 669t**
red algae (*Rhodophyta*), **275t,** 348, **349t,** 357–61, **358,** 884 (*see also* algae)
  commercial uses of, 380
  edible, 380
  life cycle of, 360–61, **362**
  unique cellular features of, 358–60
red blood cells, 219
red buckeye, **627**
redbud, **54**
red mangrove, 603
red maple, **10**
red oak, 65, **577, 625, 655,** 662, **662, 663, 669t**
redox (oxidation-reduction) reactions, 98–99, **99**
red peppers, 833
red snow, **386**
red tides, 350, **363,** 364
reduction, 98, **99,** 109
reduction division (*see* meiosis; mitosis)
redwoods, 474, **669t**
  coast, 483
  dawn, **485,** 486, **486**
  geographic distribution of, **485**
region, nodal and internodal, **621**
region of cell division, 594, **594**
region of elongation, **594,** 595
region of maturation, **594,** 595, 596
Regnellidium diphyllum, **55, 108**
regulator genes, 216, **217**
regulatory enzymes, 104
regulatory sequences, 688, **688**
regulatory transcription factors (*see* transcription factors)
reindeer moss, 337
release factors, 214
Reoviridae, 298
replacement fossils, **239**
replication, 208–9
  bubbles, 201, **202**
  forks, 201, **202**
  origin of, 201
replication of DNA (*see* DNA (deoxyribonucleic acid), replication of)
reporter genes, 698–99
repressors, 216, 217, **217,** 218
reproduction (reproductive systems) (*see also* cell division)
  asexual (*see* asexual reproduction)
  in cellular slime molds, 356
  in conifers other than pines, 482
  in fungi, 311–12
  in lichens, 338
  mitosis and, 166
  in mushrooms, 324
  of prokaryotes, 285–86
  in red algae, 360–61
  sexual (*see* sexual reproduction)
  in vascular plants, 430–31
  vegetative (*see* asexual reproduction)
  in yeasts, 330
reproductive apex, 639
reproductive isolation (*see* genetic isolation)
reservoir, in euglenoids, 351, **351**
resin ducts, 474, 659
resonance energy transfer, 131–32, 135
respiration, 6, **94,** 108–25
  ADP "recharged" to ATP in, 32
  aerobic pathway, 113–22
  anaerobic pathways, 122–23
  fats and proteins in, 122
  germination and, 565
  glucose oxidation and, 109
  mitochondria as sites of, 52
  ripening of fruits and, 683

stages of, 109 (*see also* electron transport chain; glycolysis; Krebs cycle; oxidative phosphorylation)
resting cysts, dinoflagellate, 365
restoration ecology, 792–93
restriction enzymes (restriction endonucleases), 224–25, **225,** 696
DNA sequencing techniques and, 229
restriction fragments, 224
resurrection plant *(Selaginella lepidophylla),* **439**
retinal, **134**
retinol (vitamin A), **134**
reverse transcriptase, 226
R groups, in amino acids, 26, 27, **28,** 29, **29**
rhabdovirus, **299**
Rheum rhabarbarum, **575, 643**
Rhizanthella, **529**
rhizobia, 739, **739**
Rhizobium, **294,** 738–40
rhizoids, 310
of algae, 392, **392,** 395
of bryophytes, 403, **403,** 407, **411,** 416, **419, 422t**
of *Equisetum,* **449**
of ferns, 455, 458, 459, **460, 461**
of fungi, **313, 316**
of *Psilotum,* 443
rhizomes, **180** (*see also* stems)
of *Equisetum,* 445
of ferns, 455, **455,** 458
of *Lycopodiaceae,* 435
of *Psilotum,* 443, 444
of Trillium, **5**
tubers attached to, 642, **643**
Rhizophora mangle, 603
Rhizopus, **308t,** 317
life cycle of, **316**
stolonifer, 315, 316, **316**
rhizosphere, 597
Rhodococcus, **294**
Rhodophyta (red algae), 357–61, **362,** 884 (*see also* algae; red algae)
Rhodospirillum rubrum, **291**
Rhopalotria, 487
rhubarb, **575,** 643
Rhynia, **426**
gwynne-vaughanii, **426,** 432–33, **433**
major, 434, **434**
Rhyniophyta (rhyniophytes), 431, **432,** 432–34, **463t**
Rhytidoponera metallica, **549**
riboflavin (vitamin B_2), **116**
rib, of leaf, (midrib), **628,** 631
ribonucleic acid (*see* RNA)
ribose, **19,** 30, **30, 31**
ribosomal RNA (*see* rRNA)
ribosomes, 41, 45, 48, **48,** 52, **52,** 55–56, **71t** (*see also* polysomes)
mitochondrial, **45t,** 52
plastid, **45t,** 52
of prokaryotes, 52, 282
protein synthesis and, 208, 211, 212, **212, 213,** 214, **215**
ribozymes, 31
ribulose 1,5-bisphosphate (RuBP), 139, 140, **140, 141,** 141, 142, **142,** 143, 146
ribulose bisphosphate carboxylase/oxygenase (Rubisco), 140, 142–44, 146, **149**
Riccia, 407–08
Ricciocarpus, 407–08
rice, 146, **497,** 826, 827, **827,** 835
amino acids in, 26
Azolla-Anabaena symbiosis and, 741
bran, 563
ethylene and, 682
foolish seedling disease of, 684
Ricinus communis, **562, 566,** 567, **578**
Rick, Charles, 841
Rickettsiae, 882
Rieseberg, Loren H., 255
ring bark, 658
ring-porous woods, **662, 663,** 665
ring spots, 301
ringworm, 334
ripening of fruits
ethylene and, 683, 697, **698**
genetic engineering and, 697, **698**
RNA (ribonucleic acid), **30,** 31, **31,** 41
DNA compared to, 208, **208**
gene expression and, 208
messenger (*see* mRNA)
in nucleolus, 48
protein synthesis and, 208, 210–14
ribosomal (*see* rRNA)
transfer (*see* tRNA)
of viroids, 302
of viruses, 297, 298, 300
RNA polymerase, 208, 210, **210,** 216, **218, 688**
RNA primase, **203**
Robert, Karl-Henrik, 845
Robertson, J. David, 74
Robichaux, Robert, **251**
Robinia pseudo-acacia, **627, 656, 657,** 659, **669t**
rocks
nutrients derived from, 731
weathering of, 786, **787,** 788
rock spikemoss *(Selaginella rupestris),* **439**
rockweeds, 379, **379,** 383 (see also *Fucus*)
Rohn, Megan, **520**
rootcap, 591, 592, **592, 593,** 604
gravitropism and, 704, **705**
root hairs, 342, 585, **592, 595,** 595–97, **597,** 604, **604**
density of, **759t**
nitrogen-fixing symbiosis and, 739, **739**
water absorption by roots and, 759–62
root-hair zone, 595
root pressure, 755, **760,** 760–61, **761**
root primordia, 603, **603**
root(s), 7, **8,** 426, 566, 589–609
adaptations of, 606, **607**
adventitious, **439,** 445, 566, **568,** 678
aerial, 603
air, 605, **605**
apical meristems of, **8,** 559, 562, **570,** 591–92, 604, 605
quiescent center, 594
types of organization of, 592, **593**
balance between shoot systems and, 590–91
cortex of, 592, **593, 597,** 598–600, 602
cylindrical form of, 571
cytokinin/auxin ratio and, 681, **681**
embryonic (*see* radicle)
epidermis of, 595–97, **596, 597,** 602
evolution of, 429
extent (depth and lateral spread) of, 590
feeder, 590, 596
of ferns, 455
fibrous, 590
fleshy, 606, **607**
functions of, 590
gravitropism in, 704–5, **705**
of gymnosperms, **344**
hydrotropism in, 706
hypocotyl-root axis, 482, **482, 513,** 562, 563, **637**
lateral (branch), 566, 590, **592,** 603
mycorrhizae and (*see* mycorrhizae)
nitrogen fixation and, 738, 739, **739,** 740, **740**
nutrient uptake by, 762–64, **763**
periderm of, 602
primary (taproot), 566, 590, **591**
primary growth of, 591–95, **601, 602**
near the root tip, **594,** 594–95
primary structure of, 595–600
prop, 603, **603**
secondary growth of, 600, **602**
secondary structure of, 600–2
of seedless vascular plants, 435, 438, **439, 442, 443,** 445
soil formation and, 731, **731**
systems, 426, 590–91
tips of, 591–94
transition to shoots, 636–39, **637**
vascular cylinder of, 573, 592, **593, 596, 597,** 600, **602, 603,** 604, **604**
woody, 602
Rosaceae, 256, **502**
rose
family, 256
hips, **772**
Rosemarinus officinalis, **532**
rosemary, **532**
rosettes, 686
rots, soft, 293, **294,** 317
Roundup, 697
roundworm (*see* nematodes)
rRNA (ribosomal RNA), 208
protein synthesis and, 211–12
sequencing of, molecular systematics and, 270, **270**
rubber, 34, 834
guayule as source of, 842, **842**
rubber tree, **34**
Rubiaceae, **833**
Rubisco (RuBP carboxylase/oxygenase), 140, **140, 142,** 142–44, **143,** 146, **149**
RuBP (ribulose 1,5-bisphosphate), 139, 140, **140, 141,** 141, 142, **142,** 143, 146
ruminants, 295
runners (stolons), **180**
rusts (fungi) (see *Teliomycetes*)
rutabaga, 763
rye, 146, 335, 590, 595, 719, 762

S

Sabalites montana, **518**
Saccharomyces
carlsbergensis, 331
cerevisiae, 230, 308, 330, 331, **331,** 332, 334
Saccharum, **88,** 747
officinale, 834
officinarum, **632**
Sachs, Julius von, 65, 693
sacred mushroom, 325
S-adenosylmethionine (SAM), 682, **682**
saffron, 833
sage, purple, 779, **779**
Sagittaria, **558**
saguaro cactus, **10, 497, 539, 808**
St. Helens, Mount, 789, 792, **792**
St. Anthony's fire, 335
sake, 334
Salicaceae, 342, 546
Salicornia, 744
salicylic acid, **36,** 37, 675, **675t**
saline environments, **744**
Salix, **602, 603** (*see also* willows)
Salsola, 546, **546**
salt, extreme halophilic archaea and, 295
saltbush, 585, 744, **744**
salt-marsh grass *(Spartina),* 254, **254**
salt secretion,
salt-tolerant crops, 843, **843**
salvage pathway, in photorespiration, 142–43, **143**
Salvia leucophylla, 779, **779**
Salvinia, 462, **462,** 463
Salviniales, 454, 462, **462, 463t**
SAM (S-adenosylmethionine), **682**
samaras, 545, **545**
Sambucus canadensis, **573,** 614, 618–19, **618–19,** 650, **651–653**
sandstone, 731
sandy soils, 732, **735**
Sanguinaria canadensis, 549
Sansevieria, **54**
sap, 77
cell (vascular), 54
sieve-tube, 766, **767**
velocity of flow of, 758, **758**
wind pollination and, 530
xylem, 757, **757, 758,** 758
saponins, 549
Saprolegnia, 373, **374**
life cycle of, **374**
saprophytes, 286
fungi, 310, 312, 313
oomycetes, 373
sapwood, **655,** 665
SAR (systemic acquired resistance), 37
Sargassum, 381, **381,** 383
muticum, 381
saturated fats, 24
sauerkraut, 287
Sauromatum guttatum, 37
savannas, **10, 798–99,** 805
biome map of, **798–99**
juniper, **798–99, 810**
Saxifragaceae, **522**
Saxifraga lingulata, **761**
saxifrage, **761**
scale barks, 658, **658**
scalelike outgrowths, 439, 443 (*see also* ligules)
scale mosses, 412
scales
of algae, **349t,** 367, **372t, 387**
of ferns, 458
of liverworts, **403**
scanning electron microscope (SEM), 43
scarification, **565**
breaking of dormancy by, 718
scarlet monkey-flower, **537**
scars
bundle, **657**
leaf, 636, **657**
terminal bud-scale, **657**
Schaller, George, 844
schizocarps, 545, 546
Schizosaccharomyces octosporus, **331**
Schleiden, Matthias, 41
Schwann, Theodor, 41
scientific names, 262–63
sclereids (stone cells), **62, 575,** 575–76, **576**
in xylem, 579
sclerenchyma (cells or tissue), **571,** 575, 619, **620,** 621, **622, 623,** 632 (*see also* fibers; sclereids)
Scleroderma aurantium, **277**
sclerotia, 335, 354
Scott, D. H., 457
scouring rushes, 445
screening processes
recombinant DNA and, 225–26
scrub communities, 816–17
scutellum, **562,** 563, **564,** 686, **686**
sea lettuce *(Ulva),* **393, 394,** 394
sea moss, 412
sea-nymph, **542**
sea otters, 371
sea palm, **276**
seashore environment, photosynthetic organisms and, 6–7
sea slugs (nudibranchs), 395, **396**
seasons (*see also* dormancy)
biological clock and, 708–709
seaweeds, 350, 371, **380** (*see also* brown *and* red algae)
edible, 380
Sebdenia polydactyla, **276**
Secale cereale, 146, 335, 590, 719, 762
secondary active transport, 84, **84**
secondary cell walls, 64–65, **65**
pits in, 66
secondary consumers, 783
secondary endosymbiosis, cryptomonads and, 357
secondary growth, 8, 427
in cycads, 486
diffuse, 648
in pines and other conifers, 476, **476**
in roots, 600–602
in stems, 614, 619, 647–71
bark, 654–59
lenticels, **653,** 654
periderm, 653–54
primary body of the stem, effect on, 650–59
stages of development, **651**
in veins, 631
secondary metabolites, 32–37, 54, 334, **549,** 549–51, **551** (*see also* alkaloids; phenolics; terpenoids)
secondary phloem, 427, 470, **576, 579, 580,** 600–2, 648, 649, **649,** 650, **650,** 651, **651, 652,** 654, 655, **656,** 659, 664

  - in pines and other conifers, 476, **476**
- secondary plant body, 427
- secondary plant products (*see* secondary metabolites)
- secondary structure of proteins, 28, **28**, 29, **29**
- secondary vascular tissues (*see* secondary phloem; secondary xylem; *and* vascular cambium)
- secondary xylem, 427, 457, 470, **470**, 476, **476**, **576**, **577**, 578, 600, 601, **601**, 602, **647**, 648, 649, **649**, 650, **650–652**, 654, **655**, **656**, 659–70 (*see also* wood)
- second law of thermodynamics, 95–97, **96**
- second messengers, 87, 691, **691**, **692**
- secretion, **57**, 58, 591, **592**
- secretory vesicles from the trans-Golgi network, **57**, 58, 59, 66
- sedge family, 340
- sedimentary rocks, 731
- *Sedum*, 644
- seed banks, 718, 840, **840**
- seed coat, **8**, 470, 522, 562, 563, **564**
  - of angiosperms, 509, **513**
  - dormancy of seeds and, 565, **565**
  - germination and, 565
  - of pines, **479**, 482, **482**
  - of willows and poplars, 546
- seed ferns (pteridosperms), **456**, 457, **457**, 472
- seed leaves, 482, 562–63
- seedless vascular plants, 424–66, **463t** (*see also specific phyla*)
  - of Carboniferous period, 456–57, **456–57**
  - evolution of, 425–31 (*see also* vascular plants, evolution of)
  - extinct, 425
  - *Lycophyta* (lycophytes), 435–43, **463t**
  - main features of, **463t**
  - phyla of, 431
  - *Psilotophyta*, 443–45, **463t**
  - *Pterophyta* (ferns), 449, **452**, **453**, 453–55, 457–63, **463t**
  - reproductive systems of, 430–31
  - *Rhyniophyta* (rhyniophytes), 432–34, **463t**
  - *Sphenophyta*, 445–49, **463t**
  - *Trimerophytophyta* (trimerophytes), 443, **463t**
  - *Zosterophyllophyta* (zosterophyllophytes), 434–35, **463t**
- seedlings, 555 (*see also* germination)
  - *Arabidopsis*, **560**, **561**
    - dark-grown, 712–714, **713**
    - effects of light on, 714
    - establishment of, 566–68
    - etiolated, 713, **713**, 714
    - gravitropism and, 704
- seed pieces, 180
- seed plants, 8–9
  - evolution of, 431, 471–72
  - phyla of, 470
- seeds, 9, 510
  - cold treatment of, 717
  - of cycads, **487**
  - development of, 556
  - dispersal of, 546–49
  - dormancy of, 565, 717–18
    - abscisic acid and, 684
    - gibberellin and, 685
  - of *Ephedra*, **481**
  - evolution of, 425, 468–70
  - food storage in, 470, 482, 509–10, 563, 567
  - germination of (*see* germination)
  - of *Ginkgo*, 490, **490**
  - of *Gnetum*, **491**
  - of gymnosperms, 467
  - longevity of, 718
  - of pines, 482, **482**
  - proteins in, 26
  - viability, 718
  - of *Wollemia nobilis*, **489**
  - of yews, **484**
- seed-scale complexes, 480, 481
- segregation, **186**
- segregation, principle of, 185–87, **187**, 190
- seismonastic (thigmonastic) movements, 720–21, **720**, **721**
- *Selaginella*, **428**, 559
  - *kraussiana*, **48**, **439**
  - *lepidophylla* (resurrection plant), **439**, 439–42
  - life cycle of, **440–41**
  - protostele, **442**
  - *rupestris* (rock spikemoss), **439**
  - *willdenovii*, **439**
- *Selaginellaceae*, 439–42
- self-fertilization, 438
- self-pollination, **184**, 514
  - homozygotes and, 242
  - inhibition of, 243, **243**
- self-sterility, 194, 243
- semideserts, **798–99**
- *Senecio*, **574**
- senescence
  - of flowers, genetic engineering and, 698
  - of leaves, cytokinins and, 681–82, 698
- sepal primordia, **638**
- sepals, **499**, 501
  - development of, **638**, 639
  - of early angiosperms, 524, **524**
  - of orchids, 529
- separation (abscission) layer, 636, **637**
- septa, fungal, **310**, 313, 317, 318, **322**, 323, 333
- *Sequoia*, **485**
  - *sempervirens*, 474, 483, **669t**
- *Sequoiadendron giganteum*, 483, **485**
- serial endosymbiotic theory, 272–74, 283
- serine, **27**, **28**, 143, **143**
- sesquiterpenoids, 33, 34
- sessile leaves, 624, **625**
- setae (stalks), **405**, 406, 414, **417–419**, 420, **422**
- *Setcreasea purpurea*, 89
- sex chromosomes, in bryophytes, 404
- sex expression and ethylene, 683
- sex organs (*see* antheridia; archegonia; gametangia; oogonia)
- sexual reproduction, 169, **169**, 170 (*see also* fertilization; meiosis)
  - advantages of, 180–81
  - in algae, 360–361, **362**, 365, 367, 376, **377**, 379, 382, 383, **384**, **385**, 388, **389**, 390–93, **394**, 396, 397
  - in angiosperms, **512–13**
  - in ascomycetes, 318, **318**, 320
  - in *Basidiomycota*, **321**, **328–29**
  - in bryophytes, 404, **405**, 406, **410–11**, **418–19**
  - in chytrids, **314**
  - in cycads and *Ginkgo*, **488**
  - in eukaryotes (*see by group*)
  - in fungi, 312
  - genetic variability and, 242
  - in hymenomycetes, **321**
  - in oomycetes, 371, **372**
  - in *Pinus*, 476–82
  - in seedless vascular plants, **436–37**, **440–41**, **446–47**, **450–51**, 460–61
  - in slime molds, 352, **353**, 354, 356
  - in *Teliomycetes*, **328–29**
  - in zygomycetes, **316**, 317
- shade leaves, 636, **636**
- shagbark hickory, **554**, **627**, **647**, **658**
- shale, 731
- shaman, **325**
- shape of plants and plant parts, 571–72 (*see also* morphogenesis)
- sheaths
  - bundle, **628**, **629**, 632
    - in grasses, 632, **632**, **633**
  - of hyphae, 342, **342**
  - of leaves, 624
  - mucigel, 591, **592**
  - mucilaginous, 288, 397
  - of sclerenchyma cells, 621, **622**
- sheep, karakul, **826**
- shelf fungi, 322, 323, **323**, 324
- shepherd's purse, **557**
- shiitake, 325
- shooting star, **500**
- shoot(s), **8**, 426, 610–46 (*see also* leaves; plumule; stems)
  - apex, **436**, **440**, **446**, **450**, **460**, **479**, **566**, 610–12, **612**, 614, 635, **635**
    - transition to flowering, 639
  - apical meristems of, **8**, **556**, **557**, 557, **558**, 559, **561**, 562, **562**, 568, **570**, 611–14
  - balance between root systems and, 590–91
  - cytokinin/auxin ratio and, 681, **681**
  - determinate, indeterminate, 474, 498
  - emergence during germination, 567, 568
  - of *Equisetum*, **448**
  - of gymnosperms, 474, **475**
  - gravitropism in, 704, **704**
  - reproductive (floral), 639–41
  - root compared and, 590–91, 636, **637**, 639
  - system, 426, 590–91
  - transition to root, 636–39, **637**
- shoot-tip culture, 695
- short-day plants, 709, **710**, 711
- short shoots, 474, **483**, 488
- shotgun cloning, 226
- shrubs, 648
- Shull, G. H., 243
- shuttle vesicles, **57**, 58
- sieve areas, **579**, 579–80, **580**
- sieve cells, 476, **579**, 579–80, **587t**
- sieve elements, 427, **574**, 579–81, 618
- sieve plates, 382, **382**, 580, **580**, **581**, 582, **582**, **583**
  - in algae, 382, **382**
- sieve-plate pores, 580, **581–82**
- sieve-tube elements, 579, 580, **580**, **581**, 582, **582**, **583**, **587t**, 651
- sieve tubes, **292**, 293, 300, **300**, **382**, 522, 580–82
  - assimilate transport in, 765–67, **766**
  - sap, 766, **767**
- *Sigillaria*, **466**
- signals, chemical, 86 (*see also* hormones)
- signal transduction, 86–87, **87**
- siliceous compounds, 376, 445
- silicon, **728**
- silique, 544, **544**
- *Silphium*, **506**
- Silurian period, **424**, 428, 431, 432, 434, inside front cover
- silver maple, **625**
- silversword (*Argyroxiphium sandwicense*), 250, **250**, 251
- *Silvianthemum suecicum*, **522**
- *Simmondsia chinensis*, **841**, 841–42
- simple diffusion, **82**, 82–83
- simple-sequence repeated DNA, 222
- simple tissues, 573
- Singer, S. Jonathan, 74
- single bonds, 869
- single-copy DNA, 222
- sinigrin, **549**
- siphonosteles, 428, **428**, 429, **429**, 455, **455**, **463t**
- siphonous algae, 394–95, **395**
- sister cells, 157
- sister chromatids (*see also* daughter chromosomes)
  - in meiosis, 172, **173**
  - in mitosis, 160, **160–162**, 163
- sitosterol, 25, **25**
- size exclusion limits, 88
- *Skeletonema*, **375**
- Skoog, Folke, 680
- Slack, C. R., 144
- slash pine, 659, **807**
- slate, 731
- sleep movements in leaves, **707**, 719
- slime (*see* P-protein)
- slime bodies (*see* P-protein (slime) bodies)
- slime molds (*see* cellular slime molds; plasmodial slime molds)
- slime plugs, 582, **582**
- slime sheath (*see* mucigel)
- slugs
  - pseudoplasmodia, **354**, 355
  - sea (nudibranchs), 395, **396**
- *Smilacina racemosa*, **762**
- smog, 616
- smuts, 326 (*see also Ustomycetes*)
- snake plant, **54**
- snapdragons, 194, **194**, 639
- sneezeweed, 780
- snow algae, **386**
- sodium (Na), **76**
  - chloride (NaCl), 873, **873**
  - as essential element, 729
  - halophytes and, 744
  - hydroxide (NaOH), 874, **874**
- sodium ions (Na^+), 744
- sodium-potassium pump
  - in halophytes, 744
- soft rots, 293, **294**, 317
- softwoods, 659
- soil base, 732, **732**
- soil erosion, 743
- soils, 726, 731–35
  - agriculture and, 744–45
  - cation exchange and, 735
  - classification of, 732
  - deficiencies and toxicities, 729, **730t**
  - erosion of, 736, 838, **838**
  - formation of, 731–32
  - grassland (prairie), 810, **812**
  - horizons (layers) of, 731–32, **732**
  - inorganic components in, 731
  - loam, 732, 735
  - mycorrhizae and, 738
  - organic components of, 732, **733**
  - organisms, **733**
  - particles in, 732
  - pH of, 735
  - plant nutrition and, 735
  - pore space of, 732, 734–35
  - temperate deciduous forest, 813
  - tropical, 805
  - water in, 734–35
  - weathering of rocks and, 731
- *Solanaceae*, **830**, 831
- *Solanum*, 550
  - *aviculare*, 846
  - *cheesmaniae*, 843
  - *lycopersicum*, **60**, **191**, 675, **704**, **729**
  - *melongena*, **843**
  - *tuberosum*, **642**, **643**, 698, 830, **830**
- solar energy (radiant energy), 2, 5, 94
- solar tracking, **722**, 722–23, **723t**
- *Solidago*
  - *flexicaulis*, **762**
  - *virgaurea*, 246
- Solomon's seal, **508**
- solutes
  - definition of, 873
  - diffusion of, 78
  - osmosis of, 79
  - transport across membranes, 82–84
  - water potential and, **76**, **77**
- solutions
  - acidic, 874
  - basic (alkaline), 874
  - definition of, 873
  - hypertonic, 79
  - hypotonic, 79
  - isotonic, 79
- solvent, definition of, 873
- somatic hybrids, 693, 695
- Sonoran Desert, 10, **808**, 809
- soredia, 338, **338**
- sorghum (*Sorghum vulgare*), 146
- sori
  - of ferns, **458**, **459**
  - of fungi, 322, 326
- sorrel, wood, **707**
- sorting, polypeptide (or protein), 214, **214**, **215**
- source-to-sink pattern of assimilate movement, 764
- sourwoods, eastern, **813**
- South America, 523
- Southern leaf blight of corn, 840, **840**
- Southern woodland and scrub, **798–99**
- soybean, **51**, 698, 715, 738, 739, **739**, 740, **740**, 826, **826**
  - Biloxi, **709**, 715
  - life cycle of, **512–13**
- soy paste (miso), 334
- soy sauce (shoyu), 334
- Spanish moss, **337**, **485**
- *Spartina*, 254, **254**

alterniflora, 254, **254**
anglica, 254, **254**
maritima, 254, **254**
townsendii, 254, **254**
special creation theory, 236
speciation, 248–56
process of, 249–56
allopatric speciation, 249
allopolyploidy, **252**, 252–53
autopolyploidy, 252, **252**
recombination speciation, 255
sympatric speciation, 249, 252, 253, **255**
species, taxonomic
definition of, 239, 248
species (*see also* speciation),
names of, 262–63
origin of, 248–49
reproductive isolation of, 256
Species Plantarum (Linnaeus), 262
specific epithet, 262, 263
specific gravity of wood, 668, 670
spectrophotometer, 712
spectrum
electromagnetic, **129**
of visible (white) light, 129
sperm, **170** (*see also* gametes)
of algae, **377**, 383, **384**, **385**, 387, **387**, 392, **392**, 397
of angiosperms, 474, 503, 504, **506**, 508, **508**, 509, **512**
of bryophytes, 400, 402, 404, **405**, **410**, **418**, 420
of conifers, 474
of cycads, 487, **488**
of *Equisetum*, 449, **451**
of ferns, 459, **461**
of *Ginkgo*, **488**, 490
of gnetophytes, 492
of gymnosperms, 472, 473, **473**, 474
inheritance and, 184, **186**, **187**, 190, **190**
of *Lycopodiaceae*, **437**, 438
of *Marchantia*, **405**, **411**
of mosses, **419**, 420
of pine, **479**, 481
of *Psilotum*, 445, **447**
of seedless vascular plants, 425, 472
of *Selaginella*, **441**, 442
of *Zamia*, **488**
spermatangia, 360, **362**
spermatia, 327, 360
spermatogenous (cells or tissue), 404, **405**
spermatogenous cell (body cell), 481
spermogonia, 327, **328–29**
sperm packet, **385**
Sphacelia typhina, 335
Sphagnidae (peat mosses), 412, **414**, 414–15
Sphagnum, **278**, 414
S phase (synthesis phase), 157, **157**, 158
Sphenophyta (sphenophytes), 430–31, 445, 448–51, **463t**, 887 (*see also* horsetails; *Equisetum*)
spicebush, **762**
family, 519
spices, 832–33
spike (flower), **499**
spikelet, **541**
spinach, **52**, 710, **710**, 719
Spinacia oleracea, **710**
spindle
colchicine and, 193
meiotic, **174**, 176, 177, 311
mitotic, 158–60, **160**, 161, **162**, 163, **165**, **311**
nonpersistent, 386, **387**
persistent, 386, **387**, 393, 396
spindle pole bodies, 311, **311**
spindle tuber disease of potato, 302
spines, 642, **643**
spiral grain, 667, **667**
Spiranthes, **535**
spirilla, 284, **285**
Spirillum volutans, **285**
Spirogyra, 396, **396**
Spiroplasma citri, 292, **292**
spiroplasmas, 292, **292**
splash cups, 420, **420**
spleenwort, **800**
spongy parenchyma, **628**, **629**, 630, 631
sporangia, **6**, 311, **314**, 315, **315**, **353** (*see also* capsules; eusporangia; leptosporangia; megasporangia; microsporangia; tetrasporangia; zoosporangia; zygosporangia)
of algae, **382**, 383, **384**, **394**
of bryophytes, 407, **408**, **410**, 412, **413**, 414, **415**
of early vascular plants, 432, **432**, 433, **433**, 434, 435
of *Equisetum*, 448
evolution of plants and, 425, 429, 431, **432**
of ferns, **453**, 453–54, 458, 462
of fungi, 311, **314**, **315**, **316**
of hornworts, 412, **413**
of liverworts, 407, **408**, **410**
of *Lycopodiaceae*, **436**, **437**, 438, **438**
of mosses, 414, **415**
of oomycetes, **373**, **374**
of plasmodial slime molds, 352, **353**
plurilocular, **382**, 383
of progymnosperms, **471**
of *Psilotum*, 444, **444**, **446**, **447**
of *Selaginella*, **440**, **441**, 442
unilocular, **382**, 383
sporangiophores, 315, **315**, 448
spore dispersal
in bryophytes, **405**, 408, **413**, **414**, **415**, **421**
in fungi, **316**, **318**, 319, 324
in seedless vascular plants, 448–49, 454
spore mother cells, **446**, **453**
spores, 2, 8 (*see also* aeciospores; ascospores; basidiospores; carpospores; conidia; megaspores; microspores; monospores; oospores; teliospores; tetraspores; urediniospores; zoospores; zygospores)
of algae, **394**
asexual, 356
of bacteria, 285, **285**, 286, **286**
of bryophytes, 402, 406–7, **410**, **418**
of cellular slime molds, **354**, **355**
dormant, 316
of euglenoids, 352, **353**, 354
of eusporangia, **453**
of ferns, 458
of fungi, 308, 311–13, 315, **315**, **316**, 326
of heterosporous plants, 431, **441**, **451**, 461
of homosporous plants, 430, **437**, **447**, **461**
of hornworts, 412, **413**
of leptosporangia, **453**
of liverworts, 407, 408, **409–411**
meiospores, 172
of mosses, 414, **414**, **415**, 416, **418**, **419**, 421, **421**
of oomycetes, **374**
of plasmodial slime molds, 352, **353**
pollen grains compared to, 506
of red algae, **359**, 360, 361
resting, 316
of seedless vascular plants, 438, 444, 448, **449**
sporic meiosis, **171**, 172
in brown algae, 383, **384**
sporidia, 330
sporocarps 462, **462**
sporogenous (cells or tissue), 402, **410**, **418**, **436**, **440**, **446**, **450**, **460**, 504
sporophylls, **437**, 438, **438** (*see also* megasporophylls; microsporophylls)
of early vascular plants, **456–57**
of gymnosperms, **484**
of *Lycopodiaceae*, **438**
of *Selaginella*, 439, 442
of *Isoetes*, 442
sporophytes, 172, 360
of brown algae, 383, **384**
of bryophytes, 402, **402**, 403, 404, **405**, 406
of *Lycopodiaceae*, 435, **436**, **437**, 438, 459
of *Equisetum*, **451**
evolution of vascular plants and, 425, 431
of ferns, 430, 455, **459–461**, 462
of gymnosperms, **474–75**, **483–87**, **489–92**
of homosporous plants, 430
of hornworts, 412, **413**
of *Isoetes*, 442, **442**
of liverworts, **405**, 407, 408, **408**, **410**, **411**
of *Marchantia*, **405**, **410**
of mosses, 414, 416, **417–419**, 420, 421, **422**
of *Psilotum*, 443–45, **446**, **447**
of *Selaginella*, **439**, 442
of vascular plants, 425, 430, **430**
sporophytic self-incompatibility, 511
sporopollenin, 396, 398, 402, 406–7, 504, **505**
sporozoa, 361
sporulation, 286
spring beauty, 548
spring ephemerals, 812
spruces, 483
Engelmann, 777, 791
red, **669t**
white, **788**, **818**
spurge family, 266, **266**
spur shoots, 488, **490**
squash, 567, **576**, **581**, **582**, **600**
ethylene and sex expression in, 683
stalk cell (*see* sterile cell)
stalk (*see also* rachis)
of algae, **379**, **381**, 382, **384**
of bryophytes, **405**, 406
of fungus, 324
of ovule, 506
of slime molds, **354–55**
stamen primordia, **638**
stamens, **499**, 501, 502
of apple flowers, **503**
development of, **638**, 639
of early angiosperms, 524–25
insect pollination and, 531
of orchids, 528
outcrossing and, 510, **511**
sterilization of, 524, 525
whorls of, **638**
of wind-pollinated flowers, 540, **541**
Stamnostoma huttonense, **469**
Stanley, Wendell, 296
Stapelia schinzii, **531**
Staphylococcus, 334
starch, 20, **21**, 22, **38t**
in algae, **44**, **349t**, 365, **372t**, 383
floridean, **349t**, 358
in plastids, **44**, **45**, 49–50, **51**, **126**, 142, **149**
in seeds, **51**
starch grains, **45**, 49–50, **51**, **126**
starch-statolith hypothesis, 705
stearic acid, **23**
Steinbeck, John, **812**
steles, 428, **428**, 573 (*see also* eusteles; protosteles; siphonosteles; vascular cylinder)
of *Psaronius*, 457
stem bundles, 622
Stemonitis splendens, **353**
stems, 7 (*see also* rhizomes; shoots)
apical meristems of, 611–14
arrangement of leaves on (phyllotaxy), 624
of calamites (giant horsetails), 456
of Carboniferous seed plants, 457
development of, 611–14, **670**
of eudicots, 614–15, 618–21
food storage in, 642–43
functions of, 611
herbaceous, 619–21
modifications of, 641–44, **644**
of monocots, **621–23**
of pines and other conifers, 474, 476, **476**
primary growth of, 611–14, **614**
primary structure of, 614–21, **615**
types of organization, 614
primary tissues of, 611–14
of *Psaronius*, 457
roots compared and, 595
secondary growth in, 614, 619, 647–71 (*see also* secondary xylem)
secondary structure of, 648–56, 658–59
of seedless vascular plants, **435**, 439, **442**, 444, 445, **448**, **455**
vascular cambium of, 648–50
vascular tissues of leaves and, 622–24
water storage in, 644
woody, 618, 650, 653 (*see also* wood)
external features of, **657**
Stemonitis splendens, **353**
Stenocereus thurberi, **539**
stereoisomers, 870
sterigmata, **321**, 323, 324
sterile bract, **480**
sterile cell (stalk cell), 481
sterile jacket layer, 402
sterility, male, cytoplasmic, 196
steroids, 25–26, 846, **846**
sterols, 25, **25**, **33**, **38t**
in cellular membranes, 74
Steward, F. C., 695
sticky ends, 224, **225**
Stigeoclonium, 392, **392**
stigmas (eyespot), **184**, 388, **499**, 501, 508
in euglenoids, 351, **351**
stigmas (flower)
evolution of angiosperms and, **520**, **524**, **525**
genetic self-incompatibility and, 511
outcrossing and, 510
of wind-pollinated flowers, 540, **540**, **541**
stigmasterol, 74
stigmatic tissue, 508
stinkhorns, 323, 326, **326**
stinking smut of wheat, 330
stipe
of algae, **379**, **381**, 382, **384**
of mushrooms, 324
stipules, 624, **624**
stolons
in fungi, 315, **316**
in plants (runners), 180, 642
stomata, 7, **7**, **8**
bryophyte pores as analogous to, **402**, 403
of C_3, C_4 plants, 146
of CAM plants, 147, **149**
of hornworts, 406, 412, **413**
in leaf epidermis, **584**, 584–85, **585**, 626, 627, **628**, **629**, 630, **630**
of mosses, 406, 420, **420**
movement (opening and closing) of
abscisic acid (ABA) and, 684, **692**, 692–93
cellulose microfibrils in guard cell walls and, 752–53, **753**
environmental factors and, 753–54
guard cells and, 692, 752
transpiration and, 752
water loss and, 753
in photosynthesis, **132**, 139, **140**, 144, **148**, 149, **149**
in stem epidermis, 618
stomatal crypts, **629**
stone cells (sclereids), **62**
stonecrops, 147
stoneworts, **398**
stop codons, 214
straight grain, 667
strands
antiparallel, 199
complementary, 199
lagging, 202
leading, 202
stratification, breaking seed dormancy by, 717
strawberries, **180**, 679, **679**
dispersal of, **547**
wild, **762**
streaking, by viral infection, **302**
Strelitzia reginae, **537**
Streptococcus lactis, **285**
Streptomyces, **294**
scabies, **281**
streptomycin, 287
striate venation (parallel venation), **625**, 631
strobili (cones)
of balsam fir (*Abies*), **483**
of cycads (*Cycas*), 487
of cypresses (*Cypressus*), **483**
of *Equisetum*, 448, **448**, **450**
of gnetophytes, **491**, **492**
of gymnosperms, 473, **489**
of *Lycopodiaceae*, **438**
megasporangiate (ovulate)
of balsam fir, **483**

systemic, 300
transmission of, 298, 300
wound tumor, **301**
Vitaceae, **498**
vitamin A (retinol), **134**
vitamins, 102
Vitis, 641
vinifera, 687, **687**, 825
Vittaria, **459**
viviparous mutants, 684, **684**
volcanic eruptions, 789, 793
volva, 324
Volvox, **276**, 389, 390, **390**
carteri, 390, **390**
von Sachs, Julius, 693
voodoo lily, 37
Vorticella, endosymbiosis in, 274, **274**
Voyage of the Beagle, The (Darwin), 236

W

wake-robin (*Trillium erectum*), 219, **219**
Wald, George, 130
wall pressure, 81
walnuts, 563
black, **669t**, 779
wandering Jew, **51**
Wareing, Paul F., 683
wasps, 535
water, 18
absorption by roots, 759–62
hydraulic lift, 761, **762**
passive, 761–62
pathways, 759–60, **760**
positive pressure (root pressure), **760,** 760–61, **761**
dispersal of fruits and seeds by, 546–47
in early Earth, 3
evolutionary transition of plants to land and, 7, 8
germination and, 564–65
ionized, 874, **874**
photolysis of, 137
in photosynthesis, 5–7, 127, 128, **132**, 134, **134**, 135, **136**, 137–39
pollination via, 541–42
in soils, 734, 735, **735**
as a solvent, 873, **873,** 874
structure and properties of, 872–75
transport (movement) of (*see* water transport)
water buttercup (*Ranunculus peltatus*), 197, **197**
water-conducting tissues and cells, 36, 576 (*see also* hydroids; tracheary elements; tracheids; vessel elements; xylem)
of bryophytes, 402
evolution of, 425, 428
water cycle, 734, **734**
water ferns, **462**, 462–63
water-form leaves, 626, **626**
water hemlock, **500**
water hyacinths, **787**
water lily, **575, 629, 787**
family, 521
water loss
cuticle as barrier to, 752
prevention of, 24
in bryophytes, 403, **403**
stomatal closing and, 753
by transpiration, **751t**, 751–52
watermelon, 294
water molds, **171**, 275, 373, 884 (*see also* oomycetes)
life cycle of, **374**
waternet, **391**
water potential, **76,** 76–77
osmosis and, 79, **79,** 80
of soils, 735, **735**
water-soluble fibers, **563**
water storage, stems and leaves specialized for, 644
water table, 734
water transport (movement), 751–64 (*see also* transpiration; water loss)
across cellular membranes, 76–79
bulk flow, 77
diffusion, **77,** 77–79, **79**
imbibition, 80
osmosis, 80–81
water potential gradient, **76,** 76–78
cohesion-tension theory of, 754–58, **756, 757**
lignin and, 36, 37
in osmosis, **77,** 77–81, **79**
in roots, 759–62
stomatal movement and, 750
though xylem, 754, **755**
water vesicles, 644
Watson-Crick model, **199**
Watson, James, 197–99, **199**
wavelengths, 129
wave model of light, 129, 130
waxes, 24–25, **25,** 63
cuticular, 25
epicuticular, 25, 584, **584**
jojoba, **841,** 842
weathering of rocks, 786, **787,** 788
inorganic nutrients derived from, 731
weeds, control of, 679 (*see also* herbicides)
weevils, 487
Weinberg, G., 239, 240
Welwitschia, 149, 490, 492, 518, **519**
mirabilis, 149, **492**
Went, Frits W., 676, **702,** 703
wheat, 327, 825, 835 (see also *Triticum*)
awns of, 196
bran and germ, 563
bread, 47, 563, **835**
collenchyma cells from filament of, **574**
endosperm and embryo, 563, **564**
genetic control of color in, **195t**
hybridization of, 254
kernel of, **564**
leaves of, 633
new strains of, **839**
stinking smut of, 330
wheat germ, 563
wheat rust (*see* black stem rust of wheat)
life cycle of, **328–29**
whiplash flagella, 60, **60**
whisk fern, 443
white ash, **669t**
whitebark pine, 482
white fir, 788, **789**
whiteflies, 298
white flowering spurge, **793**
white oak, **657, 666, 667, 669t**
white pine, 340, **480,** 660, **660–662, 668, 669t**
white spruce, **788, 818**
whooping cough, 287
whorled phyllotaxy, 624
whorls, 501, 502, 527
Wielandiella, **473**
Wilde, S. A., 774
Wilkesia gymnoxiphium, **251**
Wilkins, Maurice, 198, 199
Williamsoniella coronata, **473**
willows, **36,** 37, 342, 546, **602, 603**
wilting, 81
wilts (disease), 293–94
wind pollination, 487, 530, 539–40, **540, 541**
winds, prevailing, **797**
wine-making, **123**
wintercress, **728**
wintergreen, **524**
wire grass, **591**
Wirth, Tim, 845
witch hazel, 546
Witch's hair, **337**
Wolffia, **496**
borealis, **496**
Wollemia nobilis, 488–89, **489**
wood, 21, **647, 649, 652, 655–56,** 659–70 (*see also* secondary xylem)
of angiosperms, **662, 663**
commercial uses of, **669t**
compression, **666**
of conifers, **476,** 659–61
density of, 668
diffuse-porous, **662,** 665
early, late, 665
features and identification of, 667–70
grain of, 667
growth ring (layers; increments), 662, 664–66
knots, **668**
methods of sawing, 667
reaction, **666**
ring-porous, 665
of roots, 600–2
specific gravity of, 668
strength of, 668
tension, 666
wood sorrel, **707**
woody magnoliids, 519, **520,** 521, 525
woody plants, 648 (*see also* trees)
woody stems, 618, 650, 653
external features of, **657**
wound callose, 580
wound healing and repair, 574
wound lignin, 37
wound tumor virus, 301, **301**

X

Xanthidium armatum, **397**
Xanthium, **548**
strumarium, 681, 709, **710**
Xanthomonas, 293, **294**
campestris, 294
xanthophylls, 133, **134**
Xanthosoma, 827
X-disease of peach, 293
xerophytes, leaves of, 626, 627, **629**
xylans, 63
xylem, 8, **8,** 426, **427,** 572, 576–79, **587t**
cell types in, **579t**
of early vascular plants, of gymnosperms, 476
primary, **571, 573,** 577, 578, **578**
of roots, 600, 601, **601,** 602
of seedless vascular plants, **435, 442, 444,** 445, **448, 455**
of stems, 618, **619, 620**
vascular cylinder and, 600
secondary (*see* secondary xylem *and* wood)
tracheary elements of, 427, 428, **428,** 576, 577–78, **578**
embolized, **757**
of stems, 618, **619**
transfer cells and, 574
of veins, 631
water tranport through, 754, **755**
air bubbles and, **756,** 756–58, **757**
xyloglucans, **22,** 63
xylose, **22**

Y

Yabuta, T., 684
yams, 834
wild, 550, 846, **846**
yeasts, 317, 330–32, 883
classification of, 330
commercial uses of, 331, **331**
yellow pine, **480**
yellow poplar (tulip tree), 519, **662, 669t**
yellow prairie-cornflower, **793**
Yellowstone National Park
1988 fire in, **790,** 790–91, **791**
yews, **579**
arils of, 548
European, 34
family, 483, **484**
Pacific, 34
yogurt, 287
yuca, 831
Yucca bevifolia, **808**
yucca moth, **536**

Z

Zamia, 487
pumila, 486, **487, 488**
Zea
diploperennis (perennial teosinte), 829
mays ssp. *mays* (maize), 829, **829**
mays ssp. *parviglumis* (annual teosinte), 829, **829**
Zea mays (maize), **567**
grain, **562, 568**
leaf, **145, 584, 625, 630**
root, **593, 594, 603**
stem, **581, 597, 621, 622–23**
zeatin, 680, **680**
zeaxanthin, **134**
zebra, 10
Zigadenus fremontii, **535**
zinc (Z), as essential element, 727, **729, 730t**
Zingiber officinale, 832
Zinnia elegans, 676
zooplankton, 348, **785,** 786
zoosporangia, **374**
zoospores
of algae, 383, **384, 389, 391**
of chytrids, **314**
of oomycetes, 371, **373, 374**
Zooxanthellae, 365, **365**
Zosterophyllophyta (zosterophyllophytes), 431, **432, 433,** 434–35, **463t**
Zosterophyllum, **426, 433**
Zygocactus truncatus, **299**
zygomycetes (*Zygomycota*), 275, **308t,** 309, **309,** 312, 313, 315–17
life cycle of, **316**
Zygomycota, 313, 315–17, 882
zygosporangia, 316, **316,** 317
zygospores
of algae, 389, **389**
of fungi, 316, **316,** 317
zygotes, 170, **170, 171** (*see also* auxospores; macrocysts; oospores; zygospores)
of angiosperms, 509, 510, **513**
of brown algae, **384, 385**
of bryophytes, 402, 404, 406, 407, **410, 418, 419**
of cellular slime molds, **356**
of diatoms, 376, **376**
of dinoflagellates, 365
embryogenesis and, 556, **557, 558**
of fungi, 312, **314**
of green algae, **386,** 389, **389, 390,** 391, **392, 394,** 396, **396,** 397, **397,** 398, 402
of gymnosperms, **479**
of oomycetes, 372, **374**
of plasmodial slime molds, **353**
of red algae, 360, 361, **362**
of seedless vascular plants, **437,** 438, **441, 447, 450, 461**
of yeasts, 330
zygotic meiosis, 171, **171, 379,** 389, 391, 392, 397